PIPE LINE
RULES OF THUMB
HANDBOOK

Gulf Publishing Company
Houston, Texas

PIPE LINE
RULES OF THUMB
HANDBOOK

Quick and accurate solutions to your everyday pipe line problems

E.W. McAllister, Editor

This book was originally published as a targeted collection of articles from the pages of *Pipe Line Industry* magazine now renamed:

Our new edition proudly carries on the tradition of spelling pipe line as two words.

Pipe Line Rules of Thumb Handbook
Fourth Edition

Gulf Publishing Company
Book Division
P.O. Box 2608 □ Houston, Texas 77252-2608

10 9 8 7 6 5 4

Library of Congress Cataloging-in-Publication Data

Pipe line rules of thumb handbook : a manual of quick, accurate solutions to everyday pipe line problems / E. W. McAllister, editor. —4th ed.
    p.   cm.
   Includes bibliographical references and index.
   ISBN 0-88415-671-0 (alk. paper)
   1. Pipelines—Handbooks, manuals, etc. I. McAllister, E. W.
TJ930.P535 1998
621.8′672—dc21
                                                      97-22465
                                                          CIP

Printed in the United States of America.
Printed on acid-free paper (∞).

iv

Contents

1: General Information, 1

Basic formulas :   2
Mathematics—areas   3
Mathematics—surfaces and volumes   4

Cables and Ropes   5

Estimating strength of cable   5
Find the working strength of Manila rope   5
How large should drums and sheaves be for various
    types of wire rope?   5
Find advantages of block and tackle, taking into
    account pull out friction   6
Safe loads for wire rope   6
Stress in guy wires   7
Strength and weight of popular wire rope   9
Measuring the diameter of wire rope   9
Wire rope: field troubles and their causes   9
Capacity of drums 11

Belts and Shafts 11

Determine length of a V-belt 11
Calculate stress in shaft key 12
Calculate V-belt length using simple equation 12
Estimate the horsepower that can be transmitted by
    a shaft 13

Miscellaneous 13

How to estimate length of material contained in roll . . . 13
Convenient anti-freeze chart for winterizing
    cooling systems 13
How to determine glycol requirements to bring a system
    to a desired temperature protection level 14
Weight in pounds of round steel shafting 14
Properties of shafting 15
Tap drills and clearance drills for machine screws 16
Common nails 17
Drill sizes for pipe taps 17
Carbon steel—color and approximate temperature 17
Bolting dimensions for raised face weld neck flanges . . 18
Steel fitting dimensions 19
ANSI forged steel flanges 20
Color codes for locating underground utilities 21
Approximate angle of repose for sloping sides of
    excavations 21
Trench shoring—minimum requirements 22
Reuniting separated mercury in thermometers 23
Typical wire resistance 23
How to cut odd-angle long radius elbows 24
How to read land descriptions 25
Size an air receiver for engine starting 28

2: Construction, 29

Project Scoping Data **30**

Project scoping data worksheet for major facilities 30

Right of Way **31**

How to determine the crop acreage included in a
    right-of-way strip 31
Clearing and grading right-of-way: labor/equipment
    considerations 31
Estimating manhours for removing trees 32
Estimating manhours for removing tree stumps 32
Clearing and grading right-of-way 33

Ditching **34**

How many cubic yards of excavation in
    a mile of ditch? 34
Shrinkage and expansion of excavated and
    compacted soil 34
Ditching and trenching: labor/equipment
    considerations 34

Concrete Work **35**

How to approximate sacks of cement needed to fill
    a form 35
What you should know about mixing and finishing
    concrete 35

Pipe Laying **35**

How to determine the degrees of bend in a pipe that
    must fit a ditch calling for a bend in both horizontal
    and vertical planes 35
How to bend pipe to fit ditch—sags, overbends and
    combination bends 36
Pipe bending computations made with hand-held
    calculator 36
Calculate maximum bend on cold pipe 41
Determine length of a pipe bend 42
Length of pipe in arc subtended by any angle 42
Average pipelay table—Underground 43
Average pipelay table—On supports 44
Allowable pipe span between supports 44
How engineers make pipe fit the ditch 45

Pipe Lowering **48**

How to lower an existing pipe line that is
    still in service 48

Welding **51**

When should steel be preheated before welding? 51
Welding and brazing temperatures 52
Mechanical properties of pipe welding rods 52
Lens shade selector 53

Pipe Line Welding **53**

How many welds will the average welder make
    per hour? 62
How much welding rod is required for a mile of
    schedule 40 pipe line? 62
How many pounds of electrodes are required per weld
    on line pipe? 62
Welding criteria permit safe and effective pipe
    line repair 63
Cross country pipe line—vertical down electrode
    consumption, pounds of electrode per joint 69

3: Pipe Design, 70

Steel pipe design 71
Properties of pipe 76
Length of pipe in bends 78
Calculation of pipe bends 79
Spacing of pipe supports 81
American Standard Taper Pipe Threads 83
British Standard Taper Pipe Threads 84
Normal engagement between male and female
    threads to make tight joints 85
Hand-held computer calculates pipe weight, contents,
    velocity 85
Formulas and constants of value in solving problems
    relating to tubular goods 88
How to calculate the contraction or expansion of a
    pipe line 89
Estimate weight of pipe in metric tons per kilometer . . . 89
How to find pipe weight from outside diameter and
    wall thickness 90
What is the maximum allowable length of unsupported
    line pipe? 90
Identify the schedule number of pipe by direct
    measurement 90
Determine buoyancy of bare steel pipe 91
Determine buoyancy of bare and concrete coated steel
    pipe in water and mud 91
Weights of piping materials 92
Allowable working pressure for carbon steel pipe 92
Find the stress in pipe wall due to internal pressure 93

How to calculate stress in above/below ground
   transition   94
How to identify the series number of flanged fittings . . .   97
Dimensions of three-diameter ells with tangents   97
Polypipe design data   98

4: Electrical Design, 101

Electrical design 102
Hazardous locations 103
NEMA Enclosure Types 104
Size portable electric generators 105
Typical wattages for tools and appliances 106
Knockout dimensions 106
National Electrical Code Tables 107
Electrical formulas 111
Full load currents—single phase transformers 111
Full load currents—three phase transformers 112
Motor controller sizes 112
Voltage drop on circuits using 600 V. copper
   conductors in steel conduit 113
Determine the most economical size for
   electric power conductors 113
How to find the resistance and weight
   of copper wires 114
What you should remember about
   electrical formulas 114
How to calculate microwave hops on level ground 114
For quick determination of the horsepower
   per ampere for induction motors (3 phase)
   at different voltages 115
Chart gives electric motor horsepower for
   pumping units 115
Pumping stations 116

Floodlighting Concepts **117**

Terms 117
Floodlighting calculations 117
Point-by-point method 117
Beam-lumen method 118
Design procedure 118

5: Hydrostatic Testing, 120

Hydrostatic testing for pipe lines 121

Appendix A **127**

Volume of water required to fill test section 127
Volume required at test pressure 128

Appendix B **129**

How to use charts for estimating the amount of pressure
   change for a change in test water temperature 129
Compressibility factor for water 132

6: Pipe Line Drying, 133

**Pipe Line Dewatering, Cleaning,
   and Drying** **134**

Dewatering 134
Cleaning pipe lines 135
Brush pig run with gas 135
Brush pig run with liquid 135
Internal sand blasting 135
Chemical cleaning 136
Pipe line drying 136
Moisture content of air 138
Commissioning petrochemical pipe lines 140
Vacuum drying 143

7: Control Valves, 146

Control valve sizing formulas 147
Sizing control valves for thruput 151
Control valve selection 156

**Relief Valve Sizing, Selection, Installation,
   and Testing** **158**

Variable orifice rotary control valves 162

Sizing Valves for Gas and Vapor **163**

8: Corrosion/Coatings, 168

Hand-held computer determines concrete coating
   thickness 169
NAPCA specifications 171
How much primer for a mile of pipe? 174
How much coal-tar enamel for a mile of pipe? 175
How much wrapping for a mile of pipe? 175
Estimating coating and wrapping materials required
   per mile of pipe 175
Coefficient of friction for pipe coating materials 176
Troubleshooting cathodic protection systems:
   Magnesium anode system 178
Cathodic protection for pipe lines 179

Estimate the pounds of sacrificial anode material
  required for offshore pipe lines 187
Comparison of other reference electrode potentials with
  that of copper-copper sulfate reference electrode
  at 25°C 189
Chart aids in calculating ground bed resistance and
  rectifier power costs 190
How can output of magnesium anodes be predicted? . . . 191
How to determine the efficiency of a cathodic protection
  rectifier 191
How to calculate the voltage drop in ground bed
  cable quickly 192
What is the most economical size for a
  rectifier cable? 192
How to estimate the number of magnesium anodes
  required and their spacing for a bare line or for
  a corrosion "hot spot" 193
How can resistivity of fresh water be determined from
  chemical analysis? 193
What will be the resistance to earth of a single
  graphite anode? 194
How to estimate the monthly power bill for a cathodic
  protection rectifier 194
What will be the resistance to earth of a group of
  graphite anodes, in terms of the resistance of
  a single anode? 194
How can the current output of magnesium rod used for
  the cathodic protection of heat exchanger shells be
  predicted? 194
What spacing for test leads to measure current on a
  pipe line? 194
How many magnesium anodes are needed for
  supplementary protection to a short-circuited
  bare casing? 195
Group installation of sacrificial anodes 195
How can the life of magnesium anodes be predicted? . . 196
How to find the voltage rating of a rectifier if it is to
  deliver a given amount of current through a given
  ground bed (graphite or carbon) 196
Determining current requirements for coated lines 196
Determining current requirements for coated lines when
  pipe-to-soil potential values are estimated 196
HVDC effects on pipe lines 197
Troubleshooting cathodic protection systems:
  Rectifier ground bed 201
How to control corrosion at compressor stations 202
Project leak growth 203

Advances in Pipe Line Protection **204**

Methods of locating coating defects 205
Case histories 208

Estimate the number of squares of tape for pipe coating
  (machine applied) 209
Estimate the amount of primer required for tape 210
Tape requirements for fittings 210

9: Gas—General, 211

Know the gas laws 212
Calculate gas properties from a gas analysis 214
Physical properties of selected hydrocarbons and other
  chemicals and gases 218
Nomograph for calculating density and specific volume
  of gases and vapors 226

**Considerations for Selecting Energy
Measurement Equipment** **227**

Facts about methane and its behavior 233
Conversion table for pure methane 237
Categories of natural gas and reserves terminology 238
Glossary of common gas industry terms 239

10: Gas—Compression, 242

Compressors 243
Performance calculations for
  reciprocating compressors 244
Estimating suction and discharge volume bottle
  sizes for pulsation control for reciprocating
  compressors 246
Compression horsepower determination 248
Generalized compressibility factor 250
Nomograph aids in diagnosing compressor
  cylinder ills 251

Centrifugal Compressor Data **252**

Centrifugal compressor performance calculations 253
Nomographs for estimating compressor performance . . 256
Estimate hp required to compress natural gas 261
Estimate compressor hp where discharge pressure is
  1,000 psi 261
Calculate brake horsepower required to compress gas . . 262
How to find the size fuel gas line for a compressor
  station 262
Estimate engine cooling water requirements 263
Estimate fuel requirements for internal combustion
  engines 263
Estimating fuel requirements for compressor
  installation 264

11: Gas—Hydraulics, 265

Gas pipe line hydraulics calculations 266
Equivalent lengths for multiple lines based on
    Panhandle A 267
Determine pressure loss for a low-pressure
    gas system 268
Nomograph for determining pipe-equivalent factors . . . 269
How much gas is contained in a given line section? 270
How to estimate equivalent length factors
    for gas lines 270
Estimating comparative capacities of gas pipe lines 271
Determination of leakage from gas line using pressure
    drop method 271
A quick way to determine size of gas gathering lines . . . 272
Energy conversion data for estimating 272
How to estimate time required to get a shut-in test on
    gas transmission lines and approximate a maximum
    acceptable pressure loss for new lines 273
How to determine the relationship of capacity increase
    to investment increase 273
Estimate pipe size requirements for increasing
    throughput volumes of natural gas 274
Calculate line loss using cross-sectional areas table
    when testing mains with air or gas 275
Flow of fuel gases in pipe lines 276
Calculate the velocity of gas in a pipe line 277
Determining throat pressure in a blow-down system . . . 277
Estimate the amount of gas blown off through a line
    puncture 278
A practical way to calculate gas flow for pipe lines 278
How to calculate the weight of gas in a pipe line 279
Estimate average pressure in gas pipe line using up
    and downstream pressures 279
Chart for determining viscosity of natural gas 280
Flow of gas 280
Multiphase flow 284
Nomograph for calculating Reynolds number for
    compressible flow friction factor for clean steel
    and wrought iron pipe 289

12: Liquids—General, 292

Determining the viscosity of crude 293
Chart gives API gravity of blends quickly 294
Liquid gravity and density conversion chart 295
Nomograph for calculating viscosities of liquid
    hydrocarbons at high pressure 295
Calculate viscosity of a blend 297
Calculate gravity of a blend 297
Convert viscosity units 297
Convert specific gravity to API gravity 297
Calculate bulk modulus 297
Viscosities of hydrocarbon liquids 298
Nomograph for calculating viscosity of slurries 299
Nomograph for calculating velocity of liquids
    in pipes 301
Nomograph for calculating velocity of compressible
    fluids in pipes 301
Nomograph for calculating velocity of liquids
    in pipes 302
Derivation of basic ultrasonic flow equations 304
How fast does oil move in a pipe line? 306
Estimate the volume of a pipe line per linear foot
    using the inside diameter 306
What is the linefill of a given pipe in barrels
    per mile? 306
Estimate leakage amount through small holes in a
    pipe line 307
Table gives velocity heads for various pipe diameters
    and different rates of discharge 308

13: Liquids—Hydraulics, 309

Marine Hose Data 310

CALM system 310
SALM system 310
Tandem system 311
Multi-point mooring system 311
Pressure loss in hose string 313
Pressure drop calculations for rubber hose 315
Examples of pressure drop calculations for
    rubber hose 315
Typical formulas used for calculating pressure drop
    and flow rates for pipe lines 315
Hydraulic gradients 317
Equivalent lengths 320
Series systems 321
Looped systems 322
Calculate pressure loss in annular sections 323
Calculate pressure and temperature loss for viscous
    crudes ≥1,000 cP 323
Determine batch injection rate 326

Pressure Loss through
Valves and Fittings 327

Representative resistance coefficients (K) for valves
    and fittings 327
Nomograph for calculating Reynolds number for flow
    of liquids and friction factor for clean steel and
    wrought iron pipe 333

Nomograph for calculating pressure drop of liquids in
lines for turbulent flow 335

Drag-Reducing Agents **339**

The basics of drag reduction 339
How drag-reducing agents can outweigh looping
or boosting 344
How active, passive drag affect DRA injections 350
How to estimate the rate of liquid discharge
from a pipe 356
Predict subsurface temperature ranges 356
Sizing pipe lines for water flow 357
How approximate throughput of a line can be
estimated from pipe size 358
Gauge liquid flow where no weir or meter
is available 358
Estimate crude gathering line throughput for a
given pipe diameter 358
How to determine head loss due to friction in
ordinary iron pipe line carrying clear water 358
How to size lines, estimate pressure drop and
optimum station spacing for crude systems 359
Estimate the optimum working pressures in crude oil
transmission lines 359
How to size crude and products lines for capacity
increases 360
How to determine the maximum surge pressure
in liquid-filled pipe line when a main valve is
closed suddenly 360
What is the hydrostatic pressure due to a column
of oil H feet in height? 360
Transient pressure analysis 361
Tank farm line sizing 370

14: Pumps, 373

Centrifugal pumps 374
Speed torque calculation 386
Estimate the discharge of a centrifugal pump at
various speeds 387
How to estimate the head for an average centrifugal
pump 387
Find the capacity of reciprocating pump 387
How to estimate the horsepower required to pump at a
given rate at a desired discharge pressure 387
Nomograph for determining reciprocating pump
capacity 388
Nomograph for determining specific speed of pumps . . 389
Nomograph for determining horsepower requirement
of pumps 390
How to select motors for field gathering pumps 390

Reciprocating Pumps 391
Understanding the basics of rotary screw pumps 399

15: Measurement, 404

Pipe line flow measurement—the new influences 405
Liquid measurement orifice plate flange taps 408
Mass measurement light hydrocarbons 412
Pipe line measurement of supercritical
carbon dioxide 413

Gas Measurement **419**

Master meter proving orifice meters in dense phase
ethylene 419
Gas or vapor flow measurement—orifice plate
flange taps 426
Properties of gas and vapors 430
Determine required orifice diameter for any required
differential when the present orifice and differential
are known in gas measurement 435
Estimate the temperature drop across a regulator 436
Estimate natural gas flow rates 436
How to estimate the average pressure differential on
the remaining meter runs of a parallel system when
one or more runs are shut off 437
Sizing a gas metering run 437
List of typical specifications for domestic and
commercial natural gas 437
Determine the number of purges for sample cylinders . . 438
Find the British thermal units (Btu) when the specific
gravity of a pipe line gas is known 438
Estimate for variations in measurement factors 438
Rules of measurement of gas by orifice meter 439
How to measure high pressure gas 439
Four ways to calculate orifice flow in field 443
Practical maintenance tips for positive displacement
meters 446
Sizing headers for meter stations 450

16: Instrumentation, 453

Types of control systems 454

Developments in Pipe Line
Instrumentation **456**

Flow measurements 457
Proving devices 459
Valves 460
Acoustic line break detectors 461

"Smart" pressure sensors 462
Densitometers 463
Pipe line samplers 464
Pipe line monitoring systems 465
Computer systems 466
SCADA systems 468
Cathodic protection 468
System design guidelines 468
Future trends 469

**Choosing the Right Technology for
Integrated SCADA Communications** **470**

17: Leak Detection, 475

Pipe Line Leak Detection Techniques **476**

Causes and economic aspects of leaks 476
Simple leak detection systems 477
Pig-based monitoring systems 478
Computer-based monitoring systems 478
Pipe line leak phenomena 479
Background philosophy of pipe line modeling 479
Basic pipe line modeling equations 480
Impact of instrument accuracy 481
System design aspects and guidelines 482
Development of pipe line monitoring systems 483

18: Tanks, 485

Charts give vapor loss from internal
    floating-roof tanks 486
Estimating the contents of horizontal
    cylindrical tanks 488
How to gauge a horizontal cylindrical tank 489
Use nomograph to find tank capacity 489
Correct the volume of light fuels from actual
    temperature to a base of 60°F 491
Volume of liquid in vertical cylindrical tanks 491
Chart gives tank's vapor formation rate 491
Hand-held calculator program simplifies dike
    computations 492

19: Maintenance, 497

How to Plan for Oil Pipe Line Spills **498**

Regulatory requirements 498
Contingency plan objectives 498
Related studies 498
Planning concepts 499

Contingency response 500
Immediate response 501
Immediate response actions 502
Flexible response actions 502
Training 503

20: Economics, 505

Rule of thumb speeds payroll estimates 506
Rule of thumb estimates optimum time to keep
    construction equipment 507
How to estimate construction costs 509
Cost estimating strategies for pipe lines, stations,
    terminals 512
Economics 519
Time value of money: concepts and formulas 524
Decision and evaluation criteria for investments and
    financial projects 531
Accounting fundamentals 540
Estimate the cost of a pipe line in the U.S. (based on
    1994 data) 545
How to compare the cost of operating an engine on
    diesel and natural gas 545
How to estimate energy costs for different pipe line
    throughputs 545
Comparing fuel costs for diesel and electric prime
    movers 546
Nomograph for calculating scale-up of equipment or
    plant costs 546
Nomograph for calculating scale-up of tank costs 548
Nomograph for determining sum-of-years
    depreciation 549
Nomograph for estimating interest rate of return on
    investment ("profitability index") 549
Nomograph for determining break-even point 551
Chart gives unit cost per brake horsepower of
    reciprocating compressors with various types of
    prime movers 552
Chart shows influence on unit cost of numbers of
    reciprocating compressor units installed in
    one station 552
Chart gives unit cost per brake horsepower of centrifugal
    compressors with various types of prime movers ... 553

21: Rehabilitation—Risk
     Evaluation, 554

When does a pipe line need revalidation? The influence
    of defect growth rates and inspection criteria on an
    operator's maintenance program 555
Risk evaluation 564

22: Conversion Factors, 567

Units of measurement from one system to another 568
Viscosity—equivalents of absolute viscosity 577
General liquid density nomograph 577
Chart gives specific gravity/temperature relationship
    for petroleum oils 579
Weight density and specific gravity of
    various liquids 579
Metric conversions—metric to English, English to
    metric 580
Temperature conversion—centigrade to Fahrenheit,
    Fahrenheit to centigrade 581
Viscosity—equivalents of kinematic viscosity 582
Viscosity—equivalents of kinematic and Saybolt
    Universal Viscosity 582
Viscosity—equivalents of kinematic and Saybolt Furol
    Viscosity at 122°F 583

Viscosity—general conversions 584
Pressure conversion chart 585
A simple method to determine square root 585
SI data 586
Energy conversion chart 587
Flow conversion chart 587
Conversions involving different types of fuel 588
Weight of water per cubic ft at various temperatures . . . 593
Engineering constants 593
Mensuration units 594
Minutes to decimal hours conversion table 594
How to compare costs of gas and alternate fuels 595
Typical characteristics of fuel oils 595

Index, 597

1: General Information

Basic formulas 2
Mathematics—areas 3
Mathematics—surfaces and volumes 4

Cables and Ropes 5
Estimating strength of cable 5
Find the working strength of Manila rope 5
How large should drums and sheaves be for various types of wire
    rope? 5
Find advantages of block and tackle, taking into account pull out
    friction 6
Safe loads for wire rope 6
Stress in guy wires 7
Strength and weight of popular wire rope 9
Measuring the diameter of wire rope 9
Wire rope: field troubles and their causes 9
Capacity of drums 11

Belts and Shafts 11
Determine length of a V-belt 11
Calculate stress in shaft key 12
Calculate V-belt length using simple equation 12
Estimate the horsepower that can be transmitted by a shaft 13

Miscellaneous 13
How to estimate length of material contained in roll 13
Convenient anti-freeze chart for winterizing cooling systems 13
How to determine glycol requirements to bring a system to a desired
    temperature protection level 14
Weight in pounds of round steel shafting 14
Properties of shafting 15
Tap drills and clearance drills for machine screws 16
Common nails 17
Drill sizes for pipe taps 17
Carbon steel—color and approximate temperature 17
Bolting dimensions for raised face weld neck flanges 18
Steel fitting dimensions 19
ANSI forged steel flanges 20
Color codes for locating underground utilities 21
Approximate angle of repose for sloping sides of excavations 21
Trench shoring—minimum requirements 22
Reuniting separated mercury in thermometers 23
Typical wire resistance 23
How to cut odd-angle long radius elbows 24
How to read land descriptions 25
Size an air receiver for engine starting 28

Basic Formulas

1. Rate of Return Formulas:

$$S = P(1 + i)^n$$

a. Single payment compound amount, SPCA. The $(1+i)^n$ factor is referred to as the compound amount of $1.00.

b. Single payment present worth, SPPW.

$$P = S\left[\frac{1}{(1+i)^n}\right]$$

The factor $\left[\dfrac{1}{(1+i)^n}\right]$ is referred to as the present worth of $1.00

c. Uniform series compound amount, USCA:

$$S = R\left[\frac{(1+i)^n - 1}{i}\right]$$

The factor $\left[\dfrac{(1+i)^n - 1}{i}\right]$ is referred to as the compound amount of $1.00 per period.

d. Sinking fund deposit, SFD:

$$R = S\left[\frac{i}{(1+i)^n - 1}\right]$$

The factor $\left[\dfrac{i}{(1+i)^n - 1}\right]$ is referred to as the uniform series, which amounts to $1.00.

e. Capital recovery, CR:

$$R = S\left[\frac{i}{(1+i)^n - 1}\right] = P\left[\frac{i(1+i)^n}{(1+i)^n - 1}\right]$$

The factor $\left[\dfrac{i(1+i)^n}{(1+i)^n - 1}\right]$ is referred to as the uniform series that $1.00 will purchase.

f. Uniform series present worth, USPW:

$$P = R\left[\frac{(1+i)^n - 1}{i(1+i)^n}\right]$$

The factor $\left[\dfrac{(1+i)^n - 1}{i(1+i)^n}\right]$ is referred to as the present worth of $1.00 per period.

where

P = a present sum of money
S = a sum of money at a specified future date
R = a uniform series of equal end-of-period payments
n = designates the number of interest periods
i = the interest rate earned at the end of each period

Mathematics—areas

Form		Method of Finding Area
Triangle		Base x 1/2 perpendicular height $\sqrt{s(s-a)(s-b(s-c)}$ s = 1/2 sum of the three sides a,b,c
Trapezium		Sum of area of the two triangles
Trapezoid		1/2 sum of parallel sides x perpendicular height
Parallelogram		Base x perpendicular height
Reg. Polygon		1/2 sum of sides x inside radius
Circle		πr^2 = 0.78540 x diam^2. = 0.07958 x circumference^2
Sector of a Circle		$\dfrac{\pi r^2 A°}{360}$ = 0.0087266r^{2}A° = arc x 1/2 radius
Segment of a circle		$r^2 \left(\dfrac{\pi A°}{180} - \sin A° \right)$
Circle of same area as a square		Diameter = side x 1.12838
Square of same area as a circle		Side = diameter x 0.88623
Ellipse		Long diameter x short diameter x 0.78540
Parabola		Base x 2/3 perpendicular height
Irregular Plane Surface		The larger the value of n, the greater the accuracy of approximation (Simpson's Rule). $A = \dfrac{h}{3}\left[(y_o + y_n) + 4(y_1 + y_3 + ... + y_{n-1}) + 2(y_2 + y_4 + ... y_{n-2})\right]$

Mathematics—surfaces and volumes

Method of Finding Surfaces and Volumes of Solids
S = lateral or convex surface   V = Volume

Parallelopiped

S = perimeter P perp. to sides x lat. length, l

V = area of base x perpendicular height, h

V = area of section A perp. to sides x lat. length, l

Prism-right, oblique, regular or irregular

S = perimeter P perp. to sides x lat length l

V = area of base x perp. height, h

V = area of section A perp. to sides x lat. length, l

Cylinder-right, oblique, circular or elliptic, etc.

S = perimeter of base x perp. height

V = area of base x perp. height

V = area of section A perp. to sides x lat. length, l

Frustum of any prism or cylinder

V = area of base x perpendicular distance h, from base to center of gravity of opposite face. For cylinder, 1/2A(l₁+l₂)

Pyramid or Cone, right or regular

S = perimeter of base x 1/2 slant height, l

V = area of base x 1/3 perp. height, h

Pyramid or Cone-right, oblique, regular or irregular

V = area of base x 1/3 perp. height, h

V = 1/3 vol. of prism or cylinder of same base and perp height

V = 1/2 vol. of hemisphere of same base and perp. height

Frustum of pyramid or cone, right and regular, parallel ends

S = (sum of perimeter of base and top) x 1/2 slant height, l

$$V = \frac{h}{3}\left(A_1 + A_2 + \sqrt{A_1 x A_2}\right)$$

Where A_1 and A_2 are the areas of the bases

Frustum of any pyramid or cone, parallel ends

$$V = \frac{h}{3}\left(A_1 + A_2 + \sqrt{A_1 x A_2}\right)$$

Where A_1 and A_2 are the areas of the bases

Sphere

S = $4\pi r^2$

V = $4/3\pi r^3$

Spherical Sector

V = $2/3\pi r^2 b$

S = $1/2\pi r(4b+c)$

Spherical Zone

S = $2\pi r b$

V = $1/24\pi b(3a^2+3c^2+4b^2)$

CABLES AND ROPES

Estimating strength of cable

Rule.

1. Change line diameter to eighths
2. Square the numerator
3. Divide by the denominator
4. Read the answer in tons

$$\text{Diameter} = \frac{1}{2} = \frac{4}{8}$$

$$\frac{4^2}{8} = \frac{16}{8} = 2$$

Example. Estimate the strength of ½-inch steel cable:

The approximate strength of ½-inch steel cable is 2 tons.

Find the working strength of Manila rope

The working strength of Manila rope is approximately $900 \times (\text{diameter})^2$:

$$W = 900\, d^2$$

where d is expressed in inches.

W is given in pounds.

Example. What is the working strength of a ¾-inch Manila rope?
The maximum recommended pull is:

$$W = 900 \times \frac{3 \times 3}{4 \times 4} = 506 \text{ lb}$$

Example. Find the maximum working pull for a 1½-inch Manila rope.

$$W = 900 \times \frac{3}{2} \times \frac{3}{2} = 2{,}025 \text{ pounds.}$$

For rope diameters greater than 2 inches, a factor lower than 900 should be used. In working with heavier rigging it is well to refer to accepted handbooks to find safe working strengths.

How large should drums and sheaves be for various types of wire rope?

The diameter of sheaves or drums should preferably fall within the table* given below for most efficient utilization of the wire rope.

Example. What size should the hoisting drum on a dragline be, if the wire rope is 6 × 19 construction, ¾ inch in diameter?

From the table, good practice calls for 30 diameters which in this instance would be 22½ inches. Loads, speeds, bends, and service conditions will also affect the life of wire rope so it is better to stay somewhere between the "good practice" and "best wear" factors in the table.

Type of Wire Rope	6 × 19	6 × 37	8 × 19	5 × 28	6 × 25	18 × 7	6 × 7
For best wear....	45	27	31	36	45	51	72
Good practice....	30	18	21	24	30	34	42
Critical.........	16	14	14	16	16	18	28

* *Construction Methods and Machinery*, by F. H. Kellogg, Prentice-Hall, Inc., 1954.

Find advantages of block and tackle, taking into account pull out friction

The efficiency of various sheaves differs. For one with roller bearings the efficiency has been estimated at 96.2%. For plain bearing sheaves a commonly used figure is 91.7%. The following formula will give close results:

$$MA = \frac{W}{w} = E\frac{1 - E^n}{1 - E}$$

where MA = Mechanical Advantage
W = Total weight to be lifted by the assembly
w = Maximum line pull at the hoist
n = Number of working parts in the tackle
E = Efficiency of individual sheaves

It is assumed that the line leaving the upper block goes directly to the hoist without additional direction change (requiring a snatch block).

Example. Find the Mechanical Advantage of a 4-part block and tackle using upper and lower blocks having journal bearings, which have an efficiency of 91.7%.

$$MA = .917\frac{1 - .917^4}{1 - .917} = .917$$

$$\left(\frac{1 - .707}{1 - .917}\right) = .917\left(\frac{.293}{.083}\right)$$

$$MA = 3.25$$

If the load weighed 3,250 pounds, what pull would be required on the lead line?

$$\frac{W}{w} = MA$$

$$\frac{3,250}{w} = 3.25$$

w = 1,000 pounds.

Safe loads for wire rope

Safe Loading Factors For Wire Rope And Chains

Dia. In.	SINGLE PART LINE			SINGLE PART-CHOKER HITCH			TWO PARTS (60°)			TWO PARTS (30°)		
	Wire Rope (Tons)	Alloy Chain (Tons)	Wrought Chain (Tons)	Wire Rope (Tons)	Alloy Chain (Tons)	Wrought Chain (Tons)	Wire Rope (Tons)	Alloy Chain (Tons)	Wrought Chain (Tons)	Wire Rope (Tons)	Alloy Chain (Tons)	Wrought Chain (Tons)
1/4	.5	....	....	.35	....	....	.6	....	....	.35	....	....
9/32	.5	1.3	....	....	1	....	....	1.8	....	....	1	....
3/8	1	2.3	.8	.75	1.8	.63	1.4	3.1	1.1	.8	1.8	.64
7/16	1	3	1.1	....	2.4	.9	....	4.2	1.5	....	2.4	.9
1/2	1.8	4	1.4	1.4	3.2	1.1	2.4	5.6	1.9	1.4	3.2	1.1
9/16	1.8	5	....	....	4	....	....	7	....	....	4	....
5/8	2.8	6	2.2	2.1	4.8	1.8	3.7	8.2	3	2.2	4.8	1.8
3/4	3.9	7.8	3.2	2.8	6.2	2.6	5.2	10.7	4.4	3	6.2	2.6
7/8	5.2	10.4	4.3	3.7	8.3	3.4	7	14.4	6	4	8.3	3.4
1	6.6	13.8	5.7	4.6	10.8	4.6	9	18.7	7.8	5.2	10.8	4.6
1 1/8	6.6	17.8	6.7	....	14.2	5.4	....	24.7	9.3	....	14.2	5.4
1 1/4	9.7	23.	8.3	6.6	18.2	6.6	13	31.9	11.5	7.5	18.4	6.6
1 3/8	9.7	....	10	....	....	8	....	....	13.9	....	....	8
1 1/2	13.2	....	12	9.2	....	9.6	17.3	....	16.6	10	....	9.6
1 5/8	13.2	....	14	....	....	11.2	....	....	19.4	....	....	11.2
1 3/4	17	....	16	12.1	....	12.8	21.5	....	22.2	12.4	....	12.8
2	20.6	....	21	15.4	....	16.8	26.7	....	29.1	15.4	....	16.8

There are various kinds of wire rope and chain available, and where it is possible strength factors should be reviewed before using any particular kind. Tables here may be used on construction jobs. They allow for a safety factor of four as based on mild plow steels, etc. Where the wire rope is made of improved plow steel, etc., it will have a greater safety factor.

Stress in guy wires

Guys are wire ropes or strands used to hold a vertical structure in position against an overturning force. The most common types of guyed structures are stacks, derricks, masts for draglines, reversible tramways and radio transmission towers.

As a general rule, stresses in guys from temperature changes are neglected, but in structures such as radio masts this is an important feature and must be subject to special analysis.

The number of guys used for any particular installation is contingent on several variable factors such as type of structure, space available, contour of the ground, etc., and is not a part of this discussion.

It is desirable to space guys uniformly whenever possible. This equalizes the pull, P, on each guy insofar as possible, particularly against forces which change in direction, as when a derrick boom swings in its circle.

It is also desirable to equalize the erection tensions on the guys. When no external force is acting on the structure, the tension in each guy should be the same. A "Tension Indicator" is sometimes used to determine the tension in guys. If this instrument is not available, the tension can be very closely approximated by measuring the deflection at the center of the span from the chord drawn from the guy anchorage to the point of support on the structure. A good average figure to use for erection tension of guys is 20% of the maximum working tension of the guy.

This discussion outlines the method for determining the stresses in guys. One of the first considerations is the location of the guy anchorages. The anchorages should be so located that the angle α, between the horizontal plane and the guy line, is the same for all guys (to equalize erection tensions). Angle α, in good practice, seldom exceeds 45 degrees with 30 degrees being commonly used. The tension in the guys decreases as angle α becomes less. The direct load on the structure is also less with smaller values of α.

To find the maximum extra tension, T, that will be applied to any single guy by the force, F; first, determine the pull, P, which is the amount required along the guys, in the same vertical plane as the force to resist the horizontal component of the force. This pull is entirely independent of the number of guys. Assume that the following are known:

Figure 1

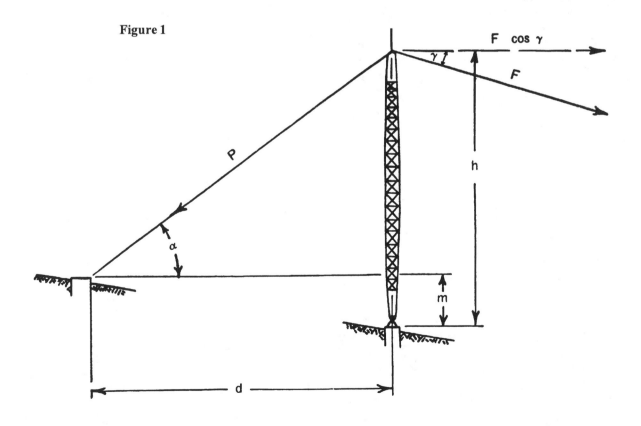

F =   The total resultant external force acting on the structure

G =   The angle between the horizonatal plane and the force F

h =   The height of the structure

d =   The horizontal distance from structure to guy anchorage

m =   The vertical height of anchorage above or below the base of the structure

The horizontal component of the force, $F, = F \cos \gamma$.

α =   The angle whose tangent is $(h \pm m) \div d$

m is plus if the anchorage is below the base of the structure and subtracted if it is above.

$$P = F \cos \gamma \div \cos \alpha$$

As $\cos \alpha$ is always less than one, P is always greater than $F \cos \gamma$, the horizontal component of force F.

It must be remembered that P represents the total pull acting along the guys at an angle, α, with the horizontal, and in the same vertical plane as the force, F.

If only one guy were used, P would represent the extra tension, T. In practice, however, a number of guys are always used and, therefore, the pull on any one guy will not be equal to P. The following table gives factors for any number of guys from 3 to 15, equally spaced about a central structure. To find the maximum exta tension, T, that will be applied to any single guy by the force, F, capable of rotating 360 degrees around a vertical axis, it is only necessary to multiply the value of P, as determined above, by the factor for the number of guys used. It must be clearly understood in using this table that the guys are uniformly spaced and under equal tension when no load is acting on the structure.

Table 1

No. of Guys	Factors*	No. of Guys	Factors*
3	1.15	10	0.45
4	1.00	11	0.40
5	0.90	12	0.37
6	0.75	13	0.35
7	0.65	14	0.32
8	0.55	15	0.30
9	0.50		

*These factors are for average conditions. If the guys are erected under accurately measured tensions of not less than 20% of the working load, the factors for five or more guys may be reduced by 10%. If the erecting tensions are low or not accurately equalized, the factors for 5 or more guys should be increased 10%.

Example–A derrick mast 90 ft high is supported by nine equally spaced guys anchored at a horizontal distance of 170 ft from the mast and the elevations of the guy anchorages are 10 ft below the base of the mast. The load on the structure is equivalent to a force of 10,000 lbs. acting on an angle of 10 degrees below the horizontal. What is the maximum pull on any single cable?

From Figure 1–

h =	90 ft
d =	170 ft
m =	10 ft
g =	10°00'
F =	10,000 lbs

$$\tan \alpha = \frac{90 + 10}{170} = \frac{100}{170} = 0.588$$

$$\alpha = 30°28'$$

$$P = \frac{F \cos \gamma}{\cos \alpha} = \frac{10,000 \times 0.985}{0.862} = 11,427 \, lbs$$

From Table 1, T = 11,427 x 0.50 = 5,714 lbs

If erection tension is 10 percent of total working tension, 5,714 is 90 percent of total working tension. Therefore, working tension = $\frac{5,714 \times 100}{90}$ = 6,349 lbs

Strength and weight of popular wire rope

The following tables give the breaking strength for wire rope of popular construction made of improved plow steel.

| SIZE | 6 x 19 | |
	Breaking Strength	Weight
¼	5,480	0.10
5/16	8,520	0.16
⅜	12,200	0.23
7/16	16,540	0.31
½	21,400	0.40
9/16	27,000	0.51
⅝	33,400	0.63
¾	47,600	0.90
⅞	64,400	1.23
1	83,600	1.60
1⅛	105,200	2.03
1¼	129,200	2.50
1⅜	155,400	3.03
1½	184,000	3.60
1⅝	214,000	4.23
1¾	248,000	4.90

| SIZE | 6 x 19 | |
	Breaking Strength	Weight
1⅞	282,000	5.63
2	320,000	6.40

Conversion Factors For Wire Rope of Other Construction

To apply the above table to wire rope of other construction, multiply by the following factors:

Wire Rope Construction	6 x 19	6 x 29	6 x 37	18 x 7
Strength Factors	1.00	0.96	0.95	0.92
Weight Factors	1.00	0.97	0.97	1.08

Example. Find the breaking strength of 6×29 improved plow steel wire rope two inches in diameter.

Strength = $320,000 \times 0.96 = 307,000$ lb

The weight can be found the same way.

Measuring the diameter of wire rope

Correct way

Incorrect way

Wire rope: field troubles and their causes

All wire rope will eventually deteriorate in operation or have to be removed simply by virtue of the loads and reversals of load applied in normal service. There are, however, many conditions of service or inadvertent abuse which will materially shorten the normal life of a wire rope of proper construction although it is properly applied. The following field troubles and their causes give some of the field conditions and practices which result in the premature replacement of wire rope. It should be borne in mind that in all cases the contributory cause of removal may be one or more of these practices or conditions.

Wire-Rope Trouble	Cause
a. Rope broken (all strands).	Overload resulting from severe impact, kinking, damage, localized wear, weakening of one or more strands, or rust-bound condition and loss of elasticity.
b. One or more whole strands parted.	Overloading, kinking, divider interference, localized wear, or rust-bound condition. Fatigue, excessive speed, slipping, or running too loosely. Concentration of vibration at dead sheave or dead-end anchor.
c. Excessive corrosion.	Lack of lubrication. Exposure to salt spray, corrosive gases, alkaline water, acid water, mud, or dirt. Period of inactivity without adequate protection.
d. Rope damage in hauling to the well or location.	Rolling reel over obstructions or dropping from car, truck, or platform. The use of chains for lashing, or the use of lever against rope instead of flange. Nailing through rope to flange.
e. Damage by improper socketing.	Improper seizing which allows slack from one or more strands to work back into rope; improper method of socketing or poor workmanship in socketing, frequently shown by rope being untwisted at socket, loose or drawn.
f. Kinks, dog legs, and other distorted places.	Kinking the rope and pulling out the loops such as in improper coiling or unreeling. Improper winding on the drum. Improper tie-down. Open-drum reels having longitudinal spokes too widely spaced. Divider interference. The addition of improperly spaced cleats to increase the drum diameter. Stressing while rope is over small sheave or obstacle.
g. Damage by hooking back slack too tightly to girt.	Operation of walking beam causing a bending action on wires at clamp and resulting in fatigue and cracking of wires, frequently before rope goes down into hole.
h. Damage or failure on a fishing job.	Rope improperly used on a fishing job, resulting in damage or failure as a result of the nature of the work.
i. Lengthening of lay and reduction of diameter.	Frequently produced by some type of overloading, such as an overload resulting in a collapse of the fiber core in swabbing lines. This may also occur in cable-tool lines as a result of concentrated pulsating or surging forces which may contribute to fiber-core collapse.
j. Premature breakage of wires.	Caused by frictional heat developed by pressure and slippage, regardless of drilling depth.
k. Excessive wear in spots.	Kinks or bends in rope due to improper handling during installation or service. Divider interference; also, wear against casing or hard shells or abrasive formations in a crooked hole. Too infrequent cut-offs on working end.
l. Spliced rope.	A splice is never as good as a continuous piece of rope, and slack is liable to work back and cause irregular wear.
m. Abrasion and broken wires in a straight line. Drawn or loosened strands. Rapid fatigue breaks.	Injury due to slipping rope through clamps.
n. Reduction in tensile strength or damage to rope.	Excessive heat due to careless exposure to fire or torch.
o. Distortion of wire rope.	Damage due to improperly attached clamps or wire-rope clips.
p. High strands.	Slipping through clamps, improper seizing, improper socketing or splicing, kinks, dog legs, and core popping.
q. Wear by abrasion.	Lack of lubrication. Slipping clamp unduly. Sandy or gritty working conditions. Rubbing against stationary object or abrasive surface. Faulty alignment. Undersized grooves and sheaves.
r. Fatigue breaks in wire.	Excessive vibration due to poor drilling conditions, i.e., high speed, rope slipping, concentration of vibration at dead sheave or dead-end anchor, undersized grooves and sheaves, and improper selection of rope construction. Prolonged bending action over spudder sheaves, such as that due to hard drilling.

s. Spiraling or curling.

Allowing rope to drag or rub over pipe, sill, or any object during installation or operation. It is recommended that a block with sheave diameter 16 times the nominal wire-rope diameter, or larger, be used during installation of the line.

t. Excessive flattening or crushing.

Heavy overload, loose winding on drum, or cross winding. Too infrequent cut-offs on working end of cable-tool lines. Improper cutoff and moving program for cable-tool lines.

u. Bird-caging or core-popping.

Sudden unloading of line such as hitting fluid with excessive speed. Improper drilling motion or jar action. Use of sheaves of too small diameter or passing line around sharp bend.

v. Whipping off of rope.

Running too loose.

w. Cutting in on drum.

Loose winding on drum. Improper cutoff and moving program for rotary drilling lines. Improper or worn drum grooving or line turn-back plate.

Capacity of Drums

The capacity of wire line drums may be figured from the following formula:

$$M = (A + B) \times A \times C \times K$$

where (see below) M = rope capacity of drum, in
A = depth of flange, in.
B = diameter of drum, in.
C = width of the drum between flanges, in.
K = constant depending on rope size shown below

Rope Size	K	Rope Size	K
3/8	1.86	1	.262
7/16	1.37	1 1/8	.207
1/2	1.05	1 1/4	.167
9/16	.828	1 3/8	.138
5/8	.672	1 1/2	.116
3/4	.465	1 5/8	.099
7/8	.342	1 3/4	.085

Figure 1. Capacity of wire line drums.

BELTS AND SHAFTS

Determine length of a V-belt

Rule: To find the nominal length of a V-belt, lay the belt on a table, and place within it a pair of circular objects of the same diameter (flanges, tin cans, or whatever may be at hand). Pull them apart until the belt is fully extended without stretching. Then measure the shortest distance between the two circles, in inches. The belt size is then twice this figure, plus 5.14 times the diameter of the circles. This value, for any standard belt, should be a whole number of inches, which is the belt size.

V-belts are made in four standard sections, classified as A, B, C, and D; the widths (at the widest part) are one-half, five-eighths, seven-eighths and 1 1/8 inches respectively. The complete designation of the belt is the letter showing the width, followed by the length in inches; thus, an A26 belt is one-half inch wide, and 26 inches long on the inside edge. The *pitch length* of the belt is measured along a median section, and corresponds to the length which runs on the pulley diameter which determines the actual speed ratio—about half of the depth of the groove. Pitch lengths for A, B, C, and D belts are greater than their nominal lengths by 1.3, 1.8, 2.9, and 3.3 inches, respectively.

Calculate Stress in Shaft Key

The shear and compressive stresses in a key are calculated using the following equations:

$$S_s = \frac{2T}{d \times W \times L} \qquad S_c = \frac{2T}{d \times h_1 \times L}$$

S_s = Shear stress in psi

S_c = Compressive stress in psi

T = Shaft torque lb-in. or $\dfrac{HP \times 63000}{RPM}$

d = shaft diameter-inches
(For taper shafts, use average diameter)

w = width key-inches
L = effective length of key-inches
h_1 = height of key in the shaft or hub that bears against the keyway—inches
$h_1 = h_2$ for square keys. For designs where unequal portions of the key are in the hub or shaft, h1 is the minimum portion.

Key material is usually AISI 1018 or AISI 1045 with the following allowable stresses:

Material	Heat Treatment	Allowable Stresses -- psi	
		Shear	Compressive
AISI 1018	None	7,500	15,000
AISI 1045	255-300 Bhn	15,000	30,000

Example: Determine key stresses for these conditions..... 300 hp @ 600 RPM; 3" dia. shaft, 3/4 x 3/4 key, 4" key engagement length.

$$T = \text{Torque} = \frac{HP \times 63000}{RPM} = \frac{300 \times 63,000}{600} = 31,500 \text{ in.-lbs}$$

$$S_s = \frac{2T}{d \times w \times L} = \frac{2 \times 31,500}{3 \times 3/4 \times 4} = 7,000 \text{ psi}$$

$$S_c = \frac{2T}{d \times h_1 \times L} = \frac{2 \times 31,500}{3 \times 3/8 \times 4} = 14,000 \text{ psi}$$

The AISI 1018 key should be used since it is within allowable stresses. NOTE: If shaft had been 2 3/4" dia. (4" long hub), the key would be 5/8 x 5/8 and S_s = 9,200 psi, S_c = 18,400 psi, and a heat treated key -- AISI 1045 would have been required.

Reprinted with permission: The Falk Corporation

Calculate V-belt length using simple equation

Rule.

L = 1.57 (D + d) + 2C

where L = Belt length, in.
        D = Diameter of larger sheave, in.
        d = Diameter of smaller sheave, in.
        C = Distance between sheave centers, in.

Estimate the horsepower that can be transmitted by a shaft

1. Where there are no stresses due to bending, weight of the shaft, pulleys, gears, or sprockets, use:

$$HP = \frac{D^3 N}{50}$$

where D = diameter of shaft, in.
N = revolutions per minute

2. For heavy duty service use:

$$HP = \frac{D^3 N}{125}$$

Example. What horsepower can be transmitted to an atmospheric cooling coil by a two-inch shaft turning at 1,800 revolutions per minute?

$$HP = \frac{(2)^3 \, (1,800)}{125}$$

$$= 115.2 \text{ horsepower}$$

MISCELLANEOUS

How to estimate length of material contained in roll

Where material of uniform thickness, like belting, is in a roll, the total length may be obtained by the following rule:

Measure the diameter of the hole in the center, and of the outside of the roll, both measurements in inches; count the number of turns; multiply the sum of the two measured diameters by the number of turns, and multiply this product by 0.13; the result is the total length of the material in feet.

Example. A roll of belting contains 24 turns. The diameter of the hole is two inches, and of the outside of the roll is 13 inches.

$(2 + 13) \times 24 \times 0.13 = 46.8$

The roll contains 46.8 feet of belting.

Note: The rule can even be applied to materials as thin as pipe line felt; counting the turns is not as difficult as might appear without a trial.

Convenient anti-freeze chart for winterizing cooling systems

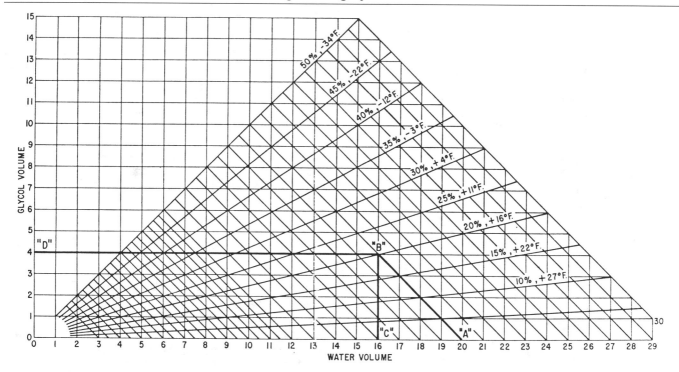

This plot of water volume versus glycol volume at various conditions of temperature and percent of glycol in the system makes winterizing field engines relatively easy.

Example. Determine the amount of glycol to be used in a 200 gallon system for protection to 16°F.

To use the chart, first find the total system capacity, 200 gallons, at Point A on the water volume axis. Point A also can represent 2, 20, 2,000, etc. Proceed along the 45° line from Point A to Point B at the intersection with the 16°F line. Then, move from Point B to Point C and read 160 on the water volume axis. This is the amount of water the system should contain. Move from Point B to Point D on the glycol axis and read 40. This is the amount of glycol that must be added to the system.

How to determine glycol requirements to bring a system to a desired temperature protection level

Solve the equation below to find the number of gallons of aqueous solution that must be removed and replaced with glycol.

$$d = \frac{D}{B}$$

where d = gallons of aqueous solution to be removed from the system and replaced with equal number of gallons of glycol

D = total gallons of glycol required for protection at desired temperature less gallons of glycol in original system

B = 1.0 − fraction of glycol in system

Example. Find the number of gallons of existing aqueous solution that must be removed and replaced with glycol in a 1,200 gallon system to give protection to − 10°F. Glycol in the system determined by Hydrometer test is 28 percent (0.28 fraction). The published figure for − 10°F protection is 519 gallons of glycol in the 1,200 gallon system.

Then: Gallons of glycol in system =
1,200 × 0.28 = 336 gallons

D = 519 − 336 = 183 gallons difference
and B = 1.0 − 0.28 = 0.72

Thus: $d = \frac{183}{0.72} = 254$ gallons

Therefore 254 gallons of existing aqueous solution need to be removed and replaced with 254 gallons of glycol for the desired protection to − 10°F.

Weight in pounds of round steel shafting

Diameter of Shaft, Inches	LENGTH IN FEET																		Weight Per Inch
	1	2	3	4	5	6	7	8	9	10	12	14	16	18	20	22	24	
15/16	2	5	7	9	12	14	16	19	21	23	28	33	37	42	47	51	56	.195
13/16	4	8	11	15	19	23	26	30	34	38	45	53	60	68	76	83	91	.314
17/16	6	11	17	22	28	33	39	44	50	55	66	77	88	99	110	121	133	.460
111/16	8	15	23	30	38	46	53	61	68	76	91	107	122	137	152	167	183	.634
115/16	10	20	30	40	50	60	70	80	90	100	120	140	161	181	201	221	241	.835
23/16	13	26	38	51	64	77	90	102	115	128	153	179	205	230	256	281	307	1.06
27/16	16	32	48	63	79	95	111	127	143	159	190	222	254	286	317	349	381	1.32
211/16	19	39	58	77	97	116	135	154	174	193	232	270	309	348	386	425	463	1.61
215/16	23	46	69	92	115	138	161	184	208	231	277	323	369	415	461	507	553	1.92
33/16	27	54	81	109	136	163	190	217	244	272	326	380	435	489	543	598	652	2.26
37/16	32	63	95	126	158	189	221	253	284	316	379	442	505	568	631	695	758	2.63
311/16	36	73	109	146	182	218	255	291	328	364	437	510	582	655	728	801	874	3.03
315/16	41	83	124	166	207	248	290	331	373	414	497	580	662	745	828	911	994	3.45
47/16	53	105	158	210	263	315	368	421	473	526	631	736	841	946	1052	1157	1262	4.38
415/16	65	130	195	260	326	391	456	521	586	651	781	911	1041	1172	1302	1432	1562	5.43
5	67	134	200	267	334	401	467	534	601	668	801	935	1068	1202	1335	1469	1602	5.56
57/16	79	158	237	316	395	474	553	632	711	790	947	1105	1263	1421	1579	1737	1894	6.58
51/2	81	162	242	323	404	485	565	646	727	808	969	1131	1292	1454	1615	1777	1938	6.73
515/16	94	188	282	377	471	565	659	753	847	941	1130	1318	1506	1695	1883	2071	2259	7.85
6	96	192	288	385	481	577	673	769	865	961	1154	1346	1538	1731	1923	2115	2307	8.01
61/2	113	226	338	451	564	677	790	903	1015	1128	1354	1579	1805	2031	2256	2482	2708	9.40
7	131	262	393	524	654	785	916	1047	1178	1309	1570	1832	2094	2355	2677	2878	3140	10.90
71/2	150	300	451	601	751	901	1051	1202	1352	1502	1803	2103	2403	2704	3004	3305	3605	12.52
8	171	342	513	684	855	1024	1196	1367	1538	1709	2051	2303	2734	3076	3418	3760	4102	14.24

Properties of shafting

Diameter of Shaft, Inches	Area of Shaft	Weight per Inch	Weight per Foot	SECTION MODULUS		MOMENT OF INERTIA		Radius of Gyration
				Bending	Torsion	Bending	Torsion	
15/16	.6903	.195	2.34	.0809	.1618	.0379	.0758	.2343
13/16	1.1075	.314	3.76	.1644	.3288	.0976	.1952	.2969
17/16	1.6230	.460	5.52	.2916	.5832	.2096	.4192	.3594
111/16	2.2365	.634	7.61	.4718	.9435	.3981	.7961	.4219
115/16	2.9483	.835	10.03	.7140	1.4281	.6971	1.3835	.4844
23/16	3.7583	1.06	12.78	1.0276	2.0553	1.1240	2.2480	.5469
27/16	4.6664	1.32	15.86	1.4218	2.8436	1.7328	3.4656	.6094
211/16	5.6727	1.61	19.29	1.9057	3.8113	2.5607	5.1215	.6719
215/16	6.7771	1.92	23.04	2.4885	4.9770	3.6549	7.3099	.7344
33/16	7.9798	2.26	27.12	3.1794	6.3589	5.0672	10.1345	.7969
37/16	9.2806	2.63	31.56	3.9878	7.9755	6.8539	13.7079	.8594
311/16	10.680	3.03	36.31	4.9226	9.8452	9.0761	18.1521	.9219
315/16	12.177	3.45	41.40	5.9932	11.9865	11.7992	23.5984	.9844
47/16	15.466	4.38	52.58	8.5786	17.1571	19.0337	38.0674	1.1094
415/16	19.147	5.43	65.10	11.8174	23.6348	29.1742	58.3483	1.2344
5	19.635	5.56	66.76	12.2718	24.5437	30.6796	61.3592	1.2500
57/16	23.221	6.58	78.96	15.7833	31.5666	42.9108	85.8217	1.3594
51/2	23.758	6.73	80.78	16.3338	32.6677	44.9180	89.8361	1.3750
515/16	27.688	7.85	94.14	20.5499	41.0999	61.0077	122.0153	1.4844
6	28.274	8.01	96.12	21.2058	42.4115	63.6172	127.2345	1.5000
61/2	33.183	9.40	112.82	26.9612	53.9225	87.6240	175.2481	1.6250
7	38.485	10.90	130.84	33.6739	67.3479	117.8588	235.7176	1.7500
71/2	44.179	12.52	150.21	41.4175	82.8350	155.3155	310.6311	1.8750
8	50.265	14.24	170.90	50.2655	100.5310	201.0619	402.1239	2.0000

Steel shaft sizes calculated for strength

$$\text{Base Formula: } d = \sqrt[3]{\frac{C.(H.P.)}{N}}$$

Base Fiber Stress = 60,000 lb. sq in.

Diameter of Shaft, Inches		H.P. / N		
		N		
		C = 50 Very Little Bending Fac. Saf. 9.4	C = 75 Average Conditions of Bending Fac. Saf. 14.1	C = 100 Very Severe Conditions of Bending Fac. Saf. 18.8
Fractional	Decimal			
1/16	0.0625	0.000004883	0.000003256	0.000002492
1/8	0.125	0.00003905	0.00002604	0.00001953
3/16	0.1875	0.0001318	0.00008788	0.00006592
1/4	0.25	0.0003124	0.0002083	0.0001563
5/16	0.3125	0.0006105	0.0004070	0.0003053
3/8	0.375	0.001054	0.0007030	0.0005272
7/16	0.4375	0.001675	0.001116	0.0008375
1/2	0.5	0.002500	0.001666	0.001250
9/16	0.5625	0.003558	0.002372	0.001779
5/8	0.625	0.004883	0.003256	0.002442

Diameter of Shaft, Inches		H.P. / N		
		N		
		C = 50 Very Little Bending Fac. Saf. 9.4	C = 75 Average Conditions of Bending Fac. Saf. 14.1	C = 100 Very Severe Conditions of Bending Fac. Saf. 18.8
Fractional	Decimal			
11/16	0.6875	0.006500	0.004333	0.003250
3/4	0.75	0.008439	0.005626	0.004220
13/16	0.8125	0.01074	0.007155	0.005367
7/8	0.875	0.01340	0.008932	0.006699
15/16	0.9375	0.01647	0.01098	0.008235
1	1.	0.02000	0.01333	0.01000
11/16	1.0625	0.02396	0.01598	0.01199
11/8	1.125	0.02846	0.01898	0.01423
13/16	1.1875	0.03353	0.02236	0.01677
11/4	1.25	0.03905	0.02604	0.01953

(table continued on next page)

Diameter of Shaft, Inches		H.P. / N			Diameter of Shaft, Inches		H.P. / N		
Fractional	Decimal	C = 50 Very Little Bending Fac. Saf. 9.4	C = 75 Average Conditions of Bending Fac. Saf. 14.1	C = 100 Very Severe Conditions of Bending Fac. Saf. 18.8	Fractional	Decimal	C = 50 Very Little Bending Fac. Saf. 9.4	C = 75 Average Conditions of Bending Fac. Saf. 14.1	C = 100 Very Severe Conditions of Bending Fac. Saf. 18.8
1 5/16	1.3125	0.04519	0.03019	0.02259	3 3/16	3.1875	0.6479	0.4319	0.3240
1 3/8	1.375	0.5199	0.03465	0.02599	3 1/4	3.25	0.6866	0.4577	0.3434
1 7/16	1.4375	0.05940	0.03960	0.02970	3 5/16	3.3125	0.7266	0.4844	0.3634
1 1/2	1.5	0.06750	0.04500	0.03375	3 3/8	3.375	0.7685	0.5123	0.3842
1 9/16	1.5625	0.07628	0.05086	0.03815	3 7/16	3.4375	0.8127	0.5418	0.4061
1 5/8	1.625	0.08578	0.05719	0.04289	3 1/2	3.5	0.8567	0.5718	0.4288
1 11/16	1.6875	0.09524	0.06409	0.04807	3 9/16	3.5625	0.9038	0.6026	0.4520
1 3/4	1.75	0.1072	0.07144	0.05358	3 5/8	3.625	0.9526	0.6351	0.4763
1 13/16	1.8125	0.1191	0.07939	0.05955	3 11/16	3.6875	1.003	0.6688	0.5017
1 7/8	1.875	0.1318	0.08788	0.06592	3 3/4	3.75	1.054	0.7030	0.5272
1 15/16	1.9375	0.1456	0.09707	0.07281	3 13/16	3.8125	1.108	0.7382	0.5538
2	2.	0.1600	0.1066	0.07998	3 7/8	3.875	1.163	0.7759	0.5820
2 1/16	2.0625	0.1755	0.1169	0.08774	3 15/16	3.9375	1.222	0.8143	0.6108
2 1/8	2.125	0.1918	0.1278	0.09592	4	4.	1.280	0.8535	0.6401
2 3/16	2.1875	0.2095	0.1397	0.1048	4 1/16	4.0625	1.340	0.8933	0.6701
2 1/4	2.25	0.2278	0.1519	0.1140	4 1/8	4.125	1.404	0.9356	0.7018
2 5/16	2.3125	0.2474	0.1649	0.1237	4 3/16	4.1875	1.469	0.9792	0.7345
2 3/8	2.375	0.2678	0.1785	0.1339	4 1/4	4.25	1.536	1.023	0.7678
2 7/16	2.4375	0.2897	0.1932	0.1449	4 5/16	4.3125	1.603	1.069	0.8019
2 1/2	2.5	0.3124	0.2083	0.1563	4 3/8	4.375	1.675	1.116	0.8375
2 9/16	2.5625	0.3363	0.2243	0.1682	4 7/16	4.4375	1.748	1.165	0.8742
2 5/8	2.625	0.3616	0.2411	0.1808	4 1/2	4.5	1.823	1.214	0.9111
2 11/16	2.6875	0.3884	0.2589	0.1942	4 9/16	4.5625	1.900	1.266	0.9497
2 3/4	2.75	0.4159	0.2772	0.2079	4 5/8	4.625	1.978	1.319	0.9893
2 13/16	2.8125	0.4449	0.2966	0.2225	4 11/16	4.6875	2.059	1.373	1.030
2 7/8	2.875	0.4754	0.3170	0.2378	4 3/4	4.75	2.143	1.429	1.072
2 15/16	2.9375	0.5074	0.3383	0.2537	4 13/16	4.8125	2.228	1.486	1.114
3	3.	0.5399	0.3683	0.2700	4 7/8	4.875	2.316	1.544	1.158
3 1/16	3.0625	0.5742	0.3836	0.2871	4 15/16	4.9375	2.407	1.604	1.203
3 1/8	3.125	0.6105	0.4070	0.3053	5	5.	2.500	1.666	1.250

Haven-Swett "Treatise on Leather Belting," American Leather Belting Association.

Tap drills and clearance drills for machine screws

Screw Size	Coarse Thread		Fine Thread		Clearance
	TPI	Drill	TPI	Drill	Drill
4	40	43	48	42	32
5	40	38	44	37	30
6	32	36	40	33	27
8	32	29	36	29	18
10	24	25	32	21	9
12	24	16	28	14	2
14	20	10	24	7	D
1/4	20	7	28	3	F
5/16	18	F	24	1	P
3/8	16	5/16	24	Q	W
7/16	14	U	20	25/64	29/64
1/2	13	27/64	20	29/64	33/64
9/16	12	31/64	18	33/64	37/64
5/8	11	17/32	18	37/64	41/64
3/4	10	21/32	16	11/16	49/64

Common Nails

Size	Length	Gauge No.	Dia. Head	Approx. No./lb
2d	1.00	15	11/64	845
3d	1.25	14	13/64	540
4d	1.50	12.5	1/4	290
5d	1.75	12.5	1/4	250
6d	2.00	11.5	17/64	265
7d	2.25	11.5	17/64	250
8d	2.50	10.25	9/32	200
9d	2.75	10.25	9/32	90

Size	Length	Gauge No.	Dia. Head	Approx. No./lb
10d	3.00	9	5/16	65
12d	3.25	9	5/16	60
16d	3.50	8	11/32	45
20d	4.00	6	13/32	30
30d	4.50	5	7/16	20
40d	5.00	4	15/32	17
50d	5.50	3	1/2	13
60d	6.00	2	17/32	10

Drill sizes for pipe taps

Tap Size (in.)	TPI	Drill Dia. (in.)
1/8	27	11/32
1/4	18	7/16
3/8	18	37/64
1/2	14	23/32
3/4	14	59/64
1	11-1/2	1-5/32
1-1/4	11-1/2	1-1/2
1-1/2	11-1/2	1-49/64

Tap Size (in.)	TPI	Drill Dia. (in.)
2	11-1/2	2-3/16
2-1/2	8	2-9/16
3	8	3-3/16
3-1/2	8	3-11/16
4	8	4-3/16
4-1/2	8	4-3/4
5	8	5-5/16
6	8	6-5/16

Carbon steel—color and approximate temperature

Color	Temperature (°F)
BLACK RED	990
DARK BLOOD RED	1050
DARK CHERRY RED	1175
MEDIUM CHERRY RED	1250
LIGHT CHERRY, SCALING	1550

Color	Temperature (°F)
SALMON, FREE SCALING	1650
LIGHT SALMON	1725
YELLOW	1825
LIGHT YELLOW	1975
WHITE	2220

Bolting Dimensions for Raised Face Weld Neck Flanges

Nom. Pipe Size	150# ANSI			300# ANSI			400# ANSI		
	Quan	Stud Dia.	Length	Quan	Stud Dia.	Length	Quan	Stud Dia.	Length
2	4	5/8	3 1/4	8	5/8	3 1/2	Note 1		
4	4	5/8	3 3/4	8	3/4	4 1/4	8	7/8	5 1/2
6	8	3/4	4	8	3/4	4 1/2	12	7/8	6
8	8	3/4	4 1/4	12	3/4	5	12	1	6 3/4
10	12	7/8	4 3/4	12	7/8	5 1/2	16	1 1/8	7 1/2
12	12	7/8	4 3/4	16	1	6 1/4	16	1 1/4	8
14	12	1	5 1/4	16	1 1/8	6 3/4	20	1 1/4	8 1/4
16	16	1	5 1/2	20	1 1/8	7	20	1 3/8	8 3/4
18	16	1 1/8	6	20	1 1/4	7 1/2	24	1 3/8	9
20	20	1 1/8	6 1/4	24	1 1/4	7 3/4	24	1 1/2	9 3/4
24	20	1 1/4	7	24	1 1/4	8 1/4	24	1 3/4	10 3/4
26	24	1 1/4	8 1/2	24	1 1/2	9 1/4	28	1 3/4	11 3/4
28	28	1 1/4	8 3/4	28	1 5/8	10 1/4	28	1 3/4	11 3/4
30	28	1 1/4	9	28	1 5/8	10 3/4	28	1 7/8	12 1/2
32	28	1 1/2	10	28	1 3/4	11 1/2	28	2	13 1/4
34	32	1 1/2	10 1/4	28	1 7/8	12 1/4	28	2	14
36	32	1 1/2	10 3/4	32	2	13	32	2	14 1/4

Nom. Pipe Size	600# ANSI			900# ANSI			1500# ANSI		
	Quan	Stud Dia.	Length	Quan	Stud Dia.	Length	Quan	Stud Dia.	Length
2	8	5/8	41/4	Note 2			8	7/8	6 1/4
3	8	3/4	5	8	7/8	5 3/4	8	1 1/8	7
4	8	7/8	5 3/4	8	1 1/8	6 3/4	8	3 3/4	7 3/4
6	12	1	6 3/4	12	1 1/8	7 3/4	12	1 3/8	10 1/4
8	12	1 1/8	7 3/4	12	1 3/8	8 3/4	12	1 5/8	11 1/2
10	16	1 1/4	8 1/2	16	1 3/8	8 3/4	12	1 7/8	13 1/2
12	20	1 1/4	8 3/4	20	1 3/8	10	16	2	15
14	20	1 3/8	9 1/4	20	1 1/2	10 3/4	16	2 1/4	16 1/4
16	20	1 1/2	10	20	1 5/8	11 1/4	16	2 1/2	17 3/4
18	20	1 5/8	10 3/4	20	1 7/8	12 3/4	16	2 3/4	19 1/2
20	24	1 5/8	11 1/2	20	2	13 1/2	16	3	21 1/4
24	24	1 7/8	13	20	2 1/2	17 1/4	16	3 1/2	24 1/4
26	28	1 7/8	13 1/2	20	2 3/4	17 3/4			
28	28	2	14	20	3	18 1/2			
30	28	2	14 1/4	20	3	19			
32	28	2 1/4	15	20	3 1/4	20 1/4			
34	28	2 1/4	15 1/4	20	3 1/2	21 1/4			
36	28	2 1/2	16	20	3 1/2	12 3/4			

Notes:

1. For 3 1/2" and smaller, use ANSI 600 Dim.

2. For 2 1/2" and smaller, use ANSI 1500 Dim.

Steel Fitting Dimensions

90° Long Radius Elbow

90° Reducing Long Radius Elbow

45° Long Radius Elbow

90° Short Radius Elbow

Cap

Lap Joint Stub End

180° Long Radius Return

180° Short Radius Return

90° 3R Elbow   45° 3R Elbow

Nom. Pipe Size	Pipe O.D. (In.)	Light Wall	SCH 20	SCH 30	STD	SCH 40	SCH 60	X STG	SCH 80	SCH 100	SCH 120	SCH 140	SCH 160	XX STG	A	B	D	E	F ANSI	F MSS	G	O	K	S	V	X	Y	Nom. Pipe Size
1/2	.84				.109	.109		.147	.147				.188	.294	1.50	.62		1.00	3.00	2.00	1.38	3.00	1.88					1/2
3/4	1.05				.113	.113		.154	.154				.219	.308	1.12	.44		1.00	3.00	2.00	1.69	2.25	1.69					3/4
1	1.32				.133	.133		.179	.179				.250	.358	1.50	.88	1.00	1.50	3.00	2.00	2.00	3.00	2.19	2.00	1.62			1
1 1/4	1.66				.140	.140		.191	.191				.250	.382	1.88	1.00	1.25	1.50	4.00	2.00	2.50	3.75	2.75	2.50	2.06			1 1/4
1 1/2	1.90				.145	.145		.200	.200				.281	.400	2.25	1.12	1.50	1.50	4.00	2.00	2.88	4.50	3.25	3.00	2.44			1 1/2
2	2.38				.154	.154		.218	.218				.344	.436	3.00	1.38	2.00	1.50	6.00	2.50	3.62	6.00	4.19	4.00	3.19	6.00	2.50	2
2 1/2	2.88				.203	.203		.276	.276				.375	.552	3.75	1.75	2.50	1.50	6.00	2.50	4.12	7.50	5.19	5.00	3.94			2 1/2
3	3.50	.188			.216	.216		.300	.300				.438	.600	4.50	2.00	3.00	2.00	6.00	2.50	5.00	9.00	6.25	6.00	4.75	9.00	3.75	3
3 1/2	4.00				.266	.266		.318	.318				.531	.674	5.25	2.25	3.50	2.50	6.00	3.00	5.50	10.50	7.25	7.00	5.50			3 1/2
4	4.50	.188			.237	.237		.337	.337		.438		.531	.674	6.00	2.50	4.00	2.50	6.00	3.00	6.19	12.00	8.25	8.00	6.25	12.00	5.00	4
5	5.56				.258	.258		.375	.375		.500		.625	.750	7.50	3.12	5.00	3.00	8.00	3.00	7.31	15.00	10.31	10.00	7.75			5
6	6.62	.219			.280	.280		.432	.432		.562		.719	.864	9.00	3.75	6.00	3.50	8.00	3.50	8.50	18.00	12.31	12.00	9.31	18.00	7.50	6
8	8.62	.219	.250	.277	.322	.322	.406	.500	.500	.594	.719	.812	.906	.875	12.00	5.00	8.00	4.00	8.00	4.00	10.62	24.00	16.31	16.00	12.31	24.00	10.00	8
10	10.75	.250	.250	.307	.365	.365	.500	.500	.594	.719	.844	1.000	1.125	1.000	15.00	6.25	10.00	5.00	10.00	5.00	12.75	30.00	20.38	20.00	15.38	30.00	12.50	10
12	12.75	.250	.250	.330	.375	.406	.562	.500	.688	.844	1.000	1.125	1.312	1.000	18.00	7.50	12.00	6.00	10.00	6.00	15.00	36.00	24.38	24.00	18.38	36.00	15.00	12
14	14.00	.250	.312	.375	.375	.438	.594	.500	.750	.938	1.094	1.250	1.406		21.00	8.75	14.00	6.50	12.00	6.00	16.25	42.00	28.00	28.00	21.00	42.00	17.50	14
16	16.00	.250	.312	.375	.375	.500	.656	.500	.844	1.031	1.219	1.438	1.594		24.00	10.00	16.00	7.00	12.00	6.00	18.50	48.00	32.00	32.00	24.00	48.00	19.88	16
18	18.00	.250	.312	.438	.375	.562	.750	.500	.938	1.156	1.375	1.562	1.781		27.00	11.25	18.00	8.00	12.00	6.00	21.00	54.00	36.00	36.00	27.00	54.00	22.38	18
20	20.00	.250	.375	.500	.375	.594	.812	.500	1.031	1.281	1.500	1.750	1.969		30.00	12.50	20.00	9.00	12.00	6.00	23.00	60.00	40.00	40.00	30.00	60.00	24.88	20
24	24.00	.250	.375	.562	.375	.688	.969	.500	1.219	1.531	1.812	2.062	2.344		36.00	15.00	24.00	10.50	12.00	6.00	27.25	72.00	48.00	48.00	36.00	72.00	29.81	24
30	30.00	.312	.500	.625	.375			.500							45.00	18.50	30.00	10.50				90.00	60.00	60.00	45.00	90.00	37.25	30
36	36.00	.312	.500	.625	.375		.750	.500							54.00	22.25	36.00	10.50					72.00		54.00	108.00	44.75	36
42	42.00				.375			.500							63.00	26.00	42.00	12.00								126.00	52.19	42
48	48.00				.375			.500							72.00	29.88	48.00	13.50								144.00	59.69	48

Straight Tee

Reducing Tee

Concentric Reducer

Eccentric Reducer

Run	Outlet	C	M	H
3/4	3/4	1.12		
3/4	1/2	1.12	1.12	1.50
1	1	1.50		
1	3/4	1.50	1.50	2.00
1	1/2	1.50	1.50	2.00
1 1/4	1 1/4	1.88		
1 1/4	1	1.88	1.88	2.00
1 1/4	3/4	1.88	1.88	2.00
1 1/4	1/2	1.88	1.88	2.00
1 1/2	1 1/2	2.25		
1 1/2	1 1/4	2.25	2.25	2.50
1 1/2	1	2.25	2.25	2.50
1 1/2	3/4	2.25	2.25	2.50
1 1/2	1/2	2.25	2.25	2.50
2	2	2.50		
2	1 1/2	2.50	2.38	3.00
2	1 1/4	2.50	2.25	3.00
2	1	2.50	2.00	3.00
2	3/4	2.50	1.75	3.00
2 1/2	2 1/2	3.00		
2 1/2	2	3.00	2.75	3.50
2 1/2	1 1/2	3.00	2.62	3.50
2 1/2	1 1/4	3.00	2.50	3.50
2 1/2	1	3.00	2.25	3.50
3	3	3.38		
3	2 1/2	3.38	3.25	3.50
3	2	3.38	3.00	3.50
3	1 1/2	3.38	2.88	3.50
3	1 1/4	3.38	2.75	3.50

Run	Outlet	C	M	H
3 1/2	3 1/2	3.75		
3 1/2	3	3.75	3.62	4.00
3 1/2	2 1/2	3.75	3.50	4.00
3 1/2	2	3.75	3.25	4.00
3 1/2	1 1/2	3.75	3.12	4.00
4	4	4.12		
4	3 1/2	4.12	4.00	4.00
4	3	4.12	3.88	4.00
4	2 1/2	4.12	3.75	4.00
4	2	4.12	3.50	4.00
4	1 1/2	4.12	3.38	4.00
5	5	4.88		
5	4	4.88	4.62	5.00
5	3 1/2	4.88	4.50	5.00
5	3	4.88	4.38	5.00
5	2 1/2	4.88	4.25	5.00
5	2	4.88	4.12	5.00
6	6	5.62		
6	5	5.62	5.38	5.50
6	4	5.62	5.12	5.50
6	3 1/2	5.62	5.00	5.50
6	3	5.62	4.88	5.50
6	2 1/2	5.62	4.75	5.50
8	8	7.00		
8	6	7.00	6.62	6.00
8	5	7.00	6.38	6.00
8	4	7.00	6.12	6.00
8	3 1/2	7.00	6.00	6.00
10	10	8.50		
10	8	8.50	8.00	7.00
10	6	8.50	7.62	7.00
10	5	8.50	7.50	7.00
10	4	8.50	7.25	7.00

Run	Outlet	C	M	H
12	12	10.00		
12	10	10.00	9.50	8.00
12	8	10.00	9.00	8.00
12	6	10.00	8.62	8.00
12	5	10.00	8.50	8.00
14	14	11.00	11.00	
14	12	11.00	10.62	13.00
14	10	11.00	10.12	13.00
14	8	11.00	9.75	13.00
14	6	11.00	9.38	13.00
16	16	12.00		
16	14	12.00	12.00	14.00
16	12	12.00	11.62	14.00
16	10	12.00	11.12	14.00
16	8	12.00	10.75	14.00
16	6	12.00	10.38	
18	18	13.50		
18	16	13.50	13.00	15.00
18	14	13.50	13.00	15.00
18	12	13.50	12.62	15.00
18	10	13.50	12.12	15.00
18	8	13.50	11.75	
20	20	15.00		
20	18	15.00	14.50	20.00
20	16	15.00	14.00	20.00
20	14	15.00	14.00	20.00
20	12	15.00	13.62	20.00
20	10	15.00	13.12	
20	8	15.00	12.75	

Run	Outlet	C	M	H
24	24	17.00		
24	22	17.00	17.00	20.00
24	20	17.00	17.00	20.00
24	18	17.00	16.50	20.00
24	16	17.00	16.00	20.00
24	14	17.00	16.00	
24	12	17.00	15.62	
24	10	17.00	15.12	
30	30	22.00		
30	24	22.00	21.00	24.00
30	20	22.00	20.00	24.00
30	18	22.00	19.50	
30	16	22.00	19.00	
30	14	22.00	19.00	
36	36	26.50		
36	30	26.50	25.00	24.00
36	24	26.50	24.00	24.00
36	20	26.50	23.00	
36	18	26.50	22.50	
36	16	26.50	22.00	
42	42	30.00		
42	36	30.00	28.00	24.00
42	30	30.00	28.00	24.00
42	24	30.00	26.00	
42	20	30.00	26.00	
48	48	35.00		
48	42	35.00	32.00	28.00
48	36	35.00	31.00	
48	30	35.00	30.00	
48	24	35.00	29.00	

NOTES:

Fittings are to ASTM A234 WPB with dimensions to ANSI B16.9 except that short radius elbows and returns are to ANSI B16.28 and 3R elbows are to MSS-SP75.

Other types, sizes and thicknesses of fittings on application.

ANSI Forged Steel Flanges

Weld Neck Flange | Slip-on Flange | Socket Type | Threaded Flange | Lap Joint Flange | Blind Flange

CLASS 150 FLANGES

Nom. Pipe Size	O	C†	Y† Weld Neck	Y† Slip-on Thrd. Socket	Y† Lap Joint	Bolt Circle	No. and Size of Holes
1/2	3.50	.44	1.88	.62	.62	2.38	4-0.62
3/4	3.88	.50	2.06	.62	.62	2.75	4-0.62
1	4.25	.56	2.19	.69	.69	3.12	4-0.62
1 1/4	4.62	.62	2.25	.81	.81	3.50	4-0.62
1 1/2	5.00	.69	2.44	.88	.88	3.88	4-0.62
2	6.00	.75	2.50	1.00	1.00	4.75	4-0.75
2 1/2	7.00	.88	2.75	1.12	1.12	5.50	4-0.75
3	7.50	.94	2.75	1.19	1.19	6.00	4-0.75
3 1/2	8.50	.94	2.81	1.25	1.25	7.00	8-0.75
4	9.00	.94	3.00	1.31	1.31	7.50	8-0.75
5	10.00	.94	3.50	1.44	1.44	8.50	8-0.88
6	11.00	1.00	3.50	1.56	1.56	9.50	8-0.88
8	13.50	1.12	4.00	1.75	1.75	11.75	8-0.88
10	16.00	1.19	4.00	1.94	1.94	14.25	12-1.00
12	19.00	1.25	4.50	2.19	2.19	17.00	12-1.00
14	21.00	1.38	5.00	2.25	USE SLIP-ON DIMS.	18.75	12-1.12
16	23.50	1.44	5.00	2.50		21.25	16-1.12
18	25.00	1.56	5.50	2.69		22.75	16-1.25
20	27.50	1.69	5.69	2.88		25.00	20-1.25
24	32.00	1.88	6.00	3.25		29.50	20-1.38
30	38.75	2.94	5.38			36.00	28-1.38
36	46.00	3.56	6.19			42.75	32-1.62
42	53.00	3.81	6.75			49.50	36-1.62
48	59.50	4.25	7.56			56.00	44-1.62
54	66.25	4.75	8.50			62.75	44-1.88
60	73.00	5.19	9.44			69.25	52-1.88

CLASS 300 FLANGES

Nom. Pipe Size	O	C†	Y† Weld Neck	Y† Slip-on Thrd. Socket	Y† Lap Joint	Bolt Circle	No. and Size of Holes
1/2	3.75	.56	2.06	.88	.88	2.62	4-0.62
3/4	4.62	.62	2.25	1.00	1.00	3.25	4-0.75
1	4.88	.69	2.44	1.06	1.06	3.50	4-0.75
1 1/4	5.25	.75	2.56	1.06	1.06	3.88	4-0.75
1 1/2	6.12	.81	2.69	1.19	1.19	4.50	4-0.88
2	6.50	.88	2.75	1.31	1.31	5.00	8-0.75
2 1/2	7.50	1.00	3.00	1.50	1.50	5.88	8-0.88
3	8.25	1.12	3.12	1.69	1.69	6.62	8-0.88
3 1/2	9.00	1.19	3.19	1.75	1.75	7.25	8-0.88
4	10.00	1.25	3.38	1.88	1.88	7.88	8-0.88
5	11.00	1.38	3.88	2.00	2.00	9.25	8-0.88
6	12.50	1.44	3.88	2.06	2.06	10.62	12-0.88
8	15.00	1.62	4.38	2.44	2.44	13.00	12-1.00
10	17.50	1.88	4.62	2.62		15.25	16-1.12
12	20.50	2.00	5.12	2.88		17.75	16-1.25
14	23.00	2.12	5.62	3.00	USE SLIP-ON DIMS.	20.25	20-1.25
16	25.50	2.25	5.75	3.25		22.50	20-1.38
18	28.00	2.38	6.25	3.50		24.75	24-1.38
20	30.50	2.50	6.38	3.75		27.00	24-1.38
24	36.00	2.75	6.62	4:19		32.00	24-1.62
30	43.00	3.62	8.25			39.25	28-1.88
36	50.00	4.12	9.50			46.00	32-1.62
42	50.75	4.69	7.88			47.50	32-1.75
48	57.75	5.25	8.81			54.00	32-2.00
54	65.25	6.00	9.94			61.00	28-2.38
60	71.25	6.44	10.75			67.00	32-2.38

CLASS 400 FLANGES

Nom. Pipe Size	O	C†	Y† Weld Neck	Y† Slip-on Thrd. Socket	Y† Lap Joint	Bolt Circle	No. and Size of Holes
1/2	3.75	.56	2.06	.88	.88	2.62	4-0.62
3/4	4.62	.62	2.25	1.00	1.00	3.25	4-0.75
1	4.88	.69	2.44	1.06	1.06	3.50	4-0.75
1 1/4	5.25	.81	2.62	1.12	1.12	3.88	4-0.75
1 1/2	6.12	.88	2.75	1.25	1.25	4.50	4-0.88
2	6.50	1.00	2.88	1.44	1.44	5.00	8-0.75
2 1/2	7.50	1.12	3.12	1.62	1.62	5.88	8-0.88
3	8.25	1.25	3.25	1.81	1.81	6.62	8-0.88
3 1/2	9.00	1.38	3.38	1.94	1.94	7.25	8-0.88
4	10.00	1.38	3.50	2.00	2.00	7.88	8-1.00
5	11.00	1.50	4.00	2.12	2.12	9.25	8-1.00
6	12.50	1.62	4.06	2.25	2.25	10.62	12-1.00
8	15.00	1.88	4.62	2.69	2.69	13.00	12-1.12
10	17.50	2.12	4.88	2.88		15.25	16-1.25
12	20.50	2.25	5.38	3.12		17.75	16-1.38
14	23.00	2.38	5.88	3.31	USE SLIP-ON DIMS.	20.25	20-1.38
16	25.50	2.50	6.00	3.69		22.50	20-1.50
18	28.00	2.62	6.50	3.88		24.75	24-1.50
20	30.50	2.75	6.62	4.00		27.00	24-1.62
24	36.00	3.00	6.88	4.50		32.00	24-1.88
30	43.00	4.00	8.62			39.25	28-2.12
36	50.00	4.50	9.88			46.00	32-2.12
42	52.00	5.25	8.81			48.25	32-2.00
48	59.50	6.00	10.12			55.25	28-2.38
54	67.00	6.69	11.38			62.25	28-2.62
60	74.25	7.31	12.56			69.00	32-2.88

CLASS 600 FLANGES

Nom. Pipe Size	O	C†	Y† Weld Neck	Y† Slip-on Thrd. Socket	Y† Lap Joint	Bolt Circle	No. and Size of Holes
1/2	3.75	.56	2.06	.88	.88	2.62	4-0.62
3/4	4.62	.62	2.25	1.00	1.00	3.25	4-0.75
1	4.88	.69	2.44	1.06	1.06	3.50	4-0.75
1 1/4	5.25	.81	2.62	1.12	1.12	3.88	4-0.75
1 1/2	6.12	.88	2.75	1.25	1.25	4.50	4-0.88
2	6.50	1.00	2.88	1.44	1.44	5.00	8-0.75
2 1/2	7.50	1.12	3.12	1.62	1.62	5.88	8-0.88
3	8.25	1.25	3.25	1.81	1.81	6.62	8-0.88
3 1/2	9.00	1.38	3.38	1.94	1.94	7.25	8-1.00
4	10.75	1.50	4.00	2.12	2.12	8.50	8-1.00
5	13.00	1.75	4.50	2.38	2.38	10.50	8-1.12
6	14.00	1.88	4.62	2.62	2.62	11.50	12-1.12
8	16.50	2.19	5.25	3.00	3.00	13.75	12-1.25
10	20.00	2.50	6.00	3.38		17.00	16-1.38
12	22.00	2.62	6.12	3.62		19.25	20-1.38
14	23.75	2.75	6.50	3.69	USE SLIP-ON DIMS.	20.75	20-1.50
16	27.00	3.00	7.00	4.19		23.75	20-1.62
18	29.25	3.25	7.25	4.62		25.75	20-1.75
20	32.00	3.50	7.50	5.00		28.50	24-1.75
24	37.00	4.00	8.00	5.50		33.00	24-2.00
30	44.50	4.50	9.75			40.25	28-2.12
36	51.75	4.88	11.12			47.00	28-2.62
42	55.25	6.62	11.00			50.50	28-2.62
48	62.75	7.44	12.44			57.50	32-2.88
54	70.00	8.25	13.75			64.25	32-3.12
60	78.50	9.19	15.31			71.75	28-3.62

CLASS 900 FLANGES

Nom. Pipe Size	O	C†	Y† Weld Neck	Y† Slip-on Thrd. Socket	Y† Lap Joint	Bolt Circle	No. and Size of Holes
1/2	4.75	.88	2.38	1.25	1.25	3.25	4-0.88
3/4	5.12	1.00	2.75	1.38	1.38	3.50	4-0.88
1	5.88	1.12	2.88	1.62	1.62	4.00	4-1.00
1 1/4	6.25	1.12	2.88	1.62	1.62	4.38	4-1.00
1 1/2	7.00	1.25	3.25	1.75	1.75	4.88	4-1.12
2	8.50	1.50	4.00	2.25	2.25	6.50	8-1.00
2 1/2	9.62	1.62	4.12	2.50	2.50	7.50	8-1.12
3	9.50	1.50	4.00	2.12	2.12	7.50	8-1.00
3 1/2							
4	11.50	1.75	4.50	2.75	2.75	9.25	8-1.25
5	13.75	2.00	5.00	3.12	3.12	11.00	8-1.38
6	15.00	2.19	5.50	3.38	3.38	12.50	12-1.25
8	18.50	2.50	6.38	4.00		15.50	12-1.50
10	21.50	2.75	7.25	4.25		18.50	16-1.50
12	24.00	3.12	7.88	4.62		21.00	20-1.50
14	25.25	3.38	8.38	5.12	USE SLIP-ON DIMS.	22.00	20-1.62
16	27.75	3.50	8.50	5.25		24.25	20-1.75
18	31.00	4.00	9.00	6.00		27.00	20-2.00
20	33.75	4.25	9.75	6.25		29.50	20-2.12
24	41.00	5.50	11.50	8.00		35.50	20-2.62
30	48.50	5.88	12.25			42.75	20-3.12
36	57.50	6.75	14.25			50.75	20-3.62
42	61.50	8.12	14.62			54.75	24-3.62
48	70.25	9.19	16.50			62.50	24-4.12

CLASS 1500 FLANGES

Nom. Pipe Size	O	C†	Y† Weld Neck	Y† Slip-on Thrd. Socket	Y† Lap Joint	Bolt Circle	No. and Size of Holes
1/2	4.75	.88	2.38	1.25	1.25	3.25	4-0.88
3/4	5.12	1.00	2.75	1.38	1.38	3.50	4-0.88
1	5.88	1.12	2.88	1.62	1.62	4.00	4-1.00
1 1/4	6.25	1.12	2.88	1.62	1.62	4.38	4-1.00
1 1/2	7.00	1.25	3.25	1.75	1.75	4.88	4-1.12
2	8.50	1.50	4.00	2.25	2.25	6.50	8-1.00
2 1/2	9.62	1.62	4.12	2.50	2.50	7.50	8-1.12
3	10.50	1.88	4.62	2.88	2.88	8.00	8-1.25
3 1/2							
4	12.25	2.12	4.88	3.56	3.56	9.50	8-1.38
5	14.75	2.88	6.12	4.12	4.12	11.50	8-1.62
6	15.50	3.25	6.75	4.69	4.69	12.50	12-1.50
8	19.00	3.62	8.38	5.62	5.62	15.50	12-1.75
10	23.00	4.25	10.00	6.25		19.00	12-2.00
12	26.50	4.88	11.12	7.12		22.50	16-2.12

CLASS 2500 FLANGES

Nom. Pipe Size	O	C†	Y† Weld Neck	Y† Slip-on Thrd. Socket	Y† Lap Joint	Bolt Circle	No. and Size of Holes
1/2	5.25	1.19	2.88	1.56	1.56	3.50	4-0.88
3/4	5.50	1.25	3.12	1.69	1.69	3.75	4-0.88
1	6.25	1.38	3.50	1.88	1.88	4.25	4-1.00
1 1/4	7.25	1.50	3.75	2.06	2.06	5.12	4-1.12
1 1/2	8.00	1.75	4.38	2.38	2.38	5.75	4-1.25
2	9.25	2.00	5.00	2.75	2.75	6.75	8-1.12
2 1/2	10.50	2.25	5.62	3.12	3.12	7.75	8-1.25
3	12.00	2.62	6.62	3.62	3.62	9.00	8-1.38
4	14.00	3.00	7.50	4.25	4.25	10.75	8-1.62
5	16.50	3.62	9.00	5.12	5.12	12.75	8-1.88
6	19.00	4.25	10.75	6.00	6.00	14.50	8-2.12
8	21.75	5.00	12.50	7.00	7.00	17.25	12-2.12
10	26.50	6.50	16.50	9.00	9.00	21.25	12-2.62
12	30.00	7.25	18.25	10.00	10.00	24.38	12-2.88

NOTES:

Flange material is ASTM A105 unless specified otherwise.

Flange dimensions are to ANSI B16.5 for sizes 24" and smaller. Sizes over 24" are to MSS-SP44.

Always specify bore or matching pipe wall when ordering.

Other types, sizes and facings on application.

† Includes .06 raised face in Class 150 and Class 300 standards. Does **NOT** include .25 raised face in Class 400 and heavier standards.

Color Codes For Locating Underground Utilities

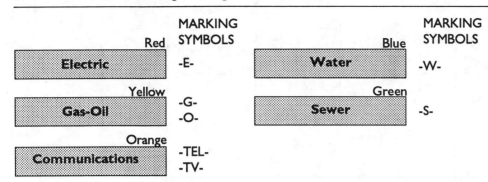

	Red	MARKING SYMBOLS
Electric		-E-
Gas-Oil	Yellow	-G- -O-
Communications	Orange	-TEL- -TV-

	Blue	MARKING SYMBOLS
Water		-W-
Sewer	Green	-S-

Approximate Angle of Repose for Sloping Sides of Excavations

Note: Clays, SIMs, Loams or Non-Hemorporus Soils Require Shoring and Bracing

The presence of ground water requires special treatment.

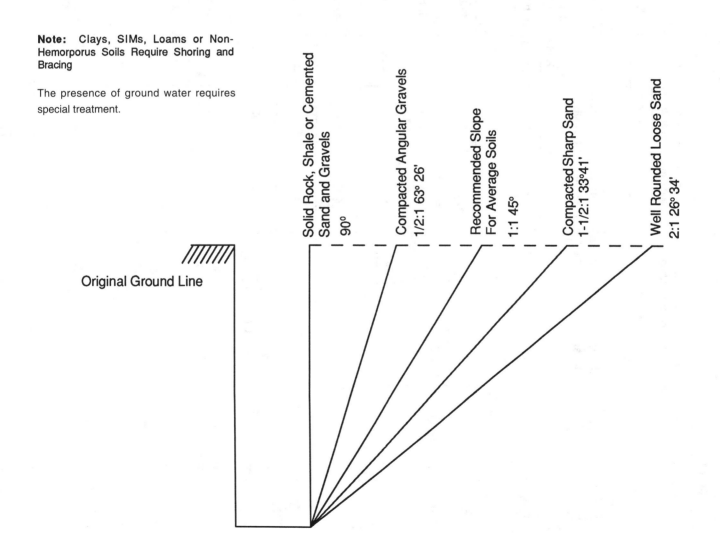

Original Ground Line

Solid Rock, Shale or Cemented Sand and Gravels 90°

Compacted Angular Gravels 1/2:1 63° 26'

Recommended Slope For Average Soils 1:1 45°

Compacted Sharp Sand 1-1/2:1 33°41'

Well Rounded Loose Sand 2:1 26° 34'

Trench Shoring - Minimum Requirements

Trench jacks may be used in lieu of, or in combination with cross braces. Shoring is not required in solid rock, hard shale, or hard slag. Where desirable, steel sheet piling and bracing of equal strength may be substituted for wood.

Depth of Trench (Feet)	Kind Or Condition of Earth	Uprights Minimum Dimensions (Inches)	Uprights Maximum Spacing (Feet)	Stringers Minimum Dimensions (Inches)	Stringers Maximum Spacing (Feet)	Cross Braces — Width of Trench: Up to 3 Feet (Inches)	3 to 6 Feet (Inches)	6 to 9 Feet (Inches)	9 to 12 Feet (Inches)	12 to 15 Feet (Inches)	Maximum Spacing Vertical (Feet)	Maximum Spacing Horizontal (Feet)
5 to 10	Hard, compact	3 x 4 or 2 x 6	6	--	--	2x6	4x4	4x6	6x6	6x8	4	6
	Likely to crack	3 x 4 or 2 x 6	3	4x6	4	2x6	4x4	4x6	6x6	6x8	4	6
	Soft, sandy, or filled	3 x 4 or 2 x 6	Close Sheeting	4x6	4	4x4	4x6	6x6	6x8	8x8	4	6
	Hydrostatic pressure	3 x 4 or 2 x 6	Close Sheeting	4x6	4	4x4	4x6	6x6	6x8	8x8	4	6
10 to 15	Hard	3 x 4 or 2 x 6	4	4x6	4	4x4	4x6	6x6	6x8	8x8	4	6
	Likely to crack	3 x 4 or 2 x 6	2	4x4	4	4x6	6x6	6x8	6x8	8x8	4	6
	Soft, sandy, or filled	3 x 4 or 2 x 6	Close Sheeting	4x6	4	4x6	6x6	6x8	8x8	8x10	4	6
	Hydrostatic pressure	3 x 6	Close Sheeting	8x10	4	4x6	6x6	6x8	8x8	8x10	4	6
15 to 20	All kinds or conditions	3 x 6	Close Sheeting	4x12	4	4x12	6x8	8x8	8x8	10x10	4	6
Over 20	All kinds or conditions	3 x 6	Close Sheeting	6x8	4	4x12	6x8	8x10	10x10	10x12	4	6

Reuniting separated mercury in thermometers

The largest single cause for failure of precision thermometers is due to separated mercury columns. This can occur in transit or in use. The mercury may be reunited by cooling the thermometer in a solution of solid CO_2 (Dry-Ice) and alcohol so that the mercury column retreats slowly into the bulb. Do not cool the stem or mercury column. Keep the bulb in the solution until the main column, as well as the separated portion retreats into the bulb. Remove and swing the thermometr in a short arc, forcing all of the mercury into the bulb.

Most mercury thermometers can be reunited using this method regardless of range (with the exception of deep immersion thermometers) provided only the bulb is immersed in the CO_2.

Caution: Do not touch the bulb until it has warmed sufficiently for the mercury to emerge from the bulb into the capillary.

Never subject the stem or mercury column to the CO_2 solution as it will freeze the mercury column in the capillary and may cause the bulb to fracture.

Typical Wire Resistance

(Stranded Copper Conductors at 59°F)

Wire Size AWG	Resistance (ohms/ft)
0000	0.00005
000	0.00006
00	0.00008
0	0.00010
1	0.00012
2	0.00016
4	0.00025
6	0.00039
8	0.00063
10	0.00098
12	0.00160

How to cut odd-angle long radius elbows

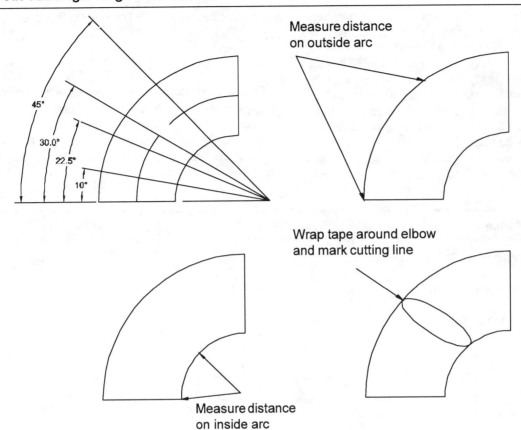

Measure distance on outside arc

Wrap tape around elbow and mark cutting line

Measure distance on inside arc

*Rounded to nearest practical fraction.

Nominal Pipe Size	Outside Arc Length*							Inside Arc Length*						
	1°	5°	10°	15°	22.5°	30°	45°	1°	5°	10°	15°	22.5°	30°	45°
2"	5/64	3/8	23/32	1-3/32	1-21/32	2-3/16	3-9/32	1/32	5/32	5/16	15/32	23/32	15/16	1-7/16
2 1/2"	3/32	7/16	29/32	1-11/32	2-1/32	2-23/32	4-1/16	3/64	3/16	13/32	19/32	29/32	1-7/32	1-13/16
3"	7/64	9/16	1-1/8	1-5/8	2-15/32	3-9/32	4-29/32	3/64	1/4	1/2	23/32	1-3/32	1-7/16	2-5/32
3 1/2"	1/8	5/8	1-9/32	1-29/32	2-27/32	3-13/16	5-11/16	1/16	9/32	9/16	27/32	1-9/32	1-11/16	2-9/16
4"	9/64	23/32	1-7/16	2-5/32	3-1/4	4-5/16	6-15/32	1/16	5/16	21/32	3-1/32	1-15/32	1 31/32	2-15/16
5"	3/16	29/32	1-25/32	2-11/16	4-1/32	5-3/8	8-1/16	5/64	13/32	13/16	1-1/4	1-27/32	2-15/32	3-23/32
6"	7/32	1-1/16	2-5/32	3-7/32	4-27/32	6-7/16	9-21/32	3/32	1/2	1	1-1/2	2-7/32	2-31/32	4-15/32
8"	9/32	1-7/16	2-27/32	4-9/32	6-13/32	8-17/32	12-13/16	1/8	11/16	1-11/32	2	3-1/32	4-1/32	6-1/32
10"	11/32	1-25/32	3-9/16	5-11/32	8	10-21/32	16	5/32	27/32	1-11/16	2-17/32	3-25/32	5-1/32	7-9/16
12"	7/16	2-1/8	4-1/4	6-3/8	9-9/16	12-3/4	19-5/32	7/32	1	2-1/32	3-1/16	4-9/16	6-3/8	9-1/8
14"	1/2	2-7/16	4-7/8	7-5/16	11	14-21/32	22	1/4	1-7/32	2-7/16	3-21/32	5-1/2	7-11/32	11
16"	9/16	2-13/16	5-19/32	8-3/8	12-9/16	16-3/4	25-1/8	9/32	1-13/32	2-13/16	4-3/16	6-9/32	8-3/8	12-5/8
18"	5/8	3-1/8	6-9/32	9-7/16	14-1/8	18-27/32	28-9/32	5/16	1-9/16	3-1/8	4-23/32	7-1/16	9-7/16	14-1/8
20"	11/16	3-1/2	7	10-15/32	15-23/32	20-15/16	31-13/32	11/32	1-3/4	3-1/2	5-1/4	7-27/32	10-15/32	15-11/16
22"	3/4	3-27/32	7-11/16	11-17/32	17-9/32	23-1/32	34-9/16	3/8	1-29/32	3-27/32	5-3/4	8-5/8	11-17/32	17-9/32
24"	27/32	4-3/16	8-3/8	12-9/16	18-27/32	25-1/8	37-11/16	13/32	2-3/32	4-3/16	6-9/32	9-7/16	12-9/16	18-27/32
26"	29/32	4-17/32	9-3/32	13-5/8	20-13/32	27-7/32	40-27/32	15/32	2-9/32	4-17/32	6-13/16	10-7/32	13-5/8	20-13/32
30"	1-1/32	5-1/4	10-15/32	15-3/4	23-9/16	31-13/32	47-1/8	17/32	2-5/8	5-1/4	7-27/32	11-25/32	15-23/32	23-9/16
36"	1-7/32	6-9/32	12-17/32	18-7/8	28-7/32	37-11/16	56-17/32	5/8	2-13/16	6-1/4	9-7/16	14-1/8	18-27/32	28-1/4

How to read land descriptions

A land description is a description of a tract of land, in legally acceptable terms, that defines exactly where the tract of land is located and how many acres it contains.

Table 1
Land Measurements

Linear Measure

1 inch	0.833 feet	16 1/2 feet	1 rod
7.92 inches	1 link	5 1/2 yards	1 rod
12 inches	1 foot	4 rods	1 link
1 vara	33 inches	66 feet	1 chain
2 3/4 feet	1 vara	80 chains	1 mile
3 feet	1 yard	320 rods	1 mile
25 links	16 1/2 feet	8000 links	1 mile
25 links	1 rod	5280 feet	1 mile
100 links	1 chain	1760 yards	1 mile

Square Measure

144 sq. in.	1 sq. ft.	43560 sq. ft.	1 acre
9 sq. ft.	1 sq. yd.	640 acres	1 sq. mile
30 1/4 sq. yds	1 sq. rod	1 sq. mile	1 section
10 sq. rods	1 sq. chain	36 sq. miles	1 township
1 sq. rod	272 1/4 sq. ft	6 miles sq.	1 township
1 sq. chain	4356 sq. ft.	208 ft. 8 in sq.	1 acre
10 sq. chains	1 acre	80 rods sq.	40 acres
160 sq. rods	1 acre	160 rods sq.	160 acres
4840 sq. yds	1 acre		

In non-rectangular land descriptions, distance is usually described in terms of either feet or rods (this is especially true in surveying today), while square measure is in terms of acres. Such descriptions are called Metes and Bounds descriptions and will be explained in detail later. In rectangular land descriptions, square measure is again in terms of acres, and the location of the land is in such terms as N 1-2 (north one-half), SE 1-4 (south east one-fourth or quarter), etc., as shown in Figures 2, 3, 4 and 5.

Meandered Water & Government Lots

A meandered lake or stream is water, next to which the adjoining land owner pays taxes on the land only. Such land is divided into divisions of land called government lots. The location, acreage and lot number of each such tract of land was determined, surveyed and platted by the original government surveyors.

The original survey of your county (complete maps of each township, meandered lakes, government lots, etc.) is in your courthouse and is the basis for all land descriptions in your county. See Figure 1.

The government lot number given to a piece of land is the legal description of that tract of land.

How can you tell whether water is meandered or privately owned? If you find government lots adjoining a body of water or stream, those waters are meandered. If there are no

Figure 1

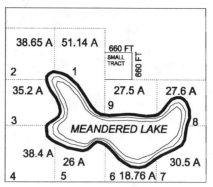

government lots surrounding water, that water is privately owned; the owner is paying taxes on the land under the water, and controls hunting, fishing, trapping rights, etc., on that water within the regulations of State and Federal laws. Note that where such water is deemed navigable, other rulings may sometimes pertain.

As a generality, meandered water is public water that the public may use for recreational purposes, fishing, hunting, trapping, etc., provided that the public can reach the waters without trespassing. There still is much litigation concerning this that will have to be settled in court.

Reading Land Descriptions

Descriptions of land always read first from either the north or the south. In Figures 2, 3, 4 and 5, notice that they all start with N (north), or S (south), such as NW, SE, etc. The are never WN (west north), ES (east south) etc.

It is simple for anyone to understand a description. The secret is to read or analyze the description from the rear, or backwards.

Example: Under Figure 4, the first description reads E 1/2, SE 1/4, SW 1/4, SW 1/4. The last part of the description reads SW 1/4, which means that the tract of land we are looking for is somewhere in that quarter (as shown in Figure 2). Next back, we find SW 1/4, which means the tract we are after is somewhere in the SW 1/4 SW 1/4 (as shown in Figure 3). Next back, we find the SE 1/4, which means that the tract is in the SE 1/4 SW 1/4 SW 1/4 (as shown in Figure 5). Next back and the last part to look up, is the E 1/2 of the above, which is the location of the tract described by the whole description (as shown in Figure 4).

Sample Sections Showing Rectangular Land Descriptions, Acreages and Distances

Figure 2

```
                       NORTH
       ┌─────────────────────┬─────────────────────┐
       │   160 RODS          │   1/2 MILE          │
       │                     │                     │
  W    │   NW 1/4            │   NE 1/4            │  E
  E    │   160 ACRES         │   160 ACRES         │  A
  S    ├─────────────────────┼─────────────────────┤  S
  T    │   830 YDS.          │   2640 FT.          │  T
       │                     │                     │
       │   SW 1/4            │   SE 1/4            │
       │   160 ACRES         │   160 ACRES         │
       │                     │                     │
       │   40 CHAINS         │   4000 LINKS        │
       └─────────────────────┴─────────────────────┘
                       SOUTH
```

Figure 3

80 RODS	1320 FEET	440 YDS	20 CHAINS
NW 1/4 NW 1/4 40	NE 1/4 NW 1/4 40	NW 1/4 NE 1/4 40	NE 1/4 NE 1/4 40
1/4 MILE SW 1/4 NW 1/4 40	SE 1/4 NW 1/4 40	SW 1/4 NE 1/4 40	SE 1/4 NE 1/4 40
NW 1/4 SW 1/4 40	NE 1/4 SW1/4 40	NW 1/4 SE 1/4 40	NE 1/4 SE 1/4 40
SW 1/4 SW 1/4 40	SE 1/4 SW 1/4 40	SW 1/4 SE 1/4 40	SE 1/4 SE 1/4 40

Figure 4

```
┌──────────────────────────────────────────────────┐
│   1 MILE - 320 RODS - 5280 FEET - 80 CHAINS       │
│                                                    │
│                   N 1/2                            │
│                   320                              │
├──────────────────────┬────┬────┬──────────────────┤
│  N 1/2 N 1/2 SW 1/4  │W1/2│E1/2│ N 1/2 NE 1/4 SE 1/4│
│         40           │NW  │NW  │        20          │
├──────────────────────┤1/4 │1/4 ├────────────────────┤
│  S 1/2 N 1/2 SW 1/4  │SE  │SE  │ S 1/2 NE 1/4 SE 1/4│
│         40           │1/4 │1/4 │        20          │
│                      │ 20 │ 20 │                    │
├────┬─────┬───────────┼────┴────┴────────────────────┤
│W1/2│NE1/4│           │                              │
│SW  │SW1/4│ SE 1/4 SW 1/4│      S 1/2 SE 1/4         │
│1/4 │ 10  │     40    │          80                  │
│ 20 │     │           │                              │
└────┴─────┴───────────┴──────────────────────────────┘
        │  └── E 1/2 SE 1/4 SW 1/4 SW 1/4 - 5 A
        └───── W 1/2 SE 1/4 SW 1/4 SW 1/4 - 5 A
```

Figure 5

```
┌──────────┬──────────┬──────────────────┐
│  W 1/2   │  E 1/2   │                  │
│  NW 1/4  │  NW 1/4  │                  │
│   80     │   80     │                  │
├──────────┴──────────┤   E 1/2          │
│                     │   320            │
│   N 1/2 SW 1/4      │                  │
│      80             │                  │
├──────────┬──────────┼───               │
│          │NW1/4 NE1/4│                  │
│N1/2SW1/4 │SE1/4 SE1/4│                  │
│SW1/4     │SW1/4 SW1/4│                  │
│  20      │ 10   10  │                  │
│       ┌──┼──────────┤                  │
│       │SE1/4│SW1/4 SE1/4│               │
│       │SE1/4│SE1/4 SE1/4│               │
│       │SW1/4│SW1/4 SW1/4│               │
│       │ 10  │ 10   10  │               │
└───────┴─────┴──────────┴──────────────┘
     │  └── S1/2SW1/4SW1/4SW1/4 - 5 A
     └───── N1/2SW1/4SW1/4SW1/4 - 5 A
```

Metes and Bounds Descriptions

A metes and bounds description is a description of land obtained by starting at a given point, running so many feet in a certain direction, so many feet in another direction, etc., back to the point of beginning.

Example: In Figure 1, notice the small tract of land outlined. A typical metes and bounds description of this tract of land would be as follows: "Begin at the center of the section, thence north 660 feet, thence east 660 feet, thence south 660 feet, thence west 660 feet, back to the point of beginning and containing 10 acres, being a part of Section No. 2."

IMPORTANT: To locate a tract of land from a metes and bounds description, start from the point of beginning and follow it out (do not read backwards as in the case of a rectangular description).

The small tract of land just located by the above metes and bounds description could also be described as the SW 1/4 SW 1/4 NE 1/4 of the section. In most cases, the same tract of land may be described in different ways. The rectangular system of describing and locating land as shown in Figures 2, 3, 4 and 5 is the simplest and is almost always used when possible.

In land descriptions, degree readings are not a measure of distance. They are combined with either north or south to show the direction a line runs from a given point.

How to Read Descriptions That Show Directions in Terms of Degrees

In Figure 6, the north-south line and the east-west line divide the circle into 4 equal parts with each part containing 90 degrees. Several different directions are shown in this diagram with the number of degrees each varies east or west from the north and south starting points (remember that all descriptions read from the north or south). North-west is a direction that is half-way between north and west. In terms of degrees, the direction north-west would read, north 45 degrees west.

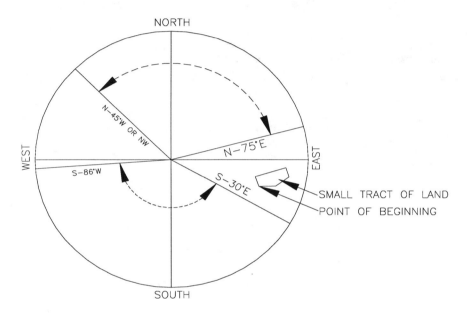

Figure 6. Land description in terms of degrees.

Notice the small tract of land in Figure 6. The following metes and bounds description will locate this tract: "Begin at the beginning point, thence N 20 degrees west—200 feet, thence N 75 degrees east—1,320 feet, thence S 30 degrees east—240 feet, thence S 45 degrees west—420 feet, thence west—900 feet back to the point of beginning, containing so many acres, etc."

Size an air receiver for engine starting

Assume that the engine starter requires 16 standard cubic feet per second, that 10 seconds is required for starting and that you want to store enough air in the receiver for three starts. Assume the initial pressure in the receiver is 215 psia and the final pressure is 115 psia.

$$V = \frac{RTM}{144(P_1 - P_2)}$$

where
V = Receiver volume, cu. ft.
R = 53.5 for air
T = Temperature (assumed constant at 520°R)
M = Weight of air required to start the engine based on density of air = 0.0763 pounds per cu. ft.
P = Pressure, pounds/sq. ft.

Example:
M = 16 cu. ft./sec. x10 sec./start x 3 starts x 0.0763 pounds per cu. ft.
  = 36.624 pounds

$$V = \frac{53.5 x 520 x 36.624}{144 x (215 - 115)}$$

$$V = 70.76 \text{ cu. ft.}$$

Assume receiver is 30" or 2.5 feet inside diameter. Find the height of the receiver.

$$V = \frac{\Pi x ID^2}{4} x H$$

where
V = Volume to be contained in receiver, cu. ft.
ID = Inside diameter of the receiver, feet
H = Height of receiver, feet
H = 14.4 feet

Use two receivers, each 7.5 feet high.

2: Construction

Project Scoping Data **30**
Project scoping data worksheet for major facilities........................... 30
Right of Way.. **31**
How to determine the crop acreage included in a right-of-way strip .. 31
Clearing and grading right-of-way: labor/equipment considerations.. 31
Estimating manhours for removing trees... 32
Estimating manhours for removing tree stumps............................. 32
Clearing and grading right-of-way 33
Ditching... **34**
How many cubic yards of excavation in a mile of ditch?.................... 34
Shrinkage and expansion of excavated and compacted soil............... 34
Ditching and trenching: labor/equipment considerations.................. 34
Concrete Work **35**
How to approximate sacks of cement needed to fill a form 35
What you should know about mixing and finishing concrete 35
Pipe Laying.. **35**
How to determine the degrees of bend in a pipe that must fit a ditch
   calling for a bend in both horizontal and vertical planes................. 35
How to bend pipe to fit ditch—sags, overbends and combination
   bends 36
Pipe bending computations made with hand-held calculator 36
Calculate maximum bend on cold pipe 41

Determine length of a pipe bend.. 42
Length of pipe in arc subtended by any angle 42
Average pipelay table—Underground.. 43
Average pipelay table—On supports.. 44
Allowable pipe span between supports.. 44
How engineers make pipe fit the ditch 45
Pipe Lowering.. **48**
How to lower an existing pipe line that is still in service.................. 48
Welding **51**
When should steel be preheated before welding?............................. 51
Welding and brazing temperatures 52
Mechanical properties of pipe welding rods................................. 52
Lens shade selector 53
Pipe Line Welding **53**
How many welds will the average welder make per hour?................ 62
How much welding rod is required for a mile of schedule 40
   pipe line?... 62
How many pounds of electrodes are required per weld on line pipe? ... 62
Welding criteria permit safe and effective pipe line repair 63
Cross country pipe line—vertical down electrode consumption,
   pounds of electrode per joint 69

PROJECT SCOPING DATA

Project Scoping Data Worksheet for Major Facilities

1.0 Type of Contract Required

1.1 Describe the type of contract the project will be based on.
1.2 What is the required completion date?
1.3 What type permits will be required and who will obtain them?
1.4 Will an environmental impact statement be required?

2.0 General Engineering

2.1 Does the client have engineering standards and specifications, or will it be necessary to develop these, or can the Engineering Contractor's specifications and standards be used?
2.2 Describe any existing above or underground obstructions at the job site.
2.3 Obtain copies of any applicable local codes dealing with air and/or water pollution, sanitary systems, electrical systems, and structures.
2.4 Are any soil bearing data available?
2.5 Describe type of soil—sandy, rocky, caliche, etc.
2.6 Does the site drain naturally? Will fill material be required for the foundations? If fill material will be required, what is the availability? Is sand-asphalt mix available?

3.0 Drafting

3.1 Will a registered engineer's stamp or signature be required on the drawings?
3.2 Determine drawing forms to be used and type of drawings.
3.3 Will client accept single line drawings or will double line drawings be required?

4.0 Structural and Architectural

4.1 Will all exposed structural steel need to be galvanized or have other protective coating?
4.2 Is sand, rock, cement, and ready-mixed concrete available locally? Price?
4.3 Requirements for building design:
Steel
Concrete/brick
How much air conditioning?
Other

5.0 Piping

5.1 How will storm water, waste water, and sewage be disposed of? If the facility is handling liquid petroleum, will a petroleum-water separator be required?
5.2 Are there any existing incoming and/or outgoing pipelines that will require connecting to?
5.3 Are there any facilities outside the job site which must be given consideration in the development of a plot plan?
5.4 What is the pressure rating for existing manifolds, suction lines, and outgoing lines?
5.5 Any special requirements for isolating the facility?
5.6 Will scraper traps be required? Incoming or outgoing or both?
5.7 Will a fire water system be required?
5.8 Is a source of fire water available? If yes, what will limit the amount of water available?
5.9 How many relief headers will be required?
5.10 Will incoming and/or outgoing relief valves be required? If yes, where will they relieve to?

6.0 Electrical

6.1 Is commercial power available?
Voltage?
Phase?
Frequency?
System capacity?
6.2 Define approximate location where incoming power will enter job site.
6.3 If commercial power is available, will the utility furnish and install the substation?
6.4 What is the short circuit capacity of the utility?
6.5 Will the utility permit across-the-line starting of large motors? What is the maximum hp they will allow to start across the line?

7.0 Instrumentation

7.1 Will a control panel be required? If yes, what type?
7.2 What type of instruments does the client prefer? Pneumatic or electronic?
7.3 Does the client have a preference for the type of final control element(s) and actuator(s) to be used?
7.4 Electrical classification for the control house.
7.5 Are communications facilities available? If yes, what type?

8.0 Equipment

8.1 Have any items of major equipment been placed on order? List items and state delivery.
8.2 Who will supply general equipment specifications?
8.3 List the design criteria for equipment sizing.
8.4 What are the spare equipment requirements?
8.5 What is the expected life of the facility?
8.6 Will secondary seals be required on storage tanks equipped with floating roofs? Tank mixers?

9.0 Material Purchasing

9.1 Obtain a copy of a preferred vendor list if one is available.
9.2 If not available, will it be necessary to develop a list for client's approval?
9.3 Describe any unusual conditions.
9.4 Determine shipping address.
9.5 Who will be responsible for receipt of material?

10.0 Special Considerations

10.1 If the site is in a remote location and is near the coast line, are there any existing piers or natural facilities available for unloading material?
10.2 What is the local availability of heavy lifting equipment? What are the lifting limitations?
10.3 What is the availability of local labor? Are they organized?
10.4 If the site is in a foreign location, is there any limit on the use of expatriates?
10.5 Are any other contractors working in the area?
10.6 What shop fabrication can be done locally?
10.7 What types of materials are available locally?
10.8 If the site is in a foreign location, who will handle clearance of material through customs?
10.9 What are the general weather conditions?
10.10 If foreign, will a construction camp be required?
10.11 If a camp is required, what is the availability of local catering?

RIGHT OF WAY

How to determine the crop acreage included in a right-of-way strip

Multiply the width of the strip in feet by the length in rods; divide this by 2,640 to obtain the acreage. If the ends of the strip are not parallel, use the length of the center line of the right-of-way.

Example. A right-of-way 35 feet wide crosses a cultivated field for a length of 14 rods; how many acres of crop were destroyed?

$35 \times 14/2,640 = 0.18$ acres, or almost one-fifth of an acre.

Example. A right-of-way 50 feet wide crosses a field for 330 rods. How many acres of crop were destroyed?

$50 \times 330/2,640 = 6.25$ acres.

The rule is exact, not an approximation.

Clearing and grading right-of-way: labor/equipment considerations

To estimate labor crew and equipment spread for clearing and grading operations, the following items, as they may apply to a given project, should be given consideration:

1. Removal of trees, brush and stumps.
2. Grubbing and removal of stumps that are in the way of the ditch.
3. Disposal of all debris, including method of disposal and length of haul.
4. Clearing area spoil a sufficient distance from the ditch line so that the spoil-bank from the ditching operations will not fall in any foreign material that might become mixed with the excavated spoil.
5. Cutting of merchantable timber into standard lengths and stacked along the right-of-way for disposition by others if specifically required by the right-of-way agreement.
6. Providing temporary walks, passageways, fences, or other structures so as not to interfere with traffic.
7. Providing sufficient and proper lighting where required.
8. Providing guards where required.
9. Preserving all trees, shrubs, hedges and lawns where required.

10. Grading irregularities where required.
11. Preserving topsoil for replacement, through all cultivated or improved fields and pastures, to its original position.
12. Proper grading of the terrain so as to allow passage of loaded trucks and equipment hauling materials and so ditching operations can be properly performed.
13. Protecting and preserving existing drainage facilities.
14. Protecting any existing structures or pipelines.
15. Protecting any telephone or utility lines and keeping them in service.
16. Cutting through fences and hedges where required and replacing these when necessary.
17. Installing gates and fences where required.

Estimating manhours for removing trees

NET MANHOURS–EACH

Average Tree Diameter in Inches	Softwood Trees		Hardwood Trees	
	Open Area	Congested Area	Open Area	Congested Area
Cross-cut Saws				
4	1.49	1.86	1.88	2.35
6	2.26	2.83	2.82	3.53
8	3.24	4.02	4.00	4.96
10	4.10	5.08	5.00	6.20
12	4.98	6.18	6.00	7.44
14	6.39	7.86	7.70	9.47
16	7.39	9.09	8.80	10.82
18	8.32	10.23	9.90	12.18
20	10.58	12.91	12.60	15.37
24	12.71	15.51	15.12	18.45
30	17.09	20.85	20.10	24.52
36	20.50	25.01	24.12	29.43

Average Tree Diameter in Inches	Softwood Trees		Hardwood Trees	
	Open Area	Congested Area	Open Area	Congested Area
Chain Saws				
4	0.37	0.46	0.47	0.59
6	0.57	0.71	0.71	0.89
8	0.81	1.00	1.00	1.24
10	1.03	1.28	1.25	1.55
12	1.25	1.55	1.50	1.86
14	1.60	1.97	1.93	2.37
16	1.85	2.28	2.20	2.71
18	2.08	2.56	2.48	3.05
20	2.65	3.23	3.15	3.84
24	3.18	3.88	3.78	4.61
30	4.28	5.22	5.03	6.14
36	5.13	6.26	6.03	7.36

Manhours include ax trimming, cutting down with cross-cut saws or chain saws, and cutting into four-feet lengths for the tree diameter sizes as listed above. Manhours are average for various heights of trees.

Manhours do not include hauling, piling and burning of trees or branches, or the removal of stumps.

Estimating manhours for removing tree stumps

NET MANHOURS – EACH

Item	Laborer	Oper. Engr.	Powder Man	Total
Grub & Removal by Hand				
8″ to 12″ diameter	6.00	–	–	6.00
14″ to 18″ diameter	7.50	–	–	7.50
20″ to 24″ diameter	9.00	–	–	9.00
26″ to 36″ diameter	11.20	–	–	11.20

NET MANHOURS – EACH

Item	Laborer	Oper. Engr.	Powder Man	Total
Blast & Pull with Tractor				
8″ to 12″ diameter	.83	.11	1.50	2.44
14″ to 18″ diameter	1.05	.23	2.33	3.61
20″ to 24″ diameter	1.50	.30	3.38	5.18
26″ to 36″ diameter	2.11	.42	4.76	7.29

Manhours include excavating and removing by hand or blasting and removing with cables and tractors. Manhours do not include burning or removal from premises.

Clearing and grading right-of-way

Equipment Spread

Equipment Description	50 Linear Feet Width				80 Linear Feet Width				100 Linear Feet Width			
	L	M	MH	H	L	M	MH	H	L	M	MH	H
D8 Tractor W/Dozer ...	1	1	1	1	1	1	2	2	1	2	3	3
D7 Tractor W/Dozer ...	0	0	1	1	1	1	1	1	1	1	2	2
Truck—2½ Ton Dump ..	2	2	2	2	2	2	2	2	2	2	3	3
Truck— Pick-up	1	1	2	2	1	2	3	3	2	3	4	4
Ripper or Brushrake .	1	1	1	1	1	1	1	1	1	2	2	2

NUMBER OF UNITS FOR

Labor Crew

Personnel Description	50 Linear Feet Width				80 Linear Feet Width				100 Linear Feet Width			
	L	M	MH	H	L	M	MH	H	L	M	MH	H
Foreman	1	1	2	2	1	2	3	3	2	3	4	4
Operator	1	1	2	2	1	2	3	3	2	4	5	5
Mechanic ...	1	1	1	1	1	1	1	1	1	1	1	1
Swamper	1	1	2	2	1	2	3	3	2	4	5	5
Truck Driver	2	2	2	2	2	2	2	2	2	2	3	3
Laborer	10	15	20	30	15	25	30	40	20	35	40	50
Total Crew ..	16	21	29	39	21	34	42	52	29	49	58	68

NUMBER OF MEN FOR

Above equipment spread should be ample for clearing and grading one mile of right-of-way per 10-hour day for the width and conditions as are outlined. Haul trucks are based on round trip haul of two miles. If brush and trees are to be burned on site, omit above dump trunks. Small tools such as saws, axes, etc., must be added as required for the individual job.

Code Description.

L = Light—light brush and grass, no trees.
M = Medium—considerable brush of larger size.
MH = Medium Heavy—large brush and small trees.
H = Heavy—much small brush, many small trees, and occasional large trees.

Above total crew should be ample for clearing and grading one mile of right-of-way per 10-hour day for the width and conditions outlined. Crew spread includes cutting, stacking or piling, loading, and hauling a round-trip distance of two miles. If burning is necessary or permitted, substitute fire tenders for dump truck drivers. See Clearing and Grading equipment spread for number of dump trucks.

Code Description.

L = Light—light brush and grass, no trees.
M = Medium—considerable brush of larger size.
MH = Medium Heavy—large brush and small trees.
H = Heavy—much small brush, many small trees, and occasional large trees.

Source

Page, J. S., *Cost Estimating Man-Hour Manual for Pipelines and Marine Structures*, Gulf Publishing Co., Houston, Texas, 1977.

DITCHING

How many cubic yards of excavation in a mile of ditch?

Multiply the width in inches by the depth in inches by 1.36; the answer is cubic yards per mile.

Example. How many cubic yards of excavation in a mile of 12-inch ditch 30-inch deep?

$12 \times 30 \times 1.36 = 490$ cubic yards per mile.

The rule is correct within about 1/5 of 1 percent; actually the errors in depth and width are much greater than this. To get the cubic yards per 1,000 feet, as in computing rock ditching, use 0.257 instead of 1.36.

Shrinkage and expansion of excavated and compacted soil

Ever notice how the spoil from a ditch occupies a greater volume than the ditch itself? There's a reason. *Excavate sand and it expands about 10 percent. Ordinary soil expands about 25 percent and clay expands about 40 percent.*

Here's a summary of the bulk you can expect from excavated soil:

Type soil	Undisturbed	Excavated	Compacted
Sand	1.00	1.11	0.95
Ordinary earth	1.00	1.25	0.90
Clay	1.00	1.43	0.90

Example. Find the volume of loose spoil from a pipe line ditch 42 inches wide, 60 inches deep. The excavation is through clay.

Volume of undisturbed clay soil:
$3.5 \times 5.0 = 17.5$ cu ft per lineal foot of ditch.

Volume of the spoil will be 143 percent of the undisturbed volume.

Volume of spoil $= 1.43 \times 17.5 = 25 +$ cu ft per lineal foot.

Through mechanical compaction this volume can be reduced to 15.75 cu ft.

Ditching and trenching: labor/equipment considerations

In determining the labor crew and equipment spread for ditching and trenching operations, the following should be given consideration should they apply to the particular project:

1. Ditching or trenching for buried pipelines should be in accordance with the following table of minimum width and coverage for all soil formations.
2. In rock, cut ditches at least six (6) inches wider.
3. If dirt-filled benches are used, ditch should be excavated deeper to obtain proper coverage.
4. Trench should be excavated to greater depth when required for proper installation of the pipe where the topography of the country warrants same.
5. Repair any damage to and maintain existing natural or other drainage facilities.
6. Do not open ditch too far in advance of pipelay crew.
7. Obtain permits for blasting.
8. When blasting, use extreme caution and protection.
9. Clean up blasted rock to prevent damage to coated pipe.

Nominal Pipe Size Inches	Minimum Width Inches	Normal Minimum Coverage Inches
4	22	30
6	26	30
8	26	30
10	26	30
12	30	30
14	32	30
16	36	30
18	38	30
20	40	30
24	44	30
30	50	30
36	52	36
42	58	36

CONCRETE WORK

How to approximate sacks of cement needed to fill a form

To obtain a close estimate of the number of sacks of cement that will be required, first determine the volume (cubic feet) to be filled in the form. Divide the volume by 4.86 to approximate the number of sacks of cement needed.

Example. How many 94-pound sacks of cement will be required to fill a form for a concrete base 10 feet by 10 feet if it is to be 6 inches thick?

$$\frac{10 \times 10 \times 0.5}{4.86} = 10.3, \text{ or } 11 \text{ sacks needed.}$$

What you should know about mixing and finishing concrete

To determine proper mix, divide the constant 44 by the sum of the parts of cement, the parts of sand, and parts of gravel to determine the number of bags of Portland cement required. Multiply the number of bags of cement as determined above by parts of sand and the constant 0.035 to calculate the number of cubic yards of sand needed. To determine cubic yards of gravel needed, multiply bags of cement by parts of gravel × 0.035.

Example. Calculate the quantities of cement, sand, and gravel required for one cubic yard of 1:2:4 concrete.

$$C = \frac{44}{1 + 2 + 4}$$

$$= 6.28 \text{ bags of cement}$$
$$S = 6.28 \times 2 \times 0.035$$
$$= 0.44 \text{ cubic yard of sand}$$
$$G = 6.28 \times 4 \times 0.035$$
$$= 0.88 \text{ cubic yard of gravel}$$

To increase or decrease slump of concrete, add or subtract one gallon of water per cubic yard of mix and subtract or add 20 pounds of aggregate to maintain yield.

To adjust from no air to air-entrained concrete and maintain strength, reduce water 1/4 gallon per sack of cement and reduce sand 10 pounds per sack of cement for each 1% of entrained air.

PIPE LAYING

How to determine the degrees of bend in a pipe that must fit a ditch calling for a bend in both horizontal and vertical planes

Rule. To find the number of degrees in the combination bend, square the side bend and the sag or overbend; add them together and extract the square root. The answer will be the number of degrees necessary to make the pipe fit the ditch.

Example. Determine the bend to make in a pipe whose ditch has a 3° overbend and a 4° sidebend. Let X = the unknown angle:

$$X^2 = 3^2 + 4^2$$
$$X^2 = 9 + 16$$

$$X^2 = 25$$
$$X = 5°$$

Example. Determine the bend to make in a pipe whose ditch has a 9° sag and 12° sidebend. Let X = the unknown angle:

$$X^2 = 9^2 + 12^2$$
$$X^2 = 225$$
$$X = 15°$$

How to bend pipe to fit ditch—sags, overbends and combination bends

To make straight sags or overbends fit the ditch (see drawing) add angles of unlike signs, subtract those of like signs.

Example. (Sta. 2 + 40)

Slope − 15° 00′
Slope − 1° 00′
(sub.) 14° 00′ overbend

In cases involving either a sag or overbend, in addition to a side-bend, the rule becomes: *Make the combination bend equal to the largest angle plus ⅓ of the smallest.*

Example. (Sta. 1 + 40)

Slope + 10° 00′
Slope − 1° 00′
(add) 11° 00′ overbend (1)
      6° 00′ side-bend left (2)

Combination bend = 11° + (⅓ × 6°) = 13° 00′ overbend left.

Note: This rule gives an error of approximately 1° for a maximum bend of 18.5°.

Pipe bending computations made with hand-held calculator

Maximum code radii for pipe cold bends requires printer for calculations

Frank E. Hangs, Sovereign Engineering, Inc., Houston

Cold bending pipe is subject to provisions of Liquid Petroleum Transportation Piping Systems, B31.4, and Gas Transmission and Distribution Piping Systems, B31.8. These codes stipulate a minimum bending radius for each size pipe in addition to requirements for thinning, flattening, etc.

The following program (Table 1), written for the Hewlett Packard 41C/CV calculator, addresses the geometry of fabricating offset bends, sloping scraper traps, connections and direction changes common to all pipe lines.

Minimum code radii are calculated for each size of oil or gas line. The vertical distance between below and above ground center lines is matched to a design distance with given tangents (5 ft is a convenient tangent for pipe bending machines). The bend angle, overall horizontal distance, and total length of pipe are calculated for each configuration.

A printer is a must. The prompting feature of this calculator asks for specific input data. The Results Recap routine prints out calculated data and inputs. Each item is identified for permanent record. Additional printouts of Results Recap may be made by XEQ "DATA."

Fig. 1 and Example 1 are a typical offset bend.

Fig. 2 and Example 2 are a conventional direction change.

Fig. 3 and Example 1 show receiver and launcher scraper trap connections, where the traps are not level.

Fig. 4 and Example 3 are a crossing under a foreign line, where offset bends are used.

The oil and gas codes differ in bending requirements. These are defined by the radius of bend/diameter ratio. The oil code, B31.4, specifies the R/D ratios as follows: for 12 in. and less, use 18; 14 in., 21; 16 in., 24; 18 in., 27; 20 in. and larger, use 30.

The gas code, B31.8, specifies for all pipe 12 in. and larger, the ratio is a constant of 38.1987. (The code states "longitudinal axis of the pipe shall not be deflected more than 1½ degrees in any length along the pipe axis equal to the diameter of the pipe.")

The program prompts: Gas pipe line greater than 12 in. Yes? or No? (Gas PL = >12Y?N?). Is line gas or oil? When this information and other data are keyed in, the proper radius is determined and the results computed and printed out.

If the scraper trap routine is desired, it should be run immediately after "Bend" while pertinent data are in storage; XEQ 15 for each receiver and each launcher.

Here again the program asks if this is a receiver (R trap) Y? or N? Note that trap angle is requested for R & L traps (they could differ), and the length of the traps may not be equal.

The direction change, Example 2 (XEQ 17), can be run at any time since it prompts for input data. If this is a code bend, then R must be determined for pipe size and service of line by—"bend" (use results of previous run for same diameter or key in data as shown on Example 1 to obtain R). Use this minimum radius or any larger radius. Points on a circle of given radius may be calculated for any angle. Subdivide angle, and for each subdivision key in R and let T = 0, thus determining points on curve.

Formulas for reverse bends

$$V = 2R(1 - \cos \angle 1) + 2T \sin \angle 1$$
$$H = 2R \sin \angle 1 + 2T \cos \angle 1 + 2T$$

$$L = 2R\frac{\angle 1}{57.2958} + 4T$$

(text continued on page 40)

TABLE 1—Examples of computations using the cold bend program

```
EXAMPLE 1
              XEQ "BEND"

 PIPE BENDS

∠1 TRIAL=?
              20.0000    RUN
D=? IN.
              12.7500    RUN
GAS PL=>12? Y?, N?
N                        RUN
V=?
              8.0000     RUN
T=?
              5.0000     RUN
∠1 INCR=?
              0.5000     RUN

RESULTS RECAP

∠1=25.5000
R/D=18.0000
R=19.1250
H=35.4929
L=37.0235
V CALC=8.0312
V GIVEN=8.0000

D IN=12.7500
D FT=1.0625
T=5.0000
∠1 INCR=0.5000
                   XEQ 15
 SCRAPERTRAPS

TRAP ∠=?
              5.0000     RUN
R TRAP? Y? N?
Y                        RUN

VR=7.5227
HR=37.1407
LR=38.6925
TRAP ∠=5.0000
LRT=?
              10.0000    RUN

VRT=6.6511
HRT=47.1027
                   XEQ 15
 SCRAPERTRAPS

TRAP ∠=?
              5.0000     RUN
R TRAP? Y? N?
N                        RUN

VL=8.3942
HL=33.8070
LL=35.3545
TRAP ∠=5.0000
```

```
EXAMPLE 1 CONT.
LLT=?
              5.0000     RUN

VLT=8.8300
HLT=38.7880

EXAMPLE 2
                   XEQ 17
 DIRECTION
   CHANGE
∠ TURN=?
              30.0000    RUN
R=?
              80.0000    RUN
T=?
              5.0000     RUN

RESULTS

V=13.2180
H=49.3301
L=51.8879

∠ TURN=30.0000
R=80.0000

EXAMPLE 3
              XEQ "BEND"

 PIPE BENDS

∠1 TRIAL=?
              8.0000     RUN
D=? IN.
              14.0000    RUN
GAS PL=>12? Y?, N?
N                        RUN
V=?
              2.5130     RUN
T=?
              5.0000     RUN
∠1 INCR=?
              0.5000     RUN

RESULTS RECAP

∠1=10.5000
R/D=21.0000
R=24.5000
H=28.7621
L=28.9797
V CALC=2.6429
V GIVEN=2.5130

D IN=14.0000
D FT=1.1667
T=5.0000
∠1 INCR=0.5000
```

(table continued on next page)

TABLE 1—Examples of computations using the cold bend program (continued)

01♦LBL "BEND"	64 38.1972	125 RCL 05	190 CLA	256 RCL 17	322 RCL 18	388 PROMPT
02 ADV	65 XEQ 13	126 COS	191 "∡1 INCR="	257 -	323 +	389 STO 32
03 SF 12		127 *	192 ARCL 11	258 RCL 18	324 STO 24	390 RCL 30
04 "PIPE "	66♦LBL 13	128 +	193 AVIEW	259 +	325 RCL 13	391 COS
05 "⊢BENDS"	67 STO 03	129 STO 12	194 CLA	260 STO 21	326 RCL 19	392 CHS
06 PRA	68 RCL 07	130 RCL 10	195 BEEP	261 RCL 13	327 +	393 1
07 ADV	69 *	131 4	196 ADV	262 RCL 19	328 STO 25	394 +
08 CF 12	70 STO 08	132 *	197 ADV	263 -	329 ADV	395 RCL 31
09 "∡1 TRIAL=?"	71 GTO 05	133 RCL 08	198 STOP	264 STO 22	330 ADV	396 *
10 PROMPT		134 2		265 ADV	331 "VR="	397 RCL 30
11 STO 05	72♦LBL 05	135 *	199♦LBL 15	266 "VL="	332 ARCL 23	398 SIN
12 "D=? IN."	73 "V=?"	136 RCL 05	200 SF 12	267 ARCL 20	333 AVIEW	399 RCL 32
13 PROMPT	74 PROMPT	137 57.2958	201 "SCRAPER"	268 AVIEW	334 CLA	400 *
14 STO 06	75 STO 09	138 /	202 "⊢TRAPS"	269 CLA	335 "HR="	401 +
15 12	76 "T=?"	139 *	203 PRA	270 "HL="	336 ARCL 24	402 STO 33
16 /	77 PROMPT	140 +	204 CF 12	271 ARCL 21	337 AVIEW	403 RCL 30
17 STO 07	78 STO 10	141 STO 13	205 ADV	272 AVIEW	338 CLA	404 SIN
18 "GAS PL=>12?"	79 "∡1 INCR=?"		206 "TRAP ∡=?"	273 CLA	339 "LR="	405 RCL 31
19 "⊢ Y?, N?"	80 PROMPT	142♦LBL "DATA"	207 PROMPT	274 "LL="	340 ARCL 25	406 *
20 AON	81 STO 11	143 ADV	208 STO 04	275 ARCL 22	341 AVIEW	407 RCL 32
21 PROMPT		144 CLA	209 COS	276 AVIEW	342 CLA	408 +
22 ASTO Y	82♦LBL 01	145 "RESULTS "	210 CHS	277 CLA	343 "TRAP ∡="	409 RCL 30
23 AOFF	83 RCL 08	146 "⊢RECAP"	211 1	278 "TRAP ∡="	344 ARCL 04	410 COS
24 "Y"	84 2	147 AVIEW	212 +	279 ARCL 04	345 AVIEW	411 RCL 32
25 ASTO X	85 *	148 CLA	213 RCL 08	280 AVIEW	346 CLA	412 *
26 X=Y?	86 RCL 05	149 ADV	214 *	281 CLA	347 "LRT=?"	413 +
27 GTO 12	87 COS	150 "∡1="	215 STO 15	282 "LLT=?"	348 PROMPT	414 STO 34
28 12	88 CHS	151 ARCL 05	216 RCL 10	283 PROMPT	349 STO 01	415 RCL 30
29 RCL 06	89 1	152 AVIEW	217 RCL 04	284 STO 02	350 RCL 04	416 57.2958
30 X<Y?	90 +	153 CLA	218 SIN	285 RCL 04	351 SIN	417 /
31 GTO 06	91 *	154 "R/D="	219 *	286 SIN	352 *	418 RCL 31
32 14	92 RCL 10	155 ARCL 03	220 STO 16	287 *	353 CHS	419 *
33 RCL 06	93 2	156 AVIEW	221 RCL 08	288 RCL 20	354 RCL 23	420 RCL 32
34 X=Y?	94 *	157 CLA	222 RCL 04	289 +	355 +	421 2
35 GTO 07	95 RCL 05	158 "R="	223 SIN	290 STO 26	356 STO 28	422 *
36 16	96 SIN	159 ARCL 08	224 *	291 RCL 02	357 RCL 24	423 +
37 RCL 06	97 *	160 AVIEW	225 STO 17	292 RCL 04	358 RCL 01	424 STO 35
38 X=Y?	98 +	161 CLA	226 RCL 10	293 COS	359 RCL 04	425 ADV
39 GTO 08	99 STO 14	162 "H="	227 RCL 04	294 *	360 COS	426 "RESULTS"
40 18	100 FS? 01	163 ARCL 12	228 COS	295 RCL 21	361 *	427 XEQ "PRA"
41 RCL 06	101 GTO 03	164 AVIEW	229 *	296 +	362 +	428 ADV
42 X=Y?	102 RCL 09	165 CLA	230 STO 18	297 STO 27	363 STO 29	429 "V="
43 GTO 09	103 X<>Y	166 "L="	231 RCL 04	298 ADV	364 ADV	430 ARCL 33
44 19.99	104 X<=Y?	167 ARCL 13	232 57.2958	299 "VLT="	365 "VRT="	431 AVIEW
45 RCL 06	105 GTO 02	168 AVIEW	233 /	300 ARCL 26	366 ARCL 28	432 CLA
46 X>Y?	106 GTO 03	169 CLA	234 RCL 08	301 AVIEW	367 AVIEW	433 "H="
47 GTO 10		170 "V CALC="	235 *	302 CLA	368 CLA	434 ARCL 34
	107♦LBL 02	171 ARCL 14	236 STO 19	303 "HLT="	369 "HRT="	435 AVIEW
48♦LBL 06	108 RCL 11	172 AVIEW	237 "R TRAP?"	304 ARCL 27	370 ARCL 29	436 CLA
49 18	109 ST+ 05	173 CLA	238 "⊢ Y? N?"	305 AVIEW	371 AVIEW	437 "L="
50 XEQ 13	110 GTO 01	174 "V GIVEN="	239 AON	306 CLA	372 CLA	438 ARCL 35
		175 ARCL 09	240 PROMPT	307 ADV	373 ADV	439 AVIEW
51♦LBL 07	111♦LBL 03	176 AVIEW	241 ASTO Y	308 ADV	374 STOP	440 CLA
52 21	112 RCL 10	177 CLA	242 AOFF	309 STOP		441 ADV
53 XEQ 13	113 2	178 ADV	243 "Y"		375♦LBL 17	442 ADV
54♦LBL 08	114 *	179 "D IN="	244 ASTO X	310♦LBL 16	376 SF 12	443 "∡ TURN="
55 24	115 RCL 08	180 ARCL 06	245 X=Y?	311 RCL 14	377 " DIRECTION "	444 ARCL 30
56 XEQ 13	116 2	181 AVIEW	246 GTO 16	312 RCL 15	378 "⊢ CHANGE"	445 AVIEW
	117 *	182 CLA	247 RCL 14	313 -	379 PRA	446 CLA
57♦LBL 09	118 RCL 05	183 "D FT="	248 RCL 15	314 RCL 16	380 CF 12	447 "R="
58 27	119 SIN	184 ARCL 07	249 -	315 -	381 "∡ TURN=?"	448 ARCL 31
59 XEQ 13	120 *	185 AVIEW	250 RCL 16	316 STO 23	382 PROMPT	449 AVIEW
	121 +	186 CLA	251 +	317 RCL 12	383 STO 30	450 CLA
60♦LBL 10	122 RCL 10	187 "T="	252 STO 20	318 RCL 10	384 "R=?"	451 ADV
61 30	123 2	188 ARCL 10	253 RCL 12	319 -	385 PROMPT	452 ADV
62 XEQ 13	124 *	189 AVIEW	254 RCL 10	320 RCL 17	386 STO 31	453 STOP
63♦LBL 12			255 -	321 +	387 "T=?"	454 .END.

Fig. 1—Cold offset bends.

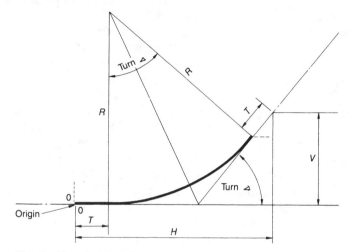

Fig. 2—Pipe direction change.

Legend & registers

Reg. no.
00	Not used	
01	*LRT*	Length of receiver trap (ft)
02	*LLT*	Length of launcher trap (ft)
03	$\frac{R}{D}$	Radius of bend/pipe diameter (ft/ft)
04	Trap ∡	Scraper trap angle in degrees
05	∡ 1	Bend angle in degrees

Note: Above ground pipe ends are modified as shown for traps not level. Face to face of valve included in *LRT* and *LLT*.

Fig. 3—Scraper traps.

Fig. 4—Crossing under foreign lines with offset bends.

06	*D*—in.	Pipe diameter (in.)
07	*D*—ft	Pipe diameter (ft)
08	*R*	Bend radius (ft)
09	*V*	Vertical distance desired (ft between center lines)
10	*T*	Tangent (ft)
11	∡ Incr.	Bend angle increment in degrees
12	*H*	Horizontal distance, overall (ft)
13	*L*	Length of pipe, (ft)
14	*V* Calc.	Vertical distance calculated (ft)
15	$R\,(1 - \cos \text{Trap } \angle)$	
16	$T \sin \text{Trap } \angle$	
17	$R \sin \text{Trap } \angle$	
18	$T \cos \text{Trap } \angle$	
	Trap ∡	
19	$\dfrac{R}{57.2958}$	
20	*VL*	Vertical distance to launcher trap connection (ft)
21	*HL*	Horizontal distance to launcher trap connection (ft)
22	*LL*	Length of pipe to launcher trap connection (ft)
23	*VR*	Vertical distance to receiver trap connection (ft)
24	*HR*	Horizontal distance to receiver trap connection (ft)
25	*LR*	Length of pipe to receiver trap connection (ft)
26	*VLT*	Vertical distance to end of launcher trap (ft)
27	*HLT*	Horizontal distance to end of launcher trap (ft)
28	*VRT*	Vertical distance to end of receiver trap (ft)
29	*HRT*	Horizontal distance to end of receiver trap (ft)
30	∡	Change direction desired in degrees
31	Radius	Change direction (ft)
32	Tangent	Change direction (ft)
33	*V*	Vertical distance calculated, change direction
34	*H*	Horizontal distance calculated, change direction
35	*L*	Length of pipe calculated, change direction.

(text continued from page 37)

Formulas for scraper trap connections

Receiver.

$$VR = V - R (1 - \cos \text{Trap} \angle) \ T \sin \text{Trap} \angle$$
$$HR = H - T + R \sin \text{Trap} \angle + T \cos \text{Trap} \angle$$

$$LR = L + R \ \frac{\text{Trap} \angle}{57.2958}$$

Launcher.

$$VL = V - R (1 - \cos \text{Trap} \angle) + T \sin \text{Trap} \angle$$
$$HL = H - T - R \sin \text{Trap} \angle + T \cos \text{Trap} \angle$$

$$LL = L - R \ \frac{\text{Trap} \angle}{57.2958}$$

Receiver trap.

$$VRT = VR - LRT \sin \text{Trap} \angle$$
$$HRT = HR + LRT \cos \text{Trap} \angle$$

Launcher trap.

$$VLT = VL + LLT \sin \text{Trap} \angle$$
$$HLT = HL + LLT \cos \text{Trap} \angle$$

Formulas for direction change

$$V = R (1 - \cos \text{Turn} \angle) + T \sin \text{Turn} \angle$$
$$H = R (\sin \text{Turn} \angle) + T (1 + \cos \text{Turn} \angle)$$

$$L = 2T + R \ \frac{\text{Turn} \angle}{57.2958}$$

User instructions

"Bend" program. Put in calculator, Size 35, XEQ "Bend." Key in prompted data and R/S each time. Key in a trial angle. Use a larger angle for smaller pipe as 15° for 12-in., 10° for 30-in. Results Recap prints out the calculated values and inputs. Each quantity is identified. Additional printouts may be made by XEQ "Data."

If one inadvertently puts in a larger trial angle than needed for a solution (i.e., calculated V approximates given V), only one calculation is made and printed out for this angle. If calculated V is too large, store smaller trial $\angle$ in 05, XEQ 01, to get proper result.

Remember to clear Flag 01 before resuming normal "Bend" calculations.

Parts of "Bend" routine may be used for small pipe in noncode work to determine bend angle for offsets. Do not XEQ "Bend" for noncode work as this determines R specified by code.

Procedure: XEQ clearing GTO "Bend."

Now select a radius (try: R = 18D: 3½ in. OD: R = 18 3.5/12 = 5.25). This can be changed if necessary. The V distance, offset, is determined by design configuration. Store R in 08, V in 09. Assume trial angle, say 20° for small pipe, store in 05. Let T = 0.5 ft, store in 10. SF 01. XEQ 01. (Only one calculation is performed.)

Inspect results. Do V given and V calc appear reasonably close? If not, take another "Fix" by changing R or angle (store new values, XEQ 01). This routine can be continued to a satisfactory solution. A point will be reached where R seems reasonable; choose an angle less than apparent solution, store in 05. CF 01, store 0.5 in 11, XEQ 01 and zero in. This returns to the iterative process for a more precise solution. T and R can be changed to suit.

"Bend" routine may be used to calculate bends for crossing under foreign lines. (See Example 3.)

The direction change routine (XEQ 17) can be used for code and noncode work. For code: determine minimum bend radius from "Bend" for pipe size and for gas or oil line. Use R from previous example or XEQ "Bend," key in data as in Example 1 for desired diameter.

For noncode work: user may employ a radius which is suitable in his judgment. Do not use "Bend" program as this calculates radius in accordance with codes. Key in radius and desired tangent. Caution: be aware that short radii for large pipes, say 8 in. and larger, soon get into hot bend category.

Example 1. Offset bend

12¾-in. oil line. Let V = 8 ft, T = 5 ft. Smaller diameters have larger bend $\angle$. Try 20°, XEQ "Bend," key in data and R/S. Flag 01 Clear. Use $\angle$ Incr. = 0.5°. Note: R/D = 18 (meets B31.4). V Calc. is close to 8 ft. Now try scraper trap routine XEQ 15. Let trap $\angle$ = 5°, T = 5 ft, receiver Y. Length of trap = 10 ft. For launcher XEQ 15, same $\angle$, T, receiver N. Length of trap = 5 ft.

Example 2. Direction change

24-in. gas line. Note from "Bend" run Min. R = 76.3944 ft. R/D = 38.1972, meets B31.8. Say 30° turn, let R = 80 ft, T = 5 ft XEQ 17. Key in data R/S.

Example 3. Crossing

A 14-in. oil line crosses under 10¾ in. foreign line. Take 2 ft, 0 in. min. clearance. Center distance = 3.0313 ft.

Min. height of bend = 2.513 ft = 'V.' Take 8° as trial ∡, T = 5 ft, ∡ Incr. = 0.5°. Note Calc. V is close to given V. This is a possible solution. Note, however, this is an alterna-

tive solution. Builders will take advantage of sag bends where possible, depending on site conditions, space available, pipe size, etc.

Calculate maximum bend on cold pipe

How much of a bend can cold straight pipe take without over stressing? In other words, in laying straight pipe cross-country, what is the smallest radius of curvature possible within the allowable stress?

The following is an analytical solution to these questions.

The deflection equation for a length of straight pipe used as a cantilever beam with a concentrated end load is:

$$\Delta y = \frac{Pl^3}{3EI} \tag{1}$$

The maximum stress in the beam is:

$$s = \frac{M}{Z} \tag{2}$$

The maximum bending moment is:

$$M = Pl \tag{3}$$

If the allowable stress is S_A, rewriting Equation 2 gives:

$$S_A = \frac{M}{Z} \tag{4}$$

$$S_A = \frac{Pl}{Z} \tag{5}$$

$$Pl = Z\,S_A \tag{6}$$

Rewriting Equation 1:

$$\Delta y = \frac{(Pl)\,l^2}{3\,EI} = \frac{Z\,S_A\,l^2}{3\,EI} \tag{7}$$

Taking the first and second derivatives of Equation 7:

$$\frac{d\,\Delta y}{d\,l} = \frac{2\,l\,ZS_A}{3\,EI} \tag{8}$$

$$\frac{d^2\,\Delta y}{d\,l^2} = \frac{2\,Z\,S_A}{3\,EI} \tag{9}$$

Radius of curvature is given by the equation:

$$R = \frac{\left[1 + \left(\dfrac{d\,\Delta y}{d\,l}\right)^2\right]^{3/2}}{\dfrac{d^2\,\Delta y}{d\,l^2}} \tag{10}$$

Substituting in Equation 10:

$$R = \frac{\left[1 + \left(\dfrac{2\,l\,ZS_A}{3\,EI}\right)^2\right]^{3/2}}{\dfrac{2\,Z\,S_A}{3\,EI}} \tag{11}$$

$$\lim_{l \to 0} R = \frac{1^{8/2}}{\dfrac{2\,Z\,S_A}{3\,EI}} = \frac{3\,EI}{2\,Z\,S_A} \tag{12}$$

Notation

Δ y deflection, in.
  l length, in.
  P load, lb
  E modulus of elasticity, lb/sq in.
  I cross-sectional moment of inertia, in⁴.
  S Stress, lb/sq in.
  S_A allowable stress, lb/sq in. from Code for Pressure Piping
  M bending moment, inch-pounds
  Z section modulus, in³.
  R Radius of curvature, in.

Example. What is the smallest radius of curvature that can be used without overstressing straight 24-in. Schedule 20 seamless steel pipe ASTM 53? The pipe is at 70°F, and is to be used for oil outside of refinery limits.

$$R = \frac{3\,EI}{2\,Z\,S_A}$$

$E = 29 \times 10^6$ lb/sq in.

$I = 1,943$ in^4.

$Z = 162$ in^3.

$S_A = 25,500$ psi (from Code for Pressure Piping)

$$R = \frac{3 \times 29 \times 10^6 \times 1,943}{2 \times 162 \times 25,500} = 20,460 \text{ in. or } 1,705 \text{ ft.}$$

The section modulus = $Z = 2I/D$, and may be simplified to $Z = I/r$ and substituted for Z in Equation 12.

I = Moment of inertia
r = Distance from center pipe
D = Outside diameter of pipe

Equation 12 then becomes:

$R = 3Er/2Sa$ (13)

$R = (21.75 \times 10^6/Sa) \times D$ (14)

Sa, the allowable stress, is determined by the design factor permitted by the design code used. Equation 14 may be further simplified by using the following table.

Minimum Radius of Curvature, R

Yield Stress (psi)	F = 0.72	F = 0.6	F = 0.5	F = 0.4
35,000	863 D	1,036 D	1,243 D	1,553 D
42,000	719 D	863 D	1,036 D	1,295 D
46,000	657 D	788 D	946 D	1,182 D
52,000	581 D	697 D	836 D	1,046 D
56,000	539 D	647 D	777 D	971 D
60,000	503 D	604 D	725 D	906 D
65,000	465 D	558 D	669 D	836 D
70,000	431 D	518 D	621 D	776 D
80,000	377 D	453 D	544 D	680 D

Example. What is the minimum radius of curvature that can be used without overstressing 42″ API 5LX X-65 line pipe at 72°F used in type C construction (gas line)? The design factor from ANSI B31.8 is 0.5; therefore use the column headed F = 0.5.

$R = 669 \times 42'' = 28,098''$ or 2,342 ft

For liquid lines designed in accordance with ANSI B31.4, use the column headed F = 0.72.

Determine length of a pipe bend

For smooth bends (other than wrinkle bends) use the following formula:

$L = R \times D \times .017 + T$

where L = length of bend
        R = radius of bend (center line of pipe)
        D = number of degrees in the bend
        T = total length of the two tangents at either end

Example. Find the length of a piece of pipe required to make a 90° bend with a radius of 60 in. There's to be a four-in. tangent at either end.

L = ?
R = 60
D = 90
T = 8
L = 60 × 90 × .01745 + 8
  = 94.23 + 8
  = 102.23 in.

Length of pipe in arc subtended by any angle

Application

To find the length of pipe subtended by an angle of 42° − 35 feet with a radius of 24 inches.

Assuming R = 1 inch,

angle of 42° subtends an arc = 0.7330″
angle of 35′ subtends an arc = $\underline{0.01018''}$
                                 0.7432″

DEGREES						MINUTES			
Degrees	Length Of Arc	Degrees	Length Of Arc	Degrees	Length Of Arc	Minutes	Length Of Arc	Minutes	Length Of Arc
1	.0175	31	.5411	61	1.0647	1	.000291	31	.009018
2	.0349	32	.5585	62	1.0821	2	.000582	32	.009308
3	.0524	33	.5760	63	1.0996	3	.000873	33	.009599
4	.0698	34	.5934	64	1.1170	4	.001164	34	.009890
5	.0873	35	.6109	65	1.1345	5	.001454	35	.010181
6	.1047	36	.6283	66	1.1519	6	.001745	36	.010472
7	.1222	37	.6458	67	1.1694	7	.002036	37	.010763
8	.1396	38	.6632	68	1.1868	8	.002327	38	.011054
9	.1571	39	.6807	69	1.2043	9	.002618	39	.011345
10	.1745	40	.6981	70	1.2217	10	.002909	40	.011636
11	.1920	41	.7156	71	1.2392	11	.003200	41	.011926
12	.2094	42	.7330	72	1.2566	12	.003491	42	.012217
13	.2269	43	.7505	73	1.2741	13	.003782	43	.012508
14	.2443	44	.7679	74	1.2915	14	.004072	44	.012799
15	.2618	45	.7854	75	1.3090	15	.004363	45	.013090
16	.2793	46	.8029	76	1.3265	16	.004654	46	.013381
17	.2967	47	.8203	77	1.3439	17	.004945	47	.013672
18	.3142	48	.8378	78	1.3614	18	.005236	48	.013963
19	.3316	49	.8552	79	1.3788	19	.005527	49	.014254
20	.3491	50	.8727	80	1.3963	20	.005818	50	.014544
21	.3665	51	.8901	81	1.4137	21	.006109	51	.014835
22	.3840	52	.9076	82	1.4312	22	.006400	52	.015126
23	.4014	53	.9250	83	1.4486	23	.006690	53	.015417
24	.4189	54	.9425	84	1.4661	24	.006981	54	.015708
25	.4363	55	.9599	85	1.4835	25	.007272	55	.015999
26	.4538	56	.9774	86	1.5010	26	.007563	56	.016290
27	.4712	57	.9948	87	1.5184	27	.007854	57	.016581
28	.4887	58	1.0123	88	1.5359	28	.008145	58	.016872
29	.5061	59	1.0297	89	1.5533	29	.008436	59	.017162
30	.5236	60	1.0472	90	1.5708	30	.008727	60	.017453

The arc subtended for any angle is directly proportional to radius:

$$\frac{1 \text{ inch}}{24 \text{ inches}} = \frac{0.7432 \text{ inches}}{\text{Arc}}$$

$$\text{Arc} = 0.7432 \times 24 = 17^{13}/_{16}''$$

Average pipelay table—Underground

Linear feet of pipelay per ten (10) hour day in trench in level, rock-free terrain

Nominal Pipe Size	PIPE WALL THICKNESS IN INCHES					
	0.000 through 0.250	0.251 through 0.375	0.376 through 0.500	0.501 through 0.625	0.626 through 0.750	0.756 through 1.000
4	10,800	10,500	10,300	—	—	—
6	9,600	9,400	9,210	—	—	—
8	9,120	8,900	8,720	8,550	—	—
10	8,900	8,760	8,580	8,410	—	—
12	8,600	8,400	8,230	8,070	7,910	—
14	8,400	8,230	8,160	8,000	7,840	—
16	8,000	7,840	7,680	7,530	7,380	7,230
18	7,800	7,640	7,500	7,350	7,200	7,050
20	7,400	7,250	7,100	6,960	6,820	6,680
24	—	6,600	6,470	6,340	6,210	6,080
30	—	6,000	5,880	5,760	5,640	5,530
36	—	—	5,400	5,300	5,200	5,090
42	—	—	4,800	4,700	4,600	4,500

Productivity will vary where different types of terrain/or rock are encountered.

Above footage is based on installing double random joints of pipe.

Average pipelay table—On supports

Linear feet of pipelay per ten (10) hour day on waist-high supports on level ground

Nominal Pipe Size	PIPE WALL THICKNESS IN INCHES					
	0.000 through 0.250	0.251 through 0.375	0.376 through 0.500	0.501 through 0.625	0.626 through 0.750	0.751 through 1.000
4	11,320	11,100	10,880	—	—	—
6	10,080	9,880	9,680	—	—	—
8	9,560	9,370	9,190	9,010	—	—
10	9,390	9,200	9,020	8,840	—	—
12	8,800	8,620	8,500	8,330	8,160	—
14	8,700	8,570	8,400	8,250	8,090	—
16	8,040	7,880	7,730	7,580	7,430	7,280
18	7,800	7,650	7,500	7,350	7,200	7,060
20	7,550	7,400	7,250	7,100	6,960	6,820
24	—	6,720	6,590	6,460	6,330	6,200
30	—	6,120	6,000	5,880	5,770	5,660
36	—	—	5,440	5,350	5,240	5,140
42	—	—	4,840	4,740	4,650	4,560

Productivity will vary with contour of terrain and accessibility.

Above footage is based on installing double random joints of pipe.

Allowable pipe span between supports

When it is desired to run pipelines on supports above the ground, the engineer must often calculate the allowable pipe span between supports from a structural viewpoint. The formula for the span length can be obtained by equating the maximum bending moment caused by the loading to the resisting moment of the pipe. This formula is:

$$L = \left(\frac{SI}{wc}\right)^{0.5}$$

where L = span length, ft
c = radius to outside of pipe, in.
I = moment of inertia of pipe, in.⁴
S = maximum allowable fiber stress of material, lb/in.²
w = uniform loading per unit length of pipe, lb/ft

The above formula applies to pipes above two inches in diameter. Reference 1 gives calculated pipe span data as well as estimated pipe loading for different pipe diameters

Figure 1. Allowable pipe span between supports.

taking into account the fluid load, dead load, and the horizontal component of the wind loading.

The design of the pipe span can be speeded up with Figure 1.

Example. Find the allowable span length for a 6-in. continuous steel pipe strung between supports. The estimated loading (sum of pipe, fluid, dead, and wind loads) is 36 lb/ft. The allowable fiber stress for steel can be taken at 20,000 lb/in.²

From any suitable handbook, c for a 6-in. pipe is 6.625/2 = 3.3175 in., I is 28.14 in.⁴ Now calculate I/c = 28.14/3.3175 = 8.5 in.³ Connect S = 20,000 lb/in.² on the S scale with 8.5 on the I/c scale to the intersection with the reference line. With the reference line as a pivot, connect the pivot point with W = 36. At the intersection with the L scale, obtain a pipe span of 68 feet.

Check the amount of deflection to make sure that it does not exceed your piping practices or applicable design code.

Source

Kuong, J. F., *Petroleum Refiner 39*, No. 10, 196 (1960).

Reference

1. Sweeney, R. J., *Chem. Eng.*, *63*, No. 3 199 (1956).

How engineers make pipe fit the ditch

By R. J. Brown

The five-men crews of engineers that keep track of the pipe and calculate side bends, overbends, sags, and combination bends can make money—or lose money—for any contractor crossing rough terrain. If they function smoothly, the bending crew and pipe gang have easy going and the lowering-in can be a breeze. But when they make mistakes, the welded-up pipe may not fit the ditch and this spells trouble for the contractor.

Bending Crew

As a rule there are two rod men, two chairmen, and one instrument man in the measurement crew. Their duties are to measure each joint of pipe, number it, and stake its position in the line. In addition they calculate degrees of bends, etc. Here is the way they perform these duties:

Measuring and Numbering Joints

(See Figure 1.) First, the crew measures all joints of pipe and numbers them from 0 to 100. (When 100 is reached, the numbering starts over.) Then, they stake the position of each joint of pipe in the line. They use stakes which are numbered according to the pipe, with the number of the joint behind on front of the stake and the joint ahead on back of the stake.

Stakes Set Back

Because the setup gang will move the pipe back half a joint's length and the pipe gang back another half, the engineers set the stakes from one to two joints length behind the corresponding numbered pipe.

Notes on Figure 1

Note that joint 25 was measured 39.9 feet, and the gap between stakes 24–25 and 25–26 was set at 29.9 feet to allow a 10-foot lap for tie-in. This leaves a pup long enough to be carried ahead and worked into the line. If the pups are distributed over sections of bends, there is a possibility of saving a cut in a full joint. In event the stakes catch up with the pipe, one joint is carried ahead of the stringing.

Calculating Bends

The method used to determine the size of overbend and sags is illustrated in Figure 2. Note that at 0 + 00 the instrument is set up on the ditch bottom at a height of 5.0 feet. The rod is set on the ditch bottom ahead of the radius of overbend. (Care should be taken to set neither transit nor rod on the arc of the change in slope.) The transit is moved behind the radius of overbend at 1 + 00 and the height of the instrument is measured to be 4.90 feet.

The rod is set just ahead of the change in radius of overbend and is read at 4.90 feet, establishing a slope of − 1°. The magnitude of bend can now be determined.

Add Unlike Angles

If the slope is positive and the tangent of the slope ahead is decreasing, an overbend is established. If the slope is pos-

Figure 1.

Figure 2.

Figure 3.

Figure 4.

Figure 5.

itive and the tangent ahead is greater positive slope, a sag is noted. The reverse holds true for change in negative slope: if the tangent ahead is increasing, the transition is an overbend; if it is decreasing, the transition is a sag. The unlike algebraic signs are added and like signs are subtracted. Figure 2, station 2 + 00, shows a positive 10° slope approaching the overbend and a negative 1° slope leaving it. Since the signs are unlike, the sum of the two angles is 11° and since the slope is positive and decreasing, the notation overbend is used.

Side Bend Angles

Side bends are determined by "eyeballing" a P.I. in the ditch and setting the instrument up over it. A back sight is taken on the ditch or preceding P.I., whichever is more easily read. The scope is flipped and a foresight is taken on the ditch ahead. Referring to Figure 2 at station 1 + 00, the deflection angle is 6° left; station 4 + 00 shows a deflection of 8° right.

Combination Bends

At points where side bends and vertical bends occur the angles are combined to make one angle rolled. The method of calculating is a rule of thumb: *the larger angle plus one-third the smaller angle.*

This rule is accurate enough to give satisfactory results in actual practice. The actual derivation of this rule comes from an approximation of the following:

$$Sin\ (combination\ angle) = \sqrt{Sin^2A + Sin^2B}$$

Angles A and B are the horizontal and vertical angles. Figure 3 shows a plot of the ⅓ rule and of the equation. From the plot it can be seen that for the maximum bend of 18.5 an error of slightly over 1° is introduced. The maximum error evidenced on the graph is due to the side and vertical bands being equal.

Marking Pipe

The final step of the measurement crew is marking the joints to be bent. After the bend is established, its position between the pipe stakes is then transferred to the pipe joint number matching the pipe stakes number. The bend size, the centerline of bend, and arrows indicating direction of construction are marked on the joint. (See Figure 5.) If a bend falls too close to the end of the pipe the joint would have to be moved, so as to leave enough room for the bend. A pup could be inserted immediately behind the joint in question to give better bend positioning.

PIPE LOWERING

How to lower an existing pipe line that is still in service

Lowering a loaded pipe line is a low-cost alternative for constructing new facilities

Marshall D. Cromwell, Senior Project Engineer, PT Caltex Pacific Indonesia, Jakarta, Indonesia

Lowering an existing line is a dirty little job, but it can have big cost benefits. The line can be lowered while remaining in service with no lost production, and the cost of lowering an existing pipe line section is relatively cheap.

No expenses are incurred for new pipe, valves, stopples, and fittings. Construction is much faster than cutting and relocating a pipe line section.

Construction of new highways, buildings, airport runways, and other facilities is often planned at locations where above-ground pipe lines are present. Relocating such lines can be extremely expensive in terms of downtime and new pipe line materials.

Long-delivery items, such as hot tap equipment, valves, and fittings can add to the pipe line construction time, delaying an already tight schedule for the new facility. The alternative is to lower the existing lines, with adequate protection, so as to eliminate the obstruction.

The following information describes how to calculate an optimum profile for smooth transitions, while keeping the pipe stresses within allowable limits, and how to physically lower the pipe line in the field to match this profile.

Figure 1. Road crossing profiles.

Precise engineering

There is more to lowering an existing line than just digging a hole and letting gravity take over. Engineering and design of a new profile must be precise so that the line is not dropped too suddenly, causing a buckle or rupture (the larger the pipe line, the more critical the profile).

For instance, in Saudi Arabia, an extensive highway modernization program resulted in scores of large-diameter pipe lines being lowered. A few line sections ruptured dur-

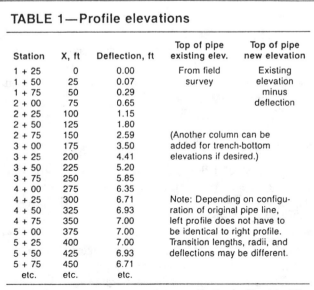

Station	X, ft	Deflection, ft	Top of pipe existing elev.	Top of pipe new elevation
1 + 25	0	0.00	From field	Existing
1 + 50	25	0.07	survey	elevation
1 + 75	50	0.29		minus
2 + 00	75	0.65		deflection
2 + 25	100	1.15		
2 + 50	125	1.80		
2 + 75	150	2.59	(Another column can be	
3 + 00	175	3.50	added for trench-bottom	
3 + 25	200	4.41	elevations if desired.)	
3 + 50	225	5.20		
3 + 75	250	5.85		
4 + 00	275	6.35		
4 + 25	300	6.71	Note: Depending on configu-	
4 + 50	325	6.93	ration of original pipe line,	
4 + 75	350	7.00	left profile does not have to	
5 + 00	375	7.00	be identical to right profile.	
5 + 25	400	7.00	Transition lengths, radii, and	
5 + 50	425	6.93	deflections may be different.	
5 + 75	450	6.71		
etc.	etc.	etc.		

After elevations are received from the field, end-points (Sta 1 + 25 and 8 + 75) should be checked by use of equations 7 and 4, if pipe line is at uphill slope.

Elevations

Station	Feet
1 + 00	7.25
1 + 25	7.50
1 + 50	8.00
1 + 50	7.43 New

$$A = \sqrt{(125-100)^2 + (7.50-7.25)^2} = 25.00125 \text{ ft}$$

$$B = \sqrt{(150-125)^2 + (7.50-7.43)^2} = 25.00010 \text{ ft}$$

$$C = \sqrt{(150-100)^2 + (7.43-7.25)^2} = 50.00032 \text{ ft}$$

$$R = 1,947.5 \text{ ft} = 23,370-\text{in.}$$

$$S = 22,340 \text{ psi}$$

No good—must come up with a new profile, with wider transition length, and check again by trial-and-error.

Figure 2. Elevation coordinates.

1. $R = \dfrac{x^2 + y^2}{2y}$

2. $x = \sqrt{2Ry - y^2}$

3. $y = R - \sqrt{R^2 - x^2}$

4. $R = \dfrac{EC^1}{S}$ or $S = \dfrac{EC^1}{R}$

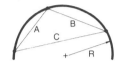

5. $y = \dfrac{EC^1}{S} - \sqrt{\left(\dfrac{EC^1}{S}\right)^2 - x^2}$

6. $x = \sqrt{2y\,\dfrac{EC^1}{S} - y^2}$

7. $R = \dfrac{A\,\mathrm{SIN}\left[90 - \mathrm{COS}^{-1}\dfrac{B^2 + C^2 - A^2}{2\,BC}\right]}{\mathrm{SIN}\left[2\,\mathrm{COS}^{-1}\dfrac{B^2 + C^2 - A^2}{2\,BC}\right]}$

(R = Feet here)

Note that formulas are independent of w.t., I, and cross-sectional area (cancelled out).

R = Radius of bend (in.)
x = Horizontal distance (in.)
y = Deflection or drop from the horizontal (in.)
S = Bending stress, psi

C^1 = One half diameter of pipeline (in.)
E = Young's modulus of elasticity
   (29×10^6 psi for steel)
A, B & C = Dimensions of triangle (ft.)

P/L profile

3 points on a circle

● = Inflection points
$x_1 = 25'$, $x_2 = 50$, etc.

Figure 3. Formulas for changes in slope.

Excavation begins. Ten-foot soil plugs remained at 50-ft centers to act as temporary pipe supports for the 36-in. line.

ing the pioneer stages due to the fact that each situation was not properly evaluated in terms of a critical engineering problem.

The key to the problem's solution is to provide a smooth profile with constant radius bends and to include a detailed list of instructions with the design package. Careful monitoring by survey equipment (levels) will assure smooth transitions in the field.

The following approach has been used effectively on numerous occasions in Saudi Arabia and Indonesia and can be recommended. Design steps to be included are:

- Determine how much soil cover there should be over the top of pipe line to assure protection from construction equipment and wheel loads (see API-RP-1102). Four feet of cover is adequate for pipe lines 12 in. and smaller. Larger pipe lines may require additional protection by way of more cover, pipe sleeves, or concrete slabs.

- Determine allowable bending stress on pipe line, based on grade of pipe. Normally, 75% of allowable stress is acceptable for cross-country pipe lines (ANSI B31.4, Section 419.6). However, 10,000-psi stress is very conservative and often preferred. If the pipe line is completely restrained (anchors on each end and no expansion loops), thermal stresses must also be considered.

- Once the total required drop of pipe line is established, calculate the minimum bending radius based on allowable bending stress. Then calculate the overall transition length from one end to the other.

- Obtain elevations on top of the pipe line every 25 ft (Table 1), marking station numbers and elevations with paint. If the line is on supports, also include grade elevations at each station.

- Starting at each end of transition, calculate the drop at 25-ft stations. Draw up a profile and fit it on existing profile. Add stations and new elevations (top of pipe and trench bottom) to profile drawing.

- Check the transition points at the beginning and end of new profile. If the pipe line was going uphill, and suddenly it needs to go downhill, the end points may be critical in terms of stress.

Since any three points not on the same line form an arc of constant radius, the stress may be approximated by Equations 7 and 4 (Figure 3). The end point should become the center of three points for calculation. If over-stressed, widen the transition length and check again.

Example 1. A new village was being built in the jungle, but access was obstructed by an above-ground, 36-in. pipe line. The village head requested that the line be buried or relocated so that bicycles, becaks, and water buffalo would have easier access.

It was determined that 4 ft of cover was required to protect the line from excessive wheel loads (people-packed buses and trucks). This means the line must be lowered 7 ft (Figure 1). The pipe is Grade B, allowable bending stress is 10,000 psi, and crossing width is 50 ft. To calculate minimum bending radius, total transition length, and deflections at 25-ft intervals, use:

$$R = EC'/S = (29,000,000)(18)/10,000 = 52,200 \text{ in.}$$
$$= 4,350 \text{ ft (min.)}$$

Total transition length = 4(X) + 50 ft
Y = 7.0 ft/2 = 3.5 ft

$$X = \sqrt{2RY - Y^2}$$

$$X = \sqrt{(2)(4,350)(3.5) - (3.5)^2} = 175 \text{ ft}$$
Total transition length = 750 ft

It is suggested that field personnel start survey (elevations) 500 ft left of crossing centerline and stop at station 10 + 00.

Deflections at 25-ft intervals are calculated from Eq. 3 (Figure 3), and a table is prepared accordingly (Table 1). Note that calculations for Y are required for 175 ft only. Since the curve is then inverted (Sta 3 + 00 to Sta 4 + 75), results are subtracted from 7.0 ft, working backwards, to arrive at true deflection or drop. For example, deflection at Sta 4 + 50 = 7.0 − 0.07 = 6.93, etc.

Example 2. Given following elevations, is pipe line being dropped too suddenly, and will it become over-stressed between stations 1 + 00 and 1 + 50 (Figure 2)?

Construction

The design package to the field should include drawing of new profiles, stations, tables of new pipe line elevations, trench elevations, and a very detailed construction sequence. The following may be used as a guide for construction:

- Starting at one end of transition, dig 40 ft of trench under the line. Leave 10 ft of soil for pipe support. Dig another 40 ft, repeat until opposite end is reached. When complete, 10-ft soil plugs are supporting the line at 50-ft centers.
- Clean the line between soil supports, take ultrasonic test (UT) readings, and install sleeve where required. Add primer and tape wrap between supports, maintaining station numbers and elevation points on the line.
- Add 12 in. of sand or oil/sand pre-mix to bottom of trench and compact to proper trench bottom elevation (profile elevation minus pipe line diameter).
- Install 6-in. × 6-in. wooden blocks every 50 ft (between soil plugs) to support the line.
- Remove soil plugs and complete line cleaning, UT readings, sleeve installations, tape wrapping, and trench preparation.
- Check elevations of new trench bottom, adjust as required. Check tape wrap with holiday detector if desired.
- Starting at one end, use six sidebooms to raise approximately 250 ft of line off its supports. Remove 6 in. (or less) of blocks on outside supports and 12 in. of blocks on all inner supports. Lower line onto supports.

Hand cleaning. The 30-in. and 36-in. crude lines were cleaned by hand to expose corroded areas and prepare for tape-wrapping.

Lowering lines. The two lines, full of crude oil, are gradually lowered with the use of six sidebooms.

Gradual lowering. A full sweep of the total transition length is made by lowering each line 6 in. or less at a time. The procedure is repeated until the lines finally reach desired profile elevations.

- Move convoy of sidebooms 250 ft, lift the line, remove 6-in. block at beginning and end supports and 12 in. of blocks at inner supports.
- Repeat until total transition length is lowered 12 in., except for several end supports, which are lowered less than 6 in. to match required deflections. Then come back and do it again, repeating several times until the pipe line is completely lowered. Note that each repeat becomes shorter. Check final pipe line elevations if desired.
- Check tape wrap for possible damage, add 12 in. of sand or pre-mix around pipe, backfill and compact.

Surprisingly, this operation goes very quickly, but must be well coordinated. One person should signal the sideboom operators to lower the pipe line concurrently, so as to eliminate excessive bends and sags. The pipe line is heavy when full of liquid, and if not handled carefully, it could easily rupture. Pressure should be reduced, and extreme caution exercised during the actual lowering process.

Strain calculations

A new profile means the pipe line has been stretched a certain amount, assuming line was horizontal originally. (If the pipe line was convex, lowering it will put the steel into compression.) Fortunately, steel is very forgiving either way, and the added strain is generally not a problem. It can be checked, however, by use of Young's Modulus of Elasticity (E = stress over strain).

The new pipe length can be calculated from the formula:

$$Arc\ Length = \pi R\theta/180$$

where $\theta = \sin^{-1} X/R$

A comparison of the new pipe length with the original pipe length will result in a figure for strain as a result of the new profile. Knowing the strain, additional longitudinal stress can be easily calculated.

In Example 1, $\theta = \sin^{-1} 175\ ft/4{,}350\ ft = 2.3056°$
Arc length (AL) $= \pi (4{,}350\ ft)(2.3056°)/180 = 175.047\ ft$
Length of new line $= 4(AL) + 50 = 750.19\ ft$

So the pipe line was actually stretched 0.19 ft or 2.28 in.

Strain $= \Delta L/L = (750.19 - 750.00)/750.00 = 0.000253$
E = Stress/Strain, so longitudinal stress = E × Strain or
LS $= 29 \times 10^6 \times 0.000253 = 7{,}347$ psi.

This is well within the 35,000-psi yield strength of Grade B pipe.

Source

Pipe Line Industry, July 1986.

WELDING

When should steel be preheated before welding?

From the chemistry of the steel determine the carbon equivalent:

$$Carbon\ Equivalent = C + \frac{Mn}{4}$$

If it exceeds 0.58 the steel may be crack sensitive and should be preheated before welding in ambient temperatures below 40°F.

Example. If steel pipe having a carbon content of 0.25 and manganese content of 0.70 is to be welded in springtime temperatures, ranging from 40°F to 80°F, is preheat necessary?

Carbon Equivalent $= .25 + \frac{.70}{4} = .25 + .175 = 0.425$

It is not necessary to preheat this particular steel before welding it.

But for another example:

Carbon = 0.20
Manganese = 1.60

Carbon Equivalent $= .20 + \frac{1.60}{4} = 0.60$

This steel should be preheated, particularly for early morning welding.

Why does preheating prevent cracking? It slows the cooling rate and reduces the amount of austenite retained as the weld cools. This prevents microcracking. Other alloying elements and pipe wall thickness may also influence when joints of high strength pipe should be preheated.

Welding and Brazing Temperatures

Carbon Steel Welding	2700-2790°F
Stainless Steel Welding	2490-2730°F
Cast Iron Welding	1920-2500°F
Copper Welding and Brazing	1980°F
Brazing Copper-Silicon with Phosphor-Bronze	1850-1900°F
Brazing Naval Bronze with Manganese Bronze	1600-1700°F
Silver Solder	1175-1600°F
Low Temperature Brazing	1175-1530°F
Soft Solder	200-730°F
Wrought Iron	2700-2750°F

Reprinted with Permission - Tube Turns, Inc.

Mechanical properties of pipe welding rods

When welded and tested in accordance with appropriate AWS or MIL specifications[1]

Electrode	AWS Class	Tensile psi	Yield psi	%Elong. in 2"	Tensile psi	Yield psi	%Elong. in 2"
Fleetweld 5P+	E6010	62 - 86,000[1]	50-75,000[1]	22-28[1]	67-78,000	51-67,000	30-34
Fleetweld 5P	E6010	62 - 75,000[1]	50-64,000[1]	22-30[1]	60-69,000	46-56,000	28-36
Shield-Arc 85	E7010-A1	70 - 78,000[1]	57-71,000[1]	22-26[1]	70-83,000	57-72,000	22-28
Shield-Arc 85P	E7010-A1	70 - 78,000[1]	57-63,000[1]	22-27[1]	70-77,000	57-68,000	22-25
Shield-Arc HYP	E7010-G	70 - 82,000[1]	60-71,000[1]	22-28[1]	72-81,000	60-72,000	25-29
Shield-Arc 70+	E8010-G	80 - 86,000[1]	67-74,000[1]	19-29[1]	76-80,000	68-72,000	23-29
Jetweld LH-70	E7018[4]	72 - 86,000	60-76,000	22-30	65-74,000	55-60,000	24-34
Jetweld LH-75MR	E8018-B2	72 - 87,000	60-74,000	22-30	70-82,000	56-72,000	29-32
Jetweld LH-90MR	E8018-C1	97 - 107,000	84-97,000	17-24	80-105,000[2]	67-93,000[2]	19-23[2]
Jet-LH8018-C1MR	E8018-C3	80 - 95,000	67-81,000	19-25	80-86,000	67-75,000	19-31
JEt-LH8018-C3MR	E8018-C3	80-90,000	68-80,000	24-30	75-84,000	66-73,000	24-32
Jetweld LH-100M1MR	MIL10018-M1	95 - 101,000	82-91,000	20-27	93-96,000[5]	80-90,000[5]	20-28[5]
Jetweld LH-110MMR	E11018-M	110 - 123,000	98-109,000	20-24	110-120,000[3]	95-107,000[3]	20-25[3]

1 Aged 48 hours @ 220°F
2 Stress Relieved @ 1275°F
3 Stress Relieved @ 1025°F
4 Imprinted identification 7018-1 to indicate
   meeting minimum CVN impact requirement of 20 ft-lbs at -50°F
5 Stress Relieved @ 1125°F

(1) Mechanical properties obtained with each type electrode also depend upon
        chemistry of the pipe
        welding procedures and
        the rate of cooling.
Since these factors vary between actual applications and testing conditions, the mechanical
properties may also vary.

Reprinted with Permission - Lincoln Electric Co.

LENS SHADE SELECTOR

Operation	Shade No.	Operation	Shade No.
Soldering	2	Nonferrous	
		Gas Tungsten-Arc Welding	
Torch Brazing	3 or 4	Gas Metal-Arc Welding	
		1/16, 3/32, 1/8, 5/32 inch electrodes	11
Oxygen Cutting			
up to 1 inch	3 or 4	Ferrous	
1 to 6 inches	4 or 5	Gas Tungsten-Arc Welding	
6 inches and over	5 or 6	Gas Metal-Arc Welding	
		1/16, 3/32, 1/8, 5/32 inch electrodes	12
Gas Welding		Shielded Metal-Arc Welding	
up to 1/8 inch	4 or 5	3/16, 7/32, 1/4 inch elctrodes	12
1/8 to 1/2 inch	5 or 6	5/16, 3/8 inch electrodes	14
1/2 inch and over	6 or 8	Atomic Hydrogen Welding	10 to 14
Shielded Metal-Arc Welding		Carbon-Arc Welding	14
1/16, 3/32, 1/8, 5/32 inch electrodes	10		

Reprinted with permission - Tube Turns, Inc.

PIPE LINE WELDING

Meeting Today's Quality Requirements For Manual Vertical Down Techniques

Pipelines are inspected more critically than ever before and today's radiographic equipment and techniques produce clearer radiographs with greater sensitivity than in the past. Although codes have not changed drastically, interpretation standards have been upgraded. The combination of more rigorous inspection, better testing methods, and high acceptability standards often approaches an attitude requiring zero defects.

This poses some serious problems because the job of welding cross-country pipelines under typical conditions has always been an extreme challenge requiring specialized and highly developed skills. Now that the demands are greater, even the best welding operators are having trouble. Rejectable defects usually require cutting out the entire weld. This is expensive and can cost competent pipeline welders their jobs.

The purpose of this bulletin is to discuss some of the more common reasons given for rejecting a pipeweld and to suggest what may be done to correct some of these conditions.

Change in Attitude

Any discussion on this subject should begin by accepting the fact that the standards are higher than they were a few years ago. Therefore, it will take a change in the attitude of the welders and everyone else involved if this new level of quality and workmanship is to be met. In turn, a commitment to new methods, equipment and theory will be required by all concerned.

Importance of Joint Preparation

It must be recognized that all joint preparation details (Figure 6) are critical and any variation could directly contribute to rejected welds.

Without good cooperation from those who prepare the pipe edges and the line-up crew, the welder has very little chance of meeting today's rigid inspection requirements. Too often the attitude exists that variations in fit-up and joint preparation are permissible and that the welder can compensate for them. This attitude cannot be tolerated. It puts too much responsibility on the welder and inevitably leads to rejects.

Recommended Procedures for Properly Cleaning Pipe

For lowest cost and highest quality, careful attention to the cleaning of pipe joint surfaces is critically important. Today's

Courtesy of *Pipe Line & Gas Industry.*

pipe welders face pipe which may be covered with a variety of coatings; these include primers, epoxy, tar, paper, varnish, rust, scale or moisture.

While joint cleanliness is important in all welding, it is especially so in the root pass of pipe. Even a thin film of contaminants, which may be difficult to see with the naked eye, may be present. Failure to recognize and properly clean joints can result in a hollow bead or other welding defects.

Follow these instructions to minimize costly defects:

1. Remove all moisture and condensations of any type prior to welding. The joint must be completely dry.

2. Clean BOTH ends of the pipe INSIDE AND OUT to remove traces of paint, rust, scale, oil, epoxy, or organic materials. The area to be cleaned should extend at least 1 inch (25 mm) from the end of the bevel (per ANSI/AWS D10.11-80 Page 1) on both the INSIDE and OUTSIDE surfaces.

3. A recommended method for cleaning in the field as described above is the use of a heavy duty straight shaft grinder with a rubber expanding wheel and a carbide coated sleeve. The small shaft and reduced overall weight allows easy access to the inside and outside surfaces of the pipe.

Internal Undercut

See Figures 1 through 5 for x-rays of the various defects that can cause rejects. Of these, one of the most common and troublesome is internal undercut (undercut on the inside of the pipe). This is understandable because it occurs on the "blind side" and the operator cannot see it happening. Consequently, he cannot immediately do anything to correct it. To make matters worse, he seldom gets a chance to see the x-ray negative or the actual weld itself. All of this makes it difficult for him to correct the situation.

Undercut may be the direct result of poor joint preparation or fit-up. If the joint does not conform to the details specified in the established procedures (see Figure 6), every reasonable effort should be made to correct it before starting to weld.

Internal undercut will tend to occur if:

1. The root face (Land) is too small.
2. The root opening is too large.
3. If a severe high-low condition exists.
4. The current is too high.

When any undesirable variation occurs in the joint preparation, the normal reaction is to compensate for it by juggling

Figure 1 — Radiographs of internal undercut. This defect may be intermittent or continuous (Figure 2) and either on one side or both sides of the weld centerline.

WHEN RADIOGRAPHS ARE INDICATIVE OF POOR WORKMANSHIP, BEST RESULTS CAN BE OBTAINED BY SHOWING THE RADIOGRAPHS TO THE WELDER SO THAT HE CAN UNDERSTAND WHAT HE IS DOING WRONG AND CORRECT IT.

Figure 2 — Radiographs of internal undercut and the lack of penetration which tends to occur at a stop and start.

Figure 3 — Radiographs of lack of penetration on the stringer bead appears as a single, straight dark line.

Figure 4 — Radiographs of unfilled wagon tracks. Distinguishing between this defect and internal undercut requires skill and experience.

Figure 5 — Radiographs of porosity.

the root spacing. Within limits, this can be fairly effective. For example:

Condition	Change in Root Spacing
Land too small.	Decrease root spacing.
Land too large.	Increase root spacing.
High-low condition.	Decrease root spacing.
Bevel too small.	Increase root spacing.

A skillful line-up crew can be helpful in juggling the root spacing to bring about the most favorable results, but this has limitations. And it is at best a poor substitute for a uniform and consistent joint preparation.

Internal Chamfer

It is important to remove any burr or overhang on the inside edge of the pipe after the root face has been machined (or

Figure 6 — Recommended joint preparation and typical procedures — operate within these tolerances to help ensure good quality welds.

Pass	Electrode[1] in. (mm)	Current[2] Amps DC(+)[5]	Alternate Stringer Bead Procedure Amps DC(−)[5]
Stringer Bead	1/8 (3.2)[3] or 5/32 (4.0)[3]	90 — 120 140 — 165[4]	85—115 115—140
Hot Pass	5/32 (4.0)	160 — 185	—
Fills and Stripper	3/16 (4.8)	160 — 190	—
Cap Pass	3/16 (4.8)	150 — 180	—

[1] Fleetweld 5P+, Shield-Arc® HYP or 70+ suggested depending upon the pipe steel requirements and contractor specifications.

[2] The ideal current should be selected from within the range shown. In general, **the best quality will be obtained by operating at the lower end of the range.** Hot pass currents up to 200 amps are often used but cause early coating breakdown and larger stub losses.

[3] Use 1/8" (3.2 mm) diameter electrode for stringer beads when the wall thickness is .250" (6.35 mm) or less and when the gap is too small to permit use of the 5/32" (4.0 mm) size [1/8" (3.2 mm) diameter electrode should also be used on other passes when the wall thickness is .250" (6.35 mm) or less; however, 5/32" (4.0 mm) diameter can also be used on fill and cap passes on .250" (6.35 mm) wall].

[4] Weld stringer bead at 23-25 arc volts and 10-16 in/min arc speed. Measure arc voltage between the pipe and electrode holder.

[5] Polarity — DC (−) (negative polarity) should be used for stringer bead welding when burn-through, internal undercut and hollow bead defects are a problem. These problems generally occur on thin wall pipe steels containing over .1% silicon. Lower currents are used with DC(−) which helps to reduce these problems. Travel speeds with DC(−) will be equal to travel speeds with DC(+).

Hot pass and all other passes should be run DC(+) (positive polarity).

Welding DC(−) (negative polarity) on the stringer bead will not be harmful to either mechanical or metallurgical properties.

General Guidelines for Power Source Adjustment (SA- and SAE Machines)
In general, the "Current Range" or "Current Control" switch (whether a tap or continuous control), should always be set as low as possible to get the current wanted, and the "Fine Current Adjustment" or "Job Selector" control (which controls open circuit voltage) should be set as high as possible. It is usually better to set this control (for OCV) at the mid-point or higher for best arc stability and fewest popouts.

When using a Lincoln SA-250 engine driven power source, it will normally be necessary to set the "Current Range Selector" tap switch one position lower than used on a Lincoln SA-200 power source to get the same operating characteristics.

ground). However, this clean-up operation should never produce an internal chamfer. A condition such as this is almost certain to result in "internal undercut" (Figure 7). See Recommended Procedures for Cleaning Pipe on the previous page.

Undercut Vs. Welding Current

Excessive current produces internal undercut. Further, the incidence of internal undercut is more prevalent in the 4-6-8 o'clock portions of the pipe. In this bottom area it is possible to use a considerably higher current than is necessary without getting burn-through, and this results in overheating and undercutting. Knowing this, it would appear to be a simple matter to specify an ideal current which would not give undercut. However, this is extremely difficult because of the many variables of the many interrelated factors.

A recommended current range is given in Figure 6, but it is necessarily rather broad to cover all reasonable variations. From a practical standpoint, the correct current should be determined by starting within the recommended range and then "fine tuning" to get precisely what "feels good" and produces the required results. This final current value will vary slightly from one operator to another and with different pipe materials, diameters, wall thickness, etc.

Because of inaccuracies in ammeters, the effect of arc voltage on current, the inability to accurately read a bobbing ammeter, etc., it is impractical to hold to an arbitrary current value. For the accomplished stringer bead welder, the selection of the ideal current is not too difficult — but he will be doing it primarily by "feel and touch."

Keyhole Size Vs. Current and Undercut

To get a good inside bead it is highly desirable to maintain a small, visible keyhole at all times. If the keyhole is allowed to get too big, it will result in internal undercut and/or windows (Figure 8).

Figure 7 — An internal chamfer will tend to leave an unfilled area on one or both sides of the stringer bead. On the x-ray this will be interpreted as internal undercut.

Assuming that the joint preparation is correct, the keyhole size is a function of current, electrode angle and pressure. The current should be "fine tuned" to produce a small keyhole when the electrode is dragged lightly using the normal electrode angle. If minor variations occur in the keyhole size (for any reason), the electrode angle and pressure can be manipulated to maintain the proper keyhole size (Figure 9).

In general, a keyhole which is about $1/8''$ (3.2 mm) in length is ideal; if it becomes $5/32''$ (4.0 mm) or longer, undercut and windows are imminent.

Frequently, a good inside bead is obtained without having a visible keyhole. For example, at a tight spot, the keyhole may disappear and the arc goes "inside the pipe." With the proper manipulative skill, this condition is tolerable and a good inside bead will be obtained. But, when it is done in this manner, the welder is largely dependent on the sound of the arc inside the pipe. A small, visible keyhole is easier to work with and is a much more controllable condition.

Significance of Arc Voltage

It is recommended that meters be used since a fairly good correlation can be established between the arc voltage and the keyhole size. For example, under controlled conditions, an

SMALL KEYHOLE
GOOD PROCEDURE

LARGE KEYHOLE
EXCESSIVE CURRENT
POOR PROCEDURE

Figure 8 — Maintaining a small keyhole will help assure a good inside bead. The larger keyhole at the right is almost certain to give internal undercut and/or windows.

ONE OF THE PRIME OBJECTIVES OF THE STRINGER BEAD CREW SHOULD BE TO PROVIDE A GOOD, STOUT STRINGER WITH AS MUCH THROAT AS POSSIBLE.

Normal Angle

Angle varied to <u>increase</u> penetration
Electrode pressure may also be increased and speed decreased.

Angle varied to <u>decrease</u> penetration
Electrode pressure may be decreased and speed increased.

Figure 9 — The penetration and keyhole size can be effectively controlled by varying the electrode angle as shown — if the joint preparation is uniform.

ideal keyhole size is consistently accompanied by an arc voltage of no more than 25 volts. When keyhole size is increased by increasing current, increasing the gap, changing the electrode angle, etc., the arc voltage increases to 26-28 volts. This is reasonable because the arc voltage reflects arc length. At the higher arc voltages, internal undercut and wagon tracks may occur.

The same arc voltage numbers quoted above may not occur in all instances. However, it is readily apparent that there is a dependable correlation between keyhole size and arc voltage. A maximum tolerable arc voltage can be determined by experimentation.

It should be further noted that this arc voltage is determined while welding and it should not be confused with the open circuit voltage. Also, it can be affected by the welding technique; changes in electrode angles and drag techniques should not be allowed to cloud the issue.

Varying Current/Varying Gap

Greater root penetration naturally occurs at the top and bottom portions of the pipe and least penetration tends to occur at the sides. This being the case, it would be desirable to alter the root spacing accordingly but this obviously is not practical. It should also be noted that a condition which permits maximum spacing on top and bottom and minimum spacing on the sides is not tolerable.

Assuming a uniform gap all the way around, the ideal condition would be to vary the current as the weld progresses from 12 to 6 o'clock to compensate for the changes in penetration due to position. Instead, penetration changes are usually controlled by the manipulative skills of the welder (see Figure 9). In an

extreme case, a second pair of hands may be required to adjust the current up or down on a signal from the welder. This requires a good communication system and a well-coordinated effort to avoid overshooting or undershooting the current. In some cases, this becomes a survival technique to make up for other undesirable variations.

Undercut—Real or Imaginary?

If visual inspection were practical for determining internal undercut, it would be easy to accurately determine its presence and severity. However, this is not the case. Radiography is the judge of internal conditions and this has led to many questionable interpretations. If internal undercut is present, it will show up on the film as a very narrow, dark line immediately adjacent to the stringer bead — on one or both sides. The darkness of the line will vary with the depth of the undercut (Figures 1 and 2).

Proper identification of undercut is sometimes difficult because its location is directly in line with the "wagon track" area (Figure 4). Distinguishing one from the other may be difficult.

Correct interpretation of internal undercut may be further complicated by the fact that a "stout" inside bead will stand out clearly (lighter on the film) against the adjacent area which is thinner and therefore darker. The thinner, darker area will be further darkened by the presence of any fairly deep widely spaced surface ripples in the cap pass. The net effect is to produce a dark shading in the area where internal undercut might occur.

Although this darker area is considerably wider than any undercut, it has been mistaken for undercut and resulted in cutting out good welds. Normally, a "stout" inside bead is considered good, but in this instance, the stouter the bead, the darker the adjacent area would appear (Figure 10).

Figure 10 — Any deep surface ripples located directly above a point adjacent to the inside bead will contribute to a dark shading of the film at this area. This has been misinterpreted by some as internal undercut.

Figure 11 — "Wagon tracks" are located on either side of the stringer bead.

Wagon Tracks

It is not possible to completely eliminate the sidewall undercut produced by the stringer bead. This condition is generally referred to as "wagon tracks" (Figure 4 and 11).

It is the responsibility of the stringer bead crew to minimize wagon tracks and the responsibility of the hot pass crew to burn them out. This is normally done with 5/32" (4.0 mm) electrode at about 175-180 amps. The secret in melting out wagon tracks lies primarily in the skill of the hot pass crew and the penetration factor of the electrode. The majority of the hot pass men use a rapid stitching technique which exposes the wagon track momentarily and permits them to direct the arc into the undercut area and melt them out.

The depth of the wagon tracks may be affected by the following:

Condition	Results
Bevel too small (Figure 12).	Increases depth of W.T.
Root Spacing (Gap) too small.	Increases depth of W.T.
Current and/or speed too high.	Increases depth of W.T.

In extremes, it may be necessary to use a grinder to open up the sidewalls to minimize deep "wagon tracks" or, if they do occur, to grind the stringer to eliminate the high peaked center (Figure 13). In all cases a 5/32" (4.0 mm) thickness disc grinder should be used to grind root beads on all pipe from 12" (304.8 mm) up irrespective of wall thickness [a 1/8" (3.2 mm) disc grinder will grind the center and roll the sides over on the wagon track unless side pressure is applied]. This will make it easier to melt them out.

If the stringer bead tends to wander to one side or the other, it will leave a deep undercut on the shallow side. This condition should be corrected immediately by changing the electrode angle (Figure 14).

Figure 12 — The depth of the wagon tracks will vary inversely with the bevel angle.

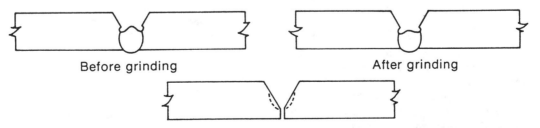

Before grinding   After grinding

A narrow bevel or tight gap may require grinding as per dotted line.

Figure 13 — A disc grinder, if used with restraint, can be helpful in correcting the condition shown above.

The electrode should be centered in the joint and not changed unless the bead wanders off to one side or the other. Changing the angle will bring it back.

Figure 14 — The correct angle can only be determined by observing the shape and location of the bead. If the bead drifts to one side for any reason, it will leave a deep undercut on the opposite sidewall. Tilting the electrode just a few degrees toward the undercut side will straighten the bead up. This change must take place rapidly if the correct bead placement is to be maintained.

The Hollow Bead

The stringer bead defect shown in this sketch is known by several names including the hollow bead, vermicular porosity and wormhole porosity. Its length varies from a fraction of an inch to several inches. Radiography exposes the presence of the problem clearly.

Vermicular porosity occurs most readily when welding high silicon pipe — generally above .10% silicon. It is aggravated by excessive travel speeds and high currents. Welding the stringer bead with 5/32″ (4.0 mm) electrode at 130-165 amps DC(+) and 12 to 14 in/min (0.3 to 0.4 m/min) travel speed minimizes its occurrence. DC(−) (negative polarity) should be used for stringer bead welding when burn-through, internal undercut and hollow bead defects are a problem. These problems generally occur on thin wall pipe and on pipe steels containing over .1% silicon. Lower currents can be used with DC(−) which helps to reduce these problems. Travel speed with DC(−) will be equal to travel speeds with DC(+).

Hot pass and all other passes should be run DC(+) (positive polarity).

Welding DC(−) (negative polarity) on the stringer bead will not be harmful to either mechanical or metallurgical properties.

Hollow bead or wormhole porosity may also be caused by poor joint preparation and/or cleanliness. See Recommended Procedures for Properly Cleaning Pipe.

Filler, Stripper and Cap

In general, there is little fault found with the fillers, strippers and caps. Most of this welding today is being done with 3/16″ (4.8 mm) Shield-Arc® HYP or 70+ at 160-180 amps and the results have been excellent.

A reasonably competent crew of firing line welders armed with FW5P+, HYP or 70+ can do a very fine job of finishing the weld if the stringer and hot pass crew have been doing their work properly.

The size and consistency of the final weld will have its influence on the x-ray radiograph. Thus, the firing line welders should be encouraged to produce a cap pass which is as uniform as possible with neatly stitched close ripples and as much reinforcement as required.

Welding Cracks

Since the advent of the higher tensile pipe steels (5LX 52, 60, 65, 70 etc.), it has been necessary to exercise better procedural control to eliminate the possibility of weld and heat-affected zone (HAZ) cracks. To do this effectively, all of the following factors must be controlled.

1. Joint preparation and root spacing must be as specified in the approved procedure. (See Figure 6.)

2. Hi-Low conditions must be held to a minimum.

3. Weld 5LX70 with Shield-Arc® 70+. Weld 5LX65 and lower with Shield-Arc HYP or Fleetweld 5P+. Fleetweld 5P+ is recommended for all stringer beads when lower hardness is a consideration. Lower stringer bead hardness will result in improved resistance to heat affected zone cracking.

4. The need for preheat varies considerably between applications. Cracking tendencies increase with higher carbon and alloy content, thicker pipe wall and diameter (faster quench rate), and lower ambient temperature. Preheat cold pipe to at least 70°F (21°C). Preheat to as much as 300°F (149°C) may be required to slow the quench rate and prevent cracking when welding high strength, large diameter or heavy wall pipe. Specific preheat requirements must be determined for each situation based on these considerations.

5. Ideally the line-up clamp should not be removed nor should any movement of the pipe take place until the stringer bead is 100% completed. Any movement of the pipe could cause a partially completed stringer bead to be overstressed and to crack. The tendency for such cracking varies with the chemistry of the pipe, its diameter and wall thickness and its temperature. Under favorable conditions it may be possible to remove the line-up clamp with as little as 60% of the stringer bead completed, but this should be done only when it has been clearly demonstrated that this practice does not cause cracks to occur.

6. After removal of the line-up clamp, the pipe must be gently and carefully lowered on skids.

7. Use only enough current on the stringer to get a good inside bead and travel slowly to get the maximum weld cross section.

8. Restrict lack of penetration on the inside bead at tie-ins to ¼″ (6.4 mm) or less. Use a disc grinder to improve this situation on starts and stops only.

9. Remove slag from each bead by power wire brushing. Grinding should not be necessary except possibly to clean up a lumpy start, humped up center or perhaps to improve a crater condition. **NOTE:** Grinding of the stringer bead should be done with a ⁵/₃₂″ (4.0 mm) disc. Excessive grinding can be detrimental.

10. Weld stringer beads with two or more persons welding on opposite sides to equalize stress. Use 3 welders on 20-30″ (508-762 mm) pipe; 4 welders on larger pipe.

Recommended pass sequence

Start

Stop

11. Start the hot pass immediately after completion of the stringer — always within five minutes. At least two hot pass welders should be used on each joint, and to put this pass in as soon as possible after the stringer bead it may require a total of four hot pass welders leap frogging each other to keep up.

12. Minimize the wagon tracks — this area is highly vulnerable to cracking until the hot pass is completed (Figure 15).

Figure 15 — Cracks tend to occur at the area indicated on one side or the other of the stringer bead. This could eventually propagate up through the weld. A properly controlled procedure and good workmanship can eliminate this condition.

The Lincoln Electric Company strongly recommends for weldments intended for sour gas, sour crude, or other critical service applications that the customer verifies that both the level of hardness and variation in hardness of the weld, heat affected zone, and base plate are within acceptable limits.

How many welds will the average welder make per hour?

Where stringer beads have been run, the average welder can complete about 140 inches of weld per hour on ordinary one-fourth-inch wall line pipe. To find the average number of welds per hour, divide the circumference of the pipe into 140.

$N = \dfrac{140}{C}$ where circumference (C) = pipe diameter in inches times 3.14

Example. How many welds per hour will a welder complete on quarter-inch wall $10^{3/4}$-inch line pipe, if the stringer beads have been run?

$N = \dfrac{140}{(10.75)(3.14)}$

$N = \dfrac{140}{33.76}$

N = 4.15 welds per hour or about 42 welds per 10-hour day.

On large diameter pipe where wall thickness has been increased, welders average about 100 inches of completed weld per hour.

Example. How many welds per hour will a welder complete on 30-inch line pipe if the stringer beads have been run?

$N = \dfrac{100}{C}$

$N = \dfrac{100}{(30)(3.14)} = \dfrac{100}{94.2} = 1.1$ welds per hour or 11 welds per 10 hour day

How much welding rod is required for a mile of schedule 40 pipe line?

Rule. For 40-foot joints of schedule 40 pipe, multiply the nominal pipe diameter by $22^{1/2}$; the answer is pounds of electrode per mile of pipe line.

Example. 6-inch pipe:

$6 \times 22^{1/2} = 135$ pounds per mile.

Example. 10-inch pipe, 30-foot joints:

$10 \times 22^{1/2} = 225$ pounds per mile

For 40-foot joints:

$225 \times \dfrac{40}{30} = 300$ pounds per mile for 30-foot joints.

The rule is based on standard weight (schedule 40) pipe, and the usual number of passes, which varies from two for small pipe up to four for 16 inches; larger pipe is usually thinner wall; the rule includes an allowance for wastage.

How many pounds of electrodes are required per weld on line pipe?

Divide the nominal pipe size by two and multiply the result by one-fourth pound.

Pounds of electrode $= \dfrac{N}{2} \times .25$

Example. How many pounds of electrode will be used per weld on 10-inch line pipe?

Pounds of electrode $= \dfrac{10}{2} \times .25 = 1.25$ pounds per weld.

Example. How many pounds of electrode will be used per weld on 24-inch line pipe?

Solution. Pounds of electrode = $12 \times .25$ or 3 pounds per weld.

Welding criteria permit safe and effective pipe line repair

Battelle Laboratories studies show weld sleeves and deposited weld metal can be applied without removing defective lines from service

J. F. Kiefner, Senior Research Engineer, Battelle Houston Operations

Research conducted by the American Gas Association (AGA)[1-5] and others[6-8] showed that pipe line defects can be repaired without removing the lines from service. Two repair methods evaluated were full-encirclement sleeves and direct deposition of weld metal on defects.* To ensure minimum risk when using these methods, criteria for repairing pipe lines in service have been formulated (Table 1).

The criteria in the table are intended as a repair guide for lines that operate at 40 percent of their specified minimum yield strength (SMYS) or more. The criteria may be unnecessarily restrictive for lines which operate at or below 40 percent of SMYS.

Note that defects in pressurized pipe lines, especially pneumatically pressurized lines, can cause sudden catastrophic ruptures. When the procedures are carefully planned, risks are minimized. Nevertheless, repair errors could reduce the safety margin of these criteria. Consequently, users of Table 1 are urged to exercise caution when performing repairs.

Choice of repair method

The repair method options included in Table 1 are:

* Type A sleeve
* Type B sleeve
* Deposited weld metal
* Removal
* Removal by hot-tapping

Type A sleeve repair consists of placing a reinforcing band concentric with the pipe while leaving the band ends unsealed. The sleeve strengthens a defective area by restraining bulging that would otherwise occur when the weakened pipe shell is pressurized. Such a sleeve may also carry a small portion of the hoop stress, but in no case can it become the sole pressure containing element since its ends are not sealed. As a result, it cannot be used to repair a leaking defect. If a gap exists over the defect, filling of the gap is required. Two classes of Type A sleeve exist: no filler; with filler.

Type B sleeve repair requires a pressure-tight reinforcing concentric band. Its ends are sealed to the carrier pipe and it can contain pressure. It can be used to repair leaking defects. In addition, a Type B can also be used as a Type A sleeve. Three classes of Type B sleeves exist: no filler; with filler; pressurized.

Deposited weld metal completely fills and eliminates clearly visible defects such as corrosion. Access to an entire defect is required so that weld metal can penetrate and bond to sound metal. It may be possible to enlarge pits, laps, or undercut by grinding a groove wide enough to permit access. Deposited weld metal can be selected when requirements to prevent burn-through are met. These requirements are discussed later.

Removal of a damaged pipe section is another way of eliminating a defect. The line must be taken out of service, purged, removed and replaced with a sound tie-in piece. Removal is often a poor economic choice in terms of wasted product and interrupted service. Thus, removal is recommended in Table 1 with the understanding that it may not be practical in certain instances.

Removal by hot-tapping takes advantage of a widely accepted method for making branch connections on live pipe lines. The coupon removed from the line must contain the entire defect. Limitations of this technique will be discussed later.

Types of defects

Classes of defects listed in Table 1 are:

* Manufacturing defects—cracks, undercut, lack of fusion, and lack of penetration in seam welds; laps, pits, cracks, and rolled-in slugs in the pipe body; hard spots.
* Environmentally caused defects—selective corrosion and hydrogen stress cracking in electric resistance welded or flash welded seams; general pipe corrosion; pitting corrosion; stress corrosion cracks; hydrogen stress cracking in hard spots.
* Defects caused by outside forces which include dents and gouges.
* Construction defects.

Submerged arc-welded (SAW) seam defects, both straight seam and spiral, include undercut, incomplete fusion, incomplete penetration, and cracks. Repair methods recommended for these defects are shown in Table 1. No filler is required with sleeves because after the weld reinforcement is ground flush, no gap between the pipe and sleeve should exist. Removal by hot-tapping is not recommended because the tap would involve cutting through the seam, a practice considered unacceptable by many companies. Repair by deposited weld metal is applicable only for undercuts and is subject to special requirements given in footnotes of Table 1.

Type B pressurized sleeves used on nonleaking defects can be pressurized only by using a tapping nipple. Since other means of sleeve repair are entirely adequate when used as directed, the intentional pressurization of a sleeve when no leak or near leak is present is not necessary.

Inside or interior defects require special consideration since they are not readily visible. Removal may be the best alternative unless one can be reasonably certain of the extent of the defects. Hot-tapping is not recommended because of the uncertainty of the extent of an inside or interior defect.

Electric resistance welded (ERW) or flash welded (FW) seam defects include: upturned fiber imperfections; incomplete fusion; penetrators; cold welds; cracks. These can only be repaired with pressurized sleeves and removal since the welds are susceptible to brittle fracture or low-resistance ductile fracture initiation. The value of restraint of bulging from either type of nonpressurized sleeve is uncertain. Hence they are not recommended for these defects. Hot-tapping also is not recommended because of the involvement of the seam weld. It is not recommended that weld metal be deposited because of possible low ductility. Required grinding before such welding would involve creating or enlarging a defect in a potentially low-toughness material. Only removal or use of a Type B pressurized sleeve which stress-relieves the defect is recommended.

Other seam defects include lap welds and furnace butt welds. These must be dealt with on a case by case basis. Such welds usually appear in older or smaller and lower pressure pipe lines where fracture resistance requirements are less stringent. On the other hand, some of these older materials can be quite susceptible to low-toughness fracture behavior. It is probably best to treat these materials with the same caution as ERW and FW seams as they are being operated at stress levels exceeding 40 percent of SMYS.

Laps, pits, seams, cracks, and rolled-in plugs should be repaired subject to these limitations: special requirements of I.D. and interior defects and leaks must be observed; hot-tapping is acceptable on nonleaking, O.D. defects as long as the entire defect is removed with the coupon; deposited weld metal may be used to repair pits or laps if they can be entirely exposed by grinding.

Hard spots created in the plate by accidental quenching on the run-out table may become flat spots in the pipe since they do not yield when the plate is formed to pipe. Such hard spots do not fail spontaneously unless they contain quench cracks, or unless they undergo hydrogen stress cracking. Such spots should be repaired if they are not cracked. Type A or B sleeves without filler would not be acceptable since they would not restrain flat spots from bulging.

Hard spots can be remedied by several methods if they occur as outside nonleaking defects. Sleeve repair methods with filler not only provide strengthening but shield the hard spot from the environment, which may cause it to become cracked.* Hard spots at the I.D. surface should not be repaired with Type A sleeves since they may be susceptible to cracking in certain types of product service. Such a sleeve would not prevent a leak if the resulting crack grew through the wall.

Selective corrosion and hydrogen stress cracking in ERW and FW seams are subject to low toughness fracture behavior. It is recommended that they be removed or repaired with pressurized Type B sleeves to relieve stress.

General or widespread corrosion is that which covers too wide an area to be repairable by means of hot-tapping or deposited weld metal. Any other repair means are acceptable if they cover the critically affected area. When internal corrosion is present, its extent must be reasonably well known and further corrosion must be prevented. Removal may sometimes be the only suitable choice.

Pitting corrosion may involve isolated pits or groups of pits. Generally, isolated single pits will not require repair if further corrosion can be halted. ASME guidelines can be used to determine whether or not a corroded region requires repair. Groups of interacting pits can substantially lower remaining strength. If strength has been lowered, the pipe should be repaired. Any of the methods of Table 1 are suitable if the noted precautions are observed.

Stress-corrosion cracks occur in clusters and often cover large areas. It is difficult to repair these cracks by hot-tapping or deposited weld metal. Repair methods for stress-corrosion cracks are confined to those which can strengthen the entire affected area.

Hydrogen stress cracking appears in hard spots attacked by hydrogen emitted from bacteria external to the pipe and from cathodic protection. Internally the hard spots can be attacked by certain types of products—especially those containing hydrogen sulfide. Because flatness often occurs near hard spots, nonfilled sleeves are not recommended. Deposited weld metal is not recommended because grinding of a hydrogen stress crack in a hard—and usually brittle—spot while the pipe is under pressure is not safe. Interior hydrogen stress cracking should not be repaired by a Type A sleeve since such a sleeve cannot prevent leaks if the crack grows through the wall.

Dents and gouges and a combination gouge-in-dent result from external encroachment by mechanical excavating

---

* One method of protecting hard spots from hydrogen stress cracking which, so far, has proven adequate, is that of using a concentric band of sheet metal spaced away from the pipe by rubber seals. The annular space is filled with coal tar to exclude ground water, and the metal band (which is itself coated) shields the hard spot from hydrogen generating cathodic protection current.

Table 1
Criteria for selecting an appropriate repair method
(X indicates an acceptable repair)

Repair method	Type of defect								
	Manufacturing defect				Environmentally caused defect in ERW or FW Seam weld in body, SAW seam or girth weld				
	In seam weld		In the body of the pipe						
	SAW Undercut, Incomplete fusion, Incomplete penetration, Crack	ERW and FW Upturned fiber imperfection, Incomplete fusion, Penetrator, Cold weld, Crack	Lap, Pit, Seam, Crack Rolled-in Slug	Hard spot exceeding $R_c>35$ and 2 in. in extent or more	Selective corrosion, Hydrogen stress cracking in weld zone	General corrosion	Pitting corrosion	Stress-corrosion cracking	Hydrogen stress cracking in hard spot
Nonleaking O.D. defect[a]									
Type A sleeve No filler ...	X	......	X	......	......	X	X	X	......
Type A sleeve With filler ..	......	......	......	X[g]	......	X	X	......	X
Type B sleeve No filler ...	X	......	X	......	......	X	X	X	......
Type B sleeve With filler ..	......	......	......	X	......	X	X	......	X
Type B sleeve pressurized	X[b]	X[b]	X[b]	X[b]	X[b]	X[b]	X[b]	X[b]	X[b]
Deposited weld metal[h] ...	X[c]	......	X[c]	......	......	......	X[c]	......	......
Removal	X	X	X	X	X	X	X	X	X
Removal by hot tap	......	......	X[e]	X[e]	......	......	X[e]	......	X[e]
Nonleaking interior or I.D. defect[a]									
Type A sleeve No filler ...	X	......	X	......	......	X	X	......	......
Type B sleeve No filler ...	X	......	X	X	......	X	X	......	X
Type B sleeve pressurized	X[b]	X[b]	X[b]	X[b]	X[b]	X[b]	X[b]	......	X[b]
Removal	X	X	X	X	X	X	X	......	X
Leaking defects									
Type B sleeve pressurized	X	X	X	X	X	X	X	X	X
Removal	X	X	X	X	X	X	X	X	X

[a] Corrosion or other defect which is known to exceed 80 percent of the wall thickness in depth should be treated as though it were a leak.

[b] Pressurization to be accomplished upon completion of repair by drilling hole through carrier pipe through a small tapping nipple.

[c] Only for undercut, laps, and pits where the defect can be safely enlarged to a weldable groove and subject to remaining wall thickness rules.

[e] Only if branch coupon entirely contains and removes defect.

[f] Only with humped sleeve or if flat sleeve is used, girth weld must be ground flush. Do not use if extreme high-low condition exists.

equipment, other kinds of equipment, or rocks. Plain dents are usually innocuous unless they are quite large*—2 percent or more of the diameter—or unless they involve a seam or girth weld. When the dent is large or welds are included in the dent or gouge-in-dent, the repair should prevent its outward movement. Sleeves with fillers or pressurized sleeves are required. If the included weld is an ERW or FW seam, the defect should be removed or the stress relieved with a pressurized sleeve.

Gouges without dents in the body of the pipe or in SAW seams or girth welds may be repaired by any means except deposited weld metal. Gouges-in-dents may be repaired only by means which prevent outward movement of the dent. Sleeves without fillers are not acceptable.

When dents are involved, hot-tapping is not recommended since it may not remove the dent entirely. For any gouge or gouge-in-dent, repair by deposited weld metal is not recommended since concealed cracks may exist. When a dent is present, weld metal may not have sufficient ductility to withstand the severe strains that accompany outward dent movement.

Construction defects in girth weld include undercut, incomplete fusion, incomplete penetration, and cracks. Deposited weld metal may be one of the best ways to repair undercut or other externally connected defects in girth welds which can be ground for access. Sleeves, if used,

---

* Present requirements of federal regulations Part 192 dictate removal of dents extending over 2 percent of the pipe diameter.

| | Defect caused by outside force | | | | Construction defect |
| In seam or other weld | | In body of pipe | | | In girth weld |
Plain dent or gouge-in-dent, SAW Seam or Girth weld	Plain dent, gouge or gouge-in-dent ERW or FW seam	Plain dent greater than 2% of pipe diameter	Gouge (Also Including gouges in SAW seams or girth welds)	Gouge-in-dent	Undercut, Incomplete, fusion Incomplete penetration, Crack
......	......	......	......	......	......
X	......	X	......	X	......
......	......	......	X	......	X(f)
X	......	X	......	X	......
X(b)	X(b)	X(b)	X(b)	X(b)	X(b,f)
......	......	......	X(c)	......	X(c)
X	X	X	X	X	X
......	......	......	X(e)	......	......
......	......	......	......	......	......
......	......	......	......	......	X(f)
......	......	......	......	......	X(b,f)
......	......	......	......	......	X
X	X	X	X	X	X(f)
X	X	X	X	X	X

(g) One method of protecting hard spots from hydrogen stress cracking which, so far, has proven adequate is that of using a concentric band of sheet metal spaced away from the pipe by rubber seals. The annular space is filled with coal tar to exclude ground water, and the metal band (which is itself coated) shields the hard spot from hydrogen-generating cathodic protection current. The role of this band is merely to shield the pipe from the cathodic protection and not to strengthen the pipe.

(h) Use of low hydrogen electrodes is recommended.

should have a special shape such as a central hump to avoid interference with the girth weld reinforcement. Only sleeves with welded ends are recommended. Such sleeves tend to strengthen the defective girth joint, whereas those with nonweld ends do not.

Burn-through

Burn-through into a live pipe line would defeat the purpose of a repair, would probably cause the line to be shut down, and might create a serious safety hazard to the repair crew.

An AGA study revealed that the following parameters control burn-through: remaining wall-thickness; welding heat input; nature and state of the pressurization media in the pipe. Consequently, it is possible to prescribe safe limits for avoiding burn-through while depositing weld metal on a defect (Table 2).

Table 2 shows a relationship between I.D. surface temperature during welding and remaining wall thickness for various pressure and flow rates of natural gas. These values were taken from AGA curves. The table is based on: heat input characteristics of the $3/32$-inch or $1/8$-inch low hydrogen electrode; 80 to 100 amps welding current at 20 volts DC; an average electrode travel speed of 4 to 5 in./min.

Energy input should not exceed these limits. Low hydrogen electrodes are recommended for reasons explained later. Under these conditions, repairs can safely be made in the flat, horizontal or overhead positions provided flow or

TABLE 2
Limitations on remaining wall-thickness for repairing without a burn-through

Values are the minimum recommended thicknesses in inches with natural gas as the pressurizing medium at the pressures and flows shown
Maximum welding voltage, 20 volts
Maximum welding current, 100 amps

Pressure, psia	Gas flow rate, feet/second			
	0	5	10	20
15	0.320	—	—	—
500	0.300	0.270	0.240	0.205
900	0.280	0.235	0.190	0.150

pressure is controlled. Analytic results predict that the I.D. wall temperature will not exceed 2,000°F and a burn-through is highly unlikely when repairs are made in this manner.

Values shown in Table 2 are believed to be quite conservative. Repairs were made without burn-throughs on 0.180-inch remaining wall with air at ambient pressure and no flow inside the pipe. This conservatism provides an extra safety margin and allows linear extrapolations for values between those shown in the table. The minimum wall-thickness of 0.150-inch is established on the basis of the experimental results. This thickness is also recommended by the British Gas Corp. Within these limits, repairs may be made by deposited weld metal (Table 1).

Underbead cracking

Underbead cracking can be minimized by making repairs with low hydrogen electrodes in a manner that avoids hard weld heat-affected zones.

One way to help assure that extremely hard weld zones are not formed during repair welding is to limit repairs to well known carbon equivalent ranges. Unfortunately, chemistries of specific samples needing repairs will seldom be known. Low hydrogen electrodes prevent hydrogen from being present in the welding atmosphere but they do not prevent formation of hard weld zones when adverse chemistries are present. Making of crack-free weldments requires careful control of heat input to avoid rapid quenching or post-repair treatments to assure that extreme hardness does not remain.

In the case of sleeves, the critical area in which cracking most often occurs is the fillet-weld at the ends. At least two procedures for making these welds, with a minimum risk of cracking, are available.[1,4,5,8]

In the case of deposited weld metal repairs, heat input must be kept low to avoid burn-through during the first or second passes. During later passes, however, higher heat inputs can be used to soften the resulting repair metal microstructure and the heat-affected-zone of base metal. A high heat input final pass can be made with a nonfusing tungsten electrode as suggested by the British Gas Corp.[8] An alternative procedure is to make an extra pass with high heat input using a conventional electrode. This pass can be ground off since its purpose is merely to soften the heat-affected microstructure of previous passes.

(Based on a paper, "Criteria for Repairing Pipelines in Service Using Sleeves and Deposited Weld Metal" presented by the author at the AGA Transmission Conference at Montreal, Que., May 8–10, 1978.)

Source

Pipe Line Industry, January 1980.

References

1. Kiefner, J. F., and Duffy, A. R., *A Study of Two Methods for Repairing Defects in Line Pipe*, Pipeline Research Committee of the American Gas Association, 1974, Catalog No. L22275.
2. Kiefner, J. F., and Duffy, A. R., Bunn, J. S., and Hanna, L. E., "Feasibility and Methods of Repairing Corroded Line Pipe," Materials Protection and Performance, Vol. III, No. 10, Oct., 1972.
3. Kiefner, J. F., "Corroded Pipe: Strength and Repair Methods," *Fifth Symposium on Line Pipe Research*, Houston, Texas, 1974, American Gas Association, Catalog No. L30174.
4. Kiefner, J. F., "Repair of Line Pipe Defects by Full-Encirclement Sleeves," *Welding Journal*, Vol. 56, No. 6, June, 1977.
5. Kiefner, J. F., Whitacre, G. R., and Eiber, R. J., "Further Studies of Two Methods for Repairing Defects in Line Pipe," to Pipeline Research Committee of the American Gas Association, NG-18 Report No. 112, March 2, 1978.
6. Cassie, B. A., "The Welding of Hot-Tap Connections to High Pressure Gas Pipelines," J. W. Jones Memorial Lecture, Pipeline Industries Guild, British Gas Corp., October, 1974.
7. Morel, R. D., *Welded Repairs on API 5LX-52 Pipe*, 13th Annual Petroleum Mechanical Engineering Conference, Denver, Colorado, September 24, 1958.

TERMINOLOGY[9-11]

Defect parameters

Defect: A crack, pit, crevice, gouge, dent, metallurgical anomaly, or combination of two or more of the above which is known or suspected to reduce the effective pipe strength level to less than 100 percent of its specified minimum yield strength (SMYS).

O.D. defect: A defect emanating at and extending radially inward from the outside surface but not entirely through the wall of the pipe.

I.D. defect: A defect emanating at and extending radially outward from the inside surface but not entirely through the wall of the pipe.

Interior defect: A defect emanating in the interior of the pipe wall but not of sufficient radial extent to be connected with either the inside or the outside surface.

Leaking O.D. defect: A defect which was initially an O.D. defect, but which has grown through the wall to become a leak.

Leaking I.D. defect: A defect which was initially an I.D. defect, but which has grown through the wall to become a leak.

Superficial defect: A lap, crevice, pit, group of pits, metallurgical anomaly, or plain dent (i.e., without scratches, gouges, or cracks) which is of insufficient extent to reduce the effective strength level of the pipe below 100 percent of SMYS.

Kinds of defects

* (Definitions marked by asterisk are from API Bulletin 5T1 *Nondestructive Testing Terminology,* Third Edition, April, 1972.)

Not every conceivable kind of defect in pipe is covered by this list. The list is limited to those which are likely to be encountered in an in-service pipe line.

Defects originating from pipe manufacture

(a) Defects not necessarily in the seam weld (primarily in the body of the pipe)
  *Lap: Fold of metal which has been rolled or otherwise worked against the surface of rolled metal, but has not fused into sound metal.
  *Pit: A depression resulting from the removal of foreign material rolled into the surface during manufacture.
  *Rolled-in slugs: A foreign metallic body rolled into the metal surface, usually not fused.
  *Seam: Crevice in rolled metal which has been more

or less closed by rolling or other work but has not been fused into sound metal.
  *Hard spot: An area in the pipe with a hardness level considerably higher than that of surrounding metal; usually caused by localized quenching.
  *Crack: A stress-induced separation of the metal which, without any other influence, is insufficient in extent to cause complete rupture of the material.

(b) Defects in the seam weld
  *Incomplete fusion: Lack of complete coalescence of some portion of the metal in a weld joint.
  *Incomplete penetration: A condition where weld metal does not continue through the full thickness of the joint.
  *Under-cut: Under-cutting on submerged-arc-welded pipe is the reduction in thickness of the pipe wall adjacent to the weld where it is fused to the surface of the pipe.
  *Weld area crack: A crack that occurs in the weld deposit, the fusion line, or the heat-affected-zone. (Crack: A stress-induced separation of the metal which, without any other influence, is sufficient in extent to cause complete rupture of the material.)
  *Upturned fiber imperfection: Metal separation, resulting from imperfections at the edge of the plate or skelp, parallel to the surface, which turn toward the I.D. or O.D. pipe surface when the edges are upset during welding.
  *Penetrator: A localized spot of incomplete fusion.
  *Cold weld: A metallurgically inexact term generally indicating a lack of adequate weld bonding strength of the abutting edges, due to insufficient heat and/or pressure. A cold weld may or may not have separation in the weld line. Other more definitive terms should be used whenever possible.

Defects originating from external or internal environmental degeneration of the pipe

(a) Seam weld defects
  Selective corrosion: Preferential corrosion in the fusion line of an electric resistance welded or flash welded longitudinal seam.
  Hydrogen stress cracking: Environmentally stimulated cracking of the weld metal or heat-affected-zone of the longitudinal seam.
(b) Defects not in the body of the pipe (possibly in the seam weld—but not specifically because of the seam weld)

8. Christian, J. R., and Cassie, B. A., "Analysis of a Live Welded Repair on An Artificial Defect," ERS.C.96, October, 1975.
9. Smith, R. B., and Eiber, R. J., "Field Failure Survey and Investigations," *Fourth Symposium on Line Pipe Research,* American Gas Association, 1969, Catalog No. L30075.
10. Eiber, R. J., "Field Failure Investigations," *Fifth Symposium on Line Pipe Research,* American Gas Association, 1974, Catalog No. L30174.
11. Wright, R. R., "Proper Inspection Methods Minimize Pipeline Failures," *Oil and Gas Journal,* May 23, 1977, pp. 51–56.

Cross country pipe line—vertical down electrode consumption, pounds of electrode per joint*

Wall Thickness	5/16"			3/8"			1/2"			5/8"			3/4"		
Elec. Dia.	5/32"	3/16"	Total	5/32"	3/16"	Total	5/32"	3/16"	Total	5/32"	3/16"	Total	5/32"	3/16"	Total
Nom. Pipe Dia., in.			lb			lb			lb			lb			lb
6	0.34	0.54	0.88	0.34	0.86	1.2									
8	0.45	0.71	1.2	0.45	1.1	1.6	0.45	2.2	2.7						
10	0.56	0.88	1.4	0.56	1.4	2.0	0.56	2.7	3.3						
12	0.67	1.0	1.7	0.67	1.7	2.4	0.67	3.2	3.9						
14	0.73	1.1	1.8	0.73	1.8	2.5	0.73	3.5	4.2	0.73	5.7	6.4	0.73	8.3	9.0
16	0.84	1.3	2.1	0.84	2.1	2.9	0.84	4.0	4.8	0.84	6.5	7.3	0.84	9.4	10.2
18	0.94	1.5	2.4	0.94	2.3	3.2	0.94	4.5	5.4	0.94	7.3	8.2	0.94	10.6	11.5
20	1.10	1.6	2.7	1.10	2.6	3.7	1.1	5.0	6.1	1.1	8.1	9.2	1.1	11.8	12.9
22	1.20	1.8	3.0	1.20	2.9	4.1	1.2	5.5	6.7	1.2	8.9	10.1	1.2	13.0	14.2
24	1.30	2.0	3.3	1.30	3.1	4.4	1.3	6.0	7.3	1.3	9.7	11.0	1.3	14.2	15.5
26	1.40	2.1	3.5	1.40	3.4	4.8	1.4	6.5	7.9	1.4	10.5	11.9	1.4	15.3	16.7
28	1.50	2.3	3.8	1.50	3.7	5.2	1.5	7.0	8.5	1.5	11.3	12.8	1.5	16.5	18.0
30	1.60	2.5	4.1	1.60	3.9	5.5	1.6	7.5	9.1	1.6	12.1	13.7	1.6	17.7	19.3
32	1.70	2.6	4.3	1.70	4.2	5.9	1.7	8.0	9.7	1.7	13.0	14.7	1.7	18.9	20.6
34	1.80	2.8	4.6	1.80	4.4	6.2	1.8	8.6	10.4	1.8	13.8	15.6	1.8	20.1	21.9
36	1.90	2.9	4.8	1.90	4.7	6.6	1.9	9.1	11.0	1.9	14.6	16.5	1.9	21.3	23.2
38	2.00	3.1	5.1	2.00	5.0	7.0	2.0	9.6	11.6	2.0	15.4	17.4	2.0	22.4	24.4
40	2.10	3.3	5.4	2.10	5.2	7.3	2.1	10.1	12.2	2.1	16.2	18.3	2.1	23.6	25.7
42				2.20	5.5	7.7	2.2	10.6	12.8	2.2	17.0	19.2	2.2	24.8	27.0
44				2.30	5.7	8.0	2.3	11.1	13.4	2.3	17.8	20.1	2.3	26.0	28.3
46				2.40	6.0	8.4	2.4	11.6	14.0	2.4	18.6	21.0	2.4	27.2	29.6
48				2.50	6.3	8.8	2.5	12.1	14.6	2.5	19.4	21.9	2.5	28.3	30.8

* "Electrode required." Figures in the table include 4-in. stub lengths. These figures will vary with different stub loss practices.

Quantities required for the 5/32" size will vary based on travel speeds of the stringer bead and hot pass. Slow travel speeds may increase these quantities by up to 50%.

Reprinted courtesy of the Lincoln Electric Company.

3: Pipe Design

Steel pipe design 71
Properties of pipe 76
Length of pipe in bends 78
Calculation of pipe bends 79
Spacing of pipe supports 81
American Standard Taper Pipe Threads 83
British Standard Taper Pipe Threads 84
Normal engagement between male and female threads to make
  tight joints 85
Hand-held computer calculates pipe weight, contents, velocity 85
Formulas and constants of value in solving problems relating to
  tubular goods 88
How to calculate the contraction or expansion of a pipe line 89
Estimate weight of pipe in metric tons per kilometer 89
How to find pipe weight from outside diameter and wall thickness ... 90
What is the maximum allowable length of unsupported line pipe? 90
Identify the schedule number of pipe by direct measurement 90
Determine buoyancy of bare steel pipe 91
Determine buoyancy of bare and concrete coated steel pipe in water
  and mud 91
Weights of piping materials 92
Allowable working pressure for carbon steel pipe 92
Find the stress in pipe wall due to internal pressure 93
How to calculate stress in above/below ground transition 94
How to identify the series number of flanged fittings 97
Dimensions of three-diameter ells with tangents 97
Polypipe design data 98

Steel Pipe Design

The maximum allowable design pressure stress will depend upon the intended service for the pipe line.

Pipe lines to be used for transporting liquid petroleum are covered by ANSI/ASME B31.4—"Liquid Petroleum Transportation Piping Systems." Pipe lines used for transporting gas are covered by ANSI/ASME B31.8—"Gas Transmission and Distribution Systems."

Pipe lines which must be operated in compliance with the Federal Pipeline Safety Regulations will also need to comply with the applicable parts of these regulations: Part 192 for gas transportation systems, and Part 195 for liquid transportation systems.

Gas Pipe Lines—ANSI/ASME B31.8

The maximum allowable pressure is calculated by the following equation:

$$P = (2St/D) \times F \times E \times T$$

where P = Design pressure, lb/in.^{2}
S = Specified minimum yield strength, lb/in.^{2} (see Table 1)
t = Nominal wall thickness, in.
D = Nominal outside diameter, in.
F = Design factor (see Table 2)
E = Longitudinal joint factor (see Table 3)
T = Temperature derating factor (see Table 4)

Class location definitions may be obtained from p. 192.111 of Part 192 of the Federal Pipeline Safety Regulations.

A typical calculation is as follows:

Pipe: 16″ OD × 0.250″ wt API 5LX X52 ERW
Location: Class 1, therefore F = 0.72 (see Table 2)
Temperature: 90°F, Temp. factor T = 1 (see Table 4)
Joint Factor: E = 1.0 (see Table 3)

$$P = (2 \times 52{,}000 \times 0.250/16.0) \times 0.72 \times 1 \times 1$$
$$P = 1{,}170 \text{ lb/in.}^2 \text{ gauge}$$

Table 1
Specified Minimum Yield Strength for Steel and Iron Pipe Commonly Used in Piping Systems

Specification	Grade	Type[1]	SMYS, psi
API 5L	A25	BW, ERW, S	25,000
API 5L	A	ERW, FW, S, DSA	30,000
API 5L	B	ERW, FW, S, DSA	35,000
API 5LS (Note (2))	A	ERW, DSA	30,000
API 5LS	B	ERW, DSA	35,000
API 5LS	X42	ERW, DSA	42,000
API 5LS	X46	ERW, DSA	46,000
API 5LS	X52	ERW, DSA	52,000
API 5LS	X56	ERW, DSA	56,000
API 5LS	X60	ERW, DSA	60,000
API 5LS	X65	ERW, DSA	65,000
API 5LS	X70	ERW, DSA	70,000
API 5LX (Note (2))	X42	ERW, FW, S, DSA	42,000
API 5LX	X46	ERW, FW, S, DSA	46,000
API 5LX	X52	ERW, FW, S, DSA	52,000
API 5LX	X56	ERW, FW, S, DSA	56,000
API 5LX	X60	ERW, FW, S, DSA	60,000
API 5LX	X65	ERW, FW, S, DSA	65,000
API 5LX	X70	ERW, FW, S, DSA	70,000
ASTM A53	Open Hrth. Bas. Oxy., Elec. Furn.	BW	25,000
ASTM A53	Bessemer	BW	30,000

Reproduced from ANSI/ASME Code B31-8-1982, Appendix D. Reprinted courtesy of The American Society of Mechanical Engineers.

Table 1 (Continued)
Specified Minimum Yield Strength for Steel and Iron Pipe Commonly Used in Piping Systems

Specification	Grade	Type[1]	SMYS, psi
ASTM A53	A	ERW, S	30,000
ASTM A53	B	ERW, S	35,000
ASTM A106	A	S	30,000
ASTM A106	B	S	35,000
ASTM A106	C	S	40,000
ASTM A134	–	EFW	[Note (3)]
ASTM A135	A	ERW	30,000
ASTM A135	B	ERW	35,000
ASTM A139	A	ERW	30,000
ASTM A139	B	ERW	35,000
ASTM A333	1	S, ERW	30,000
ASTM A333	3	S, ERW	35,000
ASTM A333	4	S	35,000
ASTM A333	6	S, ERW	35,000
ASTM A333	7	S, ERW	35,000
ASTM A333	8	S, ERW	75,000

Reproduced from ANSI/ASME Code B31-8-1982, Appendix D. Reprinted courtesy of The American Society of Mechanical Engineers.

Table 2
Values of Design Factor F

Construction Type (See 841.151)	Design Factor F
Type A	0.72
Type B	0.60
Type C	0.50
Type D	0.40

Reproduced from ANSI/ASME Code B31-8-1982, Table 841.1A. Reprinted courtesy of The American Society of Mechanical Engineers.

Table 4
Temperature Derating Factor T For Steel Pipe

Temperature, °F	Temperature Derating Factor T
250 or less	1.000
300	0.967
350	0.933
400	0.900
450	0.867

NOTE: For intermediate temperatures, interpolate for derating factor.

Reproduced from ANSI/ASME Code B31-8-1982, Table 841.1C. Reprinted courtesy of The American Society of Mechanical Engineers.

Table 3
Longitudinal Joint Factor E

Spec. Number	Pipe Class	E Factor
ASTM A53	Seamless	1.00
	Electric Resistance Welded	1.00
	Furnace Welded	0.60
ASTM A106	Seamless	1.00
ASTM A134	Electric Fusion Arc Welded	0.80
ASTM A135	Electric Resistance Welded	1.00
ASTM A139	Electric Fusion Welded	0.80
ASTM A211	Spiral Welded Steel Pipe	0.80
ASTM A381	Double Submerged-Arc-Welded	1.00
ASTM A671	Electric Fusion Welded	1.00*
ASTM A672	Electric Fusion Welded	1.00*
API 5L	Seamless	1.00
	Electric Resistance Welded	1.00
	Electric Flash Welded	1.00
	Submerged Arc Welded	1.00
	Furnace Butt Welded	0.60
API 5LX	Seamless	1.00
	Electric Resistance Welded	1.00
	Electric Flash Welded	1.00
	Submerged Arc Welded	1.00
API 5LS	Electric Resistance Welded	1.00
	Submerged Welded	1.00

NOTE: Definitions for the various classes of welded pipe are given in 804.243

*Includes Classes 12, 22, 32, 42, and 52 only.

Reproduced from ANSI/ASME Code B31-8-1982, Table 841.1B. Reprinted courtesy of The American Society of Mechanical Engineers.

Liquid Pipe Lines—ANSI/ASME B31.4

The internal design pressure is determined by using the following formula:

$$P = (2St/D) \times E \times F$$

where
P = Internal design pressure, lb/in.² gauge
S = Specified minimum yield strength, lb/in.² (see Table 5)
t = Nominal wall thickness, in.
D = Nominal outside diameter of the pipe, in.
E = Weld joint factor (see Table 6)
F = Design factor of 0.72

Note: Refer to p. 195.106 of Part 195 Federal Pipeline Safety Regulations for design factors to be used on offshore risers and platform piping and cold worked pipe.

A typical calculation of the internal design pressure is as follows:

Pipe: 26″ OD × 0.3125″ wt API 5LX X52 ERW
Weld joint factor E = 1.0 (see Table 6)
Design factor F = 0.72

$$P = (2 \times 52{,}000 \times 0.3125/26) \times 1 \times 0.72$$
$$P = 900 \text{ lb/in.}^2 \text{ gauge}$$

Table 5
Tabulation of Examples of Allowable Stresses for
Reference Use in Piping Systems

Allowable stress values (S) shown in this Table are equal to 0.72 × E (weld joint factor) × specified minimum yield strength of the pipe.

Allowable stress values shown are for new pipe of known specification. Allowable stress values for new pipe of unknown specification, ASTM A 120 specification or used (reclaimed) pipe shall be determined in accordance with 402.3.1.

For some Code computations, particularly with regard to branch connections [see 404.3.1 (d) (3)] and expansion, flexibility, structural attachments, supports, and restraints (Chapter II, Part 5), the weld joint factor E need not be considered.

For specified minimum yield strength of other grades in approved specifications, refer to that particular specification.

Allowable stress value for cold worked pipe subsequently heated to 600 F (300 C) or higher (welding excepted) shall be 75 percent of value listed in Table.

Definitions for the various types of pipe are given in 400.2.

(Metric Stress Levels are given in MPa [1 Megapascal = 1 million pascals])

Specification	Grade	Specified Min Yield Strength psi (MPa)	Notes	(E) Weld Joint Factor	(S) Allowable Stress Value −20 F to 250 F (−30 C to 120 C) psi (MPa)
Seamless					
API 5L	A25	25,000 (172)	(1)	1.00	18,000 (124)
API 5L, ASTM A53, ASTM A106	A	30,000 (207)	(1) (2)	1.00	21,600 (149)
API 5L, ASTM A53, ASTM A106	B	35,000 (241)	(1) (2)	1.00	25,200 (174)
ASTM A106	C	40,000 (278)	(1) (2)	1.00	28,800 (199)
ASTM A524	I	35,000 (241)	(1)	1.00	25,200 (174)
ASTM A524	II	30,000 (207)	(1)	1.00	21,600 (149)
API 5LU	U80	80,000 (551)	(1) (4)	1.00	57,600 (397)
API 5LU	U100	100,000 (689)	(1) (4)	1.00	72,000 (496)
API 5LX	X42	42,000 (289)	(1) (2) (4)	1.00	30,250 (208)
API 5LX	X46	46,000 (317)	(1) (2) (4)	1.00	33,100 (228)
API 5LX	X52	52,000 (358)	(1) (2) (4)	1.00	37,450 (258)
API 5LX	X56	56,000 (386)	(1) (4)	1.00	40,300 (278)
API 5LX	X60	60,000 (413)	(1) (4)	1.00	43,200 (298)
API 5LX	X65	65,000 (448)	(1) (4)	1.00	46,800 (323)
API 5LX	X70	70,000 (482)	(1) (4)	1.00	50,400 (347)
Furnace Welded-Butt Welded					
ASTM A53		25,000 (172)	(1) (2)	0.60	10,800 (74)
API 5L Class I & Class II	A25	25,000 (172)	(1) (2) (3)	0.60	10,800 (74)
API 5L (Bessemer), ASTM A53 (Bessemer)		30,000 (207)	(1) (2) (5)	0.60	12,950 (89)

Reproduced from ANSI/ASME Code B31-4-1979, Table 402.3.1(a). Reprinted courtesy of The American Society of Mechanical Engineers.

Table 5 (Continued)

Specification	Grade	Specified Min Yield Strength psi (MPa)	Notes	(E) Weld Joint Factor	(S) Allowable Stress Value −20F to 250 F (−30 C to 120 C) psi (MPa)
Furnace Welded-Lap Welded					
API 5L Class I		25,000 (172)	(1) (2) (6)	0.80	14,400 (99)
API 5L Class II		28,000 (193)	(1) (2) (6)	0.80	16,150 (111)
API 5L (Bessemer)		30,000 (207)	(1) (2) (6)	0.80	17,300 (119)
API 5L Electric Furnace		25,000 (172)	(1) (2) (6)	0.80	14,400 (99)
Electric Resistance Welded and Electric Flash Welded					
API 5L	A25	25,000 (172)	(1) (7)	1.00	18,000 (124)
API 5L, ASTM A53, ASTM A135	A	30,000 (207)	(2)	0.85	18,360 (127)
API 5L, API 5LS, ASTM A53, ASTM A135	A	30,000 (207)	(1)	1.00	21,600 (149)
API 5L, ASTM A53, ASTM A135	B	35,000 (241)	(2)	0.85	21,420 (148)
API 5L, API 5LS, ASTM A53, ASTM A135	B	35,000 (241)	(1)	1.00	25,200 (174)
API 5LS, API 5LX	X42	42,000 (289)	(1) (2) (4)	1.00	30,250 (208)
API 5LS, API 5LX	X46	46,000 (317)	(1) (2) (4)	1.00	33,100 (228)
API 5LS, API 5LX	X52	52,000 (358)	(1) (2) (4)	1.00	37,450 (258)
API 5LS, API 5LX	X56	56,000 (386)	(1) (4)	1.00	40,300 (279)
API 5LS, API 5LX	X60	60,000 (413)	(1) (4)	1.00	43,200 (297)
API 5LS, API 5LX	X65	65,000 (448)	(1) (4)	1.00	46,800 (323)
API 5LS, API 5LX	X70	70,000 (482)	(1) (4)	1.00	50,400 (347)
API 5LU	U80	80,000 (551)	(1) (4)	1.00	57,600 (397)
API 5LU	U100	100,000 (689)	(1) (4)	1.00	72,000 (496)
Electric Fusion Welded					
ASTM A 134	–	–		0.80	–
ASTM A 139	A	30,000 (207)	(1) (2)	0.80	17,300 (119)
ASTM A 139	B	35,000 (241)	(1) (2)	0.80	20,150 (139)
ASTM A 155	–	–	(2) (8)	0.90	–
ASTM A 155	–	–	(1) (8)	1.00	–
Submerged Arc Welded					
API 5L, API 5LS	A	30,000 (207)	(1)	1.00	21,600 (149)
API 5L, API 5LS	B	35,000 (241)	(1)	1.00	25,200 (174)
API 5LS, API 5LX	X42	42,000 (289)	(1) (2) (4)	1.00	30,250 (208)
API 5LS, API 5LX	X46	46,000 (317)	(1) (2) (4)	1.00	33,100 (228)
API 5LS, API 5LX	X52	52,000 (358)	(1) (2) (4)	1.00	37,450 (258)
API 5LS, API 5LX	X56	56,000 (386)	(1) (4)	1.00	40,300 (278)
API 5LS, API 5LX	X60	60,000 (413)	(1) (4)	1.00	43,200 (298)
API 5LS, API 5LX	X65	65,000 (448)	(1) (4)	1.00	46,800 (323)
API 5LS, API 5LX	X70	70,000 (482)	(1) (4)	1.00	50,400 (347)
API 5LU	U80	80,000 (551)	(1) (4)	1.00	57,600 (397)
API 5LU	U100	100,000 (689)	(1) (4)	1.00	72,000 (496)
ASTM A 381	Y35	35,000 (241)	(1) (2)	1.00	25,200 (174)
ASTM A 381	Y42	42,000 (290)	(1) (2)	1.00	30,250 (209)
ASTM A 381	Y46	46,000 (317)	(1) (2)	1.00	33,100 (228)
ASTM A 381	Y48	48,000 (331)	(1) (2)	1.00	34,550 (238)
ASTM A 381	Y50	50,000 (345)	(1)	1.00	36,000 (248)
ASTM A 381	Y52	52,000 (358)	(1)	1.00	37,450 (258)
ASTM A 381	Y60	60,000 (413)	(1)	1.00	43,200 (298)
ASTM A 381	Y65	65,000 (448)	(1)	1.00	46,800 (323)

NOTES (1) Weld joint factor E (see Table 402.4.3) and allowable stress value are applicable to pipe manufactured after 1958.

(2) Weld joint factor E (see Table 402.4.3) and allowable stress value are applicable to pipe manufactured before 1959.

(3) Class II produced under API 5L 23rd Edition, 1968, or earlier has a specified minimum yield strength of 28,000 psi (193 MPa).

(4) Other grades provided for in API 5LS, API 5LU, and API 5LX not precluded.

(5) Manufacture was discontinued and process deleted from API 5L in 1969.

(6) Manufacture was discontinued and process deleted from API 5L in 1962.

(7) A25 is not produced in electric flash weld.

(8) See applicable plate specification for yield point and refer to 402.3.1 for calculation of (S).

Table 6
Weld Joint Factor

Specification Number	Pipe Type (1)	Weld Joint Factor E	
		Pipe Mfrd. Before 1959	Pipe Mfrd. After 1958
ASTM A53	Seamless	1.00	1.00
	Electric-Resistance-Welded	0.85 (2)	1.00
	Furnace Lap-Welded	0.80	0.80
	Furnace Butt-Welded	0.60	0.60
ASTM A106	Seamless	1.00	1.00
ASTM A134	Electric-Fusion (Arc)-Welded single or double pass	0.80	0.80
ASTM A135	Electric-Resistance-Welded	0.85 (2)	1.00
ASTM A139	Electric-Fusion-Welded single or double pass	0.80	0.80
ASTM A155	Electric-Fusion-Welded	0.90	1.00
ASTM A381	Electric-Fusion-Welded, Double Submerged Arc-Welded	–	1.00
API 5L	Seamless	1.00	1.00
	Electric-Resistance-Welded	0.85 (2)	1.00
	Electric-Flash-Welded	0.85 (2)	1.00
	Electric-Induction-Welded	–	1.00
	Submerged Arc-Welded	–	1.00
	Furnace Lap-Welded	0.80	0.80 (3)
	Furnace Butt-Welded	0.60	0.60
API 5LS	Electric-Resistance-Welded	–	1.00
	Submerged Arc-Welded	–	1.00
API 5LX	Seamless	1.00	1.00
	Electric-Resistance-Welded	1.00	1.00
	Electric-Flash-Welded	1.00	1.00
	Electric-Induction-Welded	–	1.00
	Submerged Arc-Welded	1.00	1.00
API 5LU	Seamless	–	1.00
	Electric-Resistance-Welded	–	1.00
	Electric-Flash-Welded	–	1.00
	Electric-Induction-Welded	–	1.00
	Submerged Arc-Welded	–	1.00
Known	Known	(4)	(5)
Unknown	Seamless	1.00 (6)	1.00 (6)
Unknown	Electric-Resistance or Flash-Welded	0.85 (6)	1.00 (6)
Unknown	Electric-Fusion-Welded	0.80 (6)	0.80 (6)
Unknown	Furnace Lap-Welded or over NPS 4	0.80 (7)	0.80 (7)
Unknown	Furnace Butt-Welded or NPS 4 and smaller	0.60 (8)	0.60 (8)

NOTES: (1) Definitions for the various pipe types (weld joints) are given in 400.2.
(2) A weld joint factor of 1.0 may be used for electric-resistance-welded or electric-flash-welded pipe manufactured prior to 1959 where (a) pipe furnished under this classification has been subjected to supplemental tests and/or heat treatments as agreed to by the supplier and the purchaser, and such supplemental tests and/or heat treatment demonstrate the strength characteristics of the weld to be equal to the minimum tensile strength specified for the pipe, or (b) pipe has been tested as required for a new pipeline in accordance with 437.4.1.
(3) Manufacture was discontinued and process deleted from API 5L in 1962.
(4) Factors shown above for pipe manufactured before 1959 apply for new or used (reclaimed) pipe if pipe specification and pipe type are known and it is known that pipe was manufactured before 1959 or not known whether manufactured after 1958.
(5) Factors shown above for pipe manufactured after 1958 apply for new or used (reclaimed) pipe if pipe specification and pipe type are known and it is known that pipe was manufactured after 1958.
(6) Factor applies for new or used pipe of unknown specification and ASTM A120 if type of weld joint is known.
(7) Factor applies for new or used pipe of unknown specification and ASTM A120 if type of weld joint is known to be furnace lap-welded, or for pipe over NPS 4 if type of joint is unknown.
(8) Factor applies for new or used pipe of unknown specification and ASTM A120 if type of weld joint is known to be furnace butt-welded, or for pipe NPS 4 and smaller if type of joint is unknown.

Properties of Pipe

D	t	d	W	A_o	V_m	A_f	I	Z
4.500	0.125	4.250	5.840	1.178097	92.644911	0.098516	4.12	1.83
	0.156	4.188	7.240		89.961575	0.095662	5.03	2.24
	0.172	4.156	7.950		88.592057	0.094206	5.49	2.44
	0.188	4.124	8.660		87.233042	0.092761	5.93	2.64
	0.219	4.062	10.010		84.629845	0.089993	6.77	3.01
STD	0.237	4.026	10.790		83.136406	0.088405	7.56	3.36
	0.250	4.000	11.350		82.066080	0.087266	7.56	3.36
6.625	0.188	6.249	12.93	1.734421	200.292532	0.212985	19.7	6.0
	0.219	6.187	14.99		196.337808	0.208779	22.6	6.8
	0.250	6.125	17.02		192.422518	0.204616	25.5	7.7
STD	0.280	6.065	18.98		188.671072	0.200627	28.1	8.5
	0.312	6.001	21.04		184.710235	0.196415	30.9	9.3
	0.375	5.875	25.03		177.035128	0.188254	36.1	10.9
8.625	0.188	8.249	16.94	2.258020	349.016785	0.371133	44.4	10.3
	0.219	8.187	19.66		343.790038	0.365575	51.1	11.9
	0.250	8.125	22.36		338.602723	0.360059	57.7	13.4
	0.277	8.071	24.70		334.116868	0.355289	63.4	14.7
	0.312	8.001	27.70		328.346391	0.349153	70.5	16.3
STD	0.322	7.981	28.56		326.706916	0.347410	72.5	16.8
	0.344	7.937	30.43		323.114514	0.343590	76.9	17.8
	0.375	7.875	33.05		318.086203	0.338243	82.9	19.2
10.750	0.219	10.312	24.63	2.814343	545.418061	0.579980	87.0	16.2
	0.250	10.250	28.04		538.879221	0.573027	113.7	21.2
	0.279	10.192	31.20		532.797939	0.566560	125.9	23.4
	0.307	10.136	34.24		526.959102	0.560352	137.5	25.6
STD	0.365	10.020	40.49		514.966704	0.547599	160.8	29.9
12.750	0.219	12.312	29.31	3.337942	777.500935	0.826770	169.3	26.6
	0.250	12.250	33.38		769.690071	0.818464	191.9	30.1
	0.281	12.188	37.43		761.918639	0.810200	214.1	33.6
	0.312	12.126	41.45		754.186639	0.801978	236.0	37.0
STD	0.375	12.000	49.57		738.594720	0.785398	279.0	43.8
	0.438	11.874	57.60		723.165661	0.768991	321.0	50.4
	0.500	11.750	65.42		708.140511	0.753014	363.0	56.7
14.000	0.219	13.562	32.24	3.665191	943.389822	1.003171	226.0	32.3
	0.250	13.500	36.72		934.783943	0.994020	255.0	36.5
	0.281	13.438	41.18		926.217495	0.984910	285.1	40.7
	0.312	13.376	45.62		917.690481	0.975843	315.0	45.0
	0.344	13.312	50.18		908.929763	0.966527	344.3	49.2
STD	0.375	13.250	54.58		900.482886	0.957545	373.0	53.3
	0.438	13.124	63.45		883.438151	0.939420	429.0	61.4
	0.500	13.000	72.10		866.822970	0.921752	484.0	69.1

D = Outside diameter, in.
t = Wall thickness, in.
d = Inside diameter, in.
W = Pipe weight #/ft.
A_o = Outside surface area sq. ft./ft.
V_m = Displacement, bbls/mile
A_f = Flow area, sq. ft.

I = Moment of Inertia, in.^{4}
Z = Section modulus, in.^{3}
Gals/ft. = $d^2 \times 0.0408$
Gals/mile = $d^2 \times 215.4240$
Bbls/mile = $d^2 \times 5.1291$
$I = 0.0491 (D^4 - d^4)$
$Z = 0.0982 (D^4 - d^4)/D$

D	t	d	W	A_o	V_m	A_f	I	Z
16.000	0.219	15.562	36.92	4.188790	1242.151387	1.320864	338.0	42.3
	0.250	15.500	42.06		1232.273483	1.310360	384.0	48.0
	0.281	15.438	47.18		1222.435011	1.299899	429.0	53.6
	0.312	15.376	52.28		1212.635972	1.289479	474.0	59.3
	0.344	15.312	57.53		1202.562197	1.278766	519.0	64.8
STD	0.375	15.250	62.59		1192.843296	1.268432	562.0	70.3
	0.438	15.124	72.81		1173.213479	1.247558	649.0	81.1
	0.500	15.000	82.78		1154.054250	1.227185	732.0	91.5
18.000	0.250	17.500	47.40	4.712389	1570.796063	1.670335	549.0	61.0
	0.281	17.438	53.18		1559.685567	1.658520	614.2	68.2
	0.312	17.376	58.95		1548.614504	1.646747	679.0	75.5
	0.344	17.312	64.88		1537.227671	1.634639	744.0	82.7
STD	0.375	17.250	70.60		1526.236746	1.622952	807.0	89.6
	0.438	17.124	82.16		1504.021848	1.599329	932.0	103.6
	0.500	17.000	93.46		1482.318570	1.576250	1053.0	117.0
20.000	0.250	19.500	52.74	5.235988	1950.351683	2.073942	756.6	75.7
	0.281	19.438	59.19		1937.969163	2.060775	846.5	84.6
	0.312	19.376	65.61		1925.626075	2.047650	935.5	93.5
	0.344	19.312	72.22		1912.926185	2.034145	1026.5	102.6
STD	0.375	19.250	78.61		1900.663236	2.021105	1113.8	111.4
	0.438	19.124	91.52		1875.863256	1.994733	1288.6	128.9
	0.500	19.000	104.14		1851.615930	1.968950	1457.2	145.7
24.000	0.250	23.500	63.42	6.283185	2832.562043	3.012056	1315.7	109.6
	0.281	23.438	71.19		2817.635474	2.996184	1473.1	122.8
	0.312	23.376	78.94		2802.748338	2.980354	1629.2	135.8
	0.344	23.312	86.92		2787.422334	2.964056	1789.1	149.1
STD	0.375	23.250	94.63		2772.615336	2.948311	1942.8	161.9
	0.438	23.124	110.23		2742.645193	2.916442	2251.3	187.6
	0.500	23.000	125.51		2713.309770	2.885247	2550.0	212.5
26.000	0.250	25.500	68.76	6.806784	3335.216783	3.546564	1676.8	129.0
	0.281	25.438	77.20		3319.018190	3.529339	1878.0	144.5
	0.312	25.376	85.61		3302.859030	3.512155	2077.7	159.8
	0.344	25.312	94.27		3286.219968	3.494462	2282.3	175.6
STD	0.375	25.250	102.64		3270.140946	3.477364	2479.1	190.7
	0.438	25.124	119.59		3237.585722	3.442746	2874.5	221.1
	0.500	25.000	136.19		3205.706250	3.408846	3257.8	250.6
30.000	0.250	29.500	79.44	7.853982	4463.625383	4.746477	2585.8	172.4
	0.281	29.438	89.20		4444.882741	4.726547	2897.5	193.2
	0.312	29.376	98.94		4426.179533	4.706659	3207.1	213.8
	0.344	29.312	108.97		4406.914357	4.686173	3524.7	235.0
STD	0.375	29.250	118.66		4388.291286	4.666370	3830.4	255.4
	0.438	29.124	138.30		4350.565898	4.626254	4445.7	296.4
	0.500	29.000	157.55		4313.598330	4.586943	5043.5	336.2
32.000	0.250	31.500	84.78	8.377580	5089.379243	5.411884	3143.2	196.4
	0.281	31.438	95.20		5069.364577	5.390601	3522.7	220.2
	0.312	31.376	105.60		5049.389344	5.369360	3899.9	243.7
	0.344	31.312	116.32		5028.811111	5.347478	4287.0	267.9
STD	0.375	31.250	126.67		5008.916016	5.326322	4659.7	291.2
	0.438	31.124	147.66		4968.605547	5.283457	5410.3	338.1
	0.500	31.000	168.23		4929.093930	5.241442	6140.2	383.8

Length of Pipe in Bends

Radius of Pipe Bends		90° Bends	180° Bends	270° Bends	360° Bends	540° Bends
Inches	Feet					
1		1½″	3″	4¾″	6¼″	9½″
2		3	6¼	9½	12½	18¾
3	¼	4¾	9½	14¼	18¾	28¼
4		6¼	12½	18¾	25¼	37¾
5		7¾	15¾	23½	31½	47¼
6	½	9½	18¾	28¼	37¾	56½
7		11	22	33	44	66
8		12½	25¼	37¾	50¼	75½
9	¾	14¼	28¼	42½	56½	84¾
10		15¾	31½	47¼	62¾	94¼
11		17¼	34½	51¾	69	103¾
12	1	18¾	37¾	56½	75½	113
12.5		19¾	39¼	59	78½	117¾
14		22	44	66	88	132
15	1¼	23½	47	70¾	94¼	141¼
16		25¼	50¼	75½	100½	150¾
17.5		27½	55	82½	110	165
18	1½	28½	56½	84¾	113	169¾
20		31½	62¾	94¼	125¾	188½
21	1¾	33	66	99	132	198
24	2	37¾	75½	113	150¾	226¼
25		39¼	78½	117¾	157	235½
30	2½	47¼	94¼	141¼	188½	282¾
32		50¼	100½	150¾	201	301½
36	3	56½	113	169½	226¼	339¼
40		62¾	125¾	188½	251¼	377
48	4	75½	150¾	226¼	301½	452½
50		78½	157	235½	314¼	471¼
56		88	176	264	351¾	527¾
60	5	94¼	188½	282¾	377	565½
64		100½	201	301½	402	603¼
70		110	220	329¾	439¾	659¾
72	6	113	226¼	339¼	452½	678½
80		125¾	251¼	377	502¾	754
84	7	132	263¾	395¾	527¾	791½
90	7½	141¼	282¾	424	565½	848¼
96	8	150¾	301½	452½	603	904¾
100		157	314¼	471¼	628¼	942½
108	9	169½	339¼	509	678½	1017¾
120	10	188½	377	565½	754	1131
132	11	207¼	414¾	622	829½	1244
144	12	226¼	452½	678½	904¾	1357¼
156	13	245	490	735¼	980¼	1470¼
168	14	263¾	527¾	791½	1055½	1583½
180	15	282¾	565½	848¼	1131	1696½
192	16	301½	603	904¾	1206¼	1809½
204	17	320½	640¾	961¼	1281¾	1922½
216	18	339¼	678½	1017¾	1357¼	2035¾
228	19	358	716¼	1074½	1432½	2148¾
240	20	377	754	1131	1508	2262

To find the length of pipe in a bend having a radius not given above, add together the length of pipe in bends whose combined radii equal the required radius.

Example: Find length of pipe in 90° bend of 5′ 9″ radius.

Length of pipe in 90° bend of 5′ radius = 94¼″
Length of pipe in 90° bend of 9″ radius = 14¼″

Then, length of pipe in 90° bend of 5′ 9″ radius = 108½″

Calculation of Pipe Bends

Example No. 1—**Given A, B, C, R**

$$D = B - C$$
$$E = R - A$$
$$F = \sqrt{D^2 + E^2}$$
$$\frac{E}{F} = \sin \angle G$$
$$H = \sqrt{F^2 - R^2}$$

$$\frac{R}{F} = \sin \angle P$$
$$\angle S = \angle P - \angle G$$
$$\angle K = 90° - \angle S$$
$$\angle L = \tfrac{1}{2} \angle S$$
$$M = \tan \angle L \times R$$
$$N = H + M$$
$$O = B - C - M$$

Example No. 2—**Given A, B, C, R**

$$D = B - C$$
$$E = A - R$$
$$F = \sqrt{D^2 + E^2}$$
$$\frac{E}{F} = \sin \angle G$$
$$H = \sqrt{F^2 - R^2}$$

$$\frac{R}{F} = \sin \angle P$$
$$\angle S = \angle P + \angle G$$
$$\angle K = 90° - \angle S$$
$$\angle L = \tfrac{1}{2} \angle S$$
$$M = \tan \angle L \times R$$
$$N = H + M$$
$$O = B - C - M$$

Example No. 3—**Given A, B, C, R**

$$D = B - C$$
$$E = \sqrt{A^2 + D^2}$$
$$\frac{A}{E} = \sin \angle F$$

$$\angle G = \tfrac{1}{2} \angle F$$
$$H = \tan \angle G \times R$$
$$P = C - H$$

Example No. 4—**Given A, B, C, R**

$$D = B - C$$
$$E = \sqrt{A^2 + D^2}$$

$$\frac{A}{E} = \sin \angle F$$
$$\angle G = \tfrac{1}{2} \angle F$$

$$H = \tan \angle G \times R$$
$$P = C - H$$

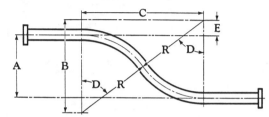

Example No. 5—**Given A, R**

$$B = 2R - A \qquad C = \sqrt{(2R)^2 - B^2} \qquad \frac{C}{2R} = \sin \angle D$$

Example No. 6—**Given A, B**

$$C = \tfrac{1}{2} B$$
$$D = \tfrac{1}{2} A$$
$$E = \sqrt{C^2 + D^2}$$

$$\frac{D}{E} = \sin \angle F$$
$$G = \tfrac{1}{2} E$$

$$\angle H = 90° - \angle F$$
$$R = \frac{A^2 + B^2}{4A}$$
$$L = 2F$$

Example No. 7—**Given A, B, C, D, R**

$$E = D - A - B$$
$$F = R - C$$
$$G = R + F$$
$$H = \sqrt{E^2 + G^2}$$

$$G/H = \sin \angle K$$
$$L = \tfrac{1}{2} H$$
$$M = \sqrt{L^2 - R^2}$$

$$M/L = \sin \angle N$$
$$\angle O = 90° - \angle K - \angle N$$
$$\angle P = \tfrac{1}{2} \angle O$$
$$S = \tan \angle P \times R$$

Calculation of Pipe Bends

Example No. 8—Given A, B, C, D, R

$$E = D - A - B$$
$$F = \sqrt{E^2 + C^2}$$
$$\frac{C}{F} = \sin \angle G$$

$$\angle H = \tfrac{1}{2} \angle G$$
$$K = \tan \angle H \times R$$
$$L = A - K$$
$$P = B - K$$
$$N = F - 2K$$

Example No. 9—Given A, B, C, D, R

$$E = D - A - B$$
$$F = 2R - C$$
$$G = \sqrt{E^2 + F^2}$$
$$\frac{F}{E} = \tan \angle H$$
$$K = \tfrac{1}{2} G$$

$$M = \sqrt{K^2 - R^2}$$
$$\frac{M}{K} = \sin \angle N$$
$$\angle P = 90° - \angle H - \angle N$$
$$\angle O = \tfrac{1}{2} \angle P$$
$$S = \tan \angle O \times R$$

Example No. 10—Given A, B, C, D, R

$$E = D - A - B$$
$$F = \sqrt{C^2 + E^2}$$
$$\frac{C}{F} = \sin \angle G$$

$$\angle H = \tfrac{1}{2} \angle G$$
$$K = \tan \angle H \times R$$
$$M = A - K$$
$$N = B - K$$
$$P = F - 2K$$

Example No. 11—Given R and 45° Angles

$$A = 3.414 \times R$$
$$B = 2.828 \times R$$
$$C = 0.828 \times R$$

$T = $ Tangent
Length of pipe in bend =
$$9.425 \times R + 2T$$

Example No. 12—Given A, R

$$C = A - 2R$$
$$D = \sqrt{(2R)^2 - C^2}$$

$$P = 2D$$
$$E = D - R$$
$$F = 2E$$

$$C/2R = \sin \angle G$$
$$\angle H = 90° + \angle G$$
$$\angle K = 180° + 2 \angle G$$

Example No. 13—Given A, B, R

$$C = \tfrac{1}{2} B$$
$$D = R + C$$
$$E = A - 2R$$
$$F = \sqrt{D^2 + E^2}$$

$$E/F = \sin \angle G$$
$$H = \tfrac{1}{2} F$$
$$K = \sqrt{H^2 - R^2}$$
$$K/H = \sin \angle L$$

$$\angle M = \angle G - \angle L$$
$$\angle N = 180° + 2 \angle M$$
$$\angle O = 90° + \angle M$$
$$P = 2D$$

Spacing of Pipe Supports

For chart on deflection of horizontal pipe lines, see next page.

When a horizontal pipe line is supported at intermediate points, sagging of the pipe occurs between these supports, the amount of sag being dependent upon the weight of the pipe, fluid, insulation, and valves or fittings which may be included in the line. If the pipe line is installed with no downward pitch, pockets will be formed in each span in which case condensation may collect if the line is transporting steam. In order to eliminate these pockets, the line must be pitched downward so that the outlet of each span is lower than the maximum sag.

Crane has conducted tests to determine the deflection of horizontal standard pipe lines filled with water, in pipe sizes ¾″ to 4″ inclusive, the results of which have indicated that for pipes larger than 2″ and with supports having center to center dimensions greater than 10 feet, the resultant deflection is less than that determined by the use of the formula for a uniformly loaded pipe fixed at both ends. For pipe sizes 2″ and smaller, the test deflection was in excess of that determined by the formula for pipe having fixed ends and approached, for the shorter spans, the deflection as determined by the use of the formula for pipe lines having unrestrained ends.

Page 83 gives the deflection of horizontal standard pipe lines filled with water, for varying spans, based upon the results obtained from tests for sizes 2″ and smaller, and upon the formula for fixed ends for the larger sizes of pipe. The deflection values given on the chart are twice those obtained from test or calculation, to compensate for any variables including weight of insulation, etc.

The formula given below indicates the vertical distance that the span must be pitched so that the outlet is lower than the maximum sag of the pipe.

$$h = \frac{144\,S^2\,y}{36\,S^2 - y^2}$$

where:
    h = Difference in elevation of span ends, inches
    S = Length of one span, feet
    y = Deflection of one span, inches

By eliminating the inconsequential term "$-y^2$" from the denominator, the formula reduces to:

$$h = 4\,y$$

The pitch of pipe spans, called the Average Gradient, is a ratio between the drop in elevation and the length of the span. This is expressed as so many inches in a certain number of feet.

$$\text{Average Gradient} = \frac{4\,y}{S}$$

The dotted lines as shown on the chart on the opposite page are plotted from the above formula and indicate average gradients of 1″ in 10′, 1″ in 15′, 1″ in 20′, 1″ in 30′, and 1″ in 40′.

Example: What is the maximum distance between supports for a 4″ Standard pipe line assuming a pitch or average gradient of 1″ in 30 feet?

Using the chart on the opposite page, find the point where the diagonal dotted line for an average gradient of 1″ in 30 feet intersects the diagonal solid line for 4″ pipe. From this point, proceed downward to the bottom line where the maximum span is noted to be approximately **22** feet.

Code for Pressure Piping: The Code for Pressure Piping, ASA B 31.1, makes the following statements relative to installations within the scope of the Code:

"**605 (g)** Supports shall be spaced so as to prevent excessive sag, bending and shear stresses in the piping, with special consideration given to those piping sections where flanges, valves, etc., impose concentrated loads. Where calculations are not made, suggested maximum spacing of hangers or supports for carbon steel piping operating at 750° F and lower are given in Table 21a (*see the table below*). "Where greater distance between supports, concentrated loads, higher temperatures, or vibration considerations are involved, special consideration should be given to effects of bending and shear stresses."

"**623 (b)** The design and spacing of supports shall be checked to assure that the sum of the longitudinal stresses due to weight, pressure, and other sustained external loading does not exceed the allowable stress (S value) in the hot condition."

Suggested Maximum Spacing Between Pipe Supports For Straight Runs of Standard Wall and Heavier Pipe (At Maximum Operating Temperature of 750 F)

Nominal Pipe Size Inches	Maximum Span Feet	Nominal Pipe Size Inches	Maximum Span Feet
1	7	8	19
1½	9	10	22
2	10	12	23
2½	11	14	25
3	12	16	27
3½	13	18	28
4	14	20	30
5	16	24	32
6	17	..	..

Notes: The values in the table do not apply where there are concentrated loads between supports such as flanges, valves, etc.

Spacing is based on a combined bending and shear stress of 1500 psi when pipe is filled with water and the pitch of the line is such that a sag of 0.1 inch between supports is permissible.

Spacing of Pipe Supports
Deflection of Horizontal Pipe Lines
(Based on Standard Pipe Filled with Water)

Distance Between Supports, Feet

American Standard Taper Pipe Threads
(NPT)

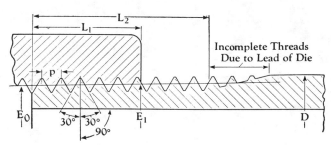

$E_0 = D - (0.050D + 1.1)p$
$E_1 = E_0 + 0.0625 L_1$
$L_2 = (0.80D + 6.8)p$

p = Pitch
Depth of thread = $0.80p$
Total Taper ¾-inch per Foot

Flush by Hand

Tolerance on Product
One turn large or small from notch on plug gauge or face of ring gauge.

Notch flush with face of fitting. If chamfered, notch flush with bottom of chamfer

Dimensions, in Inches

Nominal pipe size	D Outside diameter of pipe	E_0 Pitch diameter at end of external thread	E_1† Pitch diameter at end of internal thread	L_1¶ Normal engagement by hand between external and internal threads	L_2§ Length of effective thread	p Pitch of thread	Depth of thread	Number of threads Per Inch
⅛	0.405	0.36351	0.37476	0.180	0.2639	0.03704	0.02963	27
¼	0.540	0.47739	0.48989	0.200	0.4018	0.05556	0.04444	18
⅜	0.675	0.61201	0.62701	0.240	0.4078	0.05556	0.04444	18
½	0.840	0.75843	0.77843	0.320	0.5337	0.07143	0.05714	14
¾	1.050	0.96768	0.98887	0.339	0.5457	0.07143	0.05714	14
1	1.315	1.21363	1.23863	0.400	0.6828	0.08696	0.06957	11½
1¼	1.660	1.55713	1.58338	0.420	0.7068	0.08696	0.06957	11½
1½	1.900	1.79609	1.82234	0.420	0.7235	0.08696	0.06957	11½
2	2.375	2.26902	2.29627	0.436	0.7565	0.08696	0.06957	11½
2½	2.875	2.71953	2.76216	0.682	1.1375	0.12500	0.10000	8
3	3.500	3.34062	3.38850	0.766	1.2000	0.12500	0.10000	8
3½	4.000	3.83750	3.88881	0.821	1.2500	0.12500	0.10000	8
4	4.500	4.33438	4.38712	0.844	1.3000	0.12500	0.10000	8
5	5.563	5.39073	5.44929	0.937	1.4063	0.12500	0.10000	8
6	6.625	6.44609	6.50597	0.958	1.5125	0.12500	0.10000	8
8	8.625	8.43359	8.50003	1.063	1.7125	0.12500	0.10000	8
10	10.750	10.54531	10.62094	1.210	1.9250	0.12500	0.10000	8
12	12.750	12.53281	12.61781	1.360	2.1250	0.12500	0.10000	8
14 O.D.	14.000	13.77500	13.87262	1.562	2.2500	0.12500	0.10000	8
16 O.D.	16.000	15.76250	15.87575	1.812	2.4500	0.12500	0.10000	8
18 O.D.	18.000	17.75000	17.87500	2.000	2.6500	0.12500	0.10000	8
20 O.D.	20.000	19.73750	19.87031	2.125	2.8500	0.12500	0.10000	8
24 O.D.	24.000	23.71250	23.86094	2.375	3.2500	0.12500	0.10000	8

†Also pitch diameter at gauging notch.
§Also length of plug gauge.
¶Also length of ring gauge, and length from gauging notch to small end of plug gauge

British Standard Taper Pipe Threads

**Whitworth 55°
Form of Thread**

**Total Taper
¾-inch per Foot**

This information is taken from the British Engineering Standards Association Report No. 21-1938.

Dimensions, in Inches

Pipe Size Nominal Bore	D Approximate outside diameter of pipe	E_1 Pitch diameter at end of internal thread	L_1 Normal engagement by hand between external and internal threads	L_2 Length of effective thread	p Pitch of threads	Depth of threads	Number of threads Per Inch
⅛	13/32	0.3601	0.1563	0.2545	0.03571	0.0229	28
¼	17/32	0.4843	0.2367	0.3814	0.05263	0.0337	19
⅜	11/16	0.6223	0.2500	0.3947	0.05263	0.0337	19
½	27/32	0.7793	0.3214	0.5178	0.07143	0.0457	14
¾	1 1/16	0.9953	0.3750	0.5714	0.07143	0.0457	14
1	1 11/32	1.2508	0.4091	0.6591	0.09091	0.0582	11
1¼	1 11/16	1.5918	0.5000	0.7500	0.09091	0.0582	11
1½	1 29/32	1.8238	0.5000	0.7500	0.09091	0.0582	11
2	2⅜	2.2888	0.6250	0.9204	0.09091	0.0582	11
2½	3	2.9018	0.6875	1.0511	0.09091	0.0582	11
3	3½	3.4018	0.8125	1.1761	0.09091	0.0582	11
3½	4	3.8918	0.8750	1.2386	0.09091	0.0582	11
4	4½	4.3918	1.0000	1.4091	0.09091	0.0582	11
5	5½	5.3918	1.1250	1.5795	0.09091	0.0582	11
6	6½	6.3918	1.1250	1.5795	0.09091	0.0582	11
7	7½	7.3860	1.3750	1.9250	0.10000	0.0640	10
8	8½	8.3860	1.5000	2.0500	0.10000	0.0640	10
9	9½	9.3860	1.5000	2.0500	0.10000	0.0640	10
10	10½	10.3860	1.6250	2.1750	0.10000	0.0640	10
11	11½	11.3700	1.6250	2.3125	0.1250	0.0800	8
12	12½	12.3700	1.6250	2.3125	0.1250	0.0800	8

Normal engagement between male and female threads to make tight joints

Normal Engagement

The normal amount of engagement to make a tight joint for various types of screwed material is given in the table. These dimensions have been established from tests made under practical working conditions.

The normal engagement specified for American Standard Pipe Thread and API Line Pipe Thread joints is based on parts being threaded to the American Standard for Pipe Threads or the API Standard for Line Pipe Threads.

In order to obtain the thread engagements listed in the table it is necessary to vary the torque or power applied according to the size, metal and weight of material used. For example, it requires considerably less power to make up a screwed joint using a light bronze valve than a high pressure steel valve.

Reprinted with permission—Crane Company

American Standard and
API Line Pipe Threads

Dimensions, in Inches			
Dimensions given do not allow for variations in tapping or threading			
Size	**A**	**Size**	**A**
1/8	1/4	2 1/2	15/16
1/4	3/8	3	1
3/8	3/8	3 1/2	1 1/16
1/2	1/2	4	1 1/8
3/4	9/16	5	1 1/4
1	11/16	6	1 5/16
1 1/4	11/16	8	1 7/16
1 1/2	11/16	10	1 5/8
2	3/4	12	1 3/4

Hand-held computer calculates pipe weight, contents, velocity

Programs, written for the HP 41 CV computer equipped with printer, work with English or metric units

Frank E. Hangs, P.E., Sovereign Engineering, Inc., Houston

Pipe line engineers, accountants, and system operators are deeply involved with data concerning line pipe such as:

How much does it weigh (lb/ft, tons/mi)? What are the corresponding metric units?

How much will it contain (bbl/mi, gal/100 ft, cu m/km)?

What are velocities at various feed rates (ft/sec, mi/hr, m/hr)?

The following program, written for the Hewlett Packard HP 41 CV hand-held computer, develops a lot of useful data for steel pipe for any outside diameter and wall thickness. Either English or metric units may be entered.

A printer is essential.

Results for the WEIGHT category are calculated by an API formula and given as: Pounds (mass) per ft; kg per m. Short tons (2,000 lb) per mile; metric tons per km.

The content results are: bbl/mi, gal (U.S.)/100 ft, cu ft/mi, L/m, cu m/km.

Flowrates can be input to give: b/h or b/d; gal (U.S.)/min; or gal/hr or gal/d; cu ft/min, /hr, /d; cu m/min, /hr, /d.

Velocity results: mph; fps; km/hr; m/sec.

Figure 1. Examples of the program printed by the HP 41 CV printer.

All results, as well as the input, are printed out, so the tape becomes a permanent record.

A good feature of the program is that, for a given size pipe (after executing PWCV and A, content routine, to store data concerning that particular size pipe), velocity routine, B, can be run as often as wanted. Press B, then key in desired flowrate after the prompt, then press the RUN key.

When a different size pipe is needed: XEQ PWCV, press A, then B, as often as needed for different flowrates.

Basic data calculated by this program can be used for determining the following criteria:

- Weight: Cost of construction, hauling and stringing, investments.
- Content: Amount of water for hydrostatic test, line fill, oil stocks. Amount of one grade to be displaced for another to reach a station, shut-off valve, or terminal. When a segment must be drained, barrels or gallons to be removed.
- Velocity: Time for interface or scraper to traverse distance between two points. Time to accumulate or withdraw a given quantity of fluid. Locating head of column of a grade of oil or a scraper from given station after a certain time. Determining expected time of arrival (ETA) of interface or scraper at a given point.

List of Registers and Legend

Register	Content
00	Not used
01	Pipe OD in.
02	Pipe OD mm
03	Pipe T in.
04	Pipe T mm
05	Pipe ID in.
06	Pipe ID mm
07	LBM/FT—lb(mass)/ft
08	ST/MI—short tons/mi
09	MT/KM—metric tons/km
10	BBL/MI—bbl(42 U.S. gal) per mi
11	GAL/100 FT—gal (U.S.) per 100 ft
12	L/M—liters per m
13	M³/KM—cu m per km
14	BBL/HR—bbl (42 U.S. gal) per hr
15	GPM—gal (U.S.) per minute
16	FT³/MIN—cu ft per min
17	M³/HR—cu m per hr
18	MI/HR—mi (statute) per hr
19	KM/HR—km per hr
20	FT/S—ft per sec
21	M/S—m per sec
22	FT³/MI—cu ft per mi
23	KG/M—kg per m

LABEL	PWCV	Calls up program
LABEL	A	Calls content calculations
LABEL	B	Calls for feed rate and velocity calculations

Note. A and B are XEQ in user mode.

Numeric labels 01 to 06 are used in the program to provide necessary GTO directions. They are not stand-alone routines.

User directions. Size 25. Set user mode. Input program. If it is desired to use units as prompted, key in value and run key. If metric units are wanted, key in 0 for English units then run key.

Example. Display: Pipe OD in.? Key in 0, press run key, see pipe OD MM? Key in value and run key. See pipe T MM?

After weights are calculated: If content is wanted, XEQ "A." Calculations can be terminated at this point.

When velocities are desired, XEQ "B" input feed rate in desired units. Label "B" can be run for several feed rates as often as desired for the given size pipe. If other pipe sizes are needed, XEQ PWCV and "A," then run "B" as many times as needed for different feed rates.

Note. If feed rates other than BBL/HR are to be input, first key in 0 and run. This causes display to move up. See GPM? No. Gallons? Key in 0 and run key. See FT³/MIN? Continue to M³/HR? If you have this figure, key it in and run. Now suppose you have M³/DAY. This *does not* agree with prompt; enter M³/DAY, divide by 24, then press run key. This routine can be used for BBL/DAY, GAL/HR (divided by 60), FT³/DAY (divide by 24), etc.

Note. When results such as OD, ID, T Values, Weight KG/M are compared with printed values in API STD 5L, small discrepancies are sometimes found, usually on the order of 0.1% or less. This is evidently due to truncation differences in making conversions.

Formulas

Pipe weight LBM/FT (plain end) is calculated by formula used by API:

$$LBM/FT = 10.68(D - T)T$$

Weights in other units are found by multiplying by conversion factors:

$$BBL/MI = 5.129131(D - 2T)^2$$

Basic velocity relation:

$$\frac{BBL}{HR} \div \frac{BBL}{MI} = \frac{MI}{HR}$$

This is converted to other units.

Example 1. Size 25, Fix 3, Set user mode. Put PWCV in calculator. XEQ PWCV. Input 12.75 in. Press run key. See Pipe T in.? Input 0.375. Press run key. After calculator stops, press "A" for content. If velocities are desired, press "B." Input 4,000 BBL/HR. Press run key. Results are printed.

Now "B" can be run as often as needed.

Change feed rate units. Note zeros were put in for three feed rates, and 635.948 M³/HR input was keyed in. This is to demonstrate how other feed rates are input and the accuracy of the conversions (6 decimals). Put in feed rate 96,000 BPD. Enter, divide by 24, press run key.

Example 2. This shows how to input OD and T in MM. Key in 0 for pipe OD IN.? Then press run key. See Pipe OD MM? Key in metric units. Results are calculated and printed.

Questions and Answers

1. How many short tons are in a 55.35-mi segment of 12.75 in. × 0.375 in. WT? (Use data from Ex. 1.)

    $55.35 \times 130.843 = 7,242.160$ ST

2. How many km are in this segment above?

$55.35 \times 1.609 = 89.058$ km

3. How many bbl are in (a) 150 ft? (b) gal? (c) cu ft?

    a. $150/5,280 \times 738.595 = 20.983$ bbl
    b. $150/100 \times 587.519 = 881.279$ gal
    c. $150/5,280 \times 4,146.902 = 117.810$ cu ft

4. Suppose, in an emergency, you have to dig a pit to hold oil from 150 m of this 12-in. line. Volume? (a) cu m (b) cu ft

    a. $150/1,000 \times 72.966 = 10.945$ cu m
    b. $10.945 \times 34.315 = 386.519$ cu ft

5. How long will it take a scraper to traverse 67.85 mi of this 12-in. line at 4,000 bbl/hr?

    $67.85/5.351 = 12.680$ hrs Hours min. sec? Fix 4.

    XEQ Alpha HMS Alpha = 12 HR 40 MIN 48 SEC

Useful Conversions

Meters × 3.281 = feet
Miles × 1.609 = kilometers
Cubic ft × 7.481 = gallon (U.S.)
Kilogram × 2.205 = pounds
Cubic meters × 35.315 = cubic feet
Cubic meters × 6.290 = barrels
Cubic meters × 264.172 = gallons (U.S.)
Barrels × 5.615 = cubic feet
Barrels × 42.00 = gallons (U.S.)
Barrels × 35.00 = Imperial gallons

Source

Pipe Line Industry, May 1986.

Formulas and constants of value in solving problems relating to tubular goods

Area of Metal in Tubular Section:

Area in square inches = 3.1416 (D − t) t

where D = Outside diameter, in.
    t = Wall thickness, in.

Capacity

$B_{100} = .1237\ A = .0972\ D^2$

where B_{100} = Barrels per 100 ft
    A = Internal area of pipe, in.^{2}
    D = Internal diameter, in.

and $F_B = \dfrac{808.5}{A} = \dfrac{1,029.4}{D^2}$

where F_B = Number of feet filled by one barrel
    A = Internal area of pipe, in.^{2}
    D = Internal diameter, in.

Complete tables giving capacities of different types of tubular goods are given in chapters dealing with these types.

Steel constants* (applicable to tubular goods)
One cubic inch = 0.2833 lb
One cubic foot = 489.542 lb
Specific gravity = 7.851

Approximate coefficient of expansion for commercial casing, drill pipe, and tubing is

Coefficient = 6.9×10^{-6} per °F, over range from 0 to 400°F

Formula for expansion of pipe due to change of temperature is as follows:

---

* Engineering data Spang-Chalfant Division of the National Supply Co.

$$L_t = L_o(1 + .0000069\ t)$$

where L_o = Length at atmospheric temperature
L_t = Length after change of temperature t

Modulus: Section Modulus $\dfrac{\pi}{64}(D^4 - D_1^4)$

Polar Modulus $\dfrac{\pi}{32}(D^4 - D_1^4)$

where D = Outside diameter
D_1 = Inside diameter
π = 3.1416

Relation between twisting effort, torsional strain and stress of cylindrical shaft or tube

$$\frac{q}{r} = \frac{T}{J} = \frac{C\theta}{L}$$

How to calculate the contraction or expansion of a pipe line

Steel pipe contracts or expands approximately 0.8 inch per 100°F temperature change for each 100 feet of pipe. For construction in the U.S., slack is not normally needed in welded lines for contraction or expansion unless abnormal variations are encountered. However, slack is often provided near road crossings where the pipe may have to be lowered at some future time.

Contraction = 0.8 ×

$$\frac{\text{length of pipe (feet)}}{100} \times \frac{\text{temp. change (°F)}}{100}$$

Example. Calculate the amount a 1,000-foot section of pipe line would contract if laid at a temperature of 100°F and cooled to a low of 0°F during winter operation. (Assuming the line were free of soil or other resisting loads.)

$$\text{Contraction} = 0.8 \times \frac{1,000}{100} \times \frac{100}{100}$$

$$= 0.8 \times 10 \times 1$$
$$= 8 \text{ inches}$$

Large temperature changes may arise on exposed line or in lines downstream from compressor stations.

When slack is desired in a line, the amount of sag or rise needed may be estimated quickly. To provide for 0.8 inch of movement longitudinally along the pipe, a sag of 21 inches would be needed in a 100-foot loop. For a 150-foot loop, 31 inches are necessary. Such slack put into the line at time of construction might allow for lowering the pipe under roadbeds without putting excessive stress upon the pipe.

Location of overbends, sags, and side bends should be considered when laying slack loops. When loops are placed, parts of the line should be backfilled to serve as a tie-down. Loops should be lowered during the coolest part of the day when the line is the shortest. After lowering-in, sags should rest on the bottom of the ditch; overbends should "ride high." Side bends should rest on the bottom of the ditch and against the outside wall. The line should be lowered so that all sections of the pipe are in compression.

Estimate weight of pipe in metric tons per kilometer

To estimate the weight of pipe in metric tons per kilometer, multiply the nominal diameter by the number of sixteenths of an inch in wall thickness.

Example. Find the weight of pipe in metric tons per kilometer for 20-inch diameter pipe, wall thickness 1/4-inch.

4 × 20 = 80 tons (metric) per kilometer

Actual answer from pipe tables is 79 metric tons per kilometer.

This rule of thumb is based on a density of 490 pounds per cubic foot for steel. For larger diameter thin wall pipe, this approximation gives an answer usually about one percent low. The accompanying table gives a comparison between actual weights in metric tons per kilometer, as compared to that calculated by this rule of thumb.

PIPE DIAMETER	WALL THICKNESS	ACTUAL TONS (METRIC)	R. O. T. TONS (METRIC)
10-inch	3/16-inch	31.5	30
12-inch	5/16-inch	62	60
16-inch	7/16-inch	109	112
20-inch	1/4 -inch	79	80
20-inch	5/16-inch	99	100
20-inch	7/16-inch	137	140
24-inch	5/16-inch	118	120
24-inch	8/16-inch	189	192
28-inch	5/16-inch	139	140
30-inch	5/16-inch	148	150
32-inch	7/16-inch	221	224
36-inch	4/16-inch	143	144
36-inch	8/16-inch	285	288

How to find pipe weight from outside diameter and wall thickness

Weight (pounds per foot) = $(Dt - t^2 \times 10.68$

where D = outside diameter, in.
    t = wall thickness, in.

Example.   Outside   diameter = 4.500″,   thickness = 0.250″

$W = (4.5 \times .25 - .25^2) \times 10.68$
$= (1.125 - .0625) \times 10.68$
$= 1.0625 \times 10.68$
$= 11.35$ pounds per foot

The above equation is based on a density of 490 pounds per cubic foot for the steel. High-yield-point, thin-wall pipe may run slightly heavier than indicated.

What is the maximum allowable length of unsupported line pipe?

For Schedule 40 pipe, the maximum span between supports is given by

$S = 6.6 \sqrt{P}$

where S is the span length in feet and P is the actual outside diameter of the pipe in inches. For pipe smaller than 12-inch, the nominal size may be used, and

$S = 7 \sqrt{N}$

Example. How wide a ditch can be spanned with four-inch line pipe without the use of intermediate supports?

$S = 6.6 \sqrt{4.5} = 6.6 \times 2.12 = 13.99$ (14 feet)
$S = 7 \sqrt{4} = 7 \times 2 = 14$ feet

Note. The above rule is approximate, but conservative. It does not assume that the pipe is held fixed at either end; this condition can make longer spans safe. Also, it does not guarantee that the pipe will not be damaged by floating logs or debris! This hazard must be evaluated from the conditions encountered.

Identify the schedule number of pipe by direct measurement

Add three inches to the actual inside diameter; divide this by the wall thickness; the schedule number can then be identified by the following table:

Schedule	30	40 to 50
Schedule	40	29 to 39
Schedule	60	25 to 29
Schedule	80	20 to 23
Schedule	100	16 to 18

Schedule	120	13 to 15
Schedule	140	11 to 13
Schedule	160	9 to 11

Example. Pipe measures 12.5 inches inside diameter, wall thickness .75 inch; 15.5/.75 = 20.2; pipe is Schedule 80, 14 inches outside diameter.

The rule fails on small diameter pipe (under 6 inches), and also on schedules 10 and 20, because of the use of "standard" thicknesses in this range.

Determine buoyancy of bare steel pipe

Multiply the square of the outside diameter of the pipe, in inches, by 0.341, and subtract the pipe weight in lb/ft; the answer is the buoyancy per foot in fresh water.

Example. 12″ (12.75″ OD) 33.4 lb pipe (¼″ wall).

Buoyancy = $12.75^2 \times 0.341 - 33.4$

$= 162.56 \times 0.341 - 33.4$
$= 55.4 - 33.4 = 22$ lb/ft

Since all of the coating materials in common use are heavier than water (but not as much so as steel), coated pipe will have less buoyancy than indicated by the rule. For sea water, use 0.350 for the constant.

Determine buoyancy of bare and concrete coated steel pipe in water and mud

To find the buoyancy of bare steel and concrete coated pipe in water in lb/ft:

Buoyancy (B) $= \dfrac{D}{3}(D - 32t) + 11t^2$

for bare pipe, and

$B = \dfrac{D}{3}(D - 32t) + t_1 D \dfrac{63 - W_c}{48}$

for coated pipe

where   D = outside diameter of pipe, in.
         t = wall thickness of pipe, in.
         t_1 = thickness of concrete coating, in.
         W_c = weight of concrete, lb/ft³

To find the buoyancy of bare steel and coated pipe in mud in lb/ft:

Buoyancy (B) $= 10.7 \left(\dfrac{DW_m}{2,000} - t \right) + 11\, t^2$

for bare pipe, and

$B = 10.7 \left(\dfrac{DW_m}{2,000} - t \right) + t_1 D \left(\dfrac{W_m - W_c}{48} \right)$

for coated pipe, where W_m is weight of mud, lb/ft³.

Example 1. Find the buoyancy of steel pipe with 20-inch OD and ½-inch wall thickness in water. What thickness of concrete coating would be required to give this pipe a negative buoyancy of 100 lb/ft (W_c = 143)?

(B) $= \dfrac{20}{3}[20 - 32(½)] + 11(½)^2$

$= 26.7 + 2.8 = 29.5$ lb/ft positive buoyancy.

To give this pipe a negative buoyancy of 100 lb/ft:

$-100 = \dfrac{20}{3}[20 - 32(½)] + t_1 (20) \left(\dfrac{63 - 143}{48} \right)$

$-100 = 26.7 - 33.3\, t_1$
        $t_1 = 3.8$ in. of concrete coating.

The error introduced by this method is about 15% but would be less as the thickness of coating decreased.

Example 2. Bottom conditions at a certain crossing that require the pipe to have a negative buoyancy of at least 50 lb/ft. Let's check to see if a 10-inch pipe having a 2-inch coating of concrete (W_c = 143) has sufficient negative buoyancy.

$B = \dfrac{10}{3}(10 - 32(½)) + 2(10) \left(\dfrac{63 - 143}{48} \right)$

$B = -20 - 33.3 = -53.3$ lb/ft; so coated pipe meets design requirements.

Weights of piping materials

The weight per foot of steel pipe is subject to the following tolerances:

SPECIFICATION	TOLERANCE
A.S.T.M. A-53 A.S.T.M. A-120	STD WT + 5%, − 5% XS WT + 5%, − 5% XXS WT +10%, −10%
A.S.T.M. A-106	SCH 10–120 +6.5%, −3.5% SCH 140–160 +10%, −3.5%
A.S.T.M. A-158 A.S.T.M. A-206 A.S.T.M. A-280	12″ and under +6.5%, −3.5% over 12″ +10%, −5%
API 5L	All sizes +6.5%, −3.5%

Weight of Tube = $F \times 10.6802 \times T \times (D - T)$ lb/ft

where T = wall thickness, in.
  D = outside diameter, in.
  F = relative weight factor

The weight of tube furnished in these piping data is based on low carbon steel weighing 0.2833 lb/in.^{3}

Relative weight factor F of various metals

Aluminum	= 0.35
Brass	= 1.12
Cast Iron	= 0.91
Copper	= 1.14
Lead	= 1.44
Ferritic Stainless Steel	= 0.95
Austenitic Stainless Steel	= 1.02
Steel	= 1.00
Tin	= 0.93
Wrought Iron	= 0.98

Weight of contents of a tube = $G \times 0.3405 \times (D - 2T)^2$ lb/ft

where G = Specific gravity of contents
  T = Tube wall thickness, in.
  D = Tube outside diameter, in.

The weight of welding tees and laterals is for full size fittings. The weight of reducing fittings is approximately the same as for full size fittings.

The weight of welding reducers is for one size reduction, and is approximately correct for other reductions.

Pipe covering temperature ranges are intended as a guide only and do not constitute a recommendation for specific thickness of material.

Pipe covering thicknesses and weights indicate average conditions and include all allowance for wire, cement, canvas, bands, and paint. The listed thicknesses of combination covering is the sum of the inner and the outer layer thickness. When specific inner and outer layer thicknesses are known, add them, and use the weight for the nearest tabulated thickness.

To find the weight of covering on fittings, valves, or flanges, multiply the weight factor (light faced subscript) by the weight per foot of covering used on straight pipe. All flange weights include the proportional weight of bolts or studs required to make up all joints.

Lap joint flange weights include the weight of the lap.

Welding neck flange weights are compensated to allow for the weight of pipe displaced by the flange. Pipe should be measured from the face of the flange.

All flanged fitting weights include the proportional weight of bolts or studs required to make up all joints.

To find the approximate weight of reducing-flanged fittings, subtract the weight of a full size slip-on flange and add the weight of reduced size slip-on flange.

Weights of valves of the same type may vary because of individual Manufacturer's design. Listed valve weights are approximate only. When it is possible to obtain specific weights from the Manufacturer, such weights should be used.

To obtain the approximate weight of flanged end steel valves, add the weight of two slip-on flanges of the same size and series to the weight of the corresponding welding-end valves.

Allowable working pressure for carbon steel pipe

Based on formula No. 7 of the ASA Code for Pressure Piping, Section 3B, this table was computed for $6\frac{5}{8}$-inch OD to 36-inch OD carbon steel seamless pipe, API Specifications 5L Grade A, 5L Grade B, and API 5LX Grade X-42.

The formula is:

$$t = \frac{PD}{2(S + YP)} + u + c$$

Wall Thickness	6⅝	8⅝	10¾	12¾	14	16	18	20	22	24	26	28	30	32	34	36
0.188	895	685	548	.....	.....	.....	.....	.....	.....	.....	.....	.....	.....	.....	.....	.....
	1044	799	640	.....	.....	.....	.....	.....	.....	.....	.....	.....	.....	.....	.....	.....
	1253	959	768	.....	.....	.....	.....	.....	.....	.....	.....	.....	.....	.....	.....	.....
0.203	.....	1070	857	721	.....	.....	.....	.....	.....	.....	.....	.....	.....	.....	.....	.....
0.219	1110	849	680	572	.....	.....	.....	.....	.....	.....	.....	.....	.....	.....	.....	.....
	1295	991	793	667	.....	.....	.....	.....	.....	.....	.....	.....	.....	.....	.....	.....
	1554	1189	952	801	729	637	.....	.....	.....	.....	.....	.....	.....	.....	.....	.....
0.250	1327	1014	811	683	621	543	482	434	.....	.....	.....	.....	.....	.....	.....	.....
	1548	1184	947	797	725	633	562	506	.....	.....	.....	.....	.....	.....	.....	.....
	1858	1420	1136	956	870	760	675	607	551	505	466	433	404	378	356	336
0.281	.....	.....	.....	794	722	631	560	504	.....	.....	.....	.....	.....	.....	.....	.....
	.....	.....	.....	926	842	736	654	588	.....	.....	.....	.....	.....	.....	.....	.....
	.....	.....	.....	1111	1010	883	784	705	641	587	542	503	469	439	414	390
0.312	1765	1347	.....	905	823	719	638	574	521	478	441	.....	.....	.....	.....	.....
	2059	1572	.....	1056	961	839	745	670	608	557	514	.....	.....	.....	.....	.....
	2471	1886	.....	1267	1153	1007	894	804	730	669	617	573	534	501	471	445
0.344	1994	1520	1214	1021	928	810	720	647	587	538	.....	.....	.....	.....	.....	.....
	2326	1774	1416	1190	1083	946	839	755	685	628	.....	.....	.....	.....	.....	.....
	2791	2128	1700	1429	1299	1135	1007	906	823	753	695	645	602	564	530	501
0.375	2216	1689	.....	1133	1030	900	798	718	652	597	551	.....	.....	.....	.....	.....
	2568	1970	.....	1322	1202	1049	931	837	760	696	642	.....	.....	.....	.....	.....
	3103	2364	.....	1586	1442	1259	1118	1005	912	836	771	715	667	625	588	555
0.406	.....	.....	.....	1744	.....	.....	1228	1104	1002	918	847	786	733	687	646	610
0.438	.....	2034	1622	1362	1238	1081	959	862	782	716	661	.....	.....	.....	.....	.....
	.....	2373	1892	1589	1444	1261	1119	1005	913	836	771	.....	.....	.....	.....	.....
	.....	2848	2271	1907	1733	1513	1342	1206	1095	1003	925	858	801	750	706	666
0.500	.....	.....	.....	.....	1444	1260	1118	1004	911	834	770	.....	.....	.....	.....	.....
	.....	.....	.....	.....	1685	1470	1304	1171	1063	974	898	.....	.....	.....	.....	.....
	.....	3328	2651	2225	2022	1764	1565	1406	1276	1168	1077	1000	932	873	822	775

where   t = wall thickness, in.
P = allowable working pressure, psi
D = outside diameter, in.
S = allowable stress, psi
Y = a coefficient
u = under-thickness tolerance, in.
c = allowance for threading, mechanical strength, and corrosion, in.

In calculation of the tables, the values of Y, u, and c were:

Y = 0.4
u = 0.125t   (u = 12½% of t)
c = 0.050-inches

Substituting in formula and solving for P:

$$P = \frac{1.75\ S(t - 0.057)}{D + 0.04 - 0.7t}$$

The S-values are:

Spec. API 5L Grade A—25,500 psi
Spec. API 5L Grade B—29,750 psi
Spec. API 5LX Grade X-42—35,700 psi

Example. Find pipe specification for a 12-inch line operating at 1,078 psi: In the 12¾-inch column these conditions will be met using Grade X-42 pipe with 0.281-inch wall thickness, Grade B pipe with 0.330-inch wall thickness, or Grade A pipe with 0.375-inch wall thickness.

Find the stress in pipe wall due to internal pressure

Multiply ½ by diameter by pressure and divide by wall thickness to get stress in psi.

½ × 24 × 800 ÷ ¼ = 38,400 psi

Example. Find the wall stress due to internal pressure in a 24-inch pipe line having a ¼-inch wall thickness if the pressure gage reads 800 psi:

What wall thickness should be specified for a 30-inch discharge line where pressure runs 1,000 psi and wall stress is to be limited to 30,000 psi?

½ × 30 × 1,000 ÷ thickness = 30,000
Wall thickness = ½ inch.

This method of calculating transverse tensile stress due to internal pressure on the pipe is the Barlow formula. Actually, there are two accepted variations, and some design- ers use inside diameter of the pipe in calculating this stress. This gives a slightly lower figure for the transverse tensile stress and consequently gives less safety factor. The method used here is based on the assumption that maximum stress occurs at outside diameter of the pipe. The results give a conservative value for safety factor calculations.

How to calculate stress in above/below ground transition

Design of gas gathering/gas reinjection project in southern Iran provides solutions for deflection and anchor block forces in long underground lines

P. J. Schnackenberg, Design Engineer, Iran-Texas Engineering Co., Tehran, Iran

Stresses and deflections occur in pipe lines at the transition from the below ground (fully restrained) to the above ground (unrestrained) condition.

Analysis of the stresses and deflections in transition areas, resulting from internal pressure/temperature change, is necessary in determining anchor block requirements and design. Longitudinal deflections are used to determine whether an anchor block is required. Anchor block forces required to maintain the pipe in a fully constrained condition are then determined.

Iran-Texas Engineering Co. recently completed an analysis for a gas gathering/compression/reinjection project in southern Iran under contract to Oil Service Co. of Iran (OSCO). Each of the two fields, Bibi-Hakimeh and Rag-E-Safid have two gas/oil separation plants, and at present the gas is being flared. The broad scope of the project is as follows:

- Provision of gas gathering lines from existing and new wellhead separation to a production unit
- Design of four production units (each adjacent to the GOSP) comprising knockout facilities, compressor stations, and dehydration plant
- Provision of transfer pipe lines to send gas plus condensate as a two-phase flow to a natural gas liquid separation plant (under design by Foster Wheeler)
- Provision of reinjection system to inject lean gas back into a number of wells
- Some 9.4 MM m³/d (330 MM scfd) of gas plus 15,000 bpd condensate will be going to the NGL plant. LPG will be used as feedstock and for export.

Following is a brief review of the analysis that resulted in more accurate solutions for deflection and anchor block forces. Sample calculations are for line sizes up to 41-cm (16-in.) OD, pressure to 193 bars (2,800 psig) and temperatures to 72°C (162°F).

Case I—no anchor. Considering the pipe line as shown in Figure 1, the section up to point A is fully restrained, the

Figure 1. Area A-B depicts transition from fully restrained (left) to fully unrestrained (right).

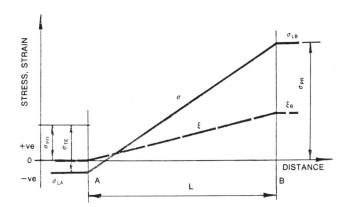

Figure 2. Transition of stress and strain between points A and B varies as linear function of length.

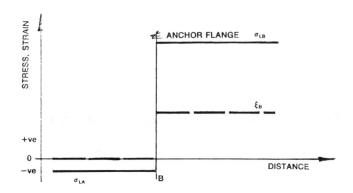

Figure 3. Distribution of stress where anchor is required to contain longitudinal deflections.

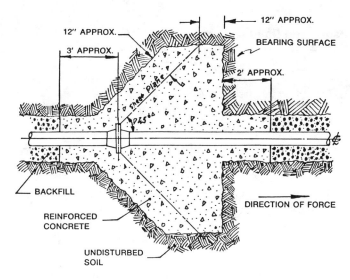

Figure 4. Sectional plan for anchor block design.

Figure 5. Sectional elevation for anchor block design.

section A-B is a transition from fully restrained to unrestrained and beyond point B is unrestrained. (Fully restrained is taken to be a condition of zero longitudinal strain.)

Analyzing a short section of fully restrained pipe, subject to internal pressure and a temperature change, there is a tensile stress due to a Poisson effect of

$$\sigma_{PO} = v\sigma_H \text{ where } \sigma_H = \frac{P\,D_i}{2t}$$

and a compressive stress due to a temperature change of

$$\sigma_{TE} = E\,\alpha\,\Delta T$$

The net longitudinal stress therefore at Point A will be

$$\sigma_{LA} = v\,\sigma_H - E\,\alpha\,\Delta T \text{ psi} \tag{1}$$

and the strain at point A, being fully restrained, will be zero

$$\varepsilon_A = 0 \tag{2}$$

In a section of unrestrained pipe at B, the longitudinal stress caused by only internal pressure will be half the hoop stress.

$$\sigma_{LB} = \frac{\sigma_H}{2} \text{ psi} \tag{3}$$

The associated strain will be made up of a

(i) Temperature effect

$$\varepsilon_{TE} = \alpha\,\Delta T$$

(ii) Pressure effect

$$\varepsilon_{PR} = \frac{\sigma_{LB}}{E} = \frac{\sigma_H}{2E}$$

(iii) Poisson effect

$$\varepsilon_{PO} = \frac{v\,\sigma_H}{E}$$

The net longitudinal strain at Point B will therefore be

$$\varepsilon_B = \alpha\Delta T + \frac{\sigma_H}{E}\left(\tfrac{1}{2} - v\right) \text{ in./in.} \tag{4}$$

The transition of stress and strain between points A and B is assumed to vary as a linear function of length as shown in Figure 2.

In order to establish the length L over which the transition occurs, the longitudinal resistance of the soil needs to be known. For this part, it is assumed that any tendency to move will be counteracted by constant and opposite soil force. Wilbur[1] has recommended a design value for average soils of

$$F_s = 80\left|\frac{D_o}{12}\right|^2 \text{ lbf./ft} \tag{5}$$

Between A and B equilibrium of forces exists and therefore,

$$F_sL = A_m(\sigma_{LB} - \sigma_{LA})$$

From which

$$L = \frac{A_m(\sigma_{LB} - \sigma_{LA})}{F_s} \text{ ft} \tag{6}$$

Total movement at B will be average strain between A and B over length L or

Table 1
Summary of results for high pressure gas transmission lines under design in Iran

No.	Pipe (All API-5LX-60)	P psig	T_0 °F	F_s lb./ft.	σ_H psi	σ_{LA} psi	σ_{LB} psi	L ft.	δ in.	F lb.
1	16″ OD × 0.312″ w.t.	1150	162	142.2	28,334	−6957	14,167	2285	9.81	324,890
2	20″ OD × 0.344″ w.t.	1000	162	222.2	28,070	−7036	14,035	2014	8.62	447,550
3	16″ OD × 0.281″ w.t.	1000	162	142.2	27,470	−7216	13,735	2045	8.71	290,800
4	16″ OD × 0.250″ w.t.	880	162	142.2	27,280	−7273	13,640	1819	7.73	258,690
5	16″ OD × 0.688″ w.t.	2800	150	142.2	29,758	−4268	14,879	4457	17.66	633,770
6	10″ NB × 0.500″ w.t.	2800	150	64.2	28,700	−4585	14,350	4728	18.60	304,850
7	6″ NB × 0.432″ w.t.	2800	150	24.4	18,670	−7594	9,335	5828	20.41	142,200

$$\delta = \frac{\varepsilon_B}{2} (12\ L)\ \text{in.} \tag{7}$$

Case II—with anchor. Where an anchor is required to contain longitudinal deflections, the stress distribution will be as shown in Figure 3. The transition from being fully restrained to unrestrained will occur at the anchor. Resultant force on the anchor will simply be the difference in stress on each side times the pipe metal area (equilibrium of forces) or

$$F = (\sigma_{LB} - \sigma_{LA})A_m\ \text{lb} \tag{8}$$

Note that for the case of an increase in wall thickness at and beyond the anchor block, the result will be essentially the same—the decrease in stress being compensated by an increase in pipe metal area.

It can be shown that this force is equal to that to produce a deflection of 2δ in Equation 7 over length L.

Table 1 is a summary of results for a number of high pressure gas transmission lines currently under design in Iran. Lines 1 through 4 are transfer lines from a number of production units to an NGL separation plant. Lines 5 through 7 are injection lines from the NGL separation plant to a number of wellheads. The following values were used for the various parameters:

Young's modulus, E, 29 × 10⁶, psi

Poisson's ratio, v, 0.3

Thermal expansion coefficient, 6.5 × 10⁻⁶, in./in. °F

Installation temperature, T_i, 80°F (minimum summer temperature when construction is in progress)

Negative stress values denote a compressive stress.

Discussion

As shown in Table 1, anchor block forces can be quite large and therefore, careful attention must be paid to the design. In order to minimize the size of the block, upper design limit values of lateral soil bearing pressure with due consideration to possible slip have been used.

For large blocks, friction between the block and the soil may also be figured in obtaining its resistance to imposed forces. The block should be reinforced concrete cast against undisturbed soil. Figures 4 and 5 indicate in broad detail the anchor block design.

Care should also be taken to ensure that connected surface piping has sufficient flexibility to absorb a degree of lateral anchor movement. Scraper traps should be installed such that they can move with the piping rather than being rigidly attached to blocks. Numerous examples of traps together with their support blocks being displaced a few inches are not uncommon.

References

1. Wilbur, W. E., *Pipe Line Industry*, February 1963.
2. Timoshenko, S., Theory of Elasticity.

Nomenclature

A_m Pipe metal area = $\pi (D_o - t)t$, sq. in.
D_o Pipe outside diameter, in.
D_i Pipe inside diameter, in.
t Pipe wall thickness, in.
E Young's modulus of elasticity, psi
F Anchor force, lb
F_s Soil resistance, lb/ft
L Transition length, ft
P Design pressure, psig
T_i Installation temperature, °F
T_o Operating temperature, °F
ΔT Temperature difference = $(T_o - T_i)$, °F
α Coefficient of thermal expansion, in./in.-°F
σ Stress, psi
ε Strain, in./in.
v Poisson's ratio
δ Deflection, in.

Subscripts (for Greek letters only)
L Longitudinal
H Hoop
A Point A, fully restrained

B Point B, unrestrained
PR Pressure
TE Temperature
PO Poisson

How to identify the series number of flanged fittings

Divide the flange thickness, expressed in sixteenths of an inch, by the nominal pipe size + 12. If the answer is less than 1.00, the flange is Series 15; if between 1.15 and 1.33, it is Series 30; if between 1.37 and 1.54, it is Series 40; and if it is between 1.60 and 1.85, it is Series 60.

Example. A six-inch flange has a thickness of $1^7/16$ inch = $^{23}/16$ inch.

$^{23}/18 = 1.28$; the flange is series 30.

The rule is not applicable to fittings below four inches in size; it also fails for a four-inch Series 60, which comes out 1.50.

Dimensions of three-diameter ells with tangents

Here are face-to-face and center-to-face dimensions of the three-diameter radius weld ells being used in natural gas compressor station piping.

Degree Ell	Outside Diameter	A (Center To Face)	A With 4″ Tan	A With 6″ Tan
30°	16″	1′-0⅞″	1′- 4⅞″	1′- 6⅞″
30°	20″	1′-4¹/₁₆″	1′- 8¹/₁₆″	1′-10¹/₁₆″
30°	24″	1′-7⁵/₁₆″	1′-11⁵/₁₆″	2′- 1⁵/₁₆″
30°	26″	1′-8⅞″	2′- 0⅞″	2′- 2⅞″
30°	30″	2′-0⅛″	2′- 4⅛″	2′- 6⅛″
45°	16″	1′-7⅞″	1′-11⅞″	2′- 1⅞″
45°	20″	2′-0⅞″	2′- 4⅞″	2′- 6⅞″
45°	24″	2′-5¹³/₁₆″	2′- 9¹³/₁₆″	2′-11¹³/₁₆″
45°	26″	2′-8⁵/₁₆″	3′- 0⁵/₁₆″	3′- 2⁵/₁₆″
45°	30″	3′-1¼″	3′- 5¼″	3′- 7¼″
90°	16″	4′-0″	4′- 4″	4′- 6″
90°	20″	5′-0″	5′- 4″	5′- 6″
90°	24″	6′-0″	6′- 4″	6′- 6″
90°	26″	6′-6″	6′-10″	7′- 0″
90°	30″	7′-6″	7′-10″	8′- 0″

Polypipe design data

Internal Pressure

Polyethylene pipe for industrial, municipal, and mining applications is manufactured to specific dimensions as prescribed by applicable American Society for Testing and Materials (ASTM) standards. Piping outside diameters meet the IPS system. Wall thicknesses are based on the Standard Dimension Ratio (SDR) system, a specific ratio of the outside diameter to the minimum specified wall thickness. Selected SDR numbers are from the ANSI Preferred Number Series. Use of the SDR number in the ISO equation, recognized as an equation depicting the relationship of pipe dimensions'—both wall and o.d.—internal pressure carrying capabilities and tensile stress, in conjunction with a suitable safety factor (F), will give the design engineer confidence that the pipe will not fail prematurely due to internal pressurization.

A pressure differential is required to move a material through a pipeline. For atmospheric systems (gravity flow), gravitational forces provide the impetus for movement of heavier-than-air mass. Internal pressure requirements must be recognized in the design stage in order to provide desired operational life. In some cases a gravity flow system must be treated comparable to the design considerations of a pressure flow system.

Polyethylene pipe under pressure is time/temperature/pressure dependent. As the temperature increases and the pressure remains constant the creep rupture strength decreases. Conversely, as the temperature decreases the strength increases. Varying these parameters, in conjunction with the SDR, can provide the desired service within the capabilities of polyethylene pipe.

Calculations for determining the pressure rating of Poly pipe is based on the ISO equation as follows:

$$P = \frac{2S}{SDR-1} \, xF$$

where
P   = Internal pressure, psi
S   = Long term hydrostatic strength, psi (1600 psi for Poly pipe)
SDR = Standard dimension ratio, (D/t)
D   = Outside diameter, inches

t = Minimum wall thickness, inches
F = Design safety factor (0.5 for water @ 73°F)

Use of other factors determined from Figures 1, 2 and 3 will provide for a more accurate performance characteristic for systems operating at other temperatures for shorter periods of time.

The ISO equation now becomes:

$$P = \frac{2S}{SDR-1} \, xFxF_1xF_2xF_3$$

F_1 = Operational life factor
F_2 = Temperature correction.
F_3 = Environmental service factor (Figure 3)

Other design factors may be necessary in the design of systems where abrasion, corrosion or chemical attack may be a factor.

Example: A horizontal process water line, 2,000 ft. in length, 125°F water, 60 psig pressure, with a desired life of five years, is proposed. Determine the standard pipe SDR for the application.

$$60 = \frac{2x1600}{SDR-1} \, x0.5x1.09x0.620$$

$$SDR - 1 = \frac{1081}{60}$$

SDR = 18+1=19
Therefore use SDR = 17.
Note: Using SDR 17 is conservative.

Figure 1
Internal Pressure Ratings (psi) (PE Pipe at 73°F for 50 years)

SDR	7.3	9	11	13.5	15.5	17	21	26	32.5
Pressure Rating	250	200	160	125	110	100	80	64	50

Figure 2
Operational Life

To use Figure 1 for determining operational life multiplier, F_1, select expected operating life (less than 50 years) on the horizontal time line, move upwards to intersect the diagonal turn line and then left to take reading at vertical axis.

Example: 5 year operational life will yield $F_1 = 1.094$.

Figure 3
Environmental Service Factor F_3

Substance	Service Factor (F_3)
Crude Oil	0.50
Wet Natural Gas	0.50
Dry Natural Gas	0.64

Figure 4 - Temperature Correction

Example: A polyethylene pipe operating at 125°F would have an F_2 factor of 0.62

Figure 5
Recommended Liquid Velocities

	Pipe Size-Inches		
	3–10	10–28	28–48
Water and Similar Viscosity Fluids			
		Velocity, fps	
Pump suction*	1 to 4	3 to 6	5 to 8
Pump discharge	3 to 5	4 to 8	6 to 12
Gravity drain system	3 to 5	4 to 8	6 to 10
Viscous liquids			
Pump suction*	0.5 to 2	1 to 4	2 to 5
Pump discharge	3 to 5	4 to 6	5 to 7

* It is important in the selection of the pipe size that the designated pump suction velocity be lower than the discharge velocity.

Pipe Diameter

Volumetric flow, Q, can be determined from the continuity equation Q = A x V by selecting a suitable velocity from Figure 5 for a given pipe size. When modified for flow in gallons per minute, the equation becomes:

$$Q = 2.449 \times V \times d^2$$

where
Q = flow rate in gpm
V = velocity if feet per second
d = inside diameter, inches

or $d = (Q \div 2.449 \times V)^{1/2}$ for pipe size determination.

Example: If the required water flow rate for a system is 2,000 gpm and the flow velocity is to be maintained below 8 fps, the pipe diameter can be found as follows:

$$d = (Q \div 2.449 \times V)^{1/2}$$

$$d = (2000 \div 2.449 \times 8)^{1/2}$$

$$d = 10.1 \text{ inches inside diameter}$$

$$OD = ID \div F \text{ (F from Figure 6)}$$

Figure 6
ID to OD Ratio

SDR	F	SDR	F
7	0.706	13.5	0.852
7.3	0.719	15.5	0.873
9	0.774	17	0.884
9.3	0.781	21	0.909
11	0.817	26	0.928
		32.5	0.944

Pressure Drop

Refer to Section 13: Liquids—Hydraulics and use the Hazen-Williams equation. Use a C of 150.

Fitting Equivalent Length

The exact pressure loss through fittings cannot readily be calculated because of the geometry of the fitting. The values shown in Figure 7 are the result of general industry consensus.

Figure 7
Equivalent Length of Fittings

Fitting Type	Equivalent length of pipe
90 deg. elbow	30 D
60 deg. elbow	25 D
45 deg. elbow	16 D
45 deg. wye	60 D
Running tee	20 D
Branch tee	50 D
Gate valve, full open	8 D
Butterfly valve (3" to 14")	40 D
Butterfly valve ≥ 14"	30 D
Swing check valve	100 D
Ball valve full bore full open	3 D

Example: A running tee has an equivalent length of 20 D. For an 18" SDR 11 line, this represents 20 x ID. From Figure 6, ID = OD x F = 18 x 0.817 = 14.7" x 20 = 294 ÷ 12 and is equal to 24.5 feet. In calculating the total pressure drop in the system, 24.5 feet of 18 inch pipe should be added to the total line length to account for the increased pressure loss due to the presence of the tee.

Published with permission—CSR™ PolyPipe

4: Electrical Design

Electrical design.. 102
Hazardous locations... 103
NEMA Enclosure Types... 104
Size portable electric generators... 105
Typical wattages for tools and appliances 106
Knockout dimensions 106
National Electrical Code Tables 107
Electrical formulas.. 111
Full load currents—single phase transformers 111
Full load currents—three phase transformers................................... 112
Motor controller sizes... 112
Voltage drop on circuits using 600 V. copper conductors in steel
    conduit 113
Determine the most economical size for electric power conductors ... 113
How to find the resistance and weight of copper wires...................... 114
What you should remember about electrical formulas........................ 114
How to calculate microwave hops on level ground............................ 114
For quick determination of the horsepower per ampere for induction
    motors (3 phase) at different voltages 115
Chart gives electric motor horsepower for pumping units 115
Pumping stations... 116
Floodlighting Concepts 117
Terms 117
Floodlighting calculations 117
Point-by-point method... 117
Beam-lumen method.. 118
Design procedure 118

Electrical Design

Most pipe line facilities use electrical equipment of some sort or another, ranging from simple power circuits of a few amperes capacity to sophisticated supervisory control and data acquisition systems. It is beyond the scope of this manual to provide a complete section on electrical information for pipe line facilities. This section is intended to provide some basic data that will prove useful to field personnel responsible for electrical installations.

Electric Motor Selection

Based on known facts and calculations, the best selection is made after a close study of the installation, operation and servicing of the motor. Basic steps in proper selection are numbered 1 through 8 and are briefly described.

1. Power Supply

(a) Voltage - NEMA has recommended the following standards.

Nominal Power System Volts	Motor Nameplate Volts
240	230
480	460
600	575

(b) Frequency - Motors rated 200 horsepower or less can vary not to exceed 5 percent above or below its rated frequency.

(c) Phases - Three-phase power supplies are found in most industrial locations; for most residential and rural areas, only single-phase power is available.

2. HP and Duty Requirements

(a) Continuous duty - means constant load for an indefinite period (about 90 percent of all motor applications).

(b) Intermittent duty - means alternate periods of load and no-load, or load and rest.

(c) Varying duty - means both the load and time operation vary to a wide degree.

3. Speeds

Single speed motors are the most common, but when a range of speeds is required multispeed motors will give 2, 3 or 4 fixed speeds.

4. Service Factors

Open, general purpose motors have a service factor depending upon the particular rating of the motor (usually between 1.15 and 1.25).

5. Selection of Motor Type

There are three main types:

(a) DC motors are designed for industrial drives requiring a controlled speed range and constant torque output on adjustable voltage control systems. Because the speed of rotation controls the flow of current in the armature, special devices must be used for starting DC motors.

(b) Single phase alternating-current motors also require some auxiliary arrangement to start rotation.

(c) Polyphase motors are alternating-current types (squirrel cage or wound rotor).

Applications - Some driven machines require a low-starting torque which gradually increases to full-load speed; others require higher-starting torque. NEMA code letters A, B, C, D, etc., on the motor nameplate designate the locked-rotor kVa per horsepower of that particular motor design. The diagram shows representative speed-torque curves for polyphase and single-phase NEMA design motors Types A through D.

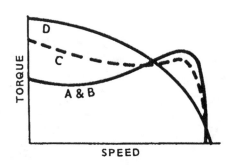

6. Torque Definitions

(a) Locked rotor torque, or "starting torque" is the minimum-torque which the motor will develop at rest for all angular positions of the rotor.

(b) Breakdown torque is the maximum-torque at rated voltage and frequency without abrupt speed drop.

(c) Full-load-torque is the torque necessary to produce rated horsepower at full-load speed.

(d) Locked-rotor current is the steady-state current of the motor with the rotor locked at rated voltage and frequency.

Code Ltr	Locked Rotor kVa/hP
A	0-3.14
B	3.15-3.54
C	3.55-3.99
D	4.0-4.49
E	4.5-4.99
F	5-5.59

7. Selection of Enclosure - The two general classifications are open, which permits passage of air over and around the windings; and totally-enclosed, which prevents exchange of air between inside and outside of frame (but is not strictly airtight).

(a) Open Drip-Proof - means that liquid or solid particles falling on the motor at an angle not greater than 15 degrees from vertical cannot enter the motor.

(b) NEMA Type 1 - A weather protected machine with its ventilating passages constructed to minimize the entrance of rain, snow and airborne particles to the electric parts and having its ventilating openings so constructed to prevent the passage of a cylindrical rod 3/4 in. in diameter.

(c) NEMA Type II - A weather protected machine which has in addition to the enclosure defined for a Type I weather-protected machine, its ventilating passages at both intake and discharge so arranged that high velocity air and airborne particles blown into the machine by storms or high winds can be discharged without entering the internal ventilating passages leading directly to the electric parts of the machine itself. The normal path of the ventilating air which enters the electric parts of the machine shall be so arranged by baffling or separate housings as to provide at least three abrupt changes in direction, none of which shall be less than 90 deg. In addition, an area of low velocity, not exceeding 600 ft/min., shall be provided in the intake air path to minimize the possibility of moisture or dirt being carried into the electric parts of the machine.

(b) Totally Enclosed Non-Vent means that a motor is not equipped for cooling by external means.

(c) Totally Enclosed Fan-Cooled means the motor has a fan integral with the motor but not external to the enclosed parts.

(d) Explosion-Proof means the enclosure is designed to withstand an explosion of a specified gas which may occur within the motor and to prevent the ignition of gas around the motor.

8. End Shield Mountings - Three types of end shields with rabbets and bolt holes for mounting are standard in the industry:

(a) Type C Face provides a male rabbet and tapped holes.

(b) Type D Flange has a male rabbet with holes for through bolts in the flange.

(c) Type P base has a female rabbet and through bolts for mounting in the flange (used for mounting vertical motors).

Hazardous Locations

Definition of Class Locations

Reprinted with permission from NFPA 70-1990, the *National Electrical Code®*, Copyright© 1989, National Fire Protection Association, Quincy, MA 02269. This reprinted material is not the complete and official position of the National Fire Protection Association, on the referenced subject which is represented only by the standard in its entirety.

The *National Electrical Code* and *NEC* are registered trademarks of the NFPA

500-5. Class I Locations. Class I locations are those in which flammable gases or vapors are or may be present in the air in quantities sufficient to produce explosive or ignitible mixtures. Class I locations shall include those specified in (a) and (b) below.

(a) Class I, Division 1. A Class I, Division 1 location is a location: (1) in which ignitible concentrations of flammable gases or vapors can exist under normal operating conditions; or (2) in which ignitible concentrations of such gases or vapors may exist frequently because of repair or maintenance operations or because of leakage; or (3) in which breakdown or faulty operation of equipment or processes might release ignitible concentrations of flammable gases or vapors, and might also cause simultaneous failure of electric equipment.

(FPN): This classification usually includes locations where volatile flammable liquids or liquefied flammable gases are transferred from one container to another; interiors of spray booths and areas in the vicinity of spraying and painting operations where volatile flammable solvents are used; locations containing open tanks or vats of volatile flammable liquids; drying rooms or compartments for the evaporation of flammable solvents; locations containing fat and oil extraction equipment using volatile flammable solvents; portions of cleaning and dyeing plants where flammable liquids are used; gas generator rooms and other portions of gas manufacturing plants where flammable gas may escape; inadequately ventilated pump rooms for flammable gas or for volatile flammable liquids; the interiors of refrigerators and freezers in which volatile flammable materials are stored in open, lightly stoppered, or easily ruptured containers; and all other locations where ignitible concentrations of flammable vapors or gases are likely to occur in the course of normal operations.

(b) Class I, Division 2. A Class I, Division 2 location is a location: (1) in which volatile flammable liquids or flammable gases are handled, processed, or used, but in which the liquids, vapors, or gases will normally be confined within closed containers or closed systems from which they can escape only in case of accidental rupture or breakdown of such containers or systems, or in the case of abnormal operation of equipment; or (2) in which ignitible concentrations of gases or vapors are normally prevented by positive mechanical ventilation, and which might become hazardous through failure or abnormal operation of the ventilating equipment; or (3) that is adjacent to a Class I, Division 1 location, and to which ignitible concentrations of gases or vapors might occasionally be communicated unless such communication is prevented by adequate positive-pressure ventilation from a source of clean air, and effective safeguards against ventilation failure are provided.

(FPN No. 1): This classification usually includes locations where volatile flammable liquids or flammable gases or vapors are used, but which, in the judgment of the authority having jurisdiction, would become hazardous only in case of an accident or of some unusual operating condition. The quantity of flammable material that might escape in case of accident, the adequacy of ventilating equipment, the total area involved, and the record of the industry or business with respect to explosions or fires are all factors that merit consideration in determining the classification and extent of each location.

(FPN No. 2): Piping without valves, checks, meters, and similar devices would not ordinarily introduce a hazardous condition even though used for flammable liquids or gases. Locations used for the storage of flammable liquid or of liquefied or compressed gases in sealed containers would not normally be considered hazardous unless subject to other hazardous conditions also.

Group A: Atmospheres containing acetylene.

Group B: Atmospheres containing hydrogen, fuel and combustible process gases containing more than 30 percent hydrogen by volume, or gases or vapors of equivalent hazard such as butadiene,* ethylene oxide,** propylene oxide,** and acrolein.**

Group C: Atmospheres such as cyclopropane, ethyl ether, ethylene, or gases or vapors of equivalent hazard.

Group D: Atmospheres such as acetone, ammonia,*** benzene, butane, ethanol, gasoline, hexane, methanol, methane, natural gas, naptha, propane or gases of vapors of equivalent hazard.

---

*Group D equipment may be used for this atmosphere if such equipment is isolated in accordance with Section 501-5(a) by sealing all conduit 1/2-inch or larger.
**Group C equipment may be used for this atmosphere if such equipment is isolated in accordance with Section 501-5(a) by sealing all conduit 1/2-inch or larger.
***For classification of areas involving ammonia atmosphere, see Safety Code for Mechanical Refrigeration, ANSI/ASHRAE 15-1978, and Safety Requirements for the Storage and Handling of Anhydrous Ammonia, ANSI/CGA G2.1-1981.

NEMA Enclosure Types[1]

Nonhazardous Locations

Type	Intended Use and Description
1	Enclosures are intended for indoor use primarily to provide a degree of protection against limited amounts of falling dirt.
2	Enclosures are intended for indoor use primarily to provide a degree of protection against limited amounts of falling water and dirt.
3	Enclosures are intended for outdoor use primarily to provide a degree of protection against rain, sleet, windblown dust, and damage from external ice formation.
3R	Enclosures are intended for outdoor use primarily to provide a degree of protection against rain, sleet, and damage from external ice formation.
3S	Enclosures are intended for outdoor use primarily to provide a degree of protection against rain, sleet, windblown dust, and to provide for operation for external mechanisms when ice laden.

4   Enclosures are intended for indoor or outdoor use primarily to provide a degree of protection against windblown dust and rain, splashing water, hose-directed water and damage from external ice formation.

4X   Enclosures are intended for indoor or out door use primarily to provide a degree of protection against corrosion, windblown dust and rain, splashing water, hose-directed water and damage from external ice formation.

5   Enclosures are intended for indoor use primarily to provide a degree of protection against settling airborne dust, falling dirt, and dripping noncorrosive liquids.

6   Enclosures are intended for indoor or out door use primarily to provide a degree of protection against hose-directed water, the entry of water during occasional temporary submersion at a limited depth, and damage from external ice formation.

6P   Enclosures are intended for indoor or out door use primarily to provide a degree of protection against hose-directed water, the entry of water during prolonged submersion at a limited depth, and damage from external ice formation.

12   Enclosures are indended for indoor use primarily to provide a degree of protection against circulating dust, falling dirt, and dripping noncorrosive liquids.

12K Enclosures with knockouts are intended for indoor use primarily to provide a degree of protection against circulating dust, falling dirt, and dripping noncorrosive liquids.

13   Enclosures are intended for indoor use primarily to provide a degree of protection against dust, spraying of water, oil, and noncorrosive coolant.

Hazardous (Classified Locations)

7   Enclosures are intended for indoor use in locations classified as Class I, Groups A, B, C, or D, as defined in the National Electrical Code.

8   Enclosures are intended for indoor use in locations classified as Class I, Groups A, B, C, or D, as defined in the National Electrical Code.

9   Enclosures are intended for indoor use in locations classified as Class II, Groups E, F, or G, as defined in the National Electrical Code.

10   Enclosures are constructed to meet the applicable requirements of the Mine Safety and Health Administration.

Reference NEMA Standards Publications/No. 250-1991
Used with permission - National Electrical Manufacturer's Association.

Size Portable Electric Generators

A power plant must be properly sized for the required load. Electric motors are particularly difficult for portable plants since they require 2 to 3 times motor nameplate amps or wattage.

A portable power plant's surge capacity is limited by engine horsepower and inertia of its rotating parts. Thus, a current surge of short duration can be supplied by a power plant, but a current demand of longer duration such as a heavily loaded motor starting a high inertia load, can overload the power plant and possibly damage the power plant and the motor. A 3450 rpm air compressor is a prime example of this type of load.

Example:

Load Requirements:

Electric Heater	1,000 watts
11 - 100 watt lamps	1,100 watts
Motor	600 watts

Total motor load = 600 x 3	= 1,800 watts
Total other load	= 2,100 watts
Total load	= 3,100 watts
Add 25% for future	= 975 watts
Total load for sizing power plant	= 4,875 watts

A generator capable of supplying 5,000 watts continuously will be adequate.

When more than one motor is connected to a power plant, always start the largest motor first. If the total load is connected to one receptacle on the power plant, be sure the ampere rating of the receptacle is not exceeded.

Typical Wattages for Tools and Appliances

Equipment	Running Watts
1/2" drill	1,000
1" drill	1,100
6" circular saw	800
10" circular saw	2,000
14" chain saw	1,100
Radio	50 to 200
Television	200 to 500
Toaster, coffeemaker	1,200
Water heater (storage type)	1,100 to 5,500
Hot plate or range (per burner)	1,000
Range oven	10,000
Skillet	1,200
Fan	50 to 200
Floodlight	1,000
Water pump	500
Vacuum cleaner	200 to 300
Refrigerator (conventional)	200 to 300
Refrigerator (freezer-combination)	250 to 600
Furnace fan (blower)	500 to 700

Knockout Dimensions

Conduit Size Ins.	Knockout Dia. Ins.		
	Min.	Nominal	Max.
1/2	0.859	0.875	0.906
3/4	1.094	1.109	1.141
1	1.359	1.375	1.406
1 1/4	1.719	1.734	1.766
1 1/2	1.958	1.984	2.016
2	2.433	2.469	2.500
2 1/2	2.938	2.969	3.000
3	3.563	3.594	3.625
3 1/2	4.063	4.125	4.156
4	4.563	4.641	4.672
5	5.625	5.719	5.750
6	6.700	6.813	6.844

Reference NEMA Standards Publications/No. 250-1991
Used with permission - National Electrical Manufacturer's Association.

National Electrical Code Tables

Table 430-150.   Full-Load Current*
Three-Phase Alternating-Current Motors

HP	Induction Type Squirrel-Cage and Wound-Rotor Amperes							Synchronous Type †Unity Power Factor Amperes			
	115V	200V	208V	230V	460V	575V	2300V	230V	460V	575V	2300V
½	4	2.3	2.2	2	1	.8					
¾	5.6	3.2	3.1	2.8	1.4	1.1					
1	7.2	4.1	4.0	3.6	1.8	1.4					
1½	10.4	6.0	5.7	5.2	2.6	2.1					
2	13.6	7.8	7.5	6.8	3.4	2.7					
3		11.0	10.6	9.6	4.8	3.9					
5		17.5	16.7	15.2	7.6	6.1					
7½		25.3	24.2	22	11	9					
10		32.2	30.8	28	14	11					
15		48.3	46.2	42	21	17					
20		62.1	59.4	54	27	22					
25		78.2	74.8	68	34	27		53	26	21	
30		92	88	80	40	32		63	32	26	
40		119.6	114.4	104	52	41		83	41	33	
50		149.5	143.0	130	65	52		104	52	42	
60		177.1	169.4	154	77	62	16	123	61	49	12
75		220.8	211.2	192	96	77	20	155	78	62	15
100		285.2	272.8	248	124	99	26	202	101	81	20
125		358.8	343.2	312	156	125	31	253	126	101	25
150		414	396.0	360	180	144	37	302	151	121	30
200		552	528.0	480	240	192	49	400	201	161	40

* These values of full-load current are for motors running at speeds usual for belted motors and motors with normal torque characteristics. Motors built for especially low speeds or high torques may require more running current, and multispeed motors will have full-load current varying with speed, in which case the nameplate current rating shall be used.

† For 90 and 80 percent power factor the above figures shall be multipled by 1.1 and 1.25 respectively.

The voltages listed are rated motor voltages. The currents listed shall be permitted for system voltage ranges of 110 to 120, 220 to 240, 440 to 480, and 550 to 600 volts.

Table 310-16.   Ampacities of Insulated Conductors
Rated 0-2000 Volts, 60° to 90°C (140° to 194°F)
Not More Than Three Conductors in Raceway or Cable or Earth
(Directly Buried), Based on Ambient Temperature of 30°C (86°F)

Size AWG kcmil	Temperature Rating of Conductor. See Table 310-13.								Size AWG kcmil
	60°C (140°F)	75°C (167°F)	85°C (185°F)	90°C (194°F)	60°C (140°F)	75°C (167°F)	85°C (185°F)	90°C (194°F)	
	TYPES †TW, †UF	TYPES †FEPW, †RH, †RHW, †THHW, †THW, †THWN, †XHHW †USE, †ZW	TYPE V	TYPES TA, TBS, SA SIS, †FEP, †FEPB, †RHH, †THHN, †THHW, †XHHW	TYPES †TW, †UF	TYPES †RH, †RHW, †THHW, †THW, †THWN, †XHHW †USE	TYPE V	TYPES TA, TBS, SA, SIS, †RHH, †THHW, †THHN, †XHHW	
	COPPER				ALUMINUM OR COPPER-CLAD ALUMINUM				
18	....	....	....	14	....	....	....	....	....
16	....	....	18	18	....	....	....	....	....
14	20†	20†	25	25†	....	....	....	....	12
12	25†	25†	30	30†	20†	20†	25	25†	10
10	30	35†	40	40†	25	30†	30	35†	8
8	40	50	55	55	30	40	40	45	
6	55	65	70	75	40	50	55	60	6
4	70	85	95	95	55	65	75	75	4
3	85	100	110	110	65	75	85	85	3
2	95	115	125	130	75	90	100	100	2
1	110	130	145	150	85	100	110	115	1
1/0	125	150	165	170	100	120	130	135	1/0
2/0	145	175	190	195	115	135	145	150	2/0
3/0	165	200	215	225	130	155	170	175	3/0
4/0	195	230	250	260	150	180	195	205	4/0
250	215	255	275	290	170	205	220	230	250
300	240	285	310	320	190	230	250	255	300
350	260	310	340	350	210	250	270	280	350
400	280	335	365	380	225	270	295	305	400
500	320	380	415	430	260	310	335	350	500
600	355	420	460	475	285	340	370	385	600
700	385	460	500	520	310	375	405	420	700
750	400	475	515	535	320	385	420	435	750
800	410	490	535	555	330	395	430	450	800
900	435	520	565	585	355	425	465	480	900
1000	455	545	590	615	375	445	485	500	1000
1250	495	590	640	665	405	485	525	545	1250
1500	520	625	680	705	435	520	565	585	1500
1750	545	650	705	735	455	545	595	615	1750
2000	560	665	725	750	470	560	610	630	2000

AMPACITY CORRECTION FACTORS

Ambient Temp. °C	For ambient temperatures other than 30°C (86°F), multiply the ampacities shown above by the appropriate factor shown below.								Ambient Temp. °F
21-25	1.08	1.05	1.04	1.04	1.08	1.05	1.04	1.04	70-77
26-30	1.00	1.00	1.00	1.00	1.00	1.00	1.00	1.00	79-86
31-35	.91	.94	.95	.96	.91	.94	.95	.96	88-95
36-40	.82	.88	.90	.91	.82	.88	.90	.91	97-104
41-45	.71	.82	.85	.87	.71	.82	.85	.87	106-113
46-50	.58	.75	.80	.82	.58	.75	.80	.82	115-122
51-55	.41	.67	.74	.76	.41	.67	.74	.76	124-131
56-60	....	.58	.67	.71	....	.58	.67	.71	133-140
61-70	....	.33	.52	.58	....	.33	.52	.58	142-158
71-80	....	....	.30	.41	....	....	.30	.41	160-176

† Unless otherwise specifically permitted elsewhere in this Code, the overcurrent protection for conductor types marked with an obelisk (†) shall not exceed 15 amperes for 14 AWG, 20 amperes for 12 AWG, and 30 amperes for 10 AWG copper; or 15 amperes for 12 AWG and 25 amperes for 10 AWG aluminum and copper-clad aluminum after any correction factors for ambient temperature and number of conductors have been applied.

Table 3A.   Maximum Number of Conductors in Trade Sizes of Conduit or Tubing
(Based on N.E.C. Table 1, Chapter 9)

Type Letters	Conductor Size AWG, kcmil	½	¾	1	1¼	1½	2	2½	3	3½	4	5	6
TW, XHHW (14 through 8)	14	9	15	25	44	60	99	142					
	12	7	12	19	35	47	78	111	171				
	10	5	9	15	26	36	60	85	131	176			
	8	2	4	7	12	17	28	40	62	84	108		
RHW and RHH (without outer covering), THW	14	6	10	16	29	40	65	93	143	192			
	12	4	8	13	24	32	53	76	117	157			
	10	4	6	11	19	26	43	61	95	127	163		
	8	1	3	5	10	13	22	32	49	66	85	133	
TW,	6	1	2	4	7	10	16	23	36	48	62	97	141
	4	1	1	3	5	7	12	17	27	36	47	73	106
THW,	3	1	1	2	4	6	10	15	23	31	40	63	91
	2	1	1	2	4	5	9	13	20	27	34	54	78
	1		1	1	3	4	6	9	14	19	25	39	57
FEPB (6 through 2), RHW and RHH (without outer covering)	1/0		1	1	2	3	5	8	12	16	21	33	49
	2/0		1	1	1	3	5	7	10	14	18	29	41
	3/0		1	1	1	2	4	6	9	12	15	24	35
	4/0			1	1	1	3	5	7	10	13	20	29
	250			1	1	1	2	4	6	8	10	16	23
	300			1	1	1	2	3	5	7	9	14	20
	350				1	1	1	3	4	6	8	12	18
	400				1	1	1	2	4	5	7	11	16
	500				1	1	1	1	3	4	6	9	14
	600					1	1	1	3	4	5	7	11
	700					1	1	1	2	3	4	7	10
	750					1	1	1	2	3	4	6	9

Note: This table is for concentric stranded conductors only. For cables with compact conductors, the dimensions in N.E.C. Table 5A shall be used.

Table 3B.   Maximum Number of Conductors in Trade Sizes of Conduit or Tubing
(Based on Table 1, Chapter 9)

Type Letters	Conductor Size AWG, kcmil	½	¾	1	1¼	1½	2	2½	3	3½	4	5	6
THWN,	14	13	24	39	69	94	154						
	12	10	18	29	51	70	114	164					
	10	6	11	18	32	44	73	104	160				
	8	3	5	9	16	22	36	51	79	106	136		
THHN, FEP (14 through 2), FEPB (14 through 8), PFA (14 through 4/0), PFAH (14 through 4/0), Z (14 through 4/0), XHHW (4 through 500 kcmil)	6	1	4	6	11	15	26	37	57	76	98	154	
	4	1	2	4	7	9	16	22	35	47	60	94	137
	3	1	1	3	6	8	13	19	29	39	51	80	116
	2	1	1	3	5	7	11	16	25	33	43	67	97
	1		1	1	3	5	8	12	18	25	32	50	72
	1/0		1	1	3	4	7	10	15	21	27	42	61
	2/0		1	1	2	3	6	8	13	17	22	35	51
	3/0		1	1	1	3	5	7	11	14	18	29	42
	4/0		1	1	1	2	4	6	9	12	15	24	35
	250			1	1	1	3	4	7	10	12	20	28
	300			1	1	1	3	4	6	8	11	17	24
	350			1	1	1	2	3	5	7	9	15	21
	400				1	1	1	3	5	6	8	13	19
	500				1	1	1	2	4	5	7	11	16
	600				1	1	1	1	3	4	5	9	13
	700					1	1	1	3	4	5	8	11
	750					1	1	1	2	3	4	7	11
XHHW	6	1	3	5	9	13	21	30	47	63	81	128	185
	600			1	1	1	1	1	3	4	5	9	13
	700				1	1	1	1	3	4	5	7	11
	750				1	1	1	1	2	3	4	7	10

Note: This table is for concentric stranded conductors only. For cables with compact conductors, the dimensions in N.E.C. Table 5A shall be used.

Table 3C. Maximum Number of Conductors in Trade Sizes of Conduit or Tubing
(Based on N.E.C. Table 1, Chapter 9)

Conduit Trade Size (Inches)		½	¾	1	1¼	1½	2	2½	3	3½	4	5	6
Type Letters	Conductor Size AWG, kcmil												
RHW,	14	3	6	10	18	25	41	58	90	121	155		
	12	3	5	9	15	21	35	50	77	103	132		
	10	2	4	7	13	18	29	41	64	86	110		
	8	1	2	4	7	9	16	22	35	47	60	94	137
RHH	6	1	1	2	5	6	11	15	24	32	41	64	93
	4	1	1	1	3	5	8	12	18	24	31	50	72
(with	3	1	1	1	3	4	7	10	16	22	28	44	63
outer	2		1	1	3	4	6	9	14	19	24	38	56
covering)	1		1	1	1	3	5	7	11	14	18	29	42
	1/0		1	1	1	2	4	6	9	12	16	25	37
	2/0			1	1	1	3	5	8	11	14	22	32
	3/0			1	1	1	3	4	7	9	12	19	28
	4/0			1	1	1	2	4	6	8	10	16	24
	250				1	1	1	3	5	6	8	13	19
	300				1	1	1	3	4	5	7	11	17
	350				1	1	1	2	4	5	6	10	15
	400				1	1	1	1	3	4	6	9	14
	500				1	1	1	1	3	4	5	8	11
	600					1	1	1	2	3	4	6	9
	700					1	1	1	1	3	3	6	8
	750						1	1	1	1	3	5	8

Note: This table is for concentric stranded conductors only. For cables with compact conductors, the dimensions in N.E.C. Table 5A shall be used.

Tables 4 and 5 give the nominal size of conductors and conduit or tubing for use in computing size of conduit or tubing for various combinations of conductors. The dimensions represent average conditions only, and variations will be found in dimensions of conductors and conduits of different manufacture.

Table 4
Dimensions and Percent Area of Conduit and of Tubing

Trade Size	Internal Diameter Inches	Total 100%	Not Lead Covered			Lead Covered				
			2 Cond. 31%	Over 2 Cond. 40%	1 Cond. 53%	1 Cond. 55%	2 Cond. 30%	3 Cond. 40%	4 Cond. 38%	Over 4 Cond. 35%
½	.622	.30	.09	.12	.16	.17	.09	.12	.11	.11
¾	.824	.53	.16	.21	.28	.29	.16	.21	.20	.19
1	1.049	.86	.27	.34	.46	.47	.26	.34	.33	.30
1¼	1.380	1.50	.47	.60	.80	.83	.45	.60	.57	.53
1½	1.610	2.04	.63	.82	1.08	1.12	.61	.82	.78	.71
2	2.067	3.36	1.04	1.34	1.78	1.85	1.01	1.34	1.28	1.18
2½	2.469	4.79	1.48	1.92	2.54	2.63	1.44	1.92	1.82	1.68
3	3.068	7.38	2.29	2.95	3.91	4.06	2.21	2.95	2.80	2.58
3½	3.548	9.90	3.07	3.96	5.25	5.44	2.97	3.96	3.76	3.47
4	4.026	12.72	3.94	5.09	6.74	7.00	3.82	5.09	4.83	4.45
5	5.047	20.00	6.20	8.00	10.60	11.00	6.00	8.00	7.60	7.00
6	6.065	28.89	8.96	11.56	15.31	15.89	8.67	11.56	10.98	10.11

Table 5
Dimensions of Rubber-Covered and
Thermoplastic-Covered Conductors

Size AWG MCM	Types RFH-2, RH, RHH,*** RHW,*** SF-2		Types TF, THW,† TW		Types TFN, THHN, THWN		Types**** FEP, FEPB, FEPW, TFE, PF, PFA, PFAH, PGF, PTF, Z, ZF, ZFF		Type XHHW, ZW††		Types KF-1, KF-2, KFF-1, KFF-2	
	Approx. Diam. Inches	Approx. Area Sq. In.	Approx. Diam. Inches	Approx. Area Sq. In.	Approx. Diam. Inches	Approx. Area Sq. In.	Approx. Diam. Inches	Approx. Area Sq. Inches	Approx. Diam. Inches	Approx. Area Sq. In.	Approx. Diam. Inches	Approx. Area Sq. In.
Col. 1	Col. 2	Col. 3	Col. 4	Col. 5	Col. 6	Col. 7	Col. 8	Col. 9	Col. 10	Col. 11	Col. 12	Col. 13
18	.146	.0167	.106	.0088	.089	.0062	.081	.0052	...	....	.065	.0033
16	.158	.0196	.118	.0109	.100	.0079	.092	.0066	...	....	.070	.0038
14	30 mils .171	.0230	.131	.0135	.105	.0087	.105 .105	.0087 .0087	...	....	.083	.0054
14	45 mils .204*	.0327*	...	....	...	....	... ...	....	.129	.0131	...	....
14	.....	....	.162†	.0206†	...	....	... ...	....	...	....	...	....
12	30 mils .188	.0278	.148	.0172	.122	.0117	.121 .121	.0115 .0115	...	....	.102	.0082
12	45 mils .221*	.0384*	...	....	...	....	... ...	....	.146	.0167	...	....
12	.....	....	.179†	.0252†	...	....	... ...	....	...	....	...	....
10	..... .242	.0460	.168	.0222	.153	.0184	.142 .142	.0158 .0158	...	....	.124	.0121
10			.199†	.0311†					.166	.0216		
8	..... .328	.0845	.245	.0471	.218	.0373	.206 .186	.0333 .0272	...	....	...	....
8			.276†	.0598†					.241	.0456		
6	.397	.1238	.323	.0819	.257	.0519	.244 .302	.0468 .0716	.282	.0625	...	....
4	.452	.1605	.372	.1087	.328	.0845	.292 .350	.0670 .0962	.328	.0845	...	....
3	.481	.1817	.401	.1263	.356	.0995	.320 .378	.0804 .1122	.356	.0995	...	....
2	.513	.2067	.433	.1473	.388	.1182	.352 .410	.0973 .1320	.388	.1182	...	....
1	.588	.2715	.508	.2027	.450	.1590	.420	.1385	.450	.1590	...	....
1/0	.629	.3107	.549	.2367	.491	.1893	.462	.1676	.491	.1893	...	....
2/0	.675	.3578	.595	.2781	.537	.2265	.498	.1948	.537	.2265	...	....
3/0	.727	.4151	.647	.3288	.588	.2715	.560	.2463	.588	.2715	...	....
4/0	.785	.4840	.705	.3904	.646	.3278	.618	.3000	.646	.3278	...	....

Size AWG MCM	Types RFH-2, RH, RHH,*** RHW,*** SF-2		Types TF, THW,† TW		Types TFN, THHN, THWN		Types**** FEP, FEPB, FEPW, TFE, PF, PFA, PFAH, PGF, PTF, Z, ZF, ZFF		Type XHHW, ZW††	
	Approx. Diam. Inches	Approx. Area Sq. In.	Approx. Diam. Inches	Approx. Area Sq. In.	Approx. Diam. Inches	Approx. Area Sq. In.	Approx. Diam. Inches	Approx. Area Sq. Inches	Approx. Diam. Inches	Approx. Area Sq. In.
Col. 1	Col. 2	Col. 3	Col. 4	Col. 5	Col. 6	Col. 7	Col. 8	Col. 9	Col. 10	Col. 11
250	.868	.5917	.788	.4877	.716	.4026	...	....	.716	.4026
300	.933	.6837	.843	.5581	.771	.4669	...	....	.771	.4669
350	.985	.7620	.895	.6291	.822	.5307	...	....	.822	.5307
400	1.032	.8365	.942	.6969	.869	.5931	...	....	.869	.5931
500	1.119	.9834	1.029	.8316	.955	.7163	...	....	.955	.7163
600	1.233	1.1940	1.143	1.0261	1.058	.8791	...	....	1.073	.9043
700	1.304	1.3355	1.214	1.1575	1.129	1.0011	...	....	1.145	1.0297
750	1.339	1.4082	1.249	1.2252	1.163	1.0623	...	....	1.180	1.0936
800	1.372	1.4784	1.282	1.2908	1.196	1.1234	...	....	1.210	1.1499
900	1.435	1.6173	1.345	1.4208	1.259	1.2449	...	....	1.270	1.2668
1000	1.494	1.7530	1.404	1.5482	1.317	1.3623	...	....	1.330	1.3893
1250	1.676	2.2062	1.577	1.9532	...	....	...	....	1.500	1.7671
1500	1.801	2.5475	1.702	2.2751	...	....	...	....	1.620	2.0612
1750	1.916	2.8832	1.817	2.5930	...	....	...	....	1.740	2.3779
2000	2.021	3.2079	1.922	2.9013	...	....	...	....	1.840	2.6590

* The dimensions of Types RHH and RHW.
† Dimensions of THW in sizes No. 14 through No. 8. No. 6 THW and larger is the same dimension as TW.
*** Dimensions of RHH and RHW without outer covering are the same as THW No. 18 through No. 10, solid; No. 8 and larger, stranded.
**** In Columns 8 and 9 the values shown for sizes No. 1 through 4/0 are for TFE and Z only. The right-hand values in Columns 8 and 9 are for FEPB, Z, ZF, and ZFF only.
†† No. 14 through No. 2.

Electrical Formulas

Single Phase

$kVA = VA/1,000$

$hp = (VA \times eff \times PF)/746$

$kw = (VA \times PF)/1,000$

$PF = kW/kVA$

Torque (ft-lb) $= hp \times 5,250/rpm$

Motor kVA $= (hp \times 0.746)/eff \times PF$

Motor kW $= (hp \times 0.746)/eff$

V = Line-to-line volts
A = Amperes

eff = Efficiency (decimal)
PF = Power Factor (decimal)
kVA = Kilovolt amperes
kW = Kilowatts
hp = Horsepower
rpm = Revolutions per minute

Three Phase

$(3^{1/2} \, V \times A)/1,000$

$(3^{1/2} \, VA \times PF \times eff)/746$

$(3^{1/2} \, VA \times PF)/1,000$

$(kW \times 1,000)/3^{1/2} \times VA$

Full load currents—single phase transformers

kVA Rating	Voltage—Volts					
	120	240	480	2,400	4,160	7,200
1.0	8.3	4.2	2.1	0.4	0.2	0.1
3.0	25.0	12.5	6.3	1.3	0.7	0.4
5.0	41.7	20.8	10.4	2.1	1.2	0.7
7.5	62.5	31.3	15.6	3.1	1.8	1.0
10.0	83.3	41.7	20.8	4.2	2.4	1.4
15.0	125.0	62.5	31.3	6.3	3.6	2.1
25.0	208.3	104.2	52.1	10.4	6.0	3.5
37.5	312.5	156.3	78.1	15.6	9.0	5.2
50.0	416.7	208.3	104.2	20.8	12.0	6.9
75.0	625.0	312.5	156.3	31.3	18.0	10.4
100.0	833.3	416.7	208.3	41.7	24.0	13.9
125.0	1,041.7	520.8	260.4	52.1	30.0	17.4
167.5	1,395.8	697.9	349.0	69.8	40.3	23.3
200.0	1,666.7	833.3	416.7	83.3	48.1	27.8
250.0	2,083.3	1,041.7	520.8	104.2	60.1	34.7
333.0	2,775.0	1,387.5	693.8	138.8	80.0	46.3
500.0	4,166.7	2,083.3	1,041.7	208.3	120.2	69.4

Full load currents—three phase transformers

kVA Rating	Voltage (line-to-line)				
	240	480	2,400	4,160	12,470
3	7.2	3.6	0.7	0.4	0.1
6	14.4	7.2	1.4	0.8	0.3
9	21.7	10.8	2.2	1.2	0.4
15	36.1	18.0	3.6	2.1	0.7
30	72.2	36.1	7.2	4.2	1.4
37.5	90.2	45.1	9.0	5.2	1.7
45	108.3	54.1	10.8	6.2	2.1
50	120.3	60.1	12.0	6.9	2.3
75	180.4	90.2	18.0	10.4	3.5
100	240.6	120.3	24.1	13.9	4.6
112.5	270.6	135.3	27.1	15.6	5.2
150	360.9	180.4	36.1	20.8	6.9
200	481.1	240.6	48.1	27.8	9.3
225	541.3	270.6	54.1	31.2	10.4
300	721.7	360.9	72.2	41.6	13.9
450	1,082.6	541.3	108.3	62.5	20.8
500	1,202.8	601.4	120.3	69.4	23.2
600	1,443.4	721.7	144.3	83.3	27.8
750	1,804.3	902.1	180.4	104.1	34.7
1,000	2,405.7	1,202.8	240.6	138.8	46.3
1,500	3,608.5	1,804.3	360.9	208.2	69.5
2,000	4,811.4	2,405.7	481.1	277.6	92.6
2,500	6,014.2	3,007.1	601.4	347.0	115.8
5,000	12,028.5	6,014.2	1,202.8	694.0	231.5
10,000	24,057.0	12,028.5	2,405.7	1,387.9	463.0

Motor controller sizes

Polyphase Motors

NEMA Size	Maximum Horsepower Full Voltage Starting	
	230 Volts	460–575 Volts
00	1.5	2
0	3	5
1	7.5	10
2	15	25
3	30	50
4	50	100
5	100	200
6	200	400
7	300	600

Single Phase Motors

NEMA Size	Maximum Horsepower Full Voltage Starting (Two Pole Contactor)	
	115 Volts	230 Volts
00	1.3	1
0	1	2
1	2	3
2	3	7.5
3	7.5	15

Voltage drop on circuits using 600 V. copper conductors in steel conduit

To determine the voltage drop, multiply the length in feet of one conductor by the current in amperes and by the number listed in the table for the type of system and power factor and divide the result by 1,000,000 to obtain the voltage loss. This table takes into consideration reactance on AC circuits and resistance of the conductor. Unless otherwise noted, the table is based on 60 Hz.

Wire Size	Direct Current	Three Phase Lagging Power Factor					Single Phase Lagging Power Factor				
		100%	90%	80%	70%	60%	100%	90%	80%	70%	60%
14	6,100	5,280	4,800	4,300	3,780	3,260	6,100	5,551	4,964	4,370	3,772
12	3,828	3,320	3,030	2,720	2,400	2,080	3,828	3,502	3,138	2,773	2,404
10	2,404	2,080	1,921	1,733	1,540	1,340	2,404	2,221	2,003	1,779	1,547
8	1,520	1,316	1,234	1,120	1,000	880	1,520	1,426	1,295	1,159	1,017
6	970	840	802	735	665	590	970	926	850	769	682
4	614	531	530	487	445	400	614	613	562	514	462
3	484	420	425	398	368	334	484	491	460	425	385
2	382	331	339	322	300	274	382	392	372	346	317
1	306	265	280	270	254	236	306	323	312	294	273
0	241	208	229	224	214	202	241	265	259	247	233
00	192	166	190	188	181	173	192	219	217	209	199
000	152	132	157	158	155	150	152	181	183	179	173
0000	121	105	131	135	134	132	121	151	156	155	152
250M	102	89	118	123	125	123	103	136	142	144	142
300M	85	74	104	111	112	113	86	120	128	130	131
350M	73	63	94	101	105	106	73	108	117	121	122
400M	64	55	87	95	98	100	64	100	110	113	116
500M	51	45	76	85	90	92	52	88	98	104	106
600M	43	38	69	79	85	87	44	80	91	98	101
700M	36	33	64	74	80	84	38	74	86	92	97
750M	34	31	62	72	79	82	36	72	83	91	95
800M	32	29	61	71	76	81	33	70	82	88	93
900M	28	26	57	68	74	78	30	66	78	85	90
1000M	26	23	55	66	72	76	27	63	76	83	88

Determine the most economical size for electric power conductors

To calculate quickly the most economical size copper wire for carrying a specified current load, use the formula:

$$A = (59)(I) \sqrt{\frac{C_e \times t}{C_c \times F}}$$

where A = Conductor cross-sectional area (circular mils)
 I = Current load (amperes)
 C_e = Cost of electrical power (cents per KWH)
 C_c = Cost of copper (cents per lb)
 t = Hours of service per year
 F = Factor for fixed charges (amortization, insurance, taxes, etc.)

Selection of the proper size electric conductor will depend on the load to be carried and the mechanical strength required, as well as economics. Once these are considered, the most economical size conductor is that for which the annual energy cost equals the copper cost. This method can be used for other metallic conductors.

Example. Determine the most economical size copper conductor for an installation operating 365 days a year, 8 hours per day, with a 100 ampere load. Energy costs are $0.00875 per KWH. Fixed charges are 23 percent, considering a five-year amortization with 3 percent for insurance and taxes. Copper costs $0.46 per lb.

$$A = (59)(100) \sqrt{\frac{0.00875 \times 365 \times 8}{0.46 \times 0.23}}$$

= 91,686 circular mils

How to find the resistance and weight of copper wires

It is easy to calculate mentally the approximate resistance and weight of standard sizes of copper wires, by remembering the following rules:

Rule 1: Number 10 wire has a resistance of 1 ohm per thousand ft. Number 2 wire weighs 200 lb per thousand ft.

Rule 2: An increase of ten numbers in the size is an increase of ten times in the resistance, and a decrease to one-tenth of the weight.

Rule 3: An increase of three numbers in the size doubles the resistance and halves the weight.

Rule 4: An increase of one number in the size multiplies the resistance by $5/4$ and the weight by $4/5$.

Always proceed from the known value (No. 10 for resistance, No. 2 for the weight) in the smallest number of jumps; that is, go by tens, then by threes, and finally by ones; the jumps do not all have to be in the same direction.

Example. What is the weight and resistance of 200 ft of No. 18? Apply the rules in order, as follows; resistance first:

Number 10 wire has a resistance of 1 ohm per thousand ft. (Rule 1)
Number 20 wire has a resistance of 10 ohms per thousand ft. (Rule 2)
Number 17 wire has a resistance of 5 ohms per thousand ft. (Rule 3)
Number 18 wire has a resistance of 6.25 ohms per thousand ft. (Rule 4)
Number 18 wire has a resistance of 1.25 ohms per 200 ft. Answer.
Number 2 wire weighs 200 lb per thousand ft. (Rule 1)
Number 12 wire weighs 20 lb per thousand ft. (Rule 2)
Number 22 wire weighs 2 lb per thousand ft. (Rule 2)
Number 19 wire weighs 4 lb per thousand ft. (Rule 3)
Number 18 wire weighs 5 lb per thousand ft. (Rule 4)
Number 18 wire weighs 1 lb per 200 ft. (Answer)

The weight determined by the above rules is that of bare copper wire. Weight of insulated wire varies widely according to the type of insulation. Resistance values are correct for bare or insulated wire, and for solid or stranded. Errors resulting from the use of the rules will rarely exceed 2 percent.

What you should remember about electrical formulas

Ohm's Law

Easily **R**emember **I**t, or $E = RI$

Power

DC Power **I**s **E**asy, or $P = IE$
It **R**eally **I**s, or $P = IRI = I^2R$

Other Facts

The number of **watts in a hp** is equal to the year Columbus discovered America divided by two, or $1492/2 = 746$ watts/hp.

The horsepower of a reciprocating engine necessary to drive a three phase, 60 cycle electric generator can be determined by multiplying KW by 1.5.

How to calculate microwave hops on level ground

Rule. Excluding obstructions such as hills, etc., the microwave (line-of-sight) distances between two towers will be twice the distance from the transmitter to the horizon. The line-of-sight distance in miles from the top of the tower to the horizon can be found by adding the square root of the tower height to the square root of the first square root (the latter of course is the fourth root).

$$x = \sqrt{h} + \sqrt{\sqrt{h}} \text{ or } \sqrt{h} + \sqrt[4]{h}$$

Example. How far apart can 100-foot-high towers be spaced to provide line-of-sight transmission if there are no obstructions between them?

$X = \sqrt{100} + \sqrt{10}$

$X = 10 + 3 +$ or 13 miles

Distance between towers can be 2X or 26 miles.

This same formula can be used to estimate distances in an airplane.

Example. How far is the line-of-sight distance to the horizon in an airplane flying at an altitude of 4,900 ft?

$(\sqrt{4,900} = 70)$

$X = \sqrt{4,900} + \sqrt{70}$

$X = 70 + 8$

Distance to horizon is 78 miles.

For quick determination of the horsepower per ampere for induction motors (3 phase) at different voltages

Voltage	hp per ampere
480	1
2,400	5
4,160	8

Chart gives electric motor horsepower for pumping units

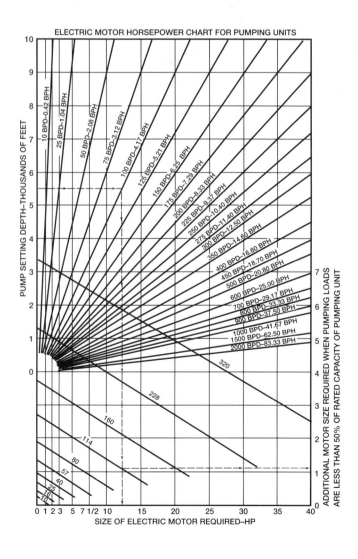

ELECTRIC MOTOR HORSEPOWER CHART FOR PUMPING UNITS

This chart provides a means of determining the power requirements for a beam pumping installation powered by an electric motor based on a fluid with a specific gravity of one, with fluid at the pump. The power requirement determined by the chart includes a mechanical efficiency factor of 0.45 and a cyclic factor of 0.75, which factors are frequently applied to motors used in sucker rod pumping service.

An arrangement is available for correcting the power requirement in the case of an underloaded pumping unit. The example shown in the chart is self-explanatory.

After the power requirements are determined from the chart the next higher size of commercially available motor is used.

Pumping stations

Table gives capacitor multipliers for kilowatt loads for different desired power factor improvements

Original Power Factor, Percent	DESIRED POWER FACTOR—Percent				
	100	95	90	85	80
50	1.732	1.403	1.248	1.112	0.982
51	1.687	1.358	1.203	1.067	0.937
52	1.643	1.314	1.159	1.023	0.893
53	1.600	1.271	1.116	0.980	0.850
54	1.559	1.230	1.075	0.939	0.809
55	1.518	1.190	1.035	0.898	0.769
56	1.480	1.151	0.996	0.860	0.730
57	1.442	1.113	0.958	0.822	0.692
58	1.405	1.076	0.921	0.785	0.655
59	1.369	1.040	0.885	0.749	0.619
60	1.333	1.004	0.849	0.713	0.583
61	1.299	0.970	0.815	0.679	0.549
62	1.266	0.937	0.782	0.646	0.516
63	1.233	0.904	0.749	0.613	0.483
64	1.201	0.872	0.717	0.581	0.451
65	1.169	0.840	0.685	0.549	0.419
66	1.138	0.809	0.654	0.518	0.388
67	1.108	0.779	0.624	0.488	0.358
68	1.078	0.749	0.594	0.458	0.328
69	1.049	0.720	0.565	0.429	0.299
70	1.020	0.691	0.536	0.400	0.270
71	0.992	0.663	0.508	0.372	0.242
72	0.964	0.635	0.480	0.344	0.214
73	0.936	0.607	0.452	0.316	0.186
74	0.909	0.580	0.425	0.289	0.159
75	0.882	0.553	0.398	0.262	0.132
76	0.855	0.526	0.371	0.235	0.105
77	0.829	0.500	0.345	0.209	0.079
78	0.802	0.473	0.318	0.182	0.052
79	0.776	0.447	0.292	0.156	0.026
80	0.750	0.421	0.266	0.130	....
81	0.724	0.395	0.240	0.104	....
82	0.698	0.369	0.214	0.078	....
83	0.672	0.343	0.188	0.052	....
84	0.646	0.317	0.162	0.026	....
85	0.620	0.291	0.136	....	.?..
86	0.593	0.264	0.109	....	....
87	0.567	0.238	0.083	....	....
88	0.540	0.211	0.056	....	....
89	0.512	0.183	0.028	....	....
90	0.484	0.155			
91	0.456	0.127			
92	0.426	0.097			
93	0.395	0.066			
94	0.363	0.034			
95	0.329	....			
96	0.292	....			
97	0.251	....			
98	0.203	....			
99	0.143	....			

Example

Assume total plant load is 100 kw at 60 percent power factor. Capacitor kvar rating necessary to improve power factor to 80 percent is found by multiplying kw (100) by multiplier in table (0.583), which gives kvar (58.3). Nearest standard rating (60 kvar) should be recommended.

FLOODLIGHTING CONCEPTS

This material is intended as a helpful guide in the evaluation of lighting requirements and is intended to serve only as a guide.

Terms

Candela or Candlepower—Light sources do not project the same amount of light in every direction. The directional characteristic of a lamp is described by the candlepower in specific directions. This directional strength of light or luminous intensity is measured in candelas.

Lumen—Light quantity, irrespective of direction, is measured in lumens. Lamp lumens are the quantity of light produced by a lamp. Average light level calculations use total lamp lumens as a basis and then adjust for all factors that lower this quantity. The amount of useful light in a floodlight beam is measured in beam lumens.

Footcandle, fc—Specifications are usually based on density of light or level of illumination which is measured in footcandles. Footcandles are the ratio of quantity of light in lumens divided by the surface area in square feet on which the lumens are falling. A density of one lumen per square foot is one footcandle. One footcandle is equal to 10.76 lux.

Lux, lx—The SI unit of illuminance. This metric measurement is based upon the density of lumens per unit surface area similar to footcandle, except one lux is one lumen per square meter. One lux is equal to 0.09 footcandle.

Light Loss Factor, LLF—These factors are used to adjust lighting calculations from laboratory test conditions to a closer approximation of expected field results. The I.E.S. Lighting Handbook, 1984 Reference Volume, defines LLF as follows: "a factor used in calculating illuminance after a given period of time and under given conditions. It takes into account temperature and voltage variations, dirt accumulations on luminaire and room surfaces, lamp depreciation, maintenance procedures and atmosphere conditions."

Floodlighting calculations

Floodlighting encompasses many variations. Since the location of the floodlight relative to the object to be lighted can be in any plane and at any distance from the source...floodlighting application is often considered the most complex and difficult of all lighting techniques.

The most commonly used systems for floodlight calculations are the point-by-point method and the beam-lumen method.

Point-by-point method

The point-by-point method permits the determination of footcandles at any point and orientation on a surface and the degree of lighting uniformity realized for any given set of conditions.

In such situations, the illumination is proportional to the candlepower of the source in a given direction, and inversely proportional to the square of the distance from the source. See Figure 1.

Footcandles on Plane (Normal to Light Ray) $= \dfrac{\text{Candlepower of Light Ray}}{\text{Distance in Feet From Source to Point-Squared}}$

$$E = I/D^2 \qquad (1)$$

Figure 1. Inverse Square Law

When the surface on which the illumination to be determined is tilted, the light will be spread over a greater area, reducing the illumination in the ratio of the area of plane A to the area of plane B as shown in Figure 2. This ratio is equal to the cosine of the angle of incidence; thus:

$$\text{Footcandles on Plane B} = \frac{\text{Candlepower of Light Ray}}{\text{Distance in Feet From Source to Point-Squared}}$$

x Cosine of Angle β

$$E = (I/D^2) \times \text{Cosine } \beta \qquad (2)$$

Then β equals the angle between the light ray and a perpendicular to the plane at that point.

Figure 2. Cosine of Angle Law

Beam-lumen method

The beam-lumen method is quite similar to the method used for interior lighting except that the utilization factors must take into consideration the fact that floodlights are not usually perpendicular to the surface and therefore not all of the useful light strikes the surface.

Beam lumens are defined as the quantity of light that is contained within the beam limits as described as "beam spread." Beam lumens equal the lamp lumens multiplied by the beam efficiency of the floodlight.

<u>Coverage</u>. It is recommended that sufficient point-by-point calculations be made for each job to check uniformity and coverage.

<u>Light</u> <u>Loss</u> <u>Factor</u> (LLF). The maintenance or light loss factor is an allowance for depreciation of lamp output with age and floodlight efficiency due to the collection of dirt on lamp, reflector, and vover glass. The total factor may vary from 0.65 to 0.85 depending on the type of lamp and luminaire used and may include losses due to lamp orientation, or "tilt."

Design procedure

Step 1—Determine the level of illumination. See Table 1 for some typical levels of illumination (fc). The basic formula is:

$$fc = \frac{N \times BL \times CBU \times LLF}{A} \qquad (3)$$

where
N   = quantity of luminaires
A   = area in square feet
BL   = beam lumens
CBU = Coefficient of beam utilization
LLF = Light loss factor

Step 2—Determine type and location of floodlights. Regardless of light source there are industry standards on beam spreads. See Table 2.

Step 3—Determine the coefficient of beam utilization. This factor, CBU, written as a decimal fraction, is expressed in the following ratio:

$$CBU = \frac{\text{Utilized Lumens}}{BL} \qquad (4)$$

Table 1
Typical Illumination Levels

Area	fc	Area	fc
Building exteriors		Storage tanks	1
Entrances		Pump areas	2
Active	5	Parking	1
Inactive	1	Substation	2
Vital locations	5	Storage yards	
Catwalks	3	Active	20
Stairs	5	Inactive	1
Barge or truck unloading	5		

Table 2
Outdoor Floodlight Luminaire Designations

Beam Spread, Degrees	NEMA Type
10 up to 18	1
>18 up to 29	2
>29 up to 46	3
>46 up to 70	4
>70 up to 100	5
>100 up to 130	6
>130 up	7

The following general principals apply in the choice of beam spread:

1. The greater the distance from the floodlight to the area to be lighted, the narrower the beam spread desired.

2. Since by definition the candlepower at the edge of a floodlight beam is 10% of the candlepower near the center of the beam, the illumination level at the edge of the beam is one-tenth or less of that at the center. To obtain reasonable uniformity of illumination the beams of individual floodlights must overlap each other as well as the edge of the surface to be lighted.

3. The percentage of beam lumens falling outside the area to be lighted is usually lower with narrow-beam units than with wide-beam units. Thus narrow-beam floodlights are preferable where they will provide the necessary degree of uniformity of illumination and the proper footcandle level.

The exact CBU can be determined graphically by projecting the outline of the area to be lighted upon the photometric data and totaling the utilized lumens. This procedure is detailed in the I.E.S. Lighting Handbook. See Table 3.

Table 3
Typical CBU's

Area	CBU	Area	fc
Baseball			
Infield	0.65	Tennis	0.75
Outfield	0.85	Large parking lots	0.80
Football	0.60	Building facades	0.65

As an approximation, the average CBU of all the floodlights in an installation should fall within the range of 0.6 to 0.9. If less than 60% of the beam lumens are utilized, a more economical lighting plan should be possible by using different locations or narrower beam floodlights. If the CBU is over 0.9, it is probable that the beam spread selected is too narrow and the resultant illumination will be spotty. An estimated CBU can be determined by experience, or by making calculations for several potential aiming points and using the average figure thus obtained.

Step 4— Determine the quantity of floodlights (N) required. Rearrange the basic formula in Step 1 as follows:

$$N = \frac{A \times fc}{BL \times CBU \times LLF} \qquad (5)$$

5: Hydrostatic Testing

Hydrostatic testing for pipe lines 121
Appendix A **127**
Volume of water required to fill test section 127
Volume required at test pressure 128
Appendix B **129**
How to use charts for estimating the amount of pressure change for
    a change in test water temperature.. 129
Compressibility factor for water.. 132

Hydrostatic testing for pipe lines

After construction of a pipe line is completed, or if pipe has been replaced or relocated, it is necessary to hydrostatically test the pipe line to demonstrate that the pipe line has the strength required to meet the design conditions, and to verify that the pipe line is leak free.

The U.S. Federal Safety Regulations for Pipelines require that pipe lines used to transport hazardous or highly volatile liquids be tested at a pressure equal to 125% of the maximum allowable operating pressure (MAOP) for at least four continuous hours and for an additional four continuous hours at a pressure equal to 110% or more of the MAOP if the line is not visually inspected for leakage during the test. A design factor of 72% of the specified minimum yield strength (SMYS) of the pipe is used to determine the maximum allowable operating pressure. The requirement to test to 125% of the MAOP will therefore cause the pipe to be tested to a pressure equal to 90% of the SMYS of the pipe. See Section 3—Pipe Design for additional information on calculating the MAOP.

The Regulations for gas lines specify design factors based on the class location of the pipe line. The class location is determined by the number of buildings in a specified area on either side of the pipe line. Refer to Part 192.111 of the Minimum Federal Safety Standards for Gas Lines for the details on how to determine the class location. See Section 3, Pipe Design for a listing of the design factors for the different class locations. The Regulations specify that the test pressure must be maintained for at least eight hours and must be equal to at least 125% of the maximum allowable operating pressure. The Regulations should be consulted for the specific testing requirements as the Regulations are subject to change.

Usually, operators will specify a test pressure range from 90% to 95% of the SMYS of the pipe. Some will allow test pressures as high as 100% of the SMYS of the pipe and some will test to slightly beyond the SMYS of the pipe. Specifying a test pressure at least equal to 90% of the SMYS of the pipe will qualify it for the maximum allowable operating pressure. In some cases, the test pressure will be based on the minimum yield strength determined from the mill test reports.

Hydrostatic testing a pipe line is certainly a major operation and should be carefully planned. Most companies have hydrostatic test manuals that detail the procedures to be followed to complete the test. Usually, this work is performed by a hydrostatic testing contractor hired by the pipe line owner, or hydrostatic testing may be included as a part of the main construction contract.

One of the first steps in planning the hydrostatic test operation is to examine the elevation gradient. The gradient, along with the location of the water source, and the pipe design data, will be used to determine the length and number of test sections. Figure 1 shows a typical pipe line elevation gradient.

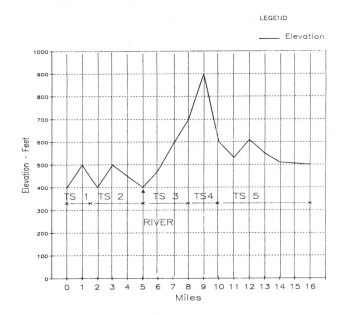

Figure 1. Pipe line elevation profile.

Where the pipe line traverses hilly terrain, the elevation gradient must be carefully considered in selecting the pipe line test segments. Different companies have differing philosophies on how to do this. Some limit the amount of elevation difference while others may specify a range of allowable percentages of the SMYS of the pipe—i.e., 90 to 95% or 90 to 100% of the SMYS of the pipe. In any case, the test gradient should be plotted to be sure the test pressure falls within the specified pressure limits.

The test gradient must be based on water head in feet, if water is used as the test medium. Water pressure may be converted to head by dividing the pressure by 0.433. If sea water is being used as the test medium, a factor of 0.445 should be used. This assumes a specific gravity of 1.02 for sea water. In any case, the factor of 0.433 should be modified according to the specific gravity of the test medium.

Let's assume that we have a line that is to be tested and the elevation at the test site is 1,000 feet and the elevation at the end of the line is 1,200 feet. Fresh water will be used as the test medium and it is desired to test the pipe to a minimum of 90% and a maximum of 96% of the SMYS of the pipe. The pipe is 30″ OD × 0.390″ wt API 5LX X60. There is a difference in elevation of 200 feet and this is equal to 86.6 psig (200 feet × 0.433 = 86.6 psig). A test pressure equal to 90% of SMYS is 1,404 psig. Since the test site is lower than the high end of the line, the 86.6 psig is added to 1,404 psig to obtain a test site pressure of 1,491 psig. The pressure at the end of the line will be 1,404 psig which equates to 90% of SMYS. The pressure at the low point equates to 96% of SMYS.

Now, let's assume a line has a high point elevation of 1,100 feet, a low elevation point of 1,000 feet, and the elevation at the test site is 1,050 feet. The test pressure at the high point will need to be 1,404 psig in order to meet the 90% of SMYS requirement. The pressure at the low point will be 1,447 psig, and the pressure at the test site will be 1,426 psig.

When testing offshore lines, the pressure at or below the water surface will be the same as at the low elevation point due to the offsetting external sub-sea pressure head. If the line previously described was laid offshore, the test pressure at the water surface would be 1,404 psig. The elevations are as follows:

Top of riser	+ 11 feet
Test site	+ 7 feet
Water surface	0 feet
Pipe depth	− 168 feet

The test pressure at the test site would be 1,404 − (7 × 0.445) or 1,401 psig. At the top of the riser, the test pressure would be 1,404 − (11 × 0.445) or 1,399 psig.

The typical profile shown in Figure 1 represents a pipe line that requires testing. The pipe line crosses a river at approximately MP 5 and river water will be used to test the line. Test sections 1 through 5 have been chosen as indicated. The length of sections 3 and 4 are limited by elevation difference. This line was designed to operate at 936 psig or 72% of the SMYS of 30″ × 0.375″ wt API 5LX X52 pipe for Class I locations, and 60% of the SMYS of 30″ × 0.390 wt API 5LX X60 pipe for Class II locations, and 50% of the SMYS of 30″ × 0.438″ wt API 5LX X65 for Class III locations.

The pipe line crosses one railroad, one highway, and the river. It also includes one main line block valve assembly. The valve assembly, river crossing, and road crossing were pre-tested before installation and are tested again after installation.

Four test manifolds were installed to facilitate filling the line and for isolating the test sections during the test operation. See Figure 2 for a typical pressure sectionalizing manifold. Two-way pigs were loaded at each of the intermediate manifolds. The pigs are moved by the fill water and are necessary to remove the air from the line.

It is generally a good idea to test the section most distant from the water source first. If it should rupture, then testing on intermediate sections can continue while repairs on the failed section are completed.

NOTE: Dimensions A thru D are based on the pipe dia. and pig length.

Figure 2. Typical pressure manifold. Courtesy Milbar Hydrotest.

Figure 3. Typical test manifold. Courtesy Milbar Hydrotest.

A - Suction Strainer
B - 6" Nominal Victaultic Pipe
C - Lowhead Pump
D - Water Filter
E - Flowmeter
F - Fill Pump
G - 6" Check Valve

Figure 4. Typical fill site. Courtesy Milbar Hydrotest.

T1– Thermometer in Thermo Well on Inlet Manifold Riser
T2– Thermometer in Backfill next to 48" Pipe line (Recording Thermometer)
T3– Thermometer in Protected Ambient Atmosphere (Recording Thermometer) – not shown

Figure 5. Test thermometer placement. Courtesy Milbar Hydrotest.

The pumps used to fill the line should have sufficient capacity to fill the line at a rate of about ¹/₂ mile per hour. Water filters should be used and are normally equipped with one hundred mesh screens. See Figure 4 for a typical fill site arrangement. Some companies may specify a finer mesh screen. The filling unit should be equipped with a flow meter to measure the amount of water pumped into the pipe line. Temperature recorders will be used to record ambient temperature, the temperature of the pipe and water, and ground temperature. See Figure 5 for typical test thermometer placement. A pressure recorder will be used to provide a recording of the pipe line test pressure. A deadweight tester will be used to calibrate the pressure recorder. A high pressure pump capable of delivering 70 to 150 gpm at a pressure exceeding the required test pressure will be used in the final pressuring operation. Small high pressure units with a capacity of 6 to 30 gpm may be used for short sections of large diameter lines and smaller diameter lines.

Deadweight pressure and temperature readings are recorded after the prescribed test pressure has been reached and the pressures and temperatures have stabilized. A pressure vs. time plot may also be made. Readings are usually made at fifteen minute intervals for the first hour and at thirty minute intervals thereafter. The procedure for accepting a leakage test will vary from company to company.

Some will accept the leakage test if there is no pressure drop in a three hour period.

A pressure-volume plot may be required, especially if the test pressures approach or exceed the SMYS of the pipe. The plot is made manually during the pressuring operation by recording pump strokes on the X-axis and pressure on the Y-axis. A straight line will be produced until the plastic range of the pipe is reached or until a leak occurs. This type of plot is also useful should a joint of lighter wall thickness or lower yield strength pipe be inadvertently placed in the pipe line.

Small leaks during the testing operation can be difficult to locate. A change in the water/pipe temperature may give the appearance of a leak. If the temperature of the pipe/water decreases, the test pressure will decrease. An increase in water/pipe temperature will cause the test pressure to increase. The effect of a temperature change may be estimated using the equations and data contained in Appendixes A and B. To achieve any degree of accuracy in these calculations, it will be necessary to have accurate temperature and pressure readings.

The pipe line test temperature may be affected by river/creek crossings along the pipe line route. Also, if cold river water is used, there will be a changing temperature gradient. Air, either trapped or entrained, will also affect the

Figure 6. Typical pressure unit set-up. Courtesy Milbar Hydrotest.

pressure-temperature calculations since the coefficient of expansion for air is not the same as for water. If the calculations are made and the decrease in pressure cannot be attributed to a temperature change, then it may be necessary to sectionalize the test section by manifolding and repressuring the shorter segments to facilitate locating the leak. Once the leak has been located and repaired, testing operations can resume.

Air compressors will be required to remove the water once the testing is complete. The maximum head for this particular project was 450 feet (195 psig). Allowance will need to be included for friction loss in the cross-over manifolds. The compressors will need to have sufficient capacity to remove the water at about the same rate as the fill rate. Most portable compressors are limited to approximately 125 psig. In this case, booster compressors will be needed for the 195 + psig. See Figure 7 for a typical dewatering connection.

Upon completion of testing, the test pressure is bled off to leave approximately 10 to 20 psig at the high point in the test section. Dewatering may be accomplished using the air compressors discussed earlier. Some companies may elect to displace the test water with the material that will normally be pumped through the line. This is usually the case with crude oil and refined product pipe lines. Pipe lines used to

transport natural gas and certain chemicals are usually dewatered using compressed air. Care must be exercised in the dewatering operation to make sure that no air is introduced into the test section, thus minimizing the possibility of air locks. An air lock is probably the most severe problem involved in dewatering. Air locks are caused by air accumulating in the downhill leg and water accumulating in the uphill leg, which creates a manometer in the pipe line. In some instances, extremely high pressures may be required to overcome the manometer. Care should be exercised to be sure the maximum allowable pipe line operating pressure is not exceeded. It may also be necessary to tap the line and vent air at the high points.

If it is necessary to pipe the water away from the right-of-way, welded pipe should be used. The pipe must be securely anchored. The use of tractors and skids will probably provide sufficient anchoring.

Always use a valve to control the amount of water being bled from the pipe line.

If there will be a delay in placing the pipe line in operation, and it is decided to leave the test water in the line until operations begin, consideration should be given to the use of a corrosion inhibitor. Long term storage of the line may necessitate the use of oxygen scavengers in addition to the inhibitor.

A– Primary Air Comp. 125 psi
B– Two stage booster comp. 125 to 900 psig
C– 3" High Pr. Flex Hose or 3" pipe w/swivel jts.
D– 2" Full Open Plug Valve
E– Pressure Gauge
F– Pressure Recorder

Figure 7. Dewatering arrangement. Courtesy Milbar Hydrotest.

FIELD PRESSURE & TEST REPORT

Page _____ of _____

Contractor or Company: _____

Pipe Line Description: _____

Section Tested: From: _____ To: _____

Test Section No.: _____ Length: _____

Type and Size of Pipe: _____ O. D. x _____ W.T. x Grade _____ Manufacturer: _____

Pressure Unit No.: _____ Gallons per Stroke: _____ Fill Unit No.: _____

Pressure Unit Location: _____ Water Source: _____

Time and Date Test Started: _____ AM / PM _____ Test Pressure (Maximum): _____ PSIG

Time and Date Test Ended: _____ AM / PM _____ Pressure Volume Plot: Yes _____ No _____

Section Accepted _____ Section Leaking _____ Section Ruptured _____

DEADWEIGHT RECORDED READING (PSIG)

TIME	PRESSURE	TEMP DEG.F	TIME	PRESSURE	TEMP DEG. F	TIME	PRESSURE	TEMP DEG. F
Date: _____								
AM PM			AM PM			AM PM		

Strokes per 10 psi: _____ Remarks: _____

_____

_____

_____

Milbar Supt. _____ Inspector _____

Witness: 1. _____ 2. _____

Gauges and Recorders Last Tested _____ Deadweight Serial No. _____

Figure 8. Typical test report. Courtesy Milbar Hydrotest.

If after dewatering, residual moisture in the pipe line poses a problem, it will be necessary to clean and dry the pipe line. See Section 6 for drying details.

A permit to discharge the water may be required, especially if there is a possibility the water may enter a stream of water. This should be determined and permits obtained, if required, in the planning phase of the testing operation. A "splash" plate or bales of hay may be used when dewatering to prevent erosion.

The most important aspect of any test operation is complete and accurate documentation, since it will become a permanent record which must be retained for as long as the facility tested remains in operation. Federal Pipeline Safety Regulations for pipe lines transporting liquids require that these records include the following information:

- The pressure recording charts;
- Test instrument calibration data;
- The names of the operator, of the person responsible for making the test, and of the test company used, if any;
- The date and time of the test;
- The minimum test pressure;

- The test medium;
- A description of the facility tested and the test apparatus;
- An explanation of any pressure discontinuities, including test failures, that appear on the pressure recording charts; and,
- Where the elevation differences in the section under test exceed 100 feet, a profile of the pipe line that shows the elevation and test sites over the entire length of the test section.

The records required by the Federal Pipeline Safety Regulations for pipe lines transporting gas are similar to those required for pipe lines used to transport liquids.

A typical test report is shown in Figure 8. If failures occur during the test, a report documenting the failure and the suspected reasons for the failure should be completed. A typical form is shown in Figure 9.

A daily operating log should be used to record activities associated with the test, operating status of the test equipment in use, engine rpm's, and any unusual circumstances that occur during the test. A typical form is shown in Figure 10.

PIPE LINE FAILURE REPORT

1. Company _____

2. Section tested_____

3. Time of failure_____AM, PM. Date_____

4. Location of failure:_____

5. Pressure, psig, at point of failure_____

6. Description of failure: Leak_____, Break_____, Length of failure_____

7. If leak, fill in blanks:_____gallons lost per hour, _____pounds lost per hour.

8. Describe any peculiarities or defects on failed part, such as corrosion or evidence of prior damage, etc.
_____

9. Possible cause of failure_____

10. Pipe size: _____OD x _____Wt., Grade_____Mfg. by_____

11. Repairs: Pipe installed_____OD x _____Wt., Grade_____

   Mfg. by_____, Length of joint or pup_____

12. Date repaired_____19 _____ By_____

13. Remarks (Damage to property, persons injured, etc.)_____
_____
_____
_____
_____

Figure 9. Pipe line failure report. Courtesy Milbar Hydrotest.

DAILY OPERATING REPORT

COMPANY & CONTRACTOR _____ DATE_____

LOCATION (City) _____ CONTRACT NO. _____

PIPE LINE DESCRIPTION: _____

_____

PERSONNEL: _____

_____

EQUIPMENT: _____

_____

_____

ACTIVITIES: _____

_____

_____

_____

_____

_____

_____

_____

_____

_____

_____

_____

_____

_____

_____

_____

_____

_____

_____

_____

_____

_____

_____

_____

_____

_____	_____	_____
COMPANY REP.	CONTRACTOR REP.	MILBAR SUPT.

Figure 10. Daily Operation Report. Courtesy Milbar Hydrotest.

APPENDIX A

VOLUME OF WATER REQUIRED TO FILL TEST SECTION:

$V = L \times (3.1416 \times d^2/4) \times 12/231$ (1)

$V = 0.0408 \times d^2 \times L$

Where: L = Length of test section, feet.
d = Inside diameter of pipe, inches.
V = Gallons required to fill @ 0 psig.

Volume Required at Test Pressure

$$V_{tp} = V \times F_{wp} \times F_{pp} \times F_{pwt} \qquad (2)$$

V_{tp} = Gallons contained in test section at test pressure P, and temperature T – °F.

F_{wp} = Factor to correct for the compressibility of water due to increasing pressure from 0 psig to test pressure P, psig

F_{pp} = Factor to correct for volume change in pipe line due to pressure increase from 0 psig to test pressure P, psig

F_{pwt} = Factor to correct for the change in water volume and pipe volume due to change in pipe and water temperature from base of 60°F to pipe and water test temperature, T – °F.

$$F_{wp} = 1/[1 - (4.5 \times 10^{-5}) \times (P/14.73)] \qquad (3)$$

P = Test pressure, psig.

$$F_{pp} = 1 + [(D/t) \times (0.91P/30 \times 10^{6})] + \\ [3.6 \times 10^{-6}(T - 60)] \qquad (4)$$

D = Outside diameter of pipe, inches.
t = Pipe wall thickness, inches.
T = Pipe temperature, °F.

$$F_{pwt} = F_{pt}/F_{wt} \qquad (5)$$

F_{pt} = Factor to correct for change in pipe volume due to thermal expansion of the pipe from base temperature of 60°F.

$$F_{pt} = 1 + [(T - 60) \times 18.2 \times 10^{-6}] \qquad (6)$$

F_{wt} = Factor to correct for thermal change in specific water volume from 60°F to test water temperature. Refer to Table 1.

SAMPLE CALCULATION:

Pipe Size: 10.750″ OD × 0.279″ w.t. X-52
Length: 5 miles
Test pressure: 2,430 psig
Temperature: 50°F

Use equation 1 to determine the initial line fill volume, V.

$$V = 0.0408 \times (10.75 - (2 \times 0.279))^2 \times 5,280 \times 5$$

$$= 111,888 \text{ gallons}$$

Volume required to achieve test pressure = $V_{tp} = V \times F_{wp} \times F_{pp} \times F_{pwt}$.

$$F_{wp} = 1/1 - [(4.5 \times 10^{-5}) \times (2430/14.73)]$$

$$= 1.007479.$$

$$F_{pp} = 1 + [(10.75/0.279) \times (0.91 \times 2430/(30 \times 10^{6}))] \\ + [3.6 \times 10^{-6} \times (50 - 60)]$$

$$= 1.002804$$

$$F_{pwt} = F_{pt}/F_{wt}$$

$$F_{pt} = 1 + [(50 - 60) \times 18.2 \times 10^{-6}]$$

$$= 0.999818$$

$F_{wt} = 0.9993061$ (from Table 1)

$$F_{pwt} = 0.999818/0.9993061$$

$$= 1.000512$$

$$V_{tp} = 111,888 \times 1.007457 \times 1.002804 \times 1.000512$$

$$= 113,099 \text{ gals.}$$

Incremental volume required to reach test pressure, P.

$$= 113,099 - 111,888 = 1,211 \text{ gals.}$$

After a period of time, the test pressure, P has decreased to 2,418 psig and the temperature of the pipe and test water has decreased to 48°F. The calculation procedure previously described may be repeated to determine if the pressure decrease is attributable to the decrease in temperature.

V = 111,888 gals.
P_1 = 2,422 psig
T_1 = 48°F.
F_{wp1} = 1.007454
F_{pp1} = 1.002796

$F_{pt1} = 0.999781$

$F_{wt1} = 0.999217$ (from Table 1)

$F_{pwt1} = 0.999781/0.999217 = 1.000565$

$V_{tp1} = 111,888 \times 1.007454 \times 1.002796 \times 1.00565$ (@ 2,422 psig and 48° F)

& 48° F.)

$V_{tp1} = 113,101$ (Volume at reduced temperature - represents volume that would be required to raise test pressure to original value at 48°F)

$V_{tp} = 113,099$ (Volume at initial test conditions)

To restore the pressure to the original test pressure, two gallons of water could be added to the test section. If more than two gallons is required to restore the test pressure, there is the possibility that a leak exists.

When making these calculations, care must be exercised to be sure that accurate temperature and pressure data is available. Where test sections are long, it is very likely that the temperature will not be uniform throughout the test section and thus affect the accuracy of the calculated results.

Table 1

Fwt—Factor to correct for the thermal change in the specific volume of water from 60°F to the test water temperature

Temp. (°F)	Fwt	Temp. (°F)	Fwt
35	0.9990777	69	1.0009290
36	0.9990590	70	1.0010364
37	0.9990458	71	1.0011696
38	0.9990375	72	1.0012832
39	0.9990340	73	1.0014229
40	0.9990357	74	1.0015420
41	0.9990421	75	1.0016883
42	0.9990536	76	1.0018130
43	0.9990694	77	1.0019657
44	0.9990903	78	1.0021222
45	0.9991150	79	1.0022552
46	0.9991451	80	1.0024178
47	0.9991791	81	1.0025561
48	0.9992168	82	1.0027251
49	0.9992599	83	1.0028684
50	0.9993061	84	1.0030435
51	0.9993615	85	1.0031919
52	0.9994112	86	1.0033730
53	0.9994715	87	1.0035573
54	0.9995322	88	1.0037133
55	0.9996046	89	1.0039034
56	0.9996683	90	1.0040642
57	0.9997488	91	1.0042601
58	0.9998191	92	1.0044357
59	0.9999074	93	1.0046271
60	1.0000000	94	1.0047972
61	1.0000803	95	1.0050043
62	1.0001805	96	1.0052142
63	1.0002671	97	1.0053915
64	1.0003746	98	1.0056067
65	1.0004674	99	1.0057884
66	1.0005823	100	1.0060090
67	1.0006811	101	1.0061949
68	1.0008031	102	1.0064207
		103	1.0066108
		104	1.0068417

APPENDIX B

How to use charts for estimating the amount of pressure change for a change in test water temperature

Example.

Pipe data	18″ OD × 0.375″ wt
Water temp. at beginning of test	70°F
Water temp. at time T	66°F
Test pressure	1,800 psig

Calculate

$D/t = 18/0.375 = 48$

where D = Pipe OD, in.
 t = Pipe wall thickness, in.

Calculate

average temperature = $(70 + 66)/2 = 68°F$

Use the chart in Figure 3. Enter at 68°F and at the intersection with D/t line representing 48, read 23 psig/°F.

Figure 1.

Figure 2.

Figure 3.

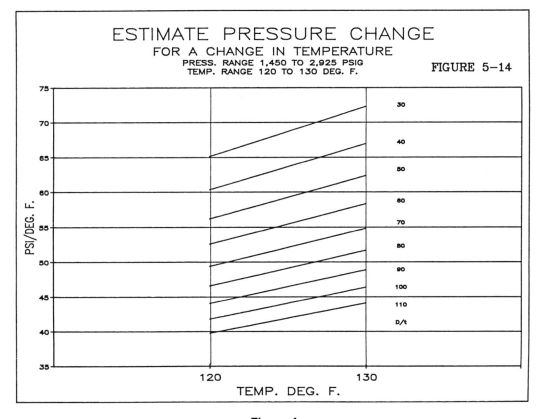

Figure 4.

Basis for chart development

$$dP = [B - 2a]/[(D \times (1 - v^2))/Et + C]$$

where dP = psig change per °C
    B = Coefficient of expansion of water
    D = Pipe OD, in.
    E = Modulus of elasticity for steel, 30×10^6
    v = Poisson's ratio, 0.3

C = Compressibility factor for water, cu in./cu in./psig
a = Coefficient of expansion for steel, 1.116×10^{-5} per °C
t = Pipe wall thickness, in.
T = Temperature, °C

$$B \times 10^6 = -64.268 + 17.0105 \times T - 0.20369 \times T^2 + 0.0016048 \times T^3$$

Compressibility factor for water

Pressure Range 1,450 psig to 2,925 psig

Temperature (°C)	Compressibility Factor (cu in./cu in./psig)
0	3.35×10^{-6}
10	3.14×10^{-6}
20	3.01×10^{-6}
50	2.89×10^{-6}

From PIPECALC™ 2.0: *Practical Pipeline Hydraulics*, Gulf Publishing Company, Houston, Texas.

6: Pipe Line Drying

Pipe Line Dewatering, Cleaning, and Drying.............134
Dewatering.. 134
Cleaning pipe lines 135
Brush pig run with gas.. 135
Brush pig run with liquid.. 135
Internal sand blasting.. 135
Chemical cleaning 136
Pipe line drying.. 136
Moisture content of air.. 138
Commissioning petrochemical pipe lines........................... 140
Vacuum drying 143

PIPE LINE DEWATERING, CLEANING, AND DRYING*

Newly constructed pipe lines are typically hydrostatically tested, using water as the test medium, upon completion of construction. Older lines already in service may be re-tested either to confirm an earlier test or to qualify the line for a higher operating pressure. Once the hydrostatic testing has been completed, it is necessary to remove the water from the line and place the line in service. Dewatering can be a simple process or, if the procedure is not properly planned, a difficult one. Pipe lines used to transport crude oil and/or refined products will probably only require removal of the test water before the line is placed in service. If the pipe line will be used to transport materials which must meet a specified dryness requirement, the pipe line will need to be dewatered, cleaned, and dried. Pipe lines used to transport natural gas will need some drying, depending on the operating pressure and the location of the line, to prevent the formation of hydrates. Other pipe lines may require drying to protect the pipe from internal corrosion caused by the formation of corrosive acids, such as carbonic acid in the case of carbon dioxide pipe lines.

* Based on a paper entitled "The State of the Art of Drying Large Diameter Gas Pipelines after Hydrotest," by Marvin D. Powers, Pipeline Dehydrators, Inc., Houston, Texas.

Dewatering

Dewatering is considered to commence with the running of the first pig after hydrostatic testing is completed and begins with the insertion of a displacer, commonly referred to as a pig, in the pipe line. The dewatering pig may be pushed through the pipe line with crude oil or other petroleum product if no drying is required. If the pipe line is to be cleaned and/or dried, the pig will be pushed by either compressed air or gas. In either case, proper precautions must be taken to be sure the test water is properly disposed of and that any required water discharge permits are obtained ahead of the dewatering operation.

Several types of pigs may be used for the dewatering phase and pig selection should depend upon the design characteristics of the pipe line and the degree of cleaning, if any, that is desired. Ideally, the pig will form a perfect seal with the inner periphery of the pipe and will not allow any of the material behind the pig to leak past the pig and comingle with the test water ahead of the pig.

Normally, the pig will move through the pipe line without difficulty. However, there are many opportunities for problems to develop, causing the pig to stick or even disintegrate.

Sticking may be caused by any or a combination of the following things:

- Pig is incorrect length and cannot negotiate check valves, tees, and bends;
- Pig is too large for heavy wall sections of the line;
- Pipe line may be equipped with reduced opening valves;
- Excess debris or construction material may be left in the pipe line;

- Air/gas may bypass the pig and create an air lock condition; and,
- Other unknown reasons.

Air locks are more likely to occur in hilly country than in flat land. Air locks occur when the accumulated static heads are greater than the available displacing pressure. In some cases, a pressure greater than the maximum allowable pipe line pressure would be required to overcome the air lock. Usually, the air lock occurs where air/gas has bypassed the pig and the downhill legs of the pipe line are filled with air and the uphill legs of the pipe line are filled with water. The air/gas can get in front of the dewatering pig in several ways such as:

- Poor filling techniques;
- Poor dewatering procedure such as draining water from the line at low points;
- Air/gas bypassing the pig because the pig is too small for the pipe line;
- Air/gas bypassing the pig before it is launched;
- Air/gas bypassing the pig in a fitting such as in a tee, or a steel shaft pig in a short radius bend; and,
- Other unknown reasons.

When an air lock condition occurs, it is necessary to either increase the displacing pressure, or remove air/gas through existing vents or other connections at high points in front of the pig.

For these reasons, it is important that the dewatering process be carefully planned, especially for pipe lines located in hilly country.

Cleaning pipe lines

It is believed there are no written specifications to define the degree of cleaning for a pipe line since there is a question of the definition of a clean pipe line and furthermore, a method for measuring the cleanliness has not been developed. It is known, however, that cleaning a pipe line does at least four good things. Cleaning the pipe line will:

- Improve flow efficiency because of a smoother pipe wall
- Reduce product contamination and formation of hydrates
- Reduce abrasive damage to pipe line appurtenances such as valves and instruments
- Facilitate pipe line drying

Internal cleaning of the pipe line may be accomplished by any of or a combination of the following methods:

- Running a brush pig with air, gas, or liquid
- Internal sand blasting
- Chemical cleaning
- Purging with air or gas followed by a liquid flush

Brush pig run with gas

One brush pig pushed by gas or air will not usually remove a significant amount of small debris from the pipe line, particularly if the debris is wet. However, the scratching action of a good brush pig will more evenly distribute the loose debris and leave a smoother pipe wall, which will improve pipe line flow efficiency.

Brush pig run with liquid

A brush pig displaced by water is a very efficient means for removing debris from the pipe line provided certain precautions are taken. Experience has shown that a pig velocity greater than three feet per second is desired. A pig, such as the Power Brush Pig, that allows some of the liquid to bypass the pig to keep the bristles clean and move the loose debris into suspension in the liquid in front of the pig is also desirable. The debris must be free of any sticky material such as oil, grease, or paraffin.

It is not necessary that the pipe line be completely filled with liquid. Short slugs of liquid batched between two pigs is usually preferable as long as there is enough water to keep the pigs apart and maintain a velocity greater than three feet per second. Short slugs will minimize the water/ debris disposal problems and will allow for higher velocities with lower pressures.

Disposal of the dirty liquid may present a problem, depending upon the location. As indicated earlier, environmental aspects should be investigated thoroughly during the pre-planning phase. Usually, rust will settle out of still water in about one hour, leaving an almost clear liquid. If the dumping of semi-clear water is acceptable, then the problem may be solved by the use of holding tanks.

The debris-holding ability of the propelling fluid is proportional to the viscosity/density of the propelling fluid. The greater the viscosity/density, the higher the holding ability. It has been reported that gels have been developed and used that are supposedly more effective than water.

Internal sand blasting

Internal sand blasting is an extremely effective method for cleaning pipe lines. This method was extensively used when natural gas was inexpensive and there were no environmental controls. Some short, small-diameter pipe lines are still being internally sand blasted by using air or nitrogen to propel the sand. This procedure, if properly executed, will leave the pipe line dry upon completion.

After the pipe line has been dewatered, clay is blown into short sections of the pipe line with a dry gas to absorb the remaining water. The correct type of sand is then blown through the dust dry section at a high velocity to remove the clay, rust, and mill scale. Caution must be taken to ensure that short sections are cleaned each time so that high velocities may be maintained; that large quantities of dry air or gas are available to sustain the high velocities; that the pipe line is dust dry; that all sand has been removed from the pipe line after the sand blasting is complete; and, that the pipe line facilities will not be damaged by the high velocity sand.

Chemical cleaning

Cleaning with chemicals is typically used for smaller diameter lines. A properly planned chemical cleaning operation can produce a very clean pipe line. Chemical cleaning will remove rust from any internal pits. Chemical cleaning is accomplished by pushing a batch of hydrochloric acid and water for rust removal, a neutralizer, and a passivator, all separated by pigs, through the pipe line. This method will leave a bright metal finish on the interior of the pipe line. If a pipe line is being converted for a different service, such as a crude oil line being converted to chemical service, a batch of detergent can be added behind the cleaning process to remove any traces of the crude oil or other product.

Typically, chemical cleaning is more involved than other cleaning processes in that some special equipment will be required to safely handle the chemicals. Disposing of the spent material will also present a problem. Care must be exercised in the selection of pigs to be sure they will not be consumed by the cleaning solution. Personnel safety must be carefully considered when choosing this method of pipe line cleaning.

Purging with gas. This method is seldom used for pipe line cleaning since the amount of debris that will be removed is small. A high gas velocity is required.

Pipe line drying

Pipe lines used to transport petrochemicals such as propylene and ethylene must be dried in order for the delivered product to meet moisture specifications. Natural gas pipe lines are usually dried to a lesser extent to prevent the formation of hydrates. It is not unusual for a petrochemical line to be dried to a dew point of $-80°F$. A typical dew point for a propylene pipe line will be $-70°F$. A carbon dioxide pipe line might be typically dried to a $-40°F$ dew point. The natural gas industry specifies dryness in pounds of water per million standard cubic feet of gas. The table at the end of this section will enable you to convert from one method to the other.

Dew point, by definition, is the temperature at which water vapor begins to condense out of a gas at atmospheric pressure. For example, at atmospheric pressure, water vapor begins to condense out of a gas that has a moisture content of seven pounds of water per million standard cubic feet at $-39°F$. Therefore, a gas that has a moisture content of seven pounds of water per million standard cubic feet has a dew point of $-39°F$.

Air at $+70°F$ with a humidity of 100% has a dew point of $+70°F$ and holds approximately 1,253 pounds of water (150 gallons) per million standard cubic feet.

The most common methods for drying pipe lines are as follows:

- Drying with super dry air
- Drying with methanol
- Drying with inert gas such as nitrogen
- Internal sand blasting
- Drying with the medium to be transported
- Vacuum drying

All of these methods may be applied to pipe line drying depending on the particular line and amount of dryness required. No single method can be considered ideal for all situations. Many times, a combination of two or more methods will be used to achieve a dry pipe line at the least cost. The first three methods are probably the most economical and technically feasible for most pipe line drying applications.

Before any type of drying operation commences, it will be necessary to clean the line using one of the previously described processes. If the rust and mill scale are not removed from the pipe wall, moisture will remain trapped and will bleed out over a long period of time. A pipe line can be dried without cleaning; however, the cost and time required will be great.

In *drying with super dry air*, soft foam pigs pushed by dry air are used to absorb any free water remaining in the pipe line after dewatering. After the line is dust dry, wire brush pigs are run to remove any water bearing debris from the pipe wall. The wire brush pigs are then followed by soft foam pigs to absorb the loosened debris. Near the end of this phase, the pigs may be weighed prior to insertion and after removal to monitor the amount of debris that is being removed. The first pigs run through the line will naturally weigh much more than their clean weight. As the cleaning and drying progress, the pig weights will approach their original weight. Pig color will also give some indication of how the debris removal is proceeding.

Dew-point readings will need to be made to determine when the line has been dried to the specified dew point.

It will be necessary to give special consideration to laterals, by-passes, and valve body cavities as any free water trapped here could affect the final dew point readings.

Drying with super dry air provides internal corrosion protection if the line is to remain out of service for some period of time before it is placed in service.

Methanol drying relies on the hygroscopic effect of the methanol. Any remaining moisture in the line will be absorbed by batches of methanol pushed through the line with either gas or dry air. Pigs are used to separate the methanol batches from the displacing medium.[1] Methanol drying usually requires fewer pig runs and, consequently, less line cleaning is accomplished. Pure methanol is expensive and sometimes a 96% methanol/water mix is used. Since the methanol mix contains water, some water will be left in the pipe line. Some of the methanol will vaporize in the pipe line and will be absorbed by the displacing medium. Towards the end of the line, the moisture content of the methanol will increase, which in turn reduces the amount of water that it can absorb. If the pressure used in the drying operation is too high, hydrate formation can occur, usually at the far end of the line. If natural gas is being used to push the methanol batches, it will probably be necessary to flare some of the line fill volume to be sure that no methanol impurities are contained in the gas.[1]

Methanol run with a dry gas will absorb most of the water and facilitate the vaporization of the remaining water. Soft swabs run through a line with a dry purge gas will ac-celerate the evaporation of remaining methanol/water solution.

Also of great concern with the methanol method of drying is the fact that explosive mixtures can easily be formed, whether gas or air is used to displace the methanol batches. It may be desirable to use an inert gas such as nitrogen to buffer the methanol batch from the air or gas used to displace the methanol batch. However, many pipe lines have been dried without the use of nitrogen buffers with no adverse results. The air/methanol mixture is also highly poisonous and corrosive.[1]

Plans will need to be made for proper disposal of the spent methanol.

If internal corrosion protection is desired, then another drying method should be considered. If the pipe line is to be used to transport sour gas, the methanol drying method should be carefully evaluated before it is used.

Drying with nitrogen is accomplished in much the same manner as when using super dry air. Nitrogen drying will cost more than super dry air by a factor of approximately 50%. Also, super dry air is plentiful and non-polluting.

Internal sand blasting will leave the pipe line clean and dry. The procedure for internal sand blasting is described earlier in this section.

Drying with natural gas requires large volumes of gas. This method is slow and not very effective unless the line is thoroughly cleaned by one of the cleaning processes previously described. If, however, the gas being used to dry with can be blended with another dry gas stream and sold or used, then this is an economical method for pipe line drying. The cost of the gas that will be used to purge the line during the drying process should be weighed against the cost of using super dry air. The Btu value of a volume of natural gas is approximately fifteen times greater than that of the fuel required to produce an equal volume of super dry air.

Vacuum drying is a slow process and all free water should be removed from the pipe line before drying begins. This method appears to be used infrequently, and perhaps only offshore.

If the pipe line has been properly cleaned by the water slug method using brush pigs run with liquid, drying can be accomplished by running soft foam pigs with dry air or gas to remove any free water left in the pipe line. This will usually produce a pipe line dry enough for natural gas operations. If additional drying is desired, it can be accomplished by using methanol or super dry air.

Reference

1. Kopp, *Pipe Line Industry*, October 1981, January 1986.

Moisture content of air

DEW POINT °F @ 14.7 PSIA	DEW POINT °C @ 14.7 PSIA	PPM BY VOLUME	PPM BY WEIGHT	LBS. WATER MMSCF DRY AIR	DEW POINT °F @ 14.7 PSIA	DEW POINT °C @ 14.7 PSIA	PPM BY VOLUME	PPM BY WEIGHT	LBS. WATER MMSCF DRY AIR	DEW POINT °F @ 14.7 PSIA	DEW POINT °C @ 14.7 PSIA	PPM BY VOLUME	PPM BY WEIGHT	LBS. WATER MMSCF DRY AIR
-110	-78.9	0.63	0.39	.032	-70	-56.7	16.6	10.3	.83	-30	-34.4	235	146	11.8
-109	-78.3	0.69	0.43	.034	-69	-56.1	17.9	11.1	.96	-29	-33.9	250	155	12.5
-108	-77.8	0.75	0.47	.038	-68	-55.6	19.2	11.9	1.0	-28	-33.3	265	165	13.3
-107	-77.2	0.82	0.51	.041	-67	-55.0	20.6	12.8	1.03	-27	-32.8	283	176	14.2
-106	-76.7	0.88	0.55	.044	-66	-54.4	22.1	13.7	1.1	-26	-32.2	300	186	15.0
-105	-76.1	1.00	0.62	.050	-65	-53.9	23.6	14.7	1.2	-25	-31.7	317	197	15.9
-104	-75.5	1.08	0.67	.054	-64	-53.3	25.6	15.9	1.3	-24	-31.1	338	210	16.9
-103	-75.0	1.18	0.73	.059	-63	-52.8	27.5	17.1	1.4	-23	-30.6	358	223	17.9
-102	-74.4	1.29	0.80	.065	-62	-52.2	29.4	18.3	1.5	-22	-30.0	378	235	18.9
-101	-73.9	1.40	0.87	.070	-61	-51.7	31.7	19.7	1.6	-21	-28.4	400	249	20.0
-100	-73.3	1.53	0.95	.077	-60	-51.1	34.0	21.1	1.7	-20	-28.9	422	212	21.1
-99	-72.8	1.66	1.03	.083	-59	-50.6	36.5	22.7	1.8	-19	-28.3	448	278	22.4
-98	-72.2	1.81	1.12	.091	-58	-50.0	39.0	24.2	2.0	-18	-27.8	475	295	23.8
-97	-71.7	1.96	1.22	.098	-57	-49.4	41.8	26.0	2.1	-17	-27.2	500	311	25.0
-96	-71.1	2.15	1.34	.108	-56	-48.9	44.6	27.7	2.2	-16	-26.7	530	329	26.5
-95	-70.6	2.35	1.46	.118	-55	-48.3	48.0	30	2.4	-15	-26.1	560	348	28.0
-94	-70	2.54	1.58	.127	-54	-47.8	51	32	2.6	-14	-25.6	590	367	29.5
-93	-69.4	2.76	1.72	.138	-53	-47.2	55	34	2.8	-13	-25.0	630	391	31.5
-92	-68.9	3.00	1.86	.150	-52	-46.7	59	37	2.9	-12	-24.4	660	410	33.0
-91	-68.3	3.28	2.04	.164	-51	-46.1	62	39	3.1	-11	-23.9	700	435	35.0
-90	-67.8	3.53	2.19	.177	-50	-45.6	67	42	3.4	-10	-22.3	740	460	37.0
-89	-67.2	3.84	2.39	.192	-49	-45.0	72	45	3.6	-9	-22.8	780	485	39.0
-88	-66.7	4.15	2.58	.221	-48	-44.4	76	47	3.8	-8	-22.2	820	509	41.0
-87	-66.1	4.50	2.80	.225	-47	-43.9	82	51	4.1	-7	-21.7	870	541	43.5
-86	-65.6	4.78	2.97	.239	-46	-43.3	87	54	4.4	-6	-21.1	920	572	46.0
-85	-65.0	5.30	3.30	.265	-45	-42.8	92	57	4.6	-5	-20.6	970	603	48.5
-84	-64.4	5.70	3.50	.285	-44	-42.2	98	61	4.9	-4	-20.0	1020	634	51.0
-83	-63.9	6.2	3.9	.31	-43	-41.7	105	65	5.3	-3	-19.4	1087	675	54.4
-82	-63.3	6.6	4.1	.33	-42	-41.1	113	70	5.7	-2	-18.9	1153	716	57.7
-81	-62.8	7.2	4.5	.36	-41	-40.6	119	74	6.0	-1	-18.3	1220	758	61.0
-80	-62.2	7.8	4.8	.39	-40	-40.0	125	80	6.4					
-79	-61.7	8.4	5.2	.42	-39	-39.4	136	85	6.8					
-78	-61.1	9.1	5.7	.46	-38	-38.9	144	89	7.2					
-77	-60.6	9.8	6.1	.49	-37	-38.3	153	95	7.7					
-76	-60.0	10.5	6.5	.53	-36	-37.9	164	102	8.2					
-75	-59.4	11.4	7.1	.57	-35	-37.2	174	108	8.7					
-74	-58.9	12.3	7.6	.62	-34	-36.7	185	115	9.3					
-73	-58.3	13.3	8.3	.67	-33	-36.1	196	122	9.8					
-72	-57.9	14.3	8.9	.72	-32	-35.6	210	131	10.5					
-71	-57.2	15.4	9.6	.77	-31	-35.0	222	140	11.1					

(table continued on next page)

DEW POINT °F @ 14.7 PSIA	DEW POINT °C @ 14.7 PSIA	PPM BY VOLUME	PPM BY WEIGHT	LBS. WATER MMSCF DRY AIR
0	-17.8	1299	807	64.9
1	-17.2	1378	856	68.9
2	-16.7	1458	906	74.3
3	-16.1	1537	955	76.9
4	-15.6	1616	1004	80.8
5	-15.0	1695	1053	84.8
6	-14.4	1775	1103	88.8
7	-13.9	1854	1152	92.7
8	-13.3	1933	1201	96.7
9	-12.8	2053	1276	103
10	-12.2	2174	1351	109
11	-11.7	2294	1425	115
12	-11.1	2415	1501	121
13	-10.6	2535	1575	127
14	-10.0	2656	1650	133
15	-9.4	2776	1725	139
16	-8.9	2897	1800	145
17	-8.3	3017	1875	151
18	-7.8	3197	1986	160
19	-7.2	3376	2098	169
20	-6.7	3556	2209	178
21	-6.1	3735	2321	187
22	-5.6	3915	2432	196
23	-5.0	4094	2544	205
24	-4.4	4274	2656	214
25	-3.9	4453	2767	223
26	-3.3	4633	2879	232
27	-2.8	4883	3034	244
28	-2.2	5133	3189	357
29	-1.7	5383	3345	269
30	-1.1	5633	3400	282
31	-0.56	5883	3655	294
32	0	6133	3811	307
33	0.56	6383	3966	319
34	1.1	6633	4121	332
35	1.7	6883	4277	344
36	2.2	7281	4524	364
37	2.8	7679	4771	384
38	3.3	8077	5019	404
39	3.9	8475	5266	424

DEW POINT °F @ 14.7 PSIA	DEW POINT °C @ 14.7 PSIA	PPM BY VOLUME	PPM BY WEIGHT	LBS. WATER MMSCF DRY AIR
40	4.4	8874	5514	444
41	5.0	9272	5761	464
42	5.6	9670	6008	484
43	6.1	10068	6256	503
44	6.7	10466	6503	522
45	7.2	10824	6725	541
46	7.8	11182	6948	559
47	8.3	11540	7170	577
48	8.9	11898	7393	595
49	9.4	12255	7614	613
50	10.0	12613	7837	631
51	10.6	12971	8059	649
52	11.1	13329	8282	666
53	11.7	13687	8504	684
54	12.2	14268	8865	713
55	12.8	14849	9226	742
56	13.3	15430	9587	772
57	13.9	16011	9948	801
58	14.4	16593	10310	830
59	15.0	17174	10671	859
60	15.6	17754	11031	889
61	16.1	18336	11393	917
62	16.7	18917	11754	946
63	17.2	19685	12231	984
64	17.8	20452	12707	1023
65	18.3	21220	13185	1061
66	18.9	21987	13661	1099
67	19.4	22755	14138	1138
68	20.0	23522	14615	1176
69	20.6	24290	15092	1215
70	21.1	25057	15569	1253
71	21.7	25825	16046	1291
72	22.2	26827	16668	1341
73	22.8	27829	17291	1391
74	23.3	28831	17914	1442
75	23.9	29833	18536	1492
76	24.4	30836	19159	1542
77	25.0	31838	19782	1592
78	25.6	32840	20405	1642
79	26.1	33842	21027	1692

DEW POINT °F @ 14.7 PSIA	DEW POINT °C @ 14.7 PSIA	PPM BY VOLUME	PPM BY WEIGHT	LBS. WATER MMSCF DRY AIR
80	26.7	34844	21650	1742
81	27.2	36138	22454	1807
82	27.8	37432	23258	1872
83	28.3	38727	24062	1936
84	28.9	40021	24866	2001
85	28.4	41315	25670	2066
86	30.0	42609	26474	2130
87	30.6	43903	27278	2195
88	31.1	45198	28083	2260
89	31.7	46492	28887	2325
90	32.2	48146	29914	2407
91	32.8	49801	30943	2490
92	33.3	51455	31971	2573
93	33.9	53109	32998	2655
94	34.4	54764	34027	2738
95	35.0	56419	35055	2821
96	35.6	58023	36083	2904
97	36.1	59728	37111	2986
98	36.7	61382	38139	3069
99	37.2	63477	39440	3174
100	37.9	65571	40741	3279
101	38.3	67666	42043	3383
102	38.9	69760	43344	3488
103	39.4	71855	44646	3593
104	40.0	73950	45948	3698
105	40.6	76045	47249	3802
106	41.1	78139	48550	3907
107	41.7	80234	49852	4012
108	42.2	83861	51484	4193
109	42.8	85489	53117	4274
110	43.3	88116	54749	4406

Commissioning Petrochemical Pipe Lines

Pipe lines carrying a variety of gaseous and volatile liquid products are extensively employed in today's industry as a safe and efficient means of transportation. New pipe lines must be placed into service initially (i.e., commissioned), and older lines occasionally are taken out of service in order to perform some desired maintenance and are then recommissioned. Among the reasons for taking a pipe line out of service (i.e., decommissioning) are: hydrostatic testing to recertify or upgrade the pipe line's ability to be used at higher operating pressures; performance of construction work on the pipe line; a change in the product transported by the pipe line. Pipe lines, or sections of a pipe line may need to be relocated because of highway work, the necessity to deepen a canal, or because of increases in the population surrounding the pipe line. It may also be necessary to replace valves, fittings or a damaged section of the pipe line or add a new connection to service a customer or supplier.

Typically, decommissioning and recommissioning a petrochemical pipe line will include the steps of decommissioning by removing the product from the pipe line, and flaring any remaining residual product. Any necessary construction, upgrading, or cleaning of the pipe line can then be performed. Usually, the pipe line is then filled with water for hydrostatic testing. After pressure testing, the water is removed, the pipe line is cleaned and dried to a specific low dewpoint (to avoid the problems of water contamination of the product) and the pipe line is inerted with nitrogen for recommissioning. Refer to the section on Pipe Line Drying for drying techniques.

In order to recommission the pipe line, the nitrogen must be displaced by the desired petrochemical. Before returning to service, product purity must be established and the line safely filled to operating pressure. The terms 'commissioning' and 'recommissioning' are used interchangeably to refer to a process whereby a first inerting gas, normally nitrogen in a pipe line is replaced with the desired product at the desired purity and pressure.

Safety and economics are two primary concerns for any proposed pipe line operation. Thus, a commissioning process which brings the purity and pressure of potentially explosive products such as ethylene or propylene up to specification quickly, but at the risk of damage to the pipe line or reduced safety to operating personnel, is not acceptable. Similarly, a process which uses large quantities of product to push nitrogen from a line results in wasted product/nitrogen mixtures that must be flared or otherwise disposed of would also be unacceptable. Improvement over known methods requires consideration of its safety and economic benefits (considering both the cost of wasted product and the cost attributed to the time the line must remain out of service) as well as its ability to bring the line back into service with product at desired pressures and purities.[1]

The main problems posed by recommissioning are (1) the fast and economical purging of nitrogen so as to obtain uncontaminated products in the line, and (2) possible damage to the pipe line due to cold temperatures to which it may be subjected during the process. The latter can be a problem when the pipe line is to be recommissioned with a product at a pressure significantly higher than that of the nitrogen inerted line. The pressure drop of a petrochemical product entering a pipe line can cause large drops in temperature and, consequently, potential damage to the carbon steel pipe line.

Due to the demands of the expanding petrochemical industry, and the aging of the present pipe line system, the need for decommissioning, repairing, cleaning, drying and recommissioning pipelines is increasing. Given the value of today's petrochemical products, a recommissioning procedure, which brings the product purity to acceptable levels as quickly as possible, is desirable. Furthermore, processes and equipment which reduce the possibility for damage to the pipe line due to cold temperature conditions which can occur during the recommissioning process are also highly desirable.[2]

Due to the unusual properties of ethylene, decommissioning and recommissioning ethylene pipelines require special consideration. Ethylene is a colorless flammable gaseous unsaturated hydrocarbon obtained by the pyrolysis of petroleum hydrocarbons. The relative density of ethylene is 0.9686 (air = 1) and its molecular weight is 28.054. The molecular weight of air is 28.964 and the molecular weight of nitrogen is 28.013.

Ethylene Decomposition

Ethylene is subject to thermal decomposition (self propagating reaction zone, explosion or flame) under certain circumstances. Thermal decomposition occurs when the temperature of a substance is raised above the value needed to cause it to begin to self-heat because of a change in molecular structure at a rate high enough to result in combustion. An external source of ignition is not required to precipitate this process nor is a source of oxygen needed. Thermal decomposition is usually initiated by a source of heat from events such as sudden compression, external source of heat, or the thermal runaway of a heater.[4] Heaters are sometimes used in ethylene measurement stations to maintain the ethylene at an optimum temperature for measurement.

Ethylene is routinely handled under conditions where decomposition may be initiated. The point at which decomposition begins is dependent upon system parameters such as line or vessel size, operating pressure, and operating temperatures. Generally, ethylene decomposition is initiated in pipe line facilities by a rapid rise in temperature associated with sudden compression in the presence of a diatomic gas such as nitrogen or oxygen, exposure of the pipe to flame, possibly hot tapping operations.

The autoignition temperature (AIT) for ethylene is 914°F at atmospheric pressure.[3] Experience of others suggests that at 1,000 to 1,500 psig, the AIT could drop to as low as 400° to 700°F.[3,4] As the concentration of oxygen increases, the decomposition initiation temperature decreases.[4] From this, it may be concluded that a mixture of low oxygen concentration and ethylene could be very hazardous. For this reason, ethylene pipelines, which have been evacuated for maintenance or repairs, are usually filled with nitrogen to displace oxygen from the pipe. Compressing nitrogen with ethylene may result in high nitrogen temperatures—especially if the compression is rapid. Also, when the pressure of the ethylene source is high, refrigeration across the throttling valve used to control the ethylene may result in extremely low temperatures and could cause the valve to fail due to excessively low temperature.

Commissioning an ethylene pipe line with warm ethylene (+/−70°F) at a low pressure (+/−50 psig) is quick, easy, inexpensive and very safe when the ethylene is heated before it is admitted to the pipe line being commissioned.

A patented process used by Pipeline Dehydrators can be used to safely commission ethylene pipelines. The process uses a shell and tube heat exchanger to heat the high pressure ethylene before the pressure is reduced to the low pressure desired for line commissioning.

The methods currently used to commission ethylene pipelines are potentially hazardous, expensive and time consuming. The two most severe problems are potential damage to the carbon steel in the pipe line due to the cold temperature of the expanded ethylene and contamination of the ethylene from the nitrogen used to inert the pipe line during commissioning.

An ethylene pipe line that is being commissioned has usually been dried to a −70°F dew point or more and inerted with nitrogen. The nitrogen pressure left on the line for commissioning varies from 10 to 1,000 psig depending on the pressure of the ethylene source and the commissioning procedure used and the pipe line owner's preference.

The source of ethylene used for commissioning an ethylene pipe line is usually over 900 psig and may be up to 2,200 psig.

If the pressure of the ethylene source used for commissioning is high and the nitrogen pressure in the pipe line being commissioned is low, the resulting temperature of the ethylene, due to the pressure drop, may be well below the −20°F design temperature of most carbon steel pipe lines.

The carbon steel in some pipelines is subject to becoming brittle at temperatures below −20°F and may fail at a very low internal pressure because of the low temperature.

At low pressures, the very cold ethylene will be much more dense than nitrogen, thereby producing an extensive product interface during commissioning or trapping pockets of the less dense nitrogen at each high place in the pipe line.

In this case, the amount of ethylene burned to produce product purity may be expensive, time consuming, create environmental problems in addition to subjecting the steel in the pipe line to sub-design temperatures.

If the pressure of the ethylene source used for commissioning is high, and the nitrogen pressure in the pipe line being commissioned is also high, cold temperatures will not be encountered. However, even at warm temperatures, high-pressure ethylene is still much more dense than nitrogen at the same pressure. Therefore, the extensive product interface and the trapped nitrogen at the high points will still occur, resulting in an expensive and time consuming purification process plus the cost of the large volume of nitrogen. Also, there is a large inventory of ethylene in a pipe line at high pressure. If a pipe line blow-down is required because of a leak, lack of purity or for whatever reason, it will be a major product loss and expense.

The Pipeline Dehydrators system utilizes a tube and shell heat exchanger and a water-glycol mix that is used as the heat transfer fluid. A pump forces the heat transfer fluid through the fired heater and then through the shell side of

the exchanger. The ethylene passes through the tubes of the heat exchanger and is safely and properly warmed by the heat transfer fluid without entering the direct fired heater, thereby eliminating the possibility of an ethylene decomposition due to heater thermal runaway.

The heater will produce ethylene at ground temperature (+/−70°F) at +/−50 psig corresponding to the conditions of the nitrogen in the pipe line. Under these conditions, ethylene and nitrogen have basically the same density; therefore, the product interface during commissioning will be insignificant, no nitrogen will be trapped at the high points in the pipe line, the steel in the pipe line will not be subjected to cold temperatures, purification will be simple, short and inexpensive and the ethylene inventory in the pipe line will be minimal in case a pipe line blow-down is required. The savings on nitrogen will also be significant.

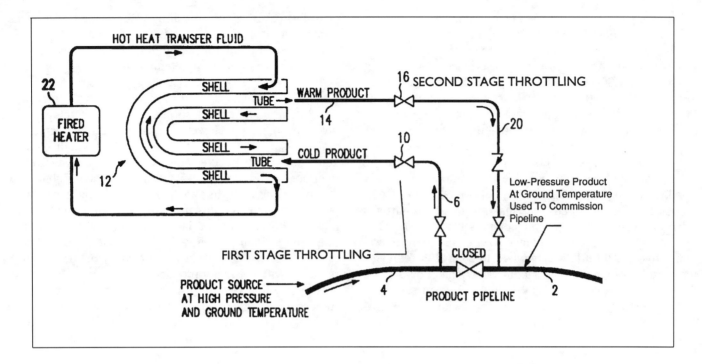

Two pressure drops will be taken on the ethylene, one on each side of the tube-shell heat exchanger. The pressure will be dropped to approximately 600 psig prior to the ethylene entering the heat exchanger. The 600 psig ethylene will be warmed to approximately 140°F. Then a second pressure drop will occur allowing the pressure and temperature of the ethylene to equalize with the pressure and temperature of the nitrogen in the pipe line (+/−50 psig @ +/−70°F). Under these conditions, PLD's 6 million BTU per hour heater can handle up to 45,000 lbs. of ethylene per hour. This high flow rate will normally be used only during the pressuring up phase of the process and not used during the purification process. The warm ethylene at low pressure, traveling about 10 mph in the small pipe lines that are typically used to transport ethylene, will produce excellent results during the purification phase of the process.

The minimum ethylene temperature encountered in this commissioning process is well above 0°F, and the maximum temperature is 160°F. This is a very safe temperature range to handle ethylene in carbon steel piping systems.

Purity can be established at 50 psig or less, with very little loss of product, and the system can be checked for leaks with ethylene in the pipe line while the pressure is low and the dollar value of ethylene in the pipe line is still small.

After purity has been established at +/−50 psig and potential leak points have been checked, the major remaining effort will be to warm enough ethylene to bring the pipe line up to approximately 600 psig. Beyond this point, the ethylene can be throttled into the pipe line without heating because the cooling due to pressure drop will no longer be great enough to cause damage to the pipe line and equipment. During this pressure-up phase of the commissioning operation, the ethylene does not have to be heated to +70°F temperature and the pipe line pressure will be increasing; both resulting in a reduction in the amount of heat required per pound of ethylene. Therefore, for this phase of the commissioning operation, PLD's new ethylene heater will handle considerably more than 45,000 lbs. of ethylene per hour.

This system provides excellent conditions for commissioning an ethylene pipe line quickly, inexpensively and safely.

This heater can also be utilized to commission other product pipelines where cold temperatures are encountered due to pressure drops such as with propylene, ethane, propane and carbon dioxide.[2]

Acknowledgement: The Editor wishes to thank Marvin D. Powers—Pipeline Dehydrators for permission to use material from paper on *Ethylene Heater For Commissioning Pipelines*.

Literature Cited:

[1]US Patent No. 5,027,842 Jul. 2, 1991 Powers

[2]Powers, Marvin D. '*Ethylene Heater For Commissioning Pipelines*' 5-89 Pipeline Dehydrators

[3]Hilado, Carlos J. and Clark, Stanley W. '*Autoignition Temperatures of Organic Chemicals*' Chemical Engineering September 4, 1972, p 76.

[4]McKay, F. F., Worrell, G. R., Thornton, B. C., and Lewis, H. L. '*How To Avoid Decomposition In Ethylene Pipelines*', Pipe Line Industry August 1978, p 80.

Vacuum Drying

The vacuum drying process is shown graphically in Figure 1 and consists of three separate phases. Corrosion is generally inhibited at relative humidity (R.H.) levels below 30% but in the presence of hygroscopic dirt (present in millscale) corrosion can occur at R.H. levels of 20%. Therefore, systems should be thoroughly drained and then vacuum dried to lower than 20% R.H.

Target Dewpoint (D.P.)

$$R.H. = \frac{\text{Water vapour pressure}}{\text{Saturated Vapour Pressure (SVP)}}$$

for a constant temperature. Assume the lowest average temperature of the system is 0°C (during the winter months) which gives

$$SVP = 0.6 \text{ KPa A}$$

The vacuum level required for water vapour at 20% R.H. is:

$$0.2 \times SVP = 0.2 \times 0.6 = 0.12 \text{ KPa A}$$

This corresponds to a dewpoint of −18°C. A safety allowance should be provided for some desorption from the pipe walls; and therefore a dewpoint of −20°C would be used.

Phase 1—Evacuation

During this phase, the pressure in the pipe line is reduced to a level where the ambient temperature of the pipe line will cause the free water to boil and change to water vapour. This pressure level corresponds to the saturated vapour pressure of the free water in the pipe line which is dependent upon the ambient temperature of the pipe line.

The approximate pressure value is calculated in advance but is easily recognized on site by a fall in the rate of pressure reduction, which is noted from the plot of pressure against time.

At some convenient point in time a "leak test" is carried out by stopping the vacuum equipment and observing the pressure, usually for a period of 4 hours. Any "air-in" leaks on flanges, fittings, or hoses are rectified at this time, although leaks are not a common occurrence.

Phase 2—Evaporation

Once the saturated vapour pressure has been reached then evaporation of the free water into water vapour will commence. During this phase, the vacuum equipment is carefully controlled to maintain the pressure at a constant level until all the free water has been converted into water vapour. This phase may take several days to complete depending on: (1) the amount of water to be evaporated; (2) the size of the vacuum equipment; and (3) the ambient temperature of the pipe line. The end of the evaporation phase will be observed on site by a noticeable decrease in pressure.

At this time it is prudent to carry out a "soak test" to ensure that all the free water has in fact evaporated. The vacuum equipment is temporarily isolated from the pipe line, usually for a period of 12 hours, and a careful note made of the pressure. If all free water has evaporated then the pressure will remain constant and the final drying phase can be commenced.

Figure 1
Vacuum Drying Pressure vs Time Curve

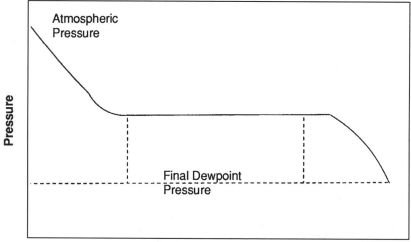

Phase 3—Final Drying

Once the free water has been converted into water vapour, the majority of it must be removed from the pipe line in order to reach the required dryness level. This is achieved by reducing the pressure in the pipe line still further which has the effect of drawing the water vapour out of the pipe line through the vacuum equipment. Obviously, the more water vapour removed, then the drier the pipe line will become.

During this phase a careful watch is kept on the slope of the final drying line to ensure that it follows the calculated value, since a shallower slope would indicate the continuing presence of some free water still remaining in the pipe line.

The necessary calculations to determine the time required for each phase of vacuum drying are detailed in Figure 2.

What is dryness?

The dryness of a pipe line is measured in terms of dewpoint, which is the temperature at which mist or dew will begin to form. A convenient method of measuring dewpoint is to use an instrument called a mirror hygrometer where the water vapour is passed across a polished surface which is slowly cooled until dew forms. The temperature at which the dew forms is the dewpoint of the water vapour and is normally expressed in degrees centigrade. The drier the air, the lower the temperature at which dew will form.

In terms of a pipe line being vacuum dried, the lower the pressure in the pipe line, the lower the dewpoint will be. For example, at a pressure level of 0.26 kPa A, the equivalent dewpoint of the pipe line would be −10°C. If the pressure were further reduced to 0.104 kPa A, then the dewpoint would be −20°C.

For gas pipelines a dewpoint level of −20°C is generally considered to be adequate and the 0.10 kPa A bar pressure level required to achieve this dewpoint is readily attainable using the portable vacuum equipment previously described.

For example, consider a 100-mile-long 36-inch-diameter pipe line which, prior to drying, contained 10,000 gallons of water as a film on the inside surface of the pipe. At a dewpoint of −10°C the quantity remaining would be reduced to only 46.5 gallons and at −20°C to 19.7 gallons. Also, this water would not be free water but rather water vapour and it would only revert back to free water if the ambient temperature of the pipe line were further reduced. This water vapour can subsequently be removed from the pipe line during the purging operation.

The relations between pressure and dewpoint are shown in Figure 3 and Table 1.

Proving the dryness

Immediately following on from the final drying phase, a dry gas purge using atmospheric air or nitrogen is carried out to prove the dryness of the pipe line. It is possible, under certain circumstances, for a small amount of free water to still remain in the pipe line. Usually this water

Figure 3
Dewpoint Temperature vs
Saturated Vapour Pressure

Table 1
Water Vapour Pressure Table

T °C	SVP (kPa A) (mb × 10⁻¹)	Vapour Density (gm m³)
−50	0.0039 (0.039 mb)	0.038
−45	0.0128	0.119
−35	0.0223	0.203
−30	0.0308	0.339
−25	0.0632	0.552
−20	0.1043	0.884
−15	0.1652	1.387
−10	0.2597	2.139
−5	0.4015	3.246
0	0.6108	4.847
5	0.8719	6.797
10	1.2270	9.399
15	1.7040	12.830
20	2.3370	17.300
25	3.1670	23.050
30	4.2430	30.380

Figure 2
Vacuum Drying Calculations

Evacuation:

$$t = \frac{V}{Se} \log e \frac{Pi - Pv}{Pf - Pv}$$

$$SE = \frac{1}{C} + \frac{1}{Sp}$$

Evaporation:

$$t = \frac{M}{SeDv}$$

Final Drying:

$$t = \frac{V}{Se} \log e \frac{Pv}{Preq}$$

Pi = Initial Pressure
Pf = Final Pressure
Pv = Vapour Pressure
Se = Effective Pump Speed
V = System Volume
Sp = Pump Speed
C = Conductance
M = Mass of Water
Dv = Vapour Density
Preq = Required Dewpoint Pressure

will have turned to ice due to the chilling effect of the vacuum drying process and may not be apparent during the final drying phase or soak test.

Nitrogen or atmospheric air is allowed to enter the pipe line through a valve at the end remote from the vacuum equipment, until the pressure has risen to the SVP equivalent of the target dewpoint.

Once this pressure level has been reached, the vacuum equipment is started and that pressure level maintained. This has the effect of drawing gas through the pipe line under vacuum at a relatively constant dewpoint equal to the final dewpoint required.

At some point in time the purge gas, now under vacuum at a dewpoint of say, −20°C., will reach the vacuum equipment and be pulled through it. The dewpoint at both ends of the pipe line is carefully monitored and compared. If there is no free water remaining in the pipe line then the dewpoint at the vacuum equipment end will be the same as the dewpoint at the remote end. However, if there is any free water present then the dry air passing through the pipe line under vacuum will absorb the water hygroscopically. The dry gas purge operation must then continue to remove the remaining free water until the dewpoint at both ends

are equal, at which time purging is discounted. The pipe line has now been vacuum dried to the required dewpoint level and the dryness proved.

Purging and commissioning

Once the dryness has been attained and proved, the pipe line is ready for commissioning. It is possible to introduce the gas directly into the vacuum or to relieve the vacuum using dry nitrogen.

Reprinted with permission Nowsco Well Service Ltd.

7: Control Valves

Control valve sizing formulas 147
Sizing control valves for thruput 151
Control valve selection 156
Relief Valve Sizing, Selection, Installation, and
Testing **158**
Variable orifice rotary control valves 162
Sizing Valves for Gas and Vapor **163**

Control valve sizing formulas

Liquid Formulas (Noncompressible Fluids).

Volume basis:

$$C_v = q \sqrt{G/\Delta P} \tag{1}$$

Weight basis:

$$C_v = W/(500 \sqrt{\Delta P \times G}) \tag{2}$$

Equations 1 and 2 exclude control valve calculations where flashing, high viscosity, two-phase flow or high pressure drops are involved.

Gas and Vapors (Other than Steam).

Volume basis:

$$C_v = (Q/1360) \sqrt{2GT/[\Delta P(P_1 + P_2)]} \tag{3}$$

Equation 3 does not take into consideration gas compressibility, mixed phase flow or high pressure drop. Compressibility is considered in Equations 12 and 13. Usable pressure drop for valve sizing calculations for gas, vapor, and steam flows is always limited to one-half of the absolute inlet pressure of the control valve. At that approximate value, critical (sonic) velocity is reached and any further reduction of the downstream pressure will not increase the velocity through a valve, thus there will be no increase in flow.

Steam Formula.

$$C_v = \frac{W(1.0 + .0007Tsh)}{2.1 \sqrt{\Delta P(P_1 + P_2)}} \tag{4}$$

Special Conditions

Equations are given below to correct for flashing, cavitation, high viscosity fluids and mixed phase flow, where the simplified equations may lead to inaccurate sizing. These additional expressions, together with the FCI equations, provide satisfactory solutions for most control valve sizing applications.

Liquids

Flashing Service. When flashing occurs, the simplified liquid sizing equation gives erroneous results. Flashing oc-

Figure 1. Typical control valve pressure gradient. The ΔP between P_1 and P_2 is much less for high recovery valves.

curs when liquids enter a valve at or near their boiling points, and they begin vaporizing as their pressures decrease, due to conversions of static pressure head to velocity head. With the advent of high recovery valves, this condition has been aggravated. High recovery valves take a much higher pressure drop from inlet to valve orifice than standard control valves (see Figure 1). When the pressure at the orifice drops to the vapor pressure of the fluid at flowing temperature, the maximum valve capacity is reached. This is true because flashing occurs across the valve orifice and a flow condition similar to the critical flow of vapors exists.

Two approaches take into account the flashing condition:

1. Hans D. Baumann, in a paper presented in *ISA Transaction*, Volume 2, Number 2 (April, 1963), introduced an equation using a C_f factor to help determine the valve capacity when flashing occurs. The equation, based also on the use of liquid vapor pressure (P_v), is

$$C_v = (Q/C_f) \sqrt{G/(P_1 - P_v)} \tag{5}$$

2. C. B. Schuder, Fisher Controls, in a paper, "How to Size High Recovery Valves Correctly," proposed an equation to solve for the maximum pressure drop to be used for liquid sizing:

$$\Delta P_m = K_m(P_1 - r_cP_v) \tag{6}$$

(See Figure 2 for values of r_c.)

a.

b.

Figure 2. These critical pressure ratios (r_c) are for use with Equation 6.

When this limiting pressure drop from Equation 6 is used in the liquid flow formula, Equation 1 gives a valve size in which flashing will not occur, unless the pressure drop increases above the calculated ΔP_m. When the ΔP exceeds ΔP_m, flashing and possible cavitation can occur. Cavitation is the effect caused by the sudden condensing of vapors when the static pressure increases to a point above the boiling point of the liquid.

When the possibility of cavitation exists, the process should be reviewed and the condition eliminated if at all possible. A high recovery valve should not be used in this application. This second approach is suggested as a good solution to the flashing problem.

Viscosity Correction. The liquid sizing formulas in general use have no factor applied to correct for viscosity. This correction is a complex function involving the Reynolds number which varies with the valve size, port area, velocity, and viscosity. No correction is required for a viscosity value below 100 SSU. Above viscosity values of 100 SSU, two equations are given to obtain correction factors:

$$R = (10,000 \times q)/\sqrt{C_v} \times cs \qquad (7)$$

This equation is used for viscosity values between 100 and 200 SSU; the viscosity must be converted to centistokes. When viscosities exceed 200 SSU, Equation 8 applies:

$$R = (46,500 \times q)/\sqrt{C_v} \times SSU \qquad (8)$$

In either equation, the following steps should be followed:

1. Solve for C_v, assuming no viscosity effects.
2. Solve for factor R using Equation 7 or 8.
3. From the viscosity correction curve, Figure 3, read the correction factor at the intercept of factor R on the curve.
4. Multiply the C_v obtained in Step 1 by the correction factor obtained in Step 3 for the corrected C_v.

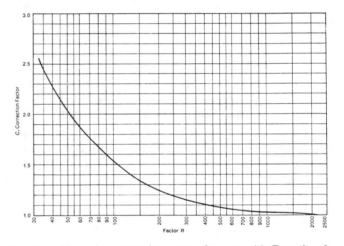

Figure 3. Viscosity correction curve for use with Equation 8.

Gases and Vapors

The simplified equation for gases and vapors (Equation 3) agrees favorably with experimental data when used for conventional globe valves, but agreement with experimental data is poor when used with high recovery valves.

A universal equation proposed by C. B. Schuder, Fisher Controls, in *Instrumentation Technology*, February, 1968, reportedly showed excellent agreement with experimental results covering a wide range of body plug configurations and pressure drop ratios.

Volume basis:

$$Q = \sqrt{\frac{520}{GT}} \, C_1 C_2 C_v P_1 \, \sin\left[\frac{3417}{C_1 C_2} \sqrt{\frac{\Delta P}{P_1}}\right] \text{Deg.} \qquad (9)$$

Weight basis:

$$W = 1.06 \sqrt{dP_1 Z}\ C_1 C_2 C_v \sin\left[\frac{3417}{C_1 C_2} \sqrt{\frac{\Delta P}{P_1}}\right] \text{Deg.} \qquad (10)$$

The C_1 factor in Equations 9 and 10 is currently available from Fisher Controls for their valves. Other companies presumably do not publish this factor. (The C_2 factor is shown in Figure 4.)

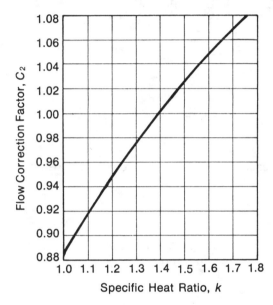

Figure 4. The C_2 factor is applicable only to Equations 9 and 10, the Fisher universal equations for gas and vapor flows.

An equation for gases and vapors proposed by the Foxboro Company (Foxboro TI Bulletin 31-4K) is identical with the FCI equation, except that P_2 is used instead of $(P_1 + P_2)/2$:

$$C_v = (Q/1360) \sqrt{GT/\Delta P(P_2)} \qquad (11)$$

A comparison of the Fisher Universal, FCI, and Foxboro equations is made in the TI Bulletin 31-4K. A plot is given showing the results of these three equations using the same data (Figure 5).

The Foxboro equation gives C_v values essentially the same as the universal equation at pressure drops up to 25% of P_1 and gives 7% larger values at a pressure drop of 50% of P_1. At the higher end of $\Delta P/P_1$ ratios, a slightly larger valve should be helpful to allow for the expanding vapors. The same agreement is claimed for steam flows although no plot is given.

To correct for deviation from the universal gas law, the compressibility factor is inserted in Equation 11, becoming:

$$C_v = (Q/1360) \sqrt{G_b TZ/(\Delta P \times P_2)} \qquad (12)$$

On a weight basis, this equation becomes:

$$C_v = (W/104) \sqrt{TZ/(\Delta P \times P_2 \times G_b)} \qquad (13)$$

Mixed Phase Flow. Mixed phase flow is generally experienced when:

1. Liquid entrainment is carried by a gaseous flow.
2. Liquid at or near its boiling point vaporizes as it flows through a line and its restrictions.

In the discussion under "Flashing Service," it is explained that flashing can be reduced or eliminated by limiting the ΔP across the valve. When two-phase flow already exists or when it occurs in the control valve, two methods are given to size the valve.

1. A rule of thumb that has been used for years is to calculate the C_v based on liquid only and increase the valve body to the next size. This is satisfactory in many cases, but it obviously does not take into account varying ratios of liquid to vapor.
2. A preferred method is to determine the amount of vapor and liquid, then calculate and add the C_v of each phase to get the overall C_v requirement. The percent vaporization can be calculated by the responsible engineer if it is not furnished by the process people.

Example.

Fluid—Propane
100 gpm @ 300° psia and 130°F
s.g. @ 130°F = 0.444
s.g. @ 60°F = 0.5077
s.g. @ P_2 (200 psia) = 0.466 @ 104°F
ΔP = 100 psi
% vaporization = 15.05%
M.W. = 44
Z = 0.8
G_b = 44/29 = 1.518
Flow = 100 gpm × 0.444 = 22,200 lb/hr
Vapor = 22,200 × 0.1505 = 3,340 lb/hr
Liquid = 22,200 − 3,340 = 18,860 lb/hr

COMPARATIVE Cv VALUES
FROM VARIOUS GAS EQUATIONS

Figure 5. These curves show the close agreement (particularly for low pressure drops) of the Foxboro, FCI, and Fisher universal equations for gas flows. Courtesy of the Foxboro Co.

Using Equation 13:

$$C_v(\text{vapor}) = (W/104) \sqrt{(T \times Z)/(\Delta P \times G_b \times P_2)}$$

$$= (3340/104) \sqrt{(564)(.8)/(100(1.518)(200)}$$

$$= 3.92$$

Using Equation 2:

$$C_v(\text{liquid}) = (W/500)/(18,860/ \sqrt{\Delta P \times G})$$

$$= 18,860/(500 \sqrt{100 \times .466})$$

$$= 5.53$$

$$C_v(\text{total}) = 3.92 + 5.53$$
$$= 9.45$$

The value of P must never exceed $P_1/2$ for sizing calculation.

Nomenclature

C_v = Valve flow coefficient
G = Specific gravity of fluid at flowing temp. (water = 1.0)

P = Pressure drop $(P_1 - P_2)$, psi
W = Flow of fluid, lb/hr
q = Flow rate, gpm
T = Flowing temp., °R
T_{sh} = Degrees of superheat, °F
C_f = Critical flow factor, from valve manufacturer
P_m = Maximum useable P (choked flow occurs), psi
K_m = Valve recovery coefficient
r_c = Theoretical critical pressure ratio at P_v
P_v = Fluid vapor pressure at flowing temp., psia
R = Factor for use with Figure 3
cs = Viscosity at flowing temp., centistokes
SSU = Viscosity at flowing temp., Saybolt Seconds Universal
C_1 = Valve sizing factor
C_2 = Specific heat correction factor (see Figure 4)
P_1 = Valve upstream pressure, psia
P_2 = Valve downstream pressure, psia
d = Density of vapors at valve inlet lb/ft³
d_1 = Density of vapors at valve outlet lb/ft³
Z = Compressibility factor
G_b = Specific gravity of gas or vapor at base conditions

Source

Applied Instrumentation in the Process Industries, Gulf Publishing Company, Houston, Texas.

Sizing control valves for thruput

These useful charts give data essential to calculation of pressure drops through regulating stations

W. G. Birkhead, Gas Measurement Engineer, United Gas Pipe Line Co., Shreveport, La.

Pressure losses through piping fittings, once given little consideration in the sizing of control valves, have become important as more efficient control valves have been developed. With the development of high capacity valves and the desire for a more exacting flow formula and efficiency, designers and users look to consideration of minor losses through elbows, tees, swages or reduced ported valves. The capacity of control valves and regulators was once calculated using the basic hydraulic head equation with an assumed efficiency. This is no longer the case.

The head loss in any closed filled conduit has been accepted as proportional to its length, diameter and fluid velocity head, and can be calculated as

$$H_L = f \frac{L}{D} \frac{V^2}{2g} \qquad (1)$$

where H_L = Head loss of flowing fluid (ft of fluid)
f = Friction factor
L = Length of pipe, ft
D = Diameter of pipe, ft
V = Average fluid velocity, ft/sec
g = Acceleration of gravity, ft/sec/sec

Sometimes L/D is referred to as the "equivalent length in diameters." Valves, fittings, etc., have often been expressed in equivalent diameters. If the friction factor f is combined with L/D, a constant K is derived for any pipe, valve or fitting. If $K = f(L/D)$, then

$$H_L = K \left(\frac{V^2}{2g}\right) \qquad (2)$$

The term K is called the "head loss factor" or "pressure drop factor."

Most control valves are rated with a capacity term called C_v. C_v is the number of gallons of water per minute that will flow through the valve with a 1 lb/in.^{2} pressure drop across the valve.

$$C_v = Q/\sqrt{\Delta P} \qquad (3)$$

Q = Quantity of water in gpm
ΔP = Pressure drop in psi

Rearranging Equation 3:

$$\Delta P = Q^2/(C_v)^2 \qquad (3a)$$

This equation may be rearranged so that C_v in terms of our K factor may be expressed as:

$$C_v = \frac{29.9 \, (d)^2}{\sqrt{K}} \qquad (4)$$

where d = inside diameter of the pipe or fitting, in.

The head loss or pressure drop of a fluid flowing through various pipes, valves, or fittings in series can be stated:

$$(\Delta P)_{Total} = (\Delta P)_1 + (\Delta P)_2 + (\Delta P)_3 + \text{---} \qquad (5)$$

Substituting Equation 3a in Equation 5:

$$\frac{1}{(C_v)^2{}_{Total}} = \frac{1}{(C_v)_1{}^2} + \frac{1}{(C_v)_2{}^2} + \frac{1}{(C_v)_3{}^2} + \text{---} \qquad (6)$$

This equation is unwieldy if several fittings are used. By equating Equation 4, substituting in Equation 6 and dividing through by $[29.9(d)^2]^2$:

$$K_{Total} = K_1 + K_2 + K_3 + \text{---} = \Sigma \, K \qquad (7)$$

Substituting in Equation 4 for K_T an over-all C_v can be derived:

$$C_v = \frac{29.9 \, (d^2)}{\sqrt{\Sigma \, K}}$$

K Factors for Various Valves and Fittings

For full open valves, pipes or fittings, the K factor is approximately equal to f(L/D) or Equation 1.

where L = Equivalent length of valve or pipe or fitting, in. (or ft)
D = Diameter of the valve, in. (or ft)
f = Friction factor

Some of the recognized K factors for various welding fittings are shown in Table 1.

Other fitting K factors can be obtained from any fluid mechanics, hydraulics or piping handbook.

Low, average and high values of K given in Table 1 point out caution necessary in selecting specific types of fittings.

Standard Welding Reducer C$_V$

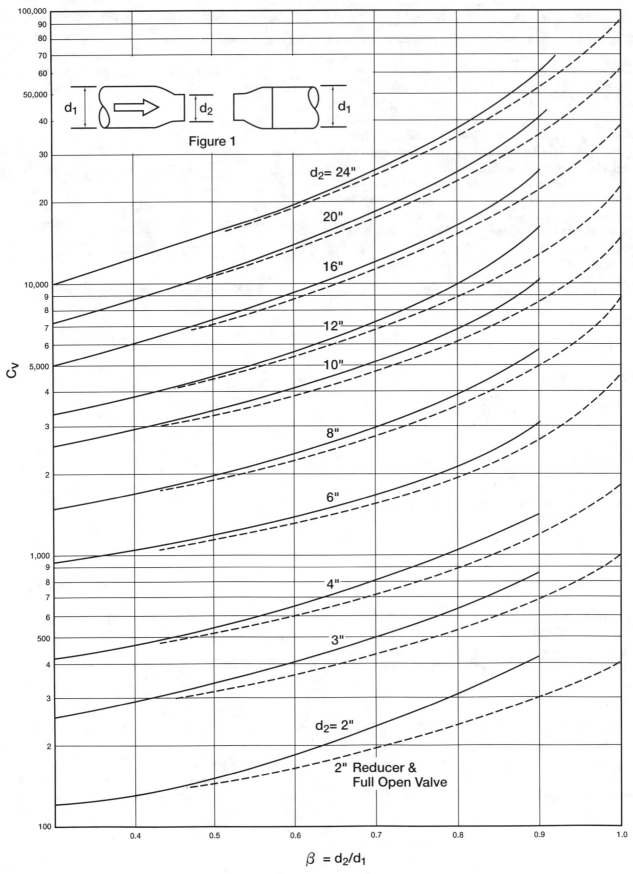

Figure 1. C$_V$ for standard welding reducers.

Cᵥ vs. K or Valve Flow Coefficient vs. Resistance Coefficient

$$C_V = \frac{29.9\, d^2}{\sqrt{K}}$$

Figure 2

Inside Diameter = d

Cᵥ= Valve Flow Coefficient

Figure 2. Cᵥ versus K values for valves.

Table 1

K Factor			
Low	Average	High	Type of Welding Fittings and Valves
0.10	0.20	0.40	90° Elbow
0.05	0.12	0.25	45° Elbow
0.25	0.50	1.00	Tee (Enter run, out branch)
0.75	1.00	1.20	Tee (Enter branch, out run)
0.05	0.10	0.15	Tee (Enter run, out run)
See Figs. 1 and 2			Standard Reducer-Increaser
0.25	0.50	1.50	Plug Valves (See manufacturer's bulletin)

Pressure losses in valves and fittings can be more accurately calculated by consulting the references.

All K factors, L/D's or C_v's must be converted to the same basis using the same diameter d when using an overall C_v for a piping run. This can be accomplished with the following equations:

$$K_a = K_b \, (f_a/f_b) \, (d_a/d_b)^4 \tag{8}$$

$$(L/D)_a = (L/D)_b \, (d_a/d_b)^4 \tag{9}$$

$$(C_v)_a = (C_v)_b \, (d_b/d_a)^2 \tag{10}$$

Examples

Refer to Figure 3 for a schematic of a typical control valve station using an old style control valve. Flow is from a header through a 12-in. by 6-in. reducer, a 6-in. isolating valve, 6-in. globe body control valve, 6-in. pipe spool, a second isolating valve, a 6-in. by 12-in. increaser, a 12-in. welding elbow and into a header. Note should be made of Figure 1 in that the C_v from the solid curve is for the complete reducer-increaser combination with regard to the ratio of diameters. In this case, the Beta ratio is 6/12 or 0.5 and C_v is 1,200 for the small diameter of 6-in. Since C_v's cannot be simply added, they must be converted to K factors. To do this, refer to Figure 2, which converts C_v to K.

Also to be considered are the isolating valves and control valves. Resistance coefficients, C_v's or pressure drop factors are published by most valve manufacturers and may be converted to a K factor by using Figure 2. In this case, 6-in. K_v is 0.05.

Pipe resistance must also be considered (Figure 5). The L/D ratio must be calculated with reference to the same units of L and D. In the example, 12 ft of 6-in. pipe has a K_p equal to 0.36.

All resistances must be converted to the same base diameter (in this case 6 in.). The entrance factor, K_e is 0.05 for 12-in. Referring to Figure 6, the 6-in. K_e is about 0.04; the 12-in. elbow coefficient converted to 6-in., 6-in. K_{el} is 0.02, and the 12-in. exit coefficient, 6-in. K_x, is 0.8. A tabulation and sum of these coefficients as determined from Figure 3 would be:

12″ Entrance	6″ K_e =	0.04
6″ Reducer—Increaser	6″ K_{ri} =	0.80
2—6″ Valves	6″ K_v =	0.10
6″ Control Valve	6″ K_v =	10.00
12-ft spool of 6″ Pipe	6″ K_p =	0.36
12″ Elbow	6″ K_{el} =	0.02
12″ Exit	6″ K_x =	0.08
	ΣK =	11.40

Refer to Figure 2 for a conversion of K to C_v. With the above values noted, the overall 6-in. C_v is 320, approximately a 6% reduction of the original C_v of 340.

Now refer to Figure 4. This is an identical control valve run with Figure 3, except that a 6-in. × 4-in. × 6-in. high-capacity ball valve has been substituted for the 6-in. single-ported globe body control valve. Note that the ΣK now is 1.93, which converted to an overall C_v is 775. This is a reduction of about 48% from the original 6-in. × 4-in. × 6-in. control valve C_v of 1,495.

If in Figure 4, the 6-in. × 4-in. × 6-in. valve were substituted with a full open 6-in. ball valve, the overall C_v is 895. This represents a reduction in C_v of about 81% from the original 6-in. ball valve C_v of 4,700.

Figure 3. Piping configuration.

Figure 4. Piping configuration.

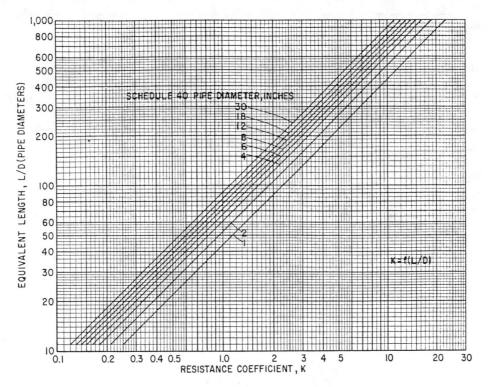

Figure 5. Resistance coefficients vs. equivalent length.

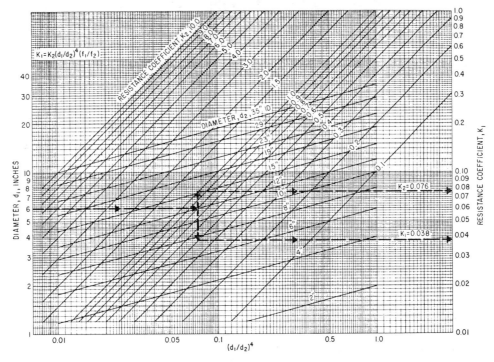

Figure 6. Equivalent resistance coefficients vs. diameter ratios.

The methods and graphs presented in this article are proposed for sizing and calculating pressure drops in a control station. The assumption of assigning values to certain un- usual piping configuration must be recognized. The author has some field data confirmation on similar control station designs.

Control valve selection

A control valve may be thought of as a section of pipe with a variable resistance. Control valves are typically used to match pumps to the pipe line system characteristics. They are not normally used with reciprocating compressors but may be used with centrifugal compressors for preventing surge.

Liquid pipe line applications usually involve controlling the flow rate, suction and discharge pressure, and other variables for centrifugal pumps. Control valves may also be used with reciprocating pumps to regulate the pump discharge pressure by recycling fluid from the pump discharge to the pump suction when the discharge pressure approaches the control point. The control valve may also be used to unload the pump during start-up by opening just prior to start-up and slowly closing after the pump driver is up to operating speed.

Other applications for control valves include:

- Pressure reduction
- Pressure relief
- Back pressure control
- Control delivery flow rates

Ideally, the pump(s) for a system will be sized such that a minimum amount of pressure throttling will be required during normal operations. System startup, system upsets, and unit startup and shutdown require that a means be provided to regulate pressures, flow rates, and in some cases motor load. This may be accomplished by using variable speed motors, variable speed couplings, or a control valve.

The control valve, when used with centrifugal pumping units, is installed in the pump discharge line. If multiple pumping units are used, only one control valve will be required.

Table 1 lists typical values of C_v for conventional and ball control valves.

When conventional globe-type control valves are used in pipe line applications for pump control, valves with equal percentage trim are the best choice. Figure 1 shows the typical control valve static characteristic curves for quick opening, linear, and equal-percentage valve trims.

Curves such as those shown in Figure 1 are based on maintaining a constant differential pressure of 1 psig across the valve for varying flow rates. In actual practice, the pressure differential across the control valve will vary according to the needs of the system. Figure 2 shows a comparison of the static and dynamic characteristics for an equal percentage control valve. The dynamic characteristic curve was developed by taking the pressure drop across a

Table 1
Typical Values of C_v for Control Valves*
(Equal Percentage Trim)

Valve Size	Double Ported Globe	Single Ported Globe	Ball Valve Full Bore†
2	48	46	406
4	195	195	1,890
6	450	400	4,580
8	750	640	8,900
10	1,160	1,000	14,600
12	1,620		22,800
14	2,000		28,700
16	2,560		38,900
20			66,000
24			98,600
26			119,800

* These values are suitable for estimating only. The valve manufacturer should be consulted for the actual C_v for the valve selected.
† Courtesy: Thunderco—Houston

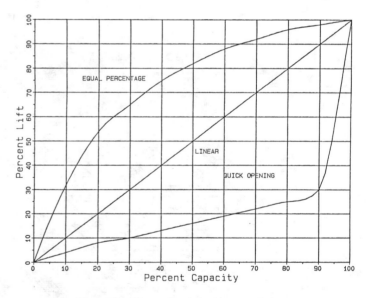

Figure 1. Control valve characteristics.

12-in. double-ported valve with equal percentage trim in series with ten miles of 12-in. pipe line. A maximum flow rate of 6,000 bbl/hr of 0.86 sp. gr. crude oil at a constant input pressure of 1,000 psig was used to calculate the pressure loss in the pipe line. The pressure drop across the control valve was calculated for various flow rates while maintaining a constant system input pressure. The C_v was then calculated for each of the flow conditions and converted to percent of valve lift.

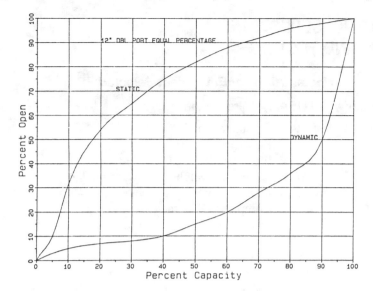

Figure 2. Control valve characteristics.

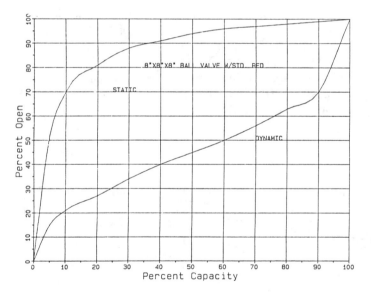

Figure 3. Control valve characteristics.

Figure 3 shows a comparison of the static and dynamic characteristics for an 8-in. × 8-in. × 8-in. ball valve under the same conditions of flow and pressure. A 6-in. × 6-in. × 6-in. ball valve with conical reducers has a C_v of 2,100, which would have satisfied the pressure drop requirements; however, the velocity through the 6-in. valve would have been approximately 45 ft/sec. Experience has shown that the valve velocity should be limited to about 25 ft/sec. The velocity through the 8-in. valve will be 27 ft/sec. The C_v for the 8-in. valve with standard welding redu-

cers is 2,600 and with conical reducers, the C_v is 4,100. The choice of using standard concentric welding reducers was made because that combination came closest to satisfying the C_v needs. When sizing and selecting valves, it is important that the valve not be oversized, as this will always result in poor system control. Reduced-size trim is available for conventional control valves and ported ball valves are available.

In this particular application, Figures 2 and 3 show that for a 10% change in system rate, the conventional globe valve must change its position by 50% (assuming that it was operating in the full open position). The ball valve would be required to change its position by 30% under the same conditions.

In pipe line applications, it is desirable to have virtually no pressure drop across a control valve when it is wide open. Any built in residual pressure drop across a control valve will result in added fuel costs; however, in certain situations, some residual pressure drop is unavoidable.

In process control, some pressure drop across the valve is desired to improve the control of the process. Ball valves have gained wide acceptance in pipe line operations because of their high C_v factor, which in turn results in very low pressure drop across the valve when wide open. As can be seen from Table 1, a 16-in. double ported valve (equal percentage trim) has a maximum C_v of 2,560. Based on a maximum pressure drop of 5 psig with the valve fully open, the maximum flow rate for 0.86 sp. gr. crude oil would be 212,000 bpd for a single valve. For flow rates greater than this, multiple valves may be used or one ball valve may be used. Very large ball valves of up to at least 36-in. in size have been successfully used for control valve applications.

The ideal control valve and control system will be quick to act when needed, and when controlling, will be stable at all control points. As shown by the dynamic characteristic curves, the valve must be capable of moving quickly from the wide open position to the control point without overshoot. Certain types of control systems may prove to be unstable on control when required to move the valve at high speeds.

High flow rates and high pressure drops across a control valve will require a valve operator with sufficient power to move the valve quickly and positively. Hydraulic systems have gained wide acceptance for this type of application. Air motors may also be used; however, the cost of the air compressor and air dryer will usually equal or exceed the cost of a hydraulic system. Other factors favoring a hydraulic system include the use of variable vane pumps, which reduces the power consumption during normal control operations. With a hydraulic system, a small hand pump may be used during electrical power outages. Control valves with air motors may be equipped with a handwheel for manual operation.

RELIEF VALVE SIZING, SELECTION, INSTALLATION AND TESTING

Effectiveness of a Safety Relief Valve Depends on Proper Sizing and Type of Equipment Selected

Gary B. Emerson, Manager SVR Products, Anderson-Greenwood USA, Bellaire, Texas

A safety-relief valve is an essential and important piece of equipment on virtually any pressured system. Required by the ASME-UPV Code, among others, it must be carefully sized to pass the maximum flow produced by emergency conditions.

After it is installed, the user hopes that it will never have to operate, which makes the safety relief valve rather unique as compared to other types of process equipment.

Sizing

The sizing formulas for vapors and gases fall into two general categories, based on the flowing pressure with respect to the discharge pressure. When the ratio of P_1 (set pressure plus allowable accumulation) to P_2 (outlet pressure) is greater than two, the flow through the relief valve is sonic; the flow reaches the speed of sound for the particular flowing medium. Once the flow becomes sonic, the velocity remains constant; it cannot go supersonic. No decrease of P_2 will increase the flowrate.

Sonic Flow

In accordance with API RP 520, Part 1, Section 3.3, the formulas used for calculating orifice areas for sonic flow are:

$$A = \frac{W\sqrt{TZ}}{CKP_1\sqrt{M}}$$

$$\text{or } A = \frac{Q\sqrt{Z}}{18.3\, K\, P_1 \left(\dfrac{C}{356}\right)\sqrt{\dfrac{520}{T}}\sqrt{\dfrac{29}{M}}}$$

where: A = Calculated orifice area (in.^{2})
W = Flow capacity (lb/h)
Q = Flow capacity (scfm)
M = Molecular weight of flowing media
T = Inlet temperature, absolute (°F + 460)
Z = Compressibility factor
C = Gas constant based on ratio of specific heats (Table 1)
K = Valve coefficient of discharge
P_1 = Inlet pressure, psia (set pressure + accumulation + atmos. pres.)

Subsonic Flow

The second general category for vapor or gas sizing is generally when P_2 is greater than half of P_1 (backpressure greater than half of inlet pressure).

Using k (ratio of specific heats) and P_1/P_2 (absolute), confirm from Fig. 1 that the subsonic ("low pressure") flow formula is required. If so, then determine "F"factor. If not, use above sonic flow formula.

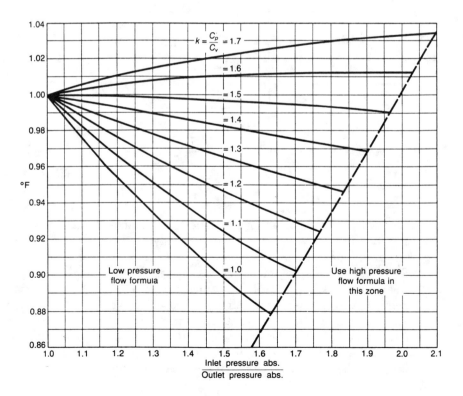

Figure 1. Low pressure and high pressure flow formulas.

$$A = \frac{Q\sqrt{GI}}{863\, K\, F\, \sqrt{(P_1 - P_2) \times P_2}}$$

where: G = Specific gravity
F = Factor obtained from Table 1
P_2 = Outlet pressure, psia (backpressure + atmos. pres.)

After determining the calculated orifice area, select the next largest standard orifice size from the relief valve manufacturer's catalog.

Selection

The fundamental selection involves the consideration of the two basic types of relief valves more commonly used; conventional spring-loaded type and pilot operated relief valves.

Advantages of the spring-loaded relief valves are:

• Competitive price at lower pressures and in smaller sizes
• Wide range of chemical compatibility
• Wide range of temperature compatibility, particularly at higher temperatures.

The disadvantages are:

• Metal-to-metal seat not tight near set pressure and usually after valve relieves
• Sensitive to conditions that can cause chatter and/or rapid cycling
• Protection against effects of backpressure is expensive, pressure limited, and creates possible, additional maintenance problems
• Testing of set pressure not easily accomplished.

Advantages of pilot operated type relief valves are:

• Seat tight to set pressure (Fig. 2)
• Ease of setting and changing set and blowdown pressures
• Can achieve short blowdown without chatter
• Pop or modulating action available
• Easy maintenance
• Easy verification of set pressure without removing relief valve from service
• Flexible—options for remote operation, backpressure protection, valve position indication.

The disadvantages are:

• Maximum temperature limitations
• Should not be used in extremely dirty or polymerizing type service.

Installation Procedures

Safety relief valve installation requires that careful consideration be given to inlet piping, discharge piping, pressure sense lines (where used) and bracing if required. Any marginal installation practice can render the safety relief valve

TABLE 1—Gas constant based on ratio of heats

Gas	Mol. Wt.	C_p/C_v	C
Acetylene	26	1.26	343
Air	29	1.40	356
Ammonia	17	1.31	348
Argon	40	1.67	378
Benzene	78	1.12	329
Butadiene	54	1.12	324
Carbon dioxide	44	1.28	345
Carbon monoxide	28	1.40	356
Ethane	30	1.19	336
Ethylene	28	1.24	341
Freon 22	86.47	1.18	335
Helium	4	1.66	377
Hexane	86	1.06	322
Hydrogen	2	1.41	357
Hydrogen sulfide	34	1.32	349
Methane	16	1.31	348
Methyl mercaptan	48.11	1.20	337
N-Butane	58	1.09	326
Natural gas (0.60)	17.4	1.27	344
Nitrogen	28	1.40	356
Oxygen	32	1.40	356
Pentane	72	1.07	323
Propane	44	1.13	330
Propylene	42	1.15	332
Sulfur dioxide	64	1.29	346
VCM	62.5	1.18	335

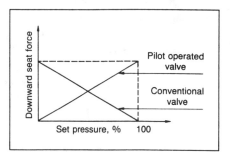

Figure 2. Setting seat pressure to set flow pressure.

inoperable, at worst, or severely restrict its design capacity at best. Either condition compromises the safety of the installation.

Inlet Piping

The proper design of inlet piping is extremely important.
Very often, safety relief valves are added to an installation at the most physically convenient location, with little regard to flow considerations. Pressure loss during flow in a pipe always occurs. Depending upon the size, geometry and inside surface condition of the pipe, the pressure loss may be large (20, 30 or 40 percent) or small (less than 5 percent).
API RP 520, Part 2, recommends a maximum inlet pipe pressure loss to a safety relief valve of 3 percent. This pressure loss shall be the sum total of the inlet loss, line loss and

when a block valve is used, the loss through it. The loss should be calculated using the maximum rated flow through the pressure relief valve.

The 3 percent maximum inlet loss is a commendable recommendation but often very difficult to achieve. If it cannot be achieved, then the effects of excessive inlet pressure should be known. These effects are rapid or short cycling with direct spring operated valves or resonant chatter with pilot operated relief valves. In addition, on pilot operated relief valves, rapid or short cycling may occur when the pilot pressure sensing line is connected to the main valve inlet. Each of these conditions results in a loss of capacity.

Pilot operated valves can tolerate higher inlet losses when the pilot senses the system pressure at a point not affected by inlet pipe pressure drop. However, even though the valve operates satisfactorily, reduced capacity will still occur because of inlet pipe pressure losses. The sizing procedure should consider the reduced flowing inlet pressure when required orifice area, "A", is calculated.

A conservative guideline to follow is to keep the equivalent L/D ratio (length/diameter) of the inlet piping to the relief valve inlet to five or less.

Rapid Cycling

Rapid valve cycling with direct spring operated valves occurs when the pressure at the valve inlet decreases excessively at the start of relief valve flow. Fig. 3 is a schematic of system pressures that is shown before and during flow. Before flow begins, the pressure is the same in the tank and at the valve inlet. During flow, the pressure at the valve inlet is reduced due to the pressure loss in the inlet pipe to the valve. Under these conditions, the valve will cycle at a rapid rate.

The valve responds to the pressure at its inlet. When that pressure decreases during flow to below the valve's reseat point, the valve will close. However, as soon as the flow stops, the inlet pipe pressure loss becomes zero and the pressure at the valve inlet rises to tank pressure once again. If the tank

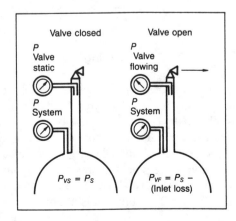

Figure 3. System pressures before and after flow.

pressure is still equal to or greater than the relief valve set pressure, the valve will open and close again. Rapid cycling reduces capacity and is sometimes destructive to the valve seat in addition to subjecting all the moving parts in the valve to excessive wear.

On pilot operated relief valves with the pilot valves sense line connected at the main valve inlet, rapid cycling can also occur for the same reasons described above; namely the pressure at the relief valve inlet is substantially reduced during flow (Fig. 4). With the pilot sense line located at the main valve inlet, the pilot valve responds to the pressure at the location and therefore closes the main valve. If the pressure loss during flow is small (less than the reseat pressure setting of the pilot) the main valve may or may not partially close depending on type of pilot design used.

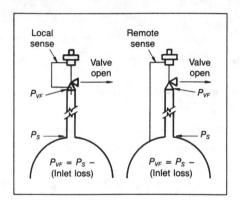

Figure 4. Schematic of rapid cycling.

In actual application, because of the long response time of a pilot operated valve compared with a direct spring operated valve, the main valve may only go into partial lift since the inlet pipe pressure loss occurs before the pilot has sufficiently vented the dome area for full piston travel.

As with direct spring operated valves, capacity is reduced but under severe inlet pressure loss conditions resonant chatter may be induced in pilot operated relief valves. This chatter is extremely destructive and may result in complete valve failure.

On existing installations the corrective action is limited. For direct spring operated valves, increasing the blowdown will minimize or eliminate rapid cycling if the blowdown can be adjusted to below the initial flowing pressure. For example:

Resonant Chatter

Resonance chatter occurs with pilot operated relief valves when the inlet piping produces excessive pressure losses at the valve inlet. The higher the set pressure or the greater the inlet pipe pressure loss, the more likely resonant chatter will occur.

Unlike the rapid cycling noted above, resonant chatter is uncontrolled; that is, once started, it cannot be stopped unless the pressure is removed from the valve inlet. In actual application, however, the valve will self-destruct before a shutdown can take place because of the magnitude of the forces involved in a resonant mode.

Discharge Piping

Discharge piping for direct spring operated valves is more critical than for pilot operated valves. As with inlet piping, pressure losses occur in discharge headers with large equivalent *L/D* ratios. Excessive backpressure will reduce the lift of a direct spring operated valve and enough backpressure (15 to 25 percent of set + overpressure) will cause the valve to reclose. As soon as the valve closes, the backpressure in the discharge header decreases and the valve opens again with the result that rapid cycling can occur.

Pilot operated relief valves with the pilot vented to the atmosphere or with a pilot balanced for backpressure are not affected by backpressure. However, if the discharge pressure can ever exceed the inlet pressure, a back flow preventer must be used.

The valve relieving capacity for either direct or pilot operated relief valves can be affected by backpressure if the flowing pressure with respect to the discharge pressure is below critical (subsonic flow).

Balanced bellows valves (direct spring operated) have limitations on maximum permissible backpressure due to the collapse pressure rating of the bellows element. This limitation will in some cases be less than the backpressure limit of a conventional valve. Manufacturer's literature should be consulted in every case. If the bellows valve is used for systems with superimposed backpressure, the additional built-up backpressure under relieving conditions must be added to arrive at maximum backpressure.

The performance of pressure relief valves, both as to operation and flow capacity, can be achieved through the following good discharge piping practices:

• Discharge piping must be at least the same size as the valve outlet connection and may be increased when necessary to larger sizes
• Flow direction changes should be minimized, but when necessary use long radius elbows and gradual transitions

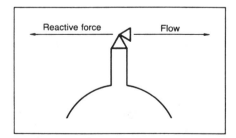

Figure 5. Reactive forces during valve relief.

• If valve has a drain port on outlet side, it should be vented to a safe area. Avoid low spots in discharge piping, preferably pitch piping away from valve outlet to avoid liquid trap at valve outlet
• Proper pipe supports to overcome thermal effects, static loads due to pipe weight and stresses that may be imposed due to reactive thrusts forces must be considered.

Reactive Force

On large orifice, high pressure valves, the reactive forces during valve relief are substantial and external bracing may be required (Fig. 5).

API RP 520, Part 2 gives the following formula for calculating this force.

$$F = \frac{Q_h}{366} \sqrt{\frac{kT}{(k+1)M}} \qquad (1)$$

where: F = Reactive force at valve outlet centerline (lb)
Q_h = Flow capacity (lb/h)
k = Ratio of specific heats (C_p/C_v)
T = Inlet temperature, absolute (°F + 460)
M = Molecular weight of flowing media.

If bracing is not feasible, a dual outlet valve (available in some pilot operated safety relief valves) can be used. The reactive forces are equal but opposite, resulting in zero force on the valve outlet, but if redirected can still impose loads that must be reacted in some manner.

An approximate method for calculating the reactive force is to multiply the following pressure (psig) by the orifice area (in.²); the reactive force obtained will be on the high side.

Example	A-G Type 22312R68 "R" Orifice (16.00 in.²)
Set pressure–	1,000 psi, 10% overpressure
Flowing media–	Natural gas, 60°F
	M = 17.4 (G = 0.60)
	C = 344 (k = 1.27)

Reactive force = 10,143 lb, using equation (1)
Reactive force = 17,600 lb, using approximate method.

Testing

Using DOT Title 49, Part 192.739 as a guideline, each valve should be inspected at least once a year to determine that it:
• Is in good mechanical condition
• Will operate at the correct set pressure
• Has adequate capacity and is operationally reliable
• Is properly installed and protected from dirt, liquids, or other conditions that might prevent proper operation.

The effectiveness of a safety relief valve installation depends greatly on proper sizing and selection of the valve type, suitable installation conditions and proper and timely testing of the valve.

Variable Orifice Rotary Control Valves

Flow Coefficient (C_v) The flow coefficient, which describes the flow capacity of a valve, is an empirically determined constant equal to the valve's flow capacity in U.S. gallons per minute of water at 60°F when a pressure drop of 1 psi exists across the valve in its most wide-open position. Table 1 lists C_v's for all sizes of DANTROL-270 valves, with and without trim.

The flow capacity of a valve in a partially open position may be calculated by selecting the appropriate C_v from Table 1 and multiplying it times the percent of maximum C_v at the desired position on the selected flow curve in Figure 2.

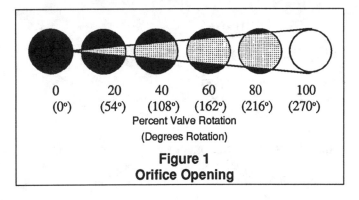

0	20	40	60	80	100
(0°)	(54°)	(108°)	(162°)	(216°)	(270°)

Percent Valve Rotation
(Degrees Rotation)

**Figure 1
Orifice Opening**

Table 1
Maximum Flow Coefficients (C_v)

Valve Size	Ball Dia.	Slot Length	No. Plates Each Slot	Cv w/Trim	Cv w/o Trim
1	4.25	7.40	8	45	53
1 1/2	5.50	9.58	16	98	120
2	6.50	11.32	16	195	220
3	8.00	13.93	18	410	480
4	9.75	16.58	20	765	860
6	13.25	23.08	25	1900	1990

Sizing Valves For Liquid Flow

These equations for single-phase Newtonian liquid flow assume that the liquid is non-viscous and neither cavitating nor flashing. (Factors and constants are subsequently outlined.)

$$C_v = \frac{Q}{N_1 F_P F_R} \sqrt{\frac{G_F}{(P_1 - P_2)}}$$

or

$$C_v = \frac{W}{N_6 F_P F_R \sqrt{(P_1 - P_2)\gamma}}$$

The Piping Geometry Factor (F_P)

The piping geometry factor (F_P) is equal to 1.0 for line-size valves. With reducers, use the value of F_P from the "Valve Reference Data" - Table 2 or 4.

The Reynold's Number Factor (F_R)

The Reynold's number factor (F_R) has a value of 1.0 for most flows in the process industry. However, sizing errors due to low Reynold's numbers (RE_v) can occur when the fluid viscosity is high or flow velocities low. In either case, RE_v can be calculated from these equations:

$$RE_v = \frac{N_3 W}{v \sqrt{F_{LP} Cv}} \left[\frac{(F_{LP} Cv)^2}{N_2 D^4} + 1 \right]^{0.25}$$

or

$$RE_v = \frac{N_4 Q}{v \sqrt{F_{LP} Cv}} \left[\frac{(F_{LP} Cv)^2}{N_2 D^4} + 1 \right]^{0.25}$$

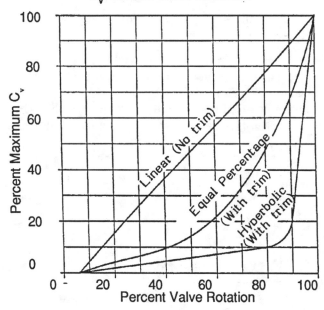

**Figure 2   Percent of Maximum
C_v versus Valve Rotation**

Using the calculated RE_v, the value of F_R can be found from the graph in Figure 3. Substitute F_L for F_{LP} with line-size valves. (Values for F_L and F_{LP} are in the "Valve Reference Data" - Table 2 or 4.)

Choked Flow

Choked flow is a limiting flow which occurs as a result of liquid vaporization when the internal valve pressure falls below the liquid's vapor pressure. Cavitation has then fully developed within the valve, and an increase in pressure drop across the valve will not produce an increase through the valve if:

$$\Delta P \geq F_{LP}^2 (P_1 - F_F P_v)$$

(Substitute F_L for F_{LP} with line-size valves.) If the valve must be operated under choked conditions, use the following choked-flow equations in place of the basic equation for valve flow capacity:

$$C_v = \frac{Q}{N_1 F_{LP}} \sqrt{\frac{G_F}{(P_1 - F_F P_v)}}$$

or

$$C_v = \frac{W}{N_6 F_{LP} \sqrt{(P_1 - F_F P_v)\gamma}}$$

Where: $F_F = 0.96 - 0.28 \dfrac{P_v}{P_c}$

(Use F_L for F_{LP} for line-size valves)

The Cavitation Index (K_c)

The cavitation index (K_c) relates to a condition just prior to flow choking where cavitation has only partially developed

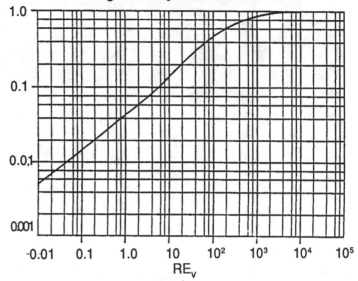

Figure 3 Reynolds Number Factor

and flow rate is no longer proportional to the square root of pressure drop. This condition occurs when:

$$\Delta P_c \geq K_c (P_1 - P_v)$$

Values for the cavitation index (K_c) can be found in the Valve Reference Data - Table 2 or 4.

Anti-Cavitation Trim

Generally, operating a valve at conditions corresponding to K_c or choked flow will result in damaging cavitation, noise and vibration.

If cavitation exists, a valve with anti-cavitation trim will prevent or lessen the effects of cavitation. When anti-cavitation trim is used, recalculate the flow-capacity coefficient (C_v) using factors for "With Trim" in the Valve Reference Data - Table . If noise is a concern, see procedure under "Hydrodynamic Noise Prediction" in manufacturer's literature.

SIZING VALVES FOR GAS AND VAPOR

Basic Valve Flow-Capacity Coefficient (Cv)

The basic valve flow coefficient (C_v) for gas or vapor flow through a variable orifice valve may be calculated using any of the following equivalent formulae for non-choked flow:

$$C_v = \frac{Q}{N_7 F_P P_1 Y} \sqrt{\frac{G_G T_1 Z}{X}}$$

or

$$C_v = \frac{W}{N_6 F_P Y \sqrt{X P_1 \gamma_1}}$$

or

$$C_v = \frac{Q}{N_9 F_P P_1 Y} \sqrt{\frac{M T_1 Z}{X}}$$

or

$$C_v = \frac{W}{N_8 F_P P_1 Y} \sqrt{\frac{T_1 Z}{X M}}$$

where: $F_k = \dfrac{K}{1.40}$ and $X = \dfrac{P_1 - P_2}{P_1}$

and $Y = 1 - \dfrac{X}{3 F_K X_{TP}} \leq 0.667*$

* See "Choked Flow" for explanation.

Substitute X_T for X_{TP} with line-size valves. Values for F_p, XT and X_{TP} are found in the "Valve Reference Data" - Table 2 or 4.

The Compressibility Factor (Z)

The compressibility factor (Z) can be found from Figure 4 after the reduced pressure (P_R) and reduced temperature (T_R) have been calculated from these equations:

$$P_R = \frac{\text{Inlet Pressure (Absolute)}}{\text{Critical Pressure (Absolute)}}$$

$$T_R = \frac{\text{Inlet Temperature (Absolute)}}{\text{Critical Temperature (Absolute)}}$$

($^\circ$R = $^\circ$F + 460, $^\circ$K = $^\circ$C + 273)

Choked Flow

Gas flow through valves can choke when the gas velocity reaches the velocity of sound. The pressure drop ratio at this point has the value:

$$X = F_K X_{TP}$$

Substitute X_T for X_{TP} with line-size valves. If this valve is to be used under choked conditions, Y is fixed at 0.667 regardless of the pressure drop. Substitute 0.667 for Y and $F_K X_T$ or $F_K X_{TP}$ for X in basic value flow-capacity coefficient equation.

Valve Outlet Velocty

Sonic velocity at the valve's outlet will cause shock waves which produce vibration and noise problems. Therefore, you should make sure the outlet velocity does not reach sonic velocity for gas and steam service. The actual valve outlet diameter must be larger than the calculated diameter from the following equations:

$$d = N_{10} \sqrt{\frac{Q \sqrt{G_G T}}{P_2}} \quad \text{for gas}$$

$$d = N_{11} \sqrt{\frac{W (1 + N_{12} T_{SH})}{P_2}} \quad \text{for steam}$$

Low-Noise Trim

When the pressure-drop ratio (X) is near or exceeds the minimum value for choked flow, check the valve for excessive noise using the procedure under "Aerodynamic Noise Prediction". If noise is excessive, use anti-noise trim; recalculate the flow-capacity coefficient (C_v) using factors from "With Trim" in the "Valve Reference Data" - Table 4. For procedures on noise level prediction, refer to the manufacturer's literature.

Reprinted with permission: Daniel Valve Company

Figure 4 Compressibility versus Reduced Pressure

Table 2
Valve Reference Data (No Trim)

	PERCENT VALVE ROTATION												
Value	10	20	30	40	50	60	70	80	85	90	95	100	D/d
F_L	0.95	0.94	0.93	0.91	0.89	0.85	0.78	0.69	0.65	0.60	0.56	0.51	
X_T	0.86	0.85	0.83	0.80	0.75	0.65	0.49	0.32	0.24	0.17	0.10	0.08	1.00
K_c	0.78	0.77	0.73	0.69	0.61	0.55	0.48	0.41	0.36	0.31	0.25	0.18	
F_P	1.0	1.0	1.0	1.0	1.0	0.99	0.98	0.95	0.91	0.86	0.78	0.62	
F_{LP}	0.95	0.94	0.93	0.91	0.88	0.83	0.76	0.64	0.59	0.52	0.45	0.35	1.25
X_{TP}	0.86	0.85	0.83	0.80	0.75	0.65	0.50	0.33	0.26	0.20	0.14	0.10	
F_P	1.0	1.0	1.0	1.0	1.0	0.99	0.97	0.92	0.89	0.83	0.74	0.56	
F_{LP}	0.95	0.94	0.93	0.91	0.88	0.82	0.75	0.64	0.57	0.50	0.42	0.34	1.33
X_{TP}	0.86	0.85	0.83	0.80	0.75	0.65	0.50	0.35	0.27	0.21	0.16	0.12	
F_P	1.0	1.0	1.0	1.0	1.0	0.99	0.96	0.91	0.86	0.76	0.64	0.50	
F_{LP}	0.95	0.94	0.93	0.91	0.88	0.82	0.74	0.63	0.56	0.49	0.40	0.32	1.50
X_{TP}	0.86	0.85	0.83	0.80	0.75	0.65	0.51	0.36	0.29	0.25	0.19	0.16	
F_P	1.0	1.0	1.0	1.0	1.0	0.99	0.94	0.85	0.78	0.68	0.56	0.44	
F_{LP}	0.95	0.94	0.93	0.91	0.88	0.81	0.73	0.62	0.55	0.46	0.38	0.29	1.67
X_{TP}	0.86	0.85	0.83	0.80	0.75	0.65	0.49	0.32	0.24	0.17	0.10	0.08	
F_P	1.0	1.0	1.0	1.0	1.0	0.97	0.93	0.83	0.76	0.65	0.50	0.40	
F_{LP}	0.95	0.94	0.93	0.91	0.88	0.80	0.72	0.60	0.53	0.45	0.36	0.27	2.00
X_{TP}	0.86	0.85	0.83	0.80	0.75	0.66	0.53	0.40	0.35	0.32	0.29	0.28	
F_P	1.0	1.0	1.0	1.0	1.0	0.95	0.89	0.75	0.64	0.46	0.38	0.33	
F_{LP}	0.95	0.94	0.93	0.91	0.88	0.80	0.72	0.60	0.53	0.45	0.36	0.24	2.67
X_{TP}	0.86	0.85	0.83	0.80	0.75	0.66	0.54	0.46	0.41	0.38	0.36	0.34	

Table 3 Constants In Liquid Sizing Equations

	N	W	Q	ΔP, P	D, d	γ	ν
N_1	0.865	-	M³/HR	BAR	-	-	-
	1.000	-	GPM	PSIA	-	-	-
N_2	0.00214	-	-	-	MM	-	-
	890	-	-	-	IN	-	-
N_3	76.0	KG/HR	-	-	MM	-	cSt
	34.5	LB/HR	-	-	IN	-	cSt
N_4	76,000	-	M³/HR	-	MM	-	cSt
	17,300	-	GPM	-	IN	-	cSt
N_6	27.3	KG/HR	-	BAR	-	KG/M³	-
	63.3	LB/HR	-	PSIA	-	LB/FT³	-

Table 5 Constants In Gas/Vapor Sizing Equations

	N	W	Q	ΔP, P	D, d	γ	T	T_{SH}
N_6	27.3	KG/HR	-	BAR	-	KG/M³	-	-
	63.3	LB/HR	-	PSIA	-	LB/FT³	-	-
N_7	417	-	M³/HR	BAR	-	-	°K	-
	1360	-	SCFH	PSIA	-	-	°R	-
N_8	94.8	KG/HR	-	BAR	-	-	°K	-
	19.3	LB/HR	-	PSIA	-	-	°R	-
N_9	2250	-	M³/HR	BAR	-	-	°K	-
	7320	-	SCFH	PSIA	-	-	°R	-
N_{10}	0.25	-	M³/HR	BAR	MM	-	°K	-
	63.3	-	SCFH	PSIA	IN	-	°R	-
N_{11}	37.1	KG/HR	-	BAR	MM	-	-	°C
	0.119	LB/HR	-	PSIA	IN	-	-	°F
N_{12}	0.00126	KG/HR	-	BAR	MM	-	-	°C
	0.0007	LB/HR	-	PSIA	IN	-	-	°F

Table 4 Valve Reference Data (With Trim)

Value	PERCENT VALVE ROTATION								D/d
	10 - 50	60	70	80	85	90	95	100	
F_L	1.0	1.0	1.0	0.92	0.85	0.77	0.67	0.54	
X_T	1.0	1.0	1.0	0.87	0.76	0.61	0.37	0.20	1.00
K_c	1.0	1.0	1.0	0.90	0.83	0.69	0.46	0.27	
F_P	1.0	1.0	1.0	0.98	0.97	0.95	0.92	0.87	
F_{LP}	1.0	1.0	1.0	0.90	0.84	0.75	0.63	0.48	1.25
X_{TP}	1.0	1.0	1.0	0.89	0.80	0.65	0.41	0.22	
F_P	1.0	1.0	1.0	0.98	0.97	0.95	0.92	0.87	
F_{LP}	1.0	1.0	1.0	0.90	0.84	0.75	0.62	0.47	1.33
X_{TP}	1.0	1.0	1.0	0.89	0.80	0.66	0.44	0.23	
F_P	1.0	0.99	0.98	0.95	0.93	0.89	0.85	0.78	
F_{LP}	1.0	1.0	1.0	0.89	0.83	0.74	0.61	0.46	1.50
X_{TP}	1.0	1.0	1.0	0.89	0.81	0.68	0.47	0.27	
F_P	1.0	0.99	0.98	0.94	0.91	0.86	0.80	0.71	
F_{LP}	1.0	1.0	1.0	0.89	0.83	0.74	0.61	0.45	1.67
X_{TP}	1.0	1.0	1.0	0.89	0.81	0.69	0.50	0.30	
F_P	1.0	0.99	0.98	0.94	0.90	0.84	0.77	0.66	
F_{LP}	1.0	1.0	1.0	0.89	0.83	0.73	0.60	0.44	2.00
X_{TP}	1.0	1.0	1.0	0.90	0.82	0.71	0.55	0.36	
F_P	1.0	0.99	0.98	0.94	0.89	0.83	0.75	0.64	
F_{LP}	1.0	1.0	1.0	0.89	0.83	0.73	0.59	0.43	2.33
X_{TP}	1.0	1.0	1.0	0.90	0.82	0.72	0.57	0.39	
F_P	1.0	0.99	0.98	0.93	0.88	0.81	0.73	0.61	
F_{LP}	1.0	1.0	1.0	0.88	0.81	0.71	0.57	0.42	2.50
X_{TP}	1.0	1.0	1.0	0.92	0.87	0.80	0.71	0.60	
F_P	1.0	0.99	0.98	0.92	0.86	0.79	0.70	0.59	
F_{LP}	1.0	1.0	1.0	0.88	0.81	0.70	0.55	0.41	2.67
X_{TP}	1.0	1.0	1.0	0.93	0.88	0.82	0.75	0.65	

8: Corrosion/Coatings

Hand-held computer determines concrete coating thickness.............. 169
NAPCA specifications.. 171
How much primer for a mile of pipe? 174
How much coal-tar enamel for a mile of pipe? 175
How much wrapping for a mile of pipe?... 175
Estimating coating and wrapping materials required per mile of
    pipe 175
Coefficient of friction for pipe coating materials 176
Trouble shooting cathodic protection systems: Magnesium anode
    system 178
Cathodic protection for pipe lines.. 179
Estimate the pounds of sacrificial anode material required for
    offshore pipe lines.. 187
Comparison of other reference electrode potentials with that of
    copper-copper sulfate reference electrode at 25°C..................... 189
Chart aids in calculating ground bed resistance and rectifier
    power costs 190
How can output of magnesium anodes be predicted? 191
How to determine the efficiency of a cathodic protection rectifier..... 191
How to calculate the voltage drop in ground bed cable quickly.......... 192
What is the most economical size for a rectifier cable? 192
How to estimate the number of magnesium anodes required and
    their spacing for a bare line or for a corrosion "hot spot" 193
How can resistivity of fresh water be determined from chemical
    analysis?... 193
What will be the resistance to earth of a single graphite anode?......... 194
How to estimate the monthly power bill for a cathodic protection
    rectifier.. 194

What will be the resistance to earth of a group of graphite anodes,
    in terms of the resistance of a single anode? 194
How can the current output of magnesium rod used for the cathodic
    protection of heat exchanger shells be predicted? 194
What spacing for test leads to measure current on a pipe line?.......... 194
How many magnesium anodes are needed for supplementary
    protection to a short-circuited bare casing?.................................. 195
Group installation of sacrificial anodes 195
How can the life of magnesium anodes be predicted? 196
How to find the voltage rating of a rectifier if it is to deliver a
    given amount of current through a given ground bed
    (graphite or carbon) 196
Determining current requirements for coated lines 196
Determining current requirements for coated lines when pipe-to-soil
    potential values are estimated.. 196
HVDC effects on pipe lines.. 197
Troubleshooting cathodic protection systems: Rectifier-ground bed 201
How to control corrosion at compressor stations 202
Project leak growth... 203
Advances in Pipe Line Protection.................................204
Methods of locating coating defects... 205
Case histories 208
Estimate the number of squares of tape for pipe coating
    (machine applied) 209
Estimate the amount of primer required for tape................................ 210
Tape requirements for fittings 210

Hand-held computer determines concrete coating thickness

Technical data for the engineer's file

Frank E. Hangs, Sovereign Engineering, Inc., Houston

Pipe line crossings, under water or in unstable soil, offer many challenges to engineers.

Pipe is buoyant and an empty line may float in water. Wet silts are like viscous fluids causing inadequately weighted pipe lines to pop up. There are various government entities that have jurisdiction over navigable rivers, bays, marshlands and offshore waters. These agencies may stipulate that pipe lines be buried at certain depths and be stabilized.

A good way to stabilize a pipe line is to use an adequate concrete weight coating. Determining the thickness of the concrete involves a process of balancing upward forces such as buoyancy of the mud and the downward forces—weights of pipe, protective coating and concrete, allowing a factor of sixty (negative buoyancy). Such computations with several variables can become involved and tedious.

Computer Program

The following program, written for the Hewlett Packard-41CV (Figure 1), calculates the thickness of concrete weight coating for submarine pipe lines expeditiously and with satisfying results.

The prompting feature (Figure 2) is employed to aid user's data input. Important calculated values and a recap of the inputs are printed out and each quantity is identified. Thus, the tape is a complete record. The program is flexible in that any inputs can be readily changed for a second run. Thus, many "what if" questions can be quickly answered. Suppose a heavier-weight pipe is used? What if the concrete density is changed? If the negative buoyancy is reduced? Etc.

```
01◆LBL "CONC"      37 RCL 15         73 *              108 STO 05        144 STO 18        177 "T="          213 "SP GR="
02 ADV             38 RCL 17         74 +              109 RCL 15        145 RCL 21        178 ARCL 05       214 ARCL 14
03 SF 12           39 +              75 X↑2            110 "D=? "        146 "T INCR=? "   179 AVIEW         215 AVIEW
04 " CONCRETE"     40 LASTX          76 .0055          111 ARCL 15       147 ARCL 21       180 CLA           216 CLA
05 "├ COATING"     41 *             77 *              112 PROMPT        148 PROMPT        181 "WPC="        217 ADV
06 PRA             42 .0218          78 STO 12         113 CLA           149 CLA           182 ARCL 06       218 "D="
07 ADV             43 *              79 RCL 20         114 STO 15        150 STO 21        183 AVIEW         219 ARCL 15
08 CF 12           44 RCL 18         80 *              115 RCL 16        151 RCL 11        184 CLA           220 AVIEW
09 "T TRIAL=?"     45 *              81 STO 09         116 "WP=? "       152 "NEG BUOY=? " 185 "WC="         221 CLA
10 PROMPT          46 STO 06         82 RCL 08         117 ARCL 16       153 ARCL 11       186 ARCL 07       222 "WP="
11 STO 05          47 RCL 15         83 X<>Y           118 PROMPT        154 PROMPT        187 AVIEW         223 ARCL 16
12 "PIPE DIAM=?"   48 RCL 17         84 -              119 CLA           155 CLA           188 CLA           224 AVIEW
13 PROMPT          49 2              85 STO 10         120 STO 16        156 STO 11        189 "WT="         225 CLA
14 STO 15          50 *              86 FS? 01         121 RCL 17        157 GTO 01        190 ARCL 08       226 "TH PC="
15 "PIPE WT=?"     51 +              87 GTO 03         122 "TH PC=? "    158◆LBL 02        191 AVIEW         227 ARCL 17
16 PROMPT          52 RCL 05         88 RCL 11         123 ARCL 17       159 RCL 21        192 CLA           228 AVIEW
17 STO 16          53 +              89 X=0?           124 PROMPT        160 ST+ 05        193 "WMD="        229 CLA
18 "PROT CT TH=?"  54 LASTX          90 GTO 04         125 CLA           161 GTO 01        194 ARCL 09       230 "DC="
19 PROMPT          55 *              91 RCL 10         126 STO 17        162◆LBL 03        195 AVIEW
20 STO 17          56 .0218          92 RCL 11         127 RCL 19        163 RCL 08        196 CLA           231 ARCL 19
21 "DEN CONC=?"    57 *              93 X<=Y?          128 "DC=? "       164 RCL 12        197 "WT-WMD="     232 AVIEW
22 PROMPT          58 RCL 19         94 GTO 03         129 ARCL 19       165 /             198 ARCL 10       233 CLA
23 STO 19          59 *              95 GTO 02         130 PROMPT        166 STO 13        199 AVIEW         234 "DM="
24 "DEN MUD=?"     60 STO 07         96◆LBL 04         131 CLA           167 62.4          200 CLA           235 ARCL 20
25 PROMPT          61 RCL 16         97 RCL 09         132 STO 19        168 /             201 "NEG BUOY="   236 AVIEW
26 STO 20          62 +              98 RCL 08         133 RCL 20        169 STO 14        202 ARCL 11       237 CLA
27 "DEN PROT CT=?" 63 RCL 06         99 X>Y?           134 "DM=? "                         203 AVIEW         238 "DPC="
28 PROMPT          64 +              100 GTO 03        135 ARCL 20       170◆LBL "DATA"    204 CLA           239 ARCL 18
29 STO 18          65 STO 08         101 GTO 02        136 PROMPT        171 ADV           205 "VT="         240 AVIEW
30 "T INCR=?"      66 RCL 15         102◆LBL 05        137 CLA           172 CLA           206 ARCL 12       241 CLA
31 PROMPT          67 RCL 17         103 RCL 05        138 STO 20        173 "RESULTS-"    207 AVIEW         242 "T INCR="
32 STO 21          68 2              104 "T=? "        139 RCL 18        174 "├RECAP."     208 CLA           243 ARCL 21
33 "NEG BUOY=?"    69 *              105 ARCL 05       140 "DPC=? "      175 AVIEW         209 "BULK DEN="   244 AVIEW
34 PROMPT          70 +              106 PROMPT        141 ARCL 18       176 ADV           210 ARCL 13       245 CLA
35 STO 11          71 RCL 05         107 CLA           142 PROMPT                          211 AVIEW         246 BEEP
36◆LBL 01          72 2                                143 CLA                             212 CLA           247 END
```

Figure 1. Hewlett Packard-41 CV computer program.

Nomenclature and Storage Registers Inputs

Reg. no.

05	T	=	Concrete thickness, in. Assume trial value. Final answer is cumulated in Reg. 05.
15	D	=	Pipe diameter, in.
16	WP	=	Weight of pipe, lb/lf.
17	TH PC	=	Thickness of protective coating. Usually 3/32 in. or 0.0938 in.
19	DC	=	Density of concrete, lb/cu ft. Usually 140, 165, 190.
20	DM	=	Density of mud or fluid, lb/cu ft. River mud is around 90.
18	DPC	=	Density of protective coating, lb/cu ft. 95 is average.
21	T INCR	=	Increment of T. Decimal of in. 0.125 is suggested for numerical precision—not that a tolerance of this order can be constantly achieved in the pipe coating world. Increment can be made smaller, if desired.
11	NEG BUOY	=	Negative buoyancy. Desired minimum value, lb/lf. Can be 0 in special circumstances—e.g., to ascertain if a given pipe will float in water, or will a certain coated pipe sink in mud?

Each value keyed in is printed on the tape as a check of inputs and the program prompting feature queries user for value of next named input (Figure 2).

Calculated Values

Reg. no.

06	WPC	=	Weight of protective coating, lb/lf.
07	WC	=	Weight of concrete coating, lb/lf.
08	WT	=	Total weight—pipe and coatings, lb/lf.
09	WMD	=	Weight of mud (fluid) displaced, lb/lf.
10	WT-WMD	=	Actual negative buoyancy, lb/lf.
12	VT	=	Total volume displaced by pipe and coatings, cu ft/lf.
13	BULK DEN	=	Bulk density, wt/vt. lb/cu ft.
14	SP GR	=	Specific gravity, bulk den/den water.

Formulas

$$WPC = (D + (THPC))(THPC)\ 0.0218\ (DPC)$$

$$WC = (D + 2(THPC) + T)\ T\ 0.0218\ (DC)$$

```
T TRIAL=?                        RESULTS-RECAP.
        3.0000    RUN            T=4.6250
PIPE DIAM=?                      WPC=4.6805
       24.0000    RUN            WC=479.3300
PIPE WT=?                        WT=578.6305
       94.6200    RUN            WMD=553.4462
PROT CT TH=?                     WT-WMD=25.1843
        0.0938    RUN            NEG BUOY=20.0000
DEN CONC=?                       VT=6.1494
      165.0000    RUN            BULK DEN=94.0954
DEN MUD=?                        SP GR=1.5079
       90.0000    RUN            D=24.0000
DEN PROT CT=?                    WP=94.6200
       95.0000    RUN            TH PC=0.0938
T INCR=?                         DC=165.0000
        0.1250    RUN            DM=90.0000
NEG BUOY=?                       DPC=95.0000
       20.0000    RUN            T INCR=0.1250
```

Figure 2. Most common questions asked when checking coating buoyancy.

$$VT = (D + 2(THPC) + 2T)^2\ 0.0055$$

$$WMD = (VT)(DM)$$

Note: User is encouraged to employ most reliable and suitable data available; such as density of protective coating, mud, concrete, etc.

Examples

Put program into calculator, XEQ Size 30. The printed RESULTS-RECAP portions of each of three examples are given. When the user XEQ "CONC" and inputs the data from the lower part of the tabulation, he will get the same calculated values above.

Figure 2 includes the most common questions asked when checking coating buoyancy. Always be sure flag 01 is clear XEQ "CONC" and put in new trial T and R/S. Also R/S after all other inputs. After all prompted data are keyed in, the processing starts and results are printed out. Notice T = 4.6250 in. has an actual negative buoyancy of 25.1843. If T = 4.5 in. the actual negative buoyancy = 18.4510 which is less than 20. Precise solution is between 4.5 in. and 4.625 in.

If user wishes to find T which produces an actual negative buoyancy nearer 20: STO 4.5 in RO 5, STO 1/16 in. (0.0625 in.), in R21, XEQ 01. This produces T = 4.5625 where actual negative buoyancy is 1.8113 greater than 20. This refinement can be found in choosing small T INCR. Note weight of concrete coating is now 471.8269 lb/lf and we have approached closer to 20.

Calculated values. Weight of protective coating and concrete, total weight and volume of pipe and coatings per linear foot will be of value in transporting coated pipe, etc.

Note: Results—RECAP of any run (which is in storage) can be reprinted as often as desired by XEQ "DATA."

If NEG BUOY is 0 and flag 01 is not set, then the program causes WT to approach WMD and calculates for this situation.

If a printer is not available, results can be obtained as follows:

1. XEQ 05, put in data and R/S after each item.
2. When beep sounds, RCL registers one by one in order given above (so as to identify values).

Figure 3 illustrates the changing of two input values: namely the pipe diameter and weight. XEQ 05, this routine calls up the contents of the input registers one at a time and questions them. Always put in a new T value. Then R/S.

When D = ? comes up, key in new value and R/S. Same with pipe WT = ?. Assume other data remain unchanged, R/S each time. (Any one value or any combination of values may be changed before rerunning the program.)

Figure 4 determines if a given pipe, 14 in. at 36.71 lb/lf and ½ in. of protective weight coating density of 135 lb/cu ft will float. T, DC, T INCR and NEG BUOY are all zero. DM becomes the density of water 62.4. Set flag 01 and XEQ 05. Key in data and R/S. When all data are in place only one computation is allowed, then printed out. WT-WMD tells us the upward forces exceed the downward forces by 19.1733 lb/lf, also the specific gravity is 0.7517 so the pipe floats. Also note BULK DEN is *less* than density of water.

```
                      XEQ 05              RESULTS-RECAP
        T=? 4.6250
                     3.0000    RUN        T=3.2500
        D=? 24.0000                       WPC=3.9034
                     20.0000   RUN        WC=273.9914
        WP=? 94.6200                      WT=375.7248
                     97.8300   RUN        WMD=352.5529
        TH PC=? 0.0938                     WT-WMD=23.1720
                               RUN        NEG BUOY=20.0000
        DC=? 165.0000                     VT=3.9173
                               RUN        BULK DEN=95.9154
        DM=? 90.0000                      SP GR=1.5371
                               RUN
        DPC=? 95.0000                     D=20.0000
                               RUN        WP=97.8300
        T INCR=? 0.1250                   TH PC=0.0938
                               RUN        DC=165.0000
        NEG BUOY=? 20.0000                DM=90.0000
                               RUN        DPC=95.0000
                                          T INCR=0.1250
```

Figure 3. Changing of two input valves.

```
            XEQ "DATA"
                                   BULK DEN=46.9065
    RESULTS-RECAP.                 SP GR=0.7517

    T=0.0000
    WPC=21.3368                    D=14.0000
    WC=0.0000                      WP=36.7100
    WT=58.0468                     TH PC=0.5000
    WMD=77.2200                    DPC=135.0000
    WT-WMD=-19.1733                DC=0.0000
    NEG BUOY=0.0000                DM=62.4000
    VT=1.2375                      T INCR=0.0000
```

Figure 4. Result of calculations to determine whether the pipe will float.

NAPCA specifications*

RECOMMENDED SPECIFICATION DESIGNATIONS FOR ENAMEL COATINGS

These are recommended enamel coating system designations for coal tar or asphalt coatings, except for specialty requirements.

SPECIFICATION SYMBOL *PART #1	**PART #2	SHOT BLAST	PRIMER	93.75 MILS MIN. ENAMEL	FIBERGLASS WRAP	#15 FELT WRAP	SEAL COAT	GLASS MAT	POLYETHYLENE WITH KRAFT	HEAVY KRAFT PAPER	ELECTRICAL INSPECTION
TF—		X	X	X		X				X	X
AF—		X	X	X		X				X	X
TG—		X	X	X	X					X	X
AG—		X	X	X	X					X	X
TGF—		X	X	X	X	X				X	X
AGF—		X	X	X	X	X				X	X
TGSF—		X	X	X	X	X	X			X	X

* NOTE: These are not all the specifications approved by NAPCA. For other specifications consult a NAPCA member.

AGSF—	X	X	X	X	X	X			X	X
TGM—	X	X	X				X		X	X
AGM—	X	X	X				X		X	X
TGMF—	X	X	X		X		X		X	X
AGMF—	X	X	X		X		X		X	X
TGMP—	X	X	X				X	X		X
AGMP—	X	X	X				X	X		X

*PART #1 SYMBOLS— ENAMEL & WRAPPERS

T—Coal Tar   G—Fiberglass
A—Asphalt   S—Seal Coat
F—#15 Felt   GM—Glass Mat
P—Polyethylene wrapper
  w/kraft paper
O—Outer wrap
  To specify Outer Wrap, use
  letter O in place of F.

**PART #2 SYMBOLS GRADES OF ENAMEL

3— Fully-Plasticized Coal Tar
  (ordered by penetration
  dictated by ambient temperature)
4— High Temperature Surface Coal Tar
7— Asphalt

EXAMPLE: 93.75 mils (minimum) fully-plasticized coal tar enamel wrapped with fiberglass, felt and kraft — the symbol would be TGF-3.

Reprinted courtesy of National Association of Pipe Coating Applicators.

STANDARD APPLIED PIPE COATING WEIGHTS FOR NAPCA COATING SPECIFICATIONS

Refer to NAPCA 1-65-83

Spec. Symbol		TF 3-4	AF-7	TG 3-4	AG-7	TGF -3-4	AGF-7
PIPE SIZE				Weight in Pounds—Per 100 Lineal Feet			
1/2"	I.D.	21	18	17	14	21	18
3/4"	"	26	22	21	17	26	22
1"	"	32	27	26	22	33	28
1 1/4"	"	40	34	33	27	41	35
1 1/2"	"	46	39	38	31	46	40
2 3/8"	O.D.	57	48	47	39	57	49
2 7/8"	"	68	58	57	47	69	59
3 1/2"	"	83	71	70	58	84	72
4"	"	95	81	80	66	96	82
4 1/2"	"	106	91	90	74	108	92
5 9/16"	"	130	111	111	91	132	113
6 5/8"	"	155	132	132	109	157	134
8 5/8"	"	201	171	171	141	204	174
10 3/4"	"	250	213	214	176	254	216
12 3/4"	"	296	252	253	209	300	256
14"	"	325	276	278	229	329	281
16"	"	372	317	318	262	377	322
18"	"	419	356	357	295	424	362
20"	"	463	393	397	327	469	400
22"	"	510	433	436	360	516	440
24"	"	555	472	476	392	563	480
26"	"	601	511	516	425	610	519
28"	"	648	550	555	458	656	559
30"	"	692	588	595	490	702	598
32"	"	739	627	634	523	749	638
34"	"	785	667	674	556	796	678
36"	"	831	706	714	589	842	717
42"	"	970	824	832	687	983	837

(continued on next page)

(table continued)

Spec. Symbol		TGM 3-4	AGMP 7	TGMP 3-4	AGM 7	TGMF 3-4	AGMF 7	TGSF 3-4	AGFS 7
PIPE SIZE		... Weight in Pounds — Per 100 Lineal Feet							
1/2"	I.D.	17	14	17	14	21	18	26	22
3/4"	"	21	17	21	17	26	22	33	28
1"	"	26	22	26	22	33	28	41	35
1 1/4"	"	33	27	33	27	41	35	51	43
1 1/2"	"	38	31	38	31	46	40	58	50
2 3/8"	O.D.	47	39	47	39	57	49	72	61
2 7/8"	"	57	47	57	47	69	59	87	74
3 1/2"	"	70	58	70	58	84	72	106	90
4"	"	80	66	80	66	96	82	121	102
4 1/2"	"	90	74	90	74	108	92	136	115
5 9/16"	"	111	91	111	91	132	113	167	141
6 5/8"	"	132	109	132	109	157	134	198	168
8 5/8"	"	171	141	171	141	204	174	258	218
10 3/4"	"	214	176	214	176	254	216	321	271
12 3/4"	"	253	209	253	209	300	256	380	321
14"	"	278	229	278	229	329	281	417	352
16"	"	318	262	318	262	377	322	477	403
18"	"	357	295	357	295	424	362	537	453
20"	"	397	327	397	327	469	400	594	501
22"	"	436	360	436	360	516	440	654	552
24"	"	476	392	476	392	563	480	713	602
26"	"	516	425	516	425	610	519	772	651
28"	"	555	458	555	458	656	559	832	702
30"	"	595	490	595	490	702	598	889	750
32"	"	634	523	634	523	749	638	949	800
34"	"	674	556	674	556	796	678	1008	850
36"	"	714	589	714	589	842	717	1067	900
42"	"	832	687	832	687	983	837	1245	1051

All weights provide for standard cutbacks.

Weights per 100 lineal feet are figured on the following material weights per square foot of pipe surface:

	Weight Lbs.—Per Sq. Ft.		Weight Lbs.—Per Sq. Ft.
Coal Tar Primer & Enamel—1/32" Min.	.2461	#15 Pipeline Felt (Tar or asphalt saturated)	.1271
Coal Tar Primer & Enamel—2/32" Min.	.4850	Outerwrap	.1225
Coal Tar Primer & Enamel—3/32" Min.	.7238	.010 . . . Fiberglass Wrap	.0110
Asphalt Primer & Enamel—1/32" Min.	.2000	Kraft Paper	.0200
Asphalt Primer & Enamel—2/32" Min.	.3950	Glass Mat	.0114
Asphalt Primer & Enamel—3/32" Min.	.5912	Polyethylene	.0450

Reprinted courtesy of National Association of Pipe Coating Applicators.

NAPCA SPECIFICATIONS PIPE LINE FELTS

PROPERTY	#15 TAR N.R.	#15 TAR R.F.	#15 ASPH. N.R.	#15 ASPH. R.F.
Saturated Felt — Not Perforated				
Weight (Min.) lbs. per 100 s.f. (ASTM-D-146-59 Sec. 1-9)	12	12	12	12
Saturation (Percent) MIN. (ASTM-D-146-59 Sec. 18 MAX. by extraction)	18 28	18 28	22 40	22 40
Ash Content (MIN).) (Percent) (ASTM-D-146-59 Sec. 21)	70	70	70	70
Caliper (Min.) (Inches) (ASTM-D-645-64T Method C)	.021	.018	.021	.018
Tensile Strength (Min.) (lbs. per inch of width) (ASTM-D-146-59 Sec. 12a) With fiber grain Across fiber grain	35 12	37 No. spec.	35 12	37 No. spec.
Tear Strength (Min.) (Grams) (ASTM-D-689-62) Across fiber grain With fiber grain	400 240	Not applicable	400 240	Not applicable

(continued on next page)

Loss on heating (Max.) (Percent) (AWWA C203-66 Sec. 2.6.6)	10	10	10	10
Pliability (AWWA C203-66 Sec. 2.6.4)	No cracking	No cracking	No cracking	No cracking
Saturated and Perforated Felt				
(1/16" perforations on 1" Staggered Centers)				
Tensile Strength (Min.) (lbs. per inch of width) (ASTM-D-146-59 Sec. 12a) With fiber grain Across fiber grain	30 10	35 No spec.	30 10	35 No. spec.
Tear Strength (MIN.) (Grams) (ASTM-D-689-62) Across fiber grain With fiber grain	348 190	Not applicable	348 190	Not applicable
Unsaturated Felt				
Asbestos Content (%) (MIN.) (ASTM-D-1918-67)	85	85	85	85

N.R. — NON REINFORCED
R.F. — REINFORCED LENGTHWISE WITH PARALLEL FIBER GLASS YARNS ON NOMINAL ¼" CENTERS.

Reprinted courtesy of National Association of Pipe Coating Applicators.

Minimum Test Voltages for Various Coating Thicknesses

Coating Thicknesses 32nd Inch	Mils	Test Volts
--	16	5,000
1	31	7,000
2	62	9,800
3	94	12,100
4	125	14,000
5	156	15,000
6	188	17,100
16	500	28,000
20	625	31,000
24	750	35,000

Reprinted courtesy of National Association of Pipe Coating Applicators.

How much primer for a mile of pipe?

Multiply pipe O.D. by 2¾ to get gallons of primer per mile.

Examples.

Pipe 24 in. in diameter will require 24 × 2¾ or 66 gallons of primer per mile;

Pipe 16 in. in diameter will require 16 × 2¾ or 44 gallons of primer per mile; and

Pipe 4½ in. in diameter will take 4½ × 2¾ or 12⅜ gallons of primer per mile (12 gallons in round figures).

The given multiplier is applicable only where the pipe is new and in good condition. For rough pipe the multiplying factor should be upped from 2¾ to 3.45. In other words, rough 20-in. pipe will require 20 × 3.45 or 69 gallons of primer per mile.

It is hard to estimate the amount of spillage and waste. Temperature and relative humidity have an effect on the way the primer goes on the pipe.

How much coal-tar enamel for a mile of pipe?

For an average cover of 3/32 in., multiply outside diameter of pipe by 0.6. The answer is in tons of enamel per mile. (Add 15% to take care of gate valves, drips, bends, etc.)

Examples. How many tons of coal-tar enamel will be required to apply a 3/32-in. coat to a 26-in. pipe line?

$26 \times 0.6 = 15.6$ tons per mile

Add 15% to take care of gate valves, bends etc.

$15.6 + 2.3 = 17.9$ or 18 tons per mile

How many tons of coal-tar enamel will be required to apply a 3/32-in. coat to a 8.625-in. pipe line?

$8.625 \times 0.6 = 5.2$ tons per mile
$5.2 \times 15\% = 0.8$
$\qquad = 6$ tons per mile

These factors strongly affect the amount of coating used per mile on pipe lines: 1. Temperature of the pipe, and ambient temperatures. (It is assumed that the temperature of the enamel can be controlled.) 2. The kind of dope machine. 3. Humidity. 4. The number of bends per mile. 5. The number of valves, crossovers, etc. 6. Condition of the pipe.

This latter point has a great effect on the amount of enamel required for a mile of pipe line. In fact, where reconditioned pipe is being coated, it is recommended that the amount of enamel be increased by about 20%.

How much wrapping for a mile of pipe?

For 14-in. to 48-in. pipe multiply diameter in in. by 15 to get number of squares (100 ft²). For pipe sizes of 12 3/4 in. and smaller add 1 (one) to nominal size of pipe and multiply by 15.

Examples. How many squares of asbestos felt would be required to wrap one mile of 30-in. pipe line?

$30 \times 15 = 450$ squares of felt

How many squares of felt would be required to wrap one mile of 4-in. pipe?

$(4 + 1) \times 15 = 75$ squares of felt

These figures are conservative and do not allow for wrapping drips, for patching holidays, and for short ends which are usually discarded. They allow for machine wrapping using 3/4-in. lap on diameters up to 10 3/4-in. and using 1-in. lap on larger diameters.

For smaller diameters of pipe this figure should be increased about 5% to take care of additional wastage and patching.

Estimating coating and wrapping materials required per mile of pipe

Primer and Coal-Tar Enamel

Gallons of primer and tons of coal-tar enamel are based on covering new pipe in good condition. For rough pipe, increase primer quantities by 25% and the coal-tar quantities by 20%. No allowance is included for spillage or waste.

Wrappings

Squares are based on machine wrapping, allowing 3/4-in. laps on pipe sizes through 10-in., and 1-in. laps on pipe sizes greater than 10 in. No allowance is included for wrapping drips, patching holidays or short ends which are usually discarded.

Nominal Pipe Size	Primer Gallons	3/32" Coal-tar Enamel Tons	Glass Wrap Squares	Kraft Paper Squares	15-Pound Felt Squares
4	11	3	75	75	75
6	17	4	105	105	105
8	22	6	135	135	135
10	28	7	165	165	165
12	33	8	195	195	195
14	39	10	210	210	210
16	44	11	240	240	240
18	50	13	270	270	270
20	55	14	300	300	300
24	66	17	360	360	360
30	83	21	450	450	450
36	99	25	540	540	540
42	116	29	630	630	630

Coefficient of friction for pipe coating materials

Tests indicate that previous values appear to be valid for thinfilm epoxies and conservative for coal tars

J. B. Ligon, Assistant Professor, Mechanical Engineering-Engineering Mechanics Dept., Michigan Technological University, Houghton, Mich., and **G. R. Mayer,** Project Engineering Manager, Bechtel, Inc., San Francisco

A major factor in the stress analysis of buried pipe lines is the movement that pipe undergoes in the presence of temperature and pressure differentials during its life. This movement is highly dependent upon friction resistance of the soil.

Although ample information is available on the static coefficient of friction for many materials, there is a lack of data on friction between soils and various coatings used in the pipe line industry. In the past, friction coefficient information was extrapolated from data in the literature that were believed to have a similarity to the external pipe coating to soil interface.

However, with the development of thinfilm epoxy resin coating systems and the increasing use of these systems in the pipe line industry, a change from a conventional coal tar felt coating to a thinfilm epoxy coating would indicate a significant change in the friction coefficient design criteria due to the extreme contrast in the surface texture of the two materials.

To evaluate the effect of the difference in surface texture on a pipe line system, test procedures were developed to determine the coefficient of friction for both coal tar felt and thinfilm epoxy to various soils and to obtain more reliable information for future pipe line designs using these coating materials.

Static friction tests were conducted to find the coefficient of friction between coal tar felt and thinfilm epoxy pipe coating and eight representative backfill soil samples from typical locations along a pipe line right of way. The results indicate that the friction coefficients are significantly larger than those previously extrapolated from literature and that coal tar has a higher friction resistance in respect to anchorage of a pipe line.

For the coal tar felt coating, the coefficient varies from 0.59 to 0.91 depending on the soil and moisture content. The thinfilm epoxy coating varies from 0.51 to 0.71 under the same conditions.

Definitions

Static coefficient of friction. The theoretical longitudinal soil force acting on the pipe surface can be calculated from the relation:

$$F = \mu \int_A p \, dA$$

where

F = longitudinal soil friction force (lb)
μ = coefficient of friction (dimensionless)
p = normal soil pressure acting on the pipe surface (psi)
dA = soil to pipe differential contact area (in.^{2})
$\int_A p \, dA$ = total normal soil force on pipe surface (lb)

The above relation is independent of the pressure per unit area of the mating surfaces as long as very high contact pressures are not encountered, which is the case for a buried pipe line.

In terms of a buried pipe line supporting a soil burden up to three pipe diameters of depth, the soil force relation becomes:

$$F = \mu \left[2D \, \gamma \left(H - \frac{D}{2} \right) + Wp \right]$$

where Wp = weight of pipe and contents (lb/in.)
D = pipe diameter (in.)
H = depth of the pipe centerline (in.)
γ = specific weight of the soil (lb/in.^{3})

The sensitivity of the soil friction forces to the coefficient of friction is readily apparent from the above relation. Since this force is inversely proportional to the active length that a pipe line moves as a result of temperature and pressure expansion, the coefficient of friction becomes a major factor in pipe line stress design.

Test system. The theoretical soil friction force acting on the surface of a coated plate can be calculated from the relation:

$$F = \mu \, N$$

where N = normal force acting on the plate surfaces (lb)
W = total weight of plate and external load (lb)

Since the coefficient of friction parameter is the same in all of the above expressions, the plate test system can be used to simulate the soil to pipe line interface.

Coal tar felt. The coal tar coating, an enamel and felt combination, was prepared to simulate an actual coated pipe line.

The surface of a steel plate was coal tar coated and then wrapped in a 15-lb felt. Therefore, the soil surface contact was predominantly to the felt material.

Thinfilm epoxy. The epoxy coating simulated a thinfilm epoxy resin pipe line coating applied by the fusion method.

Plates used in the tests were coated by the fluidized bed method. However, the surface texture was the same as that expected by coating plant production equipment.

Results

Coal tar to soils. Figure 1 illustrates that the friction coefficient varies from 0.59 to 0.91 for the various soils tested at field moisture content.

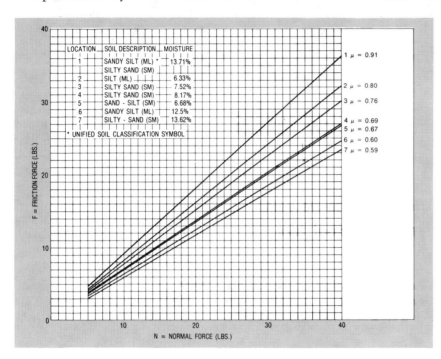

Figure 1. Coefficient of friction for coal tar coating to soil.

Figure 2. Coefficient of friction for thin film epoxy to soil.

The mechanism by which the friction coefficient changes among the various soils is not apparent from the results of these tests since the soil samples used did not vary considerably in description. However, the range over which the factor changes is considerably greater than that previously extrapolated from the literature, i.e.:

Soil description	Coefficient of friction commonly used
Silt	0.3
Sand	0.4
Gravel	0.5

The tests also indicate that the moisture content alters the friction factor to some extent, as would be expected.

In order to investigate the influence of temperature, tests were conducted with coal tar felt wrapping heated to 120°F. Only a slight softening beneath the coating surface was observed, and it is believed that temperatures up to this range will not significantly affect the magnitude of the friction coefficient.

Thinfilm epoxy to soils. As expected, results shown in Figure 2 indicate that the friction factor range of 0.51 to 0.71 for the epoxy is somewhat lower than that of the coal tar.

Again, the mechanism is not clear, but the results are fairly consistent with the friction coefficient increasing in most instances in the same order of magnitude as for coal tar. Since such information on this coating was not available in the literature, no comparison to similar tests could be made.

Temperatures in the 120°F range should have little or no effect on the coefficient of friction of epoxy to soil.

Conclusions

Although it is virtually impossible to precisely simulate the surface contact situation of a pipe line in a back-filled trench, the test procedure and apparatus reported here are a means of approximating it. The results indicate that coal tar coatings have a higher friction resistance than epoxy coatings as far as anchoring capabilities of the soil are concerned.

The selection of a coating, based on its soil friction resistance, could be of economic value in reducing those situations where extreme expansions call for reinforced fittings or elaborate culverts to overcome excessive stress levels in a pipe line system.

The test results show that previous values commonly used for the coefficient of friction were conservative for similar soils, and it is suggested that some conservatism still be incorporated in future analyses.

Since the tests indicated that the friction coefficient of the epoxies was similar to those previously used for coal tar coatings, their continued use for epoxy coated pipe lines should be valid. A more conservative approach to these values is recommended in soils where the presence of excessive moisture could change the friction factor substantially when interfaced with a smooth epoxy coating.

Troubleshooting cathodic protection systems: Magnesium anode system

The basic technique is the same as that outlined above; more measurements are required, because of the multiplicity of drain points. First, the current output of stations nearest the point of low potential should be checked; if these are satisfactory, a similar check should be extended in both directions until it is clear that the trouble must be on the line. When a given anode group shows a marked drop in current output, the cause may be drying out, shrinkage of backfill, or severed or broken lead wires. If the current is zero, the pipe-to-soil potential of the lead wire will show whether it is still connected to the pipe or the anode, and thus indicate the direction to the failure. If the current is low, there may be a loss of one or more anodes, by a severed wire; a pipe-to-soil potential survey over the anodes will show which are active, just as in the case of rectifier anodes.

Trouble indicated on the line, rather than at the anode stations, is tracked down in the same manner as that used

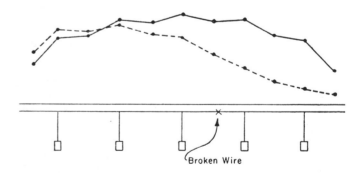

Figure 1. Locating Idle Anodes by Surface Potentials. The solid line shows the potentials found along a line of anodes when all are delivering current; the dotted line exhibits the change when there is a break in the anode lead at the point indicated. Single disconnected anodes may also be located by this method. A driven ground rod, a pipe lead, or even a rectifier terminal may be used for the reference ground; all readings should be referred to the same reference.

for line protected by rectifiers; first investigating locations where something has been done which might be responsible, next checking potentials, and finally making a line current survey.

Cathodic protection for pipe lines

Estimate the rectifier size required for an infinite line. Refer to Figure 1. If coating conductance tests have not been performed, pick a value from Table 1.

Table 1
Typical Values of Coating Conductance

Micromhos/sq ft	Coating Condition
1–10	Excellent coating—high resistivity soil
10–50	Good coating—high resistivity soil
50–100	Excellent coating—low resistivity soil
100–250	Good coating—low resistivity soil
250–500	Average coating—low resistivity soil
500–1,000	Poor coating—low resistivity soil

Use a value of -0.3 volts for ΔE_x. This is usually enough to raise the potential of coated steel to about -0.85 volts.

Use a value of 1.5 volts for ΔE at the drain point. Higher values may be used in some circumstances; however, there may be a risk of some coating disbondment at higher voltages.

Calculate I at the drain point.

$$\Delta I_A = \text{amp./in.} \times \Delta E_x/0.3 \times D$$

where D = pipe OD, in.

Example. 30-in. OD line with coating conductivity = 100 micromhos/sq ft. What is the current change at

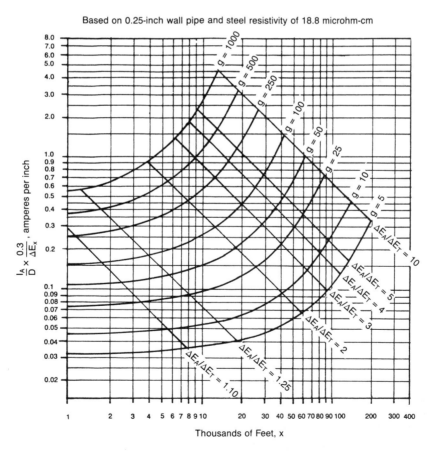

Based on 0.25-inch wall pipe and steel resistivity of 18.8 microhm-cm

Figure 1. Drainage current vs. distance for a coated infinite line.

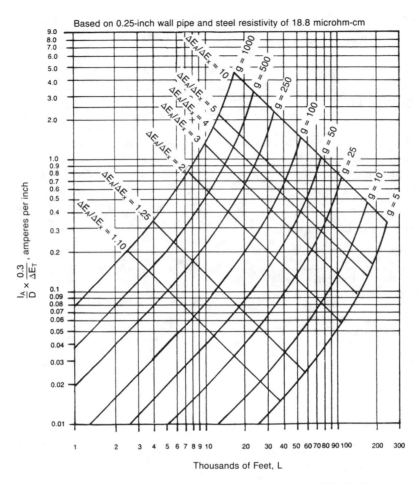

Based on 0.25-inch wall pipe and steel resistivity of 18.8 microhm-cm

Figure 2. Drainage current vs. distance for a coated finite line.

the drain point, and how far will this current protect an infinitely long pipe line?

$$\Delta E_A = 1.5$$

$$\Delta E_x = 0.3$$

$$\Delta E_A/\Delta E_x = 5.0$$

Refer to Figure 1 and read

$$L = 30,000 \text{ ft}$$

$$\Delta I_A/D \times 0.3/\Delta E_x = 0.7 \text{ (from Figure 1)}$$

$$\Delta I_A = 0.7 \times 30 = 21 \text{ amp.}$$

21 amps will protect the line for 30,000 ft in either direction from the drain point.

Estimate rectifier size for a finite line. Refer to Figure 2.

A 3½-in. OD line 20,000 ft long is protected by insulating flanges at both ends and has a coating conductivity of

500 micromhos/sq ft. What is $\Delta E_A/E_T$ and what ΔI_A is required? Assume that

$$E_T = 0.3$$

From Figure 2 read

$$\Delta E_A/E_T = 5.8$$

$$\Delta E_A = 1.74$$

$$\Delta I_A/D \times 0.3/E_T = 1.8 \text{ amp./in.}$$

$$\Delta I_A = 1.8 \times 3.5 = 6.3 \text{ amp.}$$

Estimate ground bed resistance for a rectifier installation.

Example. A ground bed for a rectifier is to be installed in 1,000 ohm-cm soil. Seven 3-in. × 60-in. vertical graphite anodes with backfill and a spacing of 10 ft will be used. What is the resistance of the ground bed?

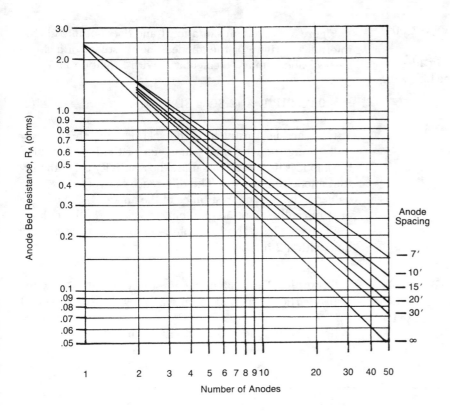

Figure 3. Anode bed resistance vs. number of anodes. 3-in. × 60-in. vertical graphite anodes in backfill.

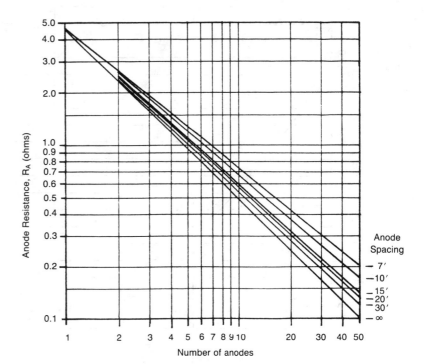

Figure 4. Anode bed resistance vs. number of anodes. 2-in. × 60-in. vertical bare anodes.

Refer to Figure 3. The resistance is 0.56 ohms.

Figure 3 is based on a soil resistivity of 1,000 ohms-cm. If the soil resistivity is different, use a ratio of the actual soil resistivity divided by 1,000 and multiply this by the reading obtained from Figure 3.

A rectifier ground bed is to be composed of 10 − 2 × 60-in. bare "Duriron" silicon anodes spaced 20 ft apart. The soil resistivity is 3,000 ohms-cm. What is the ground bed resistance? Refer to Figure 4. The resistivity from the chart is 0.55 ohms. Since the chart is based on soil resistivity of

1,000 ohms-cm, the ground bed resistivity is 3,000/ 1,000 × 0.55 or 1.65 ohms.

The resistance of multiple anodes installed vertically and connected in parallel may be calculated with the following equation:

$$R = 0.00521P/NL \times (2.3\text{Log } 8L/d - 1 + 2L/S\text{Log } 0.656N) \quad (1)$$

where   R = ground bed resistance, ohms
         P = soil resistivity, ohm-cm
         N = number of anodes
         d = diameter of anode, ft
         L = length of anode, ft
         S = anode spacing, ft

If the anode is installed with backfill such as coke breeze, use the diameter and length of the hole in which the anode is installed. If the anode is installed bare, use the actual dimensions of the anode.

Figure 5 is based on Equation 1 and does not include the internal resistivity of the anode. The resistivity of a single vertical anode may be calculated with Equation 2.

$$R = 0.00521P/L \times (2.3\text{Log } 8L/d - 1) \quad (2)$$

If the anode is installed with backfill, calculate the resistivity using the length and diameter of the hole in which the anode is installed. Calculate the resistivity using the actual anode dimensions. The difference between these two values is the internal resistance of the anode. Use the value of P, typically about 50 ohm-cm, for the backfill medium.

Figure 5 is based on 1,000 ohm-cm soil and a 7-ft × 8-in. hole with a 2-in. × 60-in. anode.

Example. Determine the resistivity of 20 anodes installed vertically in 1,500 ohm-cm soil with a spacing of 20 ft.

Read the ground bed resistivity from Figure 5.

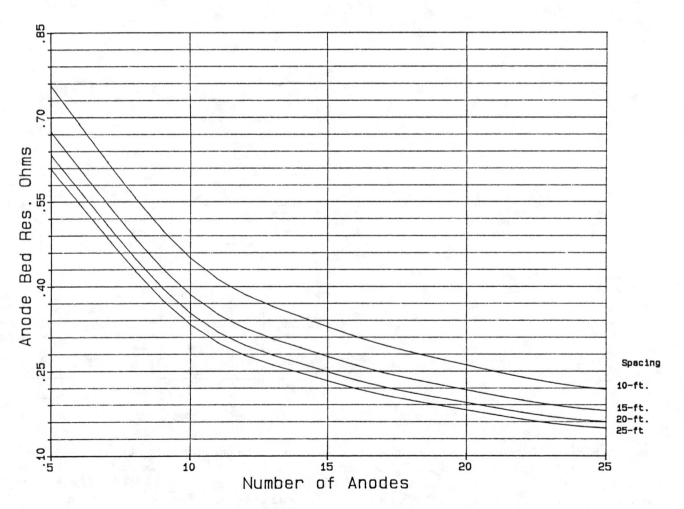

Figure 5. Anode bed resistance.

R = 0.202 ohm

Since the anodes are to be installed in 1,500 ohm-cm soil and Figure 5 is based on 1,000 ohm-cm soil, multiply R by the ratio of the actual soil resistivity to 1,000 ohm-cm.

R = 0.202 × 1,500/1,000

R = 0.303 ohm

The internal resistivity for a single 2-in. × 60-in. vertical anode installed in 50 ohm-cm backfill (7 ft × 8-in. hole) is 0.106 ohm.

Since 20 anodes will be installed in parallel, divide the resitivity for one anode by the number of anodes to obtain the internal resistivity of the anode bank.

0.106/20 = 0.005 ohm

The total resistivity of the 20 anodes installed vertically will therefore be 0.308 ohm (0.303 + 0.005).

Galvanic Anodes

Zinc and magnesium are the most commonly used materials for galvanic anodes. Magnesium is available either in standard alloy or high purity alloy. Galvanic anodes are usually pre-packaged with backfill to facilitate their installation. They may also be ordered bare if desired. Galvanic anodes offer the advantage of more uniformly distributing the cathodic protection current along the pipe line and it may be possible to protect the pipe line with a smaller amount of current than would be required with an impressed current system but not necessarily at a lower cost. Another advantage is that interference with other structures is minimized when galvanic anodes are used.

Galvanic anodes are not an economical source of cathodic protection current in areas of high soil resistivity. Their use is generally limited to soils of 3,000 ohm-cm except where small amounts of current are needed.

Magnesium is the most-used material for galvanic anodes for pipe line protection. Magnesium offers a higher solution potential than zinc and may therefore be used in areas of higher soil resistivity. A smaller amount of magnesium will generally be required for a comparable amount of current. Refer to Figure 6 for typical magnesium anode performance data. These curves are based on driving potentials of − 0.70 volts for H-1 alloy and − 0.90 volts for Galvomag working against a structure potential of − 0.85 volts referenced to copper sulfate.

The driving potential with respect to steel for zinc is less than for magnesium. The efficiency of zinc at low current levels does not decrease as rapidly as the efficiency for magnesium. The solution potential for zinc referenced to a cop-

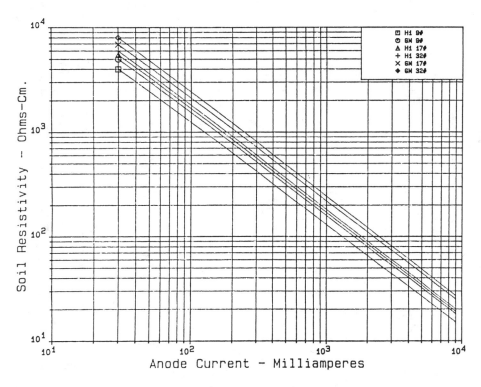

Figure 6. Magnesium anode current.

Figure 7a. Current output zinc anodes.

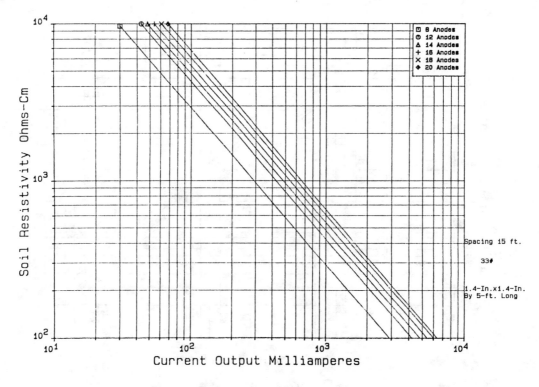

Figure 7b. Current output zinc anodes.

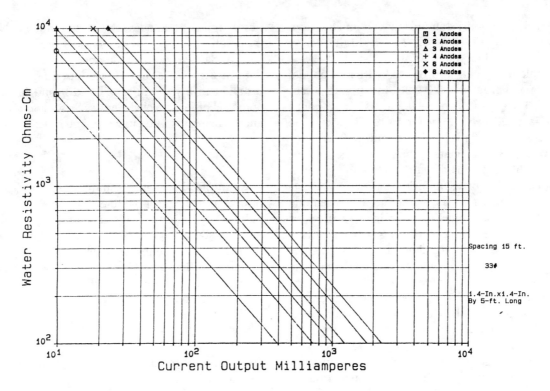

Figure 8a. Current output zinc anodes.

Figure 8b. Current output zinc anodes.

per sulfate cell is −1.1 volts; standard magnesium has a solution potential of −1.55 volts; and high purity magnesium has a solution potential of −1.8 volts.

If, for example, a pipe line is protected with zinc anodes at a polarization potential of −0.9 volts, the driving potential will be −1.1 − (−0.9) or −0.2 volts. If standard magnesium is used, the driving potential will be −1.55 − (−0.9) or −0.65 volts. The circuit resistance for magnesium will be approximately three times as great as for zinc. This would be handled by using fewer magnesium anodes, smaller anodes, or using series resistors.

If the current demands for the system are increased due to coating deterioration, contact with foreign structures, or by oxygen reaching the pipe and causing depolarization, the potential drop will be less for zinc than for magnesium anodes. With zinc anodes, the current needs could increase by as much as 50% and the pipe polarization potential would still be about 0.8 volts. The polarization potential would drop to about 0.8 volts with only a 15% increase in current needs if magnesium were used.

The current efficiency for zinc is 90% and this value holds over a wide range of current densities. Magnesium anodes have an efficiency of 50% at normal current densities. Magnesium anodes may be consumed by self corrosion if operated at very low current densities. Refer to Figures 7a, 7b, 8a, and 8b for zinc anode performance data. The data in Figures 7a and 7b are based on the anodes being installed in a gypsum-clay backfill and having a driving potential of −0.2 volts. Figures 8a and 8b are based on the anodes being installed in water and having a driving potential of −0.2 volts.[1]

Example. Estimate the number of packaged anodes required to protect a pipe line.

What is the anode resistance of a packaged magnesium anode installation consisting of nine 32 lb anodes spaced 7 ft apart in 2,000 ohm-cm soil?

Refer to Figure 9. This chart is based on 17# packaged anodes in 1,000 ohm-cm soil. For nine 32 lb anodes, the resistivity will be

$$1 \times 2,000/1,000 \times 0.9 = 1.8 \text{ ohm}$$

See Figure 10 for a table of multiplying factors for other size anodes.

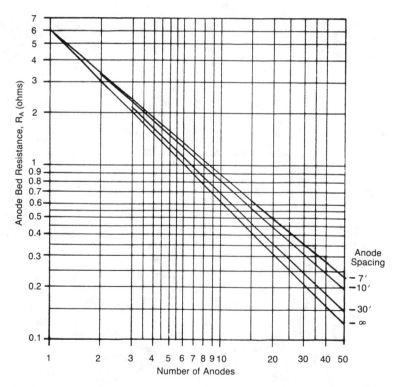

Figure 9. Anode bed resistance vs. number of anodes. 17# packaged magnesium anodes.

Chart based on 17-lb. magnesium anodes installed in 1000 ohm-cm soil in groups of 10 spaced on 10-ft. centers.

For other conditions multiply number of anodes by the following multiplying factors:

For soil resistivity: $MF = \dfrac{p}{1000}$ For 9-lb. anodes: MF = 1.25

For conventional magnesium: MF = 1.3   For 32-lb. anodes: MF = 0.9

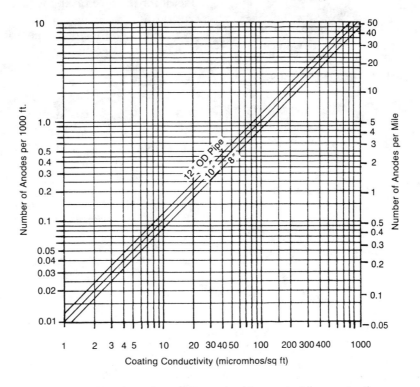

Figure 10. Number of anodes required for coated line protection.

Example. A coated pipe line has a coating conductivity of 100 micromhos/sq ft and is 10,000 ft long, and the diameter is 10³/₄-in. How many 17 lb magnesium anodes will be required to protect 1,000 ft? Refer to Figure 7 and read 2 anodes per 1,000 ft. A total of twenty 17# anodes will be required for the entire line.

Source

Pipeline Corrosion and Cathodic Protection, Gulf Publishing Company, Houston, Texas.

Reference

1. From data prepared for the American Zinc Institute.

Estimate the pounds of sacrificial anode material required for offshore pipe lines

This rule of thumb is based on the following assumptions:

- 5% of pipe line considered bare
- 3 milliamperes/sq ft required for protection
- 30-year design life

Basis for rule: sq ft of surface area of pipe line/linear ft × % of bare pipe × 3.0 m.a./sq ft × #/amp.-year × design life in years = the lb of galvanic anode material that will be required.

$$\text{Multiplier} = OD \times \text{lb/amp.-yr} \times (12\ PI/144)$$
$$\times\ 30 \times 0.05 \times 0.003 \times 1{,}000$$

or:

Pipe OD × lb/amp.-yr × 1.1781 = lb of anode material/ 1,000 linear ft of pipe line

Example. How many lb of zinc will be required for a 30-in. OD line installed in seawater and operating at 80°F? Pounds of zinc per ampere-year = 24.8 (seawater).

30 × 24.8 × 1.1781 = 877#/1,000 linear ft of pipe line

How many lb of Galvalum® III will be required for a 30-in. OD line installed in seawater and operating at 80°F? Pounds of Galvalum® III per ampere-year in seawater = 7.62 (32–95°F).

30 × 7.62 × 1.1781 = 269#/1,000 linear ft of pipe line

How many lb of Galvalum® III will be required for a 30-in. OD line installed in saline mud and operating at 80°F? Pounds of Galvalum® III per ampere-year in saline mud = 9.22 (32–95°F).

30 × 9.22 × 1.1781 = 326#/1,000 linear ft of pipe line

How many lb of Galvalum® III will be required for a 30-in. OD line installed in saline mud and operated at a temperature of 160°F? Pounds of Galvalum® III per ampere-year in saline mud = 21.9 (140–170°F).

30 × 21.9 × 1.1781 = 774#/1,000 linear ft of pipe line

It is always a good idea to check with the bracelet vendor to determine the nearest size mold available to avoid the need for a special size mold. Also, the mold size will be a factor in determining anode spacing.

These calculations assume a utilization factor of 100% and the designer should select the appropriate utilization factor for the specific application. For estimating, you may wish to use 85%. Temperatures used in the examples are assumed to be the anode operating temperature.

Generally, zinc anodes should not be used if the pipe line will be operated at temperatures above 90°F. See ASTM B418-80 and NACE Paper No. 311 by Elwood G. Haney presented at CORROSION/85.

See Table 3 for typical anode performance data. All performance data shown here are suitable for estimating. For final design purposes, the anode manufacturer or distributor should be consulted for actual anode performance data required for a specific application.

Galvalum® III is a registered trademark of Oronzio De Nora S. A., Lugano (Switzerland). Galvomag® is a registered trademark of Dow Chemical Company.

Typical Anode Specifications

Table 1
Zinc Anodes

Material	Mil. Spec. MIL-A-1800 1H	ASTM Type I	ASTM Type II
Lead	0.006 max	—	—
Iron	0.005 max	0.005 max	0.0014 max
Cadmium	0.025–0.15 Range	0.03–0.1	0.003 max
Copper	0.005 max	—	—
Aluminum	0.10–0.50 max	0.1–0.4	0.005 max
Silicon	0.125 max	—	—
Zinc	Remainder	Remainder	Remainder

Table 2*
Cast Magnesium Anodes

Material	H-1 Grade A Type III	H-1 Grade B Type II	H-1 Grade C Type I	Galvomag®
Al	5.3–6.7	6.3–6.7	5.0–7.0	0.010 max
Zn	2.5–3.5	2.5–3.5	2.0–4.0	0.05 max
Mn	0.15 min	0.15 min	0.10 min	0.5–1.3
Fe	0.003 max	0.003 max	0.003 max	0.03 max
Ni	0.002 max	0.003 max	0.003 max	0.001 max
Cu	0.02 max	0.05 max	0.10 max	0.02 max
Si	0.10 max	0.30 max	0.30 max	0.05 max
Others	0.30 max	0.30 max	0.30 max	0.05 max, each 0.30 max, total

* Reprinted courtesy of Dow Chemical U.S.A.

Table 3
Typical Anode Performance Data

Anode Type	Amp-hr/lb in Seawater	Potential Negative Volts Ag/AgCl	Approx. Current Density mA/ft²
Zinc	354	1.03	36–600
Galvalum® III	1,150–1,204 (32–95°F)	1.10	100–1,000
	in saline mud		
	950 (32–95°F)	1.10	100
	930 (104°F)	1.08	100
	600 (140°F)	1.05	100
	400 (176°F)	1.04	100
	400 (221°F)	1.02	100
	in earth		
Magnesium	400–500	1.53	36–600
Galvomag®	500–540	1.75	36–600

Table 4
Typical Environments for Sacrificial Anodes

Material	Driving Potential Volts	Environment
Magnesium	0.9	Underground Fresh water 200–10,000 ohm-cm
Aluminum	0.25	Sea water Brines Saline mud <200 ohm-cm
Zinc	0.25	Sea water Brine Underground <1,500 ohm-cm

Tables 1–4 are guideline-type information suitable for general use. The characteristics of a specific application must be considered and it may be necessary to contact the anode manufacturer or distributor to obtain detailed anode performance data before finalizing a design.

Comparison of other reference electrode potentials with that of copper-copper sulfate reference electrode at 25°C

Type of comparative reference electrode	Structure-to-comparative reference electrode reading equivalent to −0.85 volt with respect to copper sulfate reference electrode	To correct readings between structure and comparative reference electrode to equivalent readings with respect to copper sulfate reference electrode
Calomel (saturated)	−0.776 volt	Add −0.074 volt
Silver-Silver Chloride (0.1 N KCl Solution)	−0.822	Add −0.028
Silver-Silver Chloride (Silver screen with deposited silver chloride)	−0.78	Add −0.07
Pure Zinc (Special high grade)	+0.25*	Add −1.10

* Based on zinc having an open circuit potential of −1.10 volt with respect to copper sulfate reference electrode. (From A. W. Peabody, Chapter 5, p. 59, *NACE Basic Corrosion Course*, Houston, TX, 1973.)

Chart aids in calculating ground bed resistance and rectifier power cost

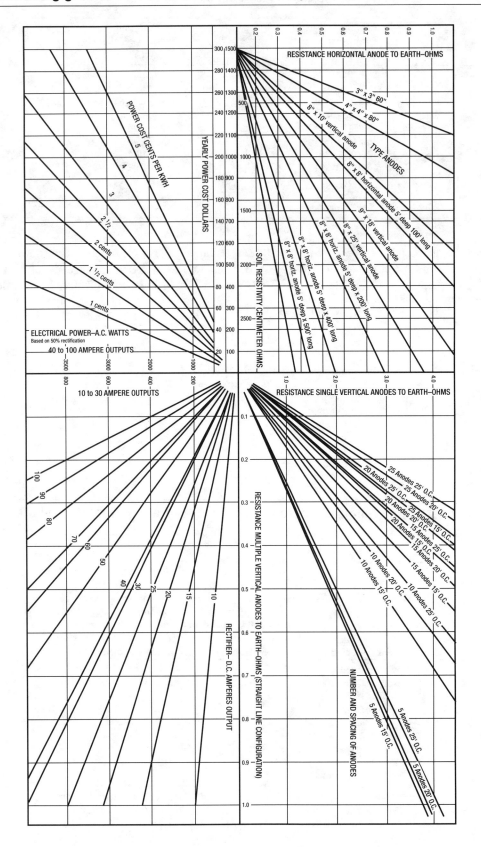

For conventional type ground beds this chart permits a rapid determination of the resistance to earth of various ground beds and the yearly power costs based on 50% efficient rectification. These curves are specifically designed for standard 3-in. by 60-in. and 4-in. by 80-in. graphite anodes centered in a vertical or horizontal excavation and backfilled to the dimensions indicated with well tamped coal coke breeze.

Example.

Soil resistivity = 1,000 ohm-cm
Power rate = 3 cents per kwh
Rectifier rating = 20-ampere

To determine the optimum size ground bed, enter the chart at the "Soil Resistivity-ohm-cm" line and extend a vertical line from the 1,000-ohm-cm point to intersect all the anode types. Proceed horizontally to the left for the resistance to earth of horizontal type anodes and horizontally to the right for vertical type anodes. Various groups of vertical anodes with 15, 20 and 25-ft spacings are represented by curves.

Resistance values for horizontal anodes obtained from the ordinate of the upper left hand section should be transferred to the abscissa of the upper right hand section to determine power consumption and annual costs from the lower right hand and lower left hand section. In obtaining power costs for the two lower sections, the ordinate shows power consumption in AC watts based on 50% efficient rectification, and the abscissa of points on the lower left hand section indicated annual power costs in dollars.

How can output of magnesium anodes be predicted?

For a single 17 lb magnesium anode, working against a fully protected pipe, divide the soil resistivity in ohm-cm units into 180,000; the answer is the current output in milliamperes.

Example. Soil resistivity 3,200 ohm-cm
180,000/3,200 = 56.25

Current output will be about 56 milliamperes.

The rule works for *single* anodes against fully protected pipe; anode groups will have lower current output per anode, and if the pipe is not fully protected, the current output will be greater. For 32 lb anodes use 195,000; for 50 lb anodes use 204,000.

How to determine the efficiency of a cathodic protection rectifier

Multiply DC voltage output times DC amperes times 100 and divide by the AC watts input. The answer is % efficiency.

Example. A rectifier has an output of 50 amperes DC at 15 volts. The wattmeter gives a direct reading of an AC input of 1155 watts.

$$\text{Efficiency} = \frac{(\text{DC volts})(\text{DC amperes})(100)}{(\text{AC watts input})}$$

$$= \frac{(15)(50)(100)}{(1,155)} = 65\%$$

On those rectifier units that do not have a wattmeter, the AC input may be obtained from the watt hour meter mounted on the pole. To do this, count the number of revolutions made by the watt hour meter in so many seconds, multiply by the meter constant stamped on the meter times 3,600 and divide by the number of seconds timed. This

gives AC input in watts. Then go ahead and use the formula as above.

Example. A rectifier unit does not have a watt meter. The wheel of the watt hour meter revolves 11 times in one minute. Meter constant is shown as 1.8. Output of the rectifier is the same as in the first case, 50 amperes at 15 volts DC.

$$\text{AC watts} = \frac{\text{K} \times \text{N} \times 3,600}{\text{T}}$$

where K is the meter constant and N is the number of meter revolutions in T seconds.

$$\text{AC watts} = \frac{(1.8)(11)(3,600)}{60} = 1,188 \text{ watts AC}$$

$$\text{Efficiency} = \frac{(15)(50)(100)}{1,188} = 63\%$$

How to calculate the voltage drop in ground bed cable quickly

Use the following formula for calculating the voltage drop in a copper ground bed cable:

$$V_T = \frac{R_{1000}}{1000} \left[I_T \, D_{GR} + D_A[(N-1)(I_T) \right.$$
$$\left. - I_A(1 + 2 + 3 + 4 \dots N - 1)] \right]$$

where V_T = total voltage drop in volts
   R_{1000} = cable resistance per 1,000 ft
   I_T = total current in amperes
   I_A = I_T/N current per anode
   D_{GB} = distance from rectifier to first anode in ft
   D_A = distance between anodes in ft
   N = number of anodes in ground bed

Example. What is the voltage drop in a copper wire ground bed cable in a cathodic protection system having a rectifier output of 225 amperes? The distance from rectifier to ground bed is 100 ft and 16 anodes are spaced on 20 ft centers. The cable used is No. 4 copper having a resistance of 0.2487 ohm per 1,000 ft.

Then R_{1000} = 0.2487 ohm
   I_T = 225 amperes
   I_A = 225/16 = 14 amperes
   D_{GB} = 100 ft
   D_A = 20 ft
   N = 16

$$V_T = \frac{.2487}{1,000} \left[(225)(100) + 20[(15)(225) \right.$$
$$\left. - (14)(120)] \right]$$

$$V_T = \frac{.2487}{1,000} (22,500 + 20[1,695])$$

$$V_T = \left(\frac{.2487}{1,000} \right) (56,400)$$

$$V_T = (.2487)(56.4)$$

$$V_T = 14.00 \text{ volts drop}$$

What is the most economical size for a rectifier cable?

Choosing the economical size for a rectifier cable involves balancing the first cost of the cable against the power losses. A rule which does not include all of the factors, but which usually gives a very close approximation, is as follows:

$$a = 2,280 \, I \, \sqrt{P}$$

where a is the area of the cable in circular mils (refer to a wire table for size), I is the current to be carried, in amperes, and P is the cost of power in cents per kilowatt-hour. This formula is based on the present prices of copper ca-

bles, and assumes the use of a good grade of direct burial cable.

Example. Suppose

I = 60 amps
P = 4 cents

$$a = (2,280)(60)(\sqrt{4})$$
$$a = 273,600 \text{ circular mils}$$

Cable is sized 250,000 and 300,000 circular mils. Use the latter.

How to estimate the number of magnesium anodes required and their spacing for a bare line or for a corrosion "hot spot"

Soil and pipe line conditions will vary widely but where these given assumptions hold, an estimate can be made for test installations using this method. First find the area of pipe to be protected, multiply by the milliamperes per sq ft necessary to protect the line and divide by design output of an anode in this type soil. This will give the number of anodes required; to find spacing, divide number of anodes into the length of line.

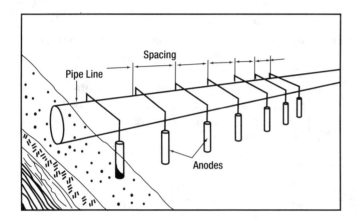

Example. Find the number of 17 lb magnesium anodes needed and spacing used to protect 10,000 ft of bare 4-in. pipe line in a corrosive soil having a resistance of 1,000 ohm per cubic centimeter. Assume a current requirement of one milliampere per sq ft, and that the anode output curve shows an output of 100 milliamperes per anode in this type soil.

$$\text{Area of pipe} = \frac{4 \times 3.14}{12} \times 10,000 = 10,500 \text{ sq ft}$$

Total current requirement = $10,500 \times 1.0 = 10,500$ milliamperes

$$\text{Anodes required} = \frac{10,500 \text{ milliamperes}}{100 \text{ ma/anode}}$$
$$= 105 \text{ anodes distributed}$$

$$\text{Spacing} = \frac{10,000}{105} = 95 \text{ ft}$$

Similarly find the number of anodes required to protect ten 100-ft long "hot spots" on a 4-in. line assuming a 100 milliampere output from a 17 lb magnesium anode. Estimate as follows:

$$\text{Area of each "hot-spot"} = \frac{4 \times 3.14}{12} = 105 \text{ sq ft}$$

$$\text{Anodes required} = \frac{105 \text{ sq ft} \times 1.0 \text{ ma per sq ft}}{100 \text{ ma per anode}} = 1.05$$

One 17 lb magnesium anode will produce approximately enough current in 1,000-ohm soil to protect one of these "hot-spots."

Ten anodes required for "hot-spots" on this section of line.

Note. For these rules the protective criterion is -0.85 volts pipe-to-soil potential using a copper-copper sulfate reference electrode; or a negative change of potential of as much as -0.25 volts. Soil and pipe line conditions vary widely, so it is well to consult an experienced corrosion engineer in the design of a complete cathodic protection system.

How can resistivity of fresh water be determined from chemical analysis?

Divide the total solids expressed in parts per million into 500,000; the result is the resistivity in ohm-centimeter units.

Example. Total solid content is 400 ppm.

$500,000 \div 400 = 1,250$ ohm centimeters restivity

This rule of thumb will give very accurate results for any naturally occurring fresh waters; severely contaminated waters may differ markedly, however, depending on the kind of contamination.

What will be the resistance to earth of a single graphite anode?

For the most common size, 3-in. by 60-in., installed in 10 ft of coke breeze backfill in an 8-in. hole, multiply the soil resistivity in ohm-cm by 0.002; this will give the resistance to earth in ohm.

Example. 1,800 ohm-cm soil

$1,800 \times 0.002 = 3.6$ ohm

This assumes that the resistivity is uniform to considerable depth; if it increases with depth, the resistance will be slightly higher.

How to estimate the monthly power bill for a cathodic protection rectifier

Multiply the rectifier output voltage and current together, and multiply this by $1\frac{1}{2}$; the result is the power consumption for one month, in kilowatt-hours. If the unit is on a meter of its own, figure the bill by the utilities rate schedule; if it represents an additional load, use the rate per kilowatt-hour for the highest bracket reached.

Example. The terminal voltage of a rectifier is 10.2 volts, and the current output is 21 amperes.

$10.2 \times 21 \times 1.5 = 321.2$

The power consumption will be about 320 kwh per month; at 3 cents this would amount to $9.60, at 1.1 cents it would be $3.52, etc.

Note that the actual current and voltage values are used; if the rating of the unit is used, the result will be the maximum power consumption within the range of the equipment. The formula is approximate; it is based on an efficiency of about 48%.

What will be the resistance to earth of a group of graphite anodes, in terms of the resistance of a single anode?

When the spacing is 20 ft (the most common value), the resistance of the group will be equal to the resistance of a single anode, divided by the square root of the number of anodes.

Example. Resistance of 4 anodes = $\frac{1}{2}$ resistance of single; resistance of 16 anodes = $\frac{1}{4}$ resistance of single, etc.

This rule is reasonably accurate, for soils which are uniform to a considerable depth; if the resistivity increases with depth, the "crowding" effect will be greater, whereas if it decreases, the individual anodes will not affect each other so much.

How can the current output of magnesium rod used for the cathodic protection of heat exchanger shells be predicted?

Divide the total dissolved solids content of the water by 13; the answer is the current output per ft of 1.315 in. (1 in. pipe size) magnesium rod to be expected.

This is an approximation to the current flow which will be found after stabilized conditions have been reached, if

protection is obtained. If insufficient magnesium is installed, the current flow will probably be greater, and if too much is installed, the current will be less.

What spacing for test leads to measure current on a pipe line?

Multiply the pipe weight, in lb/ft, by 4; the result is a spacing which will give a resistance of 0.001 ohm, or one ampere per millivolt IR drop.

Example. Pipe weight is 40.48 lb/ft.
$40.5 \times 4 = 162$
Install test leads at 162 ft spacing.

The rule works for welded lines of grade B seamless; variations in steel may introduce errors as great as 15%. It is always better to calibrate a line section if accurate results are to be had, but if the above rule is used, the spacing will not be too small for satisfactory measurement, or too great for convenience in installation.

How many magnesium anodes are needed for supplementary protection to a short-circuited bare casing?

For 17 lb anodes, use $A\rho/50{,}000$ to give the number needed for the casing alone; A is the area of the casing, in sq ft, and ρ is the soil resistivity in ohm-centimeters. The life of the anodes may be estimated from $\rho/180$. If the life in years, as given by this expression, is too short, refigure for 32-lb anodes, using $A\rho/55{,}000$ for the number and $\rho/100$ for the life. If this is still too short, figure for 50-lb anodes; number is $A\rho/60{,}000$ and life is $\rho/65$.

Such an installation, grouped on both sides of the line at each end of the casing, will enable the main line protection unit to carry beyond the casing; the values obtained above are approximate only, and should be checked by potential measurements after installation.

Example. Suppose the pipe line has shorted out to 45 ft of 24-in. bare casing. Soil resistivity is 1,100 ohm-centimeters. How many 17-lb anodes would be required for supplementary protection?

$$\text{Number of Anodes} = \frac{A\rho}{50{,}000}$$

$$A = 2 \times 3.14 \times 45 = 283$$

$$\text{Number of Anodes} = \frac{(284)(1{,}100)}{50{,}000}$$

$$= 6.21$$

Place three anodes at each end of casing. Their life expectancy will be

$$\frac{1{,}100}{180} = 6 \text{ years}$$

Group installation of sacrificial anodes

When anodes are installed in groups, a spacing correction factor must be taken into account to determine the current output. The chart below shows the correction factor for multiple anode installations.

Example. Single anode 8-in. by 60-in. in 1,500 ohm soil.

Resistance = 4.849 ohm
Potential = 0.85 volts

$$\text{Current output} = \frac{E}{R} = \frac{17.5 - 0.85}{4.849} = 0.185$$
$$\text{amps}$$

Current output of 5 anodes on 15 ft spacing under above conditions.

$$\text{Combined resistance of group} = \frac{4.849}{5} \times 1.22$$

Correction factor = 1.17 ohm

$$\text{Total current output} = \frac{1.75 - 0.85}{1.17} = 0.769 \text{ amps}$$

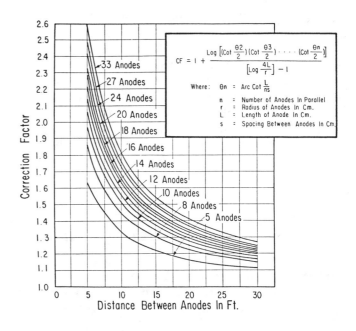

How can the life of magnesium anodes be predicted?

Multiply the anode weight in lb by 57, and divide by the current output in milliamperes.

Example. A 17-lb anode has a current output of 83 milliamperes. What is its projected life?

$17 \times 57/83 = 11.7$

The projected life is 11.7 years.

The rule is based on 50% current efficiency, and on stabilized conditions. If more anodes are to be added, or if time has not been allowed for polarization (one to four weeks), then the current output will probably fall off, and the actual life will be 10 to 40% longer than indicated by the rule.

How to find the voltage rating of a rectifier if it is to deliver a given amount of current through a given ground bed (graphite or carbon)

Multiply the desired current, in amperes, by the total resistance of the ground bed, in ohm (either calculated or measured); add two volts, to overcome the galvanic difference between graphite and protected steel.

Example. A current of 20 amperes is desired from a ground bed whose resistance is 0.9 ohm.

$(20 \times 0.9) + 2 = 20$ volts

Specify the next larger size, which will probably be (depending upon the manufacturer) 24 volts.

Note. Where the resistance is measured rather than calculated, and is the total loop resistance between pipe and ground bed, this method is quite precise; for calculated values, caution should be observed.

Determining current requirements for coated lines

Survey along the line after a polarization run of three or four hours, and locate the two points whose pipe-to-soil potentials are 1.0 volt and 0.8 volt (see Figure 1). The section of line lying between these two points will then be at an average potential of about 0.9 volt, which is approximately the average of a line protected with distributed magnesium anodes. If, then, the line current is measured at each of these two points, the difference between these two line currents will be the total amount of current picked up on the line in the section. This quantity, divided by the length of the section, will then give the current requirements of the line, in amperes per mile. Naturally this figure can only be

used for an entire line if the section chosen is truly representative; judgment must be used at this point.

Determining current requirements for coated lines when pipe-to-soil potential values are estimated

If it is not possible or convenient to find the points which have the exact values of 1.0 and 0.8, then readings can be taken which approximate these values, and both potential

and line current can then be adjusted to find the desired quantity; this is illustrated in Figure 1; readings were taken at points "A" and "B," and, by assuming a static potential

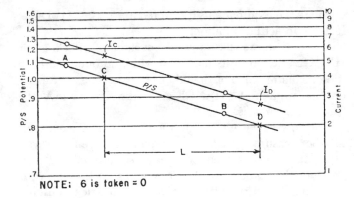

NOTE: 6 is taken = 0

of 0.6 volt and linear attentuation, it is determined that points "C" and "D" have the values of 1.0 and 0.8; since line current varies according to the same laws as potential, it is possible to determine the current values at the two latter points, and so to arrive at the desired quantity.

It is not to be expected that the estimate thus made will be highly accurate; it is intended to be merely preliminary.

HVDC effects on pipe lines

Latest methods to calculate quantitative characteristics of the earth's gradient and High Voltage Direct Current effects on pipe lines

Thomas B. Hays, George W. Silkworth, Robert D. Schilling, Cornell, Howell, Hayes & Merryfield, Inc.

A pipe line collects, conducts and discharges electric current of high voltage direct current (HVDC) origin as a result of the electric field established by the DC current transmitted through the earth and by the electrical properties of the pipe line itself.

The Electric Field

It is assumed that the electrical properties of the earth and pipe line are uniform and that the HVDC effects are isolated from other stray or induced current effects. This assumption makes possible the analysis of this portion of the total protection problem and assists in visualizing the unique effects of HVDC earth currents.

Ultimately, the total electrical current conducted by a pipe line will be the sum of all the various components. This includes protective current, telluric current, and other stray or interference currents.

Conduction of a direct current through the earth is an electrolytic phenomena, characterized by an electric field that surrounds each of the two earth electrodes and is directly proportional to the HVDC earth current (Figure 1). When the electrodes are spaced several hundred miles apart, the field appears to be concentric around each ground electrode.

The fields surrounding the opposite electrodes are of the opposite polarity. If the polarity of the ground current reverses, the surrounding electric fields will also reverse. The potential at any point in the electric field with respect to remote earth (V), due to an electrode current (I) at a distance (r) from an electrode in a homogeneous earth of the resistivity (ρ) is found by the relationship:

Figure 1. Conduction of a direct current through the earth is an electrolytic phenomena, characterized by an electric field that surrounds each of the two earth electrodes and is directly proportional to the HVDC earth current.

$$V = \frac{\rho I}{2 \pi r}$$

This relationship shows a simplified representation of what actually occurs. The earth's resistivity is not uniform at all depths or horizontal distances, which causes irregularities to the ideal voltage distribution existing around an electrode in a homogeneous earth.

To refine the earth current analysis and to account for different resistivities, computer techniques have been developed to estimate the earth's gradients existing for conditions of multi-layers of earth of different resistivities. Pipe line effects have also been predicted by this method. Reasonable agreement between these models and field data has been reported. Techniques are available to calculate the quantitative characteristics of both the earth's gradient and HVDC effects on pipe line.

HVDC Effects on a Pipe Line

Whether the HVDC component of the total current existing on the line causes a corrosion problem depends on the level of protection along the line and the effect of the added component to the level of protection. Figure 2a illustrates the HVDC component of the total conducted current along the pipe line for a given separation distance under the condition where the HVDC electrode is collecting current from the earth.

In determining protection requirements for the pipe line, the characteristic of the current transfer relationship is important (Figure 2b). In those regions of the pipe line where the current transfer is from the pipe line to the soil—along the center in this case—corrosion could occur if the HVDC component of the total transfer current to the soil is great enough to exceed any local level of protection against corrosion.

The pipe line HVDC arrangement shown in Figure 2a will serve as a basis to develop and illustrate the performance requirements for protecting a pipe line from the detrimental effects of HVDC earth currents. The hypothetical problem, as shown, is one that may be encountered in the field. In this example, the following assumptions are made:

- The pipe line is long, straight, electrically continuous, and uniform in wall thickness, diameter, material and coating.
- It is approximately 30 miles away from the nearest HVDC earth electrode.
- The earth's resistivity is uniform over the general area.
- The maximum value of the HVDC earth current in either polarity causes the pipe line effects.

While these conditions may not exist in any actual pipe line HVDC system situation, they will serve to identify and evaluate the requirements for protection against HVDC effects in an actual situation.

Figure 3 shows the range or envelope of effects as the HVDC varies from its maximum value in one polarity to the maximum value of the other polarity. The coordinates on this graph are DISTANCE along the X axis and VOLTAGE on the Y axis.

The voltage corresponds to the effect which would usually be measured on an actual pipe line: pipe-to-soil voltage measured with a copper-copper sulfate half-cell. The value of the voltage effect at any point on the line is the product of the current transfer, and the electrical resistance between the pipe material and earth over the same area.

Figure 3 also shows reasonable hypothetical scale values illustrating the various combinations of protection requirements which may exist in an actual installation. The voltage values give the change in pipe-to-soil reading that would be caused by the HVDC transfer current. If this HVDC is considered as the only source of corrosion, then a pipe-to-soil voltage reading with a copper-copper sulfate half-cell would be approximately 0.85 volts at no transfer

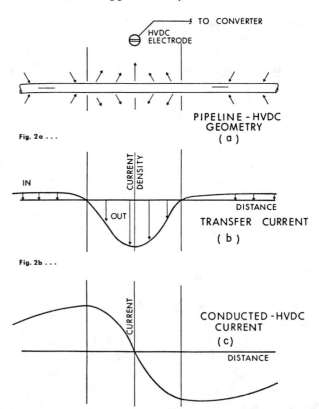

Figure 2. (a) Geometric plan between the HVDC electrode and the pipe line. Arrows indicate the direction of the HVDC component current. These would be reversed if the polarity of the electrode were reversed. (b) Curve shows characteristics of the transfer of HVDC into and out of the pipe line along its length. The transfer current is equivalent to current density. (c) Conducted current in the pipe line which results in the summation of the transfer current along the pipe.

current. This scale is shown on the left side of the figure. Corrosion occurs for any reading below this axis, since the current is leaving the pipe line and entering the soil to cause this change in the pipe-to-soil reading.

At locations on the pipe line where HVDC causes conditions which increase the volt-meter reading, current is collecting on the pipe line. The horizontal axis, therefore, represents a corrosion limit. Above this axis, a coating disbonding limit can exist where the density of the current transferred from the earth to the pipe line causes gas evolution, which may blister the coating.

In this example, a pipe-to-soil reading with a copper-copper sulfate half-cell of 1.5 volts is shown as the coating

disbonding limit. The difference between these limits gives an assumed safe operating band width of 0.65 volts. Values, other than these may be used for specific applications. These limits, however, will serve to illustrate the principles.

Protective System Limits

The requirements for protection along the line from either corrosion or disbonding is developed from examination of Figure 3 and the results are plotted in the subsequent figures. The upper limit of protection is the

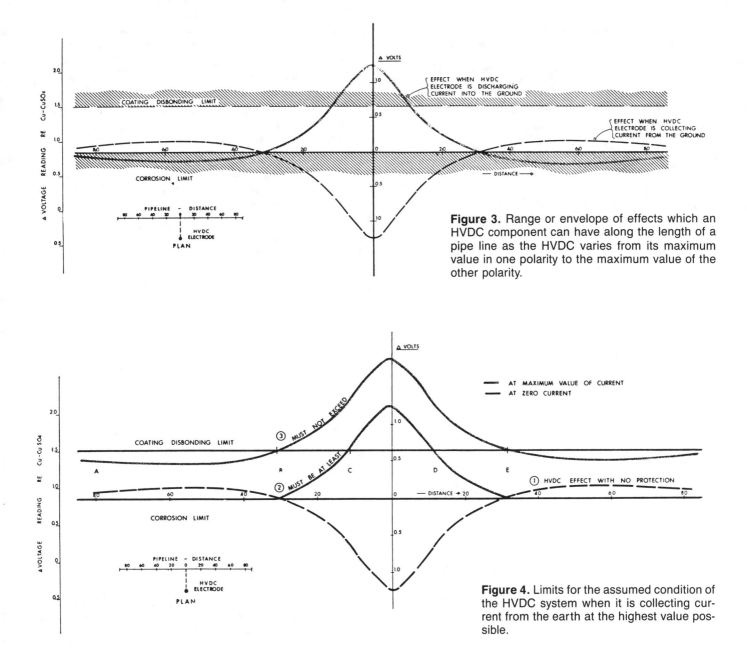

Figure 3. Range or envelope of effects which an HVDC component can have along the length of a pipe line as the HVDC varies from its maximum value in one polarity to the maximum value of the other polarity.

Figure 4. Limits for the assumed condition of the HVDC system when it is collecting current from the earth at the highest value possible.

maximum amount of current that can be collected by the pipe line without exceeding the disbonding limit. The lower limit is the minimum amount of current that can be collected by the pipe line to keep it above the corrosion limit.

At this value of HVDC (Figure 4), the uncorrected pipe line effect is shown by Curve 1. No protection is needed along either end of the pipe line in Regions A-B or E-F. Corrosion protection is needed between B and E. The minimum protection effect along the line is shown by Curve 2. The resultant of the HVDC and this protection effect lies on the corrosion limit. The maximum protective effect that can be applied along the line, without exceeding the coating disbonding limit, is shown by Curve 3. Under this maximum value and polarity of HVDC current, the region between these two protective limits, therefore, defines the range of acceptable protective system performance at any location on the pipe line.

When the HVDC effect decreases, due to a reduction in the ground current, the required level of protection on the line also decreases. Along the pipe line length from C to D, the protection *must* be reduced to avoid exceeding the disbonding limit. At zero HVDC effect, the protective system effect must fall between the corrosion and disbonding limits all along the line.

Examination of the limits for these two values of HVDC system operation shows that the protective effect must be varied in response to the HVDC effect over the length of the pipe line from C to D: automatic regulation of the protective current is needed over this length of the pipe line. With only this polarity of HVDC current, a fixed level of

protection can be applied along the pipe line from A to C and from D to F.

When the earth current can reverse polarity, as with the Pacific Northwest (PNW)-Pacific Southwest (PSW) intertie, then the same process that was used to develop the limits shown in Figure 4 can be repeated for the opposite HVDC polarity. The limits on protection along the pipe line when the HVDC electrode is discharging current, at its maximum value, into the earth are shown in Figure 5.

The combination of these two sets of limiting characteristics is shown in Figure 6. The graph shows the length along the pipe line where a fixed level of cathodic protection can be used and those regions where controlled protection is required for any value of HVDC earth current. In this particular example, variable protection is required along the pipe line for approximately 20 to 30 miles on either side of the point closest to the HVDC electrode. Cathodic protection is required when the HVDC electrode is collecting current from the earth. That is, the HVDC component of current, conducted by the pipe line, is discharging from the line near the earth electrode. The required protective current along the pipe line to counteract the HVDC effect under this condition is shown as Region 1 in Figure 6.

When the HVDC polarity is reversed and the magnitude of the ground current is increased to the maximum value, a different problem is encountered. The electric field which is established around the electrode and pipe line causes current to enter the pipe line through the coating. In this example, the disbonding limit is exceeded over the length of the pipe line from C to D at the maximum value of HVDC

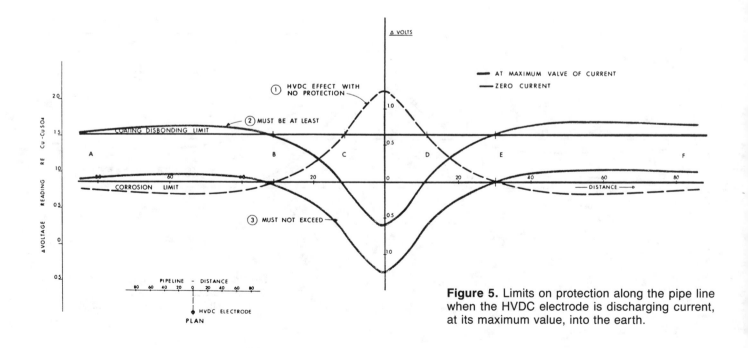

Figure 5. Limits on protection along the pipe line when the HVDC electrode is discharging current, at its maximum value, into the earth.

Figure 6. Combination of Figures 4 and 5. The length along the pipe line where a fixed level of cathodic protection can be used and those regions where controlled protection is required for any value of HVDC earth current. In this example, variable protection is required along the pipe line for approximately 20 to 30 miles on either side of the point closest to the HVDC electrodes.

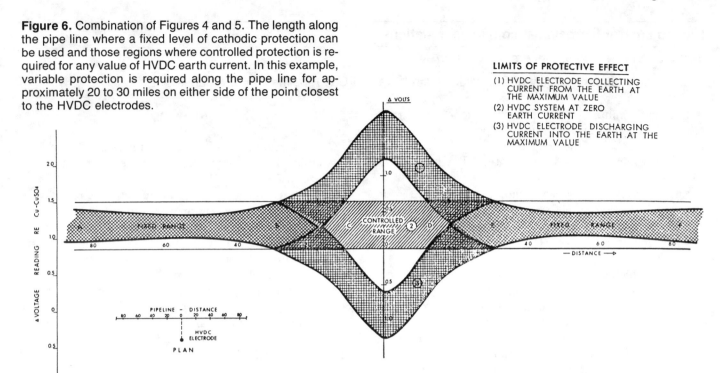

LIMITS OF PROTECTIVE EFFECT
(1) HVDC ELECTRODE COLLECTING CURRENT FROM THE EARTH AT THE MAXIMUM VALUE
(2) HVDC SYSTEM AT ZERO EARTH CURRENT
(3) HVDC ELECTRODE DISCHARGING CURRENT INTO THE EARTH AT THE MAXIMUM VALUE

earth current. The required correction for this detrimental effect is similar to the effect for the previous HVDC polarity: namely, the uncorrected electric field must be modified by an added field to give an acceptable total. Theoretically, in this situation, a portion of the HVDC earth current can be collected in a protective ground bed and conducted to the pipe line to satisfactorily modify the original field into an acceptable one.

Trouble shooting cathodic protection systems: Rectifier-ground bed

When the monitor, or a pipe-to-soil potential survey, indicates inadequacy of protection, the first place to look is at the protective unit. The curent output of the rectifier should be checked; if it is normal, the trouble is on the line itself; if it is high, and accompanied by low voltage, the trouble is certainly on the line, and is caused either by increased current demand or by a short-circuit to parasitic metal. If the current output is low, with voltage normal or high, the trouble is in the ground bed or connecting cables. A pipe-to-soil potential over the ground bed will show a peak over every anode which is working; a disconnected anode will not show at all. This test is particularly useful if comparison can be made to a similar test made at the time of the original installation. When inactive anodes are found, only digging will uncover the cause.

If the rectifier and its anode bed appear to be performing satisfactorily, the source of the low potentials must be sought on the line itself. Any area in which work has been done recently should be investigated; for example, if a new lateral has been connected, the insulation should certainly be checked. If investigation of such suspected sources discloses nothing, then a more detailed search must be made. First, the pipe-to-soil potentials should be studied, to see if the failure seems to be localized. A more thorough, but slower, approach, is that of making a detailed line current survey; find out where the drained current is coming onto the line. This will be much more easily interpreted if a similar survey, made when the line was in satisfactory condition, is on record; a comparison will often very quickly locate the offending parasite. On the other hand, if the current collected in each section is greater than on the earlier survey, with no pronounced differences, then the trouble is simply that of increased overall current requirement, probably due to coating deterioration.

How to control corrosion at compressor stations

Eight factors design engineers should consider

R. E. Hodge, Superintendent of Corrosion Control, Natural Gas Pipeline Co. of America

Corrosion failures at compressor stations result from carelessness on the part of the user, or poor choice of material/configuration by the designer.

Design engineers should consider the following:

- Coat everything underground, except ground rods, with a coating properly selected for the conditions
- Use proper coating application and inspection
- Use ground rods anodic to steel and insulated (coated) ground wires
- Avoid shielding as much as possible and allow space for corrosion control additions
- Use a minimum of insulating fittings
- Protect against water system corrosion
- Select materials carefully, with corrosion in mind as well as the more obvious considerations
- Do not retire facilities in place—take them out.

Coatings

Coatings are one of the most important considerations for control of underground corrosion. Generally, all underground metallic structures, except ground rods, should be coated. This would include gas piping, control lines, tubing, water lines, conduit, air lines, braces, etc.

There is enough buried conduit around a compressor station to soak up large amounts of cathodic protection if not coated. Non-metallic carriers can be used in some cases instead of coating.

Proper selection and use of coatings are important. The best of coatings are no good if used in the wrong place or improperly applied. The emphasis here is on the selection of coating for the temperature and environment expected.

For gas discharge lines, temperature becomes a dominant consideration. Be sure to get a coating that will withstand gas discharge temperature. Be careful, sag temperatures listed in coating literature are not maximum operating temperatures. Very few coatings on the market can withstand operating temperatures above 120° without damage. Any bare metal underground becomes a drain on the cathodic protection system and a potential shielding problem.

There are other areas for coating consideration besides underground. For instance, tank interiors and atmospheric coatings. These must also be matched to the situation.

Grounding

Grounds should be of materials anodic to steel. These might include magnesium or zinc anodes, or galvanized ground rods.

Copper ground rods should not be used. Copper is cathodic to steel. Some compressor stations have been installed with massive bare copper grounding systems. These are large galvanic cells with the steel gas piping as the anode (corroding element). Bare ground wires and rods also take cathodic protection current. This bad situation can be avoided by using galvanized rods, for example, and using copper ground wire with insulation. The junction of copper ground wire and galvanized ground rod must be coated. Grounding systems should be checked periodically for proper functioning.

Shielding

Shielding poses a major problem at compressor stations. It could be described as the blocking of cathodic protection due to geometric configuration.

Congested areas, where buried metal concentration is high, are difficult to cathodically protect. Piping, or other metal structures, will tend to shield cathodic protection currents from reaching all of the buried structures to be protected.

Cathodic protection ground beds of various styles have been used at compressor stations. Remote conventional beds and deep beds are popular for many applications. But NGPL has found that distributed anodes seem to be the best way to get adequate cathodic protection to all the shielded areas in a station. This type bed can be described as a ground bed with the anodes distributed in a non-geometrical pattern, interspersed through with piping.

Designers should allow space between lines where possible; avoid congestion. This would cause less shielding and allow space for distributed anode installations where required. It would also allow space for access to work equipment for anode installations.

Insulation

Use insulating fittings (flanges, couplings, unions, etc.) only where absolutely necessary.

Hold insulating fittings to minimum. These can often cause more problems than they solve. You might want to insulate water wells from the adjacent piping for control of cathodic protection current to the wells. This may not even be necessary if distributed ground beds are used.

Some companies insulate compressor stations from the lines entering and leaving the station. Other companies do not. This is a matter of individual company preference.

Water systems

Each water system (jacket water, piston water, boiler water, etc.) should have corrosion monitoring provisions designed into the system, such as coupons or corrosion rate probes. If there is a gas treating plant in conjunction with the compressor station, corrosion monitoring provisions should be designed into that system also.

Water treatment for corrosion control would have to be considered, depending on individual circumstances. Some treatment facilities may be designed into the systems.

Provisions should be considered for cathodic protection of the internal surfaces of storage tanks and water softeners. One problem often encountered on smaller tanks is adequate openings. Enough openings of sufficient size should be allowed on any tank for possible future work.

Water should be kept out of some systems. For instance, use air dryers on control lines.

Materials selection

Proper selection of materials is important throughout design.

Two key considerations from a corrosion standpoint are: (1) avoid the use of dissimilar metals; (2) match the material to its environment. Examples of these would include the use of a brass valve in a buried steel water line; or use of mechanically strong bolts in the wrong electrolyte, resulting in subsequent failure from stress corrosion cracking.

Select materials carefully, with corrosion in mind as well as the more obvious considerations. If design engineers do this, many subsequent failures can be eliminated.

Retirements

Many compressor station designs involve retiring something that is already there.

If something must be retired, especially underground, take it out. Many times a company pays unnecessarily to continue protection of facilities retired in place.

Even if the retired facilities can be completely disconnected from the protective systems, shielding is still a problem. Shielding is a problem by foundations also, so if they are retired they should be removed.

Thus, attacking the corrosion problems requires a truly interdisciplinary effort. In fact, corrosion offers one of the richest fields for interdisciplinary activity.

(Based on a paper, "Construction Design to Facilitate Corrosion Control—Compressor Stations," presented by the author at AGA Distribution Conference, May 8–10, 1972, Atlanta, Ga.)

Project Leak Growth

The number of leaks in older pipeline systems tend to increase exponentially. Exponential growth of leaks will fit the following equation:

$$N = N_1 e^{nA}$$

where N = number of leaks for the study area in any year considered

N_1 = number of leaks in the first year for which leak data are available

e = 2.718

n = nth year for which the number of leaks is to be estimated

A = leak growth-rate coefficient

The coefficient A can be calculated from the graph drawn of the leak data for the study period. The study period should preferable be in the range of 5 to 10 years. Any data less than this will not give enough points to accurately determine the coefficient A. Before a graph is drawn, leak report quantities must be adjusted for any changes to the system during the study period.

In other words, if a township was the study area, then pipe replaced, added, abandoned or installed should be tallied. If this is not done, errors will result. For example, if leak data were collected on a township basis, and during the study period considered, 10% of the township's pipe was renewed, leak frequency in the old area will continue to rise while in the renewed area there should be none, possibly giving an overall number of leaks less than that of the year prior to the replacement. For an accurate picture and a determination of the leak growth-rate coefficient, the number of leaks in the years follwoing the replacement should be accounted for as follows:

Let x =   percentage of replacement completed.
Let y =   actual number of leaks for the remainder of the township for the year considered

Adjusted number $N = y/(1 - x)$

The leak growth rate for some pipelines may not follow the exponential curve. For example, in a particular case, a linear increase with time may be predicted. In that case, use a linear equation instead of the exponential form.

Example:

N_1 = number of leaks in first year (i.e., 1979) = 140
N = number of leaks in year considered (1986) = 310
n = 86 - 79 = 7
$310 = 140 e^{7A}$

Hence, coefficient A = 0.113. The estimated number of leaks in 1990 will be:

$$N = 140 \, e^{11 \times 0.113} = 485 \text{ leaks}$$

Source

Ahmad, Hayat, *Gas Pipeline Renewal Insertion Technology*, 3 and 4 (1990).

ADVANCES IN PIPE LINE PROTECTION

Specialized Corrosion Surveys for Buried Pipelines: Methods and Experience

M. D. Allen, CEng, MIM, MICorrST; **N. R. Barnes,** BTech, CEng, MIM, MiCorrST
Spencer and Partners, U.K.

Synopsis

The external corrosion protection of a pipeline is commonly achieved by a system comprising an insulating coating and cathodic protection. The performance of this system is normally assessed by regular monitoring of pipe to soil potentials at selected intervals along the pipeline. Three more detailed pipeline survey procedures are available to provide additional monitoring information and the case histories described indicate the experience and value gained from their selective use.

Introduction

The external corrosion protection of a buried steel pipeline is generally achieved by a system comprising an insulating coating and cathodic protection. On modern pipelines, the coating is regarded as the primary means of protection with the cathodic protection systems providing control of corrosion where the coating has failed or has been damaged.

The performance of the cathodic protection system is usually assessed by regular monitoring of pipe to soil potentials at selected intervals along the pipeline.

From these results conclusions may be drawn concerning the level of cathodic protection being achieved and by inference the performance of the insulating coating. Under most circumstances, such measurements at selected locations will provide an acceptable indication of the overall level of corrosion protection being achieved on the pipeline.

These measurement locations however, commonly at least 1km apart, are generally selected for ease of access and thus only provide a valid assessment where pipeline coating is consistent, ensuring a constant attenuation of cathodic protection levels between each measurement point. Where the coating is variable in quality, it cannot be assumed that satisfactory levels of protection are maintained between measurement locations and a deficiency in the protection system may remain undetected allowing corrosion to occur.

Coating defect surveys have also been used to identify areas where pipeline coating quality is questionable. Historically, these surveys have almost universally taken the form of Pearson type surveys. These surveys have entailed a two-man survey team walking the length of a pipeline identifying coating defects and subsequently marking the points with pegs for further investigation.

Surveys of this type have two inherent disadvantages, namely that the survey procedures are time consuming and no indication of cathodic protection levels at the defect location is obtained.

With these shortcomings in mind two other survey procedures have been developed, the signal attenuation coating survey and the close interval potential survey. The former provides a rapid assessment and record of pipeline coating condition and more readily identifies areas where a defect or defects are present. Identified areas may then be surveyed in detail by close interval potential techniques which cannot only confirm coating defect locations, but also establish the status of the cathodic protection system at close intervals throughout the chosen areas.

Paper first presented at the 7th International Conference on the Internal and External Protection of Pipes. Held in London, England. Organized and sponsored by BHRA, The Fluid Engineering Centre © BHRA, The Fluid Engineering Centre, Cranfield, Bedford, MK43 0AJ, England 1987

Methods of locating coating defects

Pearson Survey

The Pearson survey is an above ground technique used to locate coating defects on buried pipelines. An AC signal is injected onto the pipeline from a power source connected between a cathodic protection test post or valve and an earth pin. The AC frequency selected is dependent upon the type of pipeline coating in order to minimize current loss from the pipeline through sound coating. At coating defects, the AC signal leaks to earth and flows via the soil to the earth spike to complete the circuit. The localized current flow from the pipeline at the defect creates a potential gradient through the soil and this is detected by measurement at a ground surface.

In practice, this is achieved by two surveyors walking in tandem over the entire pipeline, or section of pipeline. Earth contacts, either ski poles or foot cleats are connected to a receiver via a cable harness. As the first surveyor approaches a coating fault, an increased voltage gradient is observed, either audibly through headphones, or visually on the receiver signal level meter. As the lead surveyor passes the defect, the signal fades, and then increases again as the rear surveyors contacts pass over the defect.

The defect may be recorded on a preprepared record sheet complete with a measured distance from a fixed reference point or indicated by a marker peg or non-toxic paint. A recent development in Pearson equipment has been the introduction of a system that continuously records, through data logging equipment, the potential difference between each earth contact. In this way, data may be graphically presented.

Signal Attenuation Coating Survey

This method is similar in many respects to the Pearson survey, the essential difference being the method of defect detection. The AC current injected onto the pipeline creates a magnetic field around the pipe, the strength of this field being proportional to the amplitude of the AC signal. Measurements of the field strength are made be recording the voltage generated in a sensing coil and indicated on a dB scale, when the coil is held over the pipeline. Loss of current from a coating defect creates a change in the field strength due to a change in the pipe current and this is detected as a change in output from the sensing coil.

In practice, this is achieved by one surveyor measuring field strengths at discrete, preset intervals. At each location the field strength and pipeline chainage are recorded. A graph of distance versus field strength may then be produced, sections of pipeline having poor coating quality are indicated by the steepest gradient on the line plot. The results may also be expressed as histograms to aid interpretation. Typical results

are indicated in Figure 1. Any area indicating high signal losses may then be resurveyed using the same procedures to more closely locate the coating defects(s).

Close Interval Potential Survey

The close interval potential survey is an above ground technique used to provide a detailed profile of the pipe to soil potential with distance. This profile can be used to determine detailed information concerning the performance of the cathodic protection system, the coating system and to a lesser degree, interaction effects. The advent of portable microprocessor based measuring equipment and computer assisted data handling, together with accurate synchronized high speed current interruption devices have allowed these surveys to provide accurate, practical data and collection methods.

To eliminate errors in the pipe to soil potentials recorded, arising from the inclusion of volt drops caused by the flow of cathodic protection current in the soil, synchronized current interruptors are employed in the cathodic protection units. Pipe to soil potentials are then measured instantly after the current is interrupted, before significant chemical depolarization has occurred thus providing an 'error-free' polarized potential.

In practice, the precision current interruptors are best fitted in the DC output circuit of the cathodic protection power supplies, prior to commencing the survey to interrupt the cathodic protection current for short cyclic periods. The cycle time is chosen with regard to pipeline characteristics although a range of 'ON' and 'OFF' periods and 'ON/OFF' ratios are available.

For this survey a variety of purpose built voltmeter/microprocessors are available that record and store the pipe to soil potential data with an additional facility to store coded entries of topographical or pipeline features. This information is transferred to floppy disc storage from which graphs of ON and OFF pipe to soil potentials against pipeline chainage are produced. A typical close interval potential graph is illustrated in Figure 2.

It will be apparent from the description of the pipeline surveys that the form of results are different for each survey:

Pearson

Results are not generally presented in graphical form. Coating defects are marked in the field; in addition the detection of defects is dependent upon operator skill. Only if the recording Pearson technique is used can data be presented in graphical form. Experience of this technique is at present limited.

Figure 1. Typical signal attenuation coating graphs.

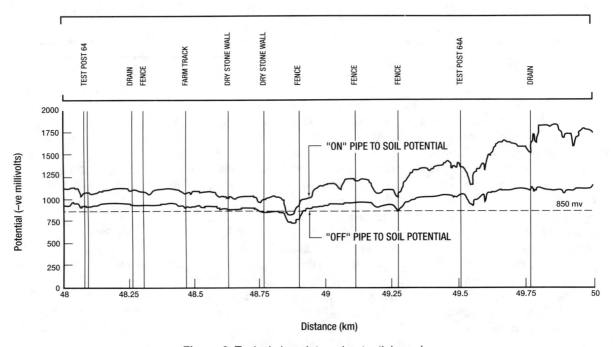

Figure 2. Typical close interval potential graph.

Signal Attenuation Coating Survey

Results are presented in graphical form although quick analysis may be performed in the field from raw data. Areas of poor coating quality may be readily identified from final graphs.

Following the production of graphs and histograms an evaluation system employed in pipeline coating quality assessment may assign values to the level of signal loss and the measurement distance interval. The sum of the two values referred to as the 'dB factor' and the 'km factor' may be used to indicate the future action required and the level of priority for each section surveyed.

The evaluation of the coating quality is illustrated in Figure 1. It should be noted, however, that gradings assigned to one particular pipeline cannot be compared to those assigned to another.

Close Interval Potential Survey

Results are shown in graphical form. Defect location is undertaken after the survey and requires experienced personnel to interpret the results.

The achievement of cathodic protection on the pipeline is indicated by the 'ON' and 'OFF' potential both being more negative than the accepted criteria.

Pipeline locations with inadequate cathodic protection will be indicated by the graph 'ON' and 'OFF' potentials shifting to a less negative level and become cause for concern if the 'OFF' potentials are more positive than say −0.85V.

It is considered that a localized coating defect would normally be indicated by a sharp positive shift of the 'ON' potential plot coupled with a plot shift of smaller amplitude of the 'OFF' potential, i.e., that the potential difference (IR drop) between the 'ON' and 'OFF' reading decreases, indicating a reduced coating resistance at that point.

The 'OFF' potential plot change would not be as large as that of the 'ON' potential, as the 'OFF' potential would be the polarized (or partially polarized) potential of the bare steel. This differs from the case where insufficient current for the achievement of polarization is available.

In all these cases, the IR drop measured between the 'ON' potentials and the 'OFF' potentials includes, along with the resistance of the coating, the resistance of the soil. The resistance of the coating is assumed to be the largest component of this resistance value.

Current is also a factor of the IR value. The current density impressed on the pipeline will increase as the cathodic protection stations are approached, giving the largest 'I' value of the IR drop near the location of the negative connection.

In order to interpret the data obtained fully, the anomalies observed in the plotted results may be classified into a number of categories relating to "point" anomalies or "general area" anomalies. For example, the following categories may be used.

Classification	Definition
Type I	Both the 'ON' and 'OFF' potentials are more positive than preset criteria. There is little or no 'IR' voltage. The potentials are unsatisfactory and the locations are not cathodically protected.
Type II	The 'OFF' potential is more positive than a preset limit, the 'ON' potential more negative. There is a reduced level of protection indicated at "point" locations which may be improved by increased output from the nearest cathodic protection station, although where this occurs over long pipeline lengths, it indicates high resistance coating or soils.
Type III	Positive change in the 'ON' and 'OFF' potential plot where both remain more negative than the selected criteria. This indicates increased current demand at these locations but that protection is maintained at a satisfactory level.
Type IV	Unusual potential variations due to outside influences, e.g., stray currents, station earthing, AC induced voltages.
Type V	A change in level of the 'ON' potential at a test post due to the change in the length of that part of the measuring circuit passing current, i.e., increasing 'IR' volt drop due to increasing pipe length in circuit.

Of the five categories listed above the first three are indicative of coating defects. Types IV and V are commonly observed on close interval potential survey results and require careful interpretation.

It is important that these considerations are borne in mind when selecting the survey or combination of surveys to be employed and in addition the choice of which survey to employ and when to employ it is dependent on a number of factors, namely,
- the purpose of the survey,
- the condition of the pipeline,
- the access over the pipeline,
- the time allowed.

As the protection system relies on both the external coating and a cathodic protection system, these may be considered as first and second lines of defense against corrosion. It is possible to take the position that concern over damaged coating is not warranted as back up from the cathodic protection system is available and therefore the integrity of the pipeline is not at risk. If this attitude is taken, then obviously close interval potential surveys would seem the most appropriate and perhaps the only survey system to consider.

Experience has shown however that the repair of significant coating defects will benefit the overall level of cathodic protection achieved, possibly permitting reduction in output levels and/or increasing the distance protected from any one protection station. In addition, the maintenance of a 'two-level' protection system also ensures the highest integrity is maintained for the pipeline. With this purpose in mind, the advantages of the coating surveys become apparent.

If we consider the requirements of a newly laid pipeline, it has been considered important to ensure as near perfect a coating as is possible at the start of the pipeline's life. It has therefore been common practice to 'Pearson' survey the pipeline after installation. As described above, this survey requires a two-man team and commonly progresses at some 4–6 km per day. This procedure may be accelerated by the use of the signal attenuation coating survey. Bearing in mind the probably high quality coating of a new pipeline, and because of the comparison basis employed by the signal attenuation coating survey, even small coating defects will become readily noticeable.

At the other extreme, with poorly coated old pipeline, a Pearson survey may have an almost continuous response making it impossible to discern individual defects or the most significant coating breakdown. The signal attenuation coating survey would, however be able to give a rapid assessment of coating condition and indicate where limited re-wrapping would be of greatest benefit.

In general, therefore, experience has shown that it is cost effective for the more detailed surveys of coating condition and cathodic protection to be undertaken in conjunction with a signal attenuation coating survey. A number of case histories are described below.

Case histories

1. Pipeline   : Multiproducts UK
   Length   : 115 kilometers
   Built   : 1982
   Coating   : Coal tar reinforced enamel with self-adhesive tape at field joints.
   Survey   : Signal attenuation coating survey
   Survey period : 9 days (4 days for disconnection of all 'temp' magnesium anodes, 5 days for survey).

Survey Results:
Survey indicated that over some 80 km of the pipeline the coating was satisfactory and further survey work was not recommended for those areas. Signal losses over these sections ranged between 1.5 dB/km and 3 dB/km. Some 9 sections, amounting to a total of some 10 km were given first priority for further investigation signal losses were as high as 17.6 dB over distances of 1.5 km. A further 2 sections, amounting to some 21 km were classified as second priority items.

It was decided that this information gathered be correlated with the results obtained from a proposed intelligent pig survey before further action.

2. Pipeline   : White Oil products, UK
   Length   : 115 kilometers
   Built   : Approx 1963
   Coating   : Coal tar enamel, tape or bitumen flood field joints
   Survey   : Signal attenuation coating survey
                  Pearson survey
                  Close interval potential survey

Survey Period : 10 days - Signal attenuation coating survey
                1 day - Pearson survey
                4 days - Close interval potential survey

Survey Results:
Initial signal attenuation coating survey identified 30 locations of significantly high signal losses, of these 12 were attributed to interpipeline bonds and sacrificial anodes, the remaining 18 were subject to more detailed investigation.

Close interval potential surveys showed that a satisfactory level of cathodic protection was being achieved at each of the locations where defects had been pinpointed using Pearson equipment.

Three particular locations were excavated, in the first a coating defect of some 800 mm associated with a welding arc burn was found within an area previously showing a signal loss of 6 dB over less than 100 m section. In the second area, showing a similar signal attenuation, a number of small coating defects were found together with a box of welding rods touching the pipe. At the third location a badly poured bitumen field joint had been indicated by a signal attenuation loss of 10 dB.

3. Pipeline   : Oil, UK
   Length   : 210 kilometers
   Built   : Approx. 1970
   Coating   : Coal tar enamel with tape wrapped field joints.
   Survey   : Close interval potential survey Recording both 'ON' and 'OFF' potentials at 5 meter intervals.
   Survey period : 40 days for 3 man team

Survey results :
Review of all the close interval potential results high-lighted some twenty-eight locations where both the 'ON' and 'OFF' potentials were less negative than the protection criteria. Of these 28 locations, 80% would not have been apparent from potential measurements recorded at adjacent test posts, thereby allowing the integrity of the pipeline to be put at risk where these unsatisfactory levels of cathodic protection coincided with coating defects.

4. Pipeline   : Aviation Fuel 6", UK
   Length   : 5 kilometers
   Built   : 1960
   Coating   : Asphalt
   Survey   : Signal attenuation survey
   Survey period : 2 days
   Survey results :
The survey showed consistently high signal losses indicating a generally deteriorated coating. Signal loss values were of the order of 40 dB per km.

Over 50% of the total signal loss was concentrated over four areas amounting to some 1,720 meters of pipe. Correlating close interval potential survey results over the same areas allowed 315 meters to be selected for rewrapping.

5. Pipeline   : Crude Oil, Middle East
   Length   : 132 kilometers
   Built   : 1986
   Coating   : Extruded polyethylene
   Survey   : Signal attenuation coating survey
   Survey period : 4 days
   Survey results :
Generally signal losses were found to average less than 1 dB per kilometer. One location was found to indicate a signal loss of 14 dB over 40 m. More detailed signal attenuation survey work located two possible coating defects with signal losses of 10 dB over 20 m and 48 dB over 20 m. A subsequent Pearson survey and two excavations revealed 5 small defects in a shrink sleeve weld joint coating for the 10 dB loss and one defect less than 100 mm^2 in the factory applied coating for the 4.8 dB.

Conclusion

It is considered that the discussion and case histories illustrated have shown that each of the above ground survey procedures offer certain advantages and disadvantages, and that many of the technical and economic penalties of choosing any one specific survey may be overcome if used in conjunction with another.

REFERENCES

J. Pearson *Petroleum Engineer* 12(10) 1941.
CCEJV "Cathodic Protection Monitoring of Buried Pipelines" —State of the Art Report by Task Group E4-2.
N. Frost "Monitoring Pipeline Corrosion." *Pipes and Pipelines International* May 1986.

C. J. Argent et al. "Coating defect location by above ground surveys"—proceedings of UK corrosion 1986.

Source
Jones, Glyn, and Thron, Joanna , *Advances in Pipeline Protection* Gulf Publishing Co. 1990

Estimate the Number of Squares of Tape for Pipe Coating (Machine Applied)

$$\text{Squares of Coating} = \frac{W \times A}{(W - O) \times 100 \text{ sq ft/square}}$$

Where: W = width of coating, inches
       A = pipe surface area, sq. ft.
       O = overlap, inches

Example:

Pipe OD   = 6 5/8"
Length   = 1 mile

Tape width   = 6"
Overlap   = 1"

Pipe surface area   = 1.73442 sq. ft/ft (Table 7 Pipe Data Section 3)

Total surface area = 1.73442 x 5280 = 9,157 sq. ft.

$$\text{Squares of Coating} = \frac{6 \times 9157}{(6 - 1) \times 100}$$

$$= 110 \text{ squares}$$

Estimate the Amount of Primer Required for Tape

Multiply pipe surface area by 0.001047 to obtain the number of gallons of primer required per mile of pipe.

Example:

| Pipe OD | = 6 5/8" |
| Length | = 1 mile |

Pipe surface area	= 1.73442 sq. ft./ft. (Table 7 Pipe Data Section 3)
Total surface area	= 1.73442 x 5280 = 9,157 sq. ft.
Gals of primer	= 9,157 x 0.001074 = 9.8 gals. (Use 10 gals.)

Tape Requirements For Fittings

Reducer Size	Tape Width	Linear Feet	Square Feet	Tee Size	Tape Width	Linear Feet	Square Feet	90° LR Ell Size	Tape Width	Linear Feet	Square Feet
3x2	2	9.2	1.5	2	2	13	2.2	2	2	9	1.5
4x3	3	8.4	2.1	3	2	21	3.5	3	2	11	1.8
6x4	4	12	4	4	2	28	4.6	4	3	14.6	3.6
8x4	6	14	17	6	3	38	9.5	6	4	26	8.6
10x6	6	18	9	8	4	45	15	8	4	40	16.5
12x8	6	25	12	10	6	34	17	10	6	48	24
14x10	6	26	13	12	6	50	25	12	6	64	32
16x12	6	38	19	14	6	60	30	14	6	81	45
18x14	6	50	25	16	6	71	35.5	16	6	101	50.5
20x16	6	66	33	18	6	85	42.5	18	6	123	61.5
				20	6	100	50	20	6	146	73
				22	6	126	63	22	6	173	86.5
				24	6	132	66	24	6	201	100.5

Tape Requirements For Fittings, Continued

90° SR Ell Size	Tape Width	Linear Feet	Square Feet	45° Ell Size	Tape Width	Linear Feet	Square Feet
2	2	7	1.5	2	2	5.6	0.93
3	2	10	1.8	3	2	9	1.5
4	3	13	3.25	4	3	13	3.2
6	4	21	7	6	4	17.3	5.75
8	4	38	13	8	4	36	12
10	6	46	23	10	6	34	17
12	6	60	30	12	6	48	22
14	6	76	38	14	6	52	26
16	6	96	48	16	6	64	32
18	6	118	59	18	6	76	38
20	6	136	68	20	6	96	48
22	6	166	83	22	6	110	55
24	6	190	95	24	6	126	63

9: Gas—General

Know the gas laws 212
Calculate gas properties from a gas analysis 214
Physical properties of selected hydrocarbons and other chemicals
    and gases... 218
Nomograph for calculating density and specific volume of gases
    and vapors.. 226
Considerations for Selecting Energy Measurement
    Equipment 227
Facts about methane and its behavior... 233
Conversion table for pure methane... 237
Categories of natural gas and reserves terminology 238
Glossary of common gas industry terms 239

Know the gas laws

The symbols that will be used in the mathematical expression of the various gas laws are:

V_1 = volume of the gas in cubic ft with original conditions
V_2 = volume of gas in cubic ft under the second set of conditions
T_1 = initial temperature of the gas in °R (°F + 460)
T_2 = temperature of gas under second conditions °R
P_1 = pressure of gas in psia under original conditions
P_2 = pressure of gas in psia under second conditions

Any other symbols used in mathematically expressing the gas laws will be explained at the time of their introduction.

Boyle's Law. If the temperature remains the same, the volume of a given quantity of gas will vary inversely as the absolute pressure. This may be expressed mathematically as:

$$\frac{V_1}{V_2} = \frac{P_2}{P_1}$$

Inasmuch as in the application of Boyle's Law we are generally interested in the volume at a second set of pressure conditions, a rearrangement of the formula more readily used is:

$$V_2 = V_1 \times \frac{P_1}{P_2} \tag{1}$$

Example. A quantity of gas at a pressure of 42 psig has a volume of 1,000 cubic ft. If the gas is compressed to 100 psig, what volume would it occupy? Assume the barometric pressure to be 14.2 psia, and the temperature to remain the same. Substituting in the second arrangement of the formula for Boyle's Law above would give:

$$V_2 = 1,000 \times \frac{42 + 14.2}{100 + 14.2}$$

$$= 492.1 \text{ cubic ft}$$

Charles' Law (sometimes called Gay-Lussac's Law). If the pressure remains the same, the volume of a given quantity of gas will vary directly as the absolute temperature.
This may be expressed mathematically as:

$$\frac{V_1}{V_2} = \frac{T_1}{T_2}$$

Again, since we are usually more interested in the volume at a second set of temperature conditions than any other information, an arrangement of the formula that would be handy is:

$$V_2 = V_1 \times \frac{T_2}{T_1} \tag{2}$$

A second part of Charles' Law is: If the volume of a quantity of gas does not change, the absolute pressure will vary directly as the absolute temperature.
Expressed mathematically this is:

$$\frac{P_1}{P_2} = \frac{T_1}{T_2}$$

In this instance, we would probably be more interested in the pressure at a second temperature condition which could be expressed as:

$$P_2 = P_1 \times \frac{T_2}{T_1} \tag{3}$$

Example. A given weight of gas has a volume of 450 cubic ft when the temperature is 45°F, and the pressure is 10 psig. If the pressure remains the same, but the temperature is changed to 90°F, what will be the volume of the gas?

TABLE 1—Molecular Weights of Hydrocarbons and Other Compounds Associated with Natural Gas

Compound	Atomic Formula	Molecular Weight
Methane	CH_4	16.043
Ethane	C_2H_6	30.070
Propane	C_3H_8	44.097
Butane	C_4H_{10}	58.124
Pentane	C_5H_{12}	72.151
Hexane	C_6H_{14}	86.178
Heptane	C_7H_{16}	100.205
Carbon Dioxide	CO_2	44.011
Nitrogen	N_2	28.016
Oxygen	O_2	32.00
Water	H_2O	18.016
Air		28.967

Note: Molecular weights based on following values of atomic weights: Hydrogen 1.008, Carbon 12.011, Nitrogen 14.008, Oxygen 16.000, and Argon 39.944. Air was assumed to consist of 78.09% nitrogen, 20.95% oxygen, 0.93% argon, and 0.03% carbon dioxide.

Substitution in the formula for the first part of the Charles' Law gives:

$$V_2 = 450 \times \frac{(90 + 460)}{(45 + 460)}$$

$$= 490 \text{ cubic ft}$$

It is desired to determine what the new pressure would be for the gas in the above example if the volume remains the same and the temperature changes from 45°F to 90°F as indicated. (Atmospheric pressure is 14.4 psia.)

Substitution in the formula gives:

$$P_2 = (10 + 14.4) \times \frac{(90 + 460)}{(45 + 460)}$$

$$= 26.6 \text{ psia or } 12.2 \text{ psig}$$

A convenient arrangement of a combination of Boyle's and Charles' laws which is easy to remember and use can be expressed mathematically as:

$$\frac{P_1 V_1}{T_1} = \frac{P_2 V_2}{T_2} \tag{4}$$

One can substitute known values in the combination formula and solve for any one unknown value. In cases where one of the parameters, such as temperature, is not to be considered, it may be treated as having the same value on both sides of the formula and consequently it can be cancelled out.

Avogadro's Law. This law states that equal volumes of all gases at the same pressure and temperature conditions contain the same number of molecules.

From this it may be seen that the weight of a given volume of gas is a function of the weights of the molecules and that there is some volume at which the gas would weigh, in lb, the numerical value of its molecular weight.

The volume at which the weight of the gas in lb is equal to the numerical value of its molecular weight (known as the "mol-volume") is 378.9 cubic ft for gases at a temperature of 60°F and a pressure of 14.73 psia. Table 1 gives the atomic formula and molecular weights for hydrocarbons and other compounds frequently associated with natural gas. Reference to the table reveals that the molecular weight for methane is 16.043. Going back to the mol-volume explanation, shows that 378.9 cubic ft of methane at 60°F and a pressure of 14.73 psia would weigh 16.043 lb.

Avogadro's law ties in closely with what is usually known as the ideal gas law.

The Ideal Gas Law. Although expressed in many slightly different arrangements, this law is most frequently expressed as:

$$pV = nRT$$

where p = pressure of the gas
V = volume of the gas
n = number of lb-mols of gas
R = the universal gas constant which varies depending upon the units of pressure, volume, and temperature employed

Since the number of lb-mols of a gas would be equal to the weight of the gas divided by the molecular weight of the gas, we can express the ideal gas law as:

$$pV = 10.722 \times \frac{W}{M} \times T \tag{5}$$

where p = pressure of the gas, psia
V = volume of the gas, cubic ft
W = weight of the gas, lb
M = molecular weight of the gas
T = temperature of the gas, °R

The constant 10.722 is based upon the generally used value for the universal gas constant of 1,544 when the pressure is expressed in lb/sq ft absolute.

This formula can be used in many arrangements. An arrangement which may be used to determine the weight of a quantity of gas is:

$$W = 0.0933 \times \frac{MVp}{T} \tag{6}$$

when the symbols and units are as above.

Example. It is desired to find the weight of a gas in a 1,000-cubic-ft container if the gas is at a pressure of 150 psig and a temperature of 90°F. The molecular weight of the gas is known to be 16.816, and the barometric pressure is 14.3 psia.

Substitution in the formula gives:

$$W = \frac{.0933 \times 16.816 \times 1,000 \times (150 + 14.3)}{(90 + 460)}$$

$$= 468.7 \text{ lb}$$

The formula above may be used when the molecular weight of a gas is known; however, at times it is desirable to determine the weight of a given volume of gas when the molecular weight is not known or cannot be readily determined. This may be accomplished if the specific gravity of the gas is known by using the formula:

$$W = 2.698 \times \frac{GVp}{T} \qquad (7)$$

where G = the specific gravity of the gas (air equals 1.000) and the other symbols and units are as previously given

Example. If the molecular weight of the gas in the preceding example was unknown, but the specific gravity was known to be 0.581, substitution in the formula would give:

$$W = 2.698 \times \frac{0.581 \times 1,000 \times (150 + 14.3)}{(90 + 460)}$$

$$= 468.3 \text{ lb}$$

Calculate gas properties from a gas analysis

In most practical applications involving gas calculations, the gas will consist of a mixture of components. Properties of the components are known; however, the properties of the mix must be determined for use in other calculations, such as compressor performance calculations. The gas molecular weight, K-value (isentropic exponent), and compressibility of the gas mix will be determined in the calculation to follow. Other data that may be needed for compressor performance determination will likely come from the process design and from the manufacturer's data.

Sample Calculation 1

Given: A gas mixture at 30 psia and 60°F consisting of the following components:

Table 1

Gas	Mol %	Mol wt	P_c psia	T_c	cp lbm-mol-°R
Ethane	5	30.070	706.5	550	12.27
Propane	80	44.097	616.0	666	17.14
n-Butane	15	58.123	550.6	766	22.96
Total	100				

Table 2
Individual Component Contributions

Gas	Mol %	Mol wt	P_c	T_c	cp
Ethane	5	1.50	35	28	0.61
Propane	80	35.28	493	533	13.71
n-Butane	15	8.72	83	115	3.44
Total	100	45.5	611	676	17.76

Therefore, the properties of the mix will be:

Molecular weight = MW = 45.5
Critical pressure = P_c = 611
Critical temperature = T_c = 676
Specific heat at constant pressure = cp = 17.76

The ratio of specific heats for the mixture are calculated from Equation 1.

$$k_{mix} = cp_{mix}/(cp_{mix} - 1.986) \qquad (1)$$

$$k_{mix} = 17.76/(17.76 - 1.986)$$
$$= 1.126$$

Determine the compressibility of the mix by calculating the reduced pressure and reduced temperature by using Equations 2 and 3.

$$P_r = P/P_c \qquad (2)$$

$$T_r = T/T_c \qquad (3)$$

$$P_r = 30/611$$
$$= 0.049$$

$$T_r = (60 + 460)/676$$
$$= 0.769$$

Refer to Figure 1 to determine the compressibility using the values of P_r and T_r. Z = 0.955.

If the mixture of components is given in mols of component or mass flow of components, these must be converted to mol % before calculating the properties of the mix.

Sample Calculation 2

Mols of components are given:

Ethane	400 mols/hr
Propane	4,800 mols/hr
n-Butane	900 mols/hr
Total Flow	6,100 mols/hr

Mol % ethane = 400/6,100 = 6.56%

Table 3

Component	Mol/hr	Mol %
Ethane	400	6.56
Propane	4,800	78.69
n-Butane	900	14.75
Totals	6,100	100.00

The mol % values listed in Table 4 can be used to calculate the properties of the mixture following the procedure detailed in Sample Calculation 1.

Sample Calculation 3

Mass flow of components is given.

Table 4

Component	Mass Flow (kg/h)	Mol wt	Mol flow (kmol/h)	Mol %
Ethane	15,000	30.070	499	11.08
Propane	150,000	44.097	3,402	75.55
n-Butane	35,000	58.123	602	13.37
Totals			4,503	100.00

Mol flow of ethane = 15,000/30.07
$$= 499$$

Mol % of ethane = 499/4,503
$$= 11.08\%$$

When a dry gas is saturated with water, it will be necessary to consider the effects of the water on the molecular weight of the mixture. The water will also affect the total mass flow of the mixture.

Sample Calculation 4

Given: Compress 1,000 lbm/min dry CO_2, which is initially water saturated. MW = 44.01, P_{inlet} = 16 psia, T_{inlet} = 100°F.

Calculate the molecular weight of the mixture and the total mass flow for the required 1,000 lbm/min of dry CO_2.

Mol fraction of water vapor = $y = P_w/P$ (4)

P_w = partial pressure of water vapor. It is exactly equal to the saturation pressure if the gas is saturated.

$$MW_{mix} = yMW_{mix} + (1 - y) \times MW_{dg} \quad (5)$$

where   w = water vapor
dg = dry gas

$$m_{total} = m_{dg}[1 + (MW_w \times P_w)/(MW_{dg} \times P_{dg})] \quad (6)$$

Saturated pressure of water vapor at 100°F is equal to 0.95 psia, and since the mixture is saturated, the partial pressure is also equal to 0.95 psia.

Calculate mol % of water vapor using Equation 4.

$$y = 0.95/16 = 5.94 \text{ mol } \% \text{ water vapor}$$

Calculate MW_{mix} using Equation 5.

$$MW_{mix} = 0.0594 (18.02) + (1 - 0.059) \times 44.01$$
$$= 42.48$$

Calculate the mass flow of the mix using Equation 6.

$$m_{total} = 1,000[1 + (18.02 \times 0.95)/(44.01 \times (16 - 0.95))]$$
$$= 1,026 \text{ lbm/min}$$

Figure 1. Compressibility chart for very low values of reduced pressure. Reproduced by permission of *Chemical Engineering*, McGraw Hill Publications Company, July 1954.

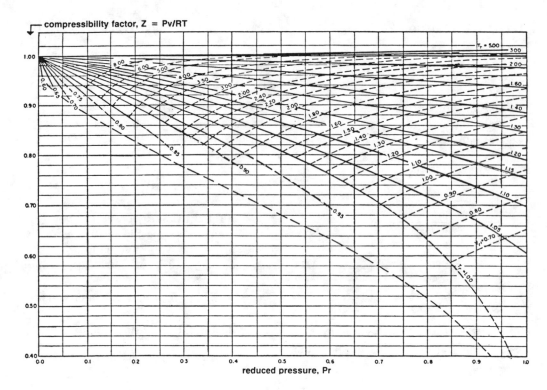

Figure 2. Compressibility chart for low values of reduced pressure. Reproduced by permission of *Chemical Engineering,* McGraw Hill Publications Company, July 1954.

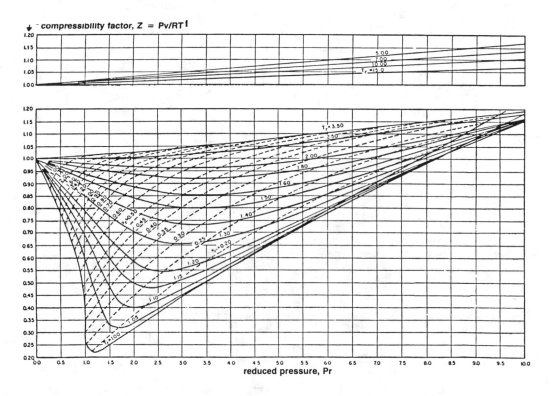

Figure 3. Compressibility chart for low to high values of reduced pressure. Reproduced by permission of *Chemical Engineering,* McGraw Hill Publications Company, July 1954.

Figure 4. Compressibility chart for low to very high values of reduced pressure. Reproduced by permission of *Chemical Engineering,* McGraw Hill Publications Company, July 1954.

Figure 5. Compressibility chart for natural gas. Reprinted by permission and courtesy of Ingersol Rand..

Physical properties of selected hydrocarbons and other chemicals and gases

Physical Constants   *See the Table of Notes and References.

Number	Compound	Formula	A. Molar mass (molecular weight)	B. Boiling point, °F 14.696 psia	Vapor pressure, psia 100°F	C. Freezing point, °F 14.696 psia	D. Refractive index, n_D 60°F	Critical constants Pressure, psia	Critical constants Temperature, °F	Critical constants Volume, ft³/lbm	Number
1	Methane	CH₄	16.043	−258.73	(5000)*	−296.44*	1.00042*	666.4	−116.67	0.0988	1
2	Ethane	C₂H₆	30.070	−127.49	(800)*	−297.04*	1.20971*	706.5	89.92	0.0783	2
3	Propane	C₃H₈	44.097	−43.75	188.64	−305.73*	1.29480*	616.0	206.06	0.0727	3
4	Isobutane	C₄H₁₀	58.123	10.78	72.581	−255.28	1.3245*	527.9	274.46	0.0714	4
5	n−Butane	C₄H₁₀	58.123	31.08	51.706	−217.05	1.33588*	550.6	305.62	0.0703	5
6	Isopentane	C₅H₁₂	72.150	82.12	20.445	−255.82	1.35631	490.4	369.10	0.0679	6
7	n−Pentane	C₅H₁₂	72.150	96.92	15.574	−201.51	1.35992	488.6	385.8	0.0675	7
8	Neopentane	C₅H₁₂	72.150	49.10	36.69	2.17	1.342*	464.0	321.13	0.0673	8
9	n−Hexane	C₆H₁₄	86.177	155.72	4.9597	−139.58	1.37708	436.9	453.6	0.0688	9
10	2−Methylpentane	C₆H₁₄	86.177	140.47	6.769	−244.62	1.37387	436.6	435.83	0.0682	10
11	3−Methylpentane	C₆H₁₄	86.177	145.89	6.103	———	1.37888	453.1	448.4	0.0682	11
12	Neohexane	C₆H₁₄	86.177	121.52	9.859	−147.72	1.37126	446.8	420.13	0.0667	12
13	2,3−Dimethylbutane	C₆H₁₄	86.177	136.36	7.406	−199.38	1.37730	453.5	440.29	0.0665	13
14	n−Heptane	C₇H₁₆	100.204	209.16	1.620	−131.05	1.38989	396.8	512.7	0.0691	14
15	2−Methylhexane	C₇H₁₆	100.204	194.09	2.272	−180.89	1.38714	396.5	495.00	0.0673	15
16	3−Methylhexane	C₇H₁₆	100.204	197.33	2.131	———	1.39091	408.1	503.80	0.0646	16
17	3−Ethylpentane	C₇H₁₆	100.204	200.25	2.013	−181.48	1.39566	419.3	513.39	0.0665	17
18	2,2−Dimethylpentane	C₇H₁₆	100.204	174.54	3.494	−190.86	1.38446	402.2	477.23	0.0665	18
19	2,4−Dimethylpentane	C₇H₁₆	100.204	176.89	3.293	−182.63	1.38379	396.9	475.95	0.0668	19
20	3,3−Dimethylpentane	C₇H₁₆	100.204	186.91	2.774	−210.01	1.38564	427.2	505.87	0.0662	20
21	Triptane	C₇H₁₆	100.204	177.58	3.375	−12.81	1.39168	428.4	496.44	0.0636	21
22	n−Octane	C₈H₁₈	114.231	258.21	0.53694	−70.18	1.39956	360.7	564.22	0.0690	22
23	Diisobutyl	C₈H₁₈	114.231	228.39	1.102	−132.11	1.39461	360.6	530.44	0.0676	23
24	Isooctane	C₈H₁₈	114.231	210.63	1.709	−161.27	1.38624	372.4	519.46	0.0656	24
25	n−Nonane	C₉H₂₀	128.258	303.47	0.17953	−64.28	1.40746	331.8	610.68	0.0684	25
26	n−Decane	C₁₀H₂₂	142.285	345.48	0.06088	−21.36	1.41385	305.2	652.0	0.0679	26
27	Cyclopentane	C₅H₁₀	70.134	120.65	9.915	−136.91	1.40896	653.8	461.2	0.0594	27
28	Methylcyclopentane	C₆H₁₂	84.161	161.25	4.503	−224.40	1.41210	548.9	499.35	0.0607	28
29	Cyclohexane	C₆H₁₂	84.161	177.29	3.266	43.77	1.42862	590.8	536.6	0.0586	29
30	Methylcyclohexane	C₇H₁₄	98.188	213.68	1.609	−195.87	1.42538	503.5	570.27	0.0600	30
31	Ethene(Ethylene)	C₂H₄	28.054	−154.73	(1400)*	−272.47*	(1.228)*	731.0	48.54	0.0746	31
32	Propene(Propylene)	C₃H₆	42.081	−53.84	227.7	−301.45*	1.3130*	668.6	197.17	0.0689	32
33	1−Butene(Butylene)	C₄H₈	56.108	20.79	62.10	−301.63*	1.3494*	583.5	295.48	0.0685	33
34	cis−2−Butene	C₄H₈	56.108	38.69	45.95	−218.06	1.3665*	612.1	324.37	0.0668	34
35	trans−2−Butene	C₄H₈	56.108	33.58	49.87	−157.96	1.3563*	587.4	311.86	0.0679	35
36	Isobutene	C₄H₈	56.108	19.59	63.02	−220.65	1.3512*	580.2	292.55	0.0682	36
37	1−Pentene	C₅H₁₀	70.134	85.93	19.12	−265.39	1.37426	511.8	376.93	0.0676	37
38	1,2−Butadiene	C₄H₆	54.092	51.53	36.53	−213.16	———	(653.)*	(340.)*	(0.065)*	38
39	1,3−Butadiene	C₄H₆	54.092	24.06	59.46	−164.02	1.3975*	627.5	305.	0.0654	39
40	Isoprene	C₅H₈	68.119	93.31	16.68	−230.73	1.42498	(558.)*	(412.)*	(0.065)*	40
41	Acetylene	C₂H₂	26.038	−120.49*	———	−114.5*	———	890.4	95.34	0.0695	41
42	Benzene	C₆H₆	78.114	176.18	3.225	41.95	1.50396	710.4	552.22	0.0531	42
43	Toluene	C₇H₈	92.141	231.13	1.033	−139.00	1.49942	595.5	605.57	0.0550	43
44	Ethylbenzene	C₈H₁₀	106.167	277.16	0.3716	−138.966	1.49826	523.0	651.29	0.0565	44
45	o−Xylene	C₈H₁₀	106.167	291.97	0.2643	−13.59	1.50767	541.6	674.92	0.0557	45
46	m−Xylene	C₈H₁₀	106.167	282.41	0.3265	−54.18	1.49951	512.9	651.02	0.0567	46
47	p−Xylene	C₈H₁₀	106.167	281.07	0.3424	55.83	1.49810	509.2	649.54	0.0570	47
48	Styrene	C₈H₈	104.152	293.25	0.2582	−23.10	1.54937	587.8	(703.)*	0.0534	48
49	Isopropylbenzene	C₉H₁₂	120.194	306.34	0.1884	−140.814	1.49372	465.4	676.3	0.0572	49
50	Methyl alcohol	CH₄O	32.042	148.44	4.629	−143.79	1.33034	1174.	463.08	0.0590	50
51	Ethyl alcohol	C₂H₆O	46.069	172.90	2.312	−173.4	1.36346	890.1	465.39	0.0581	51
52	Carbon monoxide	CO	28.010	−312.68	———	−337.00*	1.00036*	507.5	−220.43	0.0532	52
53	Carbon dioxide	CO₂	44.010	−109.257*	———	−69.83*	1.00048*	1071.	87.91	0.0344	53
54	Hydrogen sulfide	H₂S	34.08	−76.497	394.59	−121.88*	1.00060*	1300.	212.45	0.0461	54
55	Sulfur dioxide	SO₂	64.06	14.11	85.46	−103.86*	1.00062*	1143.	315.8	0.0305	55
56	Ammonia	NH₃	17.0305	−27.99	211.9	−107.88*	1.00036*	1646.	270.2	0.0681	56
57	Air	N₂+O₂	28.9625	−317.8	———	———	1.00028*	546.9	−221.31	0.0517	57
58	Hydrogen	H₂	2.0159	−422.955*	———	−435.26*	1.00013*	188.1	−399.9	0.5165	58
59	Oxygen	O₂	31.9988	−297.332*	———	−361.820*	1.00027*	731.4	−181.43	0.0367	59
60	Nitrogen	N₂	28.0134	−320.451	———	−346.00*	1.00028*	493.1	−232.51	0.0510	60
61	Chlorine	Cl₂	70.906	−29.13	157.3	−149.73*	1.3878*	1157.	290.75	0.0280	61
62	Water	H₂O	18.0153	212.000*	0.9501	32.00	1.33335	3198.8	705.16	0.0497s	62
63	Helium	He	4.0026	−452.09	———	———	1.00003*	32.99	−450.31	0.2300	63
64	Hydrogen chloride	HCl	36.461	−121.27	906.71	−173.52*	1.00042*	1205.	124.77	0.0356	64

12/12/86

NOTE: Numbers in this table do not have accuracies greater than 1 part in 1000; in some cases extra digits have been added to calculated values to achieve consistency or to permit recalculation of experimental values.

Physical Constants

*See the Table of Notes and References.

Number	E. Density of liquid 14.696 psia, 60°F			F. Temperature coefficient of density, 1/°F	G. Acentric factor, ω	H. Compressibility factor of real gas, Z 14.696 psia, 60°F	I. Ideal gas 14.696 psia, 60°F			J. Specific Heat 60°F 14.696 psia Btu/(lbm·°F)		Number
	Relative density (specific gravity) 60°F/60°F	lbm/gal.	gal./lb mole				Relative density (specific gravity) Air = 1	ft³ gas/lbm	ft³ gas/gal. liquid	C_p, Ideal gas	C_p, Liquid	
1	(0.3)*	(2.5)*	(6.4172)*	——	0.0104	0.9980	0.5539	23.654	(59.135)*	0.52669	——	1
2	0.35619*	2.9696*	10.126*	——	0.0979	0.9919	1.0382	12.620	37.476*	0.40782	0.97225	2
3	0.50699*	4.2268*	10.433*	-0.00162*	0.1522	0.9825	1.5226	8.6059	36.375*	0.38852	0.61996	3
4	0.56287*	4.6927*	12.386*	-0.00119*	0.1852	0.9711	2.0068	6.5291	30.639*	0.38669	0.57066	4
5	0.58401*	4.8690*	11.937*	-0.00106*	0.1995	0.9667	2.0068	6.5291	31.790*	0.39499	0.57272	5
6	0.62470	5.2082	13.853	-0.00090	0.2280	——	2.4912	5.2596	27.393	0.38440	0.53331	6
7	0.63112	5.2617	13.712	-0.00086	0.2514	——	2.4912	5.2596	27.674	0.38825	0.54363	7
8	0.59666*	4.9744*	14.504*	-0.00106*	0.1963	0.9582	2.4912	5.2596	26.163*	0.39038	0.55021	8
9	0.66383	5.5344	15.571	-0.00075	0.2994	——	2.9755	4.4035	24.371	0.38628	0.53327	9
10	0.65785	5.4846	15.713	-0.00076	0.2780	——	2.9755	4.4035	24.152	0.38526	0.52732	10
11	0.66901	5.5776	15.451	-0.00076	0.2732	——	2.9755	4.4035	24.561	0.37902	0.51876	11
12	0.65385	5.4512	15.809	-0.00076	0.2326	——	2.9755	4.4035	24.005	0.38231	0.51367	12
13	0.66631	5.5551	15.513	-0.00076	0.2469	——	2.9755	4.4035	24.462	0.37762	0.51308	13
14	0.68820	5.7376	17.464	-0.00068	0.3494	——	3.4598	3.7872	21.729	0.38447	0.52802	14
15	0.68310	5.6951	17.595	-0.00070	0.3298	——	3.4598	3.7872	21.568	0.38041	0.52199	15
16	0.69165	5.7664	17.377	-0.00070	0.3232	——	3.4598	3.7872	21.838	0.37882	0.51019	16
17	0.70276	5.8590	17.103	-0.00069	0.3105	——	3.4598	3.7872	22.189	0.38646	0.51410	17
18	0.67829	5.6550	17.720	-0.00070	0.2871	——	3.4598	3.7872	21.416	0.38594	0.51678	18
19	0.67733	5.6470	17.745	-0.00073	0.3026	——	3.4598	3.7872	21.386	0.39414	0.52440	19
20	0.69772	5.8170	17.226	-0.00067	0.2674	——	3.4598	3.7872	22.030	0.38306	0.50138	20
21	0.69457	5.7907	17.304	-0.00068	0.2503	——	3.4598	3.7872	21.930	0.37724	0.49920	21
22	0.70696	5.8940	19.381	-0.00064	0.3977	——	3.9441	3.3220	19.580	0.38331	0.52406	22
23	0.69793	5.8187	19.632	-0.00067	0.3564	——	3.9441	3.3220	19.330	0.37571	0.51130	23
24	0.69624	5.8046	19.679	-0.00065	0.3035	——	3.9441	3.3220	19.283	0.38222	0.48951	24
25	0.72187	6.0183	21.311	-0.00061	0.4445	——	4.4284	2.9588	17.807	0.38246	0.52244	25
26	0.73421	6.1212	23.245	-0.00057	0.4898	——	4.9127	2.6671	16.326	0.38179	0.52103	26
27	0.75050	6.2570	11.209	-0.00073	0.1950	——	2.4215	5.4110	33.856	0.27199	0.42182	27
28	0.75349	6.2819	13.397	-0.00069	0.2302	——	2.9059	4.5090	28.325	0.30100	0.44126	28
29	0.78347	6.5319	12.885	-0.00065	0.2096	——	2.9059	4.5090	29.452	0.28817	0.43584	29
30	0.77400	6.4529	15.216	-0.00062	0.2358	——	3.3902	3.8649	24.940	0.31700	0.44012	30
31	——	——	——	——	0.0865	0.9936	0.9686	13.527	——	0.35697	——	31
32	0.52095*	4.3432*	9.6889*	-0.00173*	0.1356	0.9844	1.4529	9.0179	39.167*	0.35714	0.57116	32
33	0.60107*	5.0112*	11.197*	-0.00112*	0.1941	0.9699	1.9373	6.7636	33.894*	0.35446	0.54533	33
34	0.62717*	5.2288*	10.731*	-0.00105*	0.2029	0.9665	1.9373	6.7636	35.366*	0.33754	0.52980	34
35	0.60996*	5.0853*	11.033*	-0.00106*	0.2128	0.9667	1.9373	6.7636	34.395*	0.35574	0.54215	35
36	0.60040*	5.0056*	11.209*	-0.00117*	0.1999	0.9700	1.9373	6.7636	33.856*	0.37690	0.54839	36
37	0.64571*	5.3834*	13.028	-0.00089	0.2333	——	2.4215	5.4110	29.129	0.36351	0.51782	37
38	0.65799*	5.4857*	9.8605*	-0.00101*	0.2540	(0.969)	1.8677	7.0156	38.485*	0.34347	0.54029	38
39	0.62723*	5.2293*	10.344*	-0.00110*	0.2007	(0.965)	1.8677	7.0156	36.687*	0.34120	0.53447	39
40	0.68615	5.7205	11.908	-0.00082	0.1568	——	2.3520	5.5710	31.869	0.35072	0.51933	40
41	(0.41796)	(3.4842)	(7.473)	——	0.1949	0.9930	0.8990	14.574	——	0.39754	——	41
42	0.88448	7.3740	10.593	-0.00067	0.2093	.	2.6971	4.8581	35.824	0.24296	0.40989	42
43	0.87190	7.2691	12.676	-0.00059	0.2633	——	3.1814	4.1184	29.937	0.26370	0.40095	43
44	0.87168	7.2673	14.609	-0.00056	0.3027	——	3.6657	3.5744	25.976	0.27792	0.41139	44
45	0.88467	7.3756	14.394	-0.00052	0.3942	——	3.6657	3.5744	26.363	0.28964	0.41620	45
46	0.86875	7.2429	14.658	-0.00053	0.3257	——	3.6657	3.5744	25.889	0.27427	0.40545	46
47	0.86578	7.2181	14.708	-0.00056	0.3216	——	3.6657	3.5744	25.800	0.27471	0.40255	47
48	0.91108	7.5958	13.712	-0.00053	(0.2412)	——	3.5961	3.6435	27.675	0.27110	0.41220	48
49	0.86634	7.2228	16.641	-0.00055	0.3260	——	4.1500	3.1573	22.804	0.29170	0.42053	49
50	0.79626	6.6385	4.8267	-0.00066	0.5649	——	1.1063	11.843	78.622	0.32316	0.59187	50
51	0.79399	6.6196	6.9595	-0.00058	0.6438	——	1.5906	8.2372	54.527	0.33222	0.56610	51
52	0.78939*	6.5812*	4.2561*	——	0.0484	0.9959	0.9671	13.548	89.163*	0.24847	——	52
53	0.81802*	6.8199*	6.4532*	-0.00583*	0.2667	0.9943	1.5196	8.6229	58.807*	0.19911	——	53
54	0.80144*	6.6817*	5.1005*	-0.00157*	0.0948	0.9846	1.1767	11.135	74.401*	0.23827	0.50418	54
55	1.3974*	11.650*	5.4987*	——	0.2548	0.9802	2.2118	5.9238	69.012*	0.14804	0.32460	55
56	0.61832*	5.1550*	3.3037*	——	0.2557	0.9877	0.5880	22.283	114.87*	0.49677	1.1209	56
57	0.87476*	7.2930*	3.9713*	——	——	1.0000	1.0000	13.103	95.557*	0.23988	——	57
58	0.071070*	0.59252*	3.4022*	——	-0.2202	1.0006	0.06960	188.25	111.54*	3.4038	——	58
59	1.1421*	9.5221*	3.3605*	——	0.0216	0.9992	1.1048	11.859	112.93*	0.21892	——	59
60	0.80940*	6.7481*	4.1513*	——	0.0372	0.9997	0.9672	13.546	91.413*	0.24828	——	60
61	1.4244*	11.875*	5.9710*	——	0.0878	(0.9875)	2.4482	5.3519	63.554*	0.11377	——	61
62	1.00000	8.33712	2.1609	-0.00009	0.3443	——	0.62202	21.065	175.62	0.44457	0.99974	62
63	0.12510*	1.0430*	3.8376*	——	0.	1.0006	0.1382	94.814	98.891*	1.2404	——	63
64	0.85129*	7.0973*	5.1373*	-0.00300*	0.1259	0.9923	1.2589	10.408	73.869*	0.19086	——	64

12/12/86

NOTE: Numbers in this table do not have accuracies greater than 1 part in 1000; in some cases extra digits have been added to calculated values to achieve consistency or to permit recalculation of experimental values.

Physical Constants

*See the Table of Notes and References.

Number	Compound	K. Heating value, 60°F — Net Btu/ft³ Ideal gas 14.696 psia	Net Btu/lbm Liquid	Gross Btu/ft³ Ideal gas 14.696 psia	Gross Btu/lbm Liquid	Gross Btu/gal Liquid	L. Heat of vaporization 14.696 psia at boiling point, Btu/lbm	M. Air required for combustion ideal gas ft³(air)/ft³(gas)	Flammability limits vol % in air mixture — Lower	Higher	ASTM octane number — Motor method D-357	Research method D-908	Number
1	Methane	909.4	——	1010.0	——	——	219.45	9.548	5.0	15.0	——	+1.6*	1
2	Ethane	1618.7	20277.*	1769.6	22181.*	65869.*	211.14	16.710	2.9	13.0	+0.05	+1.6*	2
3	Propane	2314.9	19757.*	2516.1	21489.*	90830.*	183.01	23.871	2.0	9.5	97.1	+1.8*	3
4	Isobutane	3000.4	19437.*	3251.9	21079.*	98917.*	157.23	31.032	1.8	8.5	97.6	+0.1*	4
5	n-Butane	3010.8	19494.*	3262.3	21136.*	102911.*	165.93	31.032	1.5	9.0	89.6*	93.8*	5
6	Isopentane	3699.0	19303.	4000.9	20891.	108805.	147.12	38.193	1.3	8.0	90.3	92.3	6
7	n-Pentane	3706.9	19335.	4008.9	20923.	110091.	153.57	38.193	1.4	8.3	62.6*	61.7*	7
8	Neopentane	3682.9	19235.*	3984.7	20822.*	103577.*	135.58	38.193	1.3	7.5	80.2	85.5	8
9	n-Hexane	4403.8	19232.	4755.9	20783.	115021.	143.94	45.355	1.1	7.7	26.0	24.8	9
10	2-Methylpentane	4395.2	19202.	4747.3	20753.	113822.	138.45	45.355	1.18	7.0	73.5	73.4	10
11	3-Methylpentane	4398.3	19213.	4750.3	20764.	115813.	140.05	45.355	1.2	7.7	74.3	74.5	11
12	Neohexane	4384.0	19163.	4736.2	20714.	112916.	131.23	45.355	1.2	7.0	93.4	91.8	12
13	2,3-Dimethylbutane	4392.9	19195.	4745.0	20746.	115246.	136.07	45.355	1.2	7.0	94.3	+0.3	13
14	n-Heptane	5100.0	19155.	5502.5	20679.	118648.	136.00	52.516	1.0	7.0	0.0	0.0	14
15	2-Methylhexane	5092.2	19133.	5494.6	20657.	117644.	131.58	52.516	1.0	7.0	46.4	42.4	15
16	3-Methylhexane	5096.0	19144.	5498.6	20671.	119197.	132.10	52.516	(1.01)	(6.6)	55.8	52.0	16
17	3-Ethylpentane	5098.3	19154.	5500.7	20679.	121158.	132.82	52.516	(1.00)	(6.5)	69.3	65.0	17
18	2,2-Dimethylpentane	5079.6	19095.	5481.9	20620.	116606.	125.12	52.516	(1.09)	(6.8)	95.6	92.8	18
19	2,4-Dimethylpentane	5084.2	19111.	5486.7	20635.	116526.	126.57	52.516	(1.08)	(6.8)	83.8	83.1	19
20	3,3-Dimethylpentane	5086.4	19119.	5488.8	20643.	120080.	127.20	52.516	(1.04)	(7.0)	86.6	80.8	20
21	Triptane	5081.2	19103.	5483.5	20628.	119451.	124.21	52.516	(1.08)	(6.8)	+0.1	+1.8	21
22	n-Octane	5796.1	19096.	6248.9	20601.	121422.	129.52	59.677	0.8	6.5			22
23	Diisobutyl	5780.5	19047.	6233.5	20552.	119586.	122.83	59.677	(0.92)	(6.3)	55.7	55.2	23
24	Isooctane	5778.8	19063.	6231.7	20568.	119389.	112.94	59.677	0.95	6.0	100.0	100.0	24
25	n-Nonane	6493.2	19054.	6996.5	20543.	123634.	124.36	66.839	0.7	5.6			25
26	n-Decane	7189.6	19018.	7742.9	20494.	125448.	119.65	74.000	0.7	5.4			26
27	Cyclopentane	3512.1	18825.	3763.7	20186.	126304.	167.33	35.806	(1.48)	(8.3)	84.9*	+0.1	27
28	Methylcyclopentane	4199.4	18771.	4501.2	20132.	126467.	148.54	42.968	1.0	8.35	80.0	91.3	28
29	Cyclohexane	4179.7	18675.	4481.7	20036.	130873.	153.03	42.968	1.2	8.35	77.2	83.0	29
30	Methylcyclohexane	4863.6	18640.	5215.9	20002.	129071.	136.30	50.129	1.1	6.7	71.1	74.8	30
31	Ethene(Ethylene)	1499.1	——	1599.8	——	——	207.41	14.323	2.7	36.0	75.6	+0.03	31
32	Propene(Propylene)	2181.8	(19858.)	2332.7	(21208.)	(92113.)	188.19	21.484	2.0	11.7	84.9	+0.2	32
33	1-Butene(Butylene)	2878.7	19309.*	3079.9	20670.*	103582.*	167.96	28.645	1.6	10.	80.8*	97.4	33
34	cis-2-Butene	2871.0	19241.*	3072.2	20602.*	107724.*	178.89	28.645	1.6	10.	83.5	100.0	34
35	trans-2-Butene	2866.8	19221.*	3068.0	20582.*	104666.*	174.37	28.645	1.6	10.			35
36	Isobutene	2859.9	19182.*	3061.1	20543.*	102830.*	169.47	28.645	1.6	10.			36
37	1-Pentene	3575.0	19184.	3826.5	20545.	110602.	154.48	35.806	1.3	10.	77.1	90.9	37
38	1,2-Butadiene	2789.0	19378.*	2939.9	20437.*	112111.*	191.88	26.258	(1.62)	(10.3)			38
39	1,3-Butadiene	2729.0	18967.*	2879.9	20025.*	104717.*	185.29	26.258	2.0	12.5			39
40	Isoprene	3410.8	18832.	3612.1	19953.	114141.	163.48	33.419	(1.12)	(8.5)	81.0	99.1	40
41	Acetylene	1423.2	(20887.)	1473.5	(21613.)	(75204.)	151.90	11.935	1.5	100.	+2.8	——	41
42	Benzene	3590.9	17256.	3741.8	17989.	132651.	169.24	35.806	1.2	8.0	+0.3	——	42
43	Toluene	4273.6	17421.	4475.0	18250.	132661.	154.83	42.968	1.2	7.1	+0.3	+5.8	43
44	Ethylbenzene	4970.5	17593.	5222.2	18492.	134387.	144.02	50.129	1.0	8.0	97.9	+0.8	44
45	o-Xylene	4958.2	17544.	5209.9	18444.	136036.	149.10	50.129	1.0	7.6	100.0	——	45
46	m-Xylene	4956.3	17541.	5207.9	18440.	133559.	147.24	50.129	1.0	7.0	+2.8	+4.0	46
47	p-Xylene	4957.1	17545.	5208.8	18444.	133131.	145.71	50.129	1.0	7.0	+1.2	+3.4	47
48	Styrene	4829.8	17414.	5031.1	18147.	137841.	152.85	47.742	1.1	8.0	+0.2	>+3.*	48
49	Isopropylbenzene	5660.9	17709.	5962.8	18662.	134792.	134.24	57.290	0.8	6.5	99.3	+2.1	49
50	Methyl alcohol	766.1	8559.	866.7	9751.	64731.	462.58	7.161	5.5	44.0	——	——	50
51	Ethyl alcohol	1448.1	11530.	1599.1	12770.	84539.	359.07	14.323	3.28	19.0	——	——	51
52	Carbon monoxide	320.5		320.5			92.77	2.387	12.50	74.20	——	——	52
53	Carbon dioxide	0.0		0.0			246.47*				——	——	53
54	Hydrogen sulfide	586.8	6337.*	637.1	6897.*	46086.*	235.63	7.161	4.30	45.50	——	——	54
55	Sulfur dioxide	0.0	——	0.0			167.22				——	——	55
56	Ammonia	359.0	——	434.4	——	——	589.48	3.581	15.50	27.00	——	——	56
57	Air	0.0	——	0.0	——	——	88.20				——	——	57
58	Hydrogen	273.8	——	324.2	——	——	192.74	2.387	4.00	74.20	——	——	58
59	Oxygen	0.0	——	0.0	——	——	91.59				——	——	59
60	Nitrogen	0.0	——	0.0	——	——	85.59				——	——	60
61	Chlorine	——	——	——	——	——	123.75				——	——	61
62	Water	0.0	*	*	0.0	0.0	970.18				——	——	62
63	Helium	0.0	——	0.0	——	——					——	——	63
64	Hydrogen chloride						190.43				——	——	64

12/12/86

NOTE: Numbers in this table do not have accuracies greater than 1 part in 1000; in some cases extra digits have been added to calculated values to achieve consistency or to permit recalculation of experimental values.

(Cont'd)
NOTES AND REFERENCES FOR THE TABLE OF PHYSICAL CONSTANTS

| | See Note No. → | | A. | B. | | C. | D. | Critical constants | | | |
Number	Compound	Formula	Molar mass (molecular weight)	Boiling point, °F 14.696 psia	Vapor pressure, psia 100 °F	Freezing point, °F 14.696 psia	Refractive index, n_D 60 °F	Pressure, psia	Temperature, °F	Volume, ft³/lbm	Number
1	Methane	CH_4	19	4	a,b	c,57	i,57	4	4	4	1
2	Ethane	C_2H_6	19	35	a,b	c,57	h,55	35	35	35	2
3	Propane	C_3H_8	19	34	34	c,57	h,55	34	34	34	3
4	Isobutane	C_4H_{10}	19	30	32	57	h,55	32	32	32	4
5	n-Butane	C_4H_{10}	19	33	33	57	h,55	33	33	33	5
6	Isopentane	C_5H_{12}	19	57	57	57	57	43	43	43	6
7	n-Pentane	C_5H_{12}	19	57	57	57	57	43	43	43	7
8	Neopentane	C_5H_{12}	19	57	57	57	h,57	43	43	43	8
9	n-Hexane	C_6H_{14}	19	57	57	57	57	43	43	43	9
10	2-Methylpentane	C_6H_{14}	19	57	57	57	57	43	43	43	10
11	3-Methylpentane	C_6H_{14}	19	57	57	f,22,29	57	43	43	43	11
12	Neohexane	C_6H_{14}	19	57	57	57	57	43	43	43	12
13	2,3-Dimethylbutane	C_6H_{14}	19	57	57	57	57	43	43	43	13
14	n-Heptane	C_7H_{16}	19	57	57	57	57	43	43	43	14
15	2-Methylhexane	C_7H_{16}	19	57	57	57	57	43	43	43	15
16	3-Methylhexane	C_7H_{16}	19	57	57	f,38	57	43	43	43	16
17	3-Ethylpentane	C_7H_{16}	19	57	57	57	57	43	43	43	17
18	2,2-Dimethylpentane	C_7H_{16}	19	57	57	57	57	43	43	43	18
19	2,4-Dimethylpentane	C_7H_{16}	19	57	57	57	57	43	43	43	19
20	3,3-Dimethylpentane	C_7H_{16}	19	57	57	57	57	43	43	43	20
21	Triptane	C_7H_{16}	19	57	57	57	57	43	43	43	21
22	n-Octane	C_8H_{18}	19	57	57	57	57	43	43	43	22
23	Diisobutyl	C_8H_{18}	19	57	57	57	57	43	43	43	23
24	Isooctane	C_8H_{18}	19	57	57	57	57	43	43	43	24
25	n-Nonane	C_9H_{20}	19	57	57	57	57	43	43	43	25
26	n-Decane	$C_{10}H_{22}$	19	57	57	57	57	43	43	43	26
27	Cyclopentane	C_5H_{10}	19	57	57	57	57	57	57	57	27
28	Methylcyclopentane	C_6H_{12}	19	57	57	57	57	57	57	57	28
29	Cyclohexane	C_6H_{12}	19	57	57	57	57	57	57	57	29
30	Methylcyclohexane	C_7H_{14}	19	57	57	57	57	57	57	57	30
31	Ethene(Ethylene)	C_2H_4	19	63	b	c,63	h,26	63	63	63	31
32	Propene(Propylene)	C_3H_6	19	5	5	5	h,26	g,5	44	57	32
33	1-Butene(Butylene)	C_4H_8	19	57	57	57	h,26	57	57	57	33
34	cis-2-Butene	C_4H_8	19	57	25	57	h,26	57	57	57	34
35	trans-2-Butene	C_4H_8	19	57	25	57	h,26	57	57	57	35
36	Isobutene	C_4H_8	19	57	57	57	h,26	57	57	57	36
37	1-Pentene	C_5H_{10}	19	57	57	57	26	57	57	57	37
38	1,2-Butadiene	C_4H_6	19	57	57	56		a	a	a	38
39	1,3-Butadiene	C_4H_6	19	57	25	56	h,26	57	57	57	39
40	Isoprene	C_5H_8	19	57	57	56		a	a	a	40
41	Acetylene	C_2H_2	19	d,1	b	c,56		57	57	57	41
42	Benzene	C_6H_6	19	14	57	14	57	14	14	14	42
43	Toluene	C_7H_8	19	57	57	15	57	57	57	57	43
44	Ethylbenzene	C_8H_{10}	19	57	57	15	57	57	57	57	44
45	o-Xylene	C_8H_{10}	19	57	57	15	57	57	57	57	45
46	m-Xylene	C_8H_{10}	19	57	57	15	57	57	57	57	46
47	p-Xylene	C_8H_{10}	19	57	57	15	57	57	57	57	47
48	Styrene	C_8H_8	19	56	13	56	57	56	a,11	56	48
49	Isopropylbenzene	C_9H_{12}	19	57	r,57	57	57	57	57	57	49
50	Methyl alcohol	CH_4O	19	58	58	58	58	58	58	58	50
51	Ethyl alcohol	C_2H_6O	19	58	58	58	58	1	58	58	51
52	Carbon monoxide	CO	19	58	b	c,58	i,58	1	1	1	52
53	Carbon dioxide	CO_2	19	d,3	b	c,3	i,58	3	3	3	53
54	Hydrogen sulfide	H_2S	19	31	31	c,31	i,58	31	31	31	54
55	Sulfur dioxide	SO_2	19	1	23	c,58	i,58	58	58	58	55
56	Ammonia	NH_3	19	1	23	c,58	i,58	1	1	1	56
57	Air	N_2+O_2	19,40	45	b		i,60	45	45	45	57
58	Hydrogen	H_2	19	p,58	b	p,c,58	i,58	p,58	p,58	p,58	58
59	Oxygen	O_2	19	u,10	b	c,u,10	i,58	46	46	46	59
60	Nitrogen	N_2	19	6	b	c,6	i,58	6	6	6	60
61	Chlorine	Cl_2	19	1	2	c,58	h,58	2	2	2	61
62	Water	H_2O	19	u,10	37	58	58	37	37	37	62
63	Helium	He	19	7	b		i,58	7	s,7	7	63
64	Hydrogen chloride	HCl	19	1	23	c,58	i,58	58	58	58	64

12/12/86

(Cont'd)

NOTES AND REFERENCES FOR THE TABLE OF PHYSICAL CONSTANTS

Number	E. Density of liquid 14.696 psia, 60°F			F. Temperature coefficient of density, 1/°F	G. Acentric factor, ω	H. Compressibility factor of real gas, Z 14.696 psia, 60°F	I. Ideal gas 14.696 psia, 60°F			J. Specific Heat 60°F 14.696 psia Btu/(lbm·°F)		Number
	Relative density (specific gravity) 60°F/60°F	lbm/gal.	gal./lb mole				Relative density (specific gravity) Air = 1	ft³ gas/lbm	ft³ gas/gal. liquid	Cp, Ideal gas	Cp, Liquid	
1	b,a	b,a,v	b,a	b,a	4	57			b,a	8	b	1
2	h,35	h,v,35	h,35	h,35	57	57			h	8	53	2
3	h,34	h,34	h,34	h,34	34	57			h	8	34	3
4	h,n,32	h,n,32	h,n,32	h,n,32	32	57			h	30	32	4
5	h,33	h,33	h,33	h,33	33	57			h	8	33	5
6	57	57	57	57	57					8	36	6
7	57	57	57	57	57					8	47	7
8	h,57	h,57	h,57	h,57	57	57			h	8	24	8
9	57	57	57	57	57					8	47	9
10	57	57	57	57	57					8	57	10
11	57	57	57	57	57					8	57	11
12	57	57	57	57	57					8	57	12
13	57	57	57	57	57					8	57	13
14	57	57	57	57	57					57	47	14
15	57	57	57	57	57					57	57	15
16	57	57	57	57	57					57	38	16
17	57	57	57	57	57					57	57	17
18	57	57	57	57	57					57	57	18
19	57	57	57	57	57					57	57	19
20	57	57	57	57	57					57	57	20
21	57	57	57	57	57					57	57	21
22	57	57	57	57	57					57	47	22
23	57	57	57	57	57					57	57	23
24	57	57	57	57	57					57	57	24
25	57	57	57	57	57					57	47	25
26	57	57	57	57	57					57	47	26
27	57	57	57	57	57					8	22,59	27
28	57	57	57	57	r,57					57	22,59	28
29	57	57	57	57	57					8	9	29
30	57	57	57	57	57					57	22,59	30
31	b	b	b	b	57	57			b	8	b	31
32	h,5	h,5	h,5	h,5	5	57			h	8	17	32
33	h	h	h	h	57	57			h	57	24	33
34	h,r	h,r	h,r	h,r	25	57			h	57	17	34
35	h,r	h,r	h,r	h,r	25	57			h	57	r,17	35
36	h,r	h,r	h,r	h,r	57	57			h	57		36
37	57	57	57	57	57					57	57	37
38	h,r	h,r	h,r	h,r	r,57	a			h	57	57	38
39	h,r	h,r	h,r	h,r	25	a			h	57	54,59	39
40	r	r	r	r	57					57	48	40
41	b	b	b	b	r,57	57			b	8		41
42	57	57	57	57	57					57	15	42
43	57	57	57	57	57					16	15	43
44	57	57	57	57	57					16	15	44
45	57	57	57	57	57					16	15	45
46	57	57	57	57	57					16	15	46
47	57	57	57	57	57					16	15	47
48	j,50	j,50	j,50	j,52	a,51					57		48
49	57	57	57	57	57					61	42	49
50	58	58	58	58	58					62	62	50
51	62	62	62	62	58					62	62	51
52	k,39	k,39	k,39		39				t	8	b	52
53	h,3	h,3	h,3	h,3					h	8		53
54	h,31	h,v,31	h,31	h,31	31	58			h	8	31	54
55	h	h	h		23	58			h	8		55
56	h	h	h		23	58			h	58		56
57	k,45	k,45	k,45						t	45		57
58	k,p,21	k,p,21	k,p,21		p,23	58			t	8	b	58
59	k,46	k,46	k,46		q,58	58			t	8	b	59
60	k,6	k,6	k,6		6	58			t	8	b	60
61	h	h	h		2	a			h	58		61
62	37	37	37	41	r,58					27	28	62
63	k	k	k		58	58			t	8	b	63
64	h	h	h	h	58	58			h	58		64

12/12/86

(Cont'd)
NOTES AND REFERENCES FOR THE TABLE OF PHYSICAL CONSTANTS

Number	Compound	K. Heating value, 60°F — Net: Btu/ft³ Ideal gas, 14.696 psia	Net: Btu/lbm Liquid	Gross: Btu/ft³ Ideal gas, 14.696 psia	Gross: Btu/lbm Liquid	Gross: Btu/gal. Liquid	L. Heat of vaporization 14.696 psia at boiling point, Btu/lbm	M. Air required for combustion ideal gas ft³(air)/ft³(gas)	Flammability limits, vol % in air mixture — Lower	Higher	ASTM octane number — Motor method D-357	Research method D-908	Number
1	Methane	57	b	57	b	b	4		x	x			1
2	Ethane	57	h,57	57	h,57	h,57	35		x	x	e	e,g	2
3	Propane	57	h,57	57	h,57	h,57	34		x	x		e,g	3
4	Isobutane	57	h,57	57	h,57	h,57	32		x	x		e,g	4
5	n-Butane	57	h,57	57	h,57	h,57	33		x	x	g	g	5
6	Isopentane	57	57	57	57	57	57		x	x			6
7	n-Pentane	57	57	57	57	57	57		x	x	g	g	7
8	Neopentane	57	h,57	57	h,57	57	57		x	x			8
9	n-Hexane	57	57	57	57	57	57		x	x			9
10	2-Methylpentane	57	57	57	57	57	57		x	x			10
11	3-Methylpentane	57	57	57	57	57	57		x	x			11
12	Neohexane	57	57	57	57	57	57		x	x			12
13	2,3-Dimethylbutane	57	57	57	57	57	57		x	x		e	13
14	n-Heptane	57	57	57	57	57	57		x	x			14
15	2-Methylhexane	57	57	57	57	57	57		x	x			15
16	3-Methylhexane	57	57	57	57	57	57		a,x	a,x			16
17	3-Ethylpentane	57	57	57	57	57	57		a,x	a,x			17
18	2,2-Dimethylpentane	57	57	57	57	57	57		a,x	a,x			18
19	2,4-Dimethylpentane	57	57	57	57	57	57		a,x	a,x			19
20	3,3-Dimethylpentane	57	57	57	57	57	57		a,x	a,x			20
21	Triptane	57	57	57	57	57	57		a,x	a,x	e	e	21
22	n-Octane	57	57	57	57	57	57		x	x			22
23	Diisobutyl	57	57	57	57	57	57		a,x	a,x			23
24	Isooctane	57	57	57	57	57	57		x	x			24
25	n-Nonane	57	57	57	57	57	57		x	x			25
26	n-Decane	57	57	57	57	57	57		a,x	a,x			26
27	Cyclopentane	57	57	57	57	57	57		a,x	a,x	g	e	27
28	Methylcyclopentane	20	20	20	20	20	57		x	x			28
29	Cyclohexane	57	57	57	57	57	57		x	x			29
30	Methylcyclohexane	20	20	20	20	20	57		x	x			30
31	Ethene(Ethylene)	57	b	57	b	b	63		x	x		e	31
32	Propene(Propylene)	57	a	57	a	a	57		x	x		e	32
33	1-Butene(Butylene)	57	h,57	57	h,57	h,57	57		x	x	g		33
34	cis-2-Butene	57	h,57	57	h,57	h,57	57		a,x	a,x			34
35	trans-2-Butene	57	h,57	57	h,57	h,57	57		a,x	a,x			35
36	Isobutene	57	h,57	57	h,57	h,57	57		a,x	a,x			36
37	1-Pentene	57	57	57	57	57	57		x	x			37
38	1,2-Butadiene	57	h,57	57	h,57	h,57	56		a,x	a,x			38
39	1,3-Butadiene	57	h,57	57	h,57	h,57	56		x	x			39
40	Isoprene	57	57	57	57	57	56		a,x	a,x			40
41	Acetylene	57	a	57	a	a	56		x	x			41
42	Benzene	57	57	57	57	57	14		x	x	e		42
43	Toluene	57	57	57	57	57	57		x	x	e	e	43
44	Ethylbenzene	57	57	57	57	57	57		x	x		e	44
45	o-Xylene	57	57	57	57	57	57		x	x			45
46	m-Xylene	57	57	57	57	57	57		x	x	e	e	46
47	p-Xylene	57	57	57	57	57	57		x	x	e	e	47
48	Styrene	20	20	20	20	20	56		x	x	e		48
49	Isopropylbenzene	57	57	57	57	57	57		x	x		e	49
50	Methyl alcohol	62	62	62	62	62	58		x	x			50
51	Ethyl alcohol	62	62	62	62	62	58		x	x			51
52	Carbon monoxide	8	b	8	b	b	58		x	x			52
53	Carbon dioxide						58						53
54	Hydrogen sulfide	18	h,58	18	h,58	h,58	31		x	x			54
55	Sulfur dioxide						56						55
56	Ammonia	58		58			58		x	x			56
57	Air						45						57
58	Hydrogen	58		58			p,58		x	x			58
59	Oxygen						46						59
60	Nitrogen						58						60
61	Chlorine						56						61
62	Water	8	w,8	w,8			37						62
63	Helium												63
64	Hydrogen chloride						56						64

12/12/86

(Cont'd)

NOTES FOR THE TABLE OF PHYSICAL CONSTANTS

a. Values in parentheses are estimated values.

b. The temperature is above the critical point.

c. At saturation pressure (triple point).

d. Sublimation point.

e. The + sign and number following specify the number of cm^3 of TEL added per gallon to achieve the ASTM octane number of 100, which corresponds to that of Isooctane (2,2,4-Trimethylpentane.)

f. These compounds form a glass. An earlier investigator [12] reported a value of -244 °F for the freezing point of 3-Methylpentane.

g. Average value from octane numbers of more than one sample.

h. Saturation pressure and 60 °F.

i. Index of refraction of the gas.

j. The temperature difference correction from [51].

k. Densities of the liquid at the normal boiling point.

m. Heat of sublimation.

n. Equation 2 of the reference was refitted to give:
$a = 0.7872957; b = 0.1294083; c = 0.03439519.$

p. Normal hydrogen (25% para, 75% ortho).

q. The vapor pressure [5] at the critical temperature [43].

r. An extrapolated value.

s. The value for the critical point of helium depends upon the temperature scale chosen. See reference 7.

t. Gas at 60 °F and the liquid at the normal boiling point.

u. Fixed points on the 1968 International Practical Temperature Scale (IPTS-68).

v. Densities at the normal boiling point are: Ethane, 4.540 [34]; Propane, 4.848 [33]; Propene, 5.083 [5]; Hydrogen Chloride, 9.948 [56]; Hydrogen Sulfide, 7.919 [30]; Ammonia, 5.688 [56]; Sulfur Dioxide, 12.20 [56].

w. Technically, water has a heating value in two cases: net (-1060. Btu/lbm) when water is liquid in the reactants, and gross (+50.313 Btu/ft^3) when water is gas in the reactants. The value is the ideal heat of vaporization (enthalpy of the ideal gas less the enthalpy of the saturated liquid at the vapor pressure). This is a matter of definition; water does not burn.

x. Extreme values of those reported by reference 24.

A. Molar mass (molecular weight) is based upon the following atomic weights: C = 12.011; H = 1.00794; O = 15.9994; N = 14.0067; S = 32.06, Cl = 35.453. The values were rounded off after calculating the molar mass using all significant figures in the atomic weights.

B. Boiling point: the temperature at equilibrium between the liquid and vapor phases at 14.696 psia.

C. Freezing point: the temperature at equilibrium between the crystalline phase and the air saturated liquid at 14.696 psia.

D. The refractive index reported refers to the liquid or gas and is measured for light of wavelength corresponding to the sodium D-line (589.26 nm).

E. The relative density (specific gravity): ρ(liquid, 60 °F)/ρ(water, 60 °F). The density of water at 60 °F is 8.33718 lbm/gal.

F. The temperature coefficient of density is related to the expansion coefficient by:

$$\frac{1}{\rho}\left(\frac{\partial \rho}{\partial T}\right)_P = -\frac{1}{V}\left(\frac{\partial V}{\partial T}\right)_P, \quad \text{in units of } 1/°F$$

G. Pitzer acentric factor:

$$\omega = -\log_{10}(P/P_c) - 1, \quad P \text{ at } T = 0.7\,T_c$$

H. Compressibility factor of the real gas, $Z = (PV)/(RT)$, is calculated using the second virial coefficient equation.

I. The density of an ideal gas relative to air is calculated by dividing the molar mass of the of the gas by 28.9625, the calculated average molar mass of air. See ref. 39 for the average composition of dry air. The specific volume of an ideal gas is calulated from the ideal gas equation. The volume ratio is: V(ideal gas)/V(liquid in vacuum).

J. The liquid value is not rigorously C_P, but rather it is C_S, the heat capacity at saturated conditions, defined by:

$$C_S = C_P - T\left(\frac{\partial V}{\partial T}\right)_P \frac{dP_S}{dT}$$

When dealing with liquids far from the critical point, $C_S \approx C_P$.

K. The heating value is the negative of the enthalpy of combustion at 60 °F and 14.696 psia in an ideal reaction (one where all gasses are ideal gasses). For an arbitrary organic compound, the combustion reaction is:

$$C_n H_m O_h S_j N_k(s, l, or\, g) + (n + \frac{m}{4} - \frac{h}{2} + j)O_2(g) \longrightarrow nCO_2(g) + \frac{m}{2}H_2O(g\, or\, l) + \frac{k}{2}N_2(g) + jSO_2(g)$$

where s, l and g denote respectively solid, liquid and ideal gas. For gross heating values, the water formed is liquid; for net heating values, the water formed is ideal gas. Values reported are on a dry basis. To account for water in the heating value, see GPA 2172. The Btu/lbm or gal. liquid column assumes a reaction with the fuel in the liquid state, while the Btu/ft^3 ideal gas column assumes the gas in the ideal gas state. Therefore, the values are not consistent if used in the same calculation, e.g. a gas plant balance.

L. The heat of vaporization is the enthalpy of the saturated vapor at the boiling point at 14.696 psia minus the enthalpy of the saturated liquid at the same pressure.

M. Air required for the combustion of ideal gas for compounds of formula $C_n H_m O_h S_j N_k$ is:

$$V(air)/V(gas) = (n + \frac{m}{4} - \frac{h}{2} + j)/0.20946$$

COMMENTS

Units - all dimensional values are reported in terms of the following three units, which are defined in terms of the corresponding SI units:

mass - pound (avdp), lbm = 0.45359337 kg

length - foot, ft = 0.3048 m

temperature - degree fahrenheit, $(t/°F) = 32 + [1.8(t/°C)]$

Temperature of the Celsius scale is defined by the International Practical Temperature of 1968 (IPTS-68).

Other derived units are:

volume - cubic foot, ft^3 = 0.02831685 m^3

gallon = 231 in^3 = 0.003785412 m^3

pressure - pound per square inch absolute = 6894.757 Pa

energy - British thermal unit (I.T.), Btu = 251.9958 cal (I.T.) = 1055.056 J

Physical constants:

gas constant, R = 1.98598 Btu(I.T.)/(°R·lb mol)
10.7316 ft^3 psia/(°R·lb mol)
8.31448 J/(K · mol)

Conversion factors:

1 ft^3 = 7.480520 gal.

1 lbm/ft^3 = 0.1336806 lbm/gal = 16.018462 kg/m^3

1 psia = 0.06804596 atm

1 atm = 14.69595 psia = 760 Torr

1 Btu (I.T.) = 252.1644 cal$_{th}$

REFERENCES FOR THE TABLE OF PHYSICAL CONSTANTS

1. Ambrose, D., National Physical Laboratory, Teddington, Middlesex, England: Feb. 1980, NPL Report Chem 107.
2. Ambrose, D.; Hall, D. J.; Lee, D. A.; Lewis, G. B.; Mash, C. J., J. Chem. Thermo., 11, 1089 (1979).
3. Angus, S.; Armstrong, B.; de Reuck, K. M., Eds. "Carbon Dioxide. International Thermodynamic Tables of the Fluid State-3"; Pergamon Press: Oxford, 1976.
4. Angus, S.; Armstrong, B.; de Reuck, K. M., Eds. "Methane. International Thermodynamic Tables of the Fluid State-5"; Pergamon Press: Oxford, 1978.
5. Angus, S.; Armstrong, B.; de Reuck, K. M., "Propylene (Propene). International Thermodynamic Tables of the Fluid State-7"; Pergamon Press: Oxford, 1980.
6. Angus, S.; de Reuck, K. M.; Armstrong, B., Eds. "Nitrogen. International Thermodynamic Tables of the Fluid State-6"; Pergamon Press: Oxford, 1979.
7. Angus, S.; de Reuck, K. M.; McCarthy, R. D., Eds. "Helium. International Thermodynamic Tables of the Fluid State-4"; Pergamon Press: Oxford, 1977.
8. Armstrong, G. T.; Jobe, T. L., "Heating Values of Natural Gas and its Components," NBSIR 82-2401, May 1982.
9. Aston, J. G.; Szasz, G. J.; Fink, H. L., J. Am. Chem. Soc., 65, 1135 (1943).
10. Barber, C. R., Metrologia 5, 35 (1969).
11. Boundy, R. H.; Boyer, R. F., (Eds.), "Styrene, Its Polymers, Copolymers and Derivatives," A.C.S. monograph No. 115, Reinholt, N.Y., 1952.
12. Bruun, J. H.; Hicks-Bruun, M. M., J. Res. NBS, 5, 933 (1930).
13. Chaiyavech, P.; Van Winkle, M., J. Chem. Eng. Data, 4, 53 (1959).
14. Chao, J., "Benzene," Key chemical data books, Thermodynamics Research Center, Texas A & M University, College Station, Texas (1978).
15. Chao, J., Hydrocarbon Processing, 59(11), 295 (1979).
16. Chao, J.; Hall, K. R., Thermochimica acta, 72, 323 (1984).
17. Chao, J.; Hall, K. R.; Yao, J., Thermochimica acta, 64, 285 (1983).
18. CODATA Task Group on Key Values for Thermodynamics, CODATA Special Report No. 7, 1978.
19. Commission on Atomic Weights and Isotopic Abundances, Pure and Appl. Chem. 49, 1102 (1983).
20. Cox, J. D.; Pilcher, G., "Thermochemistry of Organic and Organometallic Compounds," Academic Press, London, 1970.
21. Dean, J. W., "A Tabulation of the Properties of Normal Hydrogen from Low Temperature to 300 K and from 1 to 100 Atmospheres," NBS Tech. Note 120, November 1961.
22. Douslin, D. R.; Huffman, H. M., J. Am. Chem. Soc., 68, 1704 (1946).
23. Edwards, D. G., "The Vapor Pressure of 30 Inorganic Liquids Between One Atmosphere and the Critical Point," Univ. of Calif., Lawrence Radiation Laboratory, UCRL-7167. June 13, 1963.
24. Engineering Sciences Data Unit, "EDSU, Engineering Sciences Data," EDSU International Ltd., London.
25. Flebbe, J. L.; Barclay, D. A.; Manley, D. B., J. Chem. Eng. Data, 27, 405 (1982).
26. Francis, A. W., J. Chem. Eng. Data, 5, 534 (1960).
27. Friedman, A. S.; Haar, L., J. Chem. Phys. 22, 2051 (1954).
28. Ginnings, D. C.; Furukawa, G. T., J. Am. Chem. Soc. 75, 522 (1953).
29. Glasgow, A. R.; Murphy, E. T.; Willingham, C. B.; Rossini, F. D., J. Res. NBS, 37, 141 (1946).
30. Goodwin, R. D., "Isobutane: Provisional Thermodynamic Functions from 114 to 700K at Pressures to 700 Bar," NBSIR 79-1612, July 1979.
31. Goodwin, R. D., "Hydrogen Sulfide Provisional Thermochemical Properties from 188 to 700 K at Pressures to 75 MPa," NBSIR 83-1694, October 1983.
32. Goodwin, R. D.; Haynes, W. M., "Thermophysical Properties of Isobutane from 114 to 700 K at Pressures to 70 MPa," NBS Tech. Note 1051, January 1982.
33. Goodwin, R. D.; Haynes, W. M., "Thermophysical Properties of Normal Butane from 135 to 700 K at Pressures to 70 MPa," NBS Monograph 169, April 1982.
34. Goodwin, R. D.; Haynes, W. M., "Thermophysical Properties of Propane from 85 to 700 K at Pressures to 70 MPa," NBS Monograph 170, April 1982.
35. Goodwin, R. D.; Roder, H. M.; Straty, G. C.; "Thermophysical Properties of Ethane, 90 to 600 K at Pressures to 700 bar," NBS Tech. Note 684, August 1976.
36. Guthrie, G. B.; Huffman, H. M., J. Am. Chem. Soc., 65, 1139 (1943).
37. Haar, L.; Gallagher, J. S.; Kell, G. S., "NBS/NRC Steam Tables," Hemisphere Publishing Corporation, Washington, 1984.
38. Huffman, H. M.; Park, G. S.; Thomas, S. B., J. Am. Chem. Soc., 52, 3241 (1930).
39. Hust, J. G.; Stewart, R. B., "Thermodynamic Property Values for Gaseous and Liquid Carbon Monoxide from 70 to 300 Atmospheres," NBS Technical Note 202, Nov. 1963.
40. Jones, F. E., J. Res. NBS, 83, 419 (1978).
41. Kell, G. S., J. Chem. Eng. Data, 20, 97 (1975).
42. Kishimoto, K.; Suga, H.; Syuzo, S., Bull. Chem. Soc. Japan, 46, 3020 (1973).
43. Kudchadker, A. P.; Alani, G. H.; Zwolinski, B. J., Chem. Rev. 68, 659 (1968).
44. Marchman, H.; Prengle, H. W.; Motard, R. L., Ind. Eng. Chem., 41, 2658 (1949).
45. "The Matheson Unabridged Gas Data Book"; Matheson Gas Products; New York, 1974.
46. McCarty, R. D.; Weber, L. A., "Thermophysical Properties of Oxygen from the Freezing Liquid Line to 600 R for Pressures to 5000 Psia", NBS Technical Note 384, July 1971.
47. Messerly, J. F.; Guthrie, G. B.; Todd, S. S.; Finke, H. L., J. Chem. Eng. Data, 12, 338 (1967).
48. Messerly, J. F.; Todd, S. S.; Guthrie, G. B., J. Chem. Eng. Data, 15, 227 (1970).
49. Miller, A.; Scott, D. W., J. Chem. Phys., 68(3), 1317 (1978).
50. Miller, L.P.; Wachter, N. W.; Fried, V., J. Chem. Eng. Data, 20, 417 (1975).
51. Ohe, S., "Computer Aided Data Book of Vapor Pressure," Data Book Publishing Co., Tokyo, Japan, 1976.
52. Riddick, J. A.; Bunger, W. B., "Organic Solvents, Physical Properties and Methods of Purification," Wiley-Interscience: New York, 1970.
53. Roder, H. M., "Measurements of the Specific Heats, C_s, and C_v, of Dense Gaseous and Liquid Ethane," J. Res. Nat. Bur. Stand. (U.S.) 80A, 739 (1976).
54. Scott, R. B.; Meyers, C. H.; Rands, R. D.; Brickwedde, F. G.; Bekkedahl, N., J. Res. NBS, 35, 39 (1945).
55. Sliwinski, P., Z. Physik. Chem. (Frankfurt) 63, 263 (1969).
56. Stull, D. R.; Westrum, E. F.; Sinke, G. C., "The Chemical Thermodynamics of Organic Compounds," John Wiley & Sons, Inc., New York, 1969.
57. "TRC Thermodynamic Tables – Hydrocarbons," Thermodynamics Research Center, Texas A&M University System: College Station, Texas.
58. "TRC Thermodynamic Tables–Non-Hydrocarbons," Thermodynamics Research Center, Texas A&M University System: College Station, Texas.
59. Vargaftik, N. B.; "Tables on the Thermophysical Properties of Liquids and Gases," Hemisphere Publishing Corporation, Washington D.C., 1975.
60. Weast, R. C., "Handbook of Chemistry and Physics," The Chemical Rubber Co: Cleveland, Ohio, 1969.
61. Wilhoit, R. C.; Aljoe, K. L.; Adler, S. B.; Kudchadker, S. A., "Isopropylbenzene and 1-Methyl-2-, -3-, and -4-isopropylbenzene," API Monograph Series No. 722, American Petroleum Institute, 1984.
62. Wilhoit, R. C.; Zwolinski, B. J., J. Phys. Chem. Ref. Data, 2, Suppl. no. 1 (1973).
63. Younglove, B. A., J. Phys. Chem. Ref. Data, 11, Suppl. no. 1 (1982).

Nomograph for calculating density and specific volume of gases and vapors

The density of a gas or vapor can be calculated from the equation for an ideal gas:

$$\rho = \frac{144P}{RT} = \frac{MP}{10.72T} = \frac{2.70\ PSg}{T}$$

where ρ = density of gas, lb/cu ft
      P = absolute pressure, psi (psig + 14.7)
      R = individual gas constant (10.72/M)
      M = molecular weight of gas
      T = temperature, °R
      Sg = specific gravity of individual gas relative to air
          (= to ratio of gas molecular weight to that of air)

The above equation and corresponding nomograph (Figure 1) can be used to calculate the density (or specific volume) of gases at low pressures and high temperatures, where the ideal gas law holds. Note that the pressure scale is calibrated in psi gauge so that the correction of 14.7 psi is not needed when using the nomograph.

Example. What is the density of dry methane if its temperature is 100°F and its pressure is 15 psig?

Connect	With	Mark or Read
M = 16	t = 100°F	Index
Index Mark	P = 15 psig	ρ = 0.08 lb/cu ft

Source

"Fluid Flow Through Pipe, Fittings and Valves," *Technical Paper No. 410*, A-11, Crane Company, Chicago, Illinois (1957).

Figure 1. Weight density and specific volume of gases and vapors.

CONSIDERATIONS FOR SELECTING ENERGY MEASUREMENT EQUIPMENT

Keeping Track of Gas Heating Value is Important Phase of Today's Market

Paul E. Kizer, Applied Automation, Inc., Houston, Texas

With natural gas energy determination becoming an integral part of the transmission and distribution business in today's market, companies are placing increased emphasis on choosing the correct measuring device. Most current contracts have a Btu specification and many others use MMBtu instead of gas volume.

Here is what happens at the burner tip. The following is the exothermic reaction of methane and oxygen (80 to 95% of natural gas):

$$CH_4 + 2O_2 \rightarrow CO_2 + 2H_2O + \text{Heat (1010 Btu/cf)}[1]$$

for the case of ethane (up to 3% of natural gas):

$$2C_2H_6 + 7O_2 \rightarrow 4CO_2 + 6H_2O + \text{Heat (1769 Btu/cf)}[1]$$

and for the case of propane (usually up to 0.5%, except where propane/air peak shaving is used):

$$C_3H_8 + 5O_2 \rightarrow 3CO_2 + 4H_2O + \text{Heat (2516 Btu/cf)}[1]$$

Since natural gas is a mixture of these three hydrocarbons plus small amounts of other compounds, these three chemical reactions account for 90 to 99% of the heat generated when the gas is burned. However, one mole (a mole is a standard number of molecules, $(6.023 \times 10^{23}$, Avogadro's number))[17] of propane will generate over twice the heat of one mole of methane (CH_4 = 1010 Btu/cf vrs C^3H^8 = 2516 Btu/cf at 60°F and 4 14.696 psia). A small error in the heavier or longer chain hydrocarbon components of natural gas leads to a larger error in the Btu calculation. Consequently, a small shift in the composition of any component can affect the heating value of a cubic foot of carefully measured gas.

The term MMBtu first came into wide use after the 1978 Natural Gas Policy Act changed the method of custody transfer from Mcf to MMBtu. This method was upheld in a court case involving FERC Order 356 (Jan. 1984).[2]

How much is it worth to keep track of the Btu? Postulate a ±5% error between doing a lab determination of the heating value on a spot sample of gas and an on-line (nearly continuous) monitor of heating value, which results in a ±50 Btu difference at a 1,000 Btu/cf level. A thousand Btu/cf is fairly average for natural gas.

On a station that has 50 MMcf/d at $2.50 per Mcf or MMBtu, this is $125,000 worth of gas per day. Five percent of this is $6,250 per day. If a process chromatograph, $50,000 installed cost, is used to determine the energy content, a payout of less than 10 days is obtained on a 50 MMcf/d station. This explains why most major interconnects have Btu measurement on-line today.

General Methods

First, consider the direct methods of energy determination, since this is the oldest method in use today. All direct methods involve the complete oxidation (burning) of gas to determine its energy content. There are two basic classes of direct measurement devices, calorimeter and thermtitrator.

Early on, the term representing 100,000 Btu's was one therm of energy. One dekatherm (10 therms) is equal to 1,000,000 Btu or 1 MMBtu.[21] Perhaps this is where the name thermtitrator came from. However, as discussed earlier, MMBtu[2] has supplanted the therm unit of measurement for now.

The calorimeter was adapted from a laboratory physical chemistry method called "bomb calorimetry." These were laboratory instruments which yield precise results when used under controlled conditions. However, each determination required 4 to 8 hours to complete. This is fine for a research project, but in a production setting it is easy to see how gas sample cylinders would start piling up. The American calorimeter[3] provided laboratory type results in less time, but the lab conditions had to be tightly controlled with continually manned equipment. For on-line energy content measurement, the Cutler-Hammer Recording Calorimeter stood alone for decades.[3,4]

While this device still requires rigid environmental conditions, it runs unattended for long periods of time (weeks). Also it provides a recorder output, which allows the correlation of the heating value with the volume of gas. The recording calorimeter's stated precision is ±5 Btu out of 1,000 Btus, although many operators can do better.

With the introduction of solid-state electronics, micro calorimetry devices have been tried, but not with much commercial success. Hart Scientific, Provo, Utah, has used a catalyst bed to perform a controlled isothermal oxidation on a very precise volume of gas. Initial results seem to show ± 0.5 Btu at 1,000 Btu precision.[5]

The thermtitrator devices utilize a more elegant chemical method, stoichiometry—that is air to gas ratios. One instrument, the PMI thermtitrator, uses the temperature difference between two flames to maintain the gas flow ratios at stoichiometric conditions.[6,7]

Accuracy is around ±1 Btu at 1,000 Btu and compares favorably with the recording calorimeter.[8] The other stoichiometric device is the Honeywell HVT. It uses a zirconium oxide oxygen sensor to determine stoichiometric conditions and automatically adjusts air/gas ratios to minimize excess O. Both devices yield rapid reaction to step changes in Btu content and are useful in process control applications such as boiler feed forward controls.

There are two classes of indirect Btu or energy measurement, both chromatographic devices, relying on the separation, identification and quantification of all the components that make up the natural gas mixture.[9] The two classes are: laboratory chromatographs and on-line or process chromatographs. Through the years a number of different people have compared calorimetry to chromatography with varied results.[10,11]

The laboratory chromatograph (Fig. 1) is just that. It is located in a laboratory. Somehow a gas sample must be obtained and transported in a cylinder to the lab to be analyzed.

How the sample is obtained is the main difference between lab and on-line gas chromatographs (GCs). The natural gas samples analyzed in a lab today usually are called spot or grab samples. A cleaned vessel (cylinder), sometimes N_2 filled, is purged and filled at least three times for reliable sampling site. The pressured-up bottle is carried to a lab to be analyzed. More and more samples today are being collected using a continuous sampler. This device takes a small aliquot of gas at a predetermined time interval so the bottle is filled at the end of the sampling cycle required by the gas contract.

With recent developments, these devices have been tied to electronic flow computers to give flow-weighted composite samples rather than time-weighted ones. An example of such an installation is in Fig. 2.

Flow weighting generally gives even better results than time-weighted samples if there are variations in the flowrate at the sampled station. Great care should be exercised when analyzing a sample cylinder of natural gas obtained by any of the above methods. For instance, the sample bottle should be heated to 120°F for at least two hours before the analysis is made to ensure that all the heavier components are analyzed. A complete discussion of the problems associated with the recommendations for bottle sample analysis have been published previously.[12,9]

The process or on-line gas chromatograph (GC) (Fig. 3) is designed to continually purge a small amount of the gas flowing through the station or pipe line to provide a representative sample at the time the GC is ready to make a new analysis. Typically, this is once every six minutes.

The main advantage of continuous sampling and analysis is the documentation of the duration and the amount of deviation of the energy content that has occurred during the measurement period. It is possible to assign a heating value to a specific volume of gas if the on-line GC is tied directly to an electronic flow computer. This, of course, allows actual billing on an MMBtu custody transfer basis.

Gas chromatographs—lab or on-line—using the computing power of today's modern microprocessor are capable of repeatability values of ±0.5 Btu at 1,000 Btu/cf. While the pre-

Figure 1. Laboratory chromatograph used to measure gas samples.

Figure 3. Process sampler continually purges gas moving through a station or pipe line.

Figure 2. Example of on-line sampler connected to electronic flow computers.

cision of a GC makes it well-suited for custody transfer applications, the 6 to 10 minute analysis time makes it difficult to use for burner feed forward control. Generally, the on-line analyzers do not require much environmental protection at all. Most require protection from rain.

How Chromatographs Work

The way a modern microprocessor controlled on-line chromatograph works has been thoroughly discussed in previous papers.[13,14,15] However, it would be helpful to review briefly just how a GC works. Again, except for the sample conditioning system, this discussion generally applies to lab GCs as well as on-line GCs. For purposes of this discussion, the on-line GC has been divided into four sections. These are:

- Sample conditioning system
- Gas chromatograph oven
- Chromatograph controller
- Input/output chassis.

The sequence also is the route the sample takes from the pipe line to the printer. This makes it a logical flow for discussion as well.

Sample Conditioning System

To obtain a representative sample of gas, a probe should be inserted into the middle third of the pipe line. This procedure is as essential for good sampling for on-line GCs as it is for spot or composite samples. A major difference, however, is the gas pressure is reduced to 15 psig at the probe for transport to the analyzer to ensure no heavy component (C^6+) is lost. With spot or composite samples, care is taken to maintain pipe line conditions. Sometimes it is useful to use a sample bypass loop to reduce sampling lag times, as shown in Fig. 4.

Note that filtering and liquid coalescing is accomplished, if needed, to protect the integrity of the analyzer columns. Sometimes double block and bleed sample switching is used to prevent cross-contamination by the standard or other sample streams.

Gas Chromatograph Oven

Once the gas enters the oven, a repeatable gas sample is measured by a sample inject valve. The chromatograph controller then blocks the sample flow and allows it to equilibrate

Figure 4. Bypass loop reduces sampling lag time.

Figure 5. Block diagram illustrates gas chromatograph oven.

Figure 6. Separated components move through cell detectors for quantifying sample.

to atmospheric pressure. The sample is then swept onto a set of carefully selected chromatographic columns by a stream of helium gas for sequential separation of each component in the natural gas mixture. A block diagram is shown in Fig. 5.

As each column does its specified separation job, it is switched out of the helium stream. The separated components make their way through the thermal conductivity cell detector where they can be quantified. An example of a detector output is shown in Fig. 6.

These peaks must be gated and the area under each peak integrated to be quantified, which is the job of the chromatograph controller.

Chromatograph Controller

All the timing functions such as sample inject, column reverse, auto-calibration and log results must be performed by the microprocessor controller board. In addition, this device must detect each peak, open and then close integration gates, identify the peak, and assign the correct response factor. The controller must then calculate the mole fraction of each component. Since these peaks are quite arbitrary in size due to equipment variability setup and environment, high-quality standards should be used to calibrate these instruments. This calibration assures accurate determination of mole fraction possible. All Btu determination methods require high quality standard samples.

On a Btu GC, the peaks also must be normalized to compensate for barometric pressure changes (which cause sample inject size variations). This allows the calculation of Btu to repeatability of ± 0.5 Btu at 1,000 Btu/cf. This processor must also calculate the specific gravity, compressibility, condensable liquids, wobbie index, as well as the Btu. Calculation results must be made available to the outside world. This is the job of the input/output chassis.

Input/Output Chassis

The device houses communications option cards such as printer interface cards, analog output cards and computer interface cards. These usually can be configured based on the customer's needs. However, caution is advised as specific computer protocols may or may not be supported by one vendor or another. One function for this input/output chassis, for example, is to hold an RS232 output card. This card would log results to an analyzer interface unit (AIU), which would then be connected to up to 30 flow computers (Fig. 7).

Figure 7. Output card logs results in an analyzer interface unit connected to at least 30 computers.

How Btu is Calculated

The methods for calculation of the natural gas Btu from analysis data are well documented. Calculations have been published previously.[2,3,9,16,18] Nevertheless, it is good to review these methods. Although they look complicated, they really are straightforward. Below is the general formula for the Btu calculation:

$$Btu/cf = X_1 * C_1 + X_2 * C_2 + X_3 * C_3 \dots X_{11} * C_{11}$$

Where:

X_1 = Mole fraction of C_1 or methane

X_2 = Mole fraction of C_2 or ethane

X_3 = Mole fraction of C_3 or propane

"   "

"   "

"   "

X_{11} = Mole fraction of C_6 + hexanes and heavier

and:

C_1 = Molar heating value of C_1, or methane (1010 Btu/cf)

C_2 = Molar heating value of C_2, or ethane (1769 Btu/cf)

"   "

"   "

"   "

C_{11} = Molar heating value of C_6+ or hexanes and heavier (4943 Btu/cf)

The calculation of specific gravity and condensable liquids is handled in the same manner. For example:

$$Specific\ Gravity = X_1 * G_1 + X_2 * G_2 + \dots X_{11} * G_{11}$$

Where: X_1, X_2, is the mole fraction of each natural gas component and: G_1, G_2, is the molar gravity contribution of each respective component.

This calculation yields gross or total heating value and is not compensated for compressibility Z. A short method for calculation of Z is:

$$Z = 1.00369 - (0.010 * S.G.) + (0.007 * X_{inert})$$

Total heating value would then be expressed by the formula:

$$\frac{Btu/cf}{Z} = Btu/_{cf}\ Dry$$

in order to convert Dry Btu to saturated:

$$Btu\ saturated = \frac{Btu\ Dry}{1120\ saturation\ factor\ (1.0177)}$$

Some process gas chromatographs calculate all the above values at a pressure base of 14.696 psia only. Others allow user selection of a different contract pressure base and all the constants are then factored by this new pressure base.

Care should be exercised when comparing Btu, SG, and others that all these values are calculated at the same temperature base and pressure base.

Conclusions

There are valid reasons for choosing one energy measurement device over the other. Thermtitrators are fast if control capability is needed. They are not suited for hazardous locations and require more rigorous attention to their operational environment. The modern microprocessors used on most chromatographs make them extremely capable in terms of computer interfaces.

Chromatographs are suitable for hazardous area installations. They are somewhat slower reacting, having six minute cycle times. However, they generally require less maintenance than calorimeters even though they seem more technically complicated. With the AGA-8 supercompressibility calculations, there will most likely be a requirement for a total analysis of the gas to do this part of the gas volume calculations. Thermtitrators cannot do this. GCs that only present the flow computer with four analog signals, proportional to Btu, SG, CO_2, and N_2, may not be much more help.

One company has an interface unit that will pick up and store hourly averages of all 12 components from a gas chromatograph. These hourly averages can be stored for up to three streams for 35 days each so that each hour's flow in the flow computer can be correlated with each hour's gas composition. This allows composition data for not only MMBtu computation, but also AGA-8 volume calculations. From an audit standpoint, this allows downstream corrections and verification that has heretofore not been available.

REFERENCES

1. GPA Standard 2145-84, p. 2.
2. GPA Reference Bulletin, Heating Value As a Basis for Custody Transfer of Natural Gas (1984 Draft).
3. AGA Gas Measurement Manual, Section 11A.2 Determination of Heating Value of Gas, p. 11A2.1,ff.
4. Installation and Operation of Recording Calorinators, A.E. Kersey ISHM #64, 1989, p. 93 ff.
5. Field Deployable Natural Gas Energy Meter, Lewis and Greefield, Sales Information 1989.
6. "Determination of Colorific Values of Natural Gas by Combustion Instruments," R. L. Howard, ISHM #64, 1989, p. 509.
7. "Fundamentals of Energy Management," R. L. Howard, GCMSC #24, 1989, p. 43.
8. "The Therm-titrator, A Comparison With You-Know-What," John Light, GPA #60, 1981.
9. "Analysis of Natural Gas by Gas Chromatography," ASTM D1945, p. 133 ff.
10. "Comparison of Calculated and Measured Heating Values of Natural Gas," Donald C. Melrose, Appalachian Gas Measurement Short Course, 1975.
11. "Measuring Gas; Chromatography Bests Calorimetry," Leisey, Potter, and McKoy, *Oil and Gas Journal,* July 1977.
12. "Sampling and Analysis of Natural Gas," Schepens, Kilmer and Bernos, GPA proceedings #61.
13. "Energy Measurement Utilizing On-Line Chromatographs," L.N. Cox, Prod. of Inter. School of Hydrocarbon Meas. #64, 1989, p. 469 ff.
14. "Btu Measurement in Natural Gas Using Process Gas Chromatography," Kizer and Sund, Proc. of Gulf Coast Meas. Soc. #24, 1989, p. 72 ff.
15. "How to Measure Btu Effectively," Staff, *Gas Industries,* August 1989.
16. "Calculations of Gross Heating Value, Relative Density and Compressibility Factor for Natural Gas Mixtures from Compositional Analysis," GPA Std. 2171-86.
17. *College Chemistry,* Bruce H. Mahan, Addison Wesley, 1966, p. 23.
18. "Gas Heating Value: What It Is and How To Measure, Calculate," D.N. McClanahan, *Oil and Gas Journal,* Feb. 20, 1967, p. 84 ff.
19. Campbell, "436 Pushes Measurement Into The Electronics Age," *American Oil and Gas Report,* May 1987, p. 15 ff.
20. Hagar, "Pipelines, Producers Cope With Problems Under FERC Order 436," *Oil and Gas Journal,* May 1987, p. 17 ff.
21. Paul, "A Multi Btu Approach to Fuels Procurement," Nymex Energy in The News, Spring, 1990, p. 4.

Facts about methane and its behavior

A study of the physical, transport and thermodynamic properties of methane

Dr. Robert O. Parker and **Dr. William H. Kapfer,** New York University, New York, NY

Interest in liquefied natural gas (LNG) requires that the physical, transport, and thermodynamic properties of methane, the principal constituent of natural gas, be readily available. This article will present values most frequently used in process calculations and in other studies associated with the storage and processing of natural gas. In addition, sources of more detailed information are listed.

Methane, CH_4, is the hydrocarbon of lowest molecular weight (16.042). The molecule is best visualized as a regular tetrahedron with the carbon atom at the geometric center and the hydrogen atoms at each of the four apices. Methane is an end product of the anaerobic (without air) decay of plants, that is, of the breakdown of certain very complicated molecules. As such, it is the major constituent (up to 97%) of natural gas. It is the firedamp of the coal mines, and can be seen as marsh gas bubbling to the surface of swamps.

The physical, transport, and thermodynamic properties required are:

Physical

1. Critical constants
2. Equation of state (PVT relationships)
3. Vapor pressure, including bp and fp
4. Vapor and liquid density

Transport

1. Viscosity of liquid and gas
2. Thermal conductivity of liquid and gas
3. Diffusivity

Thermodynamic

1. Heat capacity
2. Latent heat
3. Enthalpy
4. Entropy

They are required for process calculations: material and energy balances; heat transfer, mass-transfer and pressure loss calculations.

Critical Constants

$$T_c = -115.8°F \ (-82.1°C)[1]$$

$$P_c = 673 \text{ psia} \ (45.8 \text{ atm})[2]$$

$$d_c = 10.17 \text{ lb/ft}^3 \ (0.1630 \text{ g/cc})[2]$$

The main utility of critical constants to the engineer is related to the fact that correlations of many of the physical, transport, and thermodynamic properties of hydrocarbons are best expressed in terms of reduced temperature, T_R, and pressure, P_R

$$\left(T_R = \frac{T}{T_c}, \ P_R = \frac{P}{P_c} \right)$$

This is simply an application of the "principle of corresponding states" and is recommended only in the absence of sufficient reliable experimental data.

Equations of State
(Pressure—Volume—Temperature Behavior)

For ordinary conditions, low to moderate pressures and temperatures not too close to the normal boiling point, i.e., when inspection of a compressibility factor chart shows it to be nearly unity, the ideal gas law

$$PV = \frac{W}{M} RT \tag{1}$$

where   P = pressure, psia
   V = volume, ft^3
   W = weight of gas, lb
   M = molecular weight (here 16.042)
   R = gas constant = 10.73 psia (ft^3)/lb mole (°R)
   T = temperature, °R

is sufficient. For most other conditions Equation 1 is modified by inclusion of the compressibility factor, Z, in the right-hand member, viz:

$$PV = \frac{W}{M} ZRT \tag{1a}$$

The writers recommend compressibility factors of Nelson and Obert.[3] If an analytical equation of state is preferred to Equation 1a the writers recommend the 8 constant equation of Benedict, Webb and Rubin.[4]

$$P = RTd + (B_oRT - A_o - C_o/T^2)d^2 + (bRT - a)d^3$$

$$+ \ a \ \alpha \ d^6 + \frac{cd^3(1 + \gamma d^2)e^{-\gamma d^2}}{T^2} \tag{2}$$

where d = density, g-moles/l

in which the constants[4] for methane are:

B_o = 0.042600

A_o = 1.85500

C_o = 22,570.0

b = 0.00338004

a = 0.0494000

c = 2,545.00

γ = 0.0060000

α = 0.000124359

With these constants, R = 0.0820 liter-atm/(°K)(gram mole), P is in atmospheres, T is in °K, and e is the base of natural logarithms. Constants in this equation are available for ethane, ethylene, propane, propylene, n-butane, i-butane, i-butene, n-pentane, i-pentane, n-hexane, and n-heptane.[4] Using Equation 2 to predict the pressure corresponding to 89 data points[5,6] it was found that the average absolute and maximum deviations for methane were 0.376% and +2.25% respectively. Reference 6 discusses a generalization of the Benedict-Webb-Rubin equation.

Vapor Pressure[2]

Freezing point in air at 1 atm = −296.46°F (182.48°C)

Boiling point at 29.921-in. Hg = −258.68°F (−161.49°C)

Vapor pressures are tabulated on p. 121 of Reference 2 from 10 mm to 1,500 mm Hg and are represented by the Antoine equation.

$$\log_{10} P = A - B/(C + t) \qquad (3)$$

or:

$$t = B/(A - \log_{10}P) - C \qquad (3a)$$

with P in mm Hg and t in °C.

A = 6.61184
B = 389.93
C = 266.00

Transport Properties

Viscosity (liquid)[2]

t°C	μ, centipoises
−185	0.226 (subcooled)
−180	0.188
−175	0.161
−170	0.142
−165	0.127
−160	0.115 (at sat. pressure)

Viscosity (vapor)[7] at 1 atm pressure:

t°C	μ, centipoises
−181.6	0.00348
−78.5	0.00760
0	0.01026
20	0.01087
100	0.01331
200.5	0.01605
284	0.01813
380	0.02026
499	0.02264

Additional data[8] are available.

Thermal Conductivity (liquid)

	t°F	K, Btu/hr ft °F
	−280	0.0841
n.b.p.	−258.68	0.0790
	−200	0.0568
crit. temp	−115.8	0.029 (reference[10])

Values for liquid thermal conductivity represent estimates by the authors using the method of Viswanath[9] which seems to be quite reliable for organic liquids (halogenated compounds excepted).

Thermal Conductivity (vapor)[7]

t°F	K, Btu/hr ft °F
−300	0.00541
−200	0.00895
−100	0.01265
−40	0.0149
−20	0.0157
0	0.01645
20	0.01725
40	0.01805
60	0.0190

t°F	K, Btu/hr ft °F
80	0.01985
100	0.0208
120	0.0216
200	0.0258

Diffusivity

The transport property, mass diffusivity (D), is of importance in calculations usually entering computations in combination with viscosity and density as the Schmidt No., $\mu/\rho D$. Reliable estimates of D_{12}, the diffusivity of species 1 through species 2 are best made by the method of Hirschfelder, Bird, and Spotz.[11]

$$D_{12} = \frac{0.001858T^{3/2}[(M_1 + M_2)/M_1M_2]^{1/2}}{P\sigma_{12}^2\Omega_D} \qquad (4)$$

where
- D = diffusivity, cm^2/sec
- T = temperature, °K
- M = molecular weight, M_1 of species 1, M_2 of species 2
- P = pressure, atm
- $\sigma_{12} = (\sigma_1 + \sigma_2)/2$ = collision diameter, A°
- Ω_D = collision integral = function of KT/ϵ_{12}
- K = Boltzmann's constant
- ϵ = a force constant

$$\epsilon_{12} = \sqrt{\epsilon_1 \epsilon_2}$$

Force constants and collision diameters for air and methane are:

	ϵ/K, °K	σ, A°
Air	97	3.617
Methane	136.5	3.822

These are known as the Lennard-Jones force constants and an extensive table is given.[11] Ω_D, which depends upon the temperature through KT/ϵ, is tabulated in the reference.[12] A few values are:

KT/ϵ	Ω_D
0.30	2.662
0.50	2.066
1.00	1.439
1.50	1.198
2.00	1.075
4.00	0.8836
10	0.7424
50	0.5756
100	0.5130
400	0.4170

If the calculation requiring diffusivities is a crucial one, we recommend that the engineer consult the principal literature reference.[11]

Thermodynamic Properties

The most extensive, reliable compilation of the thermodynamic properties is the one by Matthews and Hurd.[13] In addition to a pressure-enthalpy diagram, tabulations are provided for saturated methane from $-280°F$ to $-115.8°F$, the critical temperature, and for superheated methane to 500°F and 1,500 psia. This tabulated information is reproduced in the Chemical Engineers' Handbook 3rd Ed.[14] In the 4th Ed. (1963) superheat tables are omitted.

Corresponding to the pressure-temperature pair, values are given for:

a. Ratio fugacity (where fugacity[13] refers to "escaping tendency") to pressure
b. Specific volume of liquid and vapor
c. Enthalpy of liquid, vaporization, and vapor
d. Entropy of liquid and vapor

A few numbers of immediate interest to the process engineer are:

a. Latent heat of vaporization at normal boiling point = 219.22 Btu/lb
b. Liquid density at normal boiling point = 26.6 lb/ft^3
c. Vapor density at normal boiling point = 0.113 lb/ft^3

The liquid density is required to compute the volume of a diked area to contain a spill. Latent heat is useful in estimating the time required for a spill to be vaporized. The vapor density points up the fact that, although methane has a specific gravity referred to air of about 0.55, vapors generated at a spill are about 1.5 times as heavy as ambient air; a matter of importance in various studies.

Heat Capacities

If heat capacities are required for process calculations, they can be extracted from the thermodynamic tables mentioned above by the following procedure:

$$C_p = \frac{H_1 - H_2}{T_1 - T_2}, \text{ Btu/lb °F} \qquad (5)$$

Example. Find the heat capacity for superheated methane at 15 psia and 100°F.

Figure 1. Cold box where liquiefaction of natural gas is accomplished.

From the Matthews and Hurd superheat table:[13,14]

at 15 psia

$T_1 = 110$	$H_1 = 426.6$
$T_2 = 90$	$H_2 = 415.8$

Then $C_p = \dfrac{426.6 - 415.8}{110 - 90} = 0.54$ Btu/lb °F

The procedure given above is adequate for liquid heat capacities when enthalpy values are taken from the table for saturated methane. This works because the effect of pressure on liquid heat capacity is essentially negligible.

Alternatively, instantaneous values of heat capacity may be determined.[15]

$$C_p{}^o = 0.213 + 1.11 \times 10^{-3}T - 2.59 \times 10^{-7}T^2 \text{ Btu/lb °F} \qquad (6)$$

Note carefully that Equation 6 yields the heat capacity at low pressure, 1 atm; at elevated pressures a correction may be required. If Equation 5 above is used, no pressure correction is necessary.

Heat Capacity Ratio

The ratio $C_p/C_v = \gamma$, useful in process calculations (adapted from reference 14) is, for atmospheric pressure:

t °F	$\gamma = C_p/C_v$
− 150	1.39
− 100	1.36
− 50	1.33
0	1.31
50	1.28
100	1.27
150	1.25

Flammability Characteristics

Since methane is a combustible gas, consideration of any process involving its use or storage must include studies of the associated fire hazard. Basic data are flammable limits and the autoignition temperature. For methane in air these are[16,17] at atmospheric pressure:

a. Lower Flammable, Limit @ 25°C about 5% by vol
b. Upper Flammable, Limit @ 25°C about 15% by vol
c. Autoignition Temperature 1,004°F

These figures mean that mixtures of methane and air having less than 5% methane by volume are too lean to burn, those with more than 15% are too rich. The autoignition temperature is the minimum temperature at which ignition can occur. Figures on flammable limits refer to "homogeneous" mixtures. Industrially, few mixtures are completely homogeneous. In safety computations, therefore, the engineer must establish that the "highest local concentration" is below the lower flammable limit by a suitable margin before saying there is no fire hazard. A method dealing with LNG spills has been proposed by Parker and Spata.[18]

References

1. Kobe, K. A. and R. E. Lynn, *Chem. Revs.* 52, 117 (1953).
2. American Petroleum Institute, "Selected Values of Physical and Thermodynamic Properties of Hydrocarbons and Related Compounds," Project 44, Carnegie Press, Pittsburgh, Pa. 1953 (NBS-C-461).
3. Nelson, L. C. and E. F. Obert, *Trans. ASME* 76, 1057 (1954).
4. Benedict, M., G. B. Webb, L. C. Rubin and L. Friend, *Chem. Eng. Prog.* 47, 419, 449, 571, 609 (1951).
5. Kvalnes, H. M. and V. L. Gaddy, *J. Am. Chem. Soc.* 53, 394 (1931).
6. Edmister, W. C., J. Vairogs and A. J. Klekers, *AIChE Journal* 14, No. 3, p. 479 (1968).
7. *Handbook of Chemistry and Physics, 47th Edition*, 1966–67.
8. Huang, E. T. S., G. W. Swift and F. Kurata, *AIChE Journal* 12, p. 932 (1966).
9. Viswanath, D. S., *AIChE Journal* 13, 850 (1967).
10. Gamson, B. W., *Chem. Eng. Prog.* 45, 154 (1949).
11. Hirschfelder, J. O., R. B. Bird and E. L. Spotz, *Trans. ASME* 71, 921 (1949).
12. Hirschfelder, J. O., C. F. Curtiss and R. B. Bird, *Molecular Theory of Gases and Liquids*, John Wiley and Sons, Inc., New York, 1954.
13. Matthews, C. S. and C. O. Hurd, *Trans. AIChE* 42, p. 55 (1946); ibid., pp. 78 1009.
14. Perry, J. H. (ed.) *Chemical Engineers Handbook, 3rd ed.*, McGraw-Hill, New York, 1950.
15. Reid, R. C. and T. K. Sherwood, *The Properties of Gases and Liquids*, McGraw-Hill, New York (1958).
16. Coward, H. F. and G. W. Jones, U.S. Bureau of Mines, Bulletin 503 (1952).
17. Zabetakis, M. G., U.S. Bureau of Mines, Bulletin 627 (1965).
18. Parker, R. O. and J. K. Spata, Proc. 1st Intern. Conf. on LNG, Chicago, April 7–12, 1968, Paper No. 24.

Conversion table for pure methane

		Long tons liquid	ft³ liquid	ft³ gas	m³ liquid	m³ gas	U.S. bbl liquid	Therms	Thermies
long ton liquid	=	1	84.56	52,886	2.394	1,419.7	15.06	535.2	13,491
ft³ liquid	=	0.011826	1	625.43	0.028311	16.789	0.1781	6.329	159.54
ft³ gas x 10⁶	=	18.91	1,599	1 x 10⁶	45.27	26,847	284.785	10,121	255,115
m³ liquid	=	0.4177	35.315	22,090	1	593	6.29	223.55	5,635.2
m³ gas x 10⁹	=	704,374	59.562 x 10⁶	37.252 x 10⁹	1.686 x 10⁶	1 x 10⁹	10.608 x 10⁶	0.377 x 10⁹	9.503 x 10⁹
U.S. bbl liquid	=	0.0664	5.615	3,512	0.15896	94.268	1	35.54	895.8
Therm x 10⁶	=	1,868	157,958	98.791 x 10⁶	4,472	2.652 x 10⁶	28,132	1 x 10⁶	25.201 x 10⁶
Thermie x 10⁶	=	74.12	6,268	3.92 x 10⁶	177.44	105,228	1,116.2	39,669	1 x 10⁶

Notes

1. ft³ measured at 60°F, 14.696 lbf/in², dry

2. m³ measured at 0°C, 760 mmHg, dry

3. Gross heat content of methane is 1012 Btu/ft³ in gaseous form: Specific gravity 0.554 (Air = 1)
   Boiling point −258.7°F (14.696 lbf/in²)

4. 1 x 10⁶ ft³ gas = 24.42 long tons (0.4417 b/d) fuel oil equivalent

5. The above factors are based on the properties of pure methane; LNG as shipped in practice will vary in composition according to the presence of heavier hydrocarbons (C_2, C_3, C_4) in the gas stream.

Categories of natural gas and reserves terminology

Associated and Non-Associated Natural Gas

Natural gas is found in underground structures similar to those containing crude oil. There are three types of natural gas reservoirs:

1. Structures from which only gas can be produced economically—called **non-associated gas** (or unassociated gas).
2. Condensate reservoirs which yield relatively large amounts of gas per barrel of light liquid hydrocarbons. Although many condensate reservoirs are produced primarily for gas, there are cases where gas is re-injected or "re-cycled" to improve liquid recovery, particularly if no gas market is yet available. This gas also is termed **non-associated**.
3. Reservoirs where gas is found dissolved in crude oil (solution gas) and in some cases also in contact with underlying gas-saturated crude (**gas-cap gas**). Both are called **associated gas**. (Gas-cap gas is almost never produced until most of the economically recoverable oil has been yielded.)

In such fields, gas production rates will depend on oil output, with the oil usually representing the major part in terms of energy equivalents.

Natural Gas Reserves

There are no generally accepted definitions of natural gas reserves, and wide variations both in terms and principles may be met. The following definitions are the most commonly used:

Expected Ultimate Recovery is the total volume of hydrocarbons, gas and/or oil, which may be expected to be recoverable commercially from a given area at current or anticipated prices and costs, under existing or anticipated regulatory practices and with known methods and equipment. This total can be expressed as the sum of the "Ultimate Recovery from Existing Fields" and the "Expectation from Future Discoveries."

Ultimate Recovery from Existing Fields is the sum of the "Cumulative Production" from such fields and the "Reserves from Existing Fields."

Reserves from Existing Fields are the sum of "Proven Reserves," "Discounted (i.e., 50% of) Probable Reserves" and "Discounted (i.e., 25% of) Possible Reserves."

Proven Reserves represent the quantities of crude oil and/or natural gas and natural gas liquids which geological and engineering data demonstrate with reasonable certainty to be recoverable in the future from known oil and/or gas reservoirs. They represent strictly technical judgments, and are not knowingly influenced by attitudes of conservatism or optimism.

Discounted (i.e., 50% of) **Probable Reserves** are those quantities of crude oil and/or natural gas and natural gas liquids for which there exists a probability of 50% that they will materialize. Such reserves are usually allocated to some conjectural part of a field or reservoir as yet undrilled or incompletely evaluated where, while characteristics of the productive zone and fluid content are reasonably favorable, other uncertain features may have equal chances of being favorable or unfavorable for commercial production.

Discounted (i.e., 25% of) **Possible Reserves** are those quantities of crude oil and/or natural gas and natural gas liquids thought to be potentially producible but where the chance of their being realized is thought to be on the order of 25%. Reserves in this category are usually allocated to possible extensions of existing fields where, although geological data may be favorable, uncertainties as to the characteristics of the productive zone and/or fluid content are such as to preclude a probability greater than 25%.

It should be noted that the terms "gas-in-place" and "oil-in-place" represent estimates of the total hydrocarbons present which, in conjunction with various engineering and economic criteria, provide the basis for the "reserve" estimates. They should never be confused with "reserves," which word implies that physical recovery is possible.

Terminology of Reserves

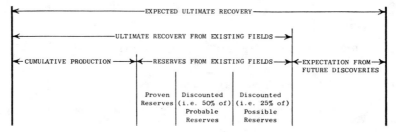

Glossary of common gas industry terms

(italics denote cross reference)

absolute pressure *gauge pressure* plus *atmospheric pressure*.

associated gas see preceding page.

atmospheric pressure the pressure of the weight of air and water vapor on the surface of the earth. Approximately 14.7 lbf/in.^{2} at sea level.

blue water gas also called "water gas." Made in a cyclic process in which an incandescent bed of coke or coal is alternately subjected to blasts of air and steam. The gas consists mainly of equal proportions of carbon monoxide and hydrogen and has a gross heat content of about 300 Btu/ft^3 (see also *carbureted water gas*).

British thermal unit the heat required to raise the temperature of one lb of water 1°F.

burner capacity the maximum Btu/hr that can be released by a burner while burning with a stable flame.

calorie the quantity of heat required to raise the temperature of one gram of water 1°C (centigrade).

calorific value the quantity of heat released by the combustion, in a calora meter at constant pressure of one atmosphere, of a unit quantity of fuel measured under given conditions. In the case of gas, the conditions of temperature, pressure, and water vapor content under which the unit volume was measured affect the calorific value.

calorimeter an instrument for determining the heating value of a fuel.

carbureted water gas (carbureted blue gas) the gas resulting from the enrichment of *blue water gas* during its manufacture by a simultaneous process of light distillate, gas oil or fuel oil gasification. The gas has a gross heat content of about 500 to 550 Btu/ft^3.

city gate a measuring station, which may also include pressure regulation, at which a distributing gas utility receives gas from a natural gas pipe line company or the transmission system.

coal gas or **coke oven gas** a *manufactured* gas made by destructive distillation ("carbonization") of bituminous coal in a gas retort or by-product coke oven. Its chief components are methane (20 to 30%) and hydrogen (about 50%). This gas generally has a gross heating value of 500 to 550 Btu/ft^3. When the process takes place in a closed oven (with gas as a by-product of coke production) it is generally designated as "Coke Oven Gas," and when produced in retorts it is called "Coal Gas."

commodity charge that portion of the charge or rate based upon the total volume of gas consumed or billed.

condensate a liquid/hydrocarbon mixture of 45 to 65°API gravity, which may be recovered at the surface from some *non-associated gas* reservoirs.

connection charge an amount to be paid by the customer in a lump sum, or in installments, for connecting the customer's facilities to the supplier's facilities.

contract demand the volume that the supplier of gas service agrees to deliver and, in general, the amount that the customer agrees to take or pay for.

cubic foot and **cubic meter** (gas) common units of measurement of gas volume. They are the amounts of gas required to fill a volume of one cubic ft or one cubic meter under stated conditions of temperature, pressure and water vapor content.

customer charge a fixed amount to be paid periodically by the customer without regard to *demand* or *commodity* consumption.

cycling or **recycling** the process by which a *non-associated gas* reservoir is produced only for the recovery of condensate. The gas, after removal of the condensate, is reinjected into the formation.

daily average send-out the total volume of gas delivered during a period of time divided by number of days in the period.

daily peak the maximum volume of gas delivered in any one day during a given period, usually a calendar year.

degree day a measure of the extent to which the mean daily temperature falls below an assumed base, say 65°F. Thus each degree by which the mean temperature for any day is less than 65°F would represent one degree day. (In Continental Europe, °C are used instead of °F, and the assumed base temperature is generally taken as 16°C equivalent to 60.8°F).

demand (load) rate of flow of gas in a specified time interval, usually expressed in cubic ft (or cubic meters) per hour, per day or per year.

demand charge that portion of the charge or rate for gas service based upon the customer's *demand* characteristics.

distribution company or **gas utility** a company which obtains the major portion of its gas operating revenues from the operation of a retail gas *distribution system* and which operates no transportation system other than incidental connections to a transportation system of another company. For purposes of American Gas Association (AGA) statistics, a distribution company obtains at least 95% of its gas operating revenues from the operation of its retail gas distribution system (see also *transmission company*).

distribution system feeders, mains, services, and equipment which carry or control the supply of gas from the point or points of local supply (usually the *city gate* station) to and including the consumer *meters*.

diversity factor the ratio of the sum of the non-coincident maximum demands of two or more loads to their coincident maximum demand for the same period.

dyne the standard centimeter-gram-second unit of force, equal to the force that produces an acceleration on one centimeter per second on a mass of one gram.

erg the centimeter-gram-second unit of work or energy, equal to the work done by a force of one dyne when its point of application moves through a distance of one centimeter in the direction of the force.

flat rate schedule a rate schedule that provides for a specified charge irrespective of the quantity of gas used or the *demand*.

gauge pressure the pressure generally shown by measuring devices. This is the pressure in excess of that exerted by the atmosphere (see *absolute pressure*).

gross calorific (heating) value the gross calorific value at constant pressure of a gaseous fuel is the number of heat units produced when unit volume of the fuel, measured under standard conditions, is burned in excess air in such a way that the materials after combustion consist of the gases carbon dioxide, sulfur dioxide, oxygen and nitrogen, water vapor equal in quantity to that in the gaseous fuel and the air before combustion, and liquid water equal in quantity to that produced during combustion, and that the pressure and temperature of the gaseous fuel, the air, and the materials after combustion are one standard atmosphere and 25°C.

inch of mercury a unit of pressure (equivalent to 0.491154 lbf/in.^{2}) by which one lb/in.^{2} equals 2.036009 in. of mercury column at 0°C.

inch of water a unit of pressure (equivalent to 0.03606 lbf/in.^{2}) by which one lb/in.^{2} equals 27.68049 in. of water column at 4°C.

interruptible gas gas made available under agreements which permit curtailment or cessation of delivery by the supplier.

joule the work done when the point of application of a force of one newton is displaced through a distance of one meter in the direction of the force. It is equal to 10^7 *ergs*, and in electrical units is the energy dissipated by one watt in a second.

kilocalorie one thousand *calories*.

kilowatt-hour a unit of energy equivalent to the energy transferred or expanded in one hour by one kilowatt of power.

line pack a method of *peak-shaving* by withdrawing gas from a section of a pipe line system in excess of the input into that section, i.e., normally the difference between the actual volume of gas in the pipe line at low flow (increased pressure) and that at normal flow.

liquefied natural gas (LNG) natural gas that has been liquefied by cooling to minus 258°F (−161°C) at atmospheric pressure.

liquefied petroleum gas (LPG) any hydrocarbon mixture, in either the liquid or gaseous state, the chief components of which consist of propane, propylene, butane, iso-butane, butylene or mixtures thereof in any ratio.

load curve a graph in which the *send-out* of a gas system, or segment of a system, is plotted against intervals of time.

load density the concentration of gas load for a given area, expressed as gas volume per unit of time and per unit of area.

load duration curve a curve of loads, plotted in descending order of magnitude, against time intervals for a specified period. The coordinates may be absolute quantities or percentages.

load factor the ratio of the average load over a designated period to the *peak load* occurring in that period. Usually expressed as a percentage.

low pressure gas distribution system a gas *distribution system*, or the mains of a segment of a distribution system, operated at low pressures of less than 15 *inches water column*.

LPG-air mixtures mixtures of *liquefied petroleum gas* and air to obtain a desired heating value and capable of being distributed through a *distribution system* also used for stand-by and *peak-shaving* purposes by gas utilities.

manufactured gas combustible gases derived from primary energy sources by processes involving chemical reaction. For instance, gas produced from coal, coke or liquid hydrocarbons (see also *town gas*).

meter (gas) a mechanical device for automatically measuring quantities of gas. Some of the more important types are:

Diaphragm Meter a gas meter in which the passage of gas through two or more chambers moves diaphragms which are geared to a volume-indicating dial. Sizes up to 20,000 cubic ft per hour and 1,000 lbf/in.^{2} working pressure are available.

Mass Flowmeter a type of meter in which the gas flow is measured in terms of mass by utilizing the velocity and density of the gas stream to calculate the mass of a gas flowing in unit time.

Orifice Meter meter for measuring the flow of fluid through a pipe or duct by measurement of the pressure differential across a plate having a precision machined hole in its center. Ratings up to very high throughputs and pressures are available.

Rotary Displacement Meter the positive-pressure blower principle in reverse is used in this meter in which gas pressure turns twin matching impellers and the quantity of gas passing through is proportional to the number of revolutions. These meters are made in the range of 3,000 to 1,000,000 cubic ft per hour and for pressures up to 1,200 lbf/in.^{2}

Turbine Flowmeter a type of meter in which there is a turbine wheel or rotor made to rotate by the flowing gas. The rate or speed at which the turbine wheel revolves is a measure of the velocity of the gas.

Venturi Meter a meter in which the flow is determined by measuring the pressure drop caused by the flow through a Venturi throat. The flow rate is proportional to the square root of the pressure drop across the throat.

minimum charge a clause which provides that the charge for a prescribed period shall not be less than a specified amount.

natural gasoline those liquid hydrocarbon mixtures containing essentially pentanes and heavier hydrocarbons which have been extracted from natural gas.

natural gas liquids a term used for those mixtures of hydrocarbon fractions that can be extracted in liquid form from natural gas. In the widest sense, they can comprise any combination of the following: ethane, propylene, propane, butylene, isobutane, n-butane, pentanes, hexanes, heptanes, and octanes.

net calorific (heating) value the net calorific value at constant pressure of a gaseous fuel is the number of heat units produced when unit volume of the fuel, measured under standard conditions, is burned in excess air in such a way that the materials after combustion consist of the gases carbon dioxide, sulfur dioxide, oxygen, nitrogen and water vapor, and that the pressure and temperature of the gaseous fuel, the air and the materials after combustion are one standard atmosphere and 25°C. This value is derived from the measured gross figure.

non-associated gas see preceding section on "reserves terminology."

off-peak the period during a day, week, month or year when the load being delivered by a gas system is not at or near the maximum volume delivered by that system.

peak or **peak load** the maximum load consumed or produced by a unit or group of units in a stated period of time.

peak shaving the practice of augmenting the normal supply of gas during peak or emergency periods from another source where gas may have either been stored during periods of low demand, or manufactured specifically to meet the peak demand.

pressure maintenance (repressing) a process in which natural gas is injected into a formation capable of producing crude petroleum to aid in maintaining pressure in an underground reservoir for the purpose of assisting in the recovery of crude.

producer gas a gas manufactured by burning coal or coke with a regulated deficiency of air, normally saturated with steam. The principal combustible component is carbon monoxide (about 30%) and the gross heat content is between 120 and 160 Btu/ft^3.

recycling see *cycling*.

reforming the process of thermal or catalytic cracking of natural gas, *liquefied petroleum gas, refinery gas* or gas from oil, resulting in the production of a gas having a different chemical composition.

refinery gas a gas resulting from oil refinery operations consisting mainly of hydrogen, methane, ethylene, propylene and the butylenes. Other gases such as nitrogen and carbon dioxide may also be present. The composition can be highly variable and the heat content can range from 1,000 to 2,000 Btu/ft^3.

reserves see preceding section.

send-out the quantity of gas delivered by a plant or system during a specified period of time.

specific gravity of gas the ratio of the density of gas to the density of dry air at the same temperature and pressure.

standard metering base standard conditions, plus agreed corrections, to which all gas volumes are corrected for purposes of comparison and payment.

summer valley the decrease which occurs in the summer months in the volume of the daily load of a gas *distribution system*.

therm 100,000 Btu's.

thermie the heat required to raise the temperature of one metric ton of water 1°C (centigrade).

town gas gas piped to consumers from a gas plant. The gas can comprise both *manufactured gas* (secondary energy) and natural gas (primary energy) used for enrichment.

transmission company a company which operates a natural gas transmission system and which either operates no retail distribution system, or, as defined by the American Gas Association (AGA) receives less than 5% of its gas operating revenues from such retail distribution system (see also *distribution company* or *gas utility*).

unaccounted for gas the difference between the total gas available from all sources and the total gas accounted for as sales, net interchange and company use. The difference includes leakage or other actual losses, discrepancies due to *meter* inaccuracies, variations of temperature, and pressure and other variants.

unassociated gas see preceding section.

utilization factor the ratio of the maximum *demand* on a system or part of a system to the rated capacity of the system or part of the system under consideration.

water gas see *blue water gas*.

watt the meter-kilogram-second unit of power, equivalent to one joule per second and equal to the power in a circuit in which a current on one ampere flows across a potential difference on one volt.

wobbe index the gross calorific value of the gas divided by the square root of the density of the gas as compared with air.

10: Gas—Compression

Compressors 243
Performance calculations for reciprocating compressors 244
Estimating suction and discharge volume bottle sizes for pulsation
    control for reciprocating compressors 246
Compression horsepower determination 248
Generalized compressibility factor 250
Nomograph aids in diagnosing compressor cylinder ills 251

Centrifugal Compressor Data 252
Centrifugal compressor performance calculations 253
Nomographs for estimating compressor performance 256
Estimate hp required to compress natural gas 261
Estimate compressor hp where discharge pressure is 1,000 psi 261
Calculate brake horsepower required to compress gas 262
How to find the size fuel gas line for a compressor station 262
Estimate engine cooling water requirements 263
Estimate fuel requirements for internal combustion engines 263
Estimating fuel requirements for compressor installation 264

Compressors

The following data are for use in the approximation of horsepower needed for compression of gas.

Definitions

The "**N value**" of gas is the ratio of the specific heat at constant pressure (C_p) to the specific heat at constant volume (C_v).

If the composition of gas is known, the value of N for a gas mixture may be determined from the molal heat capacities of the components. If only the specific gravity of gas is known an approximate N value may be obtained by using the chart in Figure 1.

Piston Displacement of a compressor cylinder is the volume swept by the piston with the proper deduction for the piston rod. The displacement is usually expressed in cubic ft per minute.

Clearance is the volume remaining in one end of a cylinder with the piston positioned at the end of the delivery stroke for this end. The clearance volume is expressed as a percentage of the volume swept by the piston in making its full delivery stroke for the end of the cylinder being considered.

Ratio of compression is the ratio of the absolute discharge pressure to the absolute inlet pressure.

Actual capacity is the quantity of gas compressed and delivered, expressed in cubic ft per minute, at the intake pressure and temperature.

Volumetric efficiency is the ratio of actual capacity, in cubic ft per minute, to the piston displacement, in cubic ft per minute, expressed in percent.

Adiabatic horsepower is the theoretical horsepower required to compress gas in a cycle in which there is no transfer of sensible heat to or from the gas during compression or expansion.

Isothermal horsepower is the theoretical horsepower required to compress gas in a cycle in which there is no change in gas temperature during compression or expansion.

Indicated horsepower is the actual horsepower required to compress gas, taking into account losses within the compressor cylinder, but not taking into account any loss in frame, gear or power transmission equipment.

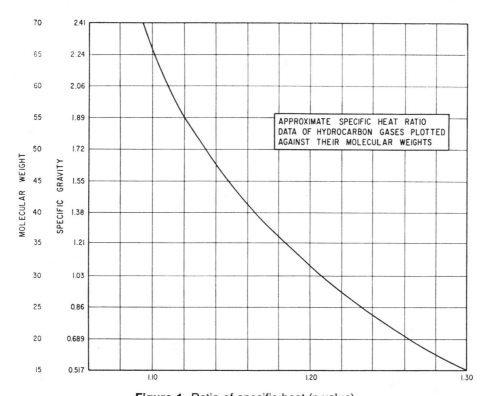

Figure 1. Ratio of specific heat (n-value).

Compression efficiency is the ratio of the theoretical horsepower to the actual indicated horsepower required to compress a definite amount of gas. The efficiency, expressed in percent, should be defined in regard to the base at which the theoretical power was calculated, whether adiabatic or isothermal.

Mechanical efficiency is the ratio of the indicated horsepower of the compressor cylinder to the brake horsepower delivered to the shaft in the case of a power driven machine. It is expressed in percent.

Overall efficiency is the product, expressed in percent, of the compression efficiency and the mechanical efficiency. It must be defined according to the base, adiabatic isother-

mal, which was used in establishing the compression efficiency.

Piston rod gas load is the varying, and usually reversing, load imposed on the piston rod and crosshead during the operation, by different gas pressures existing on the faces of the compressor piston.

The maximum piston rod gas load is determined for each compressor by the manufacturer, to limit the stresses in the frame members and the bearing loads in accordance with mechanical design. The maximum allowed piston rod gas load is affected by the ratio of compression and also by the cylinder design; i.e., whether it is single or double acting.

Performance Calculations for Reciprocating Compressors

Piston Displacement

Single acting compressor:

$$P_d = [S_t \times N \times 3.1416 \times D^2]/[4 \times 1,728] \qquad (1)$$

Double acting compressor without a tail rod:

$$P_d = [S_t \times N \times 3.1416 \times (2D^2 - d^2)]/[4 \times 1,728] \qquad (2)$$

Double acting compressor with a tail rod:

$$P_d = [S_t \times N \times 3.1416 \times 2 \times (D^2 - d^2)]/[4 \times 1,728] \qquad (3)$$

Single acting compressor compressing on frame end only:

$$P_d = [S_t \times N \times 3.1416 \times (D^2 - d^2)]/[4 \times 1,728] \qquad (4)$$

where P_d = Cylinder displacement, cu ft/min
S_t = Stroke length, in.
N = Compressor speed, number of compression strokes/min
D = Cylinder diameter, in.
d = Piston rod diameter, in.

Volumetric Efficiency

$$E_v = 0.97 - [(1/f)r_p^{1/k} - 1]C - L \qquad (5)$$

where E_v = Volumetric efficiency
f = ratio of discharge compressibility to suction compressibility Z_2/Z_1
r_p = pressure ratio
k = isentropic exponent
C = percent clearance
L = gas slippage factor

Let L = 0.3 for lubricated compressors
Let L = 0.07 for non lubricated compressors   (6)

These values are approximations and the exact value may vary by as much as an additional 0.02 to 0.03.

Note: A value of 0.97 is used in the volumetric efficiency equation rather than 1.0 since even with 0 clearance, the cylinder will not fill perfectly.

Cylinder inlet capacity

$$Q_1 = E_v \times P_d \qquad (7)$$

Piston Speed

$$PS = [2 \times S_t \times N]/12 \qquad (8)$$

Discharge Temperature

$$T_2 = T_1(r_p^{(k-1)/k}) \qquad (9)$$

where T_2 = Absolute discharge temperature °R
T_1 = Absolute suction temperature °R

Note: Even though this is an adiabatic relationship, cylinder cooling will generally offset the effect of efficiency.

Power

$$W_{cyl} = [144\, P_1 Q_1/33,000\, n_{cyl}] \times [k/(k-1)] \\ \times [r_p^{k-1/k} - 1] \qquad (10)$$

where n_{cyl} = efficiency
W_{cyl} = Cylinder horsepower

See Figure 1: for curve of efficiency vs. pressure ratio. This curve includes a 95% mechanical efficiency and a valve velocity of 3,000 ft per minute. Tables 1 and 2 permit a correction to be made to the compressor horsepower for specific gravity and low inlet pressure. While it is recognized that the efficiency is not necessarily the element affected, the desire is to modify the power required per the criteria in these figures. The efficiency correction accomplishes this. These corrections become more significant at the lower pressure ratios.

Reciprocating compressor efficiencies plotted against pressure ratio with a valve velocity of 3,000 fpm and a mechanical efficiency of 95 percent.

Figure 1. Reciprocating compressor efficiencies.

Figure 2. Volume bottle sizing.

plishes this. These corrections become more significant at the lower pressure ratios.

Inlet Valve Velocity

$$V = 288 \times P_d/A \qquad (11)$$

where V = Inlet valve velocity
      A = Product of actual lift and the valve opening periphery and is the total for inlet valves in a cylinder expressed in square in. (This is a compressor vendor furnished number.)

Example. Calculate the following:

Suction capacity
Horsepower
Discharge temperature
Piston speed

Given:

Bore = 6 in.
Stroke = 12 in.
Speed = 300 rpm
Rod diameter = 2.5 in.
Clearance = 12%
Gas = CO_2
Inlet pressure = 1,720 psia
Discharge pressure = 3,440 psia
Inlet temperature = 115°F

Calculate piston displacement using Equation 2.

$$P_d = 12 \times 300 \times 3.1416 \times [2(6)^2 - (2.5)^2]/1,728 \times 4$$
$$= 107.6 \text{ cfm}$$

Calculate volumetric efficiency using Equation 5. It will first be necessary to calculate f, which is the ratio of discharge compressibility to suction compressibility.

$$T_1 = 115 + 460 = 575°R$$

$$T_r = T/T_c$$
$$= 575/548$$
$$= 1.05$$

where T_r = Reduced temperature
      T_c = Critical temperature = 548° for CO_2
      T = Inlet temperature

$$P_r = P/P_c$$
$$= 1,720/1,071$$
$$= 1.61$$

where P_r = Reduced pressure
P_c = Critical pressure = 1,071 psia for CO_2
P = Suction pressure

From the generalized compressibility chart (page 231):

$Z_1 = 0.312$

Determine discharge compressibility. Calculate discharge temperature by using Equation 9.

$r_p = 3,440/1,720$
$= 2$

$k = cp/cv = 1.3$ for CO_2

$T_2 = 575 \times [2.0^{(1.3-1)/1.3}]$
$= 674.7°R$
$= 674.7 - 460 = 214°F$

$T_r = 674.7/548 = 1.23$

$P_r = 3,440/1,071 = 3.21$

From the generalized compressibility chart (page 231):

$Z_2 = 0.575$

$f = 0.575/1.61 = 1.843$

$E_v = 0.97 - [(1/1.843) \times 2.0^{1/1.3} - 1] \times 0.12 - 0.05$
$= 0.929$
$= 93\%$

Note: A value of 0.05 was used for L because of the high differential pressure.

Calculate suction capacity.

$Q_1 = E_v \times P_d$
$= 0.93 \times 107.6$
$= 100.1$ cfm

Calculate piston speed.

$PS = [2 \times S_t \times N]/12$
$= [2 \times 12 \times 300]/12$
$= 600$ ft/min

Table 1
Efficiency Multiplier for Low Pressure

	Pressure Psia							
r_p	10	14.7	20	40	60	80	100	150
3.0	.990	1.00	1.00	1.00	1.00	1.00	1.00	1.00
2.5	.980	.985	.990	.995	1.00	1.00	1.00	1.00
2.0	.960	.965	.970	.980	.990	1.00	1.00	1.00
1.5	.890	.900	.920	.940	.960	.980	.990	1.00

Source: Modified courtesy of the Gas Processors Suppliers Association and Ingersoll-Rand.

Table 2
Efficiency Multiplier for Specific Gravity

	SG				
r_p	1.5	1.3	1.0	0.8	0.6
2.0	0.99	1.0	1.0	1.0	1.01
1.75	0.97	0.99	1.0	1.01	1.02
1.5	0.94	0.97	1.0	1.02	1.04

Source: Modified courtesy of the Gas Processors Suppliers Association.

Estimating suction and discharge volume bottle sizes for pulsation control for reciprocating compressors

Pressure surges are created as a result of the cessation of flow at the end of the compressor's discharge and suction stroke. As long as the compressor speed is constant, the pressure pulses will also be constant. A low pressure compressor will likely require little if any treatment for pulsation control; however, the same machine with increased gas density, pressure, or other operational changes may develop a problem with pressure pulses. Dealing with pulsa-

tion becomes more complex when multiple cylinders are connected to one header or when multiple stages are used.

API Standard 618 should be reviewed in detail when planning a compressor installation. The pulsation level at the outlet side of any pulsation control device, regardless of type, should be no more than 2% peak-to-peak of the line pressure of the value given by the following equation, whichever is less.

$$P_\% = 10/P_{line}^{1/3} \qquad (12)$$

Where a detailed pulsation analysis is required, several approaches may be followed. An analog analysis may be performed on the Southern Gas Association dynamic compressor simulator, or the analysis may be made a part of the compressor purchase contract. Regardless of who makes the analysis, a detailed drawing of the piping in the compressor area will be needed.

The following equations are intended as an aid in estimating bottle sizes or for checking sizes proposed by a vendor for simple installations—i.e., single cylinder connected to a header without the interaction of multiple cylinders. The bottle type is the simple unbaffled type.

Example. Determine the approximate size of suction and discharge volume bottles for a single-stage, single-acting, lubricated compressor in natural gas service.

Cylinder bore = 9 in.
Cylinder stroke = 5 in.
Rod diameter = 2.25 in.
Suction temp = 80°F
Discharge temp = 141°F
Suction pressure = 514 psia
Discharge pressure = 831 psia
Isentropic exponent, k = 1.28
Specific gravity = 0.6
Percent clearance = 25.7%

Step 1. Determine suction and discharge volumetric efficiencies using Equations 5 and 13.

$$E_{vd} = [E_v \times f]/r_p^{1/k} \qquad (13)$$

$$r_p = 831/514$$
$$= 1.617$$

$Z_1 = 0.93$ (from Figure 4, page 199)

$Z_2 = 0.93$ (from Figure 4, page 199)

$$f = 0.93/0.93$$
$$= 1.0$$

Calculate suction volumetric efficiency using Equation 5:

$$E_v = 0.97 - [(1/1) \times (1.617)^{1/1.28} - 1] \times 0.257 - 0.03$$
$$= 0.823$$

Calculate discharge volumetric efficiency using Equation 13:

$$E_{vd} = 1 \times [0.823]/[1.617^{1/1.28}]$$
$$= 0.565$$

Calculate volume displaced per revolution using Equation 1:

$$P_d/N = S_t \times 3.1416 \times D^2/[1,728 \times 4]$$
$$= [5 \times 3.1416 \times 9^2]/[1,728 \times 4]$$
$$= 0.184 \text{ cu ft or } 318 \text{ cu in.}$$

Refer to Figure 3, volume bottle sizing, using volumetric efficiencies previously calculated, and determine the multipliers.

Suction multiplier = 13.5
Discharge multiplier = 10.4

$$\text{Discharge volume} = 318 \times 13.5$$
$$= 3,308 \text{ cu in.}$$

$$\text{Suction volume} = 318 \times 10.4$$
$$= 4,294 \text{ cu in.}$$

Calculate bottle dimensions. For elliptical heads, use Equation 14.

$$\text{Bottle diameter } d_b = 0.86 \times \text{volume}^{1/3} \qquad (14)$$

Volume = suction or discharge volume

$$\text{Suction bottle diameter} = 0.86 \times 4,294^{1/3}$$
$$= 13.98 \text{ in.}$$

$$\text{Discharge bottle diameter} = 0.86 \times 3,308^{1/3}$$
$$= 12.81 \text{ in.}$$

$$\text{Bottle length} = L_b = 2 \times d_b \qquad (15)$$

$$\text{Suction bottle length} = 2 \times 13.98$$
$$= 27.96 \text{ in.}$$

$$\text{Discharge bottle length} = 2 \times 12.81$$
$$= 25.62 \text{ in.}$$

Source

Compressors—Selection & Sizing, Gulf Publishing Company, Houston, Texas.

Compression horsepower determination

The method outlined below permits determination of approximate horsepower requirements for compression of gas.

1. From Figure 1, determine the atmospheric pressure in psia for the altitude above sea level at which the compressor is to operate.
2. Determine intake pressure (P_s) and discharge pressure (P_d) by adding the atmospheric pressure to the corresponding gage pressure for the conditions of compression.
3. Determine total compression ratio $R = P_d/P_s$. If ratio R is more than 5 to 1, two or more compressor stages will be required. Allow for a pressure loss of approximately 5 psi between stages. Use the same ratio for each stage. The ratio per stage, so that each stage has the same ratio, can be approximated by finding the nth root of the total ratio, when n = number of stages. The exact ratio can be found by trial and error, accounting for the 5 psi interstage pressure losses.
4. Determine the N value of gas (Chart, 1st ed. p. 53).

5. Figure 3 gives horsepower requirements for compression of one million cu ft per day for the compression ratios and N values commonly encountered in oil producing operations.
6. If the suction temperature is not 60°F, correct the curve horsepower figure in proportion to absolute temperature. This is done as follows:

$$HP \times \frac{460° + T_s}{460° + 60°F} = hp \text{ (corrected for suction temperature)}$$

where T_s is suction temperature in °F.

7. Add together the horsepower loads determined for each stage to secure the total compression horsepower load. For altitudes greater than 1,500 ft above sea level apply a multiplier derived from the following table to determine the nominal sea level horsepower rating of the internal combustion engine driver.

Figure 1. Atmospheres at various atmospheric pressures. *From Modern Gas Lift Practices and Principles,* Merla Tool Corp.

Figure 2. Ratio of specific heat (n-value).

N = RATIO OF SPECIFIC HEATS Cp/Cv
PS = SUCTION PRESSURE IN P.S.I.A.
PD = DISCHARGE PRESSURE IN P.S.I.A.
R = COMPRESSION RATIO Pd/Ps

NOTE:
FOR "R" GREATER THAN 4.5 TO 5 USE
MULTI-STAGE COMPRESSION, ALLOWING
FOR PRESSURE DROP BETWEEN STAGES,
USUALLY 5 P.S.I.. MAKE CORRECTION
FOR INTERSTAGE SUCTION TEMPERATURE.

Figure 3. Brake horsepower required for compressing natural gas.

Altitude—Multiplier		Altitude—Multiplier	
1,500 ft	1.000	4,000 ft	1.12
2,000 ft	1.03	4,500 ft	1.14
2,500 ft	1.05	5,000 ft	1.17
3,000 ft	1.07	5,500 ft	1.20
3,500 ft	1.10	6,000 ft	1.22

8. For a portable unit with a fan cooler and pump driven from the compressor unit, increase the horsepower figure by $7\frac{1}{2}\%$.

The resulting figure is sufficiently accurate for all purposes. The nearest commercially available size of compressor is then selected.

The method does not take into consideration the supercompressibility of gas and is applicable for pressures up to 1,000 psi. In the region of high pressures, neglecting the deviation of behavior of gas from that of the perfect gas may lead to substantial errors in calculating the compression horsepower requirements. The enthalpy-entropy charts may be used conveniently in such cases. The procedures are given in references 1 and 2.

Example. What is the nominal size of a portable compressor unit required for compressing 1,600,000 standard cubic ft of gas per 24 hours at a temperature of 85°F from 40 psig pressure to 600 psig pressure? The altitude above sea level is 2,500 ft. The N value of gas is 1.28. The suction temperature of stages, other than the first stage, is 130°F.

Solution.

$P_s = 53.41$ psia
$P_d = 613.41$ psia
$R_t = P_d/P_s = 11.5 = (3.4)^2 = (2.26)^3$

Try solution using 3.44 ratio and 2 stages.

1st stage: 53.41 psia × 3.44 = 183.5 psia discharge
2nd stage: 178.5 psia × 3.44 = 614 psia discharge

Horsepower from curve, Figure 3 = 77 hp for 3.44 ratio

$$\frac{77 \text{ hp} \times 1,600,000}{1,000,000} = 123.1 \text{ (for } 60°\text{F suction temp.)}$$

1st stage: $123.1 \text{ hp} \times \dfrac{460 + 85°}{460 + 60°} = 129.1 \text{ hp}$

2nd stage: $123.1 \text{ hp} \times \dfrac{460 + 130°}{460 + 60°} = 139.7 \text{ hp}$

1.05 (129.1 hp + 139.7 hp) = 282 hp

1.075 × 282 hp = 303 hp

Nearest nominal size compressor is 300 hp.

Centrifugal compressors

The centrifugal compressors are inherently high volume machines. They have extensive application in gas transmission systems. Their use in producing operations is very limited.

References

1. *Engineering Data Book*, Natural Gasoline Supply Men's Association, 1957.
2. Dr. George Granger Brown: "A Series of Enthalpy-entropy Charts for Natural Gas," *Petroleum Development and Technology*, Petroleum Division AIME, 1945.

Generalized compressibility factor

The nomogram (Figure 1) is based on a generalized compressibility chart.[1] It is based on data for 26 gases, excluding helium, hydrogen, water, and ammonia. The accuracy is about one percent for gases other than those mentioned.

To use the nomogram, the values of the reduced temperature (T/T_c) and reduced pressure (P/P_c) must be calculated first.

where   T = temperature in consistent units
$\quad\quad$ T_c = critical temperature in consistent units
$\quad\quad$ P = pressure in consistent units
$\quad\quad$ P_c = critical pressure in consistent units

Example. $P_r = 0.078$, $T_r = 0.84$, what is the compressibility factor, z? Connect P_r with T_r and read z = 0.948.

Source

Davis, D. S., *Petroleum Refiner*, 37, No. 11, (1961).

Reference

1. Nelson, L. C., and Obert, E. F., *Chem. Engr.*, 203 (1954).

Figure 1. Generalized compressibility factor. (Reproduced by permission *Petroleum Refiner,* Vol. 37, No. 11, copyright 1961, Gulf Publishing Co., Houston.)

Nomograph aids in diagnosing compressor cylinder ills

B. W. Robertson, Tennessee Gas Pipeline Co., Houston

Compressor cylinders, like people, develop a fever when they are sick. Unlike humans, however, the normal temperature (suction) is not constant. The trick of using temperatures to determine malfunctions lies in determining the correct starting position.

To determine the starting position, it is necessary to know the true k value of the product being pumped. The k value may be obtained from measured pressures and temperatures according to the formula:

$$T2/T1 = (P2/P1)^{\frac{k-I}{k}}$$

where the subfix 2 refers to discharge conditions and the 1 to suction, and all values of P and T are absolute.[1]

The preferred time to obtain this k would be with new equipment, or from data obtained when the equipment was new. The next best time would be with newly overhauled equipment. Lacking either of these data, any values of k must be considered to contain some malfunctions, and any unnecessarily high numbers should be discarded.

Nomograph 1 has been successfully used for a number of years, although it is slowly giving way to electronic equipment. Nomograph 2 may be finished by completing the suction and discharge temperature scales according to the following formulas:

Suction: $t_i = 8.557 \ \rho \ \text{Log} \ (t_s + 460) - 23.135 \ \rho + 2.51$ in. above base line

Discharge: $t = 23.067 \ \rho \ \text{Log} \ (t_d + 460) - 62.65 \ \rho$ in. above base line

where $\rho = \dfrac{k}{k-1}$

and Logs are to the base 10.

Reference

1. Virgil Moring Faires, *Applied Thermodynamics*, 11th Printing, 1946, p. 43.

Nomograph 1—To determine starting position, it is necessary to know the true "k" value of the product being pumped. The "k" value may be obtained from measured pressure and temperatures according to the formula given.

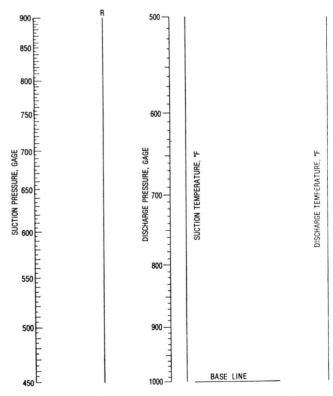

Nomograph 2—This nomograph may be finished by completing the suction and discharge temperature scales according to the formulas given.

CENTRIFUGAL COMPRESSOR DATA

Centrifugal compressor performance calculations

Centrifugal compressors are versatile, compact, and generally used in the range of 1,000 to 100,000 inlet cubic ft per minute (ICFM) for process and pipe line compression applications.

Centrifugal compressors can use either a horizontal or a vertical split case. The type of case used will depend on the pressure rating with vertical split casings generally being used for the higher pressure applications. Flow arrangements include straight through, double flow, and side flow configurations.

Centrifugal compressors may be evaluated using either the adiabatic or polytropic process method. An adiabatic process is one in which no heat transfer occurs. This doesn't imply a constant temperature, only that no heat is transferred into or out of the process system. Adiabatic is normally intended to mean adiabatic isentropic. A polytropic process is a variable-entropy process in which heat transfer can take place.

When the compressor is installed in the field, the power required from the driver will be the same whether the process is called adiabatic or polytropic during design. Therefore, the work input will be the same value for either process. It will be necessary to use corresponding values when making the calculations. When using adiabatic head, use adiabatic efficiency and when using polytropic head, use polytropic efficiency. Polytropic calculations are easier to make even though the adiabatic approach appears to be simpler and quicker.

The polytropic approach offers two advantages over the adiabatic approach. The polytropic approach is independent of the thermodynamic state of the gas being compressed, whereas the adiabatic efficiency is a function of the pressure ratio and therefore is dependent upon the thermodynamic state of the gas.

If the design considers all processes to be polytropic, an impeller may be designed, its efficiency curve determined, and it can be applied without correction regardless of pressure, temperature, or molecular weight of the gas being compressed. Another advantage of the polytropic approach is that the sum of the polytropic heads for each stage of compression equals the total polytropic head required to get from state point 1 to state point 2. This is not true for adiabatic heads.

Sample Performance Calculations

Determine the compressor frame size, number of stages, rotational speed, power requirement, and discharge temperature required to compress 5,000 lbm/min of gas from 30 psia at 60°F to 100 psia. The gas mixture molar composition is as follows:

Ethane	5%
Propane	80%
n-Butane	15%

This is the same gas mixture that was considered in Sample Calculation 1, p. 214, where the gas properties were calculated. The properties of this mixture were determined to be as follows:

$MW = 45.5$
$P_c = 611$ psia
$T_c = 676°R$
$c_p = 17.76$
$k_1 = 1.126$
$Z_1 = 0.955$

Before proceeding with the compressor calculations, let's review the merits of using average values of Z and k in calculating the polytropic head.

The inlet compressibility must be used to determine the actual volume entering the compressor to approximate the size of the compressor and to communicate with the vendor via the data sheets. The maximum value of θ is of interest and will be at its maximum at the inlet to the compressor where the inlet compressibility occurs (although using the average compressibility will result in a conservative estimate of θ).

Compressibility will decrease as the gas is compressed. This would imply that using the inlet compressibility would be conservative since as the compressibility decreases, the head requirement also decreases. If the variation in compressibility is drastic, the polytropic head requirement calculated by using the inlet compressibility would be practically useless. Compressor manufacturers calculate the performance for each stage and use the inlet compressibility for each stage. An accurate approximation may be substituted for the stage-by-stage calculation by calculating the polytropic head for the overall section using the average compressibility. This technique results in over-estimating the first half of the impellers and underestimating the last half of the impellers, thereby calculating a polytropic head very near that calculated by the stage-by-stage technique.

Determine the inlet flow volume, Q_1:

$$Q_1 = m_1[(Z_1RT_1)/(144P_1)]$$

where m = mass flow
Z_1 = inlet compressibility factor
R = gas constant = 1,545/MW
T_1 = inlet temperature °R
P_1 = inlet pressure

$$Q_1 = 5,000[(0.955)(1,545)(60 + 460)/(45.5)(144)(30)]$$
$$= 19,517 \text{ ICFM}$$

Refer to Table 1 and select a compressor frame that will handle a flow rate of 19,517 ICFM. A Frame C compressor will handle a range of 13,000 to 31,000 ICFM and would have the following nominal data:

H_{pnom} = 10,000 ft-lb/lbm (nominal polytropic head)

n_p = 77% (polytropic efficiency)

N_{nom} = 5,900 rpm

Determine the pressure ratio, r_p.

$$r_p = P_2/P_1 = 100/30 = 3.33$$

Determine the approximate discharge temperature, T_2.

$$n/n - 1 = [k/k - 1]n_p$$
$$= [1.126/(1.126 - 1.000)](0.77)$$
$$= 6.88$$

$$T_2 = T_1(r_p)^{(n-1)/n}$$
$$= (60 + 460)(3.33)^{1/6.88}$$
$$= 619°R = 159°F$$

where T_1 = inlet temp

Determine the average compressibility, Z_a.

Z_1 = 0.955 (from gas properties calculation)

where Z_1 = inlet compressibility

$$(P_r)_2 = P_2/P_c$$
$$= 100/611$$
$$= 0.164$$

$$(T_r)_2 = T_2/T_c$$
$$= 619/676$$
$$= 0.916$$

Table 1

Typical Centrifugal Compressor Frame Data*

Frame	Nominal Inlet Volume Flow		Nominal Polytropic Head		Nominal Polytropic Efficiency (%)	Nominal Rotational Speed (RPM)	Nominal Impeller Diameter	
	English (ICFM)	Metric (m³/h)	English (ft-lbf/lbm)	Metric (k·Nm/kg)			English (in)	Metric (mm)
A	1,000–7,000	1,700–12,000	10,000	30	76	11,000	16	406
B	6,000–18,000	10,000–31,000	10,000	30	76	7,700	23	584
C	13,000–31,000	22,000–53,000	10,000	30	77	5,900	30	762
D	23,000–44,000	39,000–75,000	10,000	30	77	4,900	36	914
E	33,000–65,000	56,000–110,000	10,000	30	78	4,000	44	1,120
F	48,000–100,000	82,000–170,000	10,000	30	78	3,300	54	1,370

*While this table is based on a survey of currently available equipment, the instance of any machinery duplicating this table would be purely coincidental.

Figure 1. Maximum polytropic head per stage—English system.

Refer to Figure 2, p. 216 to find Z_2, discharge compressibility.

$$Z_2 = 0.925$$

$$Z_a = (Z_1 + Z_2)/2$$
$$= 0.94$$

Determine average k-value. For simplicity, the inlet value of k will be used for this calculation. The polytropic head equation is insensitive to k-value (and therefore n-value) within the limits that k normally varies during compression. This is because any errors in the $n/(n-1)$ multiplier in the polytropic head equation tend to balance corresponding errors in the $(n-1)/n$ exponent. Discharge temperature is very sensitive to k-value. Since the k-value normally decreases during compression, a discharge temperature calculated by using the inlet k-value will be conservative and the actual temperature may be several degrees higher—possibly as much as 25–50°F. Calculating the average k-value can be time-consuming, especially for mixtures containing several gases, since not only must the mol-weighted c_p of the mixture be determined at the inlet

temperature but also at the estimated discharge temperature.

The suggested approach is as follows:

1. If the k-value is felt to be highly variable, one pass should be made at estimating discharge temperature based on the inlet k-value; the average k-value should then be calculated using the estimated discharge temperature.
2. If the k-value is felt to be fairly constant, the inlet k-value can be used in the calculations.
3. If the k-value is felt to be highly variable, but sufficient time to calculate the average value is not available, the inlet k-value can be used (but be aware of the potential discrepancy in the calculated discharge temperature).

$$k_1 = k_a = 1.126$$

Determine average $n/(n-1)$ value from the average k-value. For the same reasons discussed above, use $n/(n-1) = 6.88$.

Table 2

Approximate Mechanical Losses as a Percentage of Gas Power Requirement*

Gas Power Requirement		Mechanical Losses, L_m (%)
English (hp)	Metric (kW)	
0–3,000	0–2,500	3
3,000–6,000	2,500–5,000	2.5
6,000–10,000	5,000–7,500	2
10,000+	7,500+	1.5

*There is no way to estimate mechanical losses from gas power requirements. This table will, however, ensure that mechanical losses are considered and yield useful values for estimating purposes.

Determine polytropic head, H_p:

$$H_p = Z_a RT_1(n/n - 1)[r_p^{(n-1)/n} - 1]$$
$$= (0.94)(1,545/45.5)(520)(6.88)(3.33)^{1/6.88} - 1]$$
$$= 21,800 \text{ ft-lbf/lbm}$$

Determine the required number of compressor stages, θ:

$$\theta = [(26.1 MW)/(k_1 Z_1 T_1)]^{0.5}$$
$$= [(26.1)(45.5)/(1.126)(0.955)(520)]^{0.5}$$
$$= 1.46$$

max H_p/stage from Figure 1 using $\theta = 1.46$

$$\text{Number of stages} = H_p/\text{max. } H_p/\text{stage}$$
$$= 21,800/9,700$$
$$= 2.25$$
$$= 3 \text{ stages}$$

Determine the required rotational speed:

$$N = N_{nom}[H_p/H_{pnom} \times \text{no. stages}]^{0.5}$$
$$= 5,900[21,800/(10,000)(3)]^{0.5}$$
$$= 5,030 \text{ rpm}$$

Determine the required shaft power:

$$PWR_g = mH_p/33,000n_p$$
$$= (5,000)(21,800)/(33,000)(0.77)$$
$$= 4,290 \text{ hp}$$

Mechanical losses (L_m) = 2.5% (from Table 2)
$$L_m = (0.025)(4,290)$$
$$= 107 \text{ hp}$$

$$PWR_s = PWR_g + L_m$$
$$= 4,290 + 107$$
$$= 4,397 \text{ hp}$$

Determine the actual discharge temperature:

$$T_2 = T_1(r_p)^{(n-1)/n}$$
$$= 520(3.33)^{1/6.88}$$
$$= 619°R$$
$$= 159°F$$

The discharge temperature calculated in the last step is the same as that calculated earlier only because of the decision to use the inlet k-value instead of the average k-value. Had the average k-value been used, the actual discharge temperature would have been lower.

Source

Estimating Centrifugal Compressor Performance, Gulf Publishing Company, Houston, Texas.

Nomographs for estimating compressor performance*

Convert SCFM to inlet CFM

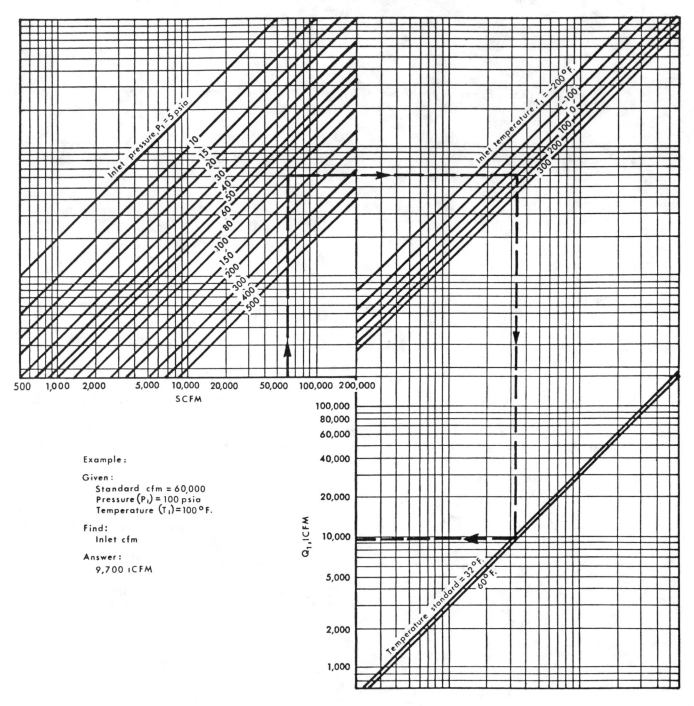

SCFM

Example:

Given:
   Standard cfm = 60,000
   Pressure (P₁) = 100 psia
   Temperature (T₁) = 100° F.

Find:
   Inlet cfm

Answer:
   9,700 ICFM

Q_1, ICFM

Inlet cubic feet per minute

* Reprinted by permission of Gas Processors Suppliers Association.

Convert weight flow to inlet CFM

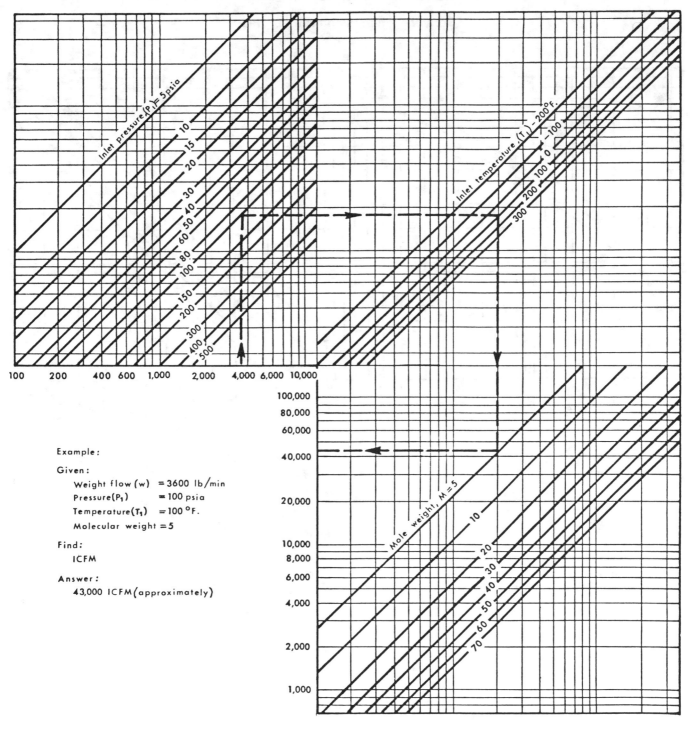

Example:

Given:
  Weight flow (w)   = 3600 lb/min
  Pressure (P_1)   = 100 psia
  Temperature (T_1) = 100 °F.
  Molecular weight = 5

Find:
  ICFM

Answer:
  43,000 ICFM (approximately)

Inlet cubic feet per minute

Convert compression ratio to head

Compression ratio, r

Head

Example:

Given:
 Compression ratio = 10
 k   = 1.15
 Inlet temperature = -50
 Molecular weight = 42

Find:
 Head

Answer:
 39,000 ft lb/lb (approximate)

Head

Convert compression ratio to discharge temperature

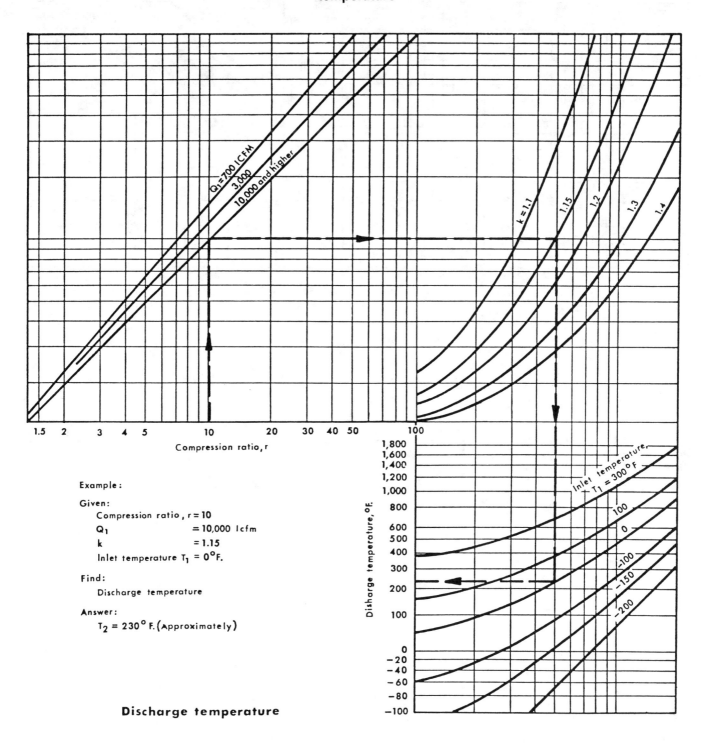

Discharge temperature

Example:

Given:
    Compression ratio, r = 10
    Q_1 = 10,000 lcfm
    k = 1.15
    Inlet temperature T_1 = 0°F.

Find:
    Discharge temperature

Answer:
    T_2 = 230° F. (Approximately)

Horsepower determination

Example: Given:

Weight flow, w = 1,000 lb/min
Head   = 70,000 ft-lb/lb

Find: Answer:

Horsepower   Hp = 3,000

Efficiency conversion

Estimate hp required to compress natural gas

To estimate the horsepower to compress a million cubic ft of gas per day, use the following formula:

$$\text{BHP/MMcfd} = \frac{R}{R + RJ}\left(\frac{5.16 + 124 \text{ Log } R}{0.97 - 0.03 R}\right)$$

where R = compression ratio. Absolute discharge pressure divided by absolute suction pressure

J = supercompressibility factor—assumed 0.022 per 100 psia suction pressure

Example. How much horsepower should be installed to raise the pressure of 10 million cubic ft of gas per day from 185.3 psi to 985.3 psi?

This gives absolute pressures of 200 and 1,000.

then $R = \dfrac{1,000}{200} = 5.0$

Substituting in the formula:

$$\text{BHP/MMcfd} = \frac{5.0}{5.0 + 5 \times 0.044} \times \frac{5.16 + 124 \times .699}{.97 - .03 \times 5}$$

= 106.5 hp = BHP for 10 MMcfd
= 1,065 hp

Where the suction pressure is about 400 psia, the brake horsepower per MMcfd can be read from the chart.

The above formula may be used to calculate horsepower requirements for various suction pressures and gas physical properties to plot a family of curves.

Estimate compressor hp where discharge pressure is 1,000 psi

Rule. Use the following formula for reciprocating compressors where the compressor ratio is between 1.20 and 1.40.

Brake horsepower per MMcfd = 11 + (42.5)(R − 1.20)

R is compression ratio.

Example. How much horsepower would have to be installed in a field compressor station where the discharge pressure would be 1,055 psia and suction would be 800 psia. The station must handle 10,000,000 cubic ft per day.

Solution. The compression ratio is

$\dfrac{1,055}{800} = 1.32$

Substituting in the formula:

Brake horsepower per MMcfd = 11 + (42.5)(1.32 − 1.20)
= 11 + 5.1
= 16.1 hp

The station would require 161 hp.

For a centrifugal compressor the formula would be:

Brake horsepower per MMcfd = 10 + 49 (R − 1.20)

Example. How much horsepower should be installed in centrifugal compressor station handling 100,000,000 cubic ft per day if the suction and discharge are the same as above?

Solution.

Brake horsepower per MMcfd = 10 + (49)(0.12) = 15.9

For 100 MMcfd the horsepower would be 1,590.

Calculate brake horsepower required to compress gas

The approximate brake horsepower per million cubic ft of gas to be compressed may be obtained from this table. The total brake horsepower required is calculated by multiplying the total million cubic ft of gas compressed. In this estimation, gas volumes are expressed at 14.7 psia and 60°F, specific gravity as 0.6 and K (the ratio of specific heats) is 1.27. The suction temperature is assumed as 80°F and interstage temperature as 130°F. The relative humidity of the air equals 100%.

Example. What is the brake horsepower required to compress 5.0 million cubic ft of gas from 200 psig suction pressure to 1,000 psig discharge pressure?

On the chart, read opposite a suction pressure of 200 and under a discharge pressure of 1,000 psig, that 100 brake horsepower is required per million cubic ft of gas compressed. Multiply 5.0 times 100 = 500 bhp.

HORSEPOWER SELECTION CHART
NATURAL GAS COMPRESSION

		DISCHARGE PRESSURE									
		250 psig	375 psig	500 psig	750 psig	1000 psig	1250 psig	1500 psig	2000 psig	3500 psig	6000 psig
SUCTION PRESSURE	0 psig	205	235	260	280	300	320	330	360	390	—
	10 psig	165	195	215	250	260	275	290	310	345	390
	25 psig	130	155	175	205	230	250	255	275	320	370
	50 psig	95	125	140	170	190	205	215	245	280	325
	75 psig	70	95	120	145	165	180	195	215	250	300
	100 psig	50	75	95	120	145	160	175	195	230	275
	200 psig	—	—	55	80	100	120	130	150	190	240
	300 psig	—	—	30	55	75	90	105	120	160	205
	500 psig	—	—	—	25	40	55	65	85	125	165
	750 psig	—	—	—	—	15	30	40	60	100	140
	1000 psig	—	—	—	—	—	—	25	40	75	125
	1500 psig	—	—	—	—	—	—	—	—	50	90
	2000 psig	—	—	—	—	—	—	—	—	—	70
	2500 psig	—	—	—	—	—	—	—	—	—	55

How to find the size fuel gas line for a compressor station

Suppose that an 8,000 hp compressor building is added to an existing compressor plant. What size fuel line should be run from the meter building 1,600 ft distant? Take-off pressure at the meter building will be 50 lb/in.², and it is desirable to maintain 25 psi pressure on the large engine fuel header in the building. Assume that the gas is heated prior to expansion so that outlet temperature is 60°F.

The first step is to estimate the ultimate compressor-building horsepower. Assume in the foreseeable future the building will be expanded to 12,000 horsepower.

Next, estimate the fuel needs.

$$Q = 12,000 \times 240 \text{ cubic ft per day}$$
$$= 2,880,000 \text{ cubic ft per day}$$

To estimate the size line which must be run for fuel use this formula:

$$Q = \frac{(36,300)\ d^3 \sqrt{P_1^2 - P_2^2}}{\sqrt{l}}$$

where Q = cubic ft of gas per 24 hours
  d = pipe ID in in.
  l = length of pipe in ft
  P_1 = psi (abs) at starting point
  P_2 = psi (abs) at ending point

Using this formula, try several pipe diameters to determine the approximate size. For instance, if 6-in. ID pipe were used, the approximate flow could be found as follows:

$$Q = \frac{36,300 \times 6^3 \sqrt{(50 + 15)^2 - (25 + 15)^2}}{\sqrt{1,600}}$$

$$Q = \frac{36,300 \times 216 \sqrt{65^2 - 40^2}}{40}$$

$$Q = \text{approximately } 10,036,000 \text{ cubic ft per 24 hours}$$

From this it is obvious that 6-in. pipe would be considerably oversized, especially if the fuel gas header in the compressor building is of large diameter.

Try substituting 4-in. diameter pipe in this formula to find an approximate size pipe.

$$Q = \frac{36,300 \times 4^3 \sqrt{(50 + 15)^2 - (25 + 15)^2}}{\sqrt{1,600}}$$

$$Q = \frac{36,300 \times 64 \sqrt{65^2 - 40^2}}{40}$$

Q = 2,973,700 cubic ft per 24 hours

Apparently 4-in. pipe will be correct size. To check it, use this more accurate formula which is based on the Weymouth:

$$Q = \frac{63,320 \times d^{8/3} \sqrt{P_1^2 - P_2^2}}{\sqrt{l}}$$

$$Q = \frac{63,320 \times 41 \sqrt{65^2 - 40^2}}{\sqrt{1,600}}$$

Four-in. pipe is adequate.

Q = 3,322,000 cubic ft per day

Note. This latter formula is based on the Weymouth formula, for gas with a specific gravity of 0.60 and a temperature of 60°F. For every 10° variation in temperature, the answer will be approximately 1% in error. For every 0.01 variation in specific gravity, the answer will be three-fourth percent in error.

Estimate engine cooling water requirements

This equation can be used for calculating engine jacket water requirements as well as lube oil cooling water requirements:

$$GPM = \frac{H \times BHP}{500\Delta t}$$

where   H = Heat dissipation in Btu's per BHP/hr. This will vary for different engines; where they are available, the manufacturers' values should be used. Otherwise, you will be safe in substituting the following values in the formula: For engines with water-cooled exhaust manifolds: Engine jacket water = 2,200 Btu's per BHP/hr. Lube oil cooling water = 600 Btu's per BHP/hr.
   For engines with dry type manifolds (so far as cooling water is concerned) use 1,500 Btu's/BHP/hr for the engine jackets and 650 Btu's/BHP/hr for lube oil cooling water requirements.

BHP = Brake Horsepower Hour
   Δt = Temperature differential across engine. Usually manufacturers recommend this not exceed 15°F; 10°F is preferable.

Example. Find the jacket water requirements for a 2,000 hp gas engine which has no water jacket around the exhaust manifold.

Solution.

$$GPM = \frac{1,500 \times 2,000}{500 \times 10}$$

$$GPM = \frac{3,000,000}{5,000} = 600 \text{ gallons per min}$$

The lube oil cooling water requirements could be calculated in like manner.

Estimate fuel requirements for internal combustion engines

When installing an internal combustion engine at a gathering station, a quick approximation of fuel consumptions could aid in selecting the type fuel used.

Using Natural Gas: Multiply the brake hp at drive by 11.5 to get cubic ft of gas per hour.

Using Butane: Multiply the brake hp at drive by 0.107 to get gallons of butane per hour.

Using Gasoline: Multiply the brake hp at drive by 0.112 to get gallons of gasoline per hour.

These approximations will give reasonably accurate figures under full load conditions.

Example. Internal combustion engine rated at 50 bhp—3 types of fuel available.

Natural Gas: $50 \times 11.5 = 575$ cubic ft of gas per hour

Butane: $50 \times 0.107 = 5.35$ gallons of butane per hour

Gasoline: $50 \times 0.112 = 5.60$ gallons of gasoline per hour

Estimating fuel requirements for compressor installation

Rule. Multiply installation horsepower by 240 to obtain cubic ft of fuel gas required per day.

Example. Estimate the amount of fuel required to supply a 1,000-hp field compressor installation.

$1,000 \times 240 = 240$ Mcfd

11: Gas—Hydraulics

Gas pipe line hydraulics calculations 266
Equivalent lengths for multiple lines based on Panhandle A 267
Determine pressure loss for a low-pressure gas system 268
Nomograph for determining pipe-equivalent factors 269
How much gas is contained in a given line section? 270
How to estimate equivalent length factors for gas lines 270
Estimating comparative capacities of gas pipe lines 271
Determination of leakage from gas line using pressure drop
    method 271
A quick way to determine size of gas gathering lines 272
Energy conversion data for estimating 272
How to estimate time required to get a shut-in test on gas
    transmission lines and approximate a maximum acceptable
    pressure loss for new lines 273
How to determine the relationship of capacity increase to
    investment increase 273
Estimate pipe size requirements for increasing throughput volumes
    of natural gas... 274
Calculate line loss using cross-sectional areas table when testing
    mains with air or gas.. 275
Flow of fuel gases in pipe lines 276
Calculate the velocity of gas in a pipe line 277
Determining throat pressure in a blow-down system 277
Estimate the amount of gas blown off through a line puncture.......... 278
A practical way to calculate gas flow for pipe lines........................... 278
How to calculate the weight of gas in a pipe line 279
Estimate average pressure in gas pipe line using up and downstream
    pressures 279
Chart for determining viscosity of natural gas 280
Flow of gas 280
Multi-phase flow.. 284
Nomograph for calculating Reynolds number for compressible flow
    friction factor for clean steel and wrought iron pipe 289

Gas pipe line hydraulics calculations

Equations commonly used for calculating hydraulic data for gas pipe lines

Panhandle A.

$$Q_b = 435.87 \times (T_b/P_b)^{1.0778} \times D^{2.6182} \times E \times$$

$$\left[\frac{P_1^2 - P_2^2 - \dfrac{(0.0375 \times G \times (h_2 - h_1) \times P_{avg}^2)}{T_{avg} \times Z_{avg}}}{G^{0.8539} \times L \times T_{avg} \times Z_{avg}}\right]^{0.5394}$$

Panhandle B.

$$Q_b = 737 \times (T_b/P_b)^{1.020} \times D^{2.53} \times E \times$$

$$\left[\frac{P_1^2 - P_2^2 - \dfrac{(0.0375 \times G \times (h_2 - h_1) \times P_{avg}^2)}{T_{avg} \times Z_{avg}}}{G^{0.961} \times L \times T_{avg} \times Z_{avg}}\right]^{0.51}$$

Weymouth.

$$Q = 433.5 \times (T_b/P_b) \times \left[\frac{P_1^2 - P_2^2}{GLTZ}\right]^{0.5} \times D^{2.667} \times E$$

$$P_{avg} = 2/3 \left[P_1 + P_2 - (P_1 \times P_2)/P_1 + P_2\right]$$

P_{avg} is used to calculate gas compressibility factor Z

Nomenclature for Panhandle equations

Q_b = flow rate, SCFD
P_b = base pressure, psia
T_b = base temperature, °R
T_{avg} = average gas temperature, °R
P_1 = inlet pressure, psia
P_2 = outlet pressure, psia
G = gas specific gravity (air = 1.0)
L = line length, miles
Z = average gas compressibility
D = pipe inside diameter, in.
h_2 = elevation at terminus of line, ft
h_1 = elevation at origin of line, ft
P_{avg} = average line pressure, psia
E = efficiency factor

   E = 1 for new pipe with no bends, fittings, or pipe diameter changes

E = 0.95 for very good operating conditions, typically through first 12–18 months
E = 0.92 for average operating conditions
E = 0.85 for unfavorable operating conditions

Nomenclature for Weymouth Equation

Q = flow rate, SCFD
T_b = base temperature, °R
P_b = base pressure, psia
G = gas specific gravity (air = 1)
L = line length, miles
T = gas temperature, °R
Z = gas compressibility factor
D = pipe inside diameter, in.
E = efficiency factor. (See Panhandle nomenclature for suggested efficiency factors)

Sample Calculations

   Q = ?
   G = 0.6
   T = 100°F
   L = 20 miles
   P_1 = 2,000 psia
   P_2 = 1,500 psia
   Elev diff. = 100 ft
   D = 4.026-in.
   T_b = 60°F
   P_b = 14.7 psia
   E = 1.0

$$P_{avg} = 2/3(2,000 + 1,500 - (2,000 \times 1,500/2,000 + 1,500))$$
$$= 1,762 \text{ psia}$$

From Figure 5, p. 217, Z at 1,762 psia and 100°F = 0.835.

Panhandle A.

$$Q_b = 435.87 \times (520/14.7)^{1.0788} \times (4.026)^{2.6182} \times 1 \times$$

$$\left[\frac{(2,000)^2 - (1,500)^2 - \dfrac{0.0375 \times 0.6 \times 100 \times (1,762)^2}{560 \times 0.835}}{(0.6)^{.8539} \times 20 \times 560 \times .835}\right]^{0.5394}$$

$$Q_b = 16,577 \text{ MCFD}$$

Panhandle B.

$Q_b = 737 \times (520/14.7)^{1.020} \times (4.026)^{2.53} \times 1 \times$

$$\left[\frac{(2,000)^2 - (1,500)^2 - \dfrac{0.0375 \times 0.6 \times 100 \times (1,762)^2}{560 \times 0.835}}{(0.6)^{.961} \times 20 \times 560 \times .835}\right]^{0.51}$$

$Q_b = 17,498$ MCFD

Weymouth.

$Q = 0.433 \times (520/14.7) \times [(2,000)^2 -$
$\quad (1,500)^2/(0.6 \times 20 \times 560 \times 0.835)]^{1/2} \times (4.026)^{2.667}$
$Q = 11,101$ MCFD

Source

Pipecalc 2.0, Gulf Publishing Company, Houston, Texas. Note: Pipecalc 2.0 will calculate the compressibility factor, minimum pipe ID, upstream pressure, downstream pressure, and flow rate for Panhandle A, Panhandle B, Weymouth, AGA, and Colebrook-White equations. The flow rates calculated in the above sample calculations will differ slightly from those calculated with Pipecalc 2.0 since the viscosity used in the examples was extracted from Figure 5, p. 147. Pipecalc uses the Dranchuk et. al. method for calculating gas compressibility.

Equivalent lengths for multiple lines based on Panhandle A

Condition I

A single pipe line which consists of two or more different diameter lines.

Let $\qquad L_E$ = equivalent length
$L_1, L_2, \ldots L_n$ = length of each diameter
$D_1, D_2, \ldots D_n$ = internal diameter of each separate line corresponding to $L_1, L_2, \ldots L_n$
D_E = equivalent internal diameter

$$L_e = L_1 \left[\frac{D_E}{D_1}\right]^{4.8539} + L_2 \left[\frac{D_E}{D_2}\right]^{4.8539} + \ldots L_n \left[\frac{D_E}{D_n}\right]^{4.8539}$$

Example. A single pipe line, 100 miles in length consists of 10 miles $10\frac{3}{4}$-in. OD; 40 miles $12\frac{3}{4}$-in. OD and 50 miles of 22-in. OD lines.

Find equivalent length (L_E) in terms of 22-in. OD pipe.

$$L_E = 50 + 40 \left[\frac{21.5}{12.25}\right]^{4.8539} + 10 \left[\frac{21.5}{10.25}\right]^{4.8539}$$

$= 50 + 614 + 364$
$= 1,028$ miles equivalent length of 22-in. OD

Condition II

A multiple pipe line system consisting of two or more parallel lines of different diameters and different length.

Let $\qquad L_E$ = equivalent length
$L_1, L_2, L_3, \ldots L_n$ = length of various looped sections

$d_1, d_2, d_3, \ldots d_n$ = internal diameter of the individual line corresponding to length L_1, L_2, L_3 & L_n

$$L_E = L_1 \left[\frac{d_E^{2.6182}}{d_1^{2.6182} + d_2^{2.6182} + d_3^{2.6182} + \ldots d_n^{2.6182}}\right]^{1.8539} + \ldots$$

$$L_n \left[\frac{d_E^{2.6182}}{d_1^{2.6182} + d_2^{2.6182} + d_3^{2.6182} + \ldots d_n^{2.6182}}\right]^{1.8539}$$

Let $\qquad L_E$ = equivalent length
L_1, L_2, L_3 & L_n = length of various looped sections
d_1, d_2, d_3 & d_n = internal diameter of individual line corresponding to lengths L_1, L_2, L_3 & L_n

$$L_E = L_1 \left[\frac{d_E^{2.6182}}{d_1^{2.6182} + d_2^{2.6182} + d_3^{2.6182} + \ldots d_n^{2.6182}}\right]^{1.8539} + \ldots$$

$$L_n \left[\frac{d_E^{2.6182}}{d_1^{2.6182} + d_2^{2.6182} + d_3^{2.6182} + \ldots d_n^{2.6182}}\right]^{1.8539}$$

when L_1 = length of unlooped section
L_2 = length of single looped section
L_3 = length of double looped section
$d_E = d_1 = d_2$

then:

$$L_E = L_1 + 0.27664\, L_2 + L_3 \left[\frac{d_1^{2.6182}}{2d_1^{2.6182} + d_3^{2.6182}}\right]^{1.8539}$$

when $d_E = d_1 = d_2 = d_3$

then $L_E = L_1 + 0.27664\ L_2 + 0.1305\ L_3$

Example. A multiple system consisting of a 15 mile section of 3–8⅝-in. OD lines and 1–10¾-in. OD line, and a 30 mile section of 2–8⅝-in. lines and 1–10¾-in. OD line.

Find the equivalent length in terms of single 12-in. ID line.

$$L_E = 15\left[\frac{12^{2.6182}}{3(7.981)^{2.6182} + 10.02^{2.6182}}\right]^{1.8539}$$

$$+ 30\left[\frac{12^{2.6182}}{2(7.981)^{2.6182} + 10.02^{2.6182}}\right]^{1.8539}$$

$$= 5.9 + 18.1$$
$$= 24.0 \text{ miles equivalent of 12-in. ID pipe}$$

Example. A multiple system consisting of a single 12-in. ID line 5 miles in length and a 30 mile section of 3–12-in. ID lines.

Find equivalent length in terms of a single 12-in. ID line.

$$L_E = 5 + 0.1305 \times 30$$
$$= 8.92 \text{ miles equivalent of single 12-in. ID line}$$

Determine pressure loss for a low-pressure gas system

Use the Spitzglass equation for systems operating at less than 1 psig.

$$Q_h = 3,550\sqrt{\frac{\Delta h_w d^5}{S_g L\left(1 + \dfrac{3.6}{d} + 0.03d\right)}}$$

where Q_h = rate of flow, in cubic feet per hour at standard conditions (14.7 psia and 60°F), scfh
   h_w = static pressure head, in inches of water
   S_g = specific gravity of gas relative to air = the ratio of the molecular weight of the gas to that of air
   d = internal diameter of the pipe, inches
   L = length of the pipe, feet

Example:

Given the following conditions, find the flow in the system:

h_w = 50 inches of water
S_g = 0.65
L = 500 feet
d = 6.187 inches

$$Q_h = 3,550\sqrt{\frac{50 \times 6.187^5}{0.65 \times 500\left(1 + \dfrac{3.6}{6.187} + 0.03 \times 6.187\right)}}$$

$$Q_h = 99,723 \text{ scfh}$$

Nomograph for determining pipe-equivalent factors

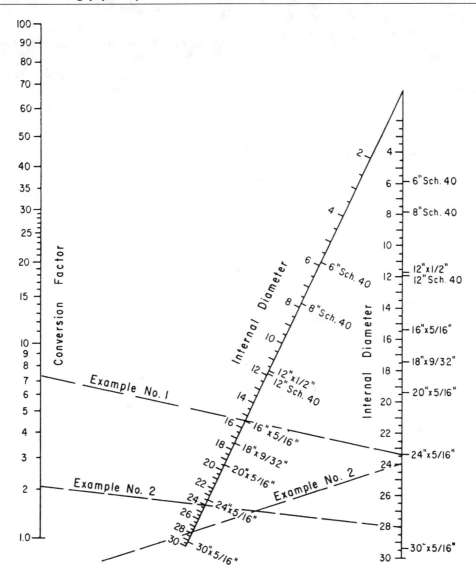

For turbulent flow, this handy nomograph will save a great deal of time in flow calculations where different sizes of pipe are connected. The advantage over a table of values is in having all sizes of pipe expressed in relation to any other size.

Example 1. 16-in. × ⁵/₁₆-in. in terms of 24-in. × ⁵/₁₆-in.

$$\left(\frac{23.375}{15.375}\right)^{4.735} = 1.5203^{4.735} = 7.26$$

Example 2. Reciprocal values can be found when chart limits are exceeded. 28-in. ID in terms of 24-in. ID falls beyond the lower limits of the chart; however, 24-in. ID in terms of 28-in. ID gives a conversion factor of 2.1.

How much gas is contained in a given line section?

Multiply the square of the inside diameter, in in., by the gauge pressure, in lb/in.²; multiply this by 0.372; the answer is the approximate number of cubic ft of gas (standard conditions) in 1,000 ft of line.

Example. How much gas in 1,000 ft of 16-in. schedule 30 pipe at 350 lb pressure?

$15.25 \times 15.25 \times 350 \times 0.372 = 30,280$ cubic ft of gas

Example. Approximately how much gas in eight miles of .250 wall 24-in. pipe if the pressure is 400 psi?

Cubic ft = $(23.5)(23.5)(400)(8)(5.28)(0.372)$
$\qquad = 3,470,000$ cubic ft

When a section of line is blown down from one pressure to another, the total gas lost may be computed by the difference in the contents at the two pressures, using the above rule.

How to estimate equivalent length factors for gas lines

Internal Diameter	Nominal Diameter	36″ 35.250	34″ 33.250	30″ 29.375	26″ 25.375	24″ 23.375	24″ 23.000	22″ 21.500	20″ 19.500	18″ 17.500	16″ 15.500	12¾″ 12.250	10¾″ 10.250	8⅝″ 8.125
35.250″	36″	1	0.758	0.416	0.204	0.137	0.126	0.092	0.056	0.034	0.019	0.006	0.002	0.0008
33.250	34	1.32	1	0.55	0.270	0.181	0.167	0.121	0.075	0.044	0.025	0.008	0.003	0.001
29.375	30	2.41	1.82	1	0.492	0.330	0.305	0.22	0.136	0.081	0.045	0.014	0.006	0.002
25.375	26	4.90	3.71	2.03	1	0.671	0.622	0.45	0.278	0.164	0.092	0.029	0.012	0.004
23.375	24	7.25	5.54	3.04	1.49	1	0.922	0.67	0.415	0.246	0.137	0.044	0.018	0.006
23.000	24	7.90	6.00	3.29	1.62	1.08	1	0.72	0.450	0.267	0.148	0.048	0.020	0.0064
21.500	22	10.93	8.31	4.55	2.24	1.50	1.37	1	0.624	0.369	0.205	0.065	0.027	0.009
19.500	20	17.60	13.33	7.31	3.59	2.41	2.22	1.60	1	0.590	0.328	0.105	0.044	0.014
17.500	18	29.8	22.5	12.35	6.08	4.075	3.78	2.72	1.69	1	0.555	0.177	0.074	0.024
15.500	16	52.8	40.5	22.2	10.9	7.345	6.75	4.88	3.04	1.80	1	0.319	0.134	0.043
12.250	12¾	168	127	69.8	34.3	23.01	21.2	15.35	9.55	5.65	3.13	1	0.420	0.136
10.250	10¾	400	303	167	81.7	54.67	50.6	36.5	22.7	13.4	7.47	2.38	1	0.324
8.125	8⅝	1235	935	513	252	168.8	156.2	112.5	70.5	41.5	23.0	7.35	3.08	1
6.125	6⅝	4860	3680	2020	992	666	615	443	276	163	90.7	29.0	12.2	3.94

This table shows the equivalent length factors for pipe lines of different diameters. It is based on the Panhandle Formula and the factors were obtained by varying the lengths of the pipe and keeping the other variables constant.

Example. Find the equivalent length of a 24-in. pipe line as compared with a 12³/4-in. line.

Enter the chart at the top—24-in.—(23.00 ID) and proceed downward to the 12.250 parameter. The equivalent length is 21.2.

This means that under the same conditions of temperatures, pressures, specific gravities, etc., one mile of 12¾-in. pipe will flow the same amount of gas that 21.2 miles of 24-in. will flow. Another way of putting it, the pressure drop in one mile of 24-in. pipe will be the same as that in .048 miles of 12¾-in. if the pressures, volumes, etc., are the same.

Where a line is composed of several sizes of pipe diameters or where the lines are looped, this chart should prove useful. In solving problems of this sort, the line is reduced to an equivalent length of some arbitrarily selected size pipe and treated as a single line of uniform size.

Estimating comparative capacities of gas pipe lines

To estimate capacity of one diameter line in terms of another read downward in the appropriate column to desired diameter line.

Example. A 24-in. (23.375-in. ID) pipe line has a capacity 160% that of a 20-in. (19.500-in. ID) line.

This chart is based on diameter factors using the Panhandle equation. Equivalent inlet and outlet pressures, lengths of line section, and uniform flow are assumed for both lines.

(Figures give capacity of one diameter pipe as percent of another.)

Internal Diameter	Nominal Diameter	36" 35.250"	34" 33.250"	30" 29.3125"	26" 25.375"	24" 2.375"	24" 23.000"	20" 19.500"	18" 17.500"	16" 15.500"	12¾" 12.250"
35.250"	36"	100	85.7	62.7	43.6	35.2	33.7	22.0	16.5	12.0	6.5
33.250"	34"	116.8	100	73.1	50.8	41.0	39.3	25.7	19.2	14.0	7.6
29.3125"	30"	159.7	137	100	69.6	56.1	53.8	35.7	26.3	19.2	10.4
25.375"	26"	229	197	149	100	80.7	77.5	50.2	37.8	27.6	14.8
23.375"	24"	285	244	178	124	100	96.0	62.3	46.9	34.2	18.5
23.000"	24"	298	255	186	129	104	100	65.1	49.0	35.7	19.2
19.500"	20"	458	392	287	199	160	154	100	75.3	35.0	29.6
17.500"	18"	608	519	379	264	213	204	132	100	73.2	39.3
15.500"	16"	835	715	522	363	293	281	182	127	100	53.8
12.250"	12¾"	1545	1322	968	673	543	521	238	254	185	100

Determination of leakage from gas line using pressure drop method

To determine the leakage in Mmcf per year at a base pressure of 14.4 psia and a temperature of 60°F, take as a basis a one mile section of line under one hour test. Use the formula:

$$Ly = 9 \times D^2 \left[\frac{P_1}{460 + t_1} - \frac{P_2}{460 + t_2} \right]$$

where D = inside diameter of the pipe, in.
 P_1 = pressure at the beginning of the test in psia
 t_1 = temperature at the beginning of the test in °F
 P_2 = pressure at end of the test
 t_2 = temperature at end of the test

The error in using this formula can be about 4%.

Example. A section of a 10-in. pipe line one mile in length is tested using the pressure drop method. ID is equal to 10.136 in. Pressure and temperature at the start of the test were 190.5 psig and 52°F. After one hour, the readings were 190.0 psig and 52°F. Calculate the leakage in Mmcf/year.

The atmospheric pressure for the area is 14.4 psia. Thus:

$P_1 = 190.5 + 14.4 = 204.9$ psia

$P_2 = 190.0 + 14.4 = 204.4$ psia

$$Ly = 9 (10.136)^2 \left[\frac{204.9}{460 + 52} - \frac{204.4}{460 + 52} \right]$$

$$= \frac{(9)(102.7)(0.5)}{512}$$

= 0.903 Mmcf per mile per year

A quick way to determine size of gas gathering lines

Here is a short-cut way to estimate gas flow in gathering lines. For small gathering lines, the answer will be within 10% of that obtained by more difficult and more accurate formulas.

$$Q = \frac{(500)(d^3)\sqrt{P_1^2 - P_2^2}}{\sqrt{L}}$$

where Q = cubic ft of gas per 24 hours
d = pipe ID in in.
P_1 = psi (abs) at starting point
P_2 = psi (abs) at ending point
L = length of line in miles

Example. How much gas would flow through four miles of 6-in. ID pipe if the pressure at the starting point is 485 psi and if the pressure at the downstream terminus is 285 psi?

Answer.

$$Q = \frac{(500)(6^3)\sqrt{(485 + 15)^2 - (285 + 15)^2}}{\sqrt{4}}$$

$$Q = \frac{(500)(216)\sqrt{500^2 - 300^2}}{2}$$

$$Q = \frac{(500)(216)(400)}{2}$$

Q = 21,600,000 cubic ft per 24 hours

Using a more accurate formula (a version of Weymouth's):

$$Q = \frac{(871)(d^{8/3})\sqrt{P_1^2 - P_2^2}}{\sqrt{L}}$$

$$Q = \frac{(871)(120)(400)}{2}$$

Q = 20,900,000 cubic ft per 24 hours

The error in this case is 700,000 cubic ft per 24 hours, or about 3%.

Energy conversion data for estimating

The accompanying table gives the high heating values of practically all commonly used fuels along with their relation one to the other.

Example. If light fuel oil will cost a potential industrial customer 18 cents per gallon and natural gas will cost him 75 cents per Mcf, which fuel will be more economical?

Solution. Assume the operator would use 10,000 gallons of light fuel oil per year at 18 cents per gallon. The cost would be $1,800. From the table, 10,000 gallons of fuel oil is equivalent to 10,000 × 134 = 1,340,000 cubic ft of natural gas. At 75 cents per Mcf this would cost 1,340 × .75 = $1,005.

The burning efficiency of both fuels is the same. The savings by using natural gas would come to $795 per year.

Fuel	Heating Value	PERCENT BURNING EFFICIENCY			Unit	NUMBER OF CUBIC FEET OF NATURAL GAS EQUIVALENT TO 1 UNIT OF OTHER FUEL		
		Domestic	Commercial	Industrial		Domestic	Commercial	Industrial
Natural Gas	1000 B.T.U. per Cubic Foot	75	75	78	Cubic Foot	1	1	1
Propane	91,600 B.T.U. per U.S. gallon	75	75	78	Gallon	91.5	91.5	91.5
Butane	102,000 B.T.U. per U.S. gallon	.....	75	78	Gallon	.....	102	102
Light Fuel Oil	134,000 B.T.U. per U.S. gallon	65–75	75	78	Gallon	117–134	134	134
Heavy Fuel Oil	146,000 B.T.U. per U.S. gallon	.....	75	78	Gallon	.....	146	146
Hydro-electricity	3,412 B.T.U. per Kilowatt Hour	100	100	100	Kilowatt Hour	4.549	4.549	4.549
*Coal	8,700–11,000 B.T.U. per pound	55–60	65	73	Pound	6.37–8.07	7.55–9.57	8.45–10.7

*Coal qualities vary considerably. The figures given are average for Western Coals; slack, run-of-the-mine; potentially available at lowest cost.

How to estimate time required to get a shut-in test on gas transmission lines and approximate a maximum acceptable pressure loss for new lines

These two rules may prove helpful for air or gas testing of gas transmission lines. They are not applicable for hydrostatic tests. Values of the constants used by different transmission companies may vary with economic considerations, line conditions and throughputs.

To determine the minimum time required to achieve a good shut-in test after the line has been charged and becomes stabilized, use the following formula:

$$H_m = \frac{3 \times D^2 \times L}{P_1}$$

where H_m = minimum time necessary to achieve an accurate test in hours
D = internal diameter of pipe in in.
L = length of section under test in miles
P_1 = initial test pressure in lb/in.² gage

Example. How long should a 14-mile section of 26-in. (25.375-in. ID) pipe line be shut in under a 1,050 psi test pressure to get a good test?

$$H_m = \frac{3 \times (25.375)^2 \times 14}{1,050}$$

$$= 25.75 \text{ hours or 25 hours and 45 minutes}$$

When a new line is being shut in for a test of this duration or longer, the following formula may be used to evaluate whether or not the line is "tight":

$$P_{ad} = \frac{H_t \times P_1}{D \times 949}$$

where P_{ad} = acceptable pressure drop in lb/in.² gage
H_t = shut-in test time in hours
D = internal diameter of pipe in in.
P_1 = initial test pressure in lb/in.² gage

Example. What would be the maximum acceptable pressure drop for a new gas transmission line under air or gas test when a 44-mile section of 30-in. (29.250-in. ID) pipe line has been shut in for 100 hours at an initial pressure of 1,100 psig?

$$P_{ad} = \frac{100 \times 1,100}{29.250 \times 949} = 4 \text{ lb/in.}^2$$

If the observed pressure drop after stabilization is less than 4 psig, the section of line would be considered "tight."

Note: Corrections must be made for the effect of temperature variations upon pressure during the test.

How to determine the relationship of capacity increase to investment increase

Example. Determine the relationship of capacity increase to investment increase when increasing pipe diameter, while keeping the factors P(in), P(out) and L constant.

Solution. The top number in each square represents the percent increase in capacity when increasing the pipe size from that size shown in the column at the left to the size shown by the column at the bottom of the graph. The lower numbers represent the percent increase in investment of the bottom size over the original pipe size. All numbers are based on pipe capable of 1,000 psi working pressure.

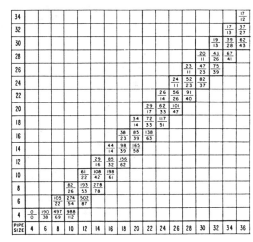

Estimate pipe size requirements for increasing throughput volumes of natural gas

Reflects the capacity to pipe size relationship when L (length), P_1 (inlet pressure) and P_2 (outlet pressure) are constant.

EXAMPLE
An 8" line will handle "X" quantity. Find pipe size required for 4.5"X". Begin at 4.5 on left scale, move horizontally to 8" curve, then down to nominal pipe size. Use next largest pipe size.

Calculate line loss using cross-sectional areas table when testing mains with air or gas

$$\text{Initial volume} = \frac{P_i}{P_B}(AL)$$

$$\text{Final volume} = \frac{P_f}{P_B}(AL)$$

where P_i = initial pressure, or pressure at start of test, psia
P_f = final pressure, or pressure at end of test, psia
P_B = base pressure, psia
A = cross-sectional area of inside of pipe in ft^2
L = length of line in ft

The loss then would be the difference of the volumes, found by the above formula, which could be written:

$$\text{Loss during test} = AL\frac{(P_i - P_f)}{P_B}$$

To facilitate calculations, Table 1 contains the cross-sectional areas in ft^2 for some of the more popular wall thickness pipe from 1-in. through 24-in. nominal size.

Example. Assume 3,562 ft of 8.125-in. inside diameter pipe is tested with air for a period of 26 hours. The initial pressure was 116 psig, the final pressure 112 psig, and the barometer was 14.4 psia in both instances. It is desired to calculate the volume of loss during the test using a pressure base of 14.65 psia.

If the area "A" is obtained from Table 1, and information as given above is substituted in the formula, we would have:

$$\text{Loss during test} = 0.360 \times 3,562 \times$$

$$\frac{[(116 + 14.4) - (112 + 14.4)]}{14.65}$$

$$= 350.1 \text{ cubic ft}$$

Table 1
Characteristics of Pipe

Nominal Pipe Size D_n	Outside Diameter	Wall Thickness	Inside Diameter	"A" Inside Area in Sq. Feet
1	1.315	.133	1.049	.006
		.179	.957	.005
1¼	1.660	.140	1.380	.010
		.191	1.278	.009
1½	1.900	.145	1.610	.014
		.200	1.500	.012
2	2.375	.154	2.067	.023
		.218	1.939	.021
2½	2.875	.203	2.469	.033
		.276	2.323	.029
3	3.500	.125	3.250	.058
		.156	3.188	.055
		.188	3.124	.053
		.216	3.068	.051
		.250	3.000	.049
3½	4.000	.125	3.750	.077
		.156	3.688	.074
		.188	3.624	.072
		.226	3.548	.069
		.250	3.500	.067
4	4.500	.125	4.250	.099
		.156	4.188	.096
		.188	4.124	.093
		.219	4.062	.090
		.237	4.026	.088
		.250	4.000	.087
5	5.563	.156	5.251	.150
		.188	5.187	.147
		.219	5.125	.143
		.258	5.047	.139
6	6.625	.188	6.249	.213
		.219	6.187	.209
		.250	6.125	.205
		.280	6.065	.201
8	8.625	.188	8.249	.371
		.219	8.187	.366
		.250	8.125	.360
		.277	8.071	.355

Nominal Pipe Size D_n	Outside Diameter	Wall Thickness	Inside Diameter	"A" Inside Area in Sq. Feet
10	10.750	.188	10.374	.587
		.219	10.312	.580
		.250	10.250	.573
		.279	10.192	.567
		.307	10.136	.560
12	12.750	.219	12.312	.827
		.250	12.250	.818
		.281	12.188	.810
		.312	12.126	.802
		.330	12.090	.797
14	14.000	.250	13.500	.994
		.281	13.438	.985
		.312	13.376	.976
		.344	13.312	.966
		.375	13.250	.958
16	16.000	.250	15.500	1.310
		.281	15.438	1.300
		.312	15.376	1.289
		.344	15.312	1.279
		.375	15.250	1.268
18	18.000	.250	17.500	1.670
		.281	17.438	1.658
		.312	17.376	1.647
		.344	17.312	1.635
		.375	17.250	1.623
20	20.000	.250	19.500	2.074
		.281	19.438	2.061
		.312	19.376	2.048
		.344	19.312	2.034
		.375	19.250	2.021
22	22.000	.312	21.376	2.492
		.344	21.312	2.477
		.375	21.250	2.463
24	24.000	.312	23.376	2.980
		.344	23.312	2.964
		.375	23.250	2.948

Flow of fuel gases in pipe lines

D. S. Davis, Special Projects, Inc., Bailey Island, Maine

The rate of flow of fuel gases at low pressures, ordinary temperatures, and under turbulent conditions in pipe lines is reflected by an equation[1] based on sound theoretical considerations:

$$q = \frac{3.185 \; \Delta h^{0.543} \; d^{2.63}}{s^{0.468}}$$

where q = rate of flow of gas, cubic ft per minute
Δh = pressure drop, in. of water per 100 ft of pipe
d = actual inner diameter of Schedule 40 steel pipe, in.
s = specific gravity of gas at the prevailing temperature and pressure relative to air at 68°F and 30 in. of mercury

The equation can be solved readily and accurately by means of the accompanying monograph constructed in accordance with methods described previously.[2]

The use of the chart is illustrated as follows. At what rate will fuel gas with a specific gravity of 0.640 flow in a 10-in. Schedule 40 pipe line if the pressure is 0.160 in. of water per 100 ft of pipe? Following the key, connect 0.160 on the Δh scale and 0.640 on the s scale with a straight line and mark the intersection with the α axis. Connect this point with 10 on the d' scale for nominal diameters and read the rate of flow at 624 cubic ft per minute on the q scale.

Some authorities give the exponent of s as 0.408 when for the data given the rate of flow as read from the chart would be decreased by 2.7% (17 cubic ft per minute) to attain 607 cubic ft per minute.

References

1. Perry, J. H., ed., *Chemical Engineers' Handbook*, 3rd ed., p. 1589, McGraw-Hill Book Co., Inc., New York (1950).
2. Davis, D. S., *Nomography and Empirical Equations*, 2nd ed., Chapter 6, Reinhold Publishing Corp., New York, 1962.

Calculate the velocity of gas in a pipe line

$$V = \frac{748.7QT}{d^2 P \times 520}$$

where

V = Velocity, feet per second
Q = Mcf/hour at standard conditions
d = Inside diameter of the pipe, inches
P = Pressure, psia
T = Average gas temperature, °R

Example: Given a pipeline having an inside diameter of 4.026 inches, a gas flow rate of 1,000Mcf/hour at a pressure of 100 psig, and an average gas temperature of 60°F, find the velocity of the gas.

Solution:

$$V = \frac{748.7 \times 1,000 \times (60 + 460)}{4.026^2 \times 100 \times 520}$$

$$V = 461.9 \text{ft / sec}$$

Determining throat pressure in a blow-down system

In designing a gas blow-down system, it is necessary to find the throat pressure in a blow-off stack of a given size with a given flow at critical velocity. The accompanying nomograph is handy to use for preliminary calculations. It was derived from the equation:

$$P_t = \frac{W}{d^2} \times \sqrt{\frac{RT_1}{K\,(K-1)} \over 11,400}$$

where P_t = pressure in the throat, psia
       W = weight of flowing gas, lb/hr
       d = actual ID of stack, in.
       R = gas constant, ft lb/°F
       K = C_p/C_v
       C_p = specific heat at constant pressure, Btu/lb/°F
       C_v = specific heat at constant volume, Btu/lb/°F
       T_1 = absolute temperature of gas flowing,
              °R = (°F + 460)

Using the following values for natural gas, R = 88.888 and K = 1.31, the above equation reduces to:

$$P_t = \frac{W}{d^2} \times .0012978 \sqrt{T_1}$$

Example. To use the accompanying nomograph, assume that the rate of flow is known to be 2.00×10^6 lb/hr, the ID of the stack is 12 in. and the flowing temperature is 480°R. Start from the abscissa at the Rate of Flow, W, and proceed upward to the ID of the pipe. From this intersection proceed to the right to the flowing temperature of 480°R. Turn upward from this intersection to the P_t scale and read the answer—395 psia.

Estimate the amount of gas blown off through a line puncture

To calculate the volume of gas lost from a puncture or blowoff, use the equation:

$$Q = D^2 P_1$$

where Q = volume of gas in Mcf/hr at a pressure of 14.9 psi, 60°F and with a specific gravity of 0.60
       D = diameter of the nipple or orifice in in.
       P = absolute pressure in lb/in.² at some nearby point upstream from the opening

Example. How much gas will be lost during a five minute blowoff through a 2-in. nipple if the upstream pressure is 1,000 psi absolute?

$$Q = D^2 P_1$$

$$Q = (2)^2 \times 1,000$$

$$Q = 4,000 \text{ Mcf/hr}$$

$$Q_{5 \text{ min}} = 4,000 \times \frac{5}{60} = 333 \text{ Mcf}$$

A practical way to calculate gas flow for pipe lines

Here is a short cut way to calculate gas flow in pipe lines. It is based on the Weymouth formula. At 60°F and specific gravity of 0.60, the answer will be accurate. For every 10° variation in temperature, the answer will be 1% error. For every 0.01 variation in specific gravity, the answer will be three-fourths percent in error.
Formula:

$$Q = \frac{(871)(d^{8/3}) \sqrt{P_1^2 - P_2^2}}{\sqrt{L}}$$

where Q = cubic ft of gas per 24 hours; 871 is a constant
       d = pipe ID in in.
       P_1 = psi (abs) at starting point
       P_2 = psi (abs) at ending point
       L = length of line in miles

Example. How much gas would flow through one mile of 6-in. ID pipe if the pressure at the starting point is 485 psi and if the pressure at the downstream terminus is 285 psi?

Solution.

$$Q = \frac{(871)(6^{8/3}) \sqrt{(485 + 15)^2 - (285 + 15)^2}}{\sqrt{1}}$$

$$= \frac{(871)(120) \sqrt{500^2 - 300^2}}{\sqrt{1}}$$

$$= (871)(120)(400)$$

$$= 41,800,000 \text{ cubic ft per 24 hours}$$

Example. What would be the flow through four miles of 10¾-in. quarter-inch wall pipe using the same pressures?

$$Q = \frac{(871)(10.25^{8/3}) \sqrt{500^2 - 300^2}}{\sqrt{4}}$$

$$Q = \frac{(871)(495.7)(400)}{2}$$

$$Q = 86,350,000 \text{ cubic ft per day}$$

How to calculate the weight of gas in a pipe line

Rule. Find the volume of the pipe in cubic ft and multiply by the weight of the gas per cubic ft. To find the latter, multiply the absolute pressure of the gas times 3 and divide by 1,000.

The basis for the latter is that gas with a specific gravity of .60 at 70°F weighs 3.06 lb/cubic ft at 1,000 psia. And everything else remaining equal, weight of the gas is proportional to absolute pressure. Thus to find the weight of gas at, say, 500 psia, multiply $3 \times 500/1,000 = 1.5$ lb. (1.53 is more accurate.)

Example. Find the weight of gas in a 1,250-ft aerial river crossing where the average pressure reads 625.3 on the gauges; the temperature of the gas is 70°F and specific gravity of gas is .60.

Solution. Change psig to psia:

$$625.3 + 14.7 = 640 \text{ psia}$$

Weight of gas per cubic ft:

$$3.06 \times \frac{640}{1,000} = 1.958 \text{ lb/cubic ft}$$

Volume of 1,250 ft of 23¼-in. ID pipe = 3,685 cubic ft

Weight of gas = 7,217 lb

This method is fairly accurate; here is the same problem calculated with the formula:

$$W = \frac{(V)(144)(P_{abs})}{R\,T}$$

where
W = weight of gas in lb
V = volume of pipe
P_{abs} = absolute pressure of gas
R = universal constant 1,544 ÷ molecular weight of gas
T = temperature of gas in °R

(To find the molecular weight of gas, multiply specific gravity × molecular weight of air, or in this case, $.6 \times 28.95 = 17.37$.)

$$W = \frac{(3,685)(144)(640)}{(1,544 \div 17.37)(70 + 460)}$$

$$W = 7,209 \text{ lb}$$

Estimate average pressure in gas pipe line using up and downstream pressures

To find the average gas pressure in a line, first divide downstream pressure by upstream pressure and look up the value of the factor X in the table shown. Then multiply X times the upstream pressure to get the average pressure.

Example. Find the average pressure in a pipe line if the upstream pressure is 775 psia and the downstream pressure is 420 psia.

$$\frac{P_d}{P_u} = \frac{420}{775} = 0.54$$

Interpolating from the table,

$$X = 0.78 + {}^{4/5}\,(0.02) = 0.796$$

$$P_{average} = 0.796 \times 775 = 617 \text{ psia}$$

This method is accurate to within 2 or 3 psi. Here is the same problem calculated with the formula:

$$P_{average} = \frac{2}{3}\left(P_u + P_d - \frac{P_u P_d}{P_u + P_d}\right)$$

$$= \frac{2}{3}\left(1,195 - \frac{325,500}{1,195}\right)$$

$$P_{average} = 615 \text{ psia}$$

P_d/P_u	X
.10	.67
.15	.68
.20	.69
.25	.70
.30	.71
.35	.725
.40	.74
.45	.76
.50	.78
.55	.80
.60	.82
.65	.84
.70	.86
.75	.88
.80	.90
.85	.925
.90	.95
.95	.975
1.00	1.00

Chart for determining viscosity of natural gas

From: Carlton Beal, "The Viscosity of Air, Water, Natural Gas, Crude Oil and Its Associated Gases at Oil Field Temperatures and Pressures," *Petroleum Development and Technology*, AIME, 1946.

Flow of gas

For gas flow problems as encountered in oil field production operations, the Fanning equation for pressure drop may be used. A modified form of this equation employing units commonly used in oil field practice is:[1,2,3,4,5,6]

$$P = \frac{FL(MCFD)^2 STZ}{20,000 \, d^5 P_{av}}$$

and for determining the friction factor F:

$$Re = \frac{20.14(MCFD)S}{d \, \mu}$$

where

P = pressure drop, psia
F = friction factor, dimensionless
L = length of pipe, ft
MCFD = gas flow at standard condition

S = specific gravity of gas
T = absolute temperature (°F + 460)
P_{av} = average flowing pressure, psia
d = internal pipe diameter, in.
Re = Reynolds number, dimensionless
μ = viscosity, centipoises

In actual practice, empirical flow formulas are used by many to solve the gas flow problems of field production operations. The Weymouth formula is the one most frequently used since results obtained by its use agree quite closely with actual values. Recent modification of the formula, by including the compressibility factor, Z, made the formula applicable for calculation of high pressure flow problems.

The modified formula is as follows:

$$Q_s = 433.45 \frac{T_s}{P_s} \times d^{2.667} \times \left(\frac{P_1^2 - P_2^2}{LSTZ}\right)^{1/2}$$

where Q_s = rate of flow of gas in cubic ft per 24 hours measured at standard conditions
d = internal diameter of pipe in in.
P_1 = initial pressure, psi absolute
P_2 = terminal pressure, psi absolute
L = length of line in miles
S = specific gravity of flowing gas (air = 1.0)
T = absolute temperature of flowing gas

Standard conditions for measurements:

Table 1
Values of $d^{2.667}$ for use in Weymouth's gas-flow formula

(Increments of tenths of an inch with certain standard pipe sizes inserted)

d	$d^{2.667}$	d	$d^{2.667}$	d	$d^{2.667}$	d	$d^{2.667}$
1.0	1.0	6.2	129.9	11.5	673.8	16.9	1881
1.1	1.3	6.3	135.4	11.6	689.5	17.0	1911
1.2	1.6	6.4	141.2	11.7	705.5	17.1	1941
1.3	2.0	6.5	147.1	11.8	721.7	17.2	1971
1.4	2.4	6.6	153.3	11.9	738.1	17.250	1988
1.5	2.9	6.7	159.5	12.000	754.8	17.3	2002
1.6	3.5	6.8	166.0	12.1	771.7	17.4	2033
1.7	4.1	6.9	172.6	12.2	788.8	17.5	2064
1.8	4.8	7.0	179.3	12.3	806.1	17.6	2096
1.9	5.5	7.1	186.2	12.4	823.7	17.7	2128
2.0	6.3	7.2	193.3	12.5	841.6	17.8	2160
2.067	6.9	7.3	200.5	12.6	859.7	17.9	2193
2.1	7.2	7.4	207.9	12.7	878.0	18.0	2225
2.2	8.2	7.5	215.5	12.8	896.5	18.1	2259
2.3	9.2	7.6	223.3	12.9	915.3	18.2	2292
2.4	10.3	7.7	231.2	13.0	934.4	18.3	2326
2.5	11.5	7.8	239.3	13.1	953.7	18.4	2360
2.6	12.8	7.9	247.6	13.2	973.2	18.5	2394
2.7	14.1	7.981	254.5	13.250	983.9	18.6	2429
2.8	15.6	8.0	256.0	13.3	993.0	18.7	2464
2.9	17.1	8.1	264.6	13.4	1013	18.8	2499
3.0	18.7	8.2	273.4	13.5	1033	18.9	2535
3.068	19.9	8.3	282.4	13.6	1054	19.0	2570
3.1	20.4	8.4	291.6	13.7	1075	19.1	2607
3.2	22.2	8.5	300.9	13.8	1096	19.182	2639
3.3	24.1	8.6	310.5	13.9	1117	19.2	2643
3.4	26.1	8.7	320.2	14.0	1138	19.250	2664
3.5	28.2	8.8	330.1	14.1	1160	19.3	2680
3.6	30.4	8.9	340.2	14.2	1182	19.4	2717
3.7	32.7	8.941	344.6	14.3	1205	19.5	2755
3.8	35.2	9.0	350.5	14.4	1227	19.6	2793
3.9	37.7	9.1	361.0	14.5	1250	19.7	2831
4.0	40.3	9.2	371.6	14.6	1273	19.8	2869
4.026	41.0	9.3	382.5	14.7	1297	19.9	2908
4.1	43.1	9.4	393.6	14.8	1321	20.0	2947
4.2	45.9	9.5	404.8	14.9	1344	20.1	2987
4.3	48.9	9.6	416.3	15.0	1368	20.2	3026
4.4	52.0	9.7	428.0	15.1	1393	20.3	3067
4.5	55.2	9.8	439.8	15.2	1418	20.4	3107
4.6	58.5	9.9	451.9	15.250	1431	20.5	3148
4.7	62.0	10.0	464.2	15.3	1443	20.6	3196
4.8	65.6	10.020	467.0	15.4	1468	20.7	3230
4.9	69.3	10.1	476.6	15.5	1494	20.8	3272
5.0	73.1	10.2	489.3	15.6	1519	20.9	3314
5.1	77.1	10.3	502.2	15.7	1545	21.0	3357
5.2	81.2	10.4	515.3	15.8	1572	21.1	3400
5.3	85.4	10.5	528.7	15.9	1599	21.1	3443
5.4	89.8	10.6	542.2	16.0	1625	21.250	3468
5.5	94.3	10.7	555.9	16.1	1653	21.3	3486
5.6	98.9	10.8	569.9	16.2	1680	21.4	3530
5.7	103.7	10.9	584.1	16.3	1708	21.5	3574
5.8	108.6	11.000	598.5	16.4	1736	21.6	3619
5.9	113.7	11.1	613.1	16.5	1765	21.7	3663
6.0	118.9	11.2	627.9	16.6	1793	21.8	3709
6.065	122.4	11.3	643.0	16.7	1822	21.9	3754
6.7	124.2	11.4	658.3	16.8	1851	22.0	3800

—The Oil & Gas Journal

T_s = standard absolute temperature
P_s = standard pressure, psi absolute
Z = compressibility factor

Values of expression $d^{2.667}$ for different inside diameter of pipe are given in Table 1.

Another formula quite commonly used is the one determined by F. H. Oliphant:

$$Q = 42a \sqrt{\frac{P_1{}^2 - P_2{}^2}{L}}$$

where Q = discharge in cubic ft per hour
       42 = a constant
       P_1 = initial pressure in lb/in.², absolute
       P_2 = final pressure in lb/in.², absolute
       L = length of line in miles
       a = diameter coefficient

This formula assumes a specific gravity of gas of 0.6. For any other specific gravity multiply the final result by:

$$\sqrt{\frac{0.6}{\text{sp. gr. gas}}}$$

The values of diameter coefficients for different sizes of pipes are given below:

¼ in.	.0317	2½ in.	10.37	8 in.	198
½ in.	.1810	3 in.	16.50	10 in.	350
¾ in.	.5012	4 in.	34.10	12 in.	556
1 in.	1.0000	5 in.	60.00	16 in.	1160
1½ in.	2.9300	5⅝ in.	81.00	18 in.	1570
2 in.	5.9200	6 in.	95.00		

For pipes greater than 12 in. in diameter the measure is taken from the outside and for pipes of ordinary thickness the corresponding inside diameters and multipliers are as follows:

Outside	Inside	Multiplier
15 in.	14¼ in.	863
16 in.	15¼ in.	1025
18 in.	17¼ in.	1410
20 in.	19¼ in.	1860

For determining gas flow rates for specific pressure drops for pipe sizes used in production operations, Tables 3, 4, and 5 may be used. The tables are calculated by the Weymouth formula for listed inside diameters of standard weight threaded line pipe for the sizes shown.

Tables 3, 4, and 5 give the hourly rates of flow of 0.70 specific gravity gas, flowing at 60°F and measured at a standard temperature of 60°F and at a standard pressure base of 14.4 psi + 4 oz. gauge pressure of 14.65 psi. For specific gravity of gas other than 0.70 and for flowing temperature other than 60°F correction can be made by use of the factors shown in Table 2. For instance, to determine the flow of gas of specific gravity 0.06 flowing at the temperature of 20°F the value obtained from the gas flow Tables 3, 4, and 5 would be multiplied by 1.12, the factor obtained from Table 2.

To find from Tables 3, 4, and 5 the volume of gas delivered through any length L, the volume found from the table should be multiplied by $1/\sqrt{L}$. L must be expressed as a multiple of the length heading the table in use. For instance if the delivery is to be determined from a 4,500 ft line, the volume found for a 1,000 ft line in Table 3 should be multiplied by $1\sqrt{4.5}$. At the bottom of each of the two tables in question multipliers are given to be used as correction factors.

Use of Tables 3, 4, and 5 may be illustrated by the following example.

Example. What is the delivery of a 2-in. gas line, 3,600 ft long with an upstream pressure of 200 psi and a downstream pressure of 50 psi? The specific gravity of gas is 0.07; the flowing temperature, 70°F.

From Table 3 the volume for 1,000 ft of line, for 0.70

Table 2
Temperature—specific gravity multipliers

Specific Gravity	FLOWING TEMPERATURE—DEGREES F.												
	0	10	20	30	40	50	60	70	80	90	100	110	129
0.60	1.15	1.14	1.12	1.11	1.10	1.09	1.08	1.07	1.06	1.05	1.04	1.03	1.02
0.70	1.06	1.05	1.04	1.03	1.02	1.01	1.00	.991	.981	.972	.964	.955	.948
0.80	.995	.984	.974	.964	.954	.945	.935	.927	.918	.910	.901	.894	.886
0.90	.938	.928	.918	.909	.899	.891	.882	.874	.965	.858	.850	.842	.835
1.00	.890	.880	.871	.862	.853	.845	.837	.829	.821	.814	.806	.799	.792

Table 3
Volume of gas delivered—thousands of cubic feet per hour
1000 foot & 2000 foot lines

0.7 SP. GR. 60° F.

GAGE PRESSURE Up Stream P.S.I.	Down Stream P.S.I.	1000 FOOT LINE — NOMINAL PIPE SIZE* (Thousands of Cu. Ft. per Hour)					2000 FOOT LINE — NOMINAL PIPE SIZE* (Thousands of Cu. Ft. per Hour)				
		2	2½	3	4	6	2	2½	3	4	6
400	10	221.5	355.8	635.0	1311.	3910.	156.6	251.6	449.0	926.8	2765.
	50	219.2	352.1	628.4	1297.	3869.	155.0	248.9	444.3	917.2	2736.
	100	213.2	342.5	611.4	1262.	3764.	150.8	242.2	432.3	892.4	2662.
	350	105.6	169.7	302.9	625.2	1865.	74.70	120.0	214.2	442.1	1319.
350	10	194.7	312.7	558.1	1152.	3436.	137.6	221.1	394.6	814.6	2430.
	50	192.0	308.5	550.6	1136.	3390.	135.8	218.1	389.3	803.6	2397.
	100	185.2	297.5	531.0	1096.	3270.	131.0	210.4	375.5	775.1	2312.
	300	98.63	158.4	282.8	583.7	1741.	69.74	112.0	200.0	412.8	1231.
300	10	167.8	269.6	481.2	993.1	2962.	118.7	190.6	340.2	702.3	2095.
	50	164.8	264.6	472.4	975.0	2908.	116.5	187.1	334.0	689.4	2056.
	100	156.8	251.9	449.5	927.9	2768.	110.9	178.1	317.9	656.1	1957.
	250	91.08	146.3	261.1	539.0	1608.	64.40	103.5	184.7	381.1	1137.
250	10	141.0	226.4	404.1	834.2	2488.	99.67	160.1	285.8	589.8	1759.
	50	137.3	220.5	393.6	812.5	2424.	97.08	155.9	278.3	574.5	1714.
	100	127.6	205.0	365.9	755.3	2253.	90.24	145.0	258.7	534.0	1583.
	200	82.84	133.1	237.5	490.3	1462.	58.58	94.09	167.9	346.7	1034.
200	10	114.0	183.2	327.0	674.9	2013.	80.64	129.5	231.2	477.2	1424.
	50	109.5	175.9	313.9	647.9	1933.	77.42	124.4	222.0	458.2	1367.
	100	97.08	155.9	278.3	574.5	1714.	68.65	110.3	196.8	406.3	1212.
	150	73.68	118.4	211.3	436.0	1301.	52.10	83.69	149.4	308.3	919.7
150	10	87.04	139.8	249.6	515.1	1537.	61.55	98.87	176.5	364.2	1086.
	25	85.45	137.3	245.0	505.7	1508.	60.42	97.06	173.2	357.6	1067.
	50	80.98	130.1	232.2	479.3	1430.	57.26	91.98	154.2	338.9	1011.
	100	63.21	101.5	181.2	374.1	1116.	44.70	71.80	128.2	264.5	789.0

GAGE PRESSURE Up Stream P.S.I.	Down Stream P.S.I.	1000 FOOT LINE — NOMINAL PIPE SIZE* (Thousands of Cu. Ft. Per Hour)					2000 FOOT LINE — NOMINAL PIPE SIZE* (Thousands of Cu. Ft. per Hour)				
		2	2½	3	4	6	2	2½	3	4	6
100	10	59.84	96.12	171.6	354.1	1056.	42.31	67.97	121.3	250.4	746.9
	25	57.50	92.36	164.9	340.3	1015.	40.66	65.31	116.6	240.6	717.7
	50	50.62	81.32	145.1	299.6	893.6	35.79	57.50	102.6	211.8	631.9
	75	38.22	61.39	109.6	226.2	674.6	27.02	43.41	77.5	159.9	477.0
80	5	49.46	79.45	141.8	292.7	873.1	34.97	56.18	100.3	207.0	617.4
	10	48.82	78.43	140.0	288.9	861.8	34.52	55.46	98.98	204.3	609.4
	30	44.50	71.64	127.9	264.0	787.3	31.54	50.66	90.42	186.6	556.7
	55	34.26	55.03	98.23	202.8	604.8	24.23	38.91	69.46	143.4	427.6
60	5	38.45	61.77	110.3	227.6	678.8	27.19	43.68	77.96	160.9	480.0
	10	37.63	60.45	107.9	222.7	664.3	26.61	42.74	76.29	157.5	469.7
	20	35.32	56.73	101.3	209.0	623.5	24.97	40.12	71.61	147.8	440.9
	40	27.17	43.65	77.91	160.8	479.7	19.21	30.86	55.09	113.7	339.2
40	5	27.21	43.71	78.01	161.0	480.3	19.24	30.91	55.16	113.9	339.6
	10	26.03	41.81	74.63	154.1	459.5	18.41	29.57	52.77	108.9	324.9
	15	24.50	39.36	70.26	145.0	432.6	17.33	27.83	49.68	102.5	305.9
	25	20.08	32.26	57.58	118.8	354.5	14.20	22.81	40.71	84.04	250.7
30	5	21.38	34.35	61.30	126.5	377.4	15.12	24.29	43.35	89.48	266.9
	10	19.86	31.90	56.94	117.5	350.6	14.04	22.56	40.26	83.11	247.9
	15	17.81	28.61	51.07	105.4	314.5	12.60	20.23	36.11	74.54	222.4
	20	15.03	24.14	43.09	88.94	265.3	10.63	17.07	30.47	62.89	187.6
20	5	15.21	24.43	43.61	90.00	268.5	10.75	17.28	30.83	63.65	189.8
	10	12.98	20.85	37.22	76.83	229.2	9.180	14.75	26.32	54.33	162.1
10	5	7.923	12.73	22.72	46.89	138.9	5.603	9.000	16.06	33.16	98.90

MULTIPLIERS TO CONVERT 1000 FOOT LINE VALUES TO LENGTHS LISTED:

Length of Line—Feet	200	400	600	800	1000	1200	1400	1600	1800	2000	2200	2400	2600	2800	3000	3200	3400	3600	3800	4000	4200	4400	4600	4800	5000	5200	Feet—Length of Line
Multipliers	2.24	1.58	1.29	1.12	1.00	.91	.85	.79	.75	.71	.67	.65	.62	.60	.58	.56	.54	.53	.51	.50	.49	.48	.47	.46	.45	.44	Multipliers

For remarks regarding use of the Table refer to page 78. Courtesy Armco Steel Corporation.

Table 4
1 mile & 2½ mile lines

0.7 SP. GR. 60° F.

GAGE PRESSURE Up Stream P.S.I.	Down Stream P.S.I.	1 MILE LINE — NOMINAL PIPE SIZE* (Thousands of Cu. Ft. per Hour)					2½ MILE LINE — NOMINAL PIPE SIZE* (Thousands of Cu. Ft. per Hour)				
		2	2½	3	4	6	2	2½	3	4	6
400	10	96.39	154.8	276.4	570.4	1701.	60.96	97.92	174.8	360.8	1076.
	50	95.38	153.2	273.5	564.5	1684.	60.32	96.90	173.0	357.0	1065.
	100	92.80	149.1	266.1	549.2	1638.	58.69	94.28	168.3	347.3	1036.
	350	45.98	73.85	131.8	272.1	811.6	29.08	46.71	83.37	172.1	513.3
350	10	84.71	136.1	242.9	501.3	1495.	53.58	86.06	153.6	317.1	945.8
	50	83.57	134.2	239.6	494.6	1475.	52.85	84.90	151.5	312.8	933.0
	100	80.61	129.5	231.1	477.1	1423.	50.98	81.90	146.2	301.7	900.0
	300	42.92	68.95	123.1	254.0	757.7	27.15	43.61	77.84	160.7	479.2
300	10	73.03	117.3	209.4	432.2	1289.	46.19	74.20	132.4	273.4	815.4
	50	71.70	115.2	205.6	424.3	1266.	45.35	72.84	130.0	268.4	800.5
	100	68.23	109.6	195.6	403.8	1204.	43.15	69.32	123.7	255.4	761.8
	250	38.64	63.67	113.6	234.6	699.7	25.07	40.27	71.88	148.4	442.5
250	10	61.34	98.53	175.9	363.0	1083.	38.80	62.32	111.2	229.6	684.8
	50	59.75	95.98	171.3	353.6	1055.	37.79	60.70	108.3	223.6	667.1
	100	55.54	89.21	159.2	328.7	980.4	35.13	56.42	100.7	207.9	620.1
	200	36.05	57.91	103.4	213.4	636.4	22.80	36.63	65.37	134.9	402.5
200	10	49.63	79.72	142.3	293.7	876.1	31.39	50.42	90.00	185.8	554.1
	50	47.65	76.54	136.6	282.0	841.1	30.13	49.41	86.40	178.3	532.0
	100	42.25	67.87	121.1	250.0	745.8	26.72	42.92	76.61	158.1	471.7
	150	32.06	51.51	91.94	189.8	566.0	20.28	32.58	58.14	120.0	358.0
150	10	37.88	60.85	108.6	224.2	668.7	23.96	38.48	68.69	141.8	422.9
	25	37.19	59.74	106.6	220.1	656.5	23.52	37.78	67.44	139.2	415.2
	50	35.24	56.61	101.0	208.6	622.1	22.29	35.80	63.91	131.9	393.5
	100	27.51	44.19	78.87	162.8	485.6	17.40	27.95	49.88	103.0	307.1

GAGE PRESSURE Up Stream P.S.I.	Down Stream P.S.I.	1 MILE LINE — NOMINAL PIPE SIZE* (Thousands of Cu. Ft. per Hour)					2½ MILE LINE — NOMINAL PIPE SIZE* (Thousands of Cu. Ft. per Hour)				
		2	2½	3	4	6	2	2½	3	4	6
100	10	26.04	41.83	74.67	154.1	459.7	16.47	26.46	47.22	97.47	290.7
	25	25.02	40.20	71.75	148.1	441.7	15.83	25.42	45.38	93.66	279.4
	50	22.03	35.39	63.16	130.4	388.9	13.93	22.38	39.95	82.46	246.0
	75	16.63	26.72	47.68	98.4	293.6	10.52	16.90	30.16	62.25	185.7
80	5	21.53	34.58	61.72	127.4	380.0	13.61	21.87	39.03	80.57	240.3
	10	21.25	34.13	60.92	125.7	375.1	13.44	21.59	38.53	79.53	237.2
	30	19.41	31.18	55.65	114.9	342.6	12.28	19.72	35.20	72.65	216.7
	55	14.91	23.95	42.75	88.24	263.2	9.430	15.15	27.04	55.81	166.5
60	5	16.74	26.88	47.98	99.04	285.4	10.58	17.00	30.35	62.64	186.8
	10	16.38	26.31	46.95	96.91	289.1	10.36	16.64	29.70	61.29	182.8
	20	15.37	24.69	44.07	90.96	271.3	9.721	15.62	27.87	57.53	171.6
	40	11.83	19.00	33.91	69.98	208.8	7.479	12.01	21.44	44.26	132.0
40	5	11.84	19.02	33.95	70.08	209.0	7.489	12.03	21.47	44.32	132.2
	10	11.33	18.20	32.48	67.04	200.0	7.165	11.51	20.54	42.40	126.5
	15	10.66	17.13	30.58	63.11	188.3	6.745	10.83	19.34	39.92	119.1
	25	8.740	14.04	25.06	51.72	154.3	5.527	8.879	15.85	32.71	97.57
30	5	9.315	14.95	26.68	55.07	164.3	5.885	9.453	16.87	34.83	103.9
	10	8.643	13.88	24.78	51.15	152.6	5.466	8.781	15.67	32.35	96.49
	15	7.752	12.45	22.23	45.88	136.8	4.903	7.875	14.06	29.02	86.55
	20	6.540	10.51	18.75	38.71	115.5	4.137	6.645	11.86	24.48	73.02
20	5	6.619	10.63	18.98	39.17	116.8	4.186	6.724	12.00	24.77	73.90
	10	5.650	9.076	16.20	33.44	99.74	3.573	5.740	10.25	21.15	63.08
10	5	3.448	5.539	9.886	20.41	60.87	2.181	3.503	6.253	12.91	38.50

MULTIPLIERS TO CONVERT 1 MILE LINE VALUES TO LENGTHS LISTED:

Length of Line—Miles	1.2	1.4	1.6	1.8	2.0	2.2	2.4	2.6	2.8	3.0	3.2	3.4	3.6	3.8	4.0	4.2	4.4	4.6	4.8	Miles—Length of Line
Multipliers	.91	.85	.79	.75	.71	.67	.65	.62	.60	.58	.56	.54	.53	.51	.50	.49	.48	.47	.46	Multipliers

For remarks regarding use of the Table refer to page 78. Courtesy of Armco Steel Corporation.

0.7 SP. GR.
60° F.

Table 5
5 mile & 10 mile lines

0.7 SP. GR.
60° F.

GAGE PRESSURE		5 MILE LINE					10 MILE LINE				
		NOMINAL PIPE SIZE▼					NOMINAL PIPE SIZE▼				
Up Stream	Down Stream	2	2½	3	4	6	2	2½	3	4	6
P.S.I.	P.S.I.	Thousands of Cu. Ft. Per Hour					Thousands of Cu. Ft. Per Hour				
400	10	43.10	69.24	123.6	255.1	760.9	30.48	48.96	87.39	180.4	538.0
	50	42.66	68.52	122.3	252.4	753.0	30.16	48.45	86.48	178.5	532.4
	100	41.50	66.67	119.0	245.6	732.6	29.35	47.14	84.14	173.7	518.0
	350	20.56	33.03	58.95	121.7	363.0	14.54	23.35	41.69	86.04	256.7
350	10	37.88	60.86	108.6	224.2	668.8	26.79	43.03	76.81	158.5	472.9
	50	37.37	60.03	107.2	221.2	659.7	26.43	42.45	75.77	156.4	466.5
	100	36.05	57.91	103.4	213.3	636.4	25.49	40.95	73.09	150.9	450.0
	300	19.20	30.84	55.04	113.6	338.9	13.57	21.80	38.92	80.33	239.6
300	10	32.66	52.46	93.64	193.3	576.6	23.09	37.10	66.22	136.7	407.7
	50	32.07	51.51	91.94	189.8	566.0	22.67	36.42	65.01	134.2	400.2
	100	30.51	49.02	87.49	180.6	538.7	21.58	34.66	61.86	127.7	380.9
	250	17.73	28.47	50.82	104.9	312.9	12.53	20.13	35.94	74.18	221.3
250	10	27.43	44.07	78.65	162.3	484.3	19.40	31.16	55.62	114.8	342.4
	50	26.72	42.92	76.61	158.1	471.7	18.89	30.35	54.17	111.8	333.5
	100	24.84	39.90	71.21	147.0	438.5	17.56	28.21	50.36	103.9	310.0
	200	16.12	25.90	46.23	95.41	284.6	11.40	18.31	32.69	67.47	201.2
200	10	22.20	35.65	63.64	131.4	391.8	15.69	25.21	45.00	92.88	277.0
	50	21.31	34.23	61.10	126.1	376.2	15.07	24.20	43.20	89.17	266.0
	100	18.89	30.35	54.17	111.8	333.5	13.36	21.46	38.31	79.07	235.8
	150	14.34	23.03	41.11	84.86	253.1	10.14	16.29	29.07	60.01	179.0
150	10	16.94	27.21	48.57	100.3	299.0	11.98	19.24	34.35	70.89	211.5
	25	16.63	26.72	47.68	98.42	293.6	11.76	18.89	33.72	69.60	207.6
	50	15.76	25.32	45.19	93.27	278.2	11.14	17.90	31.95	65.96	196.7
	100	12.30	19.76	35.27	72.81	217.2	8.699	13.97	24.94	51.48	153.6

GAGE PRESSURE		5 MILE LINE					10 MILE LINE				
		NOMINAL PIPE SIZE▼					NOMINAL PIPE SIZE▼				
Up Stream	Down Stream	2	2½	3	4	6	2	2½	3	4	6
P.S.I.	P.S.I.	Thousands of Cu. Ft. Per Hour					Thousands of Cu. Ft. Per Hour				
100	10	11.65	18.71	33.39	68.92	205.6	8.235	13.23	23.61	48.74	145.4
	25	11.19	17.98	32.09	66.23	197.5	7.913	12.71	22.69	46.83	139.7
	50	9.852	15.83	28.25	55.31	173.9	6.966	11.19	19.97	41.23	123.0
	75	7.438	11.95	21.32	44.02	131.3	5.259	8.448	15.08	31.12	92.84
80	5	9.626	15.46	27.60	56.97	169.9	6.807	10.93	19.52	40.28	120.2
	10	9.502	15.26	27.24	56.23	167.7	6.719	10.79	19.26	39.76	118.6
	30	8.680	13.94	24.89	51.37	153.2	6.138	9.860	17.60	36.32	108.4
	55	6.668	10.71	19.12	39.46	117.7	4.715	7.574	13.52	27.90	83.23
60	5	7.484	12.02	21.46	44.29	132.1	5.292	8.501	15.17	31.32	93.42
	10	7.324	11.76	21.00	43.34	129.3	5.179	8.319	14.85	30.65	91.42
	20	6.873	11.04	19.71	40.68	121.3	4.861	7.808	13.94	28.77	85.80
	40	5.288	8.495	15.16	31.30	93.36	3.740	6.007	10.72	22.13	66.01
40	5	5.296	8.507	15.18	31.34	93.48	3.745	6.015	10.74	22.16	66.10
	10	5.066	8.138	14.53	29.98	89.43	3.582	5.755	10.27	21.20	63.24
	15	4.769	7.661	13.67	28.22	84.19	3.372	5.417	9.669	19.96	59.53
	25	3.908	6.278	11.21	23.13	69.00	2.764	4.439	7.924	16.36	48.79
30	5	4.161	6.685	11.93	24.63	73.46	2.943	4.727	8.437	17.41	51.94
	10	3.865	6.209	11.08	22.87	68.23	2.733	4.390	7.836	16.17	48.25
	15	3.467	5.569	9.940	20.52	61.20	2.451	3.938	7.029	14.51	43.27
	20	2.925	4.698	8.386	17.31	51.63	2.068	3.322	5.930	12.24	36.51
20	5	2.960	4.755	8.487	17.52	52.25	2.093	3.362	6.001	12.39	36.95
	10	2.527	4.059	7.244	14.95	44.60	1.787	2.870	5.123	10.57	31.54
10	5	1.542	2.477	4.421	9.126	27.22	1.090	1.752	3.126	6.453	19.25

MULTIPLIERS TO CONVERT 10 MILE LINE VALUES TO LENGTHS LISTED.

Length of Line—Miles	5.2	5.4	5.6	5.8	6.0	6.2	6.4	6.6	6.8	7.0	7.2	7.4	7.6	7.8	8.0	8.2	8.4	8.6	8.8	9.0	9.2	9.4	9.6	9.8	Miles—Length of Line
Multipliers	1.39	1.36	1.34	1.31	1.29	1.27	1.25	1.23	1.21	1.20	1.18	1.16	1.15	1.13	1.12	1.10	1.09	1.08	1.07	1.05	1.04	1.03	1.02	1.01	Multipliers

For remarks regarding use of the Table refer to page 78.   Courtesy of Armco Steel Corporation.

specific gravity gas and 60°F flowing temperature for 200 and 50 psi up and downstream pressure, respectively, is 109.5 MCF per hour. Correction factor for the pipe length, from the same table is .53. Correction factor, from Table 2 for specific gravity and flowing temperature is .991. Therefore:

Delivery of gas = 109.5 × .53 × .991 = 57,512 cubic ft/hr

Multi-phase flow

As stated at the beginning of this chapter, the problems of the simultaneous flow of oil and gas or of oil, gas, and water through one pipe have become increasingly important in oil field production operations. The problems are complex because the properties of two or more fluids must be considered and because of the different patterns of fluid flows, depending upon the flow conditions. These patterns, in any given line, may change as the flow conditions change, and they may be coexistent at different points of the line.[3,5,7]

Different investigators recognize different flow patterns and use different nomenclature. Those generally recognized are:

1. Bubble flow—bubbles of gas flow along the upper part of the pipe with about the same velocity as the liquid.

2. Plug flow—the bubbles of gas coalesce into large bubbles and fill out the large part of the cross-sectional area of the pipe.

3. Laminar flow—the gas-liquid interface is relatively smooth with gas flowing in the upper part of the pipe.

4. Wavy flow—same as above except that waves are formed on the surface of the liquid.

5. Slug flow—the tops of some waves reach the top of the pipe. These slugs move with high velocity.

6. Annular flow—the liquid flows along the walls of the pipe and the gas moves through the center with high velocity.

7. Spray flow—the liquid is dispersed in the gas.

The above description of flow patterns has been given to emphasize the first reason why the pressure drop must be

Figure 1. Prediction of solubility from saturation pressure and gravity of crude oil. (Reference 12.)

higher for the multi-phase than the single-phase flow. In the latter the pressure drop is primarily the result of friction. In the multi-phase flow, in addition to friction, the energy losses are due to gas accelerating the fluid, due to compression of gas and formation of waves and slugs. These losses are different for different patterns, with patterns changing as a result of changes in flow conditions. The second reason is the fact that with two phases present in the pipe, less cross-sectional area is available for each of the phases. As previously discussed, the pressure drop is inversely proportional to the fifth power of pipe diameter.

The literature on multi-phase flow is quite extensive. Reference 7 gives 72 references on the subject.

Two-Phase Horizontal Flow

Over twenty correlations have been developed by different investigators for predicting the pressure drop in the two-phase flow. Under the project, sponsored jointly by the American Petroleum Institute and American Gas Association at the University of Houston, five of these correlations were tested by comparing them with 2,620 experimental measurements culled from more than 15,000 available.

The statistical part of the report on the project concludes that of the five correlations tested, the Lockhart-Martinelli correlation shows the best agreement with the experimental data, particularly for pipe sizes commonly used in oil field production operations.[8] The method may be summarized as follows:[3,4]

1. Determine the single-phase pressure drops for the liquid and the gas phase as if each one of these were flowing alone through the pipe. This is done by use of previously given equations.
2. Determine the dimensionless parameter

$$X = \sqrt{\frac{\sqrt{\Delta P_L}}{\Delta P_G}}$$

where ΔP_L and ΔP_G are single-phase pressure drops for liquid and gas respectively.
3. The method recognizes four regimes of the two-phase flow as follows:

Flow Regimes

Gas	Liquid
Turbulent	Laminar
Turbulent	Turbulent
Laminar	Laminar
Laminar	Turbulent

The type of flow of each of the phases is determined by its Reynolds number (Figure 2).
4. Determine factor Φ from a chart as a function of X for the appropriate flow regime.
5. The two-phase pressure drop is then:

$$\Delta P_{TP} = \Phi^2 P_L \text{ of G}$$

The single-phase pressure drop of either gas or liquid phase can be used. The results are the same.

For the two-phase flow problems in oil field production operations, certain simplifications can be used if only approximate estimates are desired:

1. Information which would permit determining the volume of oil and gas under flow conditions from equilibrium flash calculation is usually unavailable. The following procedure may be used:

    Gas volume is determined by use of the chart in Figure 1 which shows gas in solution for oil of differ-

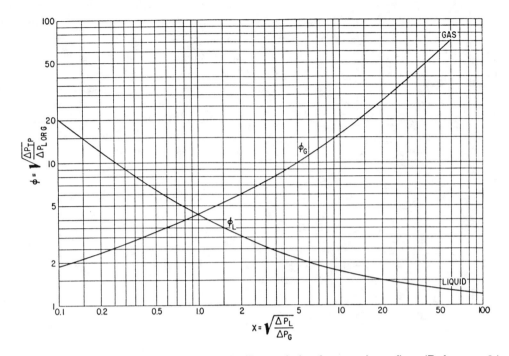

Figure 2. Parameter Φ for Lockhart-Martinelli correlation for two-phase flow. (Reference 9.)

ent gravities and different saturation pressures (in this case the flowline pressures). If the gas-oil ratio is known for one pressure the ratio for another pressure can be determined from the chart as follows:

Assume gas oil ratio of 600 cubic ft/bbl at 900 psi for a 30°API oil. Determine the gas oil ratio at 1,300 psi. From the chart, at 900 psi, the gas in solution is 205 cubic ft/bbl and at 1,300 psi is 300 cubic ft/bbl, an increase of 95 cubic ft/bbl. Therefore, at 1,300 psi the gas oil ratio will be 600 − 95 = 505 cubic ft/bbl.

Increase in volume of oil because of the increase of the dissolved gas can be obtained from charts for calculation of formation volume of bubble point liquids. However, for approximate estimates in problems involving field flowlines, this step frequently can be omitted without seriously affecting the validity of results.

2. In a majority of cases here under consideration, the flow regime is turbulent-turbulent. Therefore, the chart in Figure 2 shows the factor for only this regime.

The procedure is illustrated by the following examples.

Example 1. Determine two-phase pressure drops for different flow ratios and different flow pressures under the following conditions:

Oil	31°API
Viscosity	5 cp at 100°F
Separator pressure	900 psig
Gas oil ratio	670 cubic ft/bbl
Flow line	2 in.

Wellhead pressure of 1,300 psig was assumed and calculations were made for flow rates of 100 and 50 bbl per day. The single phase pressure drops were calculated for the gas and liquid phases, each of them flowing alone in the line, according to the Fanning equations.

Calculation of pressure drop for liquid phase:

$$H_f = \frac{f \, L \, (bbl/day)^2 \, (lb/gal)}{181,916 \, D^5}$$

where   f = friction factor
   L = length of pipe, ft
   D = inside diameter of pipe, in.

Calculation of pressure drop for gas phase: see pp. 280–281.

For the two flow rates and the two pressures, these single-phase pressure drops were calculated to be as follows:

Flow Rate bbl/day	100		50	
Pressure, psia	1,300	900	1,300	900
ΔP_G psi/100 ft	7.9×10^{-4}	2.0×10^{-3}	2.4×10^{-4}	7.7×10^{-4}
ΔP_L psi/100 ft	$2. \times 10^{-2}$	2.7×10^{-2}	1.3×10^{-2}	1.3×10^{-2}
Then $X = \sqrt{\dfrac{\Delta P_L}{\Delta P_G}}$	5.9	3.8	7.4	4.2

Figure 3. Determining two-phase flow pressure drops.

From Figure 2 ϕ	2.0	2.4	1.9	2.3
and the two-phase pressure drop				
$\Delta P_{TP} = \phi_L \times \Delta P_L$ 0.12	0.16	0.048	0.069	

With two pressure drops determined for two flow line pressures for each of the flow rates, lines can be drawn showing pressure drop for different flow line pressures for these two rates (Figure 3). From these two lines relationships can be established for pressure drops for other rates and flow line pressures. The procedure is as follows:

Start with, for instance, 800 psia line. For the rate of 50 bpd at 800 psia flow line pressure, the pressure drop from the chart is .074 psia/100 ft. From the 50 bpd rate on the scale in the upper part of the chart, go downward to the line representing .074 psia pressure drop. This gives one point of the 800 psia line.

For the rate of 100 bpd the chart indicates that at 800 psia flowline pressure, the pressure drop is .175 psia/100 ft. From the 100 bpd rate in the upper part of the chart, go downward to the .175 psia/100 ft line. This gives the second point of the 800 psia line. The line can now be drawn through these two points. Lines for other pressures can be constructed in a similar way.

The chart is used as follows: For conditions represented by this particular chart, what is the two-phase pressure drop for the rate of 60 bpd and average flow line pressure of 1,000 psia?

Start at the rate of 60 bpd on the upper scale of the chart. Go downward to the intersection with the 1,000 psia pressure, then to the left, and read the two-phase pressure drop as .075 psia/100 ft.

Example 2. If pipe diameter is to be determined for the desired pressure drop and the known length of line, flow rate, and the properties of fluids, then the problem cannot be solved directly. The procedure is as follows:

An estimate is made of the diameter of the two-phase line. The best approach is to determine, for the conditions, the diameter of the line for each of the phases flowing separately. In a majority of cases, the sum of these two diameters will be a good estimate of the diameter for the two-phase line. Calculations are then made for the two-phase pressure drop using the assumed diameter. If this drop is reasonably close to the desired one, the estimate is correct. If not, a new estimate must be made and the calculations repeated.

Inclined two-phase flow

In hilly terrain additional pressure drop can be expected in case of the two-phase flow. Little published information is available on the subject. Ovid Baker suggested the following empirical formula:[10]

$$\Delta P_{TPH} = \Delta P_{TP} + \frac{1.61\ NH\ \rho_L}{144\ V_g^{0.7}}$$

where ΔP_{TPH} = pressure drop, corrected for hilly terrain in psi
ΔP_{TP} = calculated two-phase pressure drop for horizontal pipe, psi
N = number of hills
H = average height of hills, ft
ρ_L = density of liquid phase, lb/ft^3
V_g = velocity of gas, ft/sec (based on 100% pipe area)

Horizontal three-phase flow

Very little is known about the effect on the pressure drop of addition to the gas-oil system of a third phase, an immiscible liquid, water. Formation of an emulsion results in increased viscosities. Formulas are available for approximate determination of viscosity of emulsion, if viscosity of oil and percent of water content of oil is known. The point is, however, that it is not known what portion of the water is flowing in emulsified form.

From one published reference[11] and from some unpublished data, the following conclusions appear to be well founded:

1. Within the range of water content of less than 10% or more than 90%, the flow mechanism appears to approach that of the two-phase flow.
2. Within the range of water content of from 70 to 90%, the three-phase pressure drop is considerably higher than for the two-phase flow.

For solution of three-phase flow problems of oil field flow lines, the following approach has been used. The oil and water are considered as one phase and the gas as the other phase. Previously given calculation of a two-phase drop is used. The viscosity of the oil-water mixture is determined as follows:

$$\mu_L = \frac{V_o \times \mu_o + V_w \times \mu_w}{V_o + V_w}$$

where μ_L, μ_o, and μ_w are viscosities of mixture of oil and of water respectively, and V_o and V_w represent the corresponding volumes of oil and water.

No test data are available to determine the accuracy of this approach. Until more information is developed on the subject the method may be considered for approximate estimates.

References

1. *Cameron Hydraulic Data*, 13th ed., Ingersoll-Rand Company, New York, 1962.
2. *Flow of Fluids Through Valves, Fittings and Pipe*, Technical Paper No. 410, Engineering Division, Crane Company, Chicago, 1957.
3. Streeter, Victor L., *Handbook of Fluid Dynamics*, McGraw-Hill Book Company, Inc., New York, 1961.
4. Prof. W. E. Durand's Discussions in *Handbook of the Petroleum Industry*, David T. Day; John Wiley & Sons, Inc.
5. Baker, Ovid, "Simultaneous Flow of Oil and Gas," *The Oil and Gas Journal*, July 26, 1954.
6. Campbell, Dr. John M., "Elements of Field Processing," *The Oil and Gas Journal*, December 9, 1957.
7. Campbell, Dr. John M., "Problems of Multi-phase Pipe Line Flow," in "Flow Calculations in Pipelining," *The Oil and Gas Journal*, November 4, 1959.
8. Duckler, A. E., Wicks, III, Moye, and Cleveland, R. G., "Frictional Pressure Drop in Two-Phase Flow," *AIChE Journal*, January, 1964.
9. Lockhart, R. W., and Martinelli, R. C., "Proposed Correlation of Data for Isothermal Two-Phase, Two-Component Flow in Pipes," *Chemical Engineering Progress* 45: 39–48, (1949).
10. Ovid Baker's discussion of article "How Uphill and Downhill Flow Affect Pressure Drop in Two-Phase Pipe Lines," by Brigham, W. E., Holstein, E. D., and Huntington, R. L.; *The Oil and Gas Journal*, November 11, 1957.
11. Sobocinski, D. P., and Huntington, R. D., "Concurrent Flow of Air, Gas, Oil and Water in a Horizontal Line," *Trans. ASME* 80: 1,252, 1958.
12. Beal, Carlton, "The Viscosity of Air, Water, Natural Gas, Crude Oil and Its Associated Gases at Oil Field Temperatures and Pressures," *Petroleum Development and Technology*, AIME, 1946.
13. Switzer, F. G., *Pump Fax*, Gould Pumps, Inc.
14. *Pump Talk*, The Engineering Department, Dean Brothers Pumps, Inc., Indianapolis, Ind.
15. Switzer, F. G., *The Centrifugal Pump*, Goulds Pumps, Inc.
16. Zalis, A. A., "Don't Be Confused by Rotary Pump Performance Curves," *Hydrocarbon Processing and Petroleum Refiner*, Gulf Publishing Company, September, 1961.

Nomograph for calculating Reynolds number for compressible flow friction factor for clean steel and wrought iron pipe

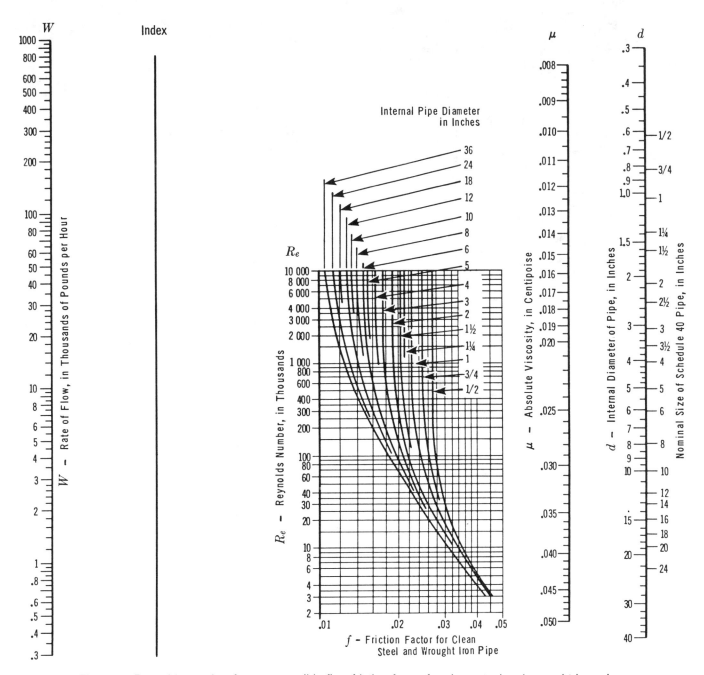

Figure 1. Reynolds number for compressible flow friction factor for clean steel and wrought iron pipe.

Background

The nomograph (Figure 1) permits calculation of the Reynolds number for compressible flow and the corresponding friction factor. It is based on the equation:

$$Re = 6.31 \frac{W}{d\mu} = 0.482 \frac{q_s \times Sg}{d\mu}$$

where W = flow rate, lb/hr
Sg = specific gravity of gas relative to air

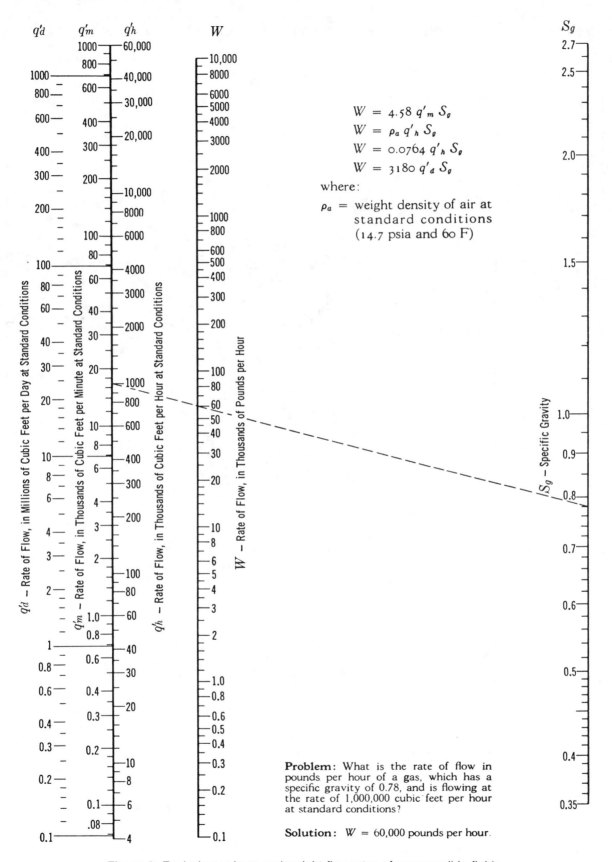

$$W = 4.58 \, q'_m \, S_g$$
$$W = \rho_a \, q'_h \, S_g$$
$$W = 0.0764 \, q'_h \, S_g$$
$$W = 3180 \, q'_d \, S_g$$

where:

ρ_a = weight density of air at standard conditions (14.7 psia and 60 F)

Problem: What is the rate of flow in pounds per hour of a gas, which has a specific gravity of 0.78, and is flowing at the rate of 1,000,000 cubic feet per hour at standard conditions?

Solution: $W = 60,000$ pounds per hour.

Figure 2. Equivalent volume and weight flow rates of compressible fluids.

q_s = volumetric flow rate, cubic ft/sec (at 14.7 psia and 60°F)

μ = fluid viscosity, centipoise

If the flow rate is given in lb/hr, the nomograph can be used directly without resorting to Figure 2 to obtain q_s. Figure 2 converts volume flow to weight flow rates if the specific gravity of the fluid is known.

Example. Natural gas at 250 psig and 60°F with a specific gravity of 0.75, flows through an 8-in. schedule 40 clean steel pipe at a rate of 1,200,000 cubic ft/hr at standard conditions.

Find the Reynolds number, and the friction factor.

At S_g = 0.75 obtain q = 69,000 from Figure 1, μ = 0.011.

Connect	With	Mark or Read
W = 69,000	μ = 0.011	Index
Index	d = 8 in.	Re = 5,000,000
Re = 5,000,000	horizontally to 8 in.	f = 0.014

Source

Flow of Fluids Through Valves, Fittings, and Pipes, Technical Paper No. 410, 3–19, Crane Company, Chicago, Illinois (1957).

12: Liquids—General

Determining the viscosity of crude... 293
Chart gives API gravity of blends quickly 294
Liquid gravity and density conversion chart 295
Nomograph for calculating viscosities of liquid hydrocarbons
    at high pressure 295
Calculate viscosity of a blend... 297
Calculate gravity of a blend.. 297
Convert viscosity units 297
Convert specific gravity to API gravity... 297
Calculate bulk modulus 297
Viscosities of hydrocarbon liquids 298
Nomograph for calculating viscosity of slurries................................ 299
Nomograph for calculating velocity of liquids in pipes 301
Nomograph for calculating velocity of compressible fluids in pipes ... 301
Nomograph for calculating velocity of liquids in pipes 302
Derivation of basic ultrasonic flow equations 304
How fast does oil move in a pipe line?... 306
Estimate the volume of a pipe line per linear foot using the inside
    diameter 306
What is the linefill of a given pipe in barrels per mile?...................... 306
Estimate leakage amount through small holes in a pipe line.............. 307
Table gives velocity heads for various pipe diameters and different
    rates of discharge 308

Determining the viscosity of crude

If the viscosity of a gas-saturated crude oil at the saturation (bubble-point) pressure is known, using this nomograph you can quickly estimate the viscosities at higher pressures.

Example. Find the viscosity at 4,200 psia for a crude oil when its viscosity is 30 cp at the saturation pressure of 1,200 psia. Notice that 4,200 psia is 3,000 psi above the saturation pressure. Connecting 3,000 on the pressure difference scale (left) with 30 on the curved scale for viscosity at the bubble-point pressure, the intersection with the scale on the right at 48 cp is the desired value.

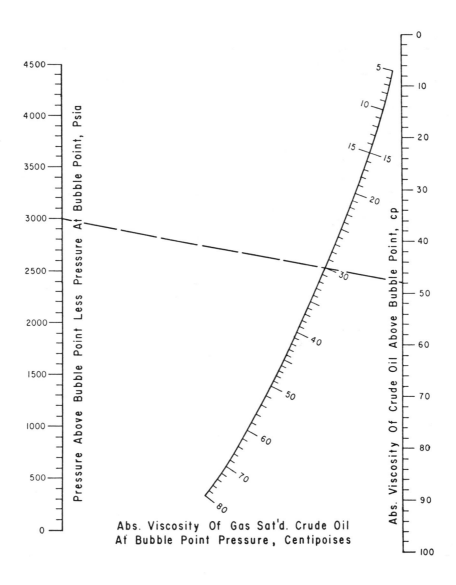

Chart gives API gravity of blends quickly

API gravity of a crude oil blend may be readily estimated from the gravity of the components and their percentage composition. It is a simple procedure to use these curves to find the resulting API gravity of the blend.

Example. If you blend a 14°API pitch (60%) with a 36°API cutting stock (40%), the resulting fuel oil has an API gravity of 22° as read from the nomograph. Calculated result from gravity tables would be 22.1°API.

Results found using this nomograph checked out within 1°API over the range of gravities and percentage of components (1) and (2) in the nomograph below. Estimates from the nomograph are based on the assumption that volumes of blends are additive and that no light components flash off in blending.

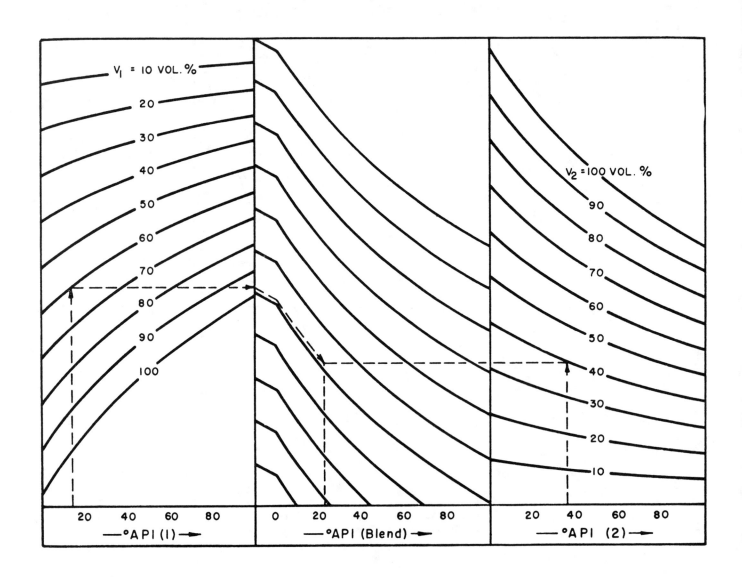

Liquid gravity and density conversion chart

This line chart provides an easy method for converting units of liquid gravity and density. Draw a horizontal line perpendicular to the scale line through a known value and read the equivalent value on any other scale.

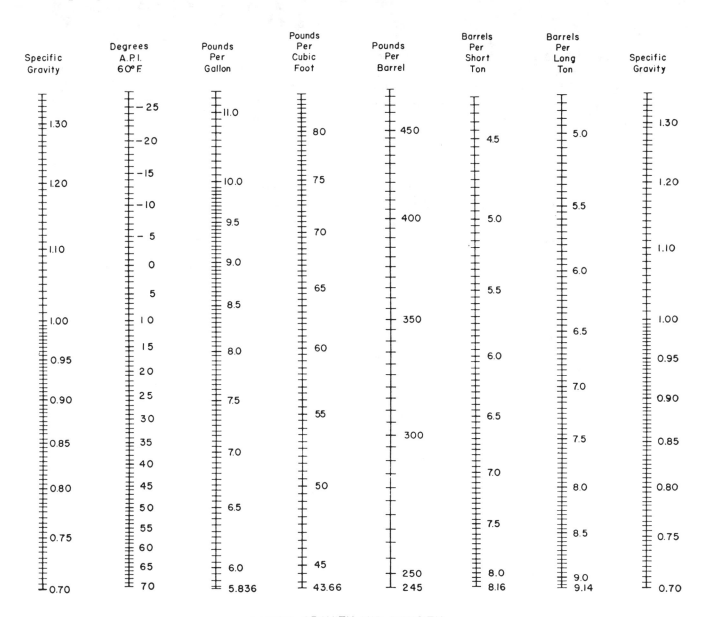

LIQUID GRAVITY AND DENSITY

Nomograph for calculating viscosities of liquid hydrocarbons at high pressure

Lockhart and Lenoir developed a graphical correlation showing the effect of pressure on viscosity of liquid hydrocarbons. This correlation is shown in Figure 1 which is based primarily on data of Hersey and Hopkins,[2] including pure hydrocarbons, lubricating oils, bright stocks, and distillates. Data from Reference 1 have also been included.

To use the nomograph, the characterization factor of Watson, K_w, and the viscosity of the liquid at atmospheric pressure are required.

The accuracy of the correlation decreases as pressure increases.

Example. What is the viscosity of an oil at 5,400 psia, if its characterization factor is 11.8 and its viscosity at atmospheric conditions is 90 centipoises?

Enter 5,400 psia in the pressure scale to the viscosity line of 90 and proceed horizontally to middle reference scale. Follow the curve lines to intersect the vertical line drawn at $K_w = 11.8$ and read the ratio of viscosity on the extreme left scale at 2.6. The viscosity of the oil is $(2.6)(90) = 234$ centipoises.

Source

Lockhart, F. J. and Lenoir, J. M., *Petroleum Refiner*, 40, No. 3, 209 (1961).

References

1. Griest, and others, *J. Chem. Physics*, 29, 711 (1958).
2. Hersey, M. D., and Hopkins, R. F., *Viscosity of Lubricants Under Pressure*, ASME, 1954.

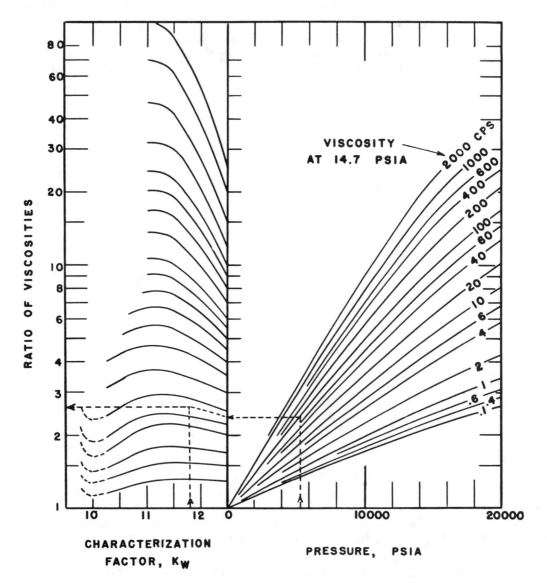

Figure 1. Isothermal effect of pressure on viscosities of liquid hydrocarbons at low reduced temperatures. (Reproduced by permission *Petroleum Refiner,* Vol. 40, No. 3 page 209, copyright 1961, Gulf Publishing Co., Houston.)

Calculate viscosity of a blend

Viscosity is not an additive function of percentage composition. The viscosity of a blend of products can be calculated with the following equation:

$$Vb = \left[\frac{G_1 + G_2 + G_3 \ldots\ldots}{\dfrac{G_1}{\sqrt{V_1}} + \dfrac{G_2}{\sqrt{V_2}} + \dfrac{G_3}{\sqrt{V_3}} \ldots\ldots} \right]^2$$

where   Vb = viscosity of blend
       G = volume (gals., gpm, b/d, etc.)
       V = viscosity in SUS

Example:

$G_1 = 10,000$ bbls $V_1 = 45$ SUS
$G_2 = 20,000$ bbls $V_2 = 50$ SUS
$G_3 = 5,000$ bbls $V_3 = 37$ SUS

$$Vb = \left[\frac{10,000 + 20,000 + 5,000}{\dfrac{10,000}{\sqrt{45}} + \dfrac{20,000}{\sqrt{50}} + \dfrac{5,000}{\sqrt{37}}} \right]^2$$

Vb = 46.4 SUS

Calculate gravity of a blend

The gravity of a blend of liquids may be calculated directly by a ratio of the gravity of each component in the blend.

Example:

Assume a blend consisting of the following:

Volume, bbls	sp. gr.	Partial sp. gr.
10,000	0.8551	0.1900
15,000	0.8337	0.2779
20,000	0.8481	0.3769
Total 45,000		0.8448

Therefore the specific gravity of the blend is 0.8448

Convert viscosity units

Convert SUS to centistokes (cS):

$$v_{cs} = 0.22\, v_{sus} \frac{-135}{v_{sus}} \quad \text{where } v_{sus} > 100\,*$$

$$v_{cs} = 0.226\, v_{sus} \frac{-195}{v_{sus}} \quad \text{where } v_{sus} \le 100\,*$$

Convert centistokes (cS) to centipoise (cP):

$$cP = v_{cs} \times \text{density, gm/cm}^3$$

$$\text{SUS} = \text{cs} \times 4.6673, \text{ if cs} > 500, \text{SUS} = \text{cs} \times 4.6347$$

Convert specific gravity to API gravity

$$\text{API} = \frac{141.5}{s_{60}} - 131.5 \quad \text{where } s = \text{specific gravity @ 60°F}$$

$$s_{60} = \frac{414.5}{\text{API}_{60} + 131.5}$$

Calculate bulk modulus

The bulk modulus, K, of a liquid is the reciprocal of its compressibility. The bulk modulus for water is K = 300,000 psi. The compressibility factors for most liquids can be found in Chapters 11.2.1 and 11.2.2 of the *API Manual of Petroleum Measurement Standards*. The following relation, known as the ARCO formula, can be used to calculate the bulk modulus for crude oil:

$$K = 1.286(10^6) + 13.55p - 4.122(10^4)T^{1/2} - 4.53(10^3)\text{API} - 10.59\text{API}^2 + 3.228T\,\text{API}*$$

where   K = adiabatic bulk moduls, psi
           p = average pressure in the line, psig
       API = API gravity
          T = absolute temperature, °R

*Reprinted with permission: Fluid Flow Consultants, Tulsa, OK.

Reprinted with Permission: Gas Processors Suppliers Association, Tenth Edition Engineering Data Book

Viscosities of Hydrocarbon Liquids

Source:
1. A.P.I. Research Project 44
2. Swift et al, AIChE Journal 6, No. 3, Sept. 1960
3. Parisot & Johnson, J. Chem. Eng. Data, 6, No. 2, Apr. 1960
4. Adapted from Heat Transfer Tables, American Locomotive Co., 1952

Temperature, °F

Viscosity, centipoise

Olefins are approximately 15% less viscous than the corresponding normal paraffin hydrocarbons

– – – Estimated

28° API Gas oil
nC_{20} – nC_{19}
nC_{18} – nC_{17}
nC_{16} – nC_{15}
nC_{14} – nC_{13}
nC_{12} – nC_{11}
nC_{10} – nC_9
n-Octane
n-Heptane
n-Hexane
n-Pentane
n-Butane
Propane
Ethane
Methane

40.2° API Kerosene
48° API (146MW)
57° API Gasoline
70° API Gasoline
75.9° API Gasoline
92° API Nat Gasoline
100° API Nat Gasoline

Bulk Modulus - psi

Based on Chapter 11.2.1 - API MPMS
Compressibility Factors for Hydrocarbons
May 21, 1992

Nomograph for calculating viscosity of slurries

The nomogram is based on the Hatschek equation for estimating the viscosity of slurries in an aqueous suspension.

$$\mu = \frac{\mu_w}{1 - x^{0.333}}$$

where μ = viscosity of slurry, centipoises
  μ_w = viscosity of water at the temperature of the slurry, centipoises
  x = volume fraction of dry solids in the slurry

The nomogram (Figure 1) shows a scale calibrated in terms of the temperature of the water instead of its viscosity. Thus, the actual water viscosity value is not needed.

Example. What is the viscosity of a slurry having a volume fraction of solids of 0.06 at a temperature of 39°C?
Connect T = 39°C with x = 0.06 and read μ_w = 1.1 centipoises on the middle scale.

Source

Davis, D. S., *Brit. Chem. Eng.*, 4, 9, 478 (1959).

Figure 1. Viscosity of slurries. (Reproduced by permission, *British Chemical Engineering*, Vol. 4, page 478, copyright 1959.)

Nomograph for calculating velocity of liquids in pipes

The mean velocity of any liquid flowing in a pipe of circular cross section can be calculated from the following formula:

$$v = 183.3 \frac{q}{d^2} = 0.408 \frac{Q}{d^2} = 0.0509 \frac{W}{d^2\rho}$$

where   v = average fluid velocity, ft/sec
d = inside diameter of pipe, in.
q = rate of flow, cubic ft/sec
Q = rate of flow, gal/min
W = rate of flow, lb/hr
ρ = fluid density, lb/cubic ft

The Nomograph for Calculating Velocity of Liquids in Pipes (see Figure 1, p. 302) can be used to calculate the liquid velocity when the rate of flow is in cubic ft, gallons or thousands of lb. Conversely, knowing the flow rate and velocity, the pipe diameter may be calculated.

Example. What is the velocity of fuel oil at 60°F flowing through a 2-in. schedule 40 pipe at a rate of 45,000 lb/hr. The oil density is 56.02 lb/gal.

Connect	With	Mark or Read
W = 45	ρ = 56.02	Q = 100 cubic ft/sec
Q = 100	d = 2-in. schedule 40	v = 10 ft/sec

Source

Flow of Fluids Through Valves, Fittings, and Pipe, Technical Paper No. 410, 3–7, Crane Company, Chicago, Illinois (1957).

Nomograph for calculating velocity of compressible fluids in pipe

The mean velocity of a compressible fluid in a pipe can be computed by the following formula, which is obtained after dividing the rate of flow in appropriate units by the cross-sectional area of the pipe:

$$V = \frac{3.06\, W\overline{V}}{d^2} = \frac{3.06\, W}{d^2\rho}$$

where   V = mean velocity of flow, ft/min
W = rate of flow, in lb/hr
$\overline{V}$ = specific volume of fluid, cubic ft/lb
d = internal diameter of pipe, in.
ρ = density of fluid, lb/cubic ft

Example. Steam at 600 psig and 850°F is to flow through a schedule 80 pipe at a rate of 30,000 lb/hr. Find the pipe size if the velocity is to be limited to 8,000 ft/min.

Connect	With	Mark or Read
850°F	600 psig line (vertically)	600 psig
600 psig	Horizontally to	$\overline{V}$ = 1.21
$\overline{V}$ = 1.21	W = 30,000	Index
Index	V = 8,000	d = 3.7, use 4-in. schedule 80 pipe

If a 4-in. schedule 80 pipe is used, the actual velocity is found by connecting the Index with 4-in. schedule 80 to get V = 7,600 ft/min.

Note: If a different fluid is involved, the value of the density (or specific volume) needed to make the calculation can be obtained from Section 9 of this book, under "Density and specific volume of gases and vapors," p. 226.

Source

Flow of Fluids Through Valves, Fittings, and Pipes, Technical Paper No. 410, 3–16, Crane Company, Chicago, Illinois (1957).

Nomograph for calculating velocity of liquids in pipes

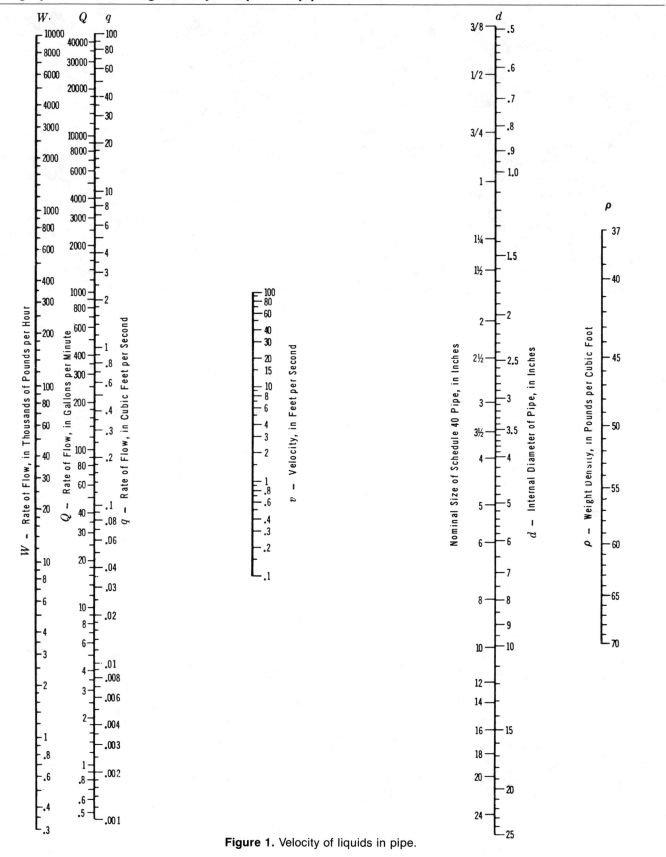

Figure 1. Velocity of liquids in pipe.

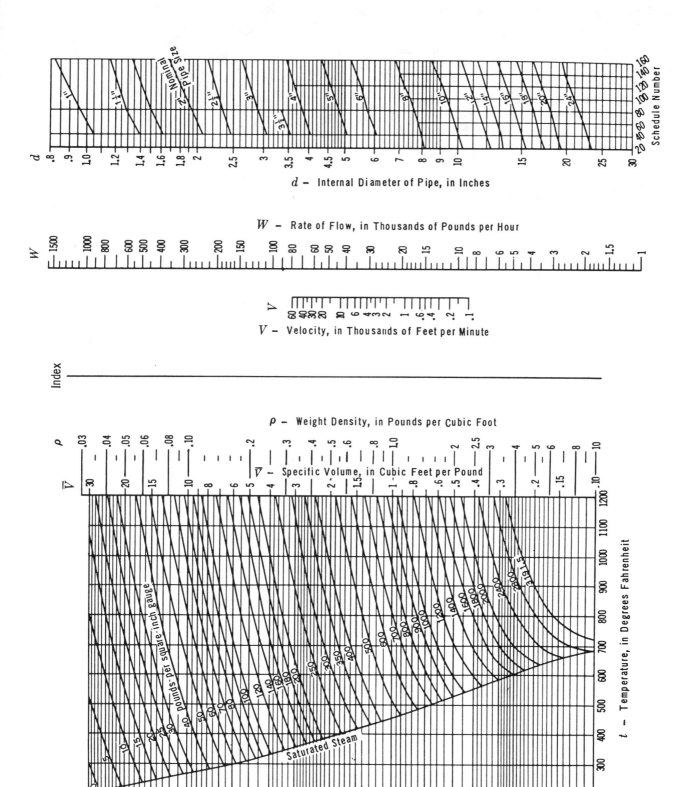

Figure 2. Velocity of Compressible Fluids in Pipes.

Derivation of basic ultrasonic flow equations

The actual measurements performed by ultrasonic flowmeters are of flow velocity, with all other variable effects due to temperature, pressure, density, viscosity, etc., being cancelled via a differential sensing technique.

Multiplication by the cross-sectional pipe area readily enables the output to be conditioned to read in volumetric flow rate scaled to engineering units. Refinement of the technique enables density and other physical properties to be measured as well, so that mass flow rates and other parameters pertaining to various process variables can be determined.

Let D = pipe diameter
     X = axial displacement of transducers
     S = path length between transducers
     Θ = angle of transducers to flow stream
     c = velocity of sound in the flowing medium
     v = velocity of flowing medium
     t_1 = flight time of upstream pulse
     t_2 = flight time of downstream pulse

then $$t_2 = \frac{S}{c + v \cos \Theta} \tag{1}$$

$$t_1 = \frac{S}{c - v \cos \Theta} \tag{2}$$

To cancel the effects of variations in the velocity of sound c in the medium itself, the circuitry is arranged to compute the reciprocal of each flight time and to subtract one from the other. This difference of the flight time reciprocals leads to the relationship:

$$\frac{1}{t_2} - \frac{1}{t_1} = \frac{2 v \cos \Theta}{S} \tag{3}$$

Solving for v

$$v = \frac{S}{2 \cos \Theta} \left(\frac{1}{t_2} - \frac{1}{t_1} \right) \tag{4}$$

but $$\cos \Theta = \frac{X}{S} \tag{5}$$

therefore $$v = \frac{S^2}{2X} \left(\frac{1}{t_2} - \frac{1}{t} \right) \tag{6}$$

The circuitry which performs the reciprocal time computation produces an output frequency since:

$$\frac{1}{t} = f \tag{7}$$

Therefore $$\frac{1}{t_2} = f_2 \tag{8}$$

and $$\frac{1}{t_1} = f_1 \tag{9}$$

and $$\frac{1}{t_2} - \frac{1}{t_1} = f_2 - f_1 = \Delta f \tag{10}$$

Substituting in Equation 6 we have the basic velocity equation:

$$v = \frac{S^2}{2X} \Delta f \tag{11}$$

or

$$v = k_1 \Delta f \tag{11a}$$

The relationship $$k_1 = \frac{S^2}{2X} \tag{12}$$

is the flow velocity scaling factor, and is usually expressed in ft per second or ft per minute. Note that it is a function of the square of the distance between the transducers. This is an important consideration when providing for re-zeroing of the meter calibration each time the transducers are removed and re-inserted.

For volumetric flow, the velocity must be multiplied by the cross-sectional area of the pipe, and the following considerations are involved:

$$A = \pi \left(\frac{D}{2} \right)^2 \tag{13}$$

Volumetric flow

$$V - v A \tag{14}$$

Therefore, volumetric flow

$$V \pi = \frac{S^2 D^2 \Delta f}{8X} \tag{15}$$

The relationship

$$k_2 = \frac{S^2 \, D^2}{8X} \, \pi \tag{16}$$

is the volumetric flow scaling factor, and is usually expressed in cubic ft per minute.

Further information on the physical properties of the flowing medium is also acquired in the process of the basic measurement. Specifically, the velocity of sound in the medium itself can be measured independent of the velocity of its motion.

Equation 3 computed the difference of the reciprocals of the upstream and downstream flight times. In the instrument this is performed in the delta computing channel. Supposing an addition of the reciprocals of the respective upstream and downstream flight times is performed in a separate computing channel, called the sigma channel. The following relationships ensue:

$$\frac{1}{t_2} + \frac{1}{t_1} = \frac{c + v \cos \Theta}{S} + \frac{c - v \cos \Theta}{S} = \frac{2c}{S} \tag{17}$$

then $c = \dfrac{S}{2} \, \Sigma \, f$ (18)

Let $k_3 \dfrac{S}{2}$ (19)

be the sound velocity scaling factor.

Then $c = k_3 \, \Sigma \, f$ (20)

The speed of sound c in a medium is a function of its bulk modulus and its density, and is expressed by the following equation:

$$c = \sqrt{\frac{B}{\rho}} \tag{21}$$

where B = bulk modulus of elasticity
       ρ = density

From this, the equation for density may be derived thus:

$$c^2 = \frac{B}{\rho} \tag{22}$$

and $\rho = \dfrac{B}{c_2}$ (23)

Substituting the value of c from Equation 20 into 23 gives

$$\rho = \frac{B}{(k_3 \, \Sigma \, f)^2} \tag{24}$$

For a given gas, B is essentially constant.

Therefore, $\dfrac{B}{k_3 2} = k_4$ (24a)

is a fair representation
Then ρ (density) $= k_4 \, \Sigma \, f^{-2}$ (25)

An interesting relationship may be derived at this point, namely, an equation for mass flow. Since

mass flow $= \rho \, V$ (26)

and since $V = k_1 \, \Delta f$ (11a)

and $\rho = \dfrac{k_4}{\Sigma \, f^2}$ (25)

then mass flow $= k_1 \, k_4 \, \dfrac{\Delta f}{\Sigma \, f^2}$ (27)

or, letting $k_1 \, k_4 = k_5$

mass flow $= k_5 \, \dfrac{\Delta f}{\Sigma \, f^2}$ (28)

Further, the density of a gas is related by the equation

$$\rho = \frac{P}{T} \tag{29}$$

where P = pressure
       T = temperature (°K)

Therefore, if the pressure is known, the temperature may be extracted from the sigma channel thusly:

From Equation 25, $\dfrac{P}{T} = k_4 \, \Sigma \, f^{-2}$ (30)

and $T = k_4 \, P^{-1} \, \Sigma \, f^{-2}$ (31)

Let us go back to Equation 24. If the chemical properties of the gas or the ratio of the mixture between two or more gases changes, then B is not a constant. If separate transducers are used to measure density, say via the pressure/temperature relationship, then the variation of B can also be measured.

Thus $\rho = B \, (k_3 \, \Sigma \, f)^{-2}$ (32)

and $B = \rho (k_3 \, \Sigma \, f)^2$ (33)

The above relationship holds for liquids or gases.
In the case of a gas, we may further say:

$$B = \frac{P}{T} \, (k_3 \, \Sigma \, f)^2 \tag{34}$$

How fast does oil move in a pipe line?

Divide the throughput, in barrels per day, by the square of the nominal diameter, in in.; divide this result by 100; and subtract 1 from the answer. The answer is the speed of the oil column in miles per hour.

Example. A 12-in. line has a throughput of 70,000 bpd

$70,000 \div 144 = 486$

$486 \div 100 = 4.86$

$4.86 - 1.00 = 3.86$

The oil moves about 4 miles per hour.

This rule is rough, but seldom is an exact answer needed for this problem. Greater accuracy can be had by using the true inside diameter of the line in the following formula:

$V = 0.0081 \times Q/d^2$

where V = speed, mph
       Q = throughput, bpd
       d = inside diameter of pipe, in.

Take the above example: If the wall thickness of the pipe is $^3/_8$ in.

$V = (0.0081)(70,000 \div 144)$

$V = 3.94$

or about 4 miles per hour.

Example. A 20-in. pipe line has a throughput of 200,000 barrels per day. How fast does the oil move inside the pipe line?

By the first method

$200,000 \div 400 = 500$

$500 \div 100 - 1 = 4$ miles per hour

By the second method

$V = (.0081)(200,000 \div 19.25^2)$

$V = 4.37$ or about 4 miles per hour

Estimate the volume of a pipe line per linear foot using the inside diameter

To find the volume of a pipe line in gallons per ft, square the inside diameter in in. and multiply by 4%. A more accurate answer may be obtained by adding 2% of this result to the first answer.

Example. Find the volume of a pipe line having an ID of 7.981 in. expressed in gallons per ft.

Volume $= (0.04) \times (7.981)^2 = 2.548$ gallons per ft

For greater accuracy add 2% of 2.548.

$2.548 + 0.051 = 2.599$ gallons per ft
Actual volume = 2.599 gallons per ft

To find the volume of the pipe line in barrels per ft, square the inside diameter in in. and divide by 1,000. For greater accuracy subtract 3% of the first answer.

Example. Find the volume of a pipe line with an ID of 7.981 in. in barrels per ft.

Volume $= \dfrac{(7.981)^2}{1,000} = 0.0637$ barrels per ft

For greater accuracy subtract 3% of 0.0637

$0.0637 - 0.0019 = 0.0618$ barrels per ft

Actual volume = 0.0619 barrels per ft

What is the linefill of a given pipe in barrels per mile?

Multiply the inside diameter of the pipe, in in., by itself, and multiply the result by 5.13; the answer is the linefill in barrels per mile.

Example. Six-in. schedule 40 pipe; inside diameter is 6.065 in.

$6.065 \times 6.065 \times 5.13 = 188.7$ barrels per mile

If the correct inside diameter is used, the rule gives the correct linefill in standard 42-gallon barrels per mile of pipe.

Estimate leakage amount through small holes in a pipe line

The amount of fluid lost through a small hole in a pipeline can be estimated using the following equation:

$$Q = 0.61A\sqrt{2gh}$$

where
 Q = flow in cu. ft./second
 A = cross sectional area, sq. ft.
 g = gravitational constant, ft/sec/sec
 h = head, feet

Example: Assume the following conditions:
Hole diameter = 0.125 inches
Pressure = 30 psig
Sp. gr. of fluid = 0.85

A = 0.0000852 sq. ft.
h = 81.5 feet
g = 32.2
$$Q = 0.61 x 0.0000852\sqrt{2 x 32.2 x 81.5}$$

$$Q = 0.003765 \text{ cu. ft./sec}$$

Table gives velocity heads for various pipe diameters and different rates of discharge

NOMINAL DIAMETER OF PIPE (Velocity Heads in Feet)

Note: In the lower portion of the table the same physical columns are re-labeled for larger pipe sizes — ⅛″ column = 8″, ¼″ column = 10″, ⅜″ column = 12″, ½″ column = 14″, ¾″ column = 15″, 1″ column = 16″, 1¼″ column = 20″, 1½″ column = 24″, 2″ column = 30″, 2½″ column = 36″, 3″ column = 42″.

Rate of Flow (GPM)	⅛″	¼″	⅜″	½″	¾″	1″	1¼″	1½″	2″	2½″	3″	4″	5″	6″	Rate of Flow (GPM)
1	.488	.148	.043	.017											1
2		.590	.175	.069	.022										2
3			.392	.155	.050	.020									3
4			.698	.276	.090	.035	.012	.006							4
5			1.089	.430	.141	.054	.018	.010							5
10				1.720	.564	.215	.071	.038	.016	.007	.003				10
15					1.265	.488	.160	.087	.036	.015	.007				15
20					2.250	.861	.286	.154	.065	.027	.013				20
25						1.345	.447	.241	.101	.041	.020				25
30						1.933	.643	.346	.146	.060	.029				30
35						2.635	.877	.472	.198	.082	.039				35
40						3.343	1.145	.618	.259	.106	.052	.016			40
45							1.548	.779	.329	.134	.065	.021			45
50							1.787	.963	.406	.166	.080	.026			50
70							3.503	1.888	.795	.326	.157	.050	.020		70
75								2.165	.912	.373	.180	.057	.023		75
100								3.851	1.621	.665	.320	.101	.041	.020	100
120								5.548	2.333	.956	.462	.146	.060	.029	120
125									2.528	1.035	.498	.158	.065	.031	125
150									3.649	1.493	.736	.229	.093	.045	150
175									4.954	2.021	.980	.309	.127	.061	175
200									6.483	2.656	1.281	.406	.166	.080	200
225										3.359	1.618	.514	.209	.101	225
	8″														
250	.040									4.131	1.992	.633	.259	.125	250
270	.047										2.333	.740	.302	.146	270
275	.048										2.429	.769	.313	.151	275
300	.057										2.883	.912	.373	.180	300
		10″													
350	.078											1.242	.509	.246	350
400	.102											1.621	.665	.320	400
450	.129	.053										2.053	.839	.406	450
			12″												
470	.141	.057										2.239	.917	.443	470
475	.144	.059										2.276	.934	.453	475
500	.159	.065	.031									2.535	1.038	.502	500
550	.193	.079	.038										1.256	.606	550
600	.229	.093	.045										1.493	.721	600
				14″											
650	.269	.109	.053	.026									1.752	.846	650
700	.312	.127	.062	.031									2.035	.980	700
750	.358	.146	.071	.035									2.337	1.123	750
800	.407	.166	.080	.040										1.281	800
850	.462	.187	.090	.045										1.548	850
					15″										
900	.516	.209	.102	.050										1.618	900
950	.574	.234	.113	.056	.043									1.813	950
1000	.637	.259	.125	.062	.048									1.999	1000
1050	.703	.296	.138	.069	.053									2.202	1050
1100	.771	.313	.152	.075	.058									2.421	1100
						16″									
1150	.841	.343	.166	.082	.064									2.648	1150
1200	.917	.373	.181	.090	.069									2.880	1200
1250	.995	.406	.196	.097	.075	.062								3.135	1250
1500	1.433	.582	.282	.145	.108	.088									1500
2000	2.547	1.035	.502	.249	.193	.158									2000
							20″								
2500		1.618	.784	.391	.301	.246									2500
3000		2.329	1.128	.562	.434	.355	.147								3000
								24″							
3500		3.224	1.540	.766	.590	.502	.199	.096							3500
									30″						
4000			2.007	1.000	.771	.633	.261	.125	.051						4000
4200			2.209	1.102	.851	.698	.288	.137	.056						4200
4500			2.539	1.265	.975	.799	.330	.158	.065						4500
5000			3.135	1.561	1.204	.985	.407	.195	.079						5000
										36″					
5500				1.884	1.460	1.196	.493	.227	.096						5500
6000				2.239	1.733	1.421	.586	.281	.115	.056					6000
6500				2.627	2.035	1.669	.690	.329	.134	.065					6500
7000				3.056	2.360	1.936	.797	.384	.156	.076					7000
7200					2.709	2.056	.846	.404	.165	.080					7200
7500					3.047	2.224	.917	.440	.180	.087					7500
8000							1.045	.498	.204	.099					8000
8500							1.177	.564	.230	.112					8500
											42″				
9000							1.319	.633	.259	.125					9000
9500							1.469	.702	.289	.140					9500
10000							1.630	.779	.319	.154	.083				10000
10500								.858	.352	.169	.092				10500
11000								.943	.387	.186	.101				11000
11500								1.033	.423	.204	.110				11500
12000								1.123	.460	.222	.120				12000
12500								1.218	.500	.240	.130				12500
13000								1.316	.541	.260	.141				13000
13500								1.421	.582	.281	.152				13500

F. G. Switzer "Pump-Fax," Goulds Pumps, Inc.

13: Liquids—Hydraulics

Marine Hose Data **310**
CALM system.. 310
SALM system.. 310
Tandem system 311
Multi-point mooring system 311
Pressure loss in hose string.. 313
Pressure drop calculations for rubber hose................. 315
Examples of pressure drop calculations for rubber hose ... 315
Typical formulas used for calculating pressure drop and flow rates
    for pipe lines 315
Hydraulic gradients... 317
Equivalent lengths 320
Series systems... 321
Looped systems 322
Calculate pressure loss in annular sections................. 323
Calculate pressure and temperature loss for viscous crudes
    ≥1,000 cP 323
Determine batch injection rate 326

Pressure Loss through Valves and Fittings.................**327**
Representative resistance coefficients (K) for valves and fittings....... 327
Nomograph for calculating Reynolds number for flow of liquids and
    friction factor for clean steel and wrought iron pipe 333
Nomograph for calculating pressure drop of liquids in lines for
    turbulent flow 335

Drag-Reducing Agents...**339**
The basics of drag reduction....................................... 339
How drag-reducing agents can outweigh looping or boosting 344
How active, passive drag affect DRA injections 350
How to estimate the rate of liquid discharge from a pipe................... 356
Predict subsurface temperature ranges 356
Sizing pipe lines for water flow... 357
How approximate throughput of a line can be estimated from
    pipe size.. 358
Gauge liquid flow where no weir or meter is available...................... 358
Estimate crude gathering line throughput for a given pipe diameter.... 358
How to determine head loss due to friction in ordinary iron pipe
    line carrying clear water 358
How to size lines, estimate pressure drop and optimum station spacing
    for crude systems 359
Estimate the optimum working pressures in crude oil
    transmission lines 359
How to size crude and products lines for capacity increases.............. 360
How to determine the maximum surge pressure in liquid-filled pipe line
    when a main valve is closed suddenly 360
What is the hydrostatic pressure due to a column of oil H feet in
    height? 360
Transient pressure analysis 361
Tank farm line sizing 370

MARINE HOSE DATA

This section presents data on how to calculate the length of hose, and pressure loss, through marine hose.

CALM system

The overall length of the floating hose string is designed as follows:

$$TL = C + R \times (1+E) + B/2 + L + H + M + G \qquad (1)$$

where

TL = required overall length
C   = diameter of buoy
R   = length of mooring hawser
E   = max. elongation of mooring hawser
B   = beam of tanker
L   = distance between bow and manifold
H   = maximum (light ballast) freeboard (approx. $0.8 \times$ tanker depth)
M   = distance between rail and manifold
G   = height of manifold from deck

SALM system

The required overall length of hose string is designed as follows:

$$\ell_1 < R$$

where

ℓ_1 = distance between hose arm* and floating hose
R   = length of mooring hawser
TW = $\sqrt{\ell_{1} + W^2}$
W   = maximum water depth at hose arm*

$$TL = HM + R \times (1+E) + TW + (B/2) + L + H + M + G + \ell_1 \quad (2)$$

where
TL = required overall length of hose string
HM = max. horizontal movement of buoy
R   = length of mooring hawser
E   = max. elongation of mooring hawser
B   = beam of tanker

L   = distance between bow and manifold
H   = maximum light (ballast) freeboard (approx. $0.8 \times$ tanker's depth)
M   = distance between rail and manifold
G   = height of manifold from deck.

*Hose arm - the horizontal (I_1) or the vertical (W) distance between the underwater hose to swivel connection and the commencement of the main floating line.

Reprinted with permission Bridgestone Engineered Products Co.

Tandem system

The required overall length of hose string is designed as follows:

$$TL = LF + HF + R + (1 + E) + LT + HT + MT + GT \quad (3)$$

where

TL = maximum water depth at hose arm
LF = distance between manifold and stern of FSV (Floating Storage Vessel)
HF = max. height of manifold of FSV from sea-surface
R = length of mooring hawser
E = max. elongation of mooring hawser
LT = distance between bow and manifold of OTT (Off-take-tanker)
HT = maximum light (ballast) freeboard of OTT (approx. 0.8 × tanker's depth)
MT = distance between rail and manifold of OTT
GT = height of manifold from deck of OTT

Floating Storage Vessel   Off-Take-Tanker

Multi-point mooring system

The required overall length of hose string is designed as follows:

$$TL = 1.5 \times L + D + H + G + M + \sqrt{(FA/2)^2 + (SW)/2)^2} \quad (4)$$

where

TL = required overall length
L = distance between PLEM and tanker hull
D = maximum water depth
H = maximum light (ballast) freeboard (approx. 0.8 × tanker's depth)
M = distance between rail and manifold
G = height of manifold from deck
FA = max. fore and aft excursion of tanker
SW = max. sideways excursion of tanker

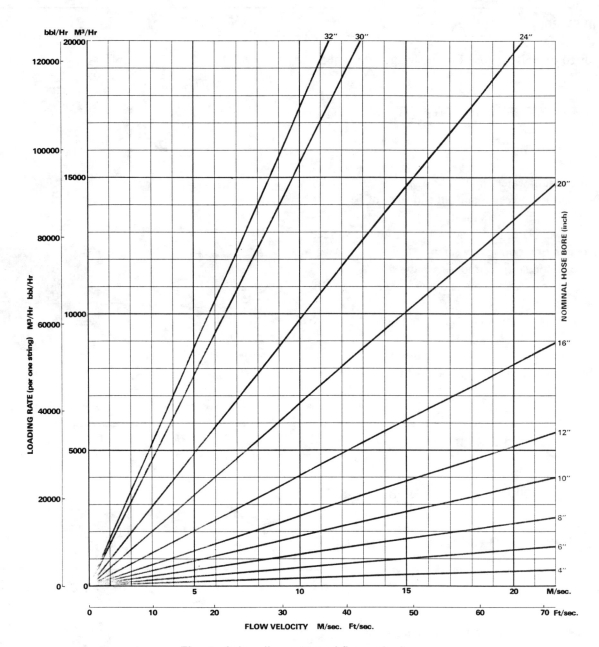

Figure 1. Loading rate and flow velocity

Pressure loss in hose string

The pressure loss in a hose string is calculated as follows:

$$\Delta P = \frac{\lambda x L x V^2 x \gamma x 10^{-4}}{2 x g x D} \tag{1}$$

where

ΔP = pressure loss (Kg/cm2)
λ = friction factor
L = length of hose string (m)
V = flow velocity (m/sec)
γ = specific weight of oil (Kg/m3)
 = $(1000 \times$ specific gravity)
g = acceleration of gravity (m/sec2)
D = inside diameter of hose (m)

λ is calculated by using the following Mises' experimental formula:

$$\lambda = 0.0096 + 5.7 x \sqrt{K/D} + 1.7 x \sqrt{1/Re} \tag{2}$$

where

K = coefficient of wall roughness by Mises' experiment (m)
K = 0.3×10^{-6} for rubber hose
Re = Reynolds number

Re is calculated as follows:

$$Re = \frac{D x V}{\nu} \tag{3}$$

where

ν = kinematic viscosity (m²/sec)

Input Data:

Viscosity and specific weight of oil depend on temperature, and the values at the operating temperature sometimes need to be converted from values at other given temperatures.

Specific weight can be converted from other units as follows:

Specific weight (Kg/m³) = 1000 x specific gravity

$$\text{Specific gravity} = \frac{141.5}{131.5 + Deg. API} \tag{4}$$

Specific gravity at operating temperature can be calculated as follows:

S. G. operating = S.G. measured - A x (Temp. operating - Temp. measured) Note: Temp: ºC. Coefficient of S.G. change "A" depends on the specific gravity of the oil and is given in Table 1.

Kinematic viscosity is converted from other units as follows:

1 cSt = 1 cP/S.G. (cP is Absolute Viscosity)
1 cSt = 0.245 x R (Redwood Sec.)
1 cSt = 7.60 x E (Engler Deg.)
1 cSt = 0.220 x S (Saybolt Universal Sec.)

Conversion of Kinematic Viscosity at a different temperature depends on the characteristics of the cargo and can be calculated from data at two different temperatures.

$$\log_{10}\log_{10}(n +0.6) = \nu - m x \log T \tag{5}$$

where
T = oK
ν = cSt

Flow velocity is determined by the internal diameter of the hose and the loading rate as follows:

$$V = \frac{4 x Q}{3600 x \Pi x D^2} \tag{6}$$

where

V = flow velocity (m/sec)
Q = loading rate (m³/hr)
D = inside diameter of hose (m)

Reprinted with permission Bridgestone Engineered Products Co.

Table 1
Change of Specific Gravity of Oil

S.G. at meas. temp.	Coeff. "A"	S.G. at meas. temp.	Coeff. "A"
0.640 - 0.643	0.00095	0.754 - 0.759	0.00078
0.644 - 0.648	0.00094	0.760 - 0.765	0.00077
0.649 - 0.654	0.00093	0.766 - 0.771	0.00076
0.655 - 0.661	0.00092	0.772 - 0.777	0.00075
0.662 - 0.667	0.00091	0.778 - 0.783	0.00074
0.668 - 0.674	0.00090	0.784 - 0.790	0.00073
0.675 - 0.681	0.00089	0.791 - 0.799	0.00072
0.682 - 0.688	0.00088	0.800 - 0.808	0.00071
0.689 - 0.696	0.00087	0.809 - 0.818	0.00070
0.697 - 0.703	0.00086	0.819 - 0828	0.00069
0.704 - 0.711	0.00085	0.829 - 0.838	0.00068
0.712 - 0.719	0.00084	0.839 - 0.852	0.00067
0.720 - 0.726	0.00083	0.853 - 0.870	0.00066
0.727 - 0.734	0.00082	0.871 - 0.890	0.00065
0.735 - 0.741	0.00081	0.891 - 0.970	0.00064
0.742 - 0.747	0.00080	0.971 - 1.000	0.00063
0.748 - 0.753	0.00079		

Example:

Conditions:

Loading Rate: 40,000 bbl/hr.
Hose size: 20" & 16"
Hose length: 100 m for 20" hose
  50 m for 16" hose
Operating Temp.: 30°C
Specific gravity of oil 34.4° API at 60°F
Kinematic viscosity: 6.90 cSt at 30°C

Calculations:

Loading Rate: Q = 40,000 x 0.15899 = 6,360 m³/hr
Flow velocity for 20" hose:

$$V_{20"} = \frac{4x6,360}{3600x\Pi x0.486^2} = 4.83(m/\sec)$$

$$V_{16"} = \frac{4x6360}{3600x\Pi x0.380^2} = 5.92(m/\sec)$$

Specific weight:

$$S.G._{60°F} = \frac{1415}{1315+34.4} = 0.853(at15.56°C)$$

$$S.G._{30°C} = 0.853 - 0.00066X(30-15.56) = 0.843$$

Specific weight = 843(Kg/m³)

Kinematic Viscosity:
$\nu = 6.9$ cSt $= 6.9$ x 10^{-6} x (m²/sec)

Reynolds Number:

$$Re = \frac{DxV}{\nu}$$

$$Re_{20"} = \frac{0.486x9.52}{6.9x10^{-6}} = 670,539$$

$$Re_{16"} = \frac{0.380x15.6}{6.9x10^{-6}} = 859,130$$

Friction factor:

$$\lambda = 0.0096 + 5.7x\sqrt{\frac{K}{D}} + 1.7x\sqrt{\frac{1}{Re}}$$

$$\lambda_{20"} = 0.0096 + 5.7x\sqrt{\frac{0.3x10^{-6}}{0.486}} + 1.7x\sqrt{\frac{1}{670539}} = 0.01615$$

$$\lambda_{16"} = 0.096 + 5.7x\sqrt{\frac{0.3x10^{-6}}{0.380}} + 1.7x\sqrt{\frac{1}{859130}} = 0.1650$$

Pressure loss:

$$\Delta P = \frac{\lambda xLxV^2 x\gamma x10^{-4}}{2xgxD}$$

$$\Delta P_{20"} = \frac{0.01615x200x9.52^2 x843x10^{-4}}{2x9.81x0.486} = 2.59(Kg/cm^2)$$

$$\Delta P_{16"} = \frac{0.01650x50x15.6^2 x843x10^{-4}}{2x9.81x0.380} = 2.27(Kg/cm^2)$$

Total pressure loss = 2.59 + 2.27 = 4.86(Kg/cm²)

Pressure drop calculations for rubber hose

Fluid Flow

The general flow formula is as follows:

$$\Delta P = \frac{0.0134 \times f \times S \times L \times Q^2}{d^5}$$

where ΔP = pressure drop, psi
  f = friction factor
  S = specific gravity
  L = length of line, ft
  Q = flow, gpm
  d = inside diameter of line, in.

$$N_R = \frac{3{,}160 \times Q}{V_C \times d}$$

where N_R = Reynolds number
  Q = flow, gpm
  V_C = fluid viscosity, centistokes
  d = inside diameter of line, in.

Friction Factor, f.

For laminar flow, $f = \dfrac{64}{N_R}$

Source

Gates Rubber Company

Examples of pressure drop calculations for rubber hose

Flow rate = Q = 1,000 gpm
Hose ID = d = 6.0 in.
Length = L = 100 ft
Fluid   = Gasoline
Specific gravity = S = 0.71
Viscosity = V_C = 0.476 centipoise/0.71 = 0.6704 centistokes

Reynolds number = N_R = $[3{,}160 \times Q]/[V_C \times d]$

  = $[3{,}160 \times 1{,}000]/[0.6704 \times 6]$
  = 785,600

From Figure 1, f = 0.012 for N_R = 785,600

P = $[0.0134 \times f \times S \times L \times Q^2]/d^5$
P = $[0.0134 \times 0.012 \times 0.71 \times 100 \times 1{,}000{,}000]/7{,}776$
P = 1.468 psi per 100 ft

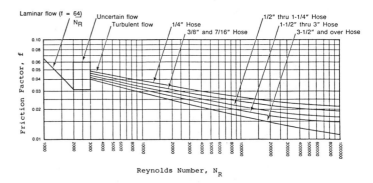

Figure 1. Friction factors for rubber hose. Reproduced with permission of Gates Rubber Company.

As a rule of thumb, loss for each set of couplings equals 5% of the loss per 100 ft of hose. For two sets of couplings,

pressure loss = $2 \times 0.05 \times 1.468 = 0.15$ psig

Typical formulas used for calculating pressure drop and flow rates for pipe lines

Shell/MIT

P = $(0.241\ fsQ^2)/d^5$

For viscous flow,

f = $0.00207 \times (1/r)$

For turbulent flow,

f = $0.0018 + 0.00662 \times (1/r)^{0.355}$

r = $0.0119Q/dv$

Viscous flow,

r = 0.1 to 0.135

Turbulent flow,

r > 0.4

Indeterminate region,

r = 0.135 to 0.4

Benjamin Miller

$B = (0.1692)(d^5 P/s)^{0.5}(\log(d^3 sP/z^2) + 4.35)$

T. R. Aude

$B = 0.871K(1/z^{0.104})((P^{0.552}d^{2.656})/s^{0.448})$

$P = ((Bz^{0.104}s^{0.448})/(0.871Kd^{2.656}))^{1.812}$

Hazen-Williams

$Q = (0.148\ d^{2.63}P^{0.54}C)/s^{0.54}$

Notation.

Q = barrels per day
B = barrels per hour
P = pressure loss, psi per mile
d = inside diameter of pipe, in.
z = absolute viscosity, centipoises
s = specific gravity
K = efficiency factor (Aude's formula)
f = friction factor
r = Reynolds number modified Re/7,742
C = Hazen-Williams coefficient (use 120 for new steel pipe; 100 for ordinary 15-year-old steel pipe; and, 150 for fiberglass pipe)
SUS = Saybolt Universal Seconds
v = kinetic viscosity, centistokes

Examples

Shell/MIT. Find the pressure loss, psi/mile, for a crude oil line where the flow rate is 30,000 barrels per day with the following conditions:

Pipe: 10³/₄-in. × 0.307-in. wt
Specific gravity = 0.85
Viscosity = 80 SUS

Convert viscosity (80 SUS) to centistokes:

$v = 0.22(SUS) - 180/SUS$
$\quad = 0.22(80) - 180/80$
$\quad = 15.35$ centistokes

$r = 0.0119Q/dv$
$\quad = (0.0119 \times 30,000)/(10.136 \times 15.35)$
$\quad = 2.295$

$f = 0.0018 + 0.00662(1/r)^{0.355}$
$\quad = 0.0018 + 0.00662(1/2.295)^{0.355}$
$\quad = 0.00673$

$P = (0.241fsQ^2)/d^5$
$\quad = (0.241 \times 0.00673 \times 0.85 \times 30,000^2)/10.136^5$
$\quad = 11.6$ psi/mile

T. R. Aude. Find the pressure loss, psi/mile, when pumping gasoline with the following conditions:

B = 2,000 barrels per hour
Pipe: 8.625-in. × 0.322-in. wt
z = 0.59 centipoise
s = 0.733
K = 0.94 (pipe roughness/efficiency factor)

$P = ((Bz^{0.104}s^{0.448})/(0.871Kd^{2.656}))^{1.812}$
$\quad = ((2,000 \times 0.59^{0.104} \times 0.733^{0.448})/(0.871 \times 0.94 \times 7.981^{2.656}))^{1.812}$
$\quad = 44.1$ psi/mile

Note: Aude's formula is based on data taken from 6-in. and 8-in. lines, and no extension data are available.

Benjamin Miller. Find the pressure loss, psi/mile, when pumping gasoline with the following conditions:

B = 4,167 barrels per hour
Pipe: 16.0-in. × 0.3125-in. wt
z = 0.59 centipoise
s = 0.733
d = 15.375-in.

$B = (0.169176)(d^5 s/P)^{0.5}(\log(d^3 sP/z^2) + 4.35)$
$4,167 = (0.169176)(15.375^5 P/0.733)^{0.5}$
$\quad\quad (\log(15.375^3 \times 0.733 \times P)/(0.59^2) + 4.35)$

Let P = 6.4 psi/mile for first try:

$4,167 = (0.169167)(15.375^5 \times 6.4)^{0.5}(\log(15.375^3 \times 0.733 \times 6.4)/(0.59^2) + 4.35)$
$4,167 = 0.169167 \times 2,738 \times (4.69 + 4.35)$
$4,167 \neq 4,187$

Let P = 6.34 psi/mile for second try

$4{,}167 = 0.169167 \times 2{,}726 \times (4.68 + 4.35)$
$4{,}167 = 4{,}167$

Therefore P = 6.34 psi/mile.

Hazen-Williams. (Typically used for water system hydraulic calculations.) Find the pressure drop, psi/mile, in a water system where the flow rate is 20,000 barrels per day.

The pipe is 4.5-in. × 0.226-in. wt new steel. Since the flowing medium is water, s = 1.0.

$20{,}000 = (0.148 \times 4.048^{2.63} \times 120 \times P^{0.54}$

$P^{0.54} = 20{,}000/702.23$
$= 28.48$

$P = 493$ psi/mile

All of the preceding examples have assumed no difference in elevation. Where a difference in elevation exists, this must be considered in determining the total pressure required for a given flow rate.

Hydraulic gradients

Some of the pressure drop formulas used to calculate pressure drop and flow rates for pipe lines are shown earlier in this section. In order to use these formulas, it is first necessary to determine the characteristics of the fluid to be transported—primarily the specific gravity, viscosity, and either the flow rate or allowable pressure drop.

Before the system design is completed, it is necessary to know the elevation of the terrain to be traversed by the pipe line. Once the elevation of the system is known and a pipe size has been selected, the pressure loss data may be used to calculate or plot a hydraulic gradient. A hydraulic gradient is the height that a column of fluid will rise to if small tubes were installed along the pipe line as shown in Figure 1. Feet of head may be converted to pressure by dividing by 2.31 and multiplying by the specific gravity. Conversely, pressure is converted to ft of head by multiplying pressure by 2.31 and dividing by the specific gravity of the fluid.

The slope of the hydraulic gradient will represent the amount of pressure drop per unit of linear distance. The pressure drop per unit of length is uniform as long as there is no change in pipe size or specific gravity, and therefore the hydraulic gradient will be a straight line with a slope proportional to the system flow rate.

Since the pipe line elevation profile is usually plotted in ft, it is helpful to plot the hydraulic gradient in ft rather than pressure. The hydraulic gradient may be plotted by converting pressure at the origin and end points to ft using the specific gravity of the fluid in the pipe line. The end point in this case will be the end of the pipe line if there are no intermediate booster stations. If the system is equipped with intermediate booster stations, then the suction of the booster station in the first segment will be the end point for that segment of the line. Don't forget to consider the suction pressure requirement for the booster station. These two points are then connected with a straight line which will represent the pressure (in ft) at any intermediate point along the pipe line. The pressure at any point may be determined by taking the difference between the gradient and pipe line elevation and converting ft to psi.

Figure 2 shows a typical elevation profile plot with two hydraulic gradients—one for no booster station and one for an intermediate booster station and no change in elevation. In Figure 3 there is an elevation change and for a hydraulically balanced system, it was necessary to locate the booster station at a point other than the geographic mid-point of the line. The location of the booster station in this example corresponds to the hydraulic mid-point of the system. A hydraulically balanced system is one in which all stations develop the same differential head. In Figure 2 where there was no elevation change, the booster station location is at the geographic mid-point of the system, which also corresponds to the hydraulic mid-point of the system. This is not meant to imply that a system must absolutely be hydraulically balanced. A booster station may be located off the hydraulic center; however, its differential head requirement will differ from that of the origin station and system throughput capacity will be affected. The amount of difference in head required, for a booster station

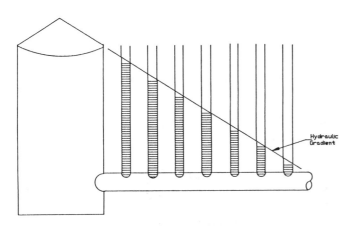

Figure 1. Schematic for hydraulic gradient.

Figure 2. Profile with no elevation change.

Figure 3. Profile with elevation change.

Figure 4. System curve. 30 mi., 12.750" x 0.250" pipe.

Figure 5. Pipe line elevation and test gradients and operating profile.

which is located off hydraulic center, will depend upon the amount of system elevation change and back pressure maintained at the end of the line.

The total system friction loss may be divided equally among an optimum number of booster stations. The discharge pressure at each station may be calculated with the following equation:

$$Pdp = (Pt + S + (N - 1) \times Ps + Pr)/N$$

where Pdp = station discharge pressure
Pt = total system friction loss
S = system static pressure

N = number of booster stations
Ps = station suction pressure
Pr = residual or back pressure at end of line

In Figure 4, three system curves are shown to illustrate how elevation will affect the capacity of a system. With a constant input pressure of 750 psig, the capacity will be 77,500 barrels per day pumping uphill (272 ft); 84,000 barrels per day with no elevation change; and 90,000 barrels per day pumping downhill (272 ft). Note that where there is a difference in elevation, the amount of displacement of the curves, 100 psi in this case, is uniform.

The hydraulic gradient may also be used to analyze a system to assure that it is operated within the pressure limits determined by the hydrostatic test and the pipe design parameters. Figure 5 shows the test and maximum allowable operating gradients for a system. The hydrostatic test point was located at MP 150 and the test pressure was 1,800 psig, which in this case corresponds to 90% of the specified minimum yield strength (SMYS) of the pipe. Water was used as the test medium and therefore 1,800 psig equates to 4,157 ft of head. The test gradient is drawn as a straight line across the profile from 8,157 ft (4,157 ft of head + 4,000 ft of pipe line elevation). This system may be operated at 80% of the test pressure since this represents 72% of the SMYS of the pipe. See Section 3 for details on pipe design and limits on maximum allowable operating pressure.

The 80% gradient (still based on water) is drawn as a straight line across the profile from 7,325 ft. The maximum allowable operating profile is now developed by taking the difference between the 80% of test pressure gradient, in ft, and the pipe line elevation and converting this difference to ft of head for the fluid to be transported. For this example a specific gravity of 0.5 was used. At MP 150, the difference between pipe line elevation and the 80% of test gradient is 3,325 ft of head (water) and for a specific gravity of 0.5, this will convert to 6,650 ft of head. Therefore, the maximum operating pressure in ft of head for a fluid with a specific gravity of 0.5 will be 10,650 ft (6,650 ft of head + 4,000 ft pipe line elevation) at MP 150.

The maximum allowable operating profile will not be a straight line but will be a displaced mirror image of the pipe line elevation profile. The remainder of the maximum allowable operating profile may be developed by following this procedure at other points along the pipe line. When fluids with different specific gravities are transported, an operating profile will need to be developed for each specific gravity. If the line is hydrostatically tested in several segments rather than in one continuous length, an operating profile will need to be developed for each of the test segments.

Once the maximum allowable operating profile has been established, the hydraulic gradient may be superimposed. If the hydraulic gradient crosses the maximum allowable operating profile, the pipe will be overpressured in the segments where the gradient crosses the profile.

Figure 6 illustrates the effect hills have on the operation of a system. In this case, a hill is located near the end of the line and, due to its elevation, it will control the system capacity since adequate back pressure must be maintained (**if** the vapor pressure of the fluid is above atmospheric pressure) at the end of the line to assure the gradient will "clear" the hill. If no back pressure is maintained at the end of the line, the gradient will intersect the side of the hill and theoretically, a siphon action will occur if the elevation

Figure 6. Pipe line elevation profile and hydraulic gradients.

difference is not too great. In reality, the system flow rate will change of its own accord to cause the gradient to intersect the top of the hill. The flow from that point to the end of the line will cascade or run away from the flow on the other side of the hill. If the vapor pressure of the fluid is very low, this condition can probably be tolerated. If segregated batches of material are handled or if the fluid vapor pressure is above atmospheric pressure, adequate back pressure must be maintained to keep the batches from mixing or forming vapor. Inadequate back pressure will cause difficulties in getting the flow in vs. flow out to balance on a short term basis as would be required for a leak detection system.

Two conditions are illustrated in Figure 6 to show the effect the flow rate will have on the amount of back pressure required to clear the hill. Some safety factor should be included when determining the amount of pressure to maintain in the system—especially when handling high vapor pressure materials. The flowing temperature will also need to be considered in making this determination.

For this example, the fluid has a specific gravity of 0.5 and its vapor pressure requires a minimum of 200 psig in the system. Therefore, it will be necessary to maintain 200 psig (924 ft) at the top of the hill located at MP 140. All gradients must be drawn through this point. If the specific gravity changes, the control point will move up or down depending upon the direction of the specific gravity change. The gradient connecting MP 0 to the end of the line passes through the control point and intersects the end of the line at 6,300 ft or 498 psig. The second condition, which assumes a booster station at MP 78, illustrates what happens to the back pressure requirement when the flow rate changes. The gradient from the discharge of the booster station must also pass through the control point and since the flow rate is greater, the slope of the gradient is steeper and intersects the end of the line at 5,800 ft or 390 psig. It follows that as the system flow rate changes, it will be necessary to adjust the amount of back pressure. If the back pressure is permanently set to the higher value, system capacity will be penalized. See Section 16: Instrumentation, for a discussion of control systems.

Equivalent lengths

Some pipe line systems are not made up of one single uniform pipe size. As the pressure in a system decreases, it may be desirable to use pipe with a thinner wall thickness to reduce the investment cost or where deliveries are made along the pipe line, it is sometimes practical to use a smaller diameter pipe for the lower flow rate. Conversely, where injections are made into a system, it may be necessary to use a larger diameter pipe for the increased flow rate.

The capacity of a system may be increased by installing loop lines. Loop lines are two or more pipe lines that are operated in parallel. Loop lines are not practical where segregated batches are pumped since there will be a problem in getting the batches to arrive simultaneously. Also, if there are injections or deliveries along such a system, it will be almost impossible to balance the amount injected or delivered between the two lines. The most practical method for operating loops where segregated batches are handled is to operate each line separately. A product system might, for example, use one line for distillates and use the other line for gasoline. If the gasoline volume is too large for one line, the excess could be batched through the other system. One other caution on loops is that the lengths of the legs may not be exactly the same.

Hydraulic systems are analogous to electrical circuits as shown in Figure 1 and this sometimes helps in analyzing a pipe line system. In a series electrical circuit, the same current flows through all of the circuit elements and the voltage drops across the elements are additive. This same logic may be applied to a pipe line system with segments of different pipe sizes connected in series. The pressure drops through the various segments are additive and the sum of the pressure drops will equal the total system pressure drop.

A loop system is analogous to a parallel electrical circuit as shown in Figure 1. Since a common pressure is applied to all the legs of the loop system, the pressure drop across each leg of a loop system must be same and the flow will divide in proportion to the friction loss in each leg.

An examination of the Shell/MIT flow formula reveals that the friction factor is dependent upon flow rate, pipe ID, and viscosity. Since the flow through each leg of a loop system will change as the total system flow changes, this will have an impact on the calculated equivalent length factor. For best accuracy, the equivalent length factor should be determined as near as possible to the expected operating parameters.

Series System

Loop System

Note: Hydraulic equivalents shown in ()
         P = pressure, Q = flow, f = friction
         V = voltage, I = current, R = resistance

Figure 1. Series and loop system analogy.

Series systems

Figure 1 shows a system curve for a pipe line system 30 miles in length and consisting of 15 miles of 12.750-in. OD × 0.250-in. pipe in series with 15 miles of 10.750-in. OD × 0.250-in. pipe. System friction curves were developed for each pipe size and for the combination of the two sizes. Inspection shows that the curve for the combination is the sum of the 12.750-in. OD curve and the 10.750-in. OD curve. Pressure drop is additive on lines of constant flow for different pipe sizes connected in series.

The step of developing the individual curves for 12.750-in. OD and 10.750-in. OD pipe may be eliminated by calculating the equivalent length of 12.750-in. OD or 10.750-in. OD pipe. A system friction loss curve may then be calculated using the equivalent length of one pipe size or the other. For this example, the fluid is crude oil with a specific gravity of 0.86 and a viscosity of 50.6 SUS.

At a flow rate of 40,000 barrels per day, the pressure drop through 12.750-in. OD × 0.250-in. pipe is 6.618 psi/mile and the pressure drop through 10.750-in. OD × 0.250-in. pipe is 15.469 psi/mile. The ratio of these two pressure drops is 6.618/15.469 or 0.428. This ratio times the length of the 12.750-in. OD pipe equals the equivalent amount of 10.750-in. OD pipe that will be required to yield the same pressure drop as 15 miles of 12.750-in. OD pipe. Therefore the equivalent length of 10.750-in. OD pipe is 15 + (0.428 × 15) or 21.41 equivalent miles of 10.750-in. OD pipe. If it is desired to use 12.750-in. OD pipe as the base size, the same procedure may be followed except that the pressure ratio will be re-versed and the reciprocal of 0.428 or 2.3364 multiplied by 15 miles of 10.750-in. OD pipe to obtain 35.046 equivalent miles of 12.750-in. OD pipe. The total equivalent length will be 15 miles + 35.046 miles, or 50.046 equivalent miles of 12.750-in. OD pipe.

This may be checked as follows:

At a flow rate of 40,000 b/d, the pressure drop through 15 miles of 10.750-in. OD is

$$15.469 \times 15 = 232 \text{ psig}$$

The pressure drop through 15 miles of 12.750-in. OD is

$$6.618 \times 15 = 99 \text{ psig}$$

The total pressure drop will be

$$232 + 99 = 331 \text{ psig}$$

The pressure drop through 21.42 equivalent miles of 10.750-in. OD is

$$21.42 \times 15.469 = 331 \text{ psig}$$

The pressure drop through 50.046 equivalent miles of 12.750-in. OD is

$$50.046 \times 6.618 = 331 \text{ psig}$$

Figure 1. System curve. 15 mi., 12.750″ × 0.250″ pipe in series with 15 mi., 10.750″ × 0.250″ pipe.

Looped systems

The ratio of pressure drops may also be used for calculating the equivalent lengths of pipe line loops; however, the procedure is slightly different from that used for series systems.

Figure 1 shows the friction curves for a looped system which is shown schematically in Figure 2. For reference, individual curves are shown for each pipe size and the combination of the two lines in parallel. The total friction loss for the system may be determined graphically using a system of curves such as shown in Figure 1 or the ratio of pressure drops may be used to determine an equivalent length for the looped system.

To use the ratio of pressure drops to convert the loop to a single line of equivalent size, determine the pressure drop through the larger size pipe (if this is to be the base size) using a chosen flow rate. In this example, a flow rate of 60,000 b/d of crude oil with a specific gravity of 0.86 and a viscosity of 50.6 SUS is used. The pressure drop through the 12.750-in. OD pipe at 60,000 b/d is 13.54 psi/mile. Next, find a flow rate through the 10.750-in. OD pipe that will yield the same pressure drop of 13.54 psi/mile. The required flow rate through the 10.750-in. OD pipe is 37,100 b/d for a drop of 13.54 psi/mile. The sum of these two rates is 97,100 b/d. Use this rate, 97,100 b/d, and determine the pressure drop/mile through the 12.750-in. OD pipe. The pressure drop/mile is 31.91 psi/mile. The factor for converting to an equivalent length of 12.750-in. OD × 0.250-in. pipe is 13.54/31.91 or 0.4243. The equivalent length of

$$Q_c = Q_a + Q_b = 80,000 \text{ B/D}$$

Figure 2. Looped pipe lines.

the loop combination will be 15 miles × 0.4243 or 6.3645 equivalent miles of 12.750-in. OD × 0.250-in. pipe.

If it is desired to convert the loop combination to a base diameter of 10.750-in. OD, select a flow rate and determine the pressure drop in the 10.750-in. OD leg. At 30,000 b/d, the pressure drop in the 10.750-in. OD leg will be 9.32 psi/mile. Determine the flow rate through the 12.750-in. OD leg to produce a pressure drop of 9.32 psi/mile. The rate will be 48,560 b/d. The sum of these two rates is 78,560 b/d. This rate through the 10.750-in. OD leg will produce a pressure drop of 51.29 psi/mile. The equivalent length factor for 10.750-in. OD pipe is 9.32/51.29 = 0.1817. The equivalent length will be 0.1817 × 15 or 2.7257 equivalent miles of 10.750-in. OD pipe.

This may be checked as follows:

The pressure drop through the loop combination at a flow rate of 80,000 b/d is

9.62 psi/mile × 15 miles = 144 psig

The flow through the 10.750-in. OD leg will be 30,550 b/d and 49,450 b/d through the 12.750-in. OD leg.

The pressure drop for 6.3645 equivalent miles of 12.750-in. OD pipe at a flow rate of 80,000 b/d is

6.3645 equivalent miles × 22.584 psi/mile = 144 psig

The pressure drop for 2.7257 equivalent miles of 10.750-in. OD pipe at a flow rate of 80,000 b/d is

2.7257 equivalent miles × 52.982 psi/mile = 144 psig

The ratio-of-pressures method will work in conjunction with any of the flow formulas listed in this section.

Figure 1. Friction curves. 10.750″ × 0.250″ and 12.750″ × 0.250″ and looped combination.

Calculate pressure loss in annular sections

$P = 0.000027 \, flpv^2/M$

Where:

f   = friction factor (from Figure 2, p. 333)
l   = length of section, ft
p   = density of fluid, #/ft³
v   = average velocity, fps
M = mean hydraulic radius, ft
P   = pressure loss of length, psi
u   = absolute viscosity, ft/lb-sec. (0.000672 * centipoise)
R_n = Reynolds number
Q = volume, cubic ft per second
e/d = relative roughness

$v = Q/0.7854 \, d_A^2$

$M = (area/wetted \; perimeter) = (d_1 - d_2)/4$

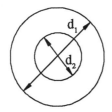

Example:

Throughput   60,000 Barrels per day = 6.44 fps

Specific gravity 0.68

Viscosity 0.50 centipoise

$d_1 = 12.615"$

$d_2 = 7.000"$

$l = 2,000$ ft

$d_A = (d_1^2 - d_2^2)^{0.5} = (12.615^2 - 7.0^2)^{0.5} = 10.495"$

$M = (d_1 - d_2)/4 = (1.051 - 0.583)/4 = 0.117$ ft.

v   = 6.44 fps
p   = 62.4 x 0.68 = 42.432 #/ft³
u   = 0.000672 x 0.50 = 0.000336 ft. lb. - sec.
Rn = (4 x 0.1178 x 6.44 x 42.432) ÷ 0.0000336 = 383,200
e/d = 0.00017
f   = 0.015 (from Figure 2, page 333)

P   = (0.000027 x 0.015 x 2000 x 42.432 x 6.44²) ÷ 0.1178
P   = 12.1 psi

Calculate pressure and temperature loss for viscous crudes ≥ 1,000 cP

When calculating pressure and temperature losses for highly viscous Newtonian crude oils, accurate flowing temperature data is required. Usually, crudes of this nature are flowing in the laminar region and a change of a few degrees in temperature can change the viscosity by as much as 50% or more and thus cause a significant change in the friction factor.

Figure 1 illustates a typical pressure and temperature profile for an uninsulated pipeline.

The data typically required for determining a pressure/temperature profile for a viscous crude oil pipeline is shown in Table 1.

Once the data has been tabulated, a computer program such as BATCH[1] can be used tocalculate the temperature/pressure profile of the pipeline. The program performs a rigorous energy balance , i.e., static head, friction and kinetic mechanical energy, plus friction, compression, stored and transferred heat are included. Thermal resistances of the inside film, pipe, any insulation and the outside environment (soil, air or water) are calculated.

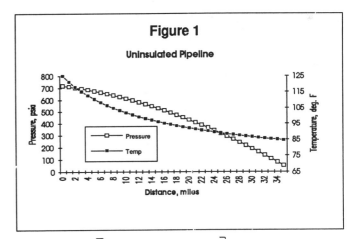

$$P_2 = P_1 - s \, \Delta x \left[\frac{C_Q \, Q^2 f}{d^5} + \frac{C_H (e_2 - e_1)}{\Delta x} \right]$$

Where:
p_1 = pressure at the upstream end of the calculation segment
p_2 = pressure at the downstream end of the calculation segment

Table 1

Data required for pressure/temperature loss calculation

Length	Heat transfer coefficient (fluid)	Inlet temperature
Size, OD	Insulation thickness	Inlet pressure
Wall thickness	Thermal conductivity of insul.	Throughput
Pipe wall roughness	Thermal conductivity of soil	Sp. gr. at two temp.
Thermal conductivity of pipe	Pipe burial depth	Viscosity at two temp.
Upstream elevation	Soil temperature	Specific heat of liquid at two temp.
Downstream elevation	Vapor pressure at two temp.	Thermal conductivity of liquid at two temp.
Number of calculation segments	Soil K factor	

Δx = length of the calculation segment
f = Darcy-Weisbach friction factor
d = inside diameter of the pipe
s = liquid specific gravity
e_1 = elevation at the upstream end of the calculation segment
e_2 = elevation at the downstream end of the calculation segment
C_Q = flow conversion constant
C_H = head conversion constant

An energy balance is performed on each calculation segment and its surroundings resulting in the next equation:

$$t_2 = [wc_{p1}t_1 - UA\,(LMTD) + w(h_1 - h_2)\,/\,J] \div (wC_{p2})$$

Where:

t_1 = flowing temp. at the upstream end of the calculation segment
t_2 = flowing temp. at the downstream end of the calculation segment
h_1 = head at the upstream end of the calculation segment
h_2 = head at the downstream end of the calculation segment
C_{p1} = heat capacity at constant pressure evaluated at t_1
C_{p2} = heat capacity at constant pressure evaluated at t_2
w = total mass flow rate
U = overall heat transfer coefficient
A = surface area of the pipe in calculation segment
LMTD = log mean temperature difference between the average flowing temperature and the average ambient temperature
J = mechanical equivalent of heat

Example:

Pipeline length	35 miles
Size	14.000"
Wall thickness	0.250"
Pipe wall roughness	0.0018"
Thermal conductivity (pipe)	29 Btu/hr-ft-F
Upstream elevation	1,500 feet
Downstream elevation	1,650 feet
Length	35 miles

Number of calculation segments	35
Heat transfer coefficient (calculated by program Btu/hr-ft2-F)	
Insulation thickness	0" (Case 1)
Insulation thermal conductivity	0 Btu/hr-ft-F
Burial depth	36"
Soil temperature	50°F
Soil K factor	1.2
Inlet temperature	125°F
Inlet pressure	calculate
Outlet pressure	50 psia
Throughput	8,000 bpd

Liquid properties:

	Value	Temp.
Specific gravity	0.95	77
	0.90	210
Viscosity, cP	3260	80
	90	180
Specific heat	0.5	60
Btu/Lbm-F	0.5	80
Thermal conductivity	0.2	60
Btu/hr-ft-F	0.22	100
Vapor pressure,	0.4	100
psia	0.3	70

The pressure/temperature profile for an uninsulated line, based on the above conditions is shown in Figure 1 and in Table 3.

If the pipeline crosses areas where the parameters such as soil thermal conductivity, pipe size, etc., change, BATCH will allow the system to be segmented into additonal pipes (up to 99 pipes) with each pipe being configured with its own input data.

Assume that 2" of urethane foam insulation is added to the pipeline previously described. The new pressure/temperature profile is shown in Figure 2 and in Table 4.

Typical soil K factors are shown in Table 2. BATCH may also be used to calculate pump requirements and heater duty.

Figure 2

Insulated Pipeline - 2" Polyurethane

Table 2
Soil K Factors
Btu/hr/ft²/°F

Soil	Condition	Cover, In.	K
Sandy	Dry	24	0.25 to 0.40
Sandy	Moist	24	0.50 to 0.60
Sandy	Soaked	24	1.10 to 1.30
Clay	Dry	24	0.20 to 0.30
Clay	Moist	24	0.40 to 0.50
Clay	Moist to wet	24	0.60 to 0.90
Sandy	Dry	8	0.60 to 0.70
Sandy	Moist to wet	8	1.20 to 2.40
River	Water	60	2.00 to 2.50
Air	Exposed	---	2.0 (approx.)

Table 3
Uninsulated Pipe

Distance miles	Pressure psia	Temp deg. F
0	718	125
1	710.59	121.3
2	702.57	118.1
3	693.79	115.3
4	684.25	112.8
5	673.93	110.5
6	662.83	108.4
7	650.95	106.5
8	638.3	104.8
9	624.89	103.3
10	610.73	101.9
11	595.82	100.6
12	580.18	99.3
13	563.81	98.2
14	546.74	97.2
15	528.97	96.2
16	510.51	95.2
17	491.38	94.4
18	471.6	93.6
19	451.18	92.8
20	430.13	92.1
21	408.46	91.4
22	386.21	90.7
23	363.37	90.1
24	339.97	89.5
25	316.02	88.9
26	291.55	88.4
27	266.55	87.9
28	241.06	87.4
29	215.09	86.9
30	188.65	86.5
31	161.76	86
32	134.43	85.6
33	106.68	85.2
34	78.54	84.9
35	50	84.5

Table 4
Insulated Pipe

Distance miles	Pressure psia	Temp. deg. F
0	506.76	125
1	499.69	123.4
2	492.32	121.9
3	484.64	120.5
4	476.65	119.1
5	468.33	117.8
6	459.68	116.5
7	450.7	115.3
8	441.38	114.1
9	431.71	112.9
10	421.68	111.8
11	411.3	110.7
12	400.56	109.7
13	389.45	108.7
14	377.98	107.8
15	366.13	106.8
16	353.9	105.9
17	341.3	105.1
18	328.32	104.3
19	314.96	103.5
20	301.22	102.7
21	287.1	101.9
22	272.59	101.2
23	257.71	100.5
24	242.44	99.8
25	226.8	99.2
26	210.77	98.5
27	194.38	97.9
28	177.6	97.3
29	160.46	96.7
30	142.95	96.2
31	125.08	95.6
32	106.84	95.1
33	88.25	94.6
34	69.3	94.1
35	50	93.6

Determine batch injection rate as per enclosure

In the operation of a pipeline system, it may be necessary to inject a specific volume into a passing block of liquid. When it is desirable to maintain the same flow rate downstream of the injection and inject at the minimum rate or into most of the passing block, the following equations can be used to determine the upstream injection rates.

$$Q_U = \frac{B + H_O Q_D + Q - \left[(B + H_O Q_D + Q)^2 - 4xH_O xBxQ\right]^{0.5}}{2xH_O}$$

$$Q_I = Q_D - Q_U$$

where

Q_D = Downstream rate that is to be held constant, bph
Q_I = Injection rate to be determined, bph
Q_U = Upstream pumping rate to be determined, bph
Q = Volume to be injected, barrels
B = Volume of passing block to be injected into, barrels
H_O = Block outage in hours. This is the time that the head and tail end of the block interface is passing the injection location during which time injection is not allowed.

Example:

Q = 15,000 bbls for injection
Q_I = Injection rate, bph
B = 73,500 bbls
Q_D = 15,600 bph (to be held constant)
H_O = 0.5 hours

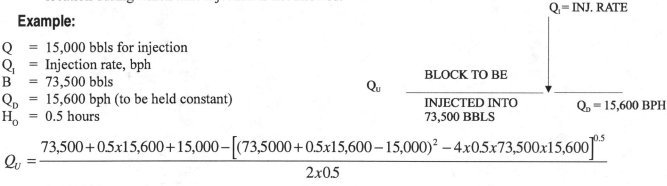

$$Q_U = \frac{73,500 + 0.5x15,600 + 15,000 - \left[(73,5000 + 0.5x15,600 - 15,000)^2 - 4x0.5x73,500x15,600\right]^{0.5}}{2x0.5}$$

$Q_U = 12,751$ bph

$Q_I = 15,600 - 12,751 = 2,849$ bph

PRESSURE LOSS THROUGH VALVES AND FITTINGS

"K" FACTOR TABLE—SHEET 1 of 4

Representative Resistance Coefficients (K) for Valves and Fittings

("K" is based on use of schedule pipe as listed on page 2-10)

PIPE FRICTION DATA FOR CLEAN COMMERCIAL STEEL PIPE
WITH FLOW IN ZONE OF COMPLETE TURBULENCE

Nominal Size	½"	¾"	1"	1¼"	1½"	2"	2½, 3"	4"	5"	6"	8-10"	12-16"	18-24"
Friction Factor (f_T)	.027	.025	.023	.022	.021	.019	.018	.017	.016	.015	.014	.013	.012

FORMULAS FOR CALCULATING "K" FACTORS*
FOR VALVES AND FITTINGS WITH REDUCED PORT
(Ref: Pages 2-11 and 3-4)

- **Formula 1**

$$K_2 = \frac{0.8\left(\sin\frac{\theta}{2}\right)(1 - \beta^2)}{\beta^4} = \frac{K_1}{\beta^4}$$

- **Formula 2**

$$K_2 = \frac{0.5(1 - \beta^2)\sqrt{\sin\frac{\theta}{2}}}{\beta^4} = \frac{K_1}{\beta^4}$$

- **Formula 3**

$$K_2 = \frac{2.6\left(\sin\frac{\theta}{2}\right)(1 - \beta^2)^2}{\beta^4} = \frac{K_1}{\beta^4}$$

- **Formula 4**

$$K_2 = \frac{(1 - \beta^2)^2}{\beta^4} = \frac{K_1}{\beta^4}$$

- **Formula 5**

$$K_2 = \frac{K_1}{\beta^4} + \text{Formula 1} + \text{Formula 3}$$

$$K_2 = \frac{K_1 + \sin\frac{\theta}{2}[0.8(1 - \beta^2) + 2.6(1 - \beta^2)^2]}{\beta^4}$$

- **Formula 6**

$$K_2 = \frac{K_1}{\beta^4} + \text{Formula 2} + \text{Formula 4}$$

$$K_2 = \frac{K_1 + 0.5\sqrt{\sin\frac{\theta}{2}}(1 - \beta^2) + (1 - \beta^2)^2}{\beta^4}$$

- **Formula 7**

$$K_2 = \frac{K_1}{\beta^4} + \beta\,(\text{Formula 2} + \text{Formula 4}) \text{ when } \theta = 180°$$

$$K_2 = \frac{K_1 + \beta\left[0.5(1 - \beta^2) + (1 - \beta^2)^2\right]}{\beta^4}$$

$$\beta = \frac{d_1}{d_2}$$

$$\beta^2 = \left(\frac{d_1}{d_2}\right)^2 = \frac{a_1}{a_2}$$

Subscript 1 defines dimensions and coefficients with reference to the smaller diameter.
Subscript 2 refers to the larger diameter.

*Use "K" furnished by valve or fitting supplier when available.

SUDDEN AND GRADUAL CONTRACTION	SUDDEN AND GRADUAL ENLARGEMENT

If: $\theta \lessgtr 45°$ .... K_2 = Formula 1

$45° < \theta \lessgtr 180°$ ... K_2 = Formula 2

If: $\theta \lessgtr 45°$ .... K_2 = Formula 3

$45° < \theta \lessgtr 180°$ ... K_2 = Formula 4

Reprinted by permission and courtesy of Crane Co. from Technical Report 410

"K" FACTOR TABLE—SHEET 2 of 4

Representative Resistance Coefficients (K) for Valves and Fittings

(for formulas and friction data, see page A-26)

("K" is based on use of schedule pipe as listed on page 2-10)

GATE VALVES
Wedge Disc, Double Disc, or Plug Type

If: $\beta = 1$, $\theta = 0$ . . . . $K_1 = 8 f_T$

$\beta < 1$ and $\theta \lesssim 45°$ . . . . $K_2 =$ Formula 5

$\beta < 1$ and $45° < \theta \lesssim 180°$ . . . $K_2 =$ Formula 6

SWING CHECK VALVES

$K = 100 f_T$ $K = 50 f_T$

Minimum pipe velocity (fps) for full disc lift $= 35 \sqrt{\overline{V}}$

Minimum pipe velocity (fps) for full disc lift $= 60 \sqrt{\overline{V}}$ except

U/L listed $= 100 \sqrt{\overline{V}}$

GLOBE AND ANGLE VALVES

If: $\beta = 1$ . . . $K_1 = 340 f_T$

If: $\beta = 1$ . . . $K_1 = 55 f_T$

If: $\beta = 1$ . . . $K_1 = 150 f_T$ If: $\beta = 1$ . . . $K_1 = 55 f_T$

All globe and angle valves,
whether reduced seat or throttled,
If: $\beta < 1$ . . . $K_2 =$ Formula 7

LIFT CHECK VALVES

If: $\beta = 1$ . . . $K_1 = 600 f_T$

$\beta < 1$ . . . $K_2 =$ Formula 7

Minimum pipe velocity (fps) for full disc lift $= 40 \ \beta^2 \sqrt{\overline{V}}$

If: $\beta = 1$ . . . $K_1 = 55 f_T$

$\beta < 1$ . . . $K_2 =$ Formula 7

Minimum pipe velocity (fps) for full disc lift $= 140 \ \beta^2 \sqrt{\overline{V}}$

TILTING DISC CHECK VALVES

	$\alpha = 5°$	$\alpha = 15°$
Sizes 2 to 8" . . . K =	$40 f_T$	$120 f_T$
Sizes 10 to 14" . . . K =	$30 f_T$	$90 f_T$
Sizes 16 to 48" . . . K =	$20 f_T$	$60 f_T$
Minimum pipe velocity (fps) for full disc lift =	$80 \sqrt{\overline{V}}$	$30 \sqrt{\overline{V}}$

Reprinted by permission and courtesy of Crane Co. from Technical Report 410

"K" FACTOR TABLE—SHEET 3 of 4

Representative Resistance Coefficients (K) for Valves and Fittings

(for formulas and friction data, see page A-26)

("K" is based on use of scheduled pipe as listed on page 2-10)

STOP-CHECK VALVES
(Globe and Angle Types)

If:
$\beta = 1 \ldots K_1 = 400\, f_T$
$\beta < 1 \ldots K_2 = $ Formula 7

Minimum pipe velocity for full disc lift
$= 55\, \beta^2 \sqrt{\overline{V}}$

If:
$\beta = 1 \ldots K_1 = 200\, f_T$
$\beta < 1 \ldots K_2 = $ Formula 7

Minimum pipe velocity for full disc lift
$= 75\, \beta^2 \sqrt{\overline{V}}$

If:
$\beta = 1 \ldots K_1 = 300\, f_T$
$\beta < 1 \ldots K_2 = $ Formula 7

If:
$\beta = 1 \ldots K_1 = 350\, f_T$
$\beta < 1 \ldots K_2 = $ Formula 7

Minimum pipe velocity (fps) for full disc lift
$= 60\, \beta^2 \sqrt{\overline{V}}$

If:
$\beta = 1 \ldots K_1 = 55\, f_T$
$\beta < 1 \ldots K_2 = $ Formula 7

If:
$\beta = 1 \ldots K_1 = 55\, f_T$
$\beta < 1 \ldots K_2 = $ Formula 7

Minimum pipe velocity (fps) for full disc lift
$= 140\, \beta^2 \sqrt{\overline{V}}$

FOOT VALVES WITH STRAINER

Poppet Disc   **Hinged Disc**

$K = 420 f_T$ $K = 75 f_T$

Minimum pipe velocity (fps) for full disc lift
$= 15 \sqrt{\overline{V}}$

Minimum pipe velocity (fps) for full disc lift
$= 35 \sqrt{\overline{V}}$

BALL VALVES

If: $\beta = 1,\ \theta = 0 \ldots \ldots \ldots \ldots K_1 = 3\, f_T$
$\beta < 1$ and $\theta \lessapprox 45° \ldots \ldots K_2 = $ Formula 5
$\beta < 1$ and $45° < \theta \lessapprox 180° \ldots K_2 = $ Formula 6

BUTTERFLY VALVES

Sizes 2 to 8" $\ldots K = 45\, f_T$
Sizes 10 to 14" $\ldots K = 35\, f_T$
Sizes 16 to 24" $\ldots K = 25\, f_T$

Reprinted by permission and courtesy of Crane Co. from Technical Report 410

Representative Resistance Coefficients (K) for Valves and Fittings
(for formulas and friction data, see page A-26)
("K" is based on use of schedule pipe as listed on page 2-10)

PLUG VALVES AND COCKS

Straight-Way   **3-Way**

View X—X

If: $\beta = 1$,   If: $\beta = 1$,   If: $\beta = 1$,
$K_1 = 18 f_T$ $K_1 = 30 f_T$ $K_1 = 90 f_T$

If: $\beta < 1 \ldots K_2 =$ Formula 6

MITRE BENDS

$\propto$	K
0°	$2 f_T$
15°	$4 f_T$
30°	$8 f_T$
45°	$15 f_T$
60°	$25 f_T$
75°	$40 f_T$
90°	$60 f_T$

90° PIPE BENDS AND FLANGED OR BUTT-WELDING 90° ELBOWS

r/d	K	r/d	K
1	$20 f_T$	8	$24 f_T$
1.5	$14 f_T$	10	$30 f_T$
2	$12 f_T$	12	$34 f_T$
3	$12 f_T$	14	$38 f_T$
4	$14 f_T$	16	$42 f_T$
6	$17 f_T$	20	$50 f_T$

The resistance coefficient, K_B, for pipe bends other than 90° may be determined as follows:

$$K_B = (n - 1) \left(0.25 \ \pi \ f_T \frac{r}{d} + 0.5 \ K \right) + K$$

n = number of 90° bends
K = resistance coefficient for one 90° bend (per table)

CLOSE PATTERN RETURN BENDS

$K = 50 f_T$

STANDARD ELBOWS

90°   45°

$K = 30 f_T$ $K = 16 f_T$

STANDARD TEES

Flow thru run . . . . $K = 20 f_T$
Flow thru branch . . . . $K = 60 f_T$

PIPE ENTRANCE

Inward Projecting   **Flush**

r/d	K
0.00*	0.5
0.02	0.28
0.04	0.24
0.06	0.15
0.10	0.09
0.15 & up	0.04

*Sharp-edged

$K = 0.78$ For K, see table

PIPE EXIT

Projecting   **Sharp-Edged**   **Rounded**

$K = 1.0$ $K = 1.0$ $K = 1.0$

Reprinted by permission and courtesy of Crane Co. from Technical Report 410

Equivalent Lengths *L* and *L/D* and Resistance Coefficient *K*

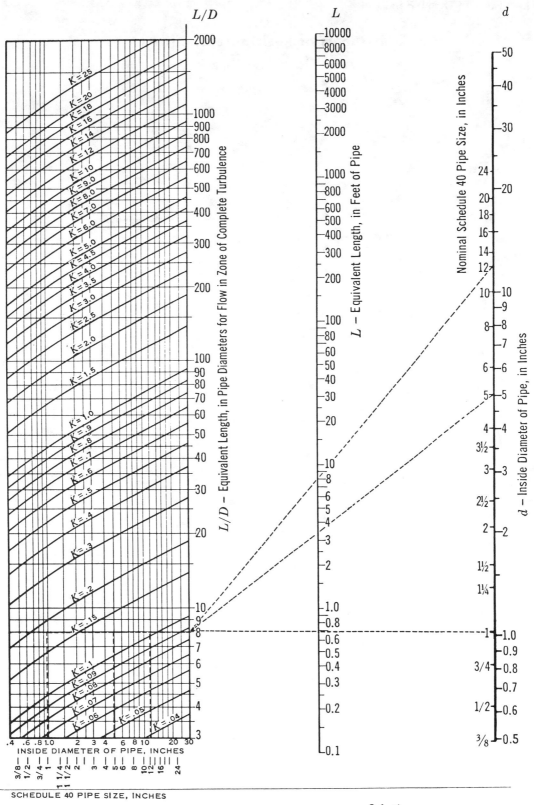

Problem: Find the equivalent length in pipe diameters and feet of Schedule 40 clean commercial steel pipe, and the resistance factor *K* for 1, 5, and 12-inch fully-opened gate valves with flow in zone of complete turbulence.

Solution

Valve Size	1"	5"	12"	*Refer to*
Equivalent length, pipe diameters	8	8	8	Page A-27
Equivalent length, feet of Sched. 40 pipe	0.7	3.4	7.9	Dotted lines
Resist. factor *K*, based on Sched. 40 pipe	0.18	0.13	0.10	on chart.

Reprinted by permission and courtesy of Crane Co. from Technical Report 410

Equivalents of Resistance Coefficient *K*
And Flow Coefficient C$_v$

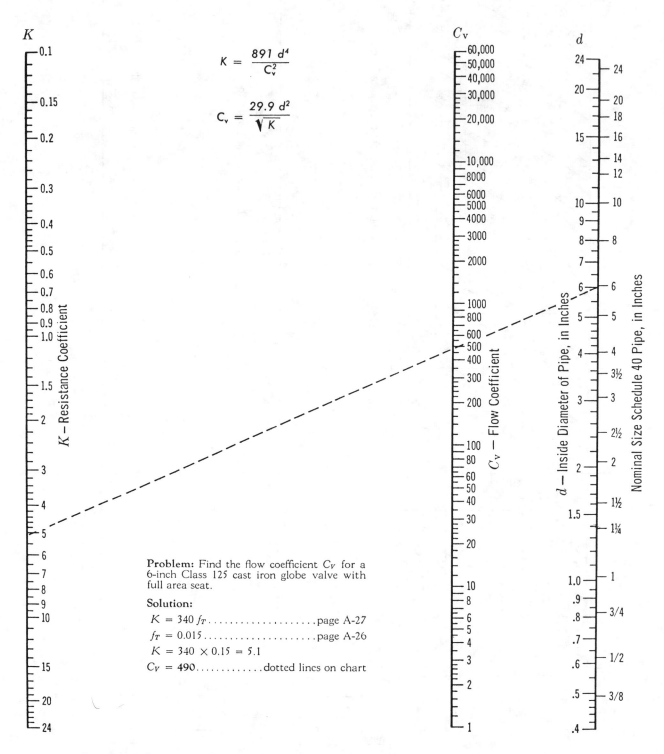

$$K = \frac{891\,d^4}{C_v^2}$$

$$C_v = \frac{29.9\,d^2}{\sqrt{K}}$$

K — Resistance Coefficient

C_v — Flow Coefficient

d — Inside Diameter of Pipe, in Inches

Nominal Size Schedule 40 Pipe, in Inches

Problem: Find the flow coefficient C_v for a 6-inch Class 125 cast iron globe valve with full area seat.

Solution:

$K = 340\, f_T$ .page A-27

$f_T = 0.015$ . . . . .page A-26

$K = 340 \times 0.15 = 5.1$

$C_v = 490$ . . . . .dotted lines on chart

Reprinted by permission and courtesy of Crane Co. from Technical Report 410

Nomograph for calculating Reynolds number for flow of liquids and friction factor for clean steel and wrought iron pipe

The nomograph (Figure 1) permits calculation of the Reynolds number for liquids and the corresponding friction factor for clean steel and wrought iron pipe.

The Reynolds number is defined as:

$$Re = 22,700 \frac{q\rho}{d\mu} = 50.6 \frac{Q\rho}{d\mu} = 6.31 \frac{W}{d\mu} = \frac{dv\rho}{\mu a}$$

where Re = Reynolds number
d = inside pipe diameter, in.

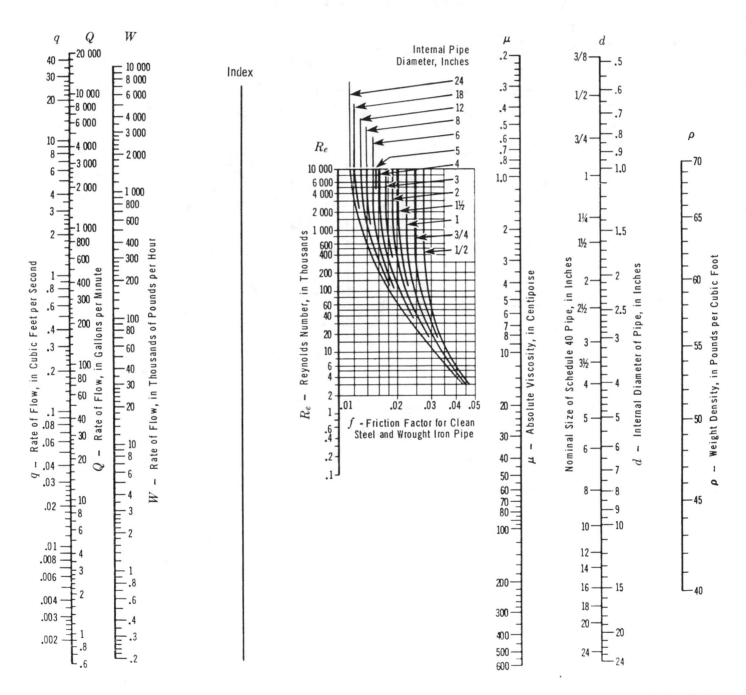

Figure 1. Reynolds number for liquid flow friction factor for clean steel and wrought iron pipe (Reproduced by permission, *Tech. Paper, 410*, Crane Co., copyright 1957).

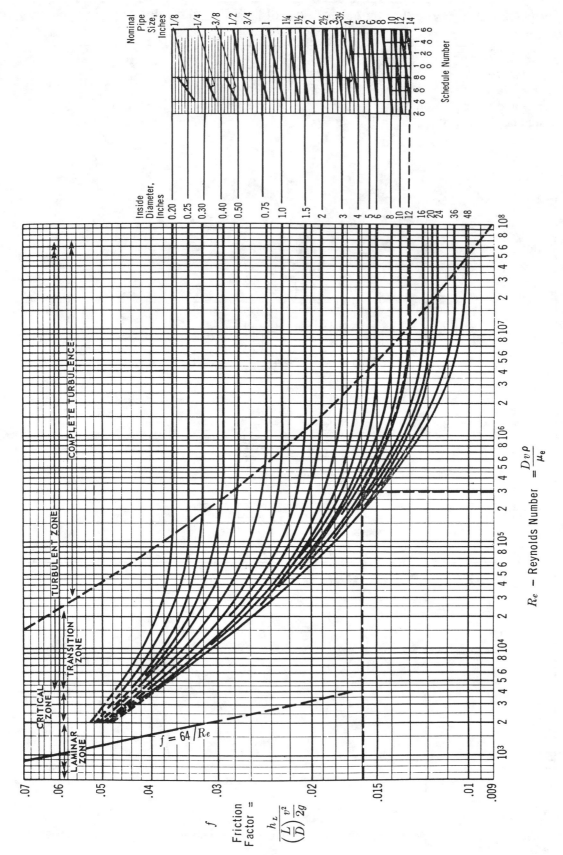

Figure 2. Friction factors for clean commercial steel and wrought iron pipe. (Reproduced by permission, *Tech. Paper, 410,* Crane Co., 1957.)

q = volumetric flow rate, cubic ft/sec
Q = volumetric flow rate, gal/min
ρ = fluid density, lb/cubic ft
μ = fluid viscosity, centipoise
μa = fluid viscosity, lb/(ft)(sec)
v = average fluid velocity, ft/sec
w = mass rate of flow, lb/hr

Example.

Given: water at 200°F
       d = 4-in. schedule 40 steel pipe
       Q = 415 gal/min
       ρ = 60.13 lb/cubic ft
       μ = 0.3 centipoises

Obtain the Reynolds number and the friction factor.

Connect	With	Mark or Read
Q = 415	ρ = 60.13	W = 200,000
W = 200,000	d = 4-in. schedule 40	Index
Index mark with	μ = 0.3	Re = 1,000,000

At Re = 1,000,000 read f = 0.017 from the graph on the center of the nomograph.

Figure 2 is an enlarged version of the friction factor chart, based on Moody's data[1] for commercial steel and wrought iron. This chart can be used instead of the smaller version shown in Figure 1.

Reference

1. Moody, L. F., *Transactions, American Society of Mechanical Engineers*, 66, 671 (1944).

Nomograph for calculating pressure drop of liquids in lines for turbulent flow

The nomograph (Figure 1) is based on a modified version of the Fanning equation. It gives the pressure drop per 100 ft of pipe for liquids flowing through a circular pipe. The basic equation is:

$$P_{100} = \frac{0.0216\, f\rho Q^2}{d^5} = 4,350\, \frac{f\rho q^2}{d^5} = 0.000336\, \frac{fW^2}{d^5\rho}$$

$$= 0.1294\, \frac{fv^2\rho}{d}$$

where P_{100} = pressure loss per 100 ft of pipe, psi
     f = friction factor (4 × f′ in Fanning equation)
     ρ = fluid density, lb/cubic ft
     d = tube internal diameter, in.
     W = flow rate, lb/hr
     Q = flow rate, gal/min
     q = flow rate, cubic ft/sec

The Moody[1] friction factor and relative roughness generalized charts (Figures 2 and 3) are also included for use with the nomogram.

Example. The estimated Reynolds number is 100,000 when water is flowing at a rate of 100 gal/min and a temperature of 70°F. Find the pressure drop per 100 ft of 2-in. diameter pipe, the relative roughness from Figure 3 is 0.0009.

At e/d = 0.0009 and Re = 100,000, the friction factor, f, is 0.022 from Figure 2.

Using Figure 1:

Connect	With	Mark or Read
f = 0.022	ρ = 62.3 lb/cubic ft	Intersection at Index 1
Index 1	Q = 100 gal/min	Intersection at Index 2
Index 2	d = 2-in.	P = 8 psi per 100 ft of pipe

Source

Flow of Fluids Through Valves, Fittings and Pipe, Technical Paper No. 410, 3–11, Crane Company, Chicago, Illinois (1957).
Anon., *British Chemical Engineering*, 2, 3, 152 (1957).

Reference

1. Moody, L. F., *Transactions, American Society of Mechanical Engineers*, 66, 671 (1944).

Figure 1. Pressure drop in liquid lines for turbulent flow.

Figure 2. Friction factors for any type of commercial pipe.

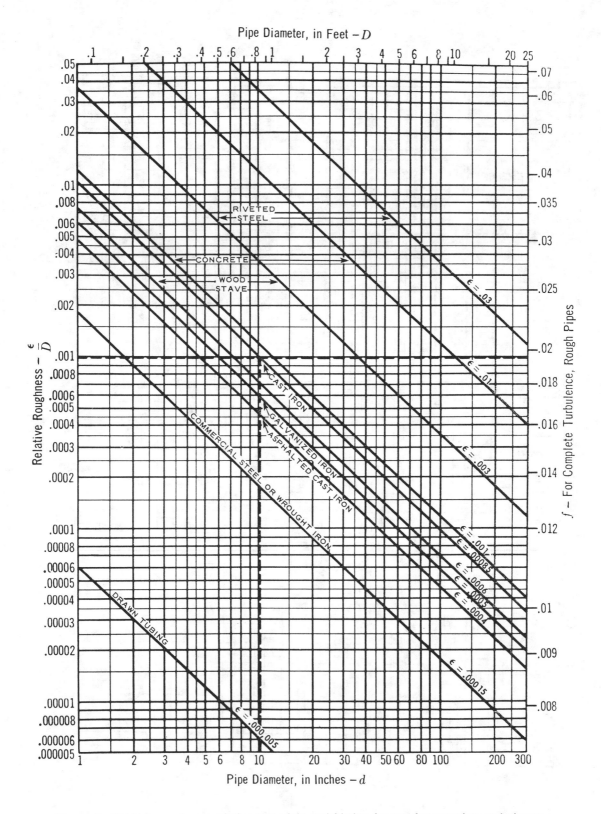

Figure 3. Relative roughness of pipe materials and friction factors for complete turbulence.

DRAG-REDUCING AGENTS

The basics of drag reduction

C. B. Lester, C. B. Lester & Associates Inc., Longview, Texas

Drag reducing agents (DRA's) can substantially reduce friction loss in most pipe lines when flowing most hydrocarbon liquids.

This article begins a series of four which will review DRA's, their fundamentals and applications, role in active and passive drag, and incidence as specific products and hardware.

When a DRA is dissolved in a solvent, it produces a solution which in laminar flow has the same pressure drop, and therefore the same Newtonian viscosity, as the solvent. But when the solution is in turbulent flow, a DRA produces a pressure drop smaller than that which would occur with untreated solvent moving at the same flow rate. The turbulent flow friction of the solution is less than that of the solvent.

DRA's are highly viscoelastic and thixotropic. They may have a viscosity at low shear rates in excess of 3,000,000 cp while having a viscosity 100–1,000 times less at higher rates. The viscosity is somewhat affected by temperature and increases with decreasing temperature. DRA-solvent solutions are viscoelastic, time-independent, shear degradeable, non-Newtonian fluids. A DRA-solvent solution handles as an ordinary hydrocarbon except in turbulent flow when the reduced friction becomes evident.

DRA's in present use in oil and products pipe lines are themselves hydrocarbons and thus should have no effect on refining processes or refined products. Millions of barrels of crude treated with DRA's on the Alyeska system have been refined without incident. Drag reduction installations need not involve large diameter pipe lines. The 200,000 b/d increase in the Alyeska system resulting from the injection of some thousands of gallons per day of DRA in this 48-in. line is dramatic.

However, the offshore operator who obtains a 20% increase in capacity in his 8-in. flowline by injecting a few dozen gallons of DRA per day is happy: Without the DRA, he would have to shut in production or loop the flowline. Either alternative would be orders of magnitude more expensive than the injection of a DRA.

And, as production decays over time, the amount of DRA injected will be reduced with the eventual result that production will match the untreated flowline capability and operations can proceed as before but with no shut in wells in past history and no sterilized investment in a loop to be carried on the books of account in future years.

Early applications

The drag reduction effect as applied to solutions was accidentally discovered by Toms in the UK in 1947.[1] It had been known for years that concentrated water-coal slurries and wood pulp-water slurries can produce a friction factor lower than that of water alone. But Dr. Toms's 1948 publication was the first indication of drag reduction in otherwise Newtonian solutions.

The first use of DRA's in the oil industry was to reduce pressure loss during pumping of fluids downhole to fracture tight formations. The quantities of DRA used are proportionally large but the savings are substantial. Data published on one DRA-kerosine solution with 600 weight ppm of DRA in kerosine showed an 80% drag reduction over untreated kerosine.

The ongoing use of DRA's to debottleneck the Alyeska 48-in. crude line has been described in several published articles. There are many other uses on commercial pipe lines which have not yet been described in published material—one considerably larger than the Alyeska example—and the list of operators using DRA is growing. Use of DRA's in multi-station pipe lines is a special application because the drag reduction effect does not survive shear stresses raised by centrifugal pumps and appears to be badly degraded going through most positive displacement pumps.

The general rule is that DRA's are always injected on the discharge side of a pump station. And if drag reduction is required all along a pipe line with N pump stations, the DRA must be injected at N points.

The effectiveness of a DRA depends to a certain extent on the viscosity of the untreated oil and decreases as viscosity increases. It varies slightly with diameter and decreases as diameter increases. And the effect varies greatly with flow velocity which, for a given oil and line diameter, means it varies greatly with Reynolds number. Drag reduction increases as Reynolds number increases. However, dilute solutions are believed to disintegrate at shear rates found in commercial pipe lines.

There are some as yet little understood practical aspects of DRA application. For instance, there seems to be some relationship between the physical characteristics of the oil and of the DRA. A certain DRA will produce good results in a given crude yet produce poor results in a seemingly similar oil.

Untreated oils which are non-Newtonian may produce non-Newtonian DRA-oil solutions. And a particular DRA may increase, rather than decrease, flowing friction. Sometimes an additional, specially compounded, additive is required to obtain the desired drag reduction. And there appears to be something in certain DRA-solvent combinations which has to do with the solubility of the DRA in the oil. Test loop studies verify that for these combinations drag reduction appears to increase asymptotically as the distance from the point of injection increases. Specially designed DRA injection nozzles help in these cases.

DRA activity

The precise mechanism describing how a DRA works to reduce friction is not established. It is believed these agents work by directly reducing turbulence or by absorbing and later returning to the flowing stream energy which otherwise would have been wasted in producing the crossflows which comprise turbulence. The absorb-and-return theory has a certain intuitive attractiveness because one of the characteristics of a viscoelastic fluid is its ability to do this on a physically large scale. Whether the effect exists at the molecular scale is unknown.

Pipe liners generally work with the rational flow formula. The friction factor therein, which is a scaling factor applied to velocity head to yield friction loss, is a macrofactor in that it lumps all of the friction-producing effects acting in a flowing system in a single numerical value. The friction factor concept does not provide an insight into how a DRA might work. It is necessary to look at the velocity distribution pattern in a pipe line in turbulent flow to understand the bases which support the concept of the f-factor and understand better how drag reduction might take place.

The concept of turbulent flow in pipe lines provides that the point flow may be directed in any direction while maintaining net pipe flow in the direction of decreasing pressure. All flow not in the direction of net pipe flow absorbs energy; the more turbulent the flow becomes, the more energy absorbed in these crossflows. In flow in hydraulically rough pipes, the f-factor does not vary with increasing flow, but the unit friction loss increases as the square of the velocity of net pipe flow.

The model of turbulent pipe flow we will use is based on the Universal Law of the Wall which applies to both smooth and rough pipe flow. It presumes two general modes of flow in a pipe: the wall layer, and the turbulent core.

The wall layer is presumed to contain all of the flow where the point velocity varies with distance of the point from the pipe wall. The turbulent core carries all of the flow where variations in point velocity are random and independent of this distance.

The wall layer itself is presumed to be comprised of three sublayers: the laminar sublayer; the buffer zone; and the turbulent sublayer.

The laminar sublayer has as its outer boundary the wall of the pipe, and flow in the entire sublayer is laminar. This definition requires that immediately at the wall lies a film of liquid which is not moving.

The increase in point velocity as the point moves away from the wall, is a linear function of the distance from the wall, and directed parallel to the wall in the direction of pipe flow. There are no crossflows in this sublayer. The contribution of the laminar sublayer to total friction arises in viscous shear generated in the sublayer.

Pipe roughness acts to inhibit laminar flow; therefore, the definition is only statistically true for rough or partially rough pipe.

In the turbulent sublayer, which has as its inner boundary the turbulent core, it is presumed that the velocity varies logarithmically with distance from the wall but that crossflows in the sublayer may be statistically large and nearly as great as in the core. The contribution of the turbulent sublayer to total friction lies in the turbulent shear generated in this sublayer.

In the buffer zone, which lies between the laminar and turbulent sublayers, variation of point velocity with point position is not established. Ideally, it should be the same as the laminar sublayer at the outer boundary and the same as the turbulent sublayer at the inner boundary. Any number of mathematical functions can be written to approximate these two requirements. Both viscosity and density contribute to shear losses in the buffer zone.

The contribution of the turbulent core to total friction loss lies entirely in turbulent shear; DRA's are presumed not to act on this component of total friction. Therefore the action of a DRA must take place in the laminar sublayer, the buffer zone, or the turbulent sublayer.

That the action of a DRA increases with increasing flowing velocity would indicate that DRA action could not be primarily in the laminar sublayer. And it sometimes would appear that this is so: Liquids treated with a DRA flowing under conditions in which the pipe is hydraulically rough and viscosity presumed to make no contribution to total friction can yield excellent drag reduction. This kind of result would lead to the presumption that a DRA works in the buffer zone.

On the other hand, large doses of a DRA have been observed to keep a liquid in laminar flow at Reynolds numbers much higher than those expected of the same liquid without DRA, which leads to the conclusion that a DRA acts in the laminar sublayer.

This conclusion suggests that a DRA may inhibit the action of crossflows, whether generated in the laminar sublayer or the buffer zone, which move inward into the turbulent sublayer and core and absorb energy which otherwise would go toward producing pipe flow. It appears this is the case.

Reduction in the energy absorbed by crossflows originating at the pipe wall can be accomplished by reducing their number, size, or velocity. Or, the energy in the crossflows may be absorbed by a DRA and later returned to the flowing stream for reuse which, in effect, means reducing their effect if not their size, number, or velocity. And it is here that theory breaks down; no one really knows which effect predominates—if either does. What is known is:

- That DRA's act to reduce turbulence.
- That to do so, they seem to act in the laminar sublayer or the buffer zone by effectively enlarging them or their effect.
- That the net effect is to reduce total friction.
- That the f-factor is less than would be calculated from the pipe liner's usual sources.

Experimental data taken on liquids flowing without a DRA do not translate directly to liquids which have been dosed with a DRA. The rational flow formula still applies but the usual f-factor relationship does not.

Therefore, the concepts of drag reduction and the drag ratio have entered the pipe liner's world.

Reduction fundamentals

Use of DRA's requires introducing a few equations (p. 343). In the following discussion, we will use the terms *treated* and *untreated* to refer to flowing liquids which have or have not been dosed with a DRA.

By common consent, drag reduction (DR) is measured in terms expressed in Equation 1. By definition, since $S_{f\ untreated}$ is larger than $S_{f\ treated}$, DR is always less than unity. DR is often expressed in percent, or $DR_{pct} = 100 \times DR$. Another commonly used factor (Equation 2), and more useful to pipe liners, is the drag ratio (D/R0) which is $(1 - DR)$. A drag reduction of 0.30, or 30%, for example, yields a drag ratio of 0.70, or 70%.

Knowing D/R, one knows the expected f-factor (Equation 3). This f-factor can be used in any of the rational flow formulas.

We need now to be able to calculate drag reduction itself. The generally accepted form of the drag reduction correlation is based on an extension of the model of turbulent viscoelastic flow of Savins and Seyer as expressed by Burger in Equation 4.[2]

The characteristic time T_p is specific to a specific DRA-solvent solution. A general form of the characteristic time equation is expressed in Equation 5.

Up-to-date values for the empirical constants are generally not available in the literature for two reasons: The science of drag reduction is one of the fastest moving disciplines of industrial chemistry, and what is published today is out of date tomorrow; also, there is a natural reluctance of a producer to make available to his competitors the technical characteristics of his product.

The data used here are based on one formulation of FLO™ Pipeline Booster produced by ChemLink, Inc., a subsidiary of Atlantic Richfield Co., with permission of the producer.

Solving equations

The Savins and Seyer correlation, taking into account the time constant equation, has been programmed in several languages for several different computers.

Program CBLFLO (Figure 2) is in HP BASIC for the HP 9845 T. Aside from the use of the REPEAT-UNTIL and SELECT CASE statements, both of which can be replaced with IF-THEN-ELSE statements, the program should run with appropriate I/O statements on almost any BASIC interpreter.

It is heavily commented, uses descriptive names for variables compatible with this text, and should be readable by anyone familiar with most versions of BASIC.

Figure 1 is a plot of parts per million (ppm) of DRA against drag ratio from D/R = 0.95 to D/R = 0.50.

Using Equation 2, we can derive the ppm-vs.-drag reduction curve shown in Figure 3 for a DR of 0.05 to 0.50.

Both curves are of the form expressed in Equation 6.

Any of the curve-fitting programs will develop the constants given a minimum of five values for ppm-D/R pairs developed using CBLFLO.

The value of this representation is that it programs well in handheld calculators and runs in seconds. The Savins

Figure 1. CBLFLO computes DRA for desired reduction.

```
1000 ! ******************** Program CBLFLO ************************
1010 ! CBLA Longview                                        08 May 84
1020                          !
1030                          !
1040 ! Program calculates gpd of hydrocarbon copolymer drag reducer required
1050 ! to produce a given drag reduction in a liquid hydrocarbon pipeline.
1060                          !
1070 ! The a,b,c,d factors in the characteristic time equation must be
1080 ! entered into the program as evaluated for the drag reducer and pipeline
1090 ! fluid under study.  Otherwise erroneous answers may result.  Enter
1100 ! these factors in the four lines following label Abcd_parameters.
1110                          !
1120 ! Combined coefficients are CBLA standards so that program finds the same
1130 ! answers for Fo, Sfo and Sfp as other CBLA programs.  Conversions from
1140 ! input data units to fps use base conversion factors directly
1150 ! instead of compound conversion factors to facilitate rewriting the
1160 ! program into other systems of units.
1170                          !
1180 ! The fundamental algorithm is the Burger interpretation of the
1190 ! Savins and Seyer model of viscoelastic flow written in the form f(x)=1.
1200                          !
1210 ! CBLFLO and FLOCBL are mirror programs; output from the first program
1220 ! inputted to the second program yields the input to the first program.
1230                          !
1240                          !
1250 !          ********** INPUT ROUTINE **********
1260                          !
1270                          !
1280 INPUT "INPUT DENSITY, VIS CST, DR PERCENT, OD IN, WT IN, ROUGH IN,
     FLOW BPD   ",Den_15,Vis_cst,Dr_pct,Od_in,Wt_in,Rough_in,Q_bpd
1290                          !
1300                          !
1310 !          ********** SET UNITS TO FPS SYSTEM **********
1320                          !
1330                          !
1340 Start:                         ! entry point for re-run using a
1350                                ! new value for a variable entered
1360                                ! from the keyboard after the
1370                                ! program has been run once with
1380                                ! INPUTTED data.
1390                          !
1400 Id_in=Od_in-2*Wt_in            ! pipe id inches
1410 Id_ft=Id_in/12                 ! pipe id feet
1420 A_sqft=PI*(Id_ft/2)^2          ! pipe area sq ft
1430 Ub_fps=42*231/1728*Q_bpd/(24*3600)/A_sqft   ! bulk velocity ft/sec
1440 Vis_sqftps=Vis_cst*1.076391E-5 ! kinematic viscosity sq ft/sec
1450 Nr=Ub_fps*Id_ft/Vis_sqftps     ! Reynolds number
1460                          !
1470                          !
1480 ! ********** USE COLEBROOK-WHITE CORRELATION FOR F-FACTOR **********
1490                          !
1500                          !
1510 C2=Rough_in/(3.7*Id_in)        ! 1st factor in C-W correlation
1520 C3=2.51/Nr                     ! 2nd factor in C-W correlation
1530 T1=10                          ! initialize transmission factor
1540 REPEAT                         ! start loop for C-W
1550     T2=-2*LGT(C2+C3*T1)        ! C-W correlation
1560     T1=(T1+T2)/2               ! refine estimate for T
1570 UNTIL T1-T2<.00001             ! loop exit condition
1580 Fo=1/(T2*T2)                   ! f = 1/(transmission factor)^2
1590                          !
1600                          !
1610 ! ********** COMPUTE VARIABLES FOR BERGER CORRELATION **********
1620                          !
1630                          !
1640 Gammao=Fo*Ub_fps^2/8/Vis_sqftps   ! wall shear rate w/o drag reducer
1650                                ! = wall shear stress/viscosity
1660                                ! = (f*Ub^2/8)/vis
1670                          !
1680 Alphao=SQR(Fo/8)               ! Alphao= friction vel/bulk vel
1690                                !       = U_star/Ub
1700                                !       = sqr(f/8)
1710                          !
1720 Sigma=1-Dr_pct/100             ! Sigma = drag ratio as a decimal
1730                          !
1740                          !
1750 !          ********** SOLVE BERGER CORRELATION **********
1760                          !
1770                          !
1780 ! Burger's correlation is:
1790                          !
1800 ! Sigma*[1+Alphao*(2^1.5*LGT(Sigma)+1.454*Gammap*Thetap-0.8809)]^2=1
1810 ! However, Gammap = Sigma*Gammao, by definition, so in terms of Gammao
1820 ! Sigma*[1+Alphao*(2^1.5*LGT(Sigma)+1.454*Gammao*Thetap*Sigma-0.8809)]^2=1
1830                          !
1840 ! It can be solved for Thetap, the characteristic time of the viscoelastic
1850 ! solution comprised of the crude oil solvent and the drag reducer polymer.
1860                          !
1870 Numerator=.8809+(Sigma^(-.5)-1)/Alphao-2^1.5*LGT(Sigma)
1880 Denominator=1.454*Gammao*Sigma
1890 Thetap=Numerator/Denominator   ! characteristic time
1900                          !
1910                          !
1920 ! ********** SOLVE CHARACTERISTIC TIME EQUATION FOR PPM **********
1930                          !
1940                          !
1950 ! Given the characteristic time (Thetap) the dosage rate Phi can be found
1960 ! by solving Thetap = A*Phi^B*(Gammao*Sigma)^C*Alpha^D for Phi.
1970                          !
1980 ! A, B, C and D are set by the drag reducer manufacturer based on tests
1990 ! or other input and take into account the kind of pipeline oil and the
2000 ! characteristics of the drag reducer chemical.
2010                          !
2020 Abcd_parameters:               ! entry for inserting a,b,c,d
2030 A=1                            !
2040 B=1                            !
2050 C=1                            !
2060 D=1                            !
2070                          !
2080 Phi=(Thetap/(A*(Sigma*Gammao)^C*Alphao^D))^(1/B)  ! ppm of drag reducer
2090                          !
2100                          !
2110 !          ********** COMPUTE GPD OF DRAG REDUCER **********
2120                          !
2130                          !
2140 Gpd=Phi*Q_bpd*42*231/1728*62.371718*Den_15/6.59/1E6   ! y = f(x)
2150                                ! density of drag reducer at
2160                                ! 15 deg C taken as 6.59 lbs/gal
2170                          !
2180                          !
2190 ! ***** COMPUTE UNTREATED AND TREATED UNIT FRICTION LOSSES *****
2200                          !
2210                          !
2220 Sfo=2.647271E-2*Fo*Q_bpd^2/Id_in^5   ! Sf mlc/km or flc/kft(untreated)
2230                                ! 2.647271E-2 is CBLA combined
2240                                ! coefficient for Sf 1/1000 base
2250                          !
2260 Sfp=(1-Dr_pct/100)*Sfo         ! Sf mlc/km or flc/kft(treated)
2270                          !
2280                          !
2290 !          ********** OUTPUT ROUTINE **********
2300                          !
2310                          !
2320 PRINTER IS 0                   ! set output to printer
2330                          !
2340 IMAGE "Q    BPD  DR PCT  OD IN  WT IN  VIS CST  f  w/o FLO      MLC/KM
     PPM    GPD "
2350 PRINT USING 2340
2360 IMAGE "                                        w/o DRA  w/DRA"
2370 PRINT USING 2360
2380 PRINT
2390 IMAGE DDDDDD,XX,DDD.DD,XX,DD.DD,XX,Z.DDD,XXX,DD.DD,XXX,XZ.DDDD,XXX,DD.DDD
     ,X,DD.DDD,XX,DDD.D,XX,DDDDD
2400 PRINT USING 2390;Q_bpd,Dr_pct,Od_in,Wt_in,Vis_cst,Fo,Sfo,Sfp,Phi,Gpd
2410                          !
2420 PRINTER IS 16                  ! reset output to screen
2430                          !
2440                          !
2450 PRINT PAGE                     ! clear screen
2460                          !
2480 !          ********** EXIT ROUTINE **********
2500 PRINT "YOU MAY CHANGE ANY INPUT PARAMETER AND RE-RUN WITH THE NEW VALUE"
2510 PRINT
2530 PRINT "YOU MAY NOW QUIT THE PROGRAM OR ENTER A NEW VALUE FOR A VARIABLE"
2540 PRINT
2550 PRINT "FOR READY REFERENCE, THE VARIABLES AVAILABLE FOR CHANGE ARE THE FOL
     LOWING"
2560 PRINT
2570 PRINT
2580 PRINT "    (1)  DENSITY                Den_15"
2590 PRINT "    (2)  VIS IN CST             Vis_cst"
2600 PRINT "    (3)  DRAG REDUCTION PERCENT Dr_pct"
2610 PRINT "    (4)  OD INCHES              Od_in"
2620 PRINT "    (5)  WT INCHES              Wt_in"
2630 PRINT "    (6)  ROUGHNESS INCHES       Rough_in"
2640 PRINT "    (7)  FLOW RATE IN BPD       Q_bpd"
2650 PRINT
2660 PRINT
2670 PRINT "    TO QUIT THE PROGRAM ENTER 0,0 AND HIT CONT"
2680 PRINT
2690 PRINT "    TO CHANGE THE VALUE OF A VARIABLE ENTER ITS NUMBER FROM THE AB
     OVE TABLE"
2700 PRINT "    AND ITS NEW VALUE, SEPARATED BY A COMMA, AND HIT CONT."
2710 PRINT
2720 PRINT
2730 INPUT "MAKE YOUR CHOICE",Choice,Value
2740 IF Choice=0 THEN
2750     END
2760 ELSE
2770     SELECT Choice
2780     CASE 1
2790         Den_15=Value
2800     CASE 2
2810         Vis_cst=Value
2820     CASE 3
2830         Dr_pct=Value
2840     CASE 4
2850         Od_in=Value
2860     CASE 5
2870         Wt_in=Value
2880     CASE 6
2890         Rough_in=Value
2900     CASE 7
2910         Q_bpd=Value
2920     END SELECT
2930 END IF
2940 GOTO Start
```

Figure 2. CBLFLO program listing.

and Seyer correlation incorporating the characteristic time equation takes many minutes to run on such machines. Another representation commonly seen in published literature which programs well in handheld calculators is found in Equation 7.

Here A and B are determined by regression of the Savins and Seyer correlation output.

This kind of presentation is shown in Figure 4. The solid line represents the output of the Savins and Seyer correlation; the straight-line Equation 7 is the dashed line. The y-intercept of the straight line is commonly thought of as the (strictly) theoretical maximum drag reduction obtainable with the given DRA and solvent. In Figure 4 the intercept is about 1.55, and maximum DR is, therefore, about 0.65. The straight line yields acceptable values up to a DR of about 0.40.

These curves apply only to a pipe line of given parameters flowing crude of given characteristics dosed with a DRA of given characteristic time coefficients for a given flow rate. As yet, with the best present knowledge, the absolute accuracy of the Savins and Seyer correlation, and therefore of the curves, is not what might be desired.

Figure 3. CBLFLO computes DRA for desired ratio.

Figure 4. Effect of increased DRA.

Equations

$$DR = (S_{f\ untreated} - S_{f\ treated})/S_{f\ untreated} \qquad (1)$$

Where: S_f = Unit friction loss taken as a slope

$$D/R = (1 - DR) = (S_{f\ treated})/(S_{f\ untreated}) \qquad (2)$$

$$f_{treated} = f_{untreated} \times D/R \qquad (3)$$

$$S \times (1 + A_o \times ([2^{1/2} \times \log_{10}S + 1.454 \times G_o \times T_p \times S - 0.8809])^2 = 1 \qquad (4)$$

Where:

S = D/R
A_o = $(f_{untreated}/8)^{1/2}$
G_o = Wall shear rate$_{untreated}$
T_p = Characteristic time of the treated DRA-solvent solution

$$T_p = a \times P^b \times G_p{}^c \times A_o{}^d \qquad (5)$$

Where:

P = DRA concentration
G_p = Wall shear rate$_{treated}$
A_o = As above in equation 4
a, b, c, and d = Constants empirically determined for each DRA-solvent solution

$$ppm = A + B \times (D/R) + C \times (D/R)^2 + D \times (D/R)^3 + E \times (D/R)^4 \qquad (6)$$

$$1/ppm = A \times (1/DR) + B \qquad (7)$$

The relative accuracy, however, given two or three experimentally obtained points for a given DRA-solvent solution, is quite good. Therefore, we will proceed with the use of the data in hand and consider that the accuracy indicated is obtainable while realizing it may not be.

For the present, it is the shape of these curves which is important.

First it should be noted that the independent variable is D/R or DR; the input to the Savins and Seyer correlation is DR and the output is ppm of DRA. The inverse problem, given ppm to find D/R or DR, requires an iterative solution to the Savins and Seyer correlation.

Second, note the effect is not linear, and the effect of the DRA decreases as concentration increases. With very large doses of DRA, the effective viscosity of the solution becomes so great that drag reduction disappears and is replaced with a drag increasing effect. The Savins and Seyer correlation will not predict this effect, and the Y-intercept value of Figure 4 is purely fictitious.

Last, note that the amount of DRA required to produce a reasonable drag reduction is physically quite small: A DR of 30% requires about 24 weight ppm of DRA.

Thus, if a pipe line is flowing under the conditions used to derive the curves and carries 1,000,000 gal/day of oil which has the same density as the DRA, injecting 24 gal/day of DRA into the line will reduce the friction loss at the same 1,000,000 gal/day to 70% of what it was before.

Caution

The accuracy implied in calculations herein should not be taken as representative of practice. Pipe line hydraulics, and especially the application of DRA's, are still inexact sciences, neither accurate nor especially precise. We justify the number of significant figures used in our calculations by noting that if calculations in such detail can be produced in two or more ways—which we tried to do—the calculations themselves are probably accurate.

Source

Oil and Gas Journal, February 4, 1985.

References

1. Toms, B. A., "Some observations on the flow of linear polymer solutions through straight pipe at large Reynolds numbers," Proceedings of the International Congress on Rheology, Scheveningen, The Netherlands, Vol. 2, pp. 135–141 (1948).
2. Burger, Edward D., Munk, Werner R., and Wahl, Harry A., "Flow increase in the Trans Alaska pipeline using a polymeric drag reducing additive," SPE 9419, 1980.

How drag-reducing agents can outweigh looping or boosting

C. B. Lester, C. B. Lester & Associates Inc., Longview, Texas

There often seem to be only two kinds of oil pipe lines: those too small or those too large.

If a pipe line is too large, an operator can worry over capital funds not well utilized. But the more fearful worry—inability to handle business offered—remains with the operator of a pipe line which is too small to take advantage of the market.

Until recently, after discharge pressures had been pushed up to the absolute maximum, the operator had to face reality: To obtain any further increase in throughput capability, his system had to be looped or boosted.

Now there is a third possibility: The operator can inject a drag-reducing agent (DRA). This liquid loop is used as a tool to benefit operators who need only a little more throughput.

Fundamental relations

A too-small pipe line is one which has no more physical room to produce an increased unit friction loss in the existing pipe. That is, the discharge pressures of the existing pump stations cannot be increased nor the suction pressures decreased. If an operator needs to double the capacity of his pipe line, his obvious solution is to loop it completely.

If he needs an increase of 10–30%, he has the choice of partial loops, booster pump stations, or injecting a DRA.

We will look at problems in this general range.

It is usual to consider booster stations and loops in terms of graphical solutions. However, since a graphical solution is not really meaningful for the DRA alternative, we will consider loops and boosters in terms of analytical solutions which can be handled by any engineering calculator.

The most commonly used, though perhaps not the best, friction factor correlation is that of Colebrook and White (C-W). This was almost never used in analytical form until the programmable calculators became available but is now in wide use. Its results can also be precisely duplicated. This brings up a problem which seems more difficult with time and involves a little-understood difference between precision and accuracy.

That the C-W correlation can be reproduced by different persons using different calculators means that it is precise but not necessarily accurate.

For purposes here, we will assume it is accurate, while recognizing that it may not be. But this will cause us little trouble because we will, in the end, be comparing flow rates of an untreated fluid with flow rates of the treated fluid in the same pipe line. In this kind of comparison, absolute accuracy becomes unimportant as long as the relative accuracy—which relates to precision—is acceptable.

The C-W correlation has a severe disadvantage when used in analytical solutions because it does require iterative techniques. To overcome this deficiency for purposes here, we will use a regression of the correlation over the range we need to use.

The C-W correlation as plotted by Moody (ASME, 1944) is on log-log coordinates.

Substantial ranges of the correlation can be approximated by a straight line, the slope of which is the derivative of the correlation.

To illustrate a specific situation, we may use the example of a 20 in. OD, 0.250 in. W.T. pipe line with 0.0007 in. equivalent roughness flowing 160,000 b/d of 5 cSt oil having a density of 0.8388.

The Reynolds number (Re) for this situation is 151,000; the roughess ratio is $0.0007/19.5 = 3.6 \times 10^{-5}$, and $S_f = 4.031$ ft of liquid column (flc)/1,000 ft.

We may derive a straight-line approximation for any range of the C-W correlation, but it is best to pick a point near the lower third of the range of use and Reynolds number which yields a derivative which is a round number to two significant figures. The slope of the f-vs.-Re C-W curve for the described conditions is -0.1900 at $Re = 147,000$, so we select this point. With some manipulation we find:

$$f = 0.161715 \times Re^{-0.1900} \tag{1}$$

where   f = D'Arcy friction factor
       Re = Reynolds number

This formula yields f less than 1% from C-W with $e/D = 3.6 \times 10^{-5}$ for a range of Re from 73,500 (half the 147,000 chosen point) to 441,000 (three times the 147,000 point).

The corresponding formula for S_f, the unit friction loss in terms of flc/1,000 ft, or meters of liquid column (mlc)/kilometer (the two are numerically the same and representing hydraulic gradeline slope on a 1:1,000 basis) is the following:

$$S_f = 4.281034 \times 10^{-3} \times Re^{-0.1900} \times (Q^2/D^5) \tag{2}$$

where Q = U.S. oil, b/d
      D = ID, in.

Given S_f, one can find Q directly by the following equation:

$$Q = 74,070 \times S_f^{0.552486} \tag{3}$$

These formulas are interesting but we are, for the most part, interested in the relations between the S_{fa} and Q_a, the unit friction loss and flow rate under existing conditions a, and S_{fb} and Q_b, the unit friction loss and flow rate under some other conditions b.

Thus we can find the following:

$$S_{fb} = S_{fa} \times (Q_b/Q_a)^{1.81} \tag{4}$$

and

$$Q_b = Q_a \times (S_{fb}/S_{fa})^{0.552486} \tag{5}$$

Here we have S_f and Q at described conditions b in terms of S_f and Q at known conditions a.

Note that while the explicit equation for S_f is written in terms flc/1,000 ft or mlc/km, the equations involving ratios can use any consistent units, i.e., psig/mile, ksc/km, etc.

Similarly, although the explicit equation for Q is in U.S. barrels per day, the formulas involving ratios can be in terms of gallons per minute, cubic meters per hour, metric tons/annum, so forth, with equal validity.

We need to know how to calculate loop lengths for a desired increase in flow compared to the single-line flow and to calculate the increase in flow obtained if a specified length of loop is constructed.

These relationships can be derived from the previous two equations if it is assumed that the loop pipe is the same outside diameter, wall thickness, and roughness as the pipe in original line—which we do assume.

We find:

$$N = (1 - 0.714809 \times M)^{-0.552486}(1 > N > 2) \tag{6}$$

$$M = (N^{1.81} - 1)/(0.714809 \times N^{1.81})(0 > M > 1) \tag{7}$$

where N = flow factor, the ratio of the flow rate without a loop to the flow rate with a loop M × L long
      M = loop factor, the fractional part of the line length L which is looped

Figure 1 shows the relationship between N and M.

We need a formula for horsepower in simple terms. Substituting in the basic U.S. horsepower formula hp = PLAN/33,000 = (pressure, psf) × (volume, cubic ft)/33,000 yields:

$$hp = 1.701389 \times 10^{-5} \times (psi) \times (b/d) \tag{8}$$

Taking 100s of psi (Cpsi) and 1,000s of b/d (Mb/d) and assuming a pump efficiency of 0.85 yields:

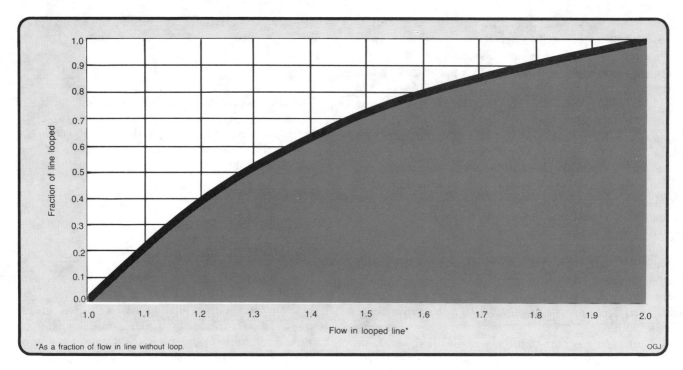

Figure 1. Flow affected by loop length.

$$hp = 2 \times Cpsi \times Mb/d \qquad (9)$$

which is simple, completely accurate at an efficiency of 0.8506, and accurate in practice for large pipe lines using pumps of modern design.

For the 0.8388 density oil selected for our example, 1 psi = 2.75 flc and

$$hp = 0.7273 \times Cflc \times Mb/d \qquad (10)$$

where Cflc = 100s flc

If the line is looped, or if a DRA is used, the horsepower required is a direct function of the flow increase. A 10% increase in flow over Q_a requires 110% of the horsepower required to pump Q_a because the discharge pressure is assumed to have been held the same at Q_b as at Q_a.

Thus the ratio of hp required at Q_b to that required at Q_a is (Q_b/Q_a) and all horsepower is installed at the existing station.

If a booster station is used, the ratio of system horsepower (the sum of horsepower at the existing station and the booster station) at Q_b to that required at Q_a is $(Q_b/Q_a)^{2.81}$. Of this, horsepower in the ratio (Q_b/Q_a), as above, is installed at the existing station and $(Q_b/Q_a)^{2.81} - (Q_b/Q_a)$ is installed at the booster station.

The ratio of the increase in system horsepower at Q_b to that required at Q_a is $[(Q_b/Q_a)^{2.81} - 1]$.

Figure 2 is a plot of the ratio of system horsepower with the booster to system horsepower without the booster-vs.-flow with the booster as a fraction of flow without the booster. We can handle drag reduction solutions in a similar manner.

Recall the following definitions of drag reduction for the same flow rate and of drag ratio:

$$DR = (S_{f\ untreated} - S_{f\ treated})/(S_{f\ untreated}) \qquad (11)$$

$$D/R = (1 - DR) = (S_{f\ treated})/(S_{f\ untreated}) \qquad (12)$$

Since one can derive ratio formulas involving S_{fa} and S_{fb} without drag reduction, formulas with drag reduction follow directly since D/R operates on S_f directly:

$$Q_{treated} = (1 - DR)^{-0.552486} \times Q_{untreated} \qquad (13)$$
$$= (D/R)^{-0.552486} \times Q_{untreated} \qquad (13a)$$

$$Q_{untreated} = (1 - DR)^{0.552486} \times Q_{treated} \qquad (14)$$
$$= (D/R)^{0.552486} \times Q_{treated} \qquad (14a)$$

Note the relationships here are the same as the fundamental relations expressed by Equation 5.

Having in hand the drag reduction required and the characteristic time coefficients for the DRA-crude oil solution, we can use the computer program of the previous part to compute amount of DRA required directly.

*Ratio of total system horsepower installed to system horsepower without booster.   †As a fraction of flow without booster.   OGJ

Figure 2. Flow affected by changes in system horsepower.

A specific application

The example we selected was a section of a 20-in. OD × 0.250-in. W.T. × 0.0007 in. roughness pipe flowing 160,000 b/d of 0.8388 density 5 cSt oil yielding a slope of 4.031 flc/1,000 ft.

To make the example more specific, we will assume the section is flat (i.e., no static heads are involved start-to-finish) and that it is constructed of API 5 LX 52 pipe having a maximum allowable operating pressure in accordance with ANSI B 31.4 of 936 psig = 2,574 flc of oil.

If the friction loss in the piping of one station:

$$H_{fs} = 2.15 \times 10^{-3} \times (Mbd)^2 = 55 \text{ flc} \qquad (18)$$

and the (reasonable) suction requirements of the pumps:

$$NPSH = 40 \text{ flc} \qquad (19)$$

then the total head available to overcome friction is:

$$H_f = (2,574 - 55 - 40)\text{flc} = 2,479 \text{ flc} \qquad (20)$$

and the indicated station spacing is:

$$L = H_f/4.031 = 615,000 \text{ ft} \qquad (21)$$

or more than 116 miles.

The mainline pumps will have to put up total dynamic head (tdh):

$$\text{tdh} = 2,479 + 55 \text{ flc} = 2,534 \text{ flc} \qquad (22)$$

We assume the operator of this pipe line wants to boost high throughput capability by 10, 20, or 30%. The degree of expansion depends on its cost and the income to be derived. The kind of expansion depends on the length of time the additional load is expected to be available.

We assume he will install pumping horsepower in the existing station to maintain the 936 psig P_d = 2,479 flc H_d under all pumping conditions.

We can first look at loops. With reference to Equation 7, we find the following:

Q = 176,000 b/d
M = 1.10
N = 0.2216
L_L = 136,300 ft

Q = 192,000 b/d
M = 1.20
N = 0.3932
L_L = 241,800 ft

Q = 208,000 b/d
M = 1.30

$N = 0.5259$
$L_L = 325,300$ ft

where L_L is the length of loop required per 615,000-ft section.

The incremental horsepower required is proportional to the increase in flow and is 295 hp, 590 hp, and 885 hp, respectively.

Next we can look at a booster. Referring to Equation 4, we find the following:

$Q = 176,000$ b/d
$S_{fb} = 4.031 \times 1.1^{1.81} = 4.790$
$H_f = 2,946$ flc

$Q = 192,000$ b/d
$S_{fb} = 4.031 \times 1.2^{1.81} = 5.607$
$H_f = 3,448$ flc

$Q = 208,000$ b/d
$S_{fb} = 4.031 \times 1.3^{1.81} = 6.481$
$H_f = 3,986$ flc

We assumed the operator would maintain pressure at the existing station and the difference between the total required head and head produced by the existing station will be picked up in the booster. For the three conditions, the differences are the following:

$$(2,946 - 2,534) + 2.15 \times 10^{-3} \times (176)^2 = 412 + 67$$
$$= 479 \text{ flc}$$

$$(3,448 - 2,534) + 2.15 \times 10^{-3} \times (192)^2 = 914 + 79$$
$$= 993 \text{ flc}$$

$$(3,986 - 2,534) + 2.15 \times 10^{-3} \times (208)^2 = 1,452 + 93$$
$$= 1,545 \text{ flc}$$

The important thing, horsepower added at the booster, is as follows:

$Q = 176,000$ b/d
$hp_{added} = 613$ hp

$Q = 192,000$ b/d
$hp_{added} = 1,387$ hp

$Q = 208,000$ b/d
$hp_{added} = 2,337$ hp

Horsepower figures were made using Equation 10.

To evaluate the DRA alternate, it is necessary to compute the dosages (in ppm) of the DRA to achieve 176,000, 192,000, and 208,000 b/d and convert these figures to gallons-per-day (b/d) requirements.

Using the $S_{untreated}$ values listed in the booster calculations, and given the coefficients a, b, c, and d of the characteristic time equation, we can find for the particular formulation of ChemLink's FLO DRA previously used as an example:

$Q = 176,000$ b/d
$D/R = 0.842$
ppm = 9

$Q = 192,000$ b/d
$D/R = 0.719$
ppm = 21

$Q = 208,000$ b/d
$D/R = 0.622$
ppm = 35

The ppm values are to the nearest ppm. We recognize that this accuracy may not be obtainable without having in hand a few experimentally taken correlation points. Since ppm are calculated in weight units, it is necessary to derive a relationship for the flowing rate of the oil in lb.

This is, for b/d of 0.8388 specific gravity oil,
lb = 1 b/d × 42 gal × 231 cubic in./1,728 cubic in./cubic ft ...
$$\times\ 62.42795 \text{ lb/cubic ft } H_2O$$
$$\times\ 0.8388 \text{ specific gravity oil} = 294 \times \text{b/d} \qquad (23)$$

The density of most DRA's is about 0.790 or 6.59 ppg. Thus the gallons-per-day requirements can be calculated by this equation:

$$\text{gpd} = 294 \times \text{b/d} \times \text{ppm}/1,000,000/6.59$$
$$= 4.46 \times 10^{-5} \times \text{b/d} \times \text{ppm} \qquad (24)$$

Using this relationship, we find the dosage rates for the three conditions to be:

$Q = 176,000$
$\text{gpd} = 4.46 \times 10^{-5} \times 176,000 \times 9 = 71$ gpd

$Q = 192,000$
$\text{gpd} = 4.46 \times 10^{-5} \times 192,000 \times 21 = 180$ gpd

$Q = 208,000$
$\text{gpd} = 4.46 \times 10^{-5} \times 208,000 \times 35 = 325$ gpd

Comparisons

Table 1 summarizes the results derived in the preceding discussion.

Table 1
Capacity Increases of 10, 20, and 30%*

	10%	20%	30%
Alternate A—partial loops			
Loop required, ft	136,300	241,800	325,300
Additional hp required			
Existing stations	295	590	885
Booster stations	0	0	0
Drag reducer required, gpd	0	0	0
Alternate B—boosters			
Loop required, ft	0	0	0
Additional hp required			
Existing stations	295	590	885
Booster stations	613	1,387	2,337
Drag reducer required, gpd	0	0	0
Alternate C—drag reducers			
Loop required, ft	0	0	0
Additional hp required			
Existing stations	295	590	885
Booster stations	0	0	0
Drag reducer required, gpd	71	180	325

Comparison of cost/ft, or of booster hp, or of gpd of DRA:

Increase, %	Flow rate, b/d	Loop, ft	Booster, hp	DRA, gpd
10	176,000	136,300	613	71
20	192,000	241,800	1,387	180
30	208,000	325,300	2,337	325

*160,000 b/d, 116.5-mile, 20-in. pipeline

Table 2
Incremental Capacity Increases of 10%*

	1st 10%	2nd 10%	3rd 10%
Alternate A—partial loops			
Loop required, ft	136,000	105,000	83,500
Additional hp required			
Existing stations	295	295	295
Booster stations	0	0	0
Drag reducer required, gpd	0	0	0
Alternate B—boosters			
Loop required, ft	0	0	0
Additional hp required			
Existing stations	295	295	295
Booster stations	613	774	950
Drag reducer required, gpd	0	0	0
Alternate C—drag reducers			
Loop required, ft	0	0	0
Additional hp required			
Existing stations	295	295	295
Booster stations	0	0	0
Drag reducer required, gpd	71	109	145

Comparisons of cost/ft, or of booster hp, or of gpd of DRA:

Increase, %	Flow rate, b/d	Loop, ft	Booster, hp	DRA, gpd
10	16,000	136,300	613	71
10	16,000	105,500	774	109
10	16,000	83,500	950	145

*160,000 b/d, 116.5-mile, 20-in. pipeline

Note that the same added horsepower is required at the existing station for each kind of solution because of our assumption that the pressure at the existing station would remain constant in each case.

For comparison purposes, therefore, we can forget the increased horsepower required at the existing station.

The comparison, then, comes down to the cost of X ft of loop, Y booster hp, or Z gpd of DRA.

Table 2 gives the same information as Table 1 except in terms of incremental throughput of 10% = 16,000 b/d each.

Instinctively, one realizes that a DRA is really a liquid loop, and the partial loop-vs.-DRA cases are directly comparable.

To make a decision for a long-term increase, one has to determine if it is better to build X ft of loop, which will have large annual fixed costs and nearly zero operating costs, or to inject Z gpd of DRA and gain the same capacity with nearly zero investment and, therefore, fixed costs while incurring the ongoing cost of DRA consumed.

Instinct fails when making long-term comparisons of boosters with either loops or DRA because boosters have not only a substantial investment, and therefore annual fixed costs, but they can have a high operating cost as well.

A kind of "reductio ad absurdum" comparison exists when comparing to DRA, in that if $365 \times Z$ gpd of DRA costs less than the annual power cost of Y booster hp, the DRA is obviously the more economic solution.

From experience, one knows that the booster station-vs.-partial loop comparison depends on the cost and time of recovery of invested capital and on operating costs. The same factors enter into long-term comparisons of the two hardware solutions with the DRA solution, and the relationships are quite complex and sometimes obscure.

As an example of this complexity, note that the efficiency of a partial loop increases as its length increases, so that it takes less loop to go from 20% to 30% increase than from 10 to 20%.

On the other hand, it takes less horsepower to increase from 10 to 20% with boosters than it does to go from 20% to 30%. And, as we have seen, DRA consumption is a higher order function of the drag ratio (D/R) and therefore has the same kind but a different order of increase with increasing capacity as the booster solution.

In every case, now that DRA's are available, DRA's provide the only quickly available, short-term solution to a capacity increase problem.

One cannot construct a substantial partial loop, or build a booster station, overnight; there is a certain minimum time required for construction.

And, usually one cannot economically construct a loop or a booster station for a 1- or 2-year increase because the cost of capital recovery over a short-term would be prohibitive.

DRA's, however, offer an almost instantaneous solution to a capacity increase problem.

Once the line is filled with treated oil, the full capacity increase is in effect.

This kind of increase is available in a matter of hours, not months or years.

A capacity increase obtained with a DRA can be an on-and-off kind of increase with no investments to salvage at a loss if the period of time the increase is required turns out to be overestimated. If a greater capacity is required this month, use it; if it isn't needed next month, don't use it. It won't go stale in the carton.

We have purposefully not included an economic evaluation of the loop-vs.-booster-vs.-DRA situation. Every operator has his own ideas as to his direct and indirect costs of construction, the rate at which he wants to recover his invested capital, and his operating costs. He can ascertain the cost of a DRA.

He also knows the worth of a capacity increase instantly available as compared to an increase obtained months or years from now.

With this information in hand an economic evaluation comes very quickly.

Source

Oil and Gas Journal, February 18, 1985.

How active, passive drag affect DRA injections

C. B. Lester, C. B. Lester & Associates Inc., Longview, Texas

This third in a series of articles on drag-reducing agents examines applications more deeply and eliminates some of the simplifying assumptions purposely made in the second part to explain fundamental functions of DRA's in pipe lines.

Line drag

Drag reduction was defined in part 1 in the following equation:

$$DR = (S_{f\ untreated} - S_{f\ treated})/S_{f\ untreated} \qquad (1)$$

Here S_f means unit friction loss in consistent units (ft of liquid column [lc] per 1,000 ft, meters of liquid column [mlc] per kilometer, psig/mile, etc.). The terms treated and untreated refer to pipe line liquids dosed or not dosed with a DRA.

A companion, sometimes more useful factor is the drag ratio:

$$D/R = (1 - DR) = (S_{f\ treated}/S_{f\ untreated}) \qquad (2)$$

DR and D/R thus computed involve only the parameters of the pipe line, the characteristics of the DRA-solvent solution, and the flow rate. Therefore, they can be related to line drag.

Line drag arises in the pipe line itself and varies with flow as a complex function of Reynolds number and pipe roughness. It also is the only drag acting in a pipe line system which can be reduced with a DRA and is therefore the active drag.

Most workers in the field of drag reduction are concerned only with active drag, thus with S_f only. This definition of drag is, for their purposes, complete and sufficient.

For a pipe liner, however, this definition is incomplete because there are other effects working in a pipe line which, while not a drag per se, act as one. These as much as any increased active drag must be overcome if a desired increase in flow is to be achieved. There are two of these passive drags: station and static.

Station drag

One of the passive drags is the turndown in the H-Q characteristic of centrifugal pumps with increasing flow, and the accompanying increase in station piping loss which increases with flow. Both reduce H_d, the pump station discharge head, as flow increases. We call this kind of loss station drag because it arises within the pump station. It varies as the square of the flow rate.

Figure 1 shows the situation for a section of a pipe line with no static head acting in the system, i.e., static heads do not act start-to-finish in the pipe line section.

We again take the example of a 20-in. × 0.250-in. × 0.0007-in. roughness pipe line flowing 160,000 b/d of 0.8388 density 5 cSt oil with S_f on a 1/1,000 basis of 4.031 flc/1,000 ft.

Under ANSI B 31.4, the pipe line is allowed to pump at 936 psig = 2,574 flc, all available to overcome friction except $2.15 \times 10^{-3} \times (1,000 \text{ b/d})^2 = 55$ flc lost in friction in the station piping and an assumed 40 flc suction requirement at the next downstream station.

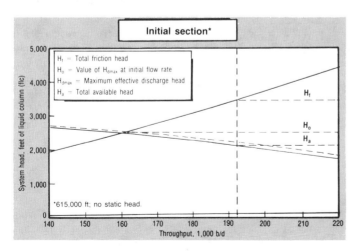

Figure 1. Example 20-in. pipe line system.

Thus at 160,000 b/d, 2,479 flc are available to overcome friction over the 615,000-ft length of the section. This value is H_{dmax}, the maximum effective discharge head of the pump station. On Figure 1, this value is the intersection of the station head curve with the 160,000 b/d flow ordinate.

The value of H_{dmax} at the 160,000-b/d initial flow rate is called H_o and is so marked on the H_d curve. H_o is the point of reference for computing active and passive drag.

To increase flow from 160,000 to 192,000 b/d, the unit friction loss, S_f, rises from 4.031 to 5.607 flc/1,000 ft and total friction head H_f rises to $615 \times 5.607 = 3,448$ flc.

This is an increase of $615 \times (5.607 - 4.031) = 969$ flc. Overcoming this increase requires a DR of

$$(H_f - H_o)/H_f = (3,448 - 2,479)/3,448 = 0.281$$

and yields a D/R of 0.609.

Note, however, that there is less effective head available at 192,000 b/d than at 160,000 b/d. Pump turndown has reduced the head available from the pumps, and the increased loss in the station piping has further reduced the head available from the station.

The dashed line pump curve in Figure 1 is the total dynamic head (tdh) curve for the pumps; the solid line below it is the station discharge head (H_d), which is the tdh curve with station losses, shown in a solid line at the bottom of the graph, subtracted. The head available (h_a) at 192,000 b/d = 2,102 flc.

The reduction in the total available head, H_a, from a pump station acts as a loss in the same way in the total flow picture as the increase in total required friction head H_f. It is a loss in the system which must be made up by reducing S_f because, other than changing out pumps or repiping the station manifold, there is no way to increase H_d. We must consider station drag as well as line drag to arrive at total drag and be able to compute the drag reduction required to achieve a given flow rate.

Solutions—1. A graphical solution for total drag in multi-station pipe lines without active static heads is handy, in a way, in that the problem is immediately seen. With $S_f = H_f/L$ and $S_a = H_a/L$, the required total drag ratio D/R can be written as follows:

$$D/R = (H_a/L)/(H_f/L) = H_a/H_f \qquad (3)$$

The corresponding required total drag reduction DR becomes the following:

$$DR = (1 - D/R) = [1 - (H_a/H_f)] \qquad (4)$$

The main problem with graphical solutions is that they don't program and are difficult to repeat. A graph yields a different answer every time it is read, and as the problems become more complex, the accuracy of the solutions suffers. Also, the pipe liner's usual way of plotting system curves does not serve well in studying drag-reduction problems. We will therefore use an analytical solution which can be applied to any kind of problem.

For the pump station design we are using, the loss in station piping has been given previously in the following form:

$$H_{fs} = kQ^2 \tag{5}$$

which we took numerically as:

$$H_{fs} = 2.15 \times 10^{-3} \times (1,000 \text{ b/d})^2 \tag{6}$$

This kind of simple velocity-squared relationship is satisfactory for any kind of problem. It may not be completely accurate but it will suffice.

The general form of the characteristic curve for a centrifugal pump of the kind used on pipe lines can be written in the following form:

$$tdh = A + BQ + CQ^2 \tag{7}$$

This representation is usually quite accurate for flow rates between $0.75 \times Q_{bep}$ and $1.25 \times Q_{bep}$, where Q_{bep} is the flow rate at the best efficiency point on the curve. A regression of the pump curve shown in Figure 1 is as follows:

$$tdh = 2,666 + 7.67 \times (1,000 \text{ b/d})$$
$$- 53.10 \times 10^{-3} \times (1,000 \text{ b/d})^2 \tag{8}$$

Inasmuch as Equation 6 subtracts directly from Equation 8 to yield H_d we may write the following:

$$H_d = 2,666 + 7.67 \times (1,000 \text{ b/d})$$
$$- 55.25 \times 10^{-3} \times (1,000 \text{ b/d})^2 \tag{9}$$

which gives us H_d in terms of the flow rate in 1,000s b/d. At 160,000 b/d

$$H_f = H_o = H_d = 2,479 \text{ flc}$$

at 192,000 b/d

$$H_f = 3,448 \text{ flc}$$

and

$$H_a = H_d = 2,102$$

Station drag and line drag cannot be added algebraically to compute total drag reduction. Note the following, with Figure 1 as reference:

For line drag:

$$DR = (H_f - H_{dmax})/H_f$$
$$= (3,448 - 2,479)/3,448$$
$$= 0.281 \tag{10}$$

And for pump drag:

$$DR = (H_{dmax} - H_a)/H_a$$
$$= (2,479 - 2,102/2,479)$$
$$= 0.152 \tag{11}$$

These yield a total of line drag + pump drag = 0.433, compared to the correct 0.390 calculated previously from the DR = H_a/H_f relationship.

However, without proof:

$$D/R_{line} = H_o/H_f$$
$$= 2,479/3,448 = 0.719 \tag{12}$$

$$D/R_{station} = H_a/H_o$$
$$= 2,102/2,479 = 0.848 \tag{13}$$

and:

$$D/R_{total} = D/R_{line} \times DR_{station}$$
$$= 0.719 \times 0.848$$
$$= 0.610 \tag{14}$$

This is the correct value for D/R.
Thus we may take the following as rules:

$$D/R_{total} = D/R_{line} \times D/R_{station} \tag{15}$$

$$DR_{total} = DR_{line} + DR_{station} - DR_{line} \times DR_{station} \tag{16}$$

Here Equation 16 follows from substitution of Equation 2 in Equation 15.

Having in hand the a, b, c, and d coefficients of the formulation of FLO produced by ChemLink Inc., a subsidiary of Atlantic Richfield Co., we can use the CBLFLO program to estimate the amount of this particular DRA required to bring the pipe line up to 192,000 b/d while taking into account the 377 flc reduction in H_d resulting from this 32,000 b/d increase; thus the following:

Q = 192,000 b/d
DR = 0.390
ppm = 38
gal/day = 326

When we assumed pump station head stayed constant as flow increased, we obtained

Q = 192,000 b/d
DR = 0.281
ppm = 21
gal/day = 180

To make up the 377 flc lost in pump drag, the pipe line operator can decide whether to spend the extra 146 gal/day of FLO or to install a pump burning 527 hp in the existing station.

Solutions—2. Previous examples have progressed from a simplistic approach to drag reduction on a single pipe line section with constant station pressures without static heads acting end-to-end in the section through an example which takes into account the reality that pump heads and station losses change with changing flow.

Now we can consider the more general problem of total drag in multi-station pipe lines with active static heads.

Static drag is a passive drag. It may be a positive drag which acts as any other drag to reduce flow, or it may be a negative drag which increases flow.

It differs from line drag and station drag in that it does not vary with flow. It does, however, directly affect the application of DRA.

For simplicity we will look at a two-station system; using more than two stations could cloud fundamentals and would not improve the example.

Section 1 is the same 160,000-b/d, 615,000-ft, 20-in. crude line without active static heads used in the previous example. Section 2 is an identical section except it is a 367,000-ft uphill section and looks at a 1,000 ft positive static head.

Figure 2 shows the system head curves for Section 2. It is important for pipe liners to understand that this set of system head curves is different from their usual kind of system curves: Instead of the positive H_s being added to H_f to obtain a curve of effective $H_t = H_s + H_f$, the H_s is subtracted from the H_d curve to yield an effective H_d curve.

This kind of representation correctly shows the relationship beween energy input and energy use in the system from the DRA standpoint but at the expense of incorrectly showing the absolute value of H_d and H_a.

The absolute value of H_d at 160,000 b/d is 2,479 flc as in Section 1, but 1,000 flc of this is used to overcome 1,000 ft of static head. The effective value of H_d in Section 2 at 160,000 b/d, therefore, is 2,479 flc − 1,000 ft = 1,479 flc and is called H_o.

Illustrated here is the most important point in handling active and passive drags in a pipe line problem: Passive drags must be removed from the problem before calculations relating to active drag are carried out.

We did this intuitively for Section 1 when we took into account station drag by using the tdh and H_{fs} equations to arrive at H_d, and this is the pipe liner's usual way. We must

now agree that, after station drags are taken into account, static drags must also be taken into account in the H_d representation by a further subtraction (or addition, if a downhill section) of H_s from H_d to yield the effective H_d curve.

In Figure 2, H_s is not shown as it would lie on the H = 1,000 flc abscissa. H_d is shown as a solid line which is the algebraic sum of the pump tdh curve, the H_{fs} curve, and the constant H_s. The curve of H_f is as shown in Figure 1 for the shorter length of Section 2.

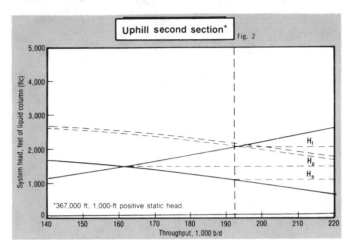

Figure 2. Example 20-in. pipe line system.

From this representation, we can take H_f = 2,057 flc, H_o = 1,479, and H_a = 1,102 flc. The following comparison results. Section 1:

D/R = H_a/H_f = 2,102/3,448 = 0.610

DR = 0.390

Section 2:

D/R = H_a/H_f = 1,102/2,057 = 0.536

DR = 0.464

Section 2 requires a greater DR to achieve 192,000 b/d than Section 1. The same pump drag must be recovered, but an additional 1,000 ft of static drag must be overcome and the active line drag on which DRA's can operate to effect drag reduction is smaller by the 366,000/615,000 ratio of the section line lengths.

The dosage of DRA works out as follows. For Section 1:

Q = 192,000 b/d
DR = 0.390
ppm = 38
gal/day = 326

for Section 2:

Q = 192,000 b/d
DR = 0.464
ppm = 57
gal/day = 488

To complete this analysis, we can assume a third 20-in. pipe line section following Section 2 which is a downhill section with a negative H_s of 1,000 ft. With the same pumps, this section would be 863,000 ft long, and the conditions working are those shown in Figure 3. Here H_f = 4,839 flc and H_o and H_a = 3,479 and 3,102 flc, respectively.

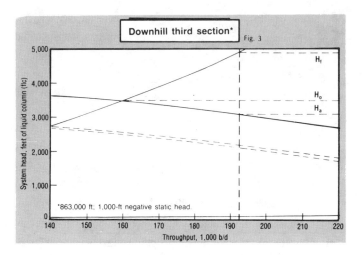

Figure 3. Example 20-in. pipe line system.

The D/R required for Section 3 is

$$H_a/H_f = 3,102/4,839 = 0.641$$

and

$$DR = (1 - D/R) = 0.359$$

To obtain this drag reduction requires 33 ppm of FLO = 283 gal/day.

We can now obtain the total cost (Table 1) of obtaining a 32,000-b/d increase in flow in the three-station, 1,835,000-ft (347.5-mi) 20-in. line.

Optimizing DRA dosage in multi-station pipe lines

The average DRA dosage rate per station is 366 gal/day, but the rates vary from 76.7% of the average to 133.3% of the average. We might suspect that there is a better way to use DRA than what we have calculated thus far. And this is sometimes true.

Remembering that the curve of ppm-vs.-D/R or DR is a higher order function of DR or D/R, we know that 100 gal/day injected in Section 2 does not produce the same incremental increase in flow rate Q as 100 gal/day in Section 1. It produces the same change in S_f in each section, but it does not produce the same change in Q. Thus the effect of passive drags is evinced.

If we could adjust the dosage so that each section receives the same dosage of DRA, we could produce a lower total dosage rate. The mathematics of the calculation are controlling.

We might suspect that we can take advantage of the flexibility of pump stations using centrifugal pumps to carry forward unused head from one section to another; i.e., to dose one section more heavily than required by the hydraulics of that section alone with the object of carrying forward the unused head to and through the succeeding downline station.

Heads thus carried forward would decrease the dosage rate of DRA in the downline section. We can do this subject to two limitations: Pump suction heads cannot drop below minimum required NPSH; and station discharge heads cannot exceed maximum working head.

The following example illustrates.

If we take the three sections together, we can find the results illustrated in Table 2. The D/R for the total line is the same as that of Section 1, something we might have suspected, since H_s for Section 2 balances out H_s for Section 3. Therefore, we should be able to inject 326 gal/day of FLO in each section and attain our objective.

But we can't. If we inject 326 gal/day at Station 1 to obtain a DR of 0.390 in Section 1, we will arrive at Station 2 with a suction head of 40 flc. But a DR of 0.390 will not be enough to overcome all of the drags working in Section 2, and we will arrive at the suction of Station 3 with something less than 40 flc which is technically unacceptable. We cannot always combine sections to reduce total DRA dosage.

Table 1
Obtaining 32,000-b/d Increase in a 20-in. Line*

Section 1	Q = 192,000 b/d	DR = 0.390	ppm = 38	gal/day =	326
Section 2	Q = 192,000 b/d	DR = 0.464	ppm = 57	gal/day =	488
Section 3	Q = 192,000 b/d	DR = 0.359	ppm = 33	gal/day =	284
Total				gal/day =	1,098
Average				gal/day =	366

* Three-station, 347.5-mi.

Table 2
Example DRA Injection by Section

		Section 1	Section 2	Section 3	Total
H_f =	*flc 192,000 b/d	3,448	2,057	4,839	10,344
H_s =	*flc 160,000 b/d	2,102	1,102	3,102	6,306
D/R =	H_s/H_f	0.610	0.536	0.641	0.610
DR =	1 − D/R	0.390	0.464	0.359	0.390

* Feet of liquid column.

Table 3
Reversed Order of DRA Injection in Example

		Section 1	Section 2	Section 3	Total
H_f =	*flc 192,000 b/d	4,839	2,057	3,448	10,344
H_s =	*flc 160,000 b/d	3,102	1,102	2,102	6,306
D/R =	H_s/H_f	0.641	0.536	0.610	0.610
DR =	1 − D/R	0.359	0.464	0.390	0.390

* Feet of liquid column.

Table 4
Calculation Sheet for Balancing DRA Requirements*

Description	Unit	Start/end	Section 1	Section 2	Section 3
Flow rate	b/d		192,000	192,000	192,000
Flowing density	gm/ml		0.8388	0.8388	0.8388
Flowing viscosity	cSt		5.0	5.0	5.0
Pipe OD	in.		20.0	20.0	20.0
Pipe W.T.	in.		0.250	0.250	0.250
Pipe roughness	in.		0.0007	0.0007	0.0007
Elevation	ft amsl†	start	1,000	0	1,000
Station suction	flc§	start	40	192	40
Pump suction	flc	start	1,040	192	1,040
Pump thd¶	flc	start	2,181	2,181	2,181
Pump discharge	ft amsl	start	3,221	2,373	3,221
Station piping loss	flc	start	79	79	79
Station discharge	ft amsl	start	3,142	2,294	3,142
S_f required	flc/1,000 ft		5.607	5.607	5.607
Drag reduction	decimal		0.3904	0.3904	0.3904
Drag ratio	decimal		0.6096	0.6096	0.6096
S_f produced	flc/1,000 ft		3.418	3.418	3.418
Length of section	ft		863,061	366,906	614,984
H_f	flc		2,950	1,254	2,102
Station suction	ft amsl	end	192	1,040	1,040
Elevation	ft amsl	end	0	1,000	1,000
Station suction	flc	end	192	40	40

* Three-station system.
† Above mean sea level.
§ Feet of liquid column.
¶ Total dynamic head.
• Station pipe loss lumped on discharge side of pumps for simplicity. For accuracy, loss should be divided between suction and discharge side to ensure pump suction does not drop below required NPSH.

We can illustrate the method by reversing the order of the sections (Table 3). These calculations, and a recommended form for making such calculations, are shown in Table 4.

Now if we start out with a DR of 0.390 in Section 1, we will come into the suction of Station 2 with a higher suction head than required, which will increase the effective H_d of Station 2. The effect will go forward ending up with the required 40 flc at the end of Section 3.

Thus we obtain the advantage of carrying drag reduction forward and our total consumption of DRA will be 3 × 326 gal/day = 978 gal/day, as compared to 1,098 gal/

day when each section is treated as a separate pipe line. The savings, 120 gal/day, is well worthwhile because while the economic effect of DRA's is sometimes dramatic they are not inexpensive.

The rule here is that when a pipe line is perfectly hydraulically balanced, i.e., when maximum allowable discharge head and minimum allowable suction head are reached simultaneously in all stations at the same flow rate (a theoretical situation usually but a good basis for making rules), drag reduction to achieve a higher flow rate can be carried forward only as long as a limiting station discharge or suction head is not reached.

For unbalanced lines, sometimes even reaching a limit in discharge or suction does not prevent unbalancing the drag-reduced system with a consequent savings in total DRA.

A long line can sometimes be divided into groups of sections and drag reduction balanced among the sections in the group. Thus while drag reduction cannot be carried forward through all the stations, it can be carried forward through a group of stations with the consequent economic advantage of reducing total DRA required to obtain a desired objective.

Source

Oil and Gas Journal, March 4, 1985.

How to estimate the rate of liquid discharge from a pipe

A quick and surprisingly accurate method of determining liquid discharge from an open ended horizontal pipe makes use of the equation:

$$gpm = \frac{2.56 \times d^2}{\sqrt{Y}}$$

where gpm = gallons per minute
d = diameter of pipe, in.
X = in.
Y = in.

Example. From the drawing:

X = 24 in.
Y = 25 in.
d = 2 in.

$$gpm = \frac{(2.56)(24)(2^2)}{\sqrt{25}}$$

$$= \frac{245.76}{5}$$

$$= 49.15$$

Predict subsurface temperature ranges

Ground temperatures are vital to pipe line design, for two reasons. First, the heat exchange between the contents of the line and the ground are greatly affected by the differential temperatures of the pipe and the soil. Second, the temperature of the pipe at times of installation will have a considerable effect on stresses, particularly where the subsequent ground temperatures will vary considerably from the installation temperatures.

For example, in the recent construction of a 34-in. line that is to operate at temperatures well over 100°F, care was exercised to limit bends when it was installed at temperatures ranging from 50 to 70°F and the underground temperature will range from about 50 to 80°F over a year's time.

Generally speaking, most areas will have somewhat similar curves from three feet on down. Of course, with depth, they approach the temperature of underground water, and the summer-winter variation ceases.

To adapt for U.S. work. As a comparison the following temperature ranges are encountered in winter and summer at depths of 3 ft in various areas in the United States.

Figure 1. The accompanying curve and table shows the range of underground temperatures you can expect in the prairie regions of Canada. At shallow depths, there is some variation and uncertainty as to temperature because of snow or sleet covering. But from three feet down, the temperature stabilizes.

California .... 51°F to 81°F
Colorado plains .... 35°F to 65°F
Texas (Panhandle) .... 42°F to 75°F
Texas (Coast) .... 60°F to 80°F
Mississippi .... 57°F to 80°F
Idaho .... 37°F to 56°F
Illinois .... 41°F to 56°F
Montana .... 33°F to 54°F
Michigan .... 38°F to 60°F
Wyoming .... 26°F to 65°F

Generally, a curve plotted for any one of the above will closely parallel the curve for Canada.

Sizing pipe lines for water flow

This form gives a reasonable degree of accuracy.

$$d^5 = \frac{(\text{gpm})^2(D)}{(10)(H)(100)}$$

where   d = inside diameter of pipe in in.
     gpm = gallons per minute
     D = equivalent distance (in ft) from standpipe (or pump) to outlet
     H = total head lost due to friction in ft

Example. What size line should be run from a water tower to a cooling tower for a minimum of 160 gpm if the equivalent distance (actual distance in ft plus allowances for valves and fittings) is 1,000 ft and the minimum head supplied by the tower will always be at least 40 ft?

Solution.

$$d^5 = \frac{(160)^2(1,000)}{(10)(40)(100)} = 640 \text{ in.}$$

Rather than extract the fifth root of 640, simply raise inside diameters of logical pipe sizes to the fifth power and compare them with 640.

If 3-in. ID pipe were used

$$d^5 = (3)(3)(3)(3)(3) = 243$$

This is obviously too small. Trying 4-in. pipe

$$d^5 = (4)(4)(4)(4)(4) = 1,024$$

How approximate throughput of a line can be estimated from pipe size

Multiply the square of the nominal pipe size by 500.

Example. The throughput of a 12-in. line is approximately 144 by 500 = 72,000 barrels per day.

Naturally, the results of this rule are at best only a very rough approximation, since no account is taken of gravity, viscosity, pressures, or even of spacing between pump stations. Nevertheless, almost all of the oil and products lines now in operation have throughput capacities which will be given by the above rule, with the constant (given as 500) ranging between 300 and 600. The rule does *not* apply to short lines, such as loading lines, or runs within a refinery yard, but only to cross-country pipe lines.

Gauge liquid flow where no weir or meter is available

Rule. Measure number of seconds required to fill five gallon pail and divide this number into the factor 10,000 and get the answer in barrels per day.

Actually the factor is 10,286, but it is so close to 10,000 that 10,000 can be used for a rule of thumb number since it is easy to remember.

Example. It requires 100 seconds to fill a 5-gallon bucket. 10,286 divided by 100 equals 102.86 bbl per day, or for practical purposes:

$$10,000 \div 100 = 100 \text{ bpd}$$

Proof: 100 seconds to fill 5-gallon pail equals 20 seconds per gallon equals .05 gallons per second.

$$60 \times 60 \times 24 = 86,400 \text{ seconds in a day}$$

$$86,400 \times .05 = 4,320 \text{ gallons per day}$$

$$4,320 \div 42 \text{ gal per bbl} = 102.86$$

Estimate crude gathering line throughput for a given pipe diameter

To select the proper size line or estimate throughput for crude oil gathering lines, an engineer with a leading oil company has devised this handy rule: **Throughput in barrels per hour is roughly equal to a constant times the square root of pressure drop per mile (psi/mile).** Constants were derived from field data for various diameter lines between 2 and 10 in.

Pipe diameter (in.)	Constant (K)
2	7.8
3	19.2
4	37.8
6	105
8	209
10	385

Example. What size gathering line would be required to move 240 barrels per hour of crude through a 10-mile line using a pump discharge pressure of 400 psi? Assuming no significant changes in head due to elevation and a pressure drop of 40 psi per mile, for the first trial calculation a 4-in. line appears to be the best choice so try the constant 37.8.

$$\text{Throughput (4-in.)} = 37.8 \sqrt{40}$$

$$= (37.8)(6.4) = 241 \text{ barrels per hour}$$

So a 4-in. gathering line will be adequate.

How to determine head loss due to friction in ordinary iron pipe line carrying clear water

The form below gives a reasonable degree of accuracy:

$$S = \frac{G^2}{10 \, d^5}$$

where S = head loss due to friction in ft per 100 ft length of pipe line
   G = rate of flow in gallons per minute
   d = nominal diameter of pipe in in.

The exact head loss due to friction in pipe lines carrying clear water is given by the Williams and Hazen formula:

$$S = \left[\frac{V}{1.318 \, C \, R^{.63}} \right] 1.85$$

where S = loss of head due to friction in ft per 100 ft
V = average velocity in ft per second
C = coefficient depending on the roughness of inside the pipe

R = the mean hydraulic radius = $\dfrac{D}{4}$

where D is the diameter in ft

The average velocity V is given by Q/a where Q is the rate of flow in ft³/sec. and a is the cross-sectional area of the pipe.

C is a function of the material and age of the pipe and the type of water. For smooth new iron pipe C = 120. For 15-year-old ordinary pipe C = 100. For 25-year-old ordinary pipe C = 90.

Example. What will be the head loss per 100 ft if a 30-in. line flows 10,000 gpm?

Solution.

$$S = \frac{(10,000)^2}{10(30)^5} = \frac{100}{243} = .41 \text{ ft}$$

Exact answer using the Williams and Hazen formula:

$$S = .34 \text{ ft}$$

Remarks. This rule of thumb could be very satisfactorily used for pipe diameters 3-in. and larger and a discharge greater than 400 gpm.

This rule applies only to wrought and cast iron pipes (20 years service). If new smooth brass or steel pipe is used, a factor should be used = .6.

For 30-year-old ordinary pipe factor = 1.2. These are empirical factors.

For pipe diameters 2-in. and less and discharges less than 200 gpm, this rule of thumb does not give satisfactory results.

How to size lines, estimate pressure drop and optimum station spacing for crude systems

This table will give "horseback" estimates as to optimum line size, pressure drop, and pump station spacing for crude oil of approximately 40° API gravity and 60 SSU viscosity.

The depreciation was assumed to be 6% straight line; return on the investment 8%, and capital and operating costs were based on Mid-Continent averages:

FOR	USE		
Throughputs (bbl per day)	Pipe diameter (in. OD)	Pressure drop (psi per mile)	Pump station spacing (miles) (for 800 psi disch. press.)
0 to 2,000	3½	—	—
2,000 to 3,000	4½	32	25
3,000 to 7,500	6⅝	16	50
7,500 to 16,500	8⅝	10.5	76
16,500 to 23,500	10¾	8.5	94
23,500 to 40,000	12¾	7	112

Estimate the optimum working pressures in crude oil transmission lines

The approximate economic optimum working pressure in psi = 1,250 − 15 × OD of line.

Example. Estimate the maximum economic optimum working pressure—which is usually station discharge pressure—for a 20-in. crude oil pipe line.

Solution. Approximate maximum working pressure = 1,250 − 15 × 20 = 950 psi.

If the line profile is approximately horizontal, the maximum working pressure occurs at the discharge of a pump station. This method is useful only for approximations.

How to size crude and products lines for capacity increases

Any new pipe line for crude or products should be sized to carry ⅔ of its ultimate capacity.

Example. Here's the way it works out. Design a line to take care of an initial capacity of 100,000 barrels per day. This could be done by laying a 16-in. line and spacing some 14 pump stations along its length. But to increase it 50% or to 150,000 barrels per day, it would be necessary to add intermediate stations—14 in all. Excessive operating costs would likely result.

However, if the line were designed so that the 100,000 barrels per day tallied up to only ⅔ its ultimate capacity, a 20-in. line with five original stations would be installed. To increase this capacity to 150,000 bpd, it would be necessary to add only five stations.

This is based on an estimate that when expanding the capacity of a line, loop it if the increase is to be greater than 50% of the original capacity. If it is to be less, simply add horsepower. This, too, often holds true.

Both of these approaches are based on statistical results, rather than on economic studies, etc.

How to determine the maximum surge pressure in liquid-filled pipe line when a main valve is closed suddenly

The maximum surge pressure that occurs in a crude or products line when a main line valve is suddenly closed may be quickly determined using the simple rule of thumb: **Line pressure equals .8 times the weight of liquid per cubic ft times its velocity in ft per second.**

This saves considerable calculation required in using the standard equation:

$$P = \frac{12\, w\, V \sqrt{\dfrac{32.2}{w}\left(\dfrac{1}{\dfrac{1}{K} + \dfrac{D}{t\,E}}\right)}}{144 \times 32.2}$$

where   P = surge pressure, psi
        w = weight of liquid per cubic ft, lb
        V = velocity change, fps
        K = bulk modulus of the liquid, psi

D = outside diameter of pipe, in.
t = pipe wall thickness, in.
E = Young's modulus of the pipe material, psi

Example. A products line transports gasoline at 4.5 ft per second in an 8-in. line. Suddenly the main line valve is closed. Assuming the weight of gasoline to be 45 lb/cubic ft, what would be the maximum surge pressure to be expected in the system?

Using the simplified equation,

P = .8wV = .8 × 45 × 4.5 = 162 psi

Using the standard equation,

P = 159 psi

Using the Rule of Thumb the error is less than 2%.

What is the hydrostatic pressure due to a column of oil H feet in height?

Multiply H by 61.41; divide this by 131.5 + °API.

Example. A column of 37° oil is 20 ft in depth; what is the pressure at the base?

P = 20 × 61.41/131.5 + 37
  = 1,228.2/168.5
  = 7.88 lb/in.²

Example. What is the pressure on a closed block gate, if the highest point on the line is 1,380 ft above the gate, and the line is standing full of 36° oil?

P = 1,380 × 61.41/131.5 + 36
  = 84,750/167.5
  = 506 lb/in.²

This rule is not an approximation, but is as accurate as the determination of specific gravity.

Transient Pressure Analysis

Transient pressure analysis is the study of a pipe line system's response to a time dependent upset condition such as a valve closing or a pump shutting down. When a valve is closed downstream of the source pressure, the pressure will increase on the upstream side of the valve while the downstream pressure decreases. The valve closure time will determine how fast the upstream pressure increases and the downstream pressure decreases. The high upstream pressure "surge" will propagate upstream while the low pressure surge moves downstream. If the downstream side is open to a tank, the downstream surge will quickly dissipate due to the surge being absorbed in the tank. Similarly, when a pump shuts down, the pressure downstream of the pump decreases and the pressure upstream of the pump will increase. Other causes of pressure surges include tank changes, manifold changes, or leaks.

Steady state vs transient conditions

Steady-state means that the pressure gradients and flow rates in the pipe line are not changing. Transient means the gradients and flows are a function of time. The event(s) that causes conditions to be time dependent is the upset condition such as valve closure or pump failure. Transient conditions must be considered when determining if the pipe line will be subjected to abnormally high operating pressures.

Fluid properties

All pressures and flows, whether steady-state or transient, depend on the fluid properties. This example deals with a pipe line filled with batches of motor gasoline and diesel fuel. The program used for the analysis is designed for almost incompressible fluids such as crude oil, products and water.

The properties required for surge analysis include: specific gravity, viscosity, bulk modulus and vapour pressure. All properties should be at the same base temperature. Multiple batches can be handled and PVT data will be needed for each batch.

Pipe

The pipe data, such as outside diameter, wall thickness, specified minimum yield strength (SMYS), Poisson's ratio and Young's modulus of elasticity are required. The length and elevation profile of the pipe line are also required.

Pipe environment

Since an isothermal surge analysis is being performed, the only environmental information required is how the pipe line is anchored. Is the pipe above ground, underwater and does it have expansion loops throughout, or is it just constrained at the source? Multiple pipes can be handled and each pipe may be constrained differently.

Equipment

The data required for a centrifugal pump includes rated flow, rated head differential, shutoff head differential (PSOH), maximum allowed discharge pressure, minimum allowed suction pressure, discharge pressure trip setting, initial speed, final speed, time it takes to go from initial to final speed (ramp-down time), time when the pump starts to ramp down, and if the pump is protected by a check valve in the pump discharge line.

The data required for valves includes valve closure time, time or pressure when the valve starts to close, closure curve shape exponent (the exponent is different for gate valves, ball valves, etc.), and the amount of valve closure.

Data required for pressure relief valves include the set pressure and the back-pressure on the relief line.

Initial conditions

Since surge is a time dependent phenomenon, we need the conditions that exist before any upset occurs. The required initial conditions are: the pressure at the delivery end of the pipe line. The program will establish the initial steady-state flow rate automatically.

Perturbations that cause surges

Any time-dependent change, such as a leak, valve closure or pump shutdown will cause surges in the system. Cascading upset events such as internal pumps tripping, may cause other pumps to trip and can be modeled.

The fluid-pipe-equipment interaction

The liquid and the pipe, according to its environment, expand and contract as the surges travel. The pumps, valves and other equipment change the pressures and/or flows as the surges reach them. The surges travel at the speed of sound in the fluidpipe system. The fluid itself is what relays the surge energy to the various pipe sections and the equipment. The fluid, pipe and equipment are inextricably linked and their behavior must be determined by a model that employs simultaneous fluid pipe and equipment solution procedures. The Method of Characteristics, which NATASHA PLUS© uses, is a simultaneous technique and therefore models "surge" phenomenon rigorously.

To illustrate the use of the program, the pipe line system shown in Figure 1 is equipped with a valve at the end of the pipe line that is caused to close at time T = 300 seconds. The valve travels from full open to full closed in 5 seconds. A pressure relief valve is connected to the system and is set to relieve at 275 psig. The system consists of six pipes configured as shown in Table 1.

Simulation

The operation of the system was simulated for 1200 seconds and the results are shown graphically in Figure 2. The program outputs a detailed tabular listing showing pressures and flow rates at various locations along the pipeline at selected time steps. A graphics file containing selected node pressures and flow rates, is also produced by by the program. Figure 2 was prepared from the graphics file produced by the program. A typical time page report is shown in Table 2. The program also prints a maximum pressure summary which is shown in Table 3. A summary of the relief valve operation is shown in Table 4 and indicates the total volume released was 855 bbls during the simulation.

The pump is not shutdown in this simulation because of the action of the relief valve. A review of the maximum pressure summary indicates that the pipeline MOP is exceeded at two locations. In order to prevent overpressure, it will be necessary

Figure 1
Sample Pipe line for Transient
Analysis Study

Product parameters:.

Product	sp. gr.	visc. cSt	bulk modulus psi
Diesel	0.79	2.15	156,495
Gasoline	0.741	0.622	121,505

Initial flow rate is 92,500 BPD
Downstream pressure is 40 psig.

Table 1

Data Required By The Program

Pipe Design Data

Pipe No.	Length Mi.	OD in.	W. T. in.	Roughness in.	SMYS psig	Design Fact.
1	2.05	12.75	0.375	0.0018	35000	72%
2	16.3	12.75	0.25	0.0018	52000	72%
3	16.3	12.75	0.25	0.0018	52000	72%
4	15.45	12.75	0.25	0.0018	52000	72%
5	16.53	12.75	0.25	0.0018	52000	72%
6	1.0	12.75	0.25	0.0018	52000	72%

Table 1 continued

Liquid Properties

Pipe No.	Sp. Gr.	Vis. cSt	Bulk Mod. psi	Vapor Press psia
1	0.741	0.622	0.122	5.7
2	0.79	2.15	0.156	10
3	0.741	0.622	0.122	5.7
4	0.79	2.15	0.156	10
5	0.741	0.622	0.122	5.7
6	0.79	2.15	0.156	10

Initial Coditions

Pipe No.	Flow BPD	Upstream Press psig
1	92500	1398.22
2	92500	1444.19
3	92500	958.58
4	92500	659.75
5	92500	363.1
6	92500	68.76

Pipe No.	Friction Factor	Seg. Length Length, mi.	Press. Drop psi/mi.	Head Drop ft/mi
1	0.014	1.03	21.5907	67.2098
2	0.0157	1.02	23.2816	67.978
3	0.014	0.91	19.4468	60.5359
4	0.0157	1.03	23.2816	67.978
5	0.014	0.92	19.4468	60.5359
6	0.0157	1.00	23.2816	67.978

PUMP PARAMETERS

PUMP AT NODE   7 THAT DISCHARGES INTO PIPE   1

RATED FLOW	=	92,500 bbl/day
ZERO FLOW HEAD DIFFERENTIAL	=	4520.00 ft-head
RATED HEAD DIFFERENTIAL	=	4228.02 ft-head
SUCTION PRESSURE	=	40.00 psig
MAXIMUM DISCHARGE CONTROLLED PRESSURE	=	1440.00 psig
DISCHARGE TRIP PRESSURE	=	1490.00 psig
INITIAL SPEED	=	3565.00 rpm
FINAL SPEED	=	3565.00 rpm
CURVE EXPONENT	=	0.5000

Table 1 continued

VALVE PARAMETERS

PRESSURE RELIEF VALVE AT NODE 5

MAX PRESS. BEFORE RELIEF	=	275.00 psig
BACK PRESSURE ON RELIEF LINE	=	0.00 psig

VALVE AT NODE 6

VALVE STARTS TO CLOSE AT T	=	300.00 sec
VALVE CLOSES IN	=	5.00 sec
VALVE STARTS TO CLOSE ABOVE P	=	99999.00 psig
FINAL FRACTIONAL AREA OPEN TO FLOW	=	0.00
CLOSURE SHAPE EXPONENT	=	1.00

Elevation Data

Pipe No.	Point	MP mi.	Elev. ft.	MOP psig
1	1	0	877	1440
	2	1.59	885	1440
	3	2.05	875	1440
2	1	0	875	1440
	2	1.13	890	1440
	3	1.27	860	1440
	4	4.37	940	1440
	5	4.96	890	1440
	6	10.96	990	1440
	7	11.98	1060	1440
	8	13.03	980	1440
	9	16.3	1000	1440
3	1	0	1000	1440
	2	1.71	1065	1440
	3	4.66	1010	1440
	4	6.42	1085	1440
3	5	9.9	930	1440
	6	10.21	1040	1440
	7	11.46	970	1440
	8	14.21	1040	1440
	9	16.3	1071	1440
4	1	0	1071	1440
	2	3.76	1050	1440
	3	5.53	1005	1440
	4	5.92	1020	1440
	5	6.32	1020	1440
	6	15.45	817	1440

Pipe No.	Point	MP mi.	Elev. ft.	MOP psig
5	1	0	817	1440
	2	4.05	770	1440
	3	5.45	875	1440
	4	5.68	810	1440
	5	6.26	915	1440
	6	7.13	805	1440
	7	8.31	930	1440
	8	12.36	755	1440
	9	16.53	746	1440
6	1	0	746	275
	2	0.28	760	275
	3	0.62	762	275
	4	1	762	275

Table 2

Typical time page report for T = 0 seconds

Pipe No.	Loc. mi.	Elev Ft.	Up Flow M bbl/day	Down Flow M bbl/day	Head Ft.	Press. psig
1	0	877	93	93	5230	1398
1	1	882	93	93	5161	1374
1	2	875	93	93	5092	1355
2	0	875	93	93	5092	1444
2	1	889	93	93	5023	1416
2	2	880	93	93	4953	1395
2	3	906	93	93	4884	1362
2	4	932	93	93	4815	1330
2	5	892	93	93	4746	1320
2	6	909	93	93	4676	1290
2	7	926	93	93	4607	1261
2	8	943	93	93	4538	1231
2	9	960	93	93	4468	1202
2	10	977	93	93	4399	1172
2	11	1007	93	93	4330	1138
2	12	1041	93	93	4261	1103
2	13	981	93	93	4191	1099
2	14	988	93	93	4122	1074
2	15	994	93	93	4053	1048
2	16	1000	93	93	3984	1022
3	0	1000	93	93	3984	959
3	1	1034	93	93	3929	930
3	2	1063	93	93	3874	903
3	3	1046	93	93	3820	891
3	4	1029	93	93	3765	879
3	5	1012	93	93	3710	867
3	5	1043	93	93	3655	839
3	6	1082	93	93	3600	809
3	7	1048	93	93	3545	802
3	8	1008	93	93	3491	798
3	9	968	93	93	3436	793
3	10	952	93	93	3381	780
3	11	1003	93	93	3326	746
3	12	978	93	93	3271	737
3	13	1001	93	93	3217	712
3	14	1024	93	93	3162	687
3	14	1044	93	93	3107	663
3	15	1058	93	93	3052	641
3	16	1071	93	93	2997	619

Table 2 continued

Typical time page report for T = 0 seconds

Pipe No.	Loc. mi.	Elev Ft.	Up Flow M bbl/day	Down Flow M bbl/day	Head Ft.	Press. psig
4	0	1071	93	93	2997	660
4	1	1065	93	93	2927	638
4	2	1059	93	93	2857	616
4	3	1054	93	93	2787	594
4	4	1041	93	93	2717	574
4	5	1015	93	93	2647	559
4	6	1020	93	93	2577	533
4	7	1000	93	93	2507	516
4	8	977	93	93	2437	500
4	9	954	93	93	2367	484
4	10	932	93	93	2297	468
4	11	909	93	93	2227	452
4	12	886	93	93	2157	435
4	13	863	93	93	2087	419
4	14	840	93	93	2017	403
4	15	817	93	93	1947	387
5	0	817	93	93	1947	363
5	1	806	93	93	1892	349
5	2	796	93	93	1836	334
5	3	785	93	93	1781	320
5	4	774	93	93	1725	305
5	5	811	93	93	1669	276
5	6	858	93	93	1614	243
5	6	894	93	93	1558	213
5	7	828	93	93	1503	217
5	8	925	93	93	1447	168
5	9	892	93	93	1391	160
5	10	853	93	93	1336	155
5	11	813	93	93	1280	150
5	12	773	93	93	1225	145
5	13	754	93	93	1169	133
5	14	752	93	93	1113	116
5	15	750	93	93	1058	99
5	16	748	93	93	1002	82
5	17	746	93	93	947	64
6	0	746	93	93	947	69
6	1	762	93	93	879	40

Figure 2

Transient Analysis

Table 3
Maximum Pressure Summary

Pipe No.	MP mi.	Elev. ft.	Hmax ft.	Pmax psig	MOP psig	Time sec.		
1	0	877	5269.51	1411.07	1440	1199.44		
1	1.03	882.16	5209.78	1390.22	1440	1200.98		
1	2.05	875	5150.04	1373.33	1440	1199.44		
2	0	875	5150.04	1464.14	1440	1199.44	X	(Exceeds MOP)
2	1.02	888.52	5089.72	1438.85	1440	1200.98		
2	2.04	879.81	5029.4	1421.18	1440	1199.44		
2	3.06	906.1	4969.08	1391.52	1440	1200.98		
2	4.07	932.39	4908.75	1361.85	1440	1199.44		
2	5.09	892.23	4848.44	1354.95	1440	1200.98		
2	6.11	909.21	4788.11	1328.47	1440	1199.44		
2	7.13	926.19	4727.8	1302	1440	1200.98		
2	8.15	943.17	4667.47	1275.53	1440	1199.44		
2	9.17	960.15	4607.16	1249.05	1440	1200.98		
2	10.19	977.13	4546.84	1222.58	1440	1199.44		
2	11.21	1006.9	4486.53	1191.73	1440	1200.98		
2	12.22	1041.33	4426.21	1159.27	1440	1199.44		
2	13.24	981.31	4365.9	1159.18	1440	1200.98		
2	14.26	987.54	4305.58	1136.38	1440	1199.44		
2	15.28	993.77	4245.27	1113.6	1440	1200.98		
2	16.3	1000	4184.95	1090.8	1440	1199.44		
3	0	1000	4184.95	1023.15	1440	1199.44		
3	0.91	1034.42	4137.43	996.82	1440	1200.98		
3	1.81	1063.11	4089.9	972.34	1440	1199.44		
3	2.72	1046.23	4042.38	962.49	1440	1200.98		
3	3.62	1029.35	3994.85	952.65	1440	1199.44		
3	4.53	1012.47	3947.33	942.81	1440	1200.98		
3	5.43	1042.95	3899.8	917.74	1440	1199.44		
3	6.34	1081.54	3852.29	890.08	1440	1200.98		
3	7.24	1048.28	3804.76	885.5	1440	1199.44		
3	8.15	1007.95	3757.25	883.2	1440	1200.98		
3	9.06	967.61	3709.72	880.89	1440	1199.44		
3	9.96	951.68	3662.21	870.74	1440	1200.98		
3	10.87	1003.23	3614.69	838.92	1440	1199.44		
3	11.77	977.95	3567.18	831.77	1440	1200.98		
3	12.68	1001	3519.66	809.1	1440	1199.44		
3	13.58	1024.05	3472.15	786.44	1440	1200.98		
3	14.49	1044.14	3424.63	764.72	1440	1199.44		
3	15.39	1057.57	3377.13	745.15	1440	1200.98		
3	16.3	1071	3329.61	725.57	1440	1199.44		
4	0	1071	3329.61	773.55	1440	1199.44		
4	1.03	1065.25	3268.67	754.64	1440	1200.98		
4	2.06	1059.49	3207.72	735.74	1440	1199.44		

Table 3
Maximum Pressure Summary

Pipe No.	MP ml.	Elev. ft.	Hmax ft.	Pmax psig	MOP psig	Time sec.		
4	3.09	1053.74	3146.78	716.84	1440	1200.98		
4	4.12	1040.85	3085.82	700.38	1440	1199.44		
4	5.15	1014.66	3024.89	688.48	1440	1200.98		
4	6.18	1020	2963.94	665.77	1440	1199.44		
4	7.21	1000.21	2903	651.68	1440	1200.98		
4	8.24	977.31	2842.06	638.65	1440	1199.44		
4	9.27	954.41	2781.13	625.63	1440	1200.98		
4	10.3	931.51	2720.18	612.6	1440	1199.44		
4	11.33	908.61	2659.25	599.57	1440	1200.98		
4	12.36	885.7	2598.31	586.55	1440	1199.44		
4	13.39	862.8	2537.39	573.52	1440	1200.98		
4	14.42	839.9	2476.45	560.5	1440	1199.44		
4	15.45	817	2415.53	547.47	1440	1200.98		
5	0	817	2415.53	513.52	1440	1200.98		
5	0.92	806.34	2367.36	501.47	1440	1199.44		
5	1.84	795.69	2319.21	489.42	1440	1200.98		
5	2.76	785.03	2271.05	477.37	1440	1199.44		
5	3.67	774.37	2222.9	465.33	1440	1200.98		
5	4.59	810.63	2174.74	438.21	1440	1199.44		
5	5.51	858.04	2126.59	407.51	1440	1200.98		
5	6.43	893.72	2078.43	380.58	1440	1199.44		
5	7.35	827.95	2030.29	386.24	1440	1200.98		
5	8.27	925.23	1982.13	339.52	1440	1199.44		
5	9.18	892.26	1933.99	334.65	1440	1200.98		
5	10.1	852.58	1885.84	331.93	1440	1199.44		
5	11.02	812.9	1837.7	329.21	1440	1200.98		
5	11.94	773.22	1789.55	326.49	1440	1199.44		
5	12.86	753.93	1741.41	317.22	1440	1200.98		
5	13.78	751.95	1693.26	302.39	1440	1199.44		
5	14.69	749.96	1645.13	287.57	1440	1200.98		
5	15.61	747.98	1596.98	272.73	1440	1199.44		
5	16.53	746	1548.9	257.93	1440	742.37		
6	0	746	1548.9	274.98	275	742.37		
6	1	762	1721.75	328.7	275	460.14	X	(Exceeds MOP)

Table 4
Pressure Relief Valve Summary

TOTAL VOLUME RELEASED AT NODE 5 = 855.4117 bbl
MAX VOLUME RATE RELEASED AT NODE 5 =   102.958 1000 bbl/day

Reprinted with permission Fluid Flow Consultants, Tulsa, OK.

to make changes to the system such as changing the relief valve setting or slowing the closing time for the valve. NATASHA-PLUS© can then be used to simulate the changes to confirm that pipeline MOP is not exceeded.

Reprinted with permission: Fluid Flow Consultants, Tulsa, OK.

Tank Farm Line Sizing

By: J. C. Stivers

The designer must locate the tanks, size the tank lines and select pumps. The designer must know the receiving flow rates, discharge flow rates, liquids to be handled along with the properties of the liquid such as viscosity, specific gravity and vapor pressure. The minimum and maximum tank levels must be determined which will set the available volume that can be pumped. Other factors include determining the number of tanks which will depend upon the type of material being stored, tank turnover frequency, system load factor (which takes into consideration system outages) and environmental considerations.

If possible, the tanks should have combination suction and fill lines. To size the lines, it is necessary to consider the fill and discharge flow rates. If a tank farm can receive from a number of small systems and has one out-going pipeline, the size of the tank lines will probably be determined by the outgoing flow rate. If a tank farm is fed by one stream and delivers to a number of smaller systems, the line sizes will probably be determined by the incoming stream rate. It is necessary to check both to confirm this generalization.

The criteria for sizing a tank line based on filling rates is somewhat arbitrary. When there are a number of sources, the operator may specify a maximum pressure of 25 to 50 psig required at the connection. The piping manifold, meters and other equipment must then be sized so the pressure will not exceed 25 to 50 psig at the maximum flow rate. If a metering system is to be used on the incoming stream and a back pressure valve is required for the meters, the tank line size may be sized to produce the minimum back pressure at the maximum flow rate. When the tank farm owner develops the pressure required to fill tanks, an economic analysis should be made to determine line sizes. Large diameter pipelines require less pressure and therefore less power cost but require greater investments than smaller diameter lines.

Determine pipeline size

The equation for determining the line size from the tanks and to fill a pump based on discharge flow rates is as follows:

NPSH available ≥ NPSH required by the pump

$$\frac{\text{Atmos. Press (In. Hg)} \times 1.135}{\text{s.g.}} - \frac{\text{Vapor Press.} \times 2.31}{\text{s.g.}} \pm \text{Static}$$

Press (Ft. of Head) - Pipe Friction Loss (Psi/mi.) x

$$\frac{\text{equiv. feet of suction line} \times 2.31}{5,280 \times \text{s.g.}} \geq \text{NPSH required by the pump.}$$

Atmospheric Pressure -- The lowest atmospheric pressure for the area should be considered. This is particularly important for tank farms where product is transferred from one tank to another using the pump. This usually occurs when the atmospheric pressure is low.

Vapor Pressure -- Vapor pressure increases with temperature and therefore reduces the NPSH available. Conversely

Figure 1
Tank Farm Line Sizing

liquid viscosity decreases with temperature increases and therefore reduces pipe friction losses. Therefore, it becomes necessary to determine the liquid temperature that produces the lowest NPSH available.

Static Pressure—This is the elevation difference between the minimum liquid level in the tank and the centerline of the pump suction. If the ground elevation in the tank farm changes, the designer can take advantage of the terrain and locate the tanks accordingly.

Pipe Friction Loss—There are many equations used to determine pipe friction loss. See Section 13.

Most tank lines are not designed to be pigged. Some crudes, particularly sour crudes, may cause a build up in the tank line, in turn reducing the pipe i.d. Any buildup will cause the friction factor to be higher and if build up is expected, the designer may wish to increase the pipe size—particularly if NPSH available to the pump is marginal.

NPSH Required By Pump

The pump manufacturer supplies the NPSH requirement for each pump. Some positive displacement pump manufacturers may not publish these requirements due to a wide range of operating speeds. They will however, provide requirements for specific conditions. If the NPSH requirements of the pump are not met, the pump will cavitate and a loss of pressure and flow will occur. Mechanical damage may also be possible.

Types of Booster Pumps

The pumps most typically used include:

Horizontal split case
Vertical Can
Vertical Inline

A horizontal split case pump may be used in some instances and will be preferred by most maintenance and operations personnel because it is easy to maintain. The NPSH requirement for a horizontal pump may be 7 to 10 feet and the suction nozzle may be 2 to 4 feet above grade. The vertical pump can have an NPSH requirement with the suction nozzle three or more feet below grade. The NPSH advantage for the can pump can range from 12 to 17 feet which may permit a size reduction of the tank lines. NPSH requirements for the vertical inline pump are similar to the horizontal pump. The advantage of the inline pump is pump and installation costs.

Design Considerations

The designer should not automatically assume that a booster pump is required. Most positive displacement pumps have low NPSH requirements and therefore a booster

pump required unless metering is installed between the tank and mainline pump(s). If the terrain permits the tanks to be located considerably higher than the main pumps, a booster pump may not be required for centrifugal pumps.

If a series of main line pumps are used and the pumping rate is determined by the number of pumps operating, it may be possible to obtain a pump with a lower NPSH requirement for the first pump and thus eliminate the need for a booster pump. This is shown by the following example.

Assume three main line pumps are connected in series.

Pump combinations produce the following flow rates

No. of Pumps Operating	Flow Rate
3	100%
2	75%
1	50%

NSPH available:

Flow Rate	NPSH Available
100%	15 feet
75%	20 feet
50%	27 feet

is seldom required unless metering is installed between the tank and mainline pump(s). If the terrain permits the tanks to be located considerably higher than the main pumps, a booster pump may not be required for centrifugal pumps.

Low NPSH requirement booster pumps are normally less efficient than mainline pumps. When this is the case, a booster pump should be selected to develop just enough head to feed the main line or delivery pumps.

If a metering system is required, it would normally be installed between the tanks and mainline pumps. In this case, a booster pump would almost certainly be required to overcome the pressure losses in the metering system. The booster pump then must develop enough head to produce the minimum back pressure required by the meters or the minimum NPSH required by the mainline pumps.

If the booster pump is used to transfer liquid from one tank to another, it will probably develop more head than required. This allows the designer to reduce the size of the transfer header and valves.

Once the minimum atmospheric pressure and vapor pressure of the liquid to be pumped is determined, there are three variables on the left side (NPSH available) and one on the right (NPSH required by the pump). The solution is trial and error and the following procedure is suggested:

Select a tentative location for the tanks and determine the minimum liquid level elevation.

Determine the equivalent pipe length of the suction line.

Locate the booster pump and assume it can operate at a specific NPSH.

This allows the solution of the equation for maximum pipe friction loss. Application of pipe pressure loss formula will give the minimum size pipe that will result in a specific NPSH. Select the next larger standard pipe size and rework the equation to determine the actual minimum NPSH available at the pump. Generally, the cost of the booster pump will decrease as the NPSH requirement increases. Higher NPSH will also allow the designer to select a different style pump. For these reasons, the equation should be solved using the next two larger pipe sizes. Pumps can now be selected for three different pipe sizes. The costs of the three systems should be estimated and an economic analysis made to select the system.

The designer must be aware that a low NPSH pump can pull a vacuum which must be considered in determining pump differential. (Pipe collapse?) Frequently, the required pump discharge pressure is used to specify pump differential head which assumes zero pressure at the pump suction. The following example assumes the booster pump is designed to operate at 7 feet NPSH and 7 feet NPSH is available. The liquid is fuel oil with a vapor pressure of 0.2 psig and 0.85 s.g. and atmospheric pressure is 14.7 psia.

$(2.31/\text{s.g.}) \times (\text{press. avail.} + \text{atmos. press} - \text{vapor press})$
$= \text{NPSH}$

Press. avail. $= [(\text{s.g.} \times \text{NPSH})/2.31] - \text{atmos. press.}$
$+ \text{vapor press.}$

Press. avail. $= [0.84 \times 7)/2.31] - 14.7 + 0.2 = 11.95$ psig

If a pump discharge pressure of 25 psig is required, the pump differential is $25 - (-11.95) = 36.95$ psig or 102 feet.

14: Pumps

Centrifugal pumps 374
Speed torque calculation.. 386
Estimate the discharge of a centrifugal pump at various speeds 387
How to estimate the head for an average centrifugal pump 387
Find the capacity of reciprocating pump 387
How to estimate the horsepower required to pump at
    a given rate at a desired discharge pressure...................................... 387
Nomograph for determining reciprocating pump capacity.................. 388
Nomograph for determining specific speed of pumps......................... 389
Nomograph for determining horsepower requirement of pumps 390
How to select motors for field gathering pumps 390
Reciprocating Pumps..391
Understanding the basics of rotary screw pumps 399

Centrifugal pumps

Centrifugal pumps are best suited for large volume applications or for smaller volumes when the ratio of volume to pressure is high. The selection of the proper pump will depend on the system throughput, viscosity, specific gravity, and head requirements for a particular application. Where a variety of products with different characteristics are handled, it may be necessary to utilize multiple pumps either in series or parallel, or a series-parallel combination, or pumps with variable speed capabilities to achieve the desired throughput at least cost. Economics will ultimately determine the type and number of pumps to be used.

The first step in selecting centrifugal pumps is to analyze the pipe line system and determine its characteristics such as the pressure required to move the desired flow rate. Initial and future conditions should be evaluated so that pumps with sufficient flexibility to handle both conditions with minor changes may be selected. Refer to Section 13—Liquid Hydraulics for data on calculating pipe line pressure drop.

Pump manufacturers publish pump performance maps similar to the one shown in Figure 1 to aid in pump selection. The performance map shows the range of differential

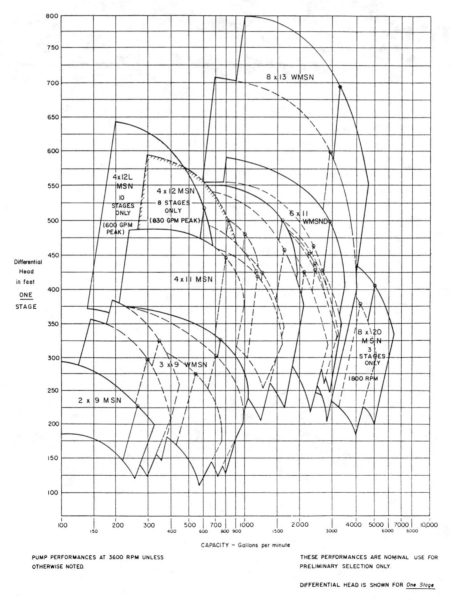

Figure 1. Typical performance map for centrifugal pumps. Courtesy United Centrifugal Pumps.

Figure 2. Typical performance curve for centrifugal pumps. Courtesy United Centrifugal Pumps.

head and capacity that will be developed by a family of pumps. Once the system data have been established, a pump may be selected from the map. The selection is further refined by referring to a curve such as that shown in Figure 2, which shows a typical pump manufacturer's catalog curve for a specific pump showing the performance for various impeller sizes as well as the brake horsepower, NPSH requirement, and pump efficiency. Pump performance curves are plotted in ft and for a specific gravity of 1.0. When pumping a nonviscous liquid, the pump will develop the same head shown on the performance curve; however, the *pressure* will change with specific gravity as will the horsepower required to drive the pump. Pressure may be calculated by multiplying the pump differential head by the specific gravity and by 0.4329. The horsepower required to drive the pump will also vary directly with the specific gravity of the fluid being pumped.

The viscosity of the fluid will influence the performance of the centrifugal pump. This is discussed in more detail later.

Centrifugal pumps are available in a number of different configurations such as horizontal, vertical, radial split, and axial split casings. The final choice will depend on the system hydraulic conditions (pressure and flow rate), the ease of maintenance desired, and amount of space available for installing the pump. The pressure rating of the case and material selection will be determined by the system requirements and the fluid to be pumped.

It is extremely important that sufficient net positive suction head (NPSH) above the vapor pressure of the liquid being pumped is available for safe and reliable pump operation. NPSH is always specified in ft of head above vapor pressure required at the center line of the impeller. The NPSH required will increase as the capacity of the pump increases, which makes it necessary for the user to make a best effort to furnish the pump manufacturer with an accurate range of flow rate conditions. The pump suction nozzle geometry, pump speed, and the impeller inlet will determine the NPSH that will be required. If too much margin of safety is included in the pump specifications, the pump will most probably be oversized to obtain a low NPSH and will not operate at its best efficiency point. Figures 8, 9, and 10 show three examples of how to calculate available NPSH. As a rule of thumb, the NPSH available should be at least 10% greater than the NPSH required by the pump. It is difficult and costly to correct NPSH prob-

Figure 3. Centrifugal pump performance data.

lems after the pump is placed in service. NPSH is discussed in detail in *Centrifugal Pumps: Design and Application* by Ross and Lobanoff, Gulf Publishing Co., and it's recommended reading for those who are involved in the selection and application of centrifugal pumps.

Table 1 gives some conversion factors and formulas that will be useful in problems dealing with pumps.

All centrifugal pumps, except those pumps that are specifically designed for low suction pressure applications such as tank farm booster pumps, should be protected by a low suction pressure shut down switch. The value of pressure to be used to set the switch will be determined by the amount of NPSH required at the maximum expected operating flow rate, the vapor pressure of the fluid being pumped, and the atmospheric pressure at the pump location. Once the maximum expected operating flow rate is known, the amount of NPSH required may be obtained from the pump performance curve. The NPSH curve is plotted in feet of head and this may be converted to psi for switch setting purposes as follows:

PSIG = NPSH × 0.433 × specific gravity
           + vapor pressure − atmospheric pressure

Figure 3 shows the performance data for a centrifugal pump that was chosen from the performance map. The impeller diameter was chosen from the curve shown in Figure 2. In this case, an impeller diameter of 11-in. was chosen.

Centrifugal pumps may be operated singly, in series, in parallel, or in a combination of series and parallel. When pumps are operated in series, the heads are additive at a given rate of flow. When pumps are operated in parallel, flow is additive at a given value of head. This is illustrated in Figure 3. When pumps are operated in a series-parallel combination, these same rules will apply to determine the total head developed by the combination of pumps. Figure 4 shows performance data for one pump and for series and parallel operation of two pumps.

Pumps connected in series have the discharge of one pump connected to the suction of the downstream pump. Pumps connected in parallel use a common suction line and common discharge line. (See Figure 4a.) Pumps that are to be operated in parallel must be matched so that both develop the same head at the same flow rate. They should also have constantly rising head curves. Pumps with a head curve that has a hump (one that develops less head at shut-off than at some other point on the head curve) should be

Table 1*
Pumps   377

Table 1*
Conversion Factors and Formulas

$gpm = .07 \times boiler\ hp$

$gpm = 449 \times cfs$

$gpm = 0.0292 \times bbl/day$

$gpm = 0.7 \times bbl/hr$

$gpm = 4.4 \times cubic\ meters/hr$

$$gpm = \frac{lb\ per\ hr}{500 \times specific\ gravity}$$

$$H = \frac{P \times 2.31}{specific\ gravity}$$

$$V = \frac{Q \times .321}{A}$$

$$U = \frac{Diameter\ (in.) \times N}{229}$$

$$h_v = \frac{V^2}{2\ g}$$

$$whp = \frac{Q \times H \times specific\ gravity}{3,960}$$

$$bhp = \frac{Q \times H \times specific\ gravity}{3,960 \times e}$$

$$bhp = \frac{Q \times P}{1,715 \times e}$$

$$T = \frac{bhp \times 5,250}{N}$$

$$N_s = \frac{N\ \sqrt{Q}}{H_1^{3/4}} = \frac{N\ \sqrt{\sqrt{H_1}} \times \sqrt{Q}}{H_1}$$

$$S = \frac{N\ \sqrt{Q}}{h_{sv}^{3/4}} = \frac{N\ \sqrt{\sqrt{h_{sv}}} \times \sqrt{Q}}{h_{sv}}$$

$$t_r = \frac{H\left(\dfrac{1}{e} - 1\right)}{780 \times C}$$

where
- N = speed in rpm
- N_s = specific speed in rpm
- S = suction specific speed in rpm
- Q = capacity in gpm
- P = pressure in psi
- H = total head in ft
- H_1 = head per stage in ft
- h_{sv} = net positive suction head in ft
- h_v = velocity head in ft
- whp = water horsepower
- bhp = brake horsepower
- U = peripheral velocity in ft per second
- g = 32.16 ft per second (acceleration of gravity)
- mgd = million gallons per day
- cfs = cubic ft per second
- bbl = barrel (42 gallons)
- C = specific heat
- psi = lb/in.^{2}
- gpm = gallons per minute
- e = pump efficiency in decimal
- D = impeller diameter in in.
- V = velocity in ft per second
- T = torque in ft-lb
- t = temperature in °F
- t_r = temperature rise in °F
- A = area in in.^{2}

* Courtesy United Centrifugal Pumps.

Figure 4. Series/parallel combinations—centrifugal pumps.

Figure 4a. Pumps connected in parallel/in series.

avoided since this could possibly preclude the second pump from pumping.

The high cost of energy will in almost all cases justify the additional cost of multiple pumping units or variable speed drivers. The use of multiple units will allow most pumping

conditions to be met without throttling and therefore wasting power. If as many as three series units can be justified, the units should be sized as follows:

One unit—full head
One unit—3/4 of the head of the full head unit
One unit—1/2 of the head of the full head unit

This will allow the following pump combinations to be used:

Units	Total head	Units	Total head
1/2	1/2	1/2, 1	1 1/2
3/4	3/4	3/4, 1	1 3/4
1	1	1/2, 3/4, 1	2 1/4
1/2, 3/4	1 1/4		

Increments of 1/4 head can be achieved up to the last 1/2 head increment. This results in maximum flexibility when constant rpm drivers are used.

Figure 5 shows the performance of series vs. parallel operation of pumps. In this particular example, series operation will provide a higher pumping rate than provided with a parallel operation and two pumps in parallel will

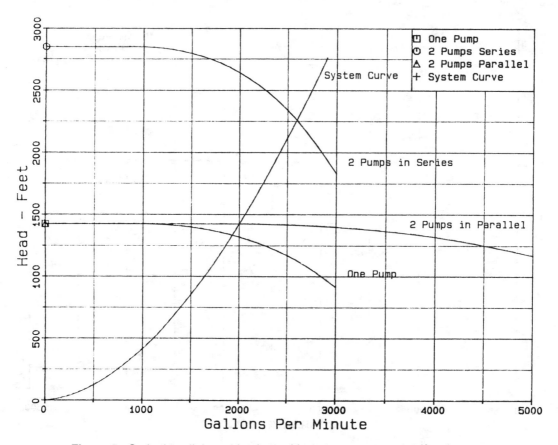

Figure 5. Series/parallel combinations with system curve—centrifugal pumps.

Figure 5a. Series and parallel operation—centrifugal pumps.

pump only slightly more than one single pump. The maximum flow rate with two pumps in series will be 2,594 gpm or 89,439 b/d. The maximum flow rate for one pump will be 1,917 gpm or 66,092 b/d.

For system rates between 66,092 b/d and 89,439 b/d, pressure throttling will be necessary to match the pump performance to the system curves, unless the pumps are equipped with variable speed drives. If throttling is necessary, it may be minimized by operating at a high flow rate for part of the day and at a lower rate for the balance of the day. If a system rate of 70,000 b/d is desired with the system shown in Figure 5, one pump operating 20 hours per day and two pumps operating 4 hours per day would achieve this rate; however, this may not be the most economical method of operation. It will be necessary to calculate and compare the cost of throttling pressure vs. the cost to operate the second unit for 4 hours. System rates less than 66,092 b/d will require pressure throttling unless the pump speed may be varied. It is not uncommon for pipe line pump stations to contain several pumps, each of which may be sized differently to increase the amount of flow rate flexibility.

The use of multiple pumps affords some protection against throughput reductions due to one or more pumps being out of service.

Parallel pump operation is usually best suited for systems where the residual static pressure is high compared to the line friction loss. Such systems usually have a flat system curve. An example of this is shown in Figure 5a.

For this system, three parallel pumps would offer the most flexibility since one unit will always be able to overcome the system friction and static and would pump almost as much as two series pumps. One could make the argument that one large pump could be selected to meet the 2 (series) condition shown in Figure 5a and a ¹/₂-head pump could be used for the maximum volume condition. This argument is valid as long as the large pump is available. When it is not available, the system will be down since the ¹/₂-head unit will not develop enough head to overcome the static and friction losses. Three pumps operating in parallel will offer the most flexibility for the system shown in Figure 5a.

Changes in system throughputs may require resizing the pump impellers to meet the new conditions. Some types of

Table 2 *
**Formulas for Recalculating Pump Performance with
Impeller Diameter or Speed Change**

Q_1, H_1, bhp_1, D_1 and N_1 = Initial Capacity, Head, Brake Horsepower, Diameter, and Speed.

Q_2, H_2, bhp_2, D_2 and N_2 = New Capacity, Head, Brake Horsepower, Diameter, and Speed.

Diameter Change Only	Speed Change Only	Diameter & Speed Change
$Q_2 = Q_1\left(\dfrac{D_2}{D_1}\right)$	$Q_2 = Q_1\left(\dfrac{N_2}{N_1}\right)$	$Q_2 = Q_1\left(\dfrac{D_2}{D_1} \times \dfrac{N_2}{N_1}\right)$
$H_2 = H_1\left(\dfrac{D_2}{D_1}\right)^2$	$H_2 = H_1\left(\dfrac{N_2}{N_1}\right)^2$	$H_2 = H_1\left(\dfrac{D_2}{D_1} \times \dfrac{N_2}{N_1}\right)^2$
$bhp_2 = bhp_1\left(\dfrac{D_2}{D_1}\right)^3$	$bhp_2 = bhp_1\left(\dfrac{N_2}{N_1}\right)^3$	$bhp_2 = bhp_1\left(\dfrac{D_2}{D_1} \times \dfrac{N_2}{N_1}\right)^3$

* Courtesy United Centrifugal Pumps.

multi-stage pumps may be destaged or upstaged by adding or taking out impellers, depending upon the number of stages in operation, and the amount of head change required. If it is contemplated that it may be necessary in the future to destage or upstage, this should be specified when the pump is purchased. It is also possible to optimize the efficiency point by chipping the pump volutes and underfiling the pump impeller(s) or adding volute inserts. The pump manufacturer should be contacted for details on how to accomplish this.

The pump affinity laws shown in Table 1 may be used to predict the performance of the pump at different operating conditions. These laws are commonly used to determine new impeller diameters required for nominal changes in flow rates or performance at different speeds. The affinity laws will allow the translation of a point on the performance curves but will not define the operation of the pump in the system. A new pump performance curve may be constructed using these laws and then applied to the system curve to determine the actual operation of the pump.

The system shown in Figure 5, with both pumps configured with 11-in. diameter impellers, will pump a maximum of 89,439 b/d. The system throughput requirements have now changed such that the maximum desired throughput is 87,931 b/d and it is desired to trim the impellers in one of the pumps to meet this new condition. The new H-Q condition for one pump will be $H_2 = 955$ ft and $Q_2 = 2,550$ gpm. In order to establish the new impeller diameter, it will be necessary to translate this point to the existing pump curve to determine H_1 and Q_1. Select flow rates on both sides of the new rate and calculate the corresponding head. These data are plotted and the resulting

Figure 6. Impeller sizing—centrifugal pumps.

line will cross the existing pump head curve as shown in Figure 6. Determine the head and flow rate at the point where this line crosses the existing head curve and these values will be H_1 and Q_1. For this calculation, flow rates were selected as follows:

Selected Q	Calculated H
2,800	1,151
2,600	993
2,400	846

The head was calculated as follows:

$$H = H_2 \times (Q/Q_2)^2$$

This line crosses the existing head curve at 2,700 gpm and 1,080 ft and these values will be the values to use for Q_1 and H_1, respectively.

Use the pump affinity law to calculate the new diameter as follows:

$$H_2/H_1 = (D_2/D_1)^2$$

$$D_2 = D_1 \times (H_2/H_1)^{1/2}$$
$$D_2 = 11.0 \times (955/1,080)^{1/2}$$
$$D_2 = 10.34 \text{ in.}$$

Since the pump affinity laws are not quite exact for a diameter change, it will be necessary to apply a correction factor to the calculated value of D_2 to determine the final diameter. The correction factor is determined by calculating the ratio of the calculated diameter to the original diameter. This value, expressed as a percentage, is then used to enter Figure 7 to determine the actual required diameter as a percentage of the original diameter.

$$D \text{ calc}/D \text{ orig} = 10.34/11 = 94\%$$

Enter the chart with 94% and read 94.5% as the actual required diameter as a percentage of the original diameter.

The new diameter will therefore be $0.945 \times 11.0 = 10.395$ in. The impellers would probably be trimmed to $10^{7}/_{16}$ in.

Once the new diameter is established, the new pump head curve may be calculated by using the affinity laws to calculate new H-Q values.

Figure 7. Impeller trim correction.

Figure 7a. Centrifugal pump performance data for speed.

The pump affinity laws are exact for a speed change. Figure 7a shows the effect of a speed change on the pump performance. The pump manufacturer should be consulted for data on the range of speeds over which the pump may be efficiently operated.

Very viscous liquids will have an impact on the performance of centrifugal pumps. Figure 11 compares pump performance when pumping water to pumping a liquid with a viscosity of 6,000 SUS. The effects of viscosity on pump performance may be predicted by the use of Figures 12 and 13. These charts are used as follows. Refer to Figure 11 and select the water capacity at bep (best efficiency point). In this example, the water capacity at bep is 23,000 gpm. Refer to Figure 12 and enter the chart with 23,000 gpm and move vertically to intersect the desired impeller diameter, 27 in., and then horizontally to intersect the 6,000 SUS line. Move vertically to intersect the line representing the water efficiency (85%) then horizontally to read the viscous efficiency 62% (not a correction factor). The previous vertical line is extended to intersect the head

Figure 8. NPSH calculation for suction lift.

Figure 9. NPSH calculation for liquid at boiling point.

Figure 10. NPSH calculation for pressured drum.

Figure 11. Effects of viscosity on centrifugal pump performance.

Figure 12. Viscosity correction chart—capacities above 10,000 GPM. Courtesy United Centrifugal Pumps.

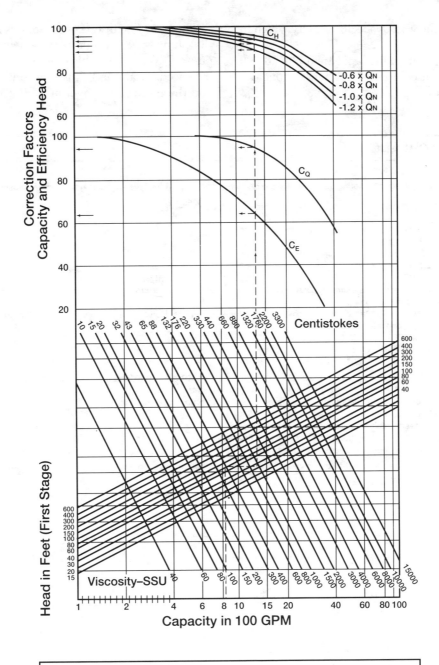

Figure 13. Viscosity correction chart—capacities up to 10,000 GPM. Courtesy United Centrifugal Pumps.

CORRECTION APPLIES TO PEAK EFFICIENCY ONLY. FOR OTHER CAP DRAW CORRECTED H.P. CURVE PARALLEL TO WATER H.P. CURVE AND CALCULATE EFFICIENCIES BASED ON ASSUMED HEAD – CAP CURVE. ADJUST AS REQUIRED TO OBTAIN A SMOOTH CURVE. USE CORRECTIONS OBTAINED FOR MAXIMUM DIAMETER FOR CUT DIAMETERS.

correction and capacity correction curves. The capacity correction factor is 93% and the head correction factor is 95%. The viscous head at shutoff will be same as the shut-off head for water. The viscous head will be 617 ft at 21,400 gpm. Plot this point and then draw the head-capacity curve through this point. The viscous brake horsepower is calculated at the bep (62% at 21,400 gpm). Draw the viscous brake horsepower curve through this point. The slope of the line will follow the slope of the water brake horsepower curve. The viscous efficiency will be 62% at 21,400 gpm and the viscous efficiency curve may now be calculated using the viscous brake horsepower, viscous head and capacity, and the viscous specific gravity which in this example is 0.9.

Figure 13 is used for pumps with capacities up to 10,000 gpm and Figure 12 is used for pumps with capacities greater than 10,000 gpm.

There may be instances where it is necessary to calculate the pump speed torque requirements to determine if the pump driver will be able to accelerate the pump to operating speed. Example 1 illustrates this calculation procedure and the results are shown in Figure 14.

SPEED TORQUE CALCULATION

The formula for calculating the speed-torque relation for a pump with a given specific gravity and viscosity is:

$$\text{Torque (lb.-ft.)} = \frac{5250 \times HP}{RPM}$$

The following data is required to plot the speed-torque curve:
1. HP @ Shutoff.
2. HP @ Open Valve (use one of the following):
   2a. HP @ 120% of peak capacity.
   2b. HP rating of driver if greater than HP @ 120% of peak capacity.
   2c. Depending on pump application maximum horsepower on performance curve which is greater than driver rated horsepower.
   NOTE: Using "2a," or "2b" will cover most of the applications. However on some applications "2c" must be considered.
3. Pump RPM.
4. Torque @ Shutoff.
5. Torque @ "2a," "2b" or "2c".

From the above data the torque with the valve closed (using HP @ Shutoff), and the torque with the valve open (using HP @ "2a," "2b" or "2c") at full speed can be computed.

The torque varies as the square of the speed; therefore to obtain the torque at:

3/4 Speed—multiply full speed torque by—0.563
1/2 Speed—multiply full speed torque by—0.250
1/4 Speed—multiply full speed torque by—0.063
1/8 Speed—multiply full speed torque by—0.016

From the figures a speed-torque curve for both closed and open valve, covering the complete speed range, can be plotted.

Example

HP @ Shutoff = 80
HP @ 120% of peak capacity = 150
Pump RPM = 3550

$$\text{Torque @ Shutoff} = \frac{5250 \times 80}{3550} = 118 \text{ lb.-ft. @ 3550 RPM}$$

$$\text{Torque @ 120\% of peak cap.} = \frac{5250 \times 150}{3550} = 222 \text{ lb.-ft. @ 3550 RPM}$$

Speed-RPM	Torque-Valve Open (150 HP)	Torque-Valve Closed (80 HP)
Full–3550	Full–222	Full–118
3/4–2660	0.563–125	0.563–67
1/2–1775	0.250–56	0.250–30
1/4–885	0.063–14	0.063–8
1/8–445	0.016–4	0.016–2
0–0	0.050–11	

Figure 14. Courtesy United Centrifugal Pumps.

Estimate the discharge of a centrifugal pump at various speeds

Within the design limits of a pump, the discharge will be approximately proportionate to the speed.

$$Q_2 = Q_1 \frac{N_2}{N_1}$$

where Q_1 is the gallons per minute at a particular speed N_1, and Q_2 is the gallons per minute at N_2.

Example. If a pump delivers 350 gpm at 1,200 rpm what will it deliver at 1,800 rpm?

$$Q_2 = 350 \times \frac{1,800}{1,200} = 525 \text{ gpm}$$

Similarly, for a given pump design head varies as the square of the impeller speed and power as the cube of the speed.

How to estimate the head for an average centrifugal pump

By slightly modifying the basic pump formula:

$$H = \left(\frac{D\,N}{1,840\,K_u}\right)^2$$

it is possible to find a means of estimating the head a pump will develop for each impeller at various speeds. The terms are defined as follows:

H = head, ft
D = diameter of the impeller, in.
N = rpm of the impeller
K_u = pump design coefficient

The values of this expression may vary from 0.95 to 1.15. For estimating purposes substitute 1,900 for 1,840 K_u. The formula becomes:

$$H = \left(\frac{D\,N}{1,900}\right)^2 \times \text{(the number of impellers)}$$

Example. What head will be developed by a well-designed centrifugal pump with four 11.75-in. impellers turning at 1,200 rpm?

$$H = 4 \times \left(\frac{11.75 \times 1,200}{1,900}\right)^2 = 4 \times 7.4^2 = 220 \text{ ft of head}$$

Find the capacity of reciprocating pump

Multiply the diameter of the plunger by itself, multiply this by the stroke, then by the number of strokes per minute, and finally by 0.11; the result is the capacity in barrels per day.

$$C = B^2 \times S \times N \times 0.11$$

Example. Four-in. bore, 6-in. stroke, 80 strokes per minute.

$$4 \times 4 \times 6 \times 80 \times 0.11 = 845 \text{ barrels per day}$$

Note. The rule is approximate only; it allows about 5% for "slip," which is actually somewhat variable. Also note that it is applicable only to single-acting pumps; for a double-acting piston type pump, N must be twice the rpm, and a reduction must be made for the area of the piston rod on the crank end of the pump.

How to estimate the hp required to pump at a given rate at a desired discharge pressure

Ignoring effects of temperature, viscosity, etc.

$$hp = psi \times \frac{\text{barrels per day}}{1,000} \times .022$$

The factor .022 is arrived at by using a pump efficiency of 80% and a gear efficiency of 97%.

Example. What horsepower would be required to pump 15,000 barrels per day with an 800 psi discharge?

$$hp = 800 \times 15 \times .022$$
$$= 264 \text{ horsepower}$$

The factor .022 was derived from the formula:

$$hp = \frac{(psi)(bpd)}{58,800 \times \text{pump efficiency} \times \text{gear efficiency}}$$

Nomograph for determining reciprocating pump capacity

The theoretical capacity of reciprocating pumps can be calculated by the following equation:

$$C = 6/55 \; D^2 \, L \, N$$

where   C = theoretical pump capacity (delivery), gal/min
        D = diameter of cylinder, ft

L = length of stroke, ft
N = number of strokes/min

The nomograph (Figure 1) is based on this equation. The diameter and length-of-stroke scales are calibrated in in. units instead of ft. The nomograph also gives the delivery in British imperial gallons.

Figure 1. Theoretical capacity of reciprocating pumps.

Example. What is the theoretical capacity of a reciprocating pump having the following characteristics:

D = 3 in.
L = 3 in.
N = 60 strokes/min

Connect	With	Mark or Read
D = 3 in.	L = 3 in.	Reference Scale
Ref. point	N = 60	C = 5 gal/min

Source

Macdonald, J. O. S., *British Chemical Engineering*, 1, 6, 301 (1956).

Nomograph for determining specific speed of pumps

The specific speed is a criterion used in the selection of pumps for a given service. It can be used to (a) avoid cavitation troubles, (b) to select the most economic pump for a given process layout, and (c) to assist in redesigning process conditions.

For single-stage, side-entering pumps, the NPSH is defined by the following equation:[1,2]

$$S = N \frac{Q^{1/2}}{H^{3/4}} \tag{1}$$

where S = specific speed, no units
 Q = pump capacity
 N = pump speed, rev/min
 H = net positive suction head above the liquid vapor pressure (NPSH) measured at the pump suction flange, ft of fluid

H (in ft of liquid) is given by:

$$H = h_a \pm h_h - h_p - h_v \tag{2}$$

where h_a = ambient pressure at the liquid surface
 h_h = static head between liquid surfaces
 h_p = combined entrance, piping and fitting friction loss
 h_v = proper pressure at pumping temperature

The specific speed of centrifugal pumps may vary between 1,100 to 15,000. Values above 12,000 are not considered practical and, where possible, the use of a specific speed below 8,500 is recommended. Specific speed limits for each type of pump are usually available from manufacturers of pumps.

The nomograph (Figure 1) solves Equation 1.

The specific speed is also defined in terms of a dimensionless number, using consistent units, by the following equation:

$$S = \frac{Q^{0.5}N}{(\Delta H \times g_c)^{0.75}} \tag{3}$$

where g_c = gravitational conversion constant, 32.2 (lb mass/lb force)(ft/sec²)

Figure 1. Pump specific speed from net positive suction head (Reproduced by permission, *Petroleum Refiner,* Vol 36, No. 10, copyright 1957).

The range for this equation is 0.03 to 0.87. Results of Equation 1 can be converted to these units by dividing by 17,200.

Example. The following data are given for a single-suction reboiler pump:

Q = 400 gal/min
N = 3,500 rev/min

The reboiler liquid level is 18 ft above the pump and friction from entrance, pipe, and fittings loss add up to 5 ft of liquid.

For a reboiler, the liquid is at its boiling point, so $h_a = h_v$, using Equation 2, H = 13 ft.

Connect	With	Mark or Read
NPSH = 13	Q = 400 gal/min	Reference Line
Ref. line pivot	N = 3,500	S = 10,200

Source

Jacobs, J. K., *Petroleum Refiner*, 36, 10, 145 (1957).

References

1. *Hydraulic Institute Standards*, 10th Edition, New York.
2. Brown, et. al., *Unit Operations*, 190, John Wiley and Sons, New York (1950).

Nomograph for determining horsepower requirement of pumps

In the selection of pumps it is necessary to calculate the horsepower requirement to handle a specified delivery rate of fluid, against the specific head. The formula for the horsepower required by a pump is as follows:

$$hp = \frac{G \times H \times \text{specific gravity}}{3{,}960 \times e}$$

where hp = pump horsepower
       G = flow rate, gal/min
       H = total dynamic head (change in head), ft
       e = pump efficiency

Figure 1 solves this equation.

Example. Given:

G = 200 U.S. gal/min
H = 60 ft
Specific gravity = 1.1
e = .80

What is the required horsepower?

Connect	With	Mark or Read
e = .80	H = 60 ft	C
C	G = 200 gal/min	hp = 0.78 horsepower

Source

Macdonald, J. O. S., *British Chemical Engineering*, 1, 6, 290 (1956).

Figure 1. Pump horsepower.

How to select motors for field gathering pumps

The following simple formulas will provide a reasonably close approximation of the motor hp necessary to drive duplex double acting pumps normally used in the field.

Rule.

1. $hp = \dfrac{bpd \times psi}{48{,}000}$

2. $hp = \dfrac{bph \times psi}{2{,}000}$

3. $hp = \dfrac{gpm \times psi}{1{,}400}$

Motors selected with these formulas will fit delivery rates used in the formula if they are within capacity of the pump. If anticipated rate is below pump capacity, a savings may be made in first cost and power bills by selecting a motor corresponding to the actual rate rather than to full pump capacity.

It should be noted that net bpd values are to be used in Formula 1. That is, if 600 barrels are to be pumped each day, but it is desired to accomplish this in only 8 hours, the rate to be used in the formula is 1,800 bpd.

Reciprocating Pumps

A. How a Reciprocating Pump Works

A reciprocating pump is a positive displacement mechanism with liquid discharge pressure being limited only by the strength of the structural parts. Liquid volume or capacity delivered is constant regardless of pressure, and is varied only by speed changes.

Characteristics of a GASO reciprocating pump are 1) positive displacement of liquid, 2) high pulsations caused by the sinusoidal motion of the piston, 3) high volumetric efficiency, and 4) low pump maintenance cost.

B. Plunger or Piston Rod Load

Plunger or piston "rod load" is an important power and design consideration for reciprocating pumps. Rod load is the force caused by the liquid pressure acting on the face of the piston or plunger. This load is transmitted directly to the power frame assembly and is normally the limiting factor in determining maximum discharge pressure ratings. This load is directly proportional to the pump gauge discharge pressure and proportional to the square of the plunger or piston diameter.

Occasionally, allowable liquid end pressures limit the allowable rod load to a value below the design rod load. IT IS IMPORTANT THAT LIQUID END PRESSURES DO NOT EXCEED PUBLISHED LIMITS.

C. Calculations of Volumetric Efficiency

Volumetric efficiency (E_v) is defined as the ratio of plunger or piston displacement to liquid displacement. The volumetric efficiency calculation depends upon the internal configuration of each individual liquid body, the piston size, and the compressibility of the liquid being pumped.

D. Tools for Liquid Pulsation Control, Inlet and Discharge

Pulsation Control Tools ("PCT", often referred to as "dampeners" or "stabilizers") are used on the inlet and discharge piping to protect the pumping mechanism and associated piping by reducing the high pulsations within the liquid caused by the motions of the slider-crank mechanism. A properly located and charged pulsation control tool may reduce the length of pipe used in the acceleration head equation to a value of 5 to 15 nominal pipe diameters. Figure 1 is a suggested piping system for power pumps. The pulsation control tools are specially required to compensate for inadequately designed or old/adapted supply and discharge systems.

E. Acceleration Head

Whenever a column of liquid is accelerated or decelerated, pressure surges exist. This condition is found on the inlet side of the pump as well as the discharge side. Not only can the surges cause vibration in the inlet line, but they can restrict and impede the flow of liquid and cause incomplete filling of the inlet valve chamber. The magnitude of the surges and how they will react in the system is impossible to predict without an extremely complex and costly analysis of the system. Since the behavior of the natural frequencies in the system is not easily predictable, as much of the surge as possible must be eliminated at the source. Proper installation of an inlet pulsation control PCT will absorb a large percentage of the surge before it can travel into the system. The function of the PCT is to absorb the "peak" of the surge and feed it back at the low part of the cycle. The best position for the PCT is in the liquid supply line as close to the pump as possible, or attached to the blind flange side of the pump inlet. In either location, the surges will be dampened and harmful vibrations reduced.

RECIPROCATING PUMPS FLOW CHARACTERISTICS

DUPLEX DOUBLE-ACTING

Average Flow — 100%
Maximum Flow — 100% + 24%
Minimum Flow — 100% − 22%
Total Flow Var. — 46%

TRIPLEX SINGLE-ACTING

Average Flow — 100%
Maximum Flow — 100% + 6%
Minimum Flow — 100% − 17%
Total Flow Var. — 23%

Figure 1. Suggested piping system for power pumps.

NOTES

1) Centerline of pump suction (inlet) to be slightly higher than centerline of suction (inlet) valve—so any air in suction system promptly reaches pump and is expelled.

2) Slope by-pass line so low point drain will fully empty relief and choke valves and all liquid in by-pass circuit.

3) Do not locate piping or other equipment in front of or above pump liquid end—preventing servicing. Refer to pump manufacturer for minimum clearances.

4) Locate charging pump at point shown—if a charging pump is

necessary. (As for volatile fluids, for example.)

5) If desired, a two-way motor-operated by-pass valve may be used rather than manual type. It should be designed to open automatically while starting or stopping.

6) Feed line and by-pass line should feed liquid into tank below minimum liquid level.

7) To remove piping strain and vibration, a flexible hose, expansion joint, or swivel joint pair should be positioned to minimize effects of piping thermal expansion, contraction, and piping weight.

8) Suction and discharge piping must be supported and anchored.

9) To protect suction system against hazard of discharge pressure entry (as when pump is idle), a small relief valve is often connected here.

10) All system components must have adequate pressure ratings for operating, starting, and upset conditions. In order to reduce potential hazards, particular attention is recommended for the surge condition that will result downstream of the relief valve when normal discharge is blocked.

QUINTUPLEX SINGLE-ACTING

Average Flow — 100%
Maximum Flow — 100% + 2%
Minimum Flow — 100% – 5%
Total Flow Var. — 7%

SEPTUPLEX SINGLE-ACTING

Average Flow — 100%
Maximum Flow — 100% + 1.2%
Minimum Flow — 100% – 2.6%
Total Flow Var. — 3.8%

NONUPLEX SINGLE-ACTING

Average Flow — 100%
Maximum Flow — 100% + 0.6%
Minimum Flow — 100% – $1.5%
Total Flow Var. — 2.1%

REQUIRED FORMULAE AND DEFINITIONS

Acceleration Head

$$h_a = \frac{LVNC}{Kg} \qquad V = \frac{GPM}{(2.45)(ID)^2}$$

Where

h_a = Acceleration head (in feet)
L = Length of liquid supply line (in line)
V = Average velocity in liquid supply line (in fps)
N = Pump speed (revolutions per minute)
C = Constant depending on the type of pump
   C = 0.200 for simplex double-acting
    = 0.200 for duplex single-acting
    = 0.115 for duplex double-acting
    = 0.066 for triplex single or double-acting
    = 0.040 for quintuplex single or double-acting
    = 0.028 for septuplex, single or double-acting
    = 0.022 for nonuplex, single or double-acting
K = Liquid compressibility factor
   K = 2.5 For relatively compressible liquids
      (ethane, hot oil)
   K = 2.0 For most other hydrocarbons
   K = 1.5 For amine, glycol and water
   K = 1.4 For liquids with almost no compressibility
      (hot water)
g = Gravitational constant = 32.2 ft/sec²
d = Inside diameter of pipe (inches)

Stroke

One complete uni-directional motion of piston or plunger. Stroke length is expressed in inches.

Pump Capacity (Q)

The capacity of a reciprocating pump is the total volume through-put per unit of time at suction conditions. It includes both liquid and any dissolved or entrained gases at the stated operating conditions. The standard unit of pump capacity is the U.S. gallon per minute.

Pump Displacement (D)

The displacement of a reciprocating pump is the volume swept by all pistons or plungers per unit time. Deduction for piston rod volume is made on double acting piston type pumps when calculating displacement. The standard unit of pump displacement is the U.S. gallon per minute.

For single-acting pumps: $D = \dfrac{Asnm}{231}$

For double-acting piston pumps:

$$D - \frac{(2A - a)\,snm}{231}$$

Where

A = Plunger or piston area, square inch
a = Piston rod cross-sectional area, square inch
   (double-acting pumps)
s = Stroke length, inch
n = RPM of crankshaft
m = Number of pistons or plungers

Plunger or Piston Speed (v)

The plunger or piston speed is the average speed of the plunger or piston. It is expressed in feet per minute.

$$v = \frac{ns}{6}$$

Pressures

The standard unit of pressure is the pound force per square inch.

Discharge Pressure (Pd)—The liquid pressure at the centerline of the pump discharge port.

Suction Pressure (Ps)—The liquid pressure at the centerline of the suction port.

Differential pressure (Ptd)—The difference between the liquid discharge pressure and suction pressure.

Net Positive Suction Head Required (NPSHR)—The amount of suction pressure, over vapor pressure, required by the pump to obtain satisfactory volumetric efficiency and prevent excessive cavitation.

The pump manufacturer determines (by test) the net positive suction head required by the pump at the specified operating conditions.

NPSHR is related to losses in the suction valves of the pump and frictional losses in the pump suction manifold and pumping chambers. Required NPSHR does not include system acceleration head, which is a system-related factor.

Slip (S)

Slip of a reciprocating pump is the loss of capacity, expressed as a fraction or percent of displacement, due to leaks past the valves (including the back-flow through the valves caused by delayed closing) and past double-acting pistons. Slip does not include fluid compressibility or leaks from the liquid end.

Power (P)

Pump Power Input (Pi)—The mechanical power delivered to a pump input shaft, at the specified operating conditions. Input horsepower may be calculated as follows:

$$Pi = \frac{Q \times Ptd}{1714 \times \eta p}$$

Pump Power Output (Po)—The hydraulic power imparted to the liquid by the pump, at the specified operating conditions. Output horsepower may be calculated as follows:

$$Po = \frac{Q \times Ptd}{1714}$$

The standard unit for power is the horsepower.

Efficiencies (n)

Pump Efficiency (ηp) (also called pump mechanical efficiency)—The ratio of the pump power output to the pump power input.

$$\eta p = \frac{Po}{Pi}$$

Volumetric Efficiency (ηv)—The ratio of the pump capacity to displacement.

$$\eta v = \frac{Q}{D}$$

Plunger Load (Single-Acting Pump)

The computed axial hydraulic load, acting upon one plunger during the discharge portion of the stroke is the plunger load. It is the product of plunger area and the gauge discharge pressure. It is expressed in pounds force.

Piston Rod Load (Double-Acting Pump)

The computed axial hydraulic load, acting upon one piston rod during the forward stroke (toward head end) is the piston rod load.

It is the product of piston area and discharge pressure, less the product of net piston area (rod area deducted) and suction pressure. It is expressed in pounds force.

Liquid pressure $\left(\frac{\text{pounds}}{\text{square inch}}\right)$ or(psi) =

Cylinder area (square inches) = (3.1416) x (Radius(inches))² = $\frac{(3.1416) \times (\text{diameters(inches)})^2}{4}$

Cylinder force (pounds)—pressure (psi) x area (square inches)

Cylinder speed or average liquid velocity through piping (feet/second) = $\frac{\text{flow rate (gpm)}}{2.448 \times (\text{inside diameter (inch)})^2}$

Reciprocating pump displacement (gpm) = $\frac{\text{rpm} \times \text{displacement (cubic in/revolution)}}{231}$

Pump input horsepower—see horsepower calculations

Shaft torque (foot-pounds) = $\frac{\text{horsepower} \times 5252}{\text{shaft speed (rpm)}}$

Electric motor speed (rpm) = $\frac{120 \times \text{frequency (Hz)}}{\text{number of poles}}$

Three phase motor horsepower (output) = $\frac{1.73 \times \text{amperes} \times \text{volts} \times \text{efficiency} \times \text{power factor}}{746}$

Static head of liquid (feet) = $\frac{2.31 \times \text{static pressure (psig)}}{\text{specific gravity}}$

Velocity head of liquid (feet) =

$$\frac{\text{liquid velocity}^2}{g = 32.2 \text{ ft/sec}^2}$$

Absolute viscosity (centipoise) =
specific gravity x Kinematic viscosity (centistrokes)

Kinematic viscosity (centistrokes) =
$$0.22 \times \text{Saybolt viscosity (ssu)} = \frac{180}{\text{Saybolt viscosity (ssu)}}$$

Absolute pressure (psia) =
Local atmospheric pressure + gauge pressure (psig)

Gallon per revolution =
$$\frac{\text{Area of plunger (sq in) x length of stroke(in) x number of plungers}}{231}$$

Barrels per day = gal/rev x pump speed (rpm) x 34.3

TABLE 1 Water Compressibility

Compressibility Factor $\beta t \times 10^{-6}$ = Contraction in Unit Volume Per Psi Pressure
Compressibility from 14.7 Psia, 32 F to 212 F and from Saturation Pressure Above 212 F

TEMPERATURE

Pressure Psia	0 C 32 F	20 C 63 F	40 C 104 F	60 C 140 F	80 C 176 F	100 C 212 F	120 C 248 F	140 C 284 F	160 C 320 F	180 C 356 F	200 C 392 F	220 C 428 F	240 C 464 F	260 C 500 F	280 C 536 F	300 C 572 F	320 C 608 F	340 C 644 F	360 C 680 F
200	3.12	3.06	3.06	3.12	3.23	3.40	3.66	4.00	4.47	5.11	6.00	7.27							
400	3.11	3.05	3.05	3.11	3.22	3.39	3.64	3.99	4.45	5.09	5.97	7.21							
600	3.10	3.05	3.05	3.10	3.21	3.39	3.63	3.97	4.44	5.07	5.93	7.15	8.95						
800	3.10	3.04	3.04	3.09	3.21	3.38	3.62	3.96	4.42	5.04	5.90	7.10	8.85	11.6					
1000	3.09	3.03	3.03	3.09	3.20	3.37	3.61	3.95	4.40	5.02	5.87	7.05	8.76	11.4	16.0				
1200	3.08	3.02	3.02	3.08	3.19	3.36	3.60	3.94	4.39	5.00	5.84	7.00	8.68	11.2	15.4				
1400	3.07	3.01	3.01	3.07	3.18	3.35	3.59	3.92	4.37	4.98	5.81	6.95	8.61	11.1	15.1	23.0			
1600	3.06	3.00	3.00	3.06	3.17	3.34	3.58	3.91	4.35	4.96	5.78	6.91	8.53	10.9	14.8	21.9			
1800	3.05	2.99	3.00	3.05	3.16	3.33	3.57	3.90	4.34	4.94	5.75	6.87	8.47	10.8	14.6	21.2	36.9		
2000	3.04	2.99	2.99	3.04	3.15	3.32	3.56	3.88	4.32	4.91	5.72	6.83	8.40	10.7	14.3	20.7	34.7		
2200	3.03	2.98	2.98	3.04	3.14	3.31	3.55	3.87	4.31	4.89	5.69	6.78	8.33	10.6	14.1	20.2	32.9	86.4	
2400	3.02	2.97	2.97	3.03	3.14	3.30	3.54	3.85	4.29	4.87	5.66	6.74	8.26	10.5	13.9	19.8	31.6	69.1	
2600	3.01	2.96	2.96	3.02	3.13	3.29	3.53	3.85	4.28	4.85	5.63	6.70	8.20	10.4	13.7	19.4	30.5	61.7	
2800	3.00	2.95	2.96	3.01	3.12	3.28	3.52	3.83	4.26	4.83	5.61	6.66	8.14	10.3	13.5	19.0	29.6	57.2	238.2
3000	3.00	2.94	2.95	3.00	3.11	3.28	3.51	3.82	4.25	4.81	5.58	6.62	8.08	10.2	13.4	18.6	28.7	53.8	193.4
3200	2.99	2.94	2.94	3.00	3.10	3.27	3.50	3.81	4.23	4.79	5.55	6.58	8.02	10.1	13.2	18.3	27.9	51.0	161.0
3400	2.98	2.93	2.93	2.99	3.09	3.26	3.49	3.80	4.22	4.78	5.53	6.54	7.96	9.98	13.0	17.9	27.1	48.6	138.1
3600	2.97	2.92	2.93	2.98	3.09	3.25	3.48	3.79	4.20	4.76	5.50	6.51	7.90	9.89	12.9	17.6	26.4	45.4	122.4
3800	2.96	2.91	2.92	2.97	3.08	3.24	3.47	3.78	4.19	4.74	5.47	6.47	7.84	9.79	12.7	17.3	25.8	44.5	110.8
4000	2.95	2.90	2.91	2.97	3.07	3.23	3.46	3.76	4.17	4.72	5.45	6.43	7.78	9.70	12.5	17.1	25.2	42.8	101.5
4200	2.95	2.90	2.90	2.96	3.06	3.22	3.45	3.75	4.16	4.70	5.42	6.40	7.73	9.62	12.4	16.8	24.6	41.3	93.9
4400	2.94	2.89	2.90	2.95	3.05	3.21	3.44	3.74	4.14	4.68	5.40	6.36	7.68	9.53	12.2	16.5	24.1	40.0	87.6
4600	2.93	2.83	2.89	2.94	3.05	3.20	3.43	3.73	4.13	4.66	5.37	6.32	7.62	9.44	12.1	16.3	23.6	38.8	82.3
4800	2.92	2.87	2.88	2.94	3.04	3.20	3.42	3.72	4.12	4.64	5.35	6.29	7.57	9.36	12.0	16.0	23.2	37.6	77.7
5000	2.91	2.87	2.87	2.93	3.03	3.10	3.41	3.71	4.10	4.63	5.32	6.25	7.52	9.28	11.8	15.8	22.7	36.6	73.9
5200	2.90	2.85	2.87	2.92	3.02	3.18	3.40	3.69	4.09	4.61	5.30	6.22	7.47	9.19	11.7	15.6	22.3	35.6	70.3
5400	2.90	2.85	2.86	2.91	3.01	3.17	3.39	3.68	4.07	4.59	5.27	6.19	7.41	9.12	11.6	15.3	21.9	34.6	66.9

EXAMPLE: Find the volumetric efficiency of a reciprocating pump with the following conditions:

Type of pump	3 in diam plunger x 5 in stroke triplex
Liquid pumped	Water
Suction pressure	Zero psig
Discharge pressure	1785 psig
Pumping temperature	140 F
c	127.42 cu in
d	35.343 cu in
S	.02

Find βt from Table of Water Compressibility (Table 1).

βt = .00000305 at 140 F and 1800 psia Calculate volumetric efficiency:

Vol. Eff. =

$$\frac{1 - \left[P_{td}\, \beta t \left(1 + \frac{c}{d} \right) \right]}{1 - P_{td}\, \beta t} - S =$$

$$\frac{1 - \left[(1785 - 0)(.00000305) \left[1 + \frac{127.42}{35.343} \right] \right]}{1 - (1785 - 0)(.00000305)} - .02$$

= .96026

= 96 per cent

Specific gravity (at 60°F) =
$$\frac{141.5}{131.5 + \text{API gravity (degree)}}$$

Bolt clamp load (lb) =
0.75 x proof strength (psi) x tensile stress area (in²)

Bolt torque (ft-lb) =
$$\frac{0.2 \text{ (or 0.15)}}{12} \times \text{nominal diameter in inches} \times \text{bolt clamp load (lb)}$$
0.2 for dry
0.15 for lubricated, plating, and hardened washers

Calculating Volumetric Efficiency for Water

The volumetric efficiency of a reciprocating pump, based on capacity at suction conditions, using table of water compressibility, shall be calculated as follows:

Vol. Eff. =

$$\frac{1 - \text{Ptd } \beta t \left(1 + \dfrac{c}{d}\right)}{1 - \text{Ptd } \beta t} - S$$

Where

βt = Compressibility factor at temperature t (degrees Fahrenheit or centigrade). (See Tables 1 and 2).

c = Liquid chamber volume in the passages of chamber between valves when plunger is at the end of discharge stroke in cubic inches

d = Volume displacement per plunger in cubic inches
Ptd = Discharge pressure minus suction pressure in psi
S = Slip, expressed in decimal value

Calculating Volumetric Efficiency For Hydrocarbons

The volumetric efficiency of a reciprocating pump based on capacity at suction conditions, using compressibility factors for hydrocarbons, shall be calculated as follows:

$$\text{Vol. Eff.} = 1 - \left[S - \frac{c}{d} \left(1 - \frac{\rho d}{\rho s}\right)\right]$$

Where

c = Fluid chamber volume in the passages of chamber between valves, when plunger is at the end of discharge strike, in cubic inches

d = Volume displacement per plunger, in cubic inches

P = Pressure in psia (P_s = suction pressure in psia; P_d = discharge pressure in psia)

P_c = Critical pressure of liquid in psia (See p. 218).

P_r = Reduced pressure
$$\frac{\text{Actual pressure in psia}}{\text{Critical pressure in psia}} = \frac{P}{P_c}$$

P_{rs} = Reduced suction pressure = $\dfrac{P_s}{P_c}$

P_{rd} = Reduced discharge pressure = $\dfrac{P_d}{P_c}$

TABLE 2 Water Compressibility

Compressibility Factor $\beta t \times 10^{-6}$ = Contraction in Unit Volume Per Psi Pressure
Compressibility from 14.7 Psia at 68 F and 212 F and from Saturation Pressure at 392 F

Pressure Psia	Temperature			Pressure Psia	Temperature		
	20 C 68 F	100 C 212 F	200 C 392 F		20 C 68 F	100 C 212 F	200 C 392 F
6000	2.84	3.14	5.20	22000	2.61	2.42	3.75
7000	2.82	3.10	5.09	23000	2.59	2.38	3.68
8000	2.80	3.05	4.97	24000	2.58	2.33	3.61
9000	2.78	3.01	4.87	25000	2.57	2.29	3.55
10000	2.76	2.96	4.76	26000	2.56	2.24	3.49
11000	2.75	2.92	4.66	27000	2.55	2.20	3.43
12000	2.73	2.87	4.57	28000	2.55	2.15	3.37
13000	2.71	2.83	4.47	29000	2.54	2.11	3.31
14000	2.70	2.78	4.38	30000	2.53	2.06	3.26
15000	2.69	2.74	4.29	31000	2.52	2.02	3.21
16000	2.67	2.69	4.21	32000	2.51	1.97	3.16
17000	2.66	2.65	4.13	33000	2.50	1.93	3.11
18000	2.65	2.60	4.05	34000	2.49	1.88	3.07
19000	2.64	2.56	3.97	35000	2.49	1.84	3.03
20000	2.63	2.51	3.89	36000	2.48	1.79	2.99
21000	2.62	2.47	3.82				

S = Slip expressed in decimal value

t = Temperature, in degrees Rankine

= Degrees F + 460 (t_s = suction temperature in degrees Rankine; t_d = discharge temperature in degrees Rankine)

T_c = Critical temperature of liquid, in degrees Rankine (See Table 3)

T_r = Reduced temperature

= $\dfrac{\text{actual temp. in degrees Rankine}}{\text{critical temp. in degrees Rankine}}$

= $\dfrac{t}{T_c}$ (See Fig. 2)

T_{rs} = Reduced suction temperature

= $\dfrac{t_s}{T_c}$

T_{rd} = Reduced discharge temperature

= $\dfrac{t_d}{T_c}$

Vol. Eff. = Volumetric efficiency expressed in decimal value.

= $\dfrac{1}{1}$ x   x 62.4 = density of liquid in lb per cu ft

s = Density in lb per cu ft at suction pressure

d = Density in lb per cu ft at discharge pressure

= Expansion factor of liquid

$\dfrac{1}{1}$ = Characteristic constant in grams per cubic centimeter for any one liquid which is established by density measurements and the corresponding values of   (See Table 3).

TABLE 3

Carbon Atoms	Name	T_c Degrees Rankine	P_c Lb Per Sq In	pl/wl Grams Per cc
1	Methane	343	673	3.679
2	Ethane	550	717	4.429
3	Propane	666	642	4.803
4	Butane	766	544	5.002
5	Pentane	847	482	5.128
6	Hexane	915	433	5.216
7	Heptane	972	394	5.285
8	Octane	1025	362	5.340
9	Nonane	1073	332	5.382
10	Decane	1114	308	5.414
12	Dodecane	1185	272	5.459
14	Tetradecane	1248	244	5.483

Example: Find volumetric efficiency of the previous reciprocating pump example with the following new conditions:

Type of Pump	3 inch dia plunger x 5 inch stroke triplex
Liquid pumped	Propane
Suction temperature	70 F
Discharge temperature	80 F
Suction pressure	242 psig
Discharge pressure	1911 psig

Find density at suction pressure:

$T_{rs} = \dfrac{t_s}{T_c} = \dfrac{460 + 70}{666} = .795$

$P_{rs} = \dfrac{P_s}{P_c} = \dfrac{257}{642} = .4$

$\dfrac{\rho_1}{\omega_1} = 4.803$ (From Table 3, propane)

$\omega = .1048$ (From Fig. 2)

$\rho_s = \dfrac{\rho_1}{\omega_1}$ x ω x 62.4

= 4.803 x .1048 x 62.4
= 31.4 lb per cu ft

Find density at discharge pressure:

$T_{rd} = \dfrac{t_d}{T_c} = \dfrac{460 + 80}{666} = .81$

$P_{rd} = \dfrac{P_d}{P_c} = \dfrac{1926}{642} = 3.0$

$\omega = .1089$ (From Fig. 2)

$\rho_d = \dfrac{\rho_1}{\omega_1}$ x ω x 62.4

= 4.803 x .1089 x 62.4
= 32.64 lb per cu ft

Therefore

Vol. Eff. = $1 - \left[S - \dfrac{c}{d}\left(1 - \dfrac{\rho_d}{\rho_s} \right) \right]$

= $1 - \left[.02 - \dfrac{127.42}{35.343}\left(1 - \dfrac{32.64}{31.4} \right) \right]$

= .8376

= 83.76 per cent

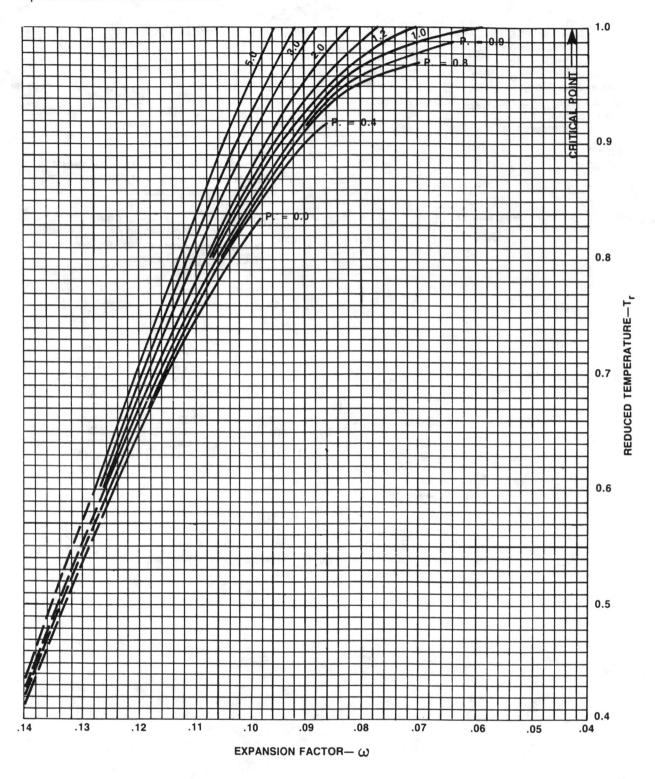

Figure 2. Thermal Expansion and Compressibility of Liquids

Understanding the basics of

Rotary Screw Pumps

. . . operating principles, application factors, and performance characteristics

By **ALBERT A. ZALIS,** Vice President Engineering, Warren Pumps, Inc.
IMO Industries, Inc.

BASICALLY only two categories of pumps need be considered when a pumping application is evaluated: centrifugal pumps and positive-displacement pumps. Centrifugal pumps depend on rotational speed to establish head and capacity characteristics and are capable of moderately high discharge heads and very high pumping capacities. Centrifugal pump design configurations include single and multistage, enclosed and open impeller, single and double suction, and horizontal and vertical.

When the application of a centrifugal pump is considered, in addition to the normal requirements of head and capacity, the pumped medium fluid characteristics must be known in detail:

- Abrasiveness
- Consistency and viscosity
- System head/capacity variations
- Net positive suction head available (NPSHA)
- Corrosiveness
- Required construction materials.

Although centrifugal pumps successfully handle all types of slurries and viscous fluids, a suitable relationship between head, capacity, and fluid characteristics must exist. A simple end suction centrifugal pump with open impeller can be used for handling both clear water and slurries, Fig. 1. Typical centrifugal pump performance curves are shown in Fig. 2. Centrifugal pump performance can be evaluated using the affinity laws: With the impeller diameter held constant, capacity Q varies as the ratio of the change in speed N

$$\frac{Q_2}{Q_1} = \frac{N_2}{N_1}$$

Head H varies as the square of the speed change ratio

$$\frac{H_2}{H_1} = \left(\frac{N_2}{N_1}\right)^2$$

Brake horsepower *bhp* varies as the cube of the speed change ratio

$$\frac{bhp_2}{bhp_1} = \left(\frac{N_2}{N_1}\right)^3$$

These laws make it simple to correct for varying speeds. However, these relationships assume that efficiency remains constant. Tests have shown that the correction for head, with regard to speed, is not precisely correct when head is reduced more than the amount calculated to result from a speed increase. Also, a speed increase yields a reduction in efficiency exceeding that calculated using the brake horsepower and head formula. However, the deviation of values is not great, and the affinity laws are a valuable tool when used with good engineering judgment.

Positive-displacement pumps in-

Fig. 1. The end suction centrifugal pump is a general-purpose pump that is widely used for pumping a broad range of liquids. Seal may be flushed with clean liquid when corrosive or abrasive slurries are pumped.

FLUSHING LIQUID SUPPLY

SEAL

IMPELLER

clude several pump designs. The major classifications are reciprocating and rotary pumps. Reciprocating pumps include power and steam driven, single and double acting, and plunger and piston.

Reciprocating pumps have high head capability, easily handle all types of slurries, and have high volumetric efficiency, permitting low-capacity/high-head relationships.

When the application of a reciprocating pump is considered in addition to the usual head and capacity data, the following conditions related to the fluid must be known:
- NPSHA
- Allowable pulsation levels
- Corrosiveness or required construction materials.

Rotary pumps cover a broad range including screw, gear, vane, and radial piston types.

Screw pumps can be further classified as single (progressive cavity), two screw (timed), and three screw (untimed). Gear pumps include external and internal gear designs.

Rotary pumps operate in generally the same pressure ranges as centrifugal pumps, except at lower capacities. Rotary pumps are not particularly suitable for handling severely abrasive fluids, except for the single-screw progressive cavity pump, which performs well in pumping abrasive slurries. Rotary pumps are generally capable of handling a wide range of fluid viscosities, at medium pressures, over a broad range of flow rates.

The remainder of this article is confined to a discussion of screw pumps, primarily the two-screw timed pump that utilizes timing gears to drive the screws and prevent driving contact between the intermeshing screws. This pump uses bearings to support the shafts and absorb some or all of the radial load, giving it a broad field of application suitable for pumping corrosive and noncorrosive fluids, lubricating and nonlubricating fluids, clean fluids and slurries, and Newtonian and non-Newtonian fluids.

Relatively high capacities, for positive-displacement pumps, can be handled by screw pumps. The primary limitation is the production of screws having the displace-

ment required to produce a high discharge rate.

Performance characteristics of a screw pump are functions of screw design, provided it is used with properly designed inlet and outlet ports. In a particular pump size, performance characteristics can be radically changed by altering screw design. When a screw pump performs poorly in a specific application, the application parameters should be rechecked and the screw design examined.

The capacity of a screw pump is a function of four critical screw dimensions, which control pump displacement, Fig. 3. For the pump to deliver maximum displacement efficiently, the profiles of the intermeshing threads must be given careful attention. The

outside diameter of the screw and the root diameter are cylindrical. Clearances of these dimensions average 0.0015 in. per in. of diameter. The flanks (thread sides) are profiled, and must be carefully controlled to obtain good pumping efficiency. One of the two flanks must have a profile defined as an epicycloid. The other flank may have an epicycloidal curve or any form of convex curve that will form a continuous seal when it is mated with the meshing screw. The total axial or flank clearance normally equals the total diametrical clearance between the screw outside diameter and the casing bore.

There must be small clearances between the screws and between the screws and casing. These clearances allow a small amount

Fig. 2. Typical performance curves for a centrifugal pump plot the brake horsepower, efficiency, and head (discharge pressure) as functions of capacity. The plots of horsepower and head are not straight lines, a fact that increases the difficulty of accurately predicting performance.

Fig. 3. Four basic dimensions establish the pump displacement characteristics. One of the flanks must have an epicycloidal profile to establish a seal with the mating screw. The mating flank may have an epicycloidal curve or any form of convex curve that will achieve a sealed running mesh.

Fig. 4. In a two-shaft, four-screw pump, the material enters at each end and discharges in the center. End thrust forces are balanced. Radial forces are functions of outlet pressure and screw dimensions, and must be carried by the bearings at each end.

Fig. 5. Pumps with two screws require timing gears to maintain constant relationship between screws. Axial forces are normally counterbalanced by hydraulic loading. The gear set may be product lubricated (a) or provided with an external lubrication system (b).

of leakage from the pump outlet cavity to the pump inlet cavity. This leakage, identified as slip, is a function of pump outlet pressure, fluid viscosity, and fluid characteristics. If the screws are designed with epicycloidal curves on both flanks, the slip will be 70 percent greater than if the design provides for one epicycloidal curve and one convex curve. However, for certain applications the double epicycloid has distinct advantages.

In a pump design utilizing four pumping screws, the arrangement usually allows the fluid to enter the screws on each end and discharge in the center, Fig. 4. When the screws are all the same diameter, axial thrust forces are equal and opposite, and therefore cancel each other. Although other arrangements for balancing radial forces are available, they are, at best, only partly effective and sometimes introduce other problems. Radial forces in screw pumps are functions of outlet pressure and screw dimensions. If the epicycloid/convex screw flank design is used, proper location of the profiles for the two screws on each shaft will partially balance radial loads.

Timed screw pumps are available in which only two pumping screws are used, Fig. 5. In these pumps, strategic location of the flank profiles has no effect on radial forces. Axial forces are of appreciable magnitude and are normally counterbalanced hydraulically.

Screw pumps are capable of discharge pressures from atmospheric to more than 3000 psig. However, the same screw design cannot be used throughout this range of pressures. In a centrifugal pump, high discharge heads are achieved either by high speeds or, more often, by staging a number of impellers to operate in series; each impeller carries a proportionate share of the discharge head. Discharge head in a screw pump is regulated by controlling the number of screw turns. The minimum number of screw turns that can be used is two. Fewer than two turns results in erratic performance and excessive slip. Unlike the centrifugal pump impeller, each screw turn does not generate a proportionate share of the outlet pressure. The

". . . lower volumetric efficiency could result in a higher mechanical efficiency."

additional screw turns assist in reducing slip, thereby increasing volumetric efficiency.

A higher volumetric efficiency can mean higher mechanical efficiency, and with today's power costs it would appear desirable to use screws having many turns to achieve a high volumetric efficiency. This is not necessarily true. The greater the number of screw turns is, the longer the screw (screw length equals the number of turns multiplied by the screw pitch) is. Screw length has a proportional bearing on the horsepower required to drive the pump at zero outlet pressure.

If a low-viscosity fluid is being pumped, this horsepower will be of small magnitude. However, the horsepower requirement increases rapidly as viscosity increases. Therefore, a slightly lower volumetric efficiency could result in a higher mechanical efficiency. Mechanical efficiency controls power costs; consequently, the application engineer, when selecting a screw pump, must consider both slip and friction horsepower to determine the best arrangement.

The remaining parts of a screw pump are subject to conventional mechanical design: Shafts must be designed for bending and torsional loads, and bearings must match the center distance between the two parallel shafts. Timing gears should be designed in accordance with American Gear Manufacturers Association standards. If spur gears are used, both shafts should be rigidly located axially. If herringbone gears are used to reduce noise level, the driven shaft should be allowed to float axially and be positioned by the apex of the gear. Velocities through the fluid inlet passages should be kept low, normally not more than 5 fps for average viscosities of 150 to 1000 SSU, and decreased as the viscosity increases.

The performance characteristics of a screw pump are similar to those of other positive-displacement pumps. Capacity, horsepower, and efficiency are plotted as functions of discharge pressure. For centrifugal pumps, horsepow-

er, efficiency, and discharge pressure are plotted as functions of capacity. With centrifugal pumps, plots of horsepower and head against capacity are not straight lines. Therefore, accurately predicting performance is difficult, and more testing and modeling are required. For positive-displacement pumps, plots of capacity and brake horsepower against pressure normally result in straight lines when coordinate paper is used. However, there are instances in which the capacity versus pressure line is slightly curved. This characteristic usually results from excessive clearance in the pump permitting deflection and eccentric operation of the rotating screw.

Performance conditions can be manipulated to reflect operation at speeds or pressures other than those shown in Fig. 6, using these fundamental relationships:

Friction horsepower is the power required to overcome losses including packing seal friction, timing gear inefficiencies, bearing friction, and pumped fluid viscous shear. All of the losses are so small when compared to the viscous drag of the pumped fluid that it can be safely assumed that friction horsepower is entirely dependent on fluid viscosity and pump speed. Even pump rotor clearances, within the limits of efficient pump operation, have such a minor influence on the power consumed that extremely sensitive instrumentation is required for taking measurements.

Fig. 7 shows the relationship between friction horsepower and viscosity at 1200 rpm. If, for example, it is necessary to determine friction horsepower fhp at 1800 rpm, the following relationship exists:

$$fhp_2 = fhp_1 \left(\frac{1800}{1200}\right)^{1.5}$$

With today's programmable calculators and computers, the curves shown in Fig. 7 may be easier to work with if they are reduced to mathematical formulas. Because the equation will take the form $x = ay^n$, by means of simultaneous equations the friction horsepower equation becomes:

$$fhp_2 = 2.9541 \times (\text{viscosity in SSU})^{0.42065}$$

Hydraulic horsepower is the measure of the total hydraulic work done by the pump. This term should not be confused with water horsepower or work horse-

Fig. 6. In contrast to performance curves for a centrifugal pump, the screw pump performance curves for capacity and brake horsepower are normally straight lines. Capacity, brake horsepower, mechanical efficiency, and volumetric efficiency are plotted as functions of discharge pressure.

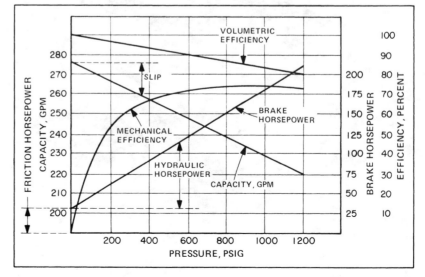

power, the measure of the useful work done by the pump.

These terms can be defined by the formulas:

Hydraulic horsepower =

$$\frac{(\text{displaced gpm})(\text{psi})}{1714}$$

Water horsepower =

$$\frac{(\text{delivered gpm})(\text{psi})}{1714}$$

Displaced gpm is the capacity at zero psi discharge pressure. Delivered gpm is the capacity at the rated or required discharge pressure, or the displaced gpm minus the slip gpm.

Referring to Fig. 6, the brake horsepower is the sum of the friction horsepower and the hydraulic horsepower. This approach yields a fairly accurate brake horsepower through a wide range of operating conditions.

Slip is the loss of capacity from the higher pressure area through the internal clearances. Slip is a function of viscosity and differential pressure. Figure 7 is a slip/viscosity curve for 500 psi operating pressure. Slip varies in direct proportion to the change in pressure. Similarly, with the friction horsepower curve, slip can be expressed by the formula:

Slip, gpm = 294.95
$$\times (\text{viscosity in SSU})^{-0.4660}$$

Mechanical efficiency is a calculated quantity that until recently received very little attention. The energy crisis has made it more important. Mechanical efficiency is calculated by:

Mechanical efficiency =

$$\frac{(\text{delivered gpm})(\text{psi})}{1714(\text{bhp})} \times 100$$

Volumetric efficiency is an infrequently used measure of internal slip and can be calculated by:

Volumetric efficiency =

$$\frac{\text{delivered gpm}}{\text{displaced gpm}} \times 100$$

It may be well to review a suggested approach to the pumping of non-Newtonian fluids. One fairly common application for screw pumps is pumping pseudoplastic polymers of various types. A pseudoplastic fluid is one whose viscosity decreases as the shear rate in the pump increases. Data on pseudoplastic fluids are usually presented as a curve plotting apparent viscosity vs. shear rate in reciprocal seconds. The determi-

nation of shear rate in a positive-displacement pump is a widely discussed topic. It is difficult, if not impossible, to arrive at one formula that is applicable to all pumps. For the timed screw pump, successful application has been made using the formula:

Shear rate, sec^{-1} =

$$\frac{\pi(\text{rotor diameter})(\text{rpm})}{(\text{radial clearance})(60)}$$

where:
rotor diameter = pumping screw diameter, in.
radial clearance = clearance between screw diameter and casing bore, in.

This formula gives consideration to only the area of maximum shear in the pump. Experience has shown that refining this formula to include all shear surfaces yields values that are grossly incorrect. The shear rate calculated from the formula will permit a satisfactory pump selection; however, care should be taken to consider the shear rate in the inlet piping. This relatively small quantity results in a much higher viscosity and thereby influences the inlet pressure requirements.

Sometimes it becomes necessary to take a surplus pump and apply it to a new application. This often happens in chemical plants where processes are frequently subject to change. Earlier discussions dealt in some detail with screw profile variations, importance of screw turns, slip, and friction horsepower. All these factors must be considered for the new application. Performance characteristics can be changed radically by installation of a new rotating element in an existing casing.

This article has not covered all there is to learn about screw pumps and their applications but has discussed the more important areas on which there is little, if any, literature published.

Fig. 7. Slip is a loss of capacity from the higher pressure area to the lower pressure area and is a function of viscosity and pressure differential. Slip varies in direct proportion to the change in pressure. At a specific pressure as the viscosity increases, slip decreases, while the friction horsepower increases.

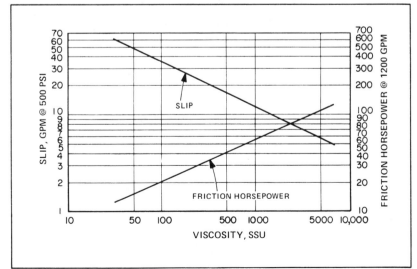

Reprinted from **PLANT ENGINEERING**
© CAHNERS PUBLISHING COMPANY

15: Measurement

Pipe line flow measurement—the new influences................................. 405
Liquid measurement orifice plate flange taps..................................... 408
Mass measurement light hydrocarbons 412
Pipe line measurement of supercritical carbon dioxide..................... 413
Gas Measurement 419
Master meter proving orifice meters in dense phase ethylene............. 419
Gas or vapor flow measurement—orifice plate flange taps 426
Properties of gas and vapors... 430
Determine required orifice diameter for any required differential
    when the present orifice and differential are known in
    gas measurement.. 435
Estimate the temperature drop across a regulator.............................. 436
Estimate natural gas flow rates... 436
How to estimate the average pressure differential on the
    remaining meter runs of a parallel system when one or more
    runs are shut off 437
Sizing a gas metering run 437
List of typical specifications for domestic and commercial
    natural gas.. 437
Determine the number of purges for sample cylinders....................... 438
Find the British thermal units (Btu) when the specific gravity of
    a pipe line gas is known... 438
Estimate for variations in measurement factors................................. 438
Rules of measurement of gas by orifice meter 439
How to measure high pressure gas 439
Four ways to calculate orifice flow in field....................................... 443
Practical maintenance tips for positive displacement meters 446
Sizing headers for meter stations... 450

Pipe line flow measurement—the new influences

E. L. Upp, Flow Measurement Consultant

Introduction

There are many considerations that influence our choice of methods and equipment in the measurement of liquids and gases in pipe lines. A review of these problems is timely, since we have changing influences that are affecting our choices to accomplish the job of flow measurement. These are accuracy, electronic equipment, economics, fluid characteristics, standards and governmental influences, and testing or proving practices. All of these factors have had importance to some degree in the past; however, their relative importance to flow measurement and more importantly to a pipe line's overall operation has changed considerably as we have seen the prices of the fluids handled make dramatic changes along with supply and demand fluctuations of considerable magnitude vs. a steady growth pattern of just a few years back.

Accuracy

Today's Demand—Better Flow Measurement Accuracy. In analyzing the economics of the flow measurement industry, we find the customers are no longer in an expansion mode. With the turndown in their economies, they are more concerned with improving efficiencies in their present installations. Better measurement accuracy is one way to achieve this. The users are asking the manufacturer to improve his present devices as well as develop new meters and new technology to do the job "better." No one can fault these desires. However, measurement is a very conservative, "tradition-following" market. Industry metering standards are slow to change, and it has been difficult in the past to get the users to adopt new meters and products.

In the gas industry, contracts are written by lawyers who are not measurement experts and who tend to follow a contract form that was successful in the past. Manufacturers tried to switch users to electronic readout devices some 20 years ago. Electronic computers and flow recorders have been available to industry all this time, though admittedly not of today's quality. Yet, today in the United States—for pipe line large-volume natural gas measurement—less than 2% of the stations installed utilize electronic computer controls and readouts. The industry has been more successful in applying electronic digital readout devices to liquid metering in conjunction with the turbine meter and are beginning to see production fields being automated for both liquid and gas measurement.

If the industry had a new meter today that offered absolute accuracy; infinite rangeability; met all the environ-mental requirements, codes, and governmental regulations; could be installed and operated by present personnel with minimal maintenance; was priced low compared to present devices on the market—it would still be difficult to sell it in sufficient quantities in the market place to make it profitable to manufacturers. The companies would at best be willing to try the meter for a year and compare it with present meters. Then and only then, would they add it to their accepted list. The manufacturers are asking the user to change his cash register. The user has to be convinced that he is on solid legal, engineering, and procedural grounds.

Costs of Accuracy. Many companies are taking a hard look at all capital expenditures in this time of high interest rates. Management people feel that the high price of the fluids they are handling allows a larger investment in order to get accurate measurement. Simple expenditure of capital does not eliminate the fact that normally the more expensive the device the "better answer" we will get. However, careful consideration must be given to what the "better answer" will mean to our operations.

Quite often the flow measurement industry is called upon to solve flow measurement problems. Sometimes the end result is not exactly what the users had in mind. A new seller of natural gas or fluids that has a meter reading low by 1% would be delighted to use the new "more accurate meter." However, a purchaser who was just as interested in his balance sheet and who had his flow figure raised by 1%, would not be too happy with a decision to use this "more accurate meter."

Many times, users have evaluated electronic flow computers, and when compared with the existing meter will read differently. There are good solid technical arguments why the computer is more accurate, but there are also political and economic reasons why customers cannot use more accuracy "if it lowers my billing." When two parties and a contract are involved, you must keep in mind that you can't have a more accurate metering device that will raise measurement for both sides of the equation. As manufacturers, the only path that can be followed is to sell the most accurate metering device they know how to make and to let the customer work with the political and economic consequences. These influences certainly affect the industry's ability to bring out new products and have them accepted based on accuracy.

Their Place in Flow Measurement. There is no denying that the high technology age is upon us. This basically refers to electronic calculating and data handling systems as far as the pipe line flow measurement problems are concerned. These tools have been available for several decades; however, we are finally seeing the beginning of the significant use of electronics to replace our previous mechanical methods of converting meter readings to custody transfer volumes. This is basically due to the availability of low cost, reliable electronic devices that will meet the field conditions where flow measurement takes place, and a growing awareness of the user on the requirements to successfully apply these devices. We have moved through stages of applications from "flow computers can do anything," to "flow computers don't operate in the field," to the stage we are in now where properly applied flow computers are a viable consideration for accomplishing the job we have done in the past, plus additional timely uses of the flow data to improve operations. However, this decision is not made without a serious look at the economic consequences.

Economics of Flow Measurement

Accuracy vs. Cost. As the cost of the fluids handled goes up, there is a demand by all parties involved in its purchase and sale to "reduce the tolerances," "get more accurate measurement," or "get more sophisticated equipment to get a better answer." Capital investment required to upgrade measurement is getting a hard look by pipe line management, as are all of their expenditures. So our desire for better measurement is tempered by available capital and the resulting savings to be expected. The definition of "more accurate answers" of various measuring devices, both in terms of theoretical and practical use, must be reviewed to improve our ability to get the best measurement for the dollars invested.

If you are producing a well at .25 MMcfd, one level of equipment and procedure investment is justified; whereas, a station delivering 25 MMcfd or 250 MMcfd may require another. Similar thinking applies to liquid measurement. Physical limitations such as available power, cost of space to install equipment, and effect on the environment must also be considered. On offshore platforms, where space costs a premium, meters that can be installed with a minimum of piping have been used because of the savings realized. A large portion of the new metering devices' electronic readouts require good quality and dependable power. In some remote locations where commercial power is not available, the cost of getting the proper power may exceed the cost of the metering. In this case, the cost of power may dictate the use of mechanical or pneumatic readout devices or some local generated source such as solar or thermoelectric. These power sources are normally low wattage devices, which may restrict the capabilities of the electronics or limit the choices of devices.

Personnel Requirements. A very definite consideration affecting the economics of flow measurement is the personnel required to accomplish a given job. This may mean at least a retraining, if not a replacing of present measurement people. Generally they will be in an upgraded position that requires both electronic and mechanical skills along with a solid understanding of flow measurement. Otherwise, you may end up with an electronics oriented person chasing the details of proper programs and round-off procedures of the measurement readout system while not recognizing that the basic meter may be improperly applied or functioning and the total system is in error. The reverse can be true if a mechanically oriented person is lacking in electronic knowledge. Fortunately, in the last few years most new workers have had good exposure to electronic devices and are more comfortable with electronic flow approaches than some of our older personnel. This problem exists at all levels of our company's operations and to a large degree will be the deciding factor of the final success of our system. It has a definite economic impact on our choice of changing our flow measurement systems.

Improving Present Flow Measurement. Since most pipe lines are in a nonexpansion mode, if not a reduction in throughput, operating people are spending a large portion of their efforts in improving efficiencies of their systems. This includes much closer attention to the efficiencies of compressor or pump stations. We are not talking about large percentage reductions in efficiencies being significant cost savings.

As an example, some gas pipe lines have as much as 80% of their total cost of operation coming from fuel costs, so even a 5% reduction relates to a reduction in operation cost of 4%. To study these efficiencies for real results requires very accurate flow measurement including individual measurement on each compressor or pump and engine. This is a case where the use of flow measurement to reduce costs is the driving factor rather than the custody transfer service that we generally worry about.

Fluid characteristics

A Changing Problem. For years we have had the luxury of handling fairly simple fluids with well-defined and/or accepted correction factors to reduce fluids at flowing conditions to fluids at base conditions. We now are being faced with pipe lines handling "dense fluids" such as carbon dioxide and ethylene and liquids such as non-stabilized crudes, light hydrocarbons operating at their vapor pressures, and refrigerated LPG. In these cases, the choice of the basic meter is less of a concern than the proper preparation of the fluid prior to passing through the meter, and the proper factors for making corrections to base conditions.

Sometimes we are trapped by our previous practices established with well-behaving fluids and can find very large errors when these practices are applied to ill-defined fluids. In these cases the problems are not in the basic flow meter, but in the flow meter system application of the correcting factors based on inadequate knowledge. The proper answer to these problems has not been worked out to give the accuracy of flow measurement we desire, but with the recognition of the problem we have been able to minimize some of the errors while continuing to seek new flow meters or methods of correcting for the fluid characteristics to solve the problem.

Standards and governmental effects on flow measurement

Outside Influences on Flow Measurement. Industry practices, standards and contracts, along with the ever increasing governmental requirements must be considered to complete the picture of present day pipe line flow measurement. It is the belief of some of the bodies that measurement can be accomplished by legislation.

All orifice plates in some countries, for example, have to be stamped by a government inspector, which makes them acceptable and correct. The manufacturer or user then can adopt the attitude that if the standards body or government finds it correct, it is not his worry and he really does not need to know much about flow measurement. This fundamentally has a serious flaw, since it removes the one item effective in obtaining accuracy—personal commitment and responsibility to accomplish the job. The manufacturers and users in countries where this is acceptable practice have passed the responsibility back to the government. They have reduced the caring as to whether or not their metering is accurate or properly installed, and whether or not it maintains this accuracy in long-term service.

As we get further into the governmental agencies controlling our business we have seen that the cost of doing the job is of secondary importance to meeting DOE, EPA or OSHA standards, as an example. Unfortunately, the governmental bodies are looking at getting further in the measurement of fluids business with every passing day. Experience tells us that present economics of fluid measurement will be less of a controlling factor as this continues to occur.

Testing or proving flow meters

Present Practices. The term "proving" in flow measurement means a throughput test and comes to us from the liquid measurement people who developed the mechanical displacement prover (now termed the "pipe prover") for proving turbine and positive displacement meters on crude oil and petroleum products. Prior to this development, we had critical flow and low flow provers used on PD meters on gas, but these had relatively limited use. On the other hand, "proving head meters" was never done. We substituted mechanical inspection of the primary devices and calibration of the secondary devices for the throughput test.

Proving is the checking of the throughput of a meter vs. a standard determination of volume from another device. The purpose of all inspection and calibration or proving is to establish a meter's performance accuracy to some tolerance. This tolerance is dependent on the basic accuracy of the standard used plus the conditions of the test. The allowable tolerance is dependent on the value and quantity of the product handled. Because of this, some metering installations cannot justify an extra few dollars in cost. In those cases we continue to use inspection and calibration to determine the meter's accuracy.

Proving meters can be quite expensive in terms of permanently installed provers. The decision is based on "cost considerations" and is balanced by the degree of "accuracy needed" and operations and maintenance money available to be spent in obtaining and maintaining the "accuracy." Here, then, are several of the influences that affect our measurement testing and proving: present practice, standards, accuracy desired, and economics of the measurement job.

Improvement of Present Practices. Present practices will be continued in concept; however, equipment development has improved our tools such as air deadweight devices for differential pressure tests and precision test gauges for static pressure for gas meters. Liquid meters have improved equipment available for sampling and fluid characteristic testing.

The standards for flow measurement have had a great deal of study done recently to improve the accuracy and application range with such programs as the American Petroleum Institute work on the fluid characteristics, the development of the small provers for liquids, and the basic flow coefficient research on the orifice by the American Petroleum Institute and the Gas Processors Association. This work of upgrading standards to improve accuracy is continuing at a rapid pace. The accuracy requirements for all flow measurement is improved because of the high cost of the measured fluids, and we anticipate that this demand will increase.

Summary

The business of pipe line flow measurement has assumed a more important role in the overall pipe line operation picture because the economics of the job to be done requires: more accuracy; better knowledge of the fluids being handled; wider application of electronics to accomplish the basic flow measurement plus other uses of these data to improve operations; the adaptation to changing standards and governmental rules and the use of new equipment and techniques to determine the accuracy of a flow meter. The people who are responsible for flow measurement have a challenge to respond to these changing influences to improve their company's flow measurement.

Liquid measurement orifice plate flange taps

This program, designed for the HP-41CV calculator, will calculate the bore required for an orifice plate for a given rate of flow, or the flow rate for a given bore size, when the following variables are entered into the program.

Initiate program by pressing **A**

Enter:

Flow rate, lb/hr	Press **R/S**
Flowing temperature, °F	Press **R/S**
Flowing specific gravity, dimensionless	Press **R/S**
Differential pressure range, in. of water	Press **R/S**
Pipe inside diameter, in.	Press **R/S**
Absolute viscosity, centipoises	Press **R/S**

Program calculates beta ratio (dimensionless) and orifice bore (in.).

If an even-sized bore is required, press **C** and enter:

Required orifice bore, in.	Press **R/S**

Program calculates beta ratio (dimensionless) and new flow rate (lb/hr).

Notes

1. The E subroutine is available to reenter data and recalculate.
2. Viscosity correction is automatic.
3. If the orifice plate material is not 316SS or 304SS, see Table 1.

Sequence of Calculations

1. Calculate the S value.

$$S = \frac{W_m}{D^2(2{,}831 + .0567T_f)\sqrt{G_f h_m}}$$

2. Calculate Reynolds number.

$$R_D = \frac{4.42\, W_m}{D\mu}$$

Table 1
Table of Equal Thermal Expansion Temperatures, °F, for Various Orifice Plate Materials

Type 316 or Type 304	Aluminum	Copper	Type 430	2% CRMO	5% CRMO	Bronze	Steel	Monel
	−298					−360		
	−230	−378				−273		
−325	−178	−266				−216		
−226	−131	−194				−163		−281
−152	−85	−132				−111		−191
−86	−41	−72				−60		−108
−24	+3	−15				−8		−33
+38	+47	+42	+24	+28	+27	+43	+31	+38
+101	+91	+99	+112	+108	+110	+95	+106	+104
+158			+202	+184	+191	+149	+181	+168
213			290	260	272	200	254	230
266			374	331	352	249	324	290
318			456	400	426	297	392	348
367			531	468	493	345	457	405
414			604	534	560	393	520	461
463			677	597	628	440	542	516
512			749	660	696		644	571
560			817	720	763		705	626
607			884	780	827		765	680
655			952	842	888		823	733
703			1020	899	948		879	785
750			1087	956	1009		934	835
796			1155	1011	1071		988	886
840			1223	1065	1133		1041	935

The program is written for 316SS and 304SS. If another material is used, move down the scale for the selected material, then across (left) to the equivalent 316SS/304SS temperature. Enter this value. For example, if the material selected for orifice plate was Monel, with a flow temperature of 405°F, enter 367°F. (The temperature is only entered in the program for orifice plate thermal expansion. It does not affect any other calculation.)

3. Iterate to find uncorrected beta ratio.

$$S = 0.598\beta^2 + .01\beta^3 + .518\beta^{6.425}$$

$$fx = 0.598\beta^2 + .01\beta^3 + .518\beta^{6.425} - S$$

$$f^1x = 1.196\beta + .03\beta^2 + 3.3282\beta^{5.425}$$

Start with $\beta = 0.5$ and calculate fx/f^1x. If greater than 0.0001, subtract fx/f^1x from current value of β and repeat until the error is less than 0.00001.

4. Calculate Reynolds number correction factor. If $R_D < 350\, \beta^{1.3}$:

HP 41CV Data Sheet

```
* * * * * * * * * *          ENTER DIFF RANGE INS H2O                    2.3391
                                       100.0000   ***
ORIFICE  LIQD              ENTER PIPE ID INS           ENTER REQD  ORIFICE INS
FLANGE    TAPS                          4.0260   ***               2.3750   ***
                          ENTER VISCOSITY cP           CALC    BETA
* * * * * * * * * *                     0.8000   ***          RATIO
                          CALC   BETA                              0.5899   ***
ENTER FLOW  PPH           RATIO
    100,000.0000   ***                  0.5810   ***   CALC    FLOW
ENTER TEMP DEG F          CALC ORIFICE                  PPH
        150.0000   ***         BORE INS                      103,700.4948   ***
ENTER SG                                2.3391   ***
          0.9800   ***      * * * * * * * * * * * *     * * * * * * * * * *
```

HP 41CV Register Loading

SIZE 080	R26= 0.056700	R53= "DIFF R"
R00= 0.000000	R27= 2,831.000000	R54= "ANGE I"
R01= 0.000000	R28= 0.000000	R55= "******"
R02= 0.000000	R29= 0.000000	R56= "CALC "
R03= 0.000000	R30= 72.450000	R57= "ENTER "
R04= 0.000000	R31= 46.470000	R58= 44.000000
R05= 0.000000	R32= 57.484900	R59= 0.000000
R06= 0.000000	R33= 57.505100	R60= "NS H2O"
R07= 0.000000	R34= 57.520000	R61= "PIPE I"
R08= 0.000000	R35= 57.535460	R62= "D INS "
R09= 0.000000	R36= 57.616200	R63= "VISCOS"
R10= 0.000000	R37= 57.636400	R64= "ITY cP"
R11= 0.000000	R38= 0.000000	R65= "BETA "
R12= 0.000000	R39= 56.656600	R66= "RATIO "
R13= 0.000000	R40= 71.676869	R67= "RIFICE"
R14= 0.000000	R41= 57.707273	R68= " BORE "
R15= 0.000000	R42= 0.000000	R69= "INS "
R16= 0.000000	R43= 56.656600	R70= "REQD "
R17= 0.000000	R44= 56.484900	R71= "CALC O"
R18= 0.000000	R45= "E LIQD"	R72= "ORIFIC"
R19= 1,200.000000	R46= "FLANGE"	R73= "E INS "
R20= 3.328200	R47= " TAPS"	R74= 0.000000
R21= 5.425000	R48= "FLOW "	R75= 0.000000
R22= 1.196000	R49= "PPH "	R76= 0.000000
R23= 0.000010	R50= "TEMP D"	R77= 0.000000
R24= 0.518040	R51= "EG F "	R78= 0.000000
R25= 6.425000	R52= "SG "	R79= 0.000000

HP 41CV Listing

SUBROUTINES
```
01*LBL "FT LIQ"
02 GTO A
```

PRINT LINE OF STARS
```
03*LBL a
04 FC? 55
05 RTN
06 55.55
07 ADV
08 SF 12
09 XEQ b
10 ADV
11 RTN
```

PRINT LINES
```
12*LBL d
13 SF 12
14 1
15 ST+ 58
16 RCL IND 58
17*LBL b
18 STO 59
19 CLA
20*LBL c
21 ARCL IND 59
22 RCL 59
23 FRC
24 100
25 *
26 STO 59
27 X≠0?
28 GTO c
29 FC? 55
30 GTO 01
31 PRA
32 CF 12
33 RTN
34*LBL 01
35 CF 12
36 AVIEW
37 RTN
```

PRINT & STORE VARIABLES
```
38*LBL e
39 1
40 ST+ 58
41 RCL IND 58
42 X=0?
43 GTO d
44 XEQ b
45 1
46 ST+ 01
47 RCL IND 01
48 TONE 9
49 FS? 00
50 GTO 01
51 STOP
```

```
52 STO IND 01
53 .5
54 STO 12
55 RCL IND 01
56◆LBL 01
57 XEQ J
58 GTO e
```

RE-ENTER VARIABLES

```
59◆LBL "REENTER"
60 ADV
61 1
62 +
63 STO 01
64 RCL IND 01
65 XEQ J
66 STOP
67 XEQ J
68 STO IND 01
69 ADV
70 SF 00
71 SF 02
72 STOP
73 GTO 02
```

BYPASS PRINTER

```
74◆LBL J
75 FC? 55
76 GTO 01
77 PRX
78 RTN
79◆LBL 01
80 VIEW X
81 RTN
```

CALC S

```
82◆LBL G
83 RCL 12
84 PSE
85 X↑2
86 .598
87 *
88 RCL 12
89 3
90 Y↑X
91 .01
92 *
93 +
94 RCL 12
95 RCL 25
96 Y↑X
97 RCL 24
98 *
99 +
100 RTN
```

ITERATE β (ENDS AT 218)

```
101◆LBL D
102 RCL 12
103 RCL 22
104 *
105 RCL 12
106 X↑2
107 .03
108 *
109 +
110 RCL 12
111 RCL 21
112 Y↑X
113 RCL 20
114 *
115 +
116 1/X
```

```
117 XEQ G
118 RCL 11
119 RCL 10
120 /
121 -
122 *
123 ST- 12
124 ABS
125 RCL 23
126 X<>Y
127 X>Y?
128 GTO D
```

CALC F_c WHEN $R_D \leq 700\beta^{1.3}$

```
129 RCL 12
130 1.3
131 Y↑X
132 700
133 *
134 STO 18
135 STO 15
136 2
137 /
138 STO 14
139 RCL 09
140 /
141 1/X
142 SQRT
143 90
144 *
145 SIN
146 RCL 12
147 .15
148 Y↑X
149 /
150 STO 17
151 RCL 09
152 RCL 14
153 X>Y?
154 GTO I
155 RCL 17
156 ABS
157 SQRT
158 STO 17
```

CALC F_c WHEN $R_D > 700\beta^{1.3}$

```
159 RCL 09
160 RCL 18
161 X>Y?
162 GTO I
163 1
164 STO 17
165 RCL 09
166 STO 18
167 STO 15
168 RCL 09
169 RCL 19
170 X>Y?
171 GTO I
172 STO 18
```

CALC F_c WHEN $R_D = 1200$

```
173◆LBL I
174 0
175 RCL 12
176 .2
177 -
178 X>Y?
179 0
180 X↑2
```

```
181 RCL 12
182 3.9
183 *
184 E↑X
185 8.7
186 *
187 RCL 15
188 -.89
189 Y↑X
190 *
191 +
192 STO 28
193 RCL 14
194 RCL 18
195 /
196 RCL 12
197 SQRT
198 SQRT
199 .12
200 -
201 Y↑X
202 *
203 1
204 RCL 18
205 RCL 19
206 /
207 -
208 SQRT
209 *
210 CHS
211 RCL 28
212 +
213 RCL 17
214 *
215 .997
216 +
217 STO 10
218 RTN
```

MAIN PROGRAM ENTER DATA

```
219◆LBL A
220 CF 02
221 1
222 STO 01
223 STO 10
224 29
225 STO 58
226 XEQ a
227 XEQ d
228 XEQ d
229 XEQ a
230 XEQ e
231 CF 00
```

CALC S

```
232◆LBL 02
233 39
234 STO 58
235 RCL 02
236 RCL 06
237 X↑2
238 /
239 RCL 27
240 RCL 26
241 RCL 03
242 *
243 +
244 /
245 RCL 04
246 RCL 05
247 *
248 SQRT
```

```
249 /
250 STO 11
```

CALC R_D

```
251 4.42
252 RCL 02
253 *
254 RCL 06
255 /
256 RCL 07
257 /
258 STO 09
```

CALC β

```
259 XEQ D
260 XEQ D
261 XEQ D
262 RCL 12
263 XEQ J
```

CALC d

```
264 XEQ d
265 RCL 06
266 RCL 12
267 *
268 STO 29
269 GTO 01
```

ENTER REQD BORI

```
270◆LBL "EVEN SZ"
271 7
272 STO 01
273 40
274 STO 58
275 XEQ e
```

CALC β

```
276 RCL 08
277 RCL 06
278 /
279 STO 12
280 XEQ J
281 XEQ G
282 STO 16
```

CALC FLOW

```
283 XEQ d
284 RCL 02
285 RCL 11
286 /
287 RCL 16
288 *
289 STO 29
290◆LBL 01
291 XEQ J
292 FS? 00
293 STOP
294 XEQ a
295 RCL 29
296 VIEW X
297 BEEP
298 ADV
299 STOP
300 .END.
```

$$F_c = \left[A - \left(A(.5^{(\beta^{.25} - .12)}) \sqrt{1 - \frac{700\beta^{1.3}}{1,200}} \right) \right]$$

$$\left[\frac{SIN \left(90 \sqrt{\frac{R_D}{350\beta^{1.3}}} \right)}{\beta^{.15}} \right] + .997$$

where $A = (\beta - .2)^2 + (8.7e^{3.9\beta})(700\beta^{1.3})^{-.89}$

when $\beta - .2 > 0$

$(\beta - .2)^2 = 0$

If $350^{1.3} < R_D < 700\beta^{1.3}$:

$$F_c = \left[A - \left(A(.5^{(\beta^{.25} - .12)}) \sqrt{\frac{1 - 700\beta^{1.3}}{1,200}} \right) \right]$$

$$\left[\sqrt{\frac{SIN \left(90 \sqrt{\frac{R_D}{350\beta^{1.3}}} \right)}{\beta^{.15}}} \right] + .997$$

If $700\beta^{1.3} < R_D < 1,200$:

$$F_c = \left[A - \left(A \frac{(350\beta^{1.3})^{(\beta^{.25} - .12)}}{R_D} \sqrt{1 - \frac{R_D}{1,200}} \right) \right] + .997$$

When $R_D > 1,200$:

$F_c = (A + .997)$

5. Divide original S by F_c for viscosity correction.
6. Repeat step 3 for new β for corrected S.
7. Repeat step 4 for final F_c for corrected β.
8. Repeat step 5 for final S.
9. Repeat step 3 for final β.
10. Calculate orifice bore.

$d = \beta D$ in.

11. Calculate flow rate for an even-sized orifice plate.

$$\beta = \frac{d}{D}$$

$$S = .598\beta^2 + .01\beta^3 + .518\beta^{6.425}$$

$$new\ flow = \frac{new\ S}{original\ S} \times original\ flow,\ lb/hr$$

Nomenclature

A	partial calculation of F_c, dimensionless
D	inside diameter of pipe, in.
d	diameter of orifice bore, in.
F_c	Reynolds number correction factor, dimensionless
G_f	specific gravity of liquid at flowing temperature, dimensionless
h_m	maximum differential range, in. of water
R_D	Reynolds number, dimensionless
S	Beta ratio index, dimensionless
T_f	flowing temperature, °F
W_m	maximum rate of flow, lb/hr
μ	absolute viscosity, centipoises
β	Beta ratio d/D, dimensionless

Source

Instrument Engineering Programs, Gulf Publishing Company, Houston, Texas.

Reference

1. Spink, L. K., *The Principles and Practice of Flowmeter Engineering*, The Foxboro Company.

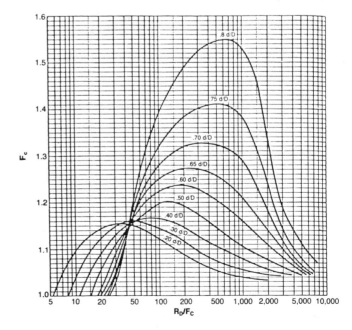

Viscosity Corrections for Concentric Orifices

Mass measurement light hydrocarbons

Streams which transport mixed natural gas liquids require the use of mass measurement to accurately account for the volume of the components which make up the mix. When natural gas liquid components are mixed, non-ideal mixing occurs. Compressibility factors are difficult to determine for mixes. Mass measurement eliminates the effects of non-ideal mixing and the need for compressibility factors.

Certain precautions must be noted in using the mass measurement method. The density and volume must be measured at the same flowing conditions. An accurate composite sample of the flowing stream must be collected and analyzed with care. On systems where there are a number of input points that will be compared to an output point, it may be desirable to have all composite samples analyzed at the same laboratory.

The following data are needed for the calculation of quantities of natural gas liquid components:

- Metered volume at operating conditions
- Meter factor, if applicable
- Density at flowing conditions
- Composite sample and sample analysis for metered volume
- Component lb mass for each component in the stream
- Density—lb mass per U.S. barrel for each component in the stream

Sample calculation

Metered volume = 25,000 barrels
Meter factor = 1.0015
Density = 0.480

Composite sample analysis

Component	mol %
Carbon dioxide	0.08
Methane	1.90
Ethane	41.30
Propane	38.00
Isobutane	5.00
Normal butane	9.30
Isopentane	1.50
Normal pentane	1.10
Hexanes +	1.82
Total	100.00

See Table 1 for the details of the calculation.

Table 1
Mass Measurement—Light Hydrocarbons

Component	Mol % 1	Mol Wt 2	Mol % × Mol Wt 3	Wt Frac. of Comp. 4	# Mass Comp. 5	Density #/bbl 6	U.S. bbl 7
CO_2	0.08	44.010	3.5208	0.000852	3,588	286.4358	13
C_1	1.90	16.043	30.4817	0.007381	31,063	105.0000	296
C_2	41.30	30.070	1,241.891	0.300739	1,265,556	124.7232	10,147
C_3	38.00	44.097	1,675.686	0.405787	1,707,617	177.5256	9,619
iC_4	5.00	58.123	290.615	0.070375	296,153	197.0934	1,503
NC_4	9.30	58.123	540.5439	0.130899	550,844	204.4980	2,694
iC_5	1.50	72.150	108.225	0.026208	110,287	218.7444	504
NC_5	1.10	72.150	79.365	0.019219	80,877	220.9914	366
$C_6{}^+$	1.82	87.436	159.1335	0.038536	162,166	244.2300	664
	100.00		4,129.461		4,208,151		25,805

Total lb mass = 25,000 × 1.0015 × 0.480 × 8.337 × 42 = 4,208,151

Column 1—From Sample Analysis
Column 2—From Table of Hydrocarbon Properties (Figure 1)
Column 3—Column 1 × Column 2
Column 4—Mol % × Mol Wt divided by sum of Column 3
Column 5—Column 4 × total lb mass
Column 6—From Table of Hydrocarbon Properties (Figure 1)
Column 7—Column 5 divided by Column 6

Reference: GPA Standard 8173-83, Method for Converting Natural Gas Liquids and Vapors to Equivalent Liquid Volumes. GPA Standard 8182-82, Tentative Standard for the Mass Measurement of Natural Gas Liquids.

Pipe line measurement of supercritical carbon dioxide

G. W. Marsden, Shell Pipe Line Corporation and **D. G. Wolter,** Shell Oil Company

Introduction

The Shell Companies have been involved in carbon dioxide systems used for enhanced oil recovery for a number of years. Pilot programs existed in the late 1970s using recovered combustion generated carbon dioxide. In 1981 a plan was in place for a large scale commercial pipe line transportation system connecting carbon dioxide producing fields in southwestern Colorado and the oil fields of the Permian Basin in West Texas (see Figure 1). The Cortez Pipeline Company was set up as the ownership company and Shell Pipe Line Corporation was established as the operator.

The 810 km (502 mi) long 760 mm (30-in.) diameter Cortez pipe line was designed to operate under supercritical conditions of 9,000 to 19,000 kPa (1,300 to 2,700 psig), 286 to 322°K (55 to 120°F), and with the nominal stream composition shown in Table 1. Design flow rates were 4.6 to 13.83 kmol/s (330 to 1,000 MMscfd). Line fill was completed in May 1984 and the system has been in continuous operation since then. There are ten receipt meters and nine delivery meters currently in operation, each with a custody transfer measurement facility.

Table 1
Stream Composition

Major Components	m^3/m^3	% Vol
CO_2, minimum	.95	95
Nitrogen, maximum	.04	4
Methane	.01–.05	1–5
Contaminants		
Hydrogen sulfide	.00002	0.002
Water	0.48 g per m^3 at standard conditions	30 lb/MMscf

Definitions

"Carbon dioxide," as referred to here and as used in the pipe line industry, is a CO_2-rich mixture normally having nitrogen and methane as principal volumetric contaminants. A minimum quality of 95% carbon dioxide purity is used by many pipe line carriers, while Cortez experience thus far is circa 98% CO_2 purity. The measured quantity is the total of all mixture constituents.

Another term used by liquid pipe line operators is "custody transfer." Liquid pipe line operators typically do not own the fluid being transported, but instead accept cus-

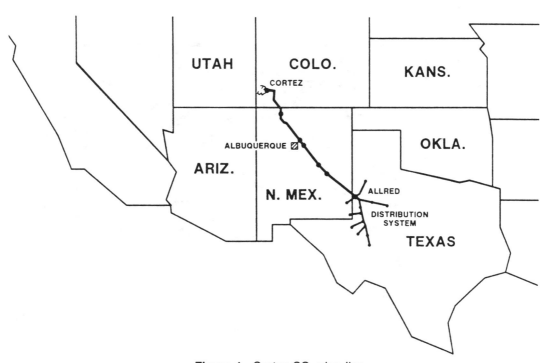

Figure 1. Cortez CO_2 pipe line.

tody at pipe line origin (receipt) and transfer that custody back to the owner or other party at the pipe line destination (delivery).

Pipe lines thus have custody transfer measurement facilities, which are comprised of generally accepted, commercially available, and accurate measurement equipment. Additionally, the operation of this equipment is accomplished in a uniform manner acceptable to all parties to the transaction: buyers, transporters, and sellers. The API Manual of Petroleum Measurement Standards is widely used in the design and operation of these facilities.

These brief definitions give some insight into the selection of custody transfer measurement methods and equipment. Let us return now to the CO_2-rich fluid we want to address.

Physical properties

CO_2, in its pure form, has a critical point of 304 K (88°F) and 7,415 kPa (1,071 psia). Within practical limits, the characteristics of pipe line quality CO_2-rich mixtures will approximate those of pure CO_2. Refer to the pressure-enthalpy diagram for CO_2 (Figure 2) which illustrates the area of normal measurement for Cortez Pipeline.

The most economical pipe line operation is in the supercritical single phase area. Fortunately, this operating area also coincides with our preferred measurement envelope. There is no pressure letdown for the delivery; in practice, the receiving party prefers the fluid with a pressure level similar to the pipe line operating pressure. This leads to measurement systems where pressure conditions are quite similar at both the inlet to and the outlet from the pipe line.

Product temperature does vary from the inlet measurement facility, which is downstream of compression equipment, to the delivery stations where the CO_2 stream approaches the ambient ground temperature. Future increases in throughput will require additional pumping/compression equipment which will tend to create measurement operating temperatures that are similar at both pipe line inlet and outlet.

The upshot of these temperature and pressure conditions is the measurement envelope shown in Figure 2. This envelope is considerably more confined during normal measurement operations than is the pipe line operating envelope. The operating envelope includes the changes in hydraulic conditions which occur along the route of the pipe line.

Looking again at the P-E diagram (Figure 2), observe that the density of the CO_2-rich stream is in the range of 640 to 960 kg/m³ (40 to 60 lb/ft³). Additionally, note that the prime determinant of a change in density will be a change in temperature. This sensitivity of change in density with temperature is of concern in the design of the measurement facilities, and it has led to special attention being given to temperature loss/gain between elements of the measuring system. There is an approximate 0.9% volume change for each °C (0.5%/°F).

The change of density with pressure change in the measurement envelope is considerably less prominent, being on the order of one tenth as severe as temperature.

Figure 2. Carbon dioxide pressure enthalpy diagram.

Measurement methods in use today

The commercial planning and development of the Cortez carbon dioxide project preceded the engineering activity, and the supply and sales contract negotiations were well advanced before custody transfer flow measurement methods were investigated. The resulting contracts, based on existing natural gas pipe line systems, dictated the use of MMscfd volumetric units for gas phase equivalent volume of carbon dioxide.

Custody transfer measurement of carbon dioxide in large volumes at high pressures presented significant technological hurdles in the methods, performance, and materials areas. Firstly, the probable stream compositions and the variations of the source field carbon dioxide were uncertain. This uncertainty of composition, and thus the physical properties, led to the selection of mass measurement techniques which would provide flexibility and operability when product conditions changed. Shell's decade long experience with mass measurement of variable mix hydrocarbon streams and with supercritical streams suggested mass measurement as one of the strongest contenders. Mass measurement, however, would not directly provide the volumetric quantities which had been established in the preceding business negotiations.

The selection of meter systems to meet the above requirements resulted in two systems which are used today in the carbon dioxide transportation industry. The orifice meter, common in the natural gas industry, and the turbine meter, similar to those used in liquid measurement, were the dominant choices. As with most other choices, there are advantages and disadvantages (see Table 2) to each method. Both are operated as mass meter systems with subsequent conversion to volumetric terms for contract compliance and accounting requirements.

Table 2
Carbon Dioxide Pipe Line Metering Options

	Orifice Meter	Turbine Meter
Advantages	Industry acceptance Proven equipment and operations Least sensitive to density determination accuracy	Accuracy $\pm 0.5\%$ Rangeability $\pm 10:1$
Disadvantages	Limited rangeability Accuracy $\pm 1\%$ Costly installation	Performance unknown with no experience in CO_2 No experience with proving in CO_2

Fortunately, Shell Pipe Line had several years of operating experience with mass measurement of petrochemical products, i.e., ethane and ethylene, as background in the orifice measurement system. The orifice meter was selected based on analysis of the factors shown in Table 2. The units conversion process, however, required additional facilities and methods for routinely and rapidly determining the composition of the streams, the calculation of molecular weight for the equivalent gas, and the conversion to volume at the selected reference pressure and temperature conditions. The resulting methods and facilities in service today are discussed below.

Orifice measurement practices

The measurement system that provided basic simplicity, reliability, wide acceptance, and the capability of handling variable mix streams without breaking new frontiers in measurement methods was resolved to be the orifice flow measuring element with online density meter and microprocessor flow computer. This basic assembly of three proven devices provides real time-mass flow output. The orifice is the key to the success of the measurement system.

The orifice plate has a fifty-year record of widespread use in the measurement of all manner of fluids. It is a static device generally inert to the measured fluid conditions, and calibration consists of simple dimensional measurement and conformance to specified physical tolerances. This, along with the periodic calibration of the associated differential pressure transmitter, completes the verification of this part of the measurement system. The orifice meter is one of the few measurement devices accepted for custody transfer which does not require in-service proving.

The second element, the density meter, has been selected to be an external unit that is easily isolated from the flowing stream for calibration, inspection, and maintenance. The relative insensitivity of CO_2 density to small changes in pressure in the primary flow to the orifice meters permits locating the density meter upstream of the meter manifold, thereby serving several meters. All installations are insulated to assure that pipe line density is maintained in the density meter.

The microprocessor flow computer, or third element of the system, is essential to achieving the advantages of integrated mass flow. This advantage comes from the ability of the computer to make computations in essentially "real time." How this advantage comes about is seen upon examination of the flow equation.

Orifice flow equation

The API Manual of Petroleum Measurement Standards, Chapter 14, Section 3 (also API 2530, AGA Report No. 3,

ANSI/API 2530-1985, and GPA 8185-85) provides the widely accepted formulation for mass flow. Mass flow is considered by this source to be the fundamental orifice meter flow equation.

Equation (38) of this publication, "Equation for Mass Flow Rate when Density of Fluid Flowing is Known" is shown in Figure 3.

The terms used in Figure 3 are:

q_m is the mass flow rate.

K and Y, the flow coefficient and the expansion factor, are the empirical elements of the orifice equation, based upon the work reported by Buckingham and Bean in 1934.

The flow coefficient, K, is in part dependent upon the viscosity of the fluid. Pipe line CO_2 viscosity ranges from approximately 0.05 to 0.08 mPa's (0.05 to 0.08 centipoise) in the measurement envelope.

The expansion factor, Y, is a function of the ratio of the orifice differential pressure to static pressure and to the specific heat ratio of the CO_2. The specific heat ratio ranges from 2.9 to 4.0 for CO_2 in the measurement envelope.

ρ_{fl} is the density output of the density meter.

ΔP is the differential pressure across the orifice plate.

Two correction factors, F_a and F_{pwl}, from Appendix E of Chapter 14, Section 3 are also utilized. F_a provides for the thermal expansion of the orifice plate, and F_{pwl} provides for local gravitational correction for deadweight testers.

Note that errors in density are taken at half their real value by integrating the density value with the orifice differential before taking the square root.

Figure 4 illustrates the accumulation of individual tolerances to arrive at the mass measurement system uncertainty tolerance. Individual element tolerances are modi-

---

14.3.5.4.3   "Equation for Mass Flow Rate when Density of Fluid Flowing is Known"

$$q_m = \frac{\pi}{4}[2g_c]^{0.5} K Y_1 d^2 [\rho_{fl} \Delta P]^{0.5}$$

Correction Factors in Accordance with Appendix E

E.3: Orifice Thermal Expansion Factor

$$F_a = 1 + [0.0000185(t_f - t_{meas})]$$

E.7: Local Gravitational Correction for Deadweight Tester

$$F_{pwl} = \left[\frac{g}{g_o}\right]^{0.5}$$

Figure 3. CO_2-rich mixture orifice measurement equation. From the *API Manual of Petroleum Measurement Standards,* 2nd Edition, September 1985.

VARIABLE	TOLER-ANCE %	EFFECT	SQ OF TOL & EFF	SUM OF SQUARES	SQ. RT. OF SUM %
Basic Orifice, F(B)	0.50	1.00	.2500	.2500	.5000
Diffntl Press, H	0.50	0.50	.0625	.3125	.5590
Density	1.00	0.50	.2500	.5625	.7500
Processor Error	0.01	1.00	.0001	.5626	.7500
Reynolds, F(R)	0.03	1.00	.0009	.5635	.7506
Expansion, Y	0.02	1.00	.0004	.5639	.7509
Orifice, F(A)	0.05	1.00	.0025	.5664	.7523

Individual factors with tolerance as above will provide
system measurement with tolerance of (plus or minus)   0.75%

Figure 4. Orifice meter system tolerance (root-sum-square method).

fied by an effect factor, then the square root of the sum of the squares of these products is computed. The resulting overall measurement uncertainty for the Cortez mass measurement system is $\pm 0.75\%$.

Pipe line operation

Full Cortez pipe line operation is accomplished using the mass flow output of the measurement system. This system also forms the basis for all custody transfer measurements in volumetric units.

It is necessary to introduce a mass-to-volume conversion step to satisfy the volumetric contracts. This conversion is accomplished using an analysis of the composite stream with a gas-liquid chromatograph maintained at each measurement area. The CO_2, N_2, and CH_4 composition of the CO_2-rich stream is determined for each measurement ticket period. An off-line equation of state computer program then calculates the appropriate standard volumetric units. A custody transfer measurement ticket (see Figure 5) is produced to document all pertinent data, including mass units, 3-component composition and the standard volumetric units.

The equation of state computer program utilized by Cortez Pipeline for this conversion was developed specifically for CO_2-rich mixtures. It is based upon the IUPAC (International Union of Pure and Applied Chemists) publication, "CO_2 International Thermodynamic Tables of the Fluid State," Volume III, compiled by Angus, Armstrong, and DeReuch. The method of extended corresponding states is utilized for the stated properties of the CO_2-rich mixture.

Turbine metering

The authors do not have first hand experience or knowledge with turbine metering of carbon dioxide and therefore will not delve into the details of the application. Turbine metering was considered in 1981 as one of the two major alternatives and rejected for a variety of reasons. Interestingly, other companies did select turbine metering and experience is now being accumulated on large scale carbon dioxide pipe lines.

Turbine meters were not selected for Cortez, or subsequent Shell systems, because the experience with mechanical displacement provers at such high pressure with compressible fluids was non-existent. Likewise, the application of liquid or gas turbine meters in supercritical carbon dioxide service was a complete unknown.

The volumetric accuracy of turbine meters was attractive, and no doubt led others to select them. However, the direct influence of density in the mass flow calculation is a disadvantage (as compared to orifice metering where density enters the equation as the square root of density and effectively halves the sensitivity of mass flow accuracy to density determination errors). (See Figure 6.)

The proving of turbine meters was a concern and has been a significant commissioning problem on systems utilizing turbine meters. High pressure provers in the 900 to 1,500 ANSI flange ratings were scarce and expensive. The seals, design, and materials were considered potential problems. Significant progress has been made by those utilizing the equipment. In the long term, turbine meters and provers will likely be more widely applied as seal systems

Figure 5. Cortez Pipeline Company. CO_2 meter ticket.

```
┌─────────────────────────────────────────────────────────────┐
│                                                               │
│               FOR ORIFICE METERING:                           │
│                                                               │
│   MASS = (CONSTANT) (DENSITY * DIFFERENTIAL) 0.5              │
│                                                               │
│                                                               │
│            FOR TURBINE OR VORTEX METERING:                    │
│                                                               │
│             MASS = (VOLUME) (DENSITY)                         │
│                                                               │
└─────────────────────────────────────────────────────────────┘
```

Figure 6. Mass flow calculation for orifice metering.

are demonstrated to be reliable and as the other operational aspects are demonstrated to be manageable.

Potential measurement methods

Two additional meter types appear attractive and are coming into use in carbon dioxide service for various applications. Cortez Pipeline now utilizes the vortex shedding meter for line integrity/leak detection purposes, and the coriolis meter is attractive due to its simplicity and apparent accuracy.

The vortex shedding meter offers advantages in higher rangeability, requiring fewer meter runs than orifice meters, and thus lower installation costs. We believe the rangeability to be between 10 and 20 to one and the off-the-shelf accuracy to be 1% or better. Operating personnel have made on-line comparisons with adjacent, though several miles distant, orifice metering systems, and observe agreement within less than 1% for extended periods. No detailed studies have been made, however.

Currently, no custody transfer applications of the vortex shedding meter are known, although consideration is becoming more common. Application will first occur on low volume service between agreeable parties or at the opportunity which assures that a proving system will be available. (Proving via master meter and master prover techniques is one possibility which has been tested on other similar products.) Once the proving methods are applied, it is expected that accuracies better than 0.5% mass basis would be achievable.

Application of the coriolis meter lags the vortex shedding meter since it is a newer development. The expected accuracy of better than 1% and rangeability of 10 to 1 are attractive. The coriolis meter also provides direct mass measurement, which removes the need for a separate density determination and should reduce both the installed and operating costs significantly. The meter currently lacks widespread industry acceptance for custody transfer measure-

ment applications, and the high pressure operation of carbon dioxide systems currently limits selection to the smaller meter sizes. No custody transfer metering applications are known at this time.

Future application of the coriolis meter in carbon dioxide service will be much the same as described above for the vortex shedding meter. Industry experience, acceptance, application of proving methods, and availability in larger sizes with higher pressure ratings will open the routes for increased use.

The future

Carbon dioxide measurement technology is advancing rapidly, and the major difficulties in accuracy and operation are now identified. One of the accuracy enhancements which is progressing is the proprietary development of a flow computer with an accurate equation of state for carbon dioxide density determination from on-line/real time pressure and temperature inputs and keypad, or possibly analog, input of stream composition.

The Shell equation of state is a proprietary unit at this time and is being used in its main frame form for conversion of mass to volumetric units on Cortez Pipeline. Industry acceptance and verification in field practice has been excellent. The equation of state is a modification of the IUPAC equation for pure carbon dioxide and incorporates the effects of nitrogen and methane impurities over a range of 80 to 100% carbon dioxide. The equation of state is considered to be accurate to $\pm 0.2\%$ for density determination in the supercritical regions. The program is being applied to a flow computer with density update calculations every 1 to 2 minutes. Pressure and temperature are real time analog inputs, while composition will be manually input via the keypad on a periodic basis dependent upon compositional stability of the fluid at each location. The planned operation region is shown on Table 3.

It is recognized that the National Bureau of Standards and others are pursuing similar equation-of-state expressions. Early investigation, however, determined that an in-

**Table 3
Region of Operation for CO_2 Equation of State Flow Computer**

Fluid	Single Phase CO_2		
		m³/m³	vol %
Composition	CO_2	0.8–1.0	80–100
	Methane	0–0.02	0–2
	Nitrogen	0–0.20	0–20
Pressure	3,400–20,500 kPa absolute (500–3,000 psia)		
Temperature	273.16–333.16 K (32–140°F)		

dustry-accepted equation of state was in the distant future and the earliest implementation could be achieved on a proprietary development with industry acceptance to follow. This decision was also supported from the view that numerous in-house uses could be made of the computers where outside acceptance was not required. These applications included process operations at the carbon dioxide production, separation, and reinjection plants where density meter operation is a considerable burden and optimized accuracy is not justified. Likewise, the computer is attractive for comparison against critical density meters and for line integrity/leak detection flow measurement facilities. We

are hoping to implement some of these applications in 1986.

In addition to the equation of state, it is our intent to remain flexible and consider the use of the potential meter types discussed above. The meter manifolds on the Cortez system have been designed with flexibility for other meter types and proving operations if these become sufficiently attractive.

The technology of carbon dioxide pipe line measurement is gaining uniformity via conferences such as this and joint industry development projects such as the equation-of-state work at NBS. In the commercial area, it is desirable that similar efforts be directed toward the use of mass measurement for operation and custody transfer purposes and to do away with the processes necessary for conversion to volumetric units. Mass units are obtained with simpler systems, at a lower cost, and more accurately. We should all work toward converting the contracts and agreements to mass units.

Source

Reprinted from *Oil & Gas Journal*, November 3, 1986 by permission and courtesy of PennWell Publishing Co.; G. W. Marsden, Shell Pipe Line Corporation; and D. G. Wolter, Shell Oil Company.

GAS MEASUREMENT

Master meter proving orifice meters in dense phase ethylene

James E. Gallagher, Shell Oil Company, Houston, Texas, USA

Abstract

Shell Pipe Line Corporation, cognizant of the importance of highly accurate orifice measurement for polymer-grade ethylene, embarked on a development program to improve the overall uncertainty of orifice metering. The program's goal was to successfully apply master meter proving techniques to dense phase ethylene orifice meter facilities operated on a mass basis.

This paper presents the techniques and results of mass proving concentric, square edged, flange tapped orifice meters. All data were obtained on commercially operated metering facilities which are in compliance with ANSI 2530 requirements.

Results indicate an overall orifice meter uncertainty tolerance of approximately $\pm 0.31\%$ is attainable when orifice meters are mass proved in situ using master meter proving techniques.

Goal of development program

During the last decade, increasing commercial activity for ethylene demanded improvement of measurement technology. Significant measurement uncertainties were being experienced on ethylene pipe line systems. In response, a development program to improve the overall uncertainty of orifice metering was initiated by Shell Pipe Line Corporation.

The program's goal was to develop an economical method for proving ethylene orifice meters under actual operating conditions. Master meter proving on a mass basis was selected as the only viable approach.

Pipe line transportation and measurement systems

An ethylene distribution system consists of a pipe line network and salt dome storage facility linking producers and consumers. Since producers and consumers are not equipped with on-site storage facilities, the pipe line(s) is designed with maximum flexibility to satisfy the continually changing demands on the operations.

Shell's ethylene systems are operated in the dense phase fluid region due to lower transportation costs. The ethylene meter stations operate in two regions depicted in Figure 1—the dense phase fluid and single phase gas regions.

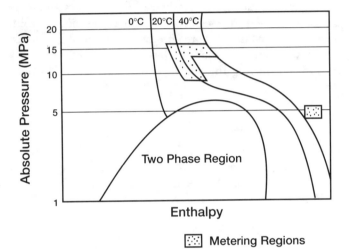

Figure 1. Ethylene pressure-enthalpy diagram.

The quantity of ethylene measured by concentric, square edged, flange tapped orifice meters, shown in Figure 3, is calculated through the combined use of the mass orifice equation and API's ethylene density equation. In the U.S., the applicable standards are API, Chapter 14, Section 3—Orifice Metering of Natural Gas and Other Related Hydrocarbon Fluids, Second Edition, September 1985 (ANSI 2530), and API Chapter 11.3.2.1—Ethylene Density.

In Shell, the concentric square edge flange tap orifice predominates as the primary element. Most installations are equipped with a dual chamber orifice fitting to allow retrieval of the orifice plate under pressure. For the most part, the orifice plates are sized to maintain a 0.3 to 0.6 beta ratio. Solid state electronic flow computers and transmitters are installed to maximize metering accuracy. Parallel meter runs and stacked "dp" transmitters maximize rangeability while maintaining a high degree of accuracy.

Historically, the industry has presumed that compliance with the reference standard piping conditions infers accuracy. No consideration was given to deviation from the empirical equation's criteria upon initial installation or further deviation over the life of the facility. The major

Velocity Profile
- **Obstructions, Deposits, Etc.**
- **Piping Configuration**
- **Pre & Post Run Lengths**
- **Pipe Condition**
  - Wall Roughness
  - Roundness
- **Plate Condition**
  - Concentric
  - Edge Sharpness
  - Nicks or Burrs
  - Flatness
  - Roughness
- **D.P. Sensing Taps**
  - Burrs

Operating Conditions
- **Pulsating Flow**
- **Widely Varying Flow**

Fluid Properties
- **Density**
  - Equation of State Uncertainty
  - Transmitters' Sensitivity
- **Expansion Factor**
- **Viscosity**

Instrumentation
- **Ambient Effects**
- **Static Pressure on "DP" Sensing**
- **Calibration Accuracy**

Figure 2. Influence quantities on orifice meters.

Figure 3. Ethylene orifice meter.

weakness of orifice metering has been the inability to prove the meter under actual operating conditions.

In 1982, the Industry Ethylene Measurement Project conducted at Mt. Belvieu, Texas, demonstrated the feasibility of gas turbine meters, conventional piston pipe provers, and API's equation of state for custody transfer. The project also demonstrated the turbine meter and pipe prover approach produced a lower uncertainty than orifice meters. The ability to accurately prove the turbine meter, shown in Figure 4, on a mass basis under actual operating conditions eliminated the reliance on reference conditions.

However, installation costs could not justify this technique for most custody transfer installations. As a result, the majority of Shell's ethylene custody transfers continues to be based on the traditional orifice meter.

Figure 4. Ethylene turbine meter.

Development program

Development of the master meter proving technique was conducted on commercially operating ethylene pipe lines by field personnel. Orifice meters associated with these systems were proven under actual operating conditions.

Fluid properties. The thermodynamic and physical properties of polymer-grade ethylene have been the subject of intensive investigations. Comprehensive treatises on the density of pure ethylene have been published by two highly reputable sources—API and the NBS. The pipe line industry in the United States presently utilizes API.

Chapter 11.3.2—Ethylene Density (API 2565). Ethylene densities are calculated using measured temperatures and pressures. The accuracy of the calculated density depends on the accuracy of the equation of state, the accuracy of the measuring devices, and the sensitivity of the calculated density to small temperature and pressure variations.

Portable proving facility. A truck-mounted 900 ANSI proving system was developed to allow maximum utilization of capital investment. The truck is equipped with a small volume prover, master turbine meter, and associated instrumentation, as shown in Figures 5 and 6.

Figure 5. Master meter method—mechanical.

Figure 6. Master meter method—electronic.

The small volume prover, SVP, operates on the principle of repeatable displacement of a known volume of fluid in the calibrated section. The mass of the fluid in the prover is determined through the combination of the prover's base volume, the fluid's pressure and temperature, and the API ethylene density equation.

The SVP was selected based on the following criteria:

- Minimize capital investment by using portable system
- Unidirectional flow design
- Honed and calibrated pipe section
- Dynamic piston seal leak detection
- Single fiber optic piston position detector
- Control of launching pressure wave
- Double chronometry pulse interpolation

Principal parts include an outer housing, a honed inner measuring cylinder, a piston displacer with an Invar extension rod, and an automatic hydraulic system. The hydraulic system returns the piston displacer to the upstream launch position after each proving run and actuates the prover bypass valve to allow the fluid to bypass the prover while the piston is being returned to the launch position.

The pressure wave that occurs when launching the prover's piston is monitored by a high speed recorder which was added during initial testing. The differential pressure, or pressure wave, across the prover is recorded to assure the piston does not act as a compressor or brake. The hydraulic system controls the piston to maintain a constant velocity through the calibrated section.

A gas turbine meter specially modified by the manufacturer for ethylene service was selected for the master turbine meter. The meter is equipped with a Lexan rotor, ball bearings, external lubricator, and conventional preamplified electronic pulse output. The master meter has a stated linearity of $\pm 0.25\%$ over a 5:1 turndown. The truck is equipped with two master turbine meters to cover normal operating flowrates—100 mm (4 in.) and 50 mm (2 in.) turbine meters.

The SVP is controlled by a microprocessor based computer. The computer monitors repeatability, piston position, and prover bypass valve position; performs dual chronometric pulse interpolation; and calculates/prints a hard copy report for each successful proving.

The computer receives input from the prover's temperature and pressure transmitters, the turbine meter's pulse train, the fiber optic gate detector, and various hydraulic controls.

Two microprocessor-based flow computers are used to calculate the turbine meter's mass flow. The first computer calculates the API ethylene density based on the turbine meter's pressure and temperature transmitters. The second computer calculates the turbine's mass flow based on the turbine meter's pulse input and contains an internal mass proving counter.

Solid state temperature transmitters, equipped with high accuracy 100 ohm platinum RTD sensors, are installed 4 to 5 pipe diameters downstream of the turbine meter and at the exit of the prover. The calibrated accuracy is $\pm 0.08°C$ $(0.15°F)$.

Solid state pressure transmitters, equipped with a sealed capacitance sensing element, are installed 4–5 pipe diameters downstream of the turbine meter and at the inlet to the prover. The calibrated accuracy is ± 6.89 kPa (1.0 psi).

Solid state differential pressure (dp) transmitters, equipped with a sealed capacitance sensing element, are installed across the prover and the piston's seal. The calibrated accuracy is ± 12 Pa (0.05 in. H_2O).

A high speed chart recorder monitors the prover's "dp" transmitters:

- To assure that the prover does not induce a pressure wave greater than 6.89 kPa (1 psi) while the piston is in the calibrated section
- To assure that the piston's seal does not leak in the dynamic state

A solid state counter, equipped to receive a contact closure from the orifice's computer and send a dry contact closure output to the turbine meter's flow computer, is installed to start and stop the master meter proving run. This device triggers the running start/stop master meter proving.

Master meter proving principle. The proving is based on the concept that a "master meter" can act as a transfer standard between the primary standard and another meter.

Meter proving by the master meter method requires two distinct steps:

- Mass prove the master turbine meter on site under operating conditions
- Master meter prove the orifice meter on a mass basis using the master turbine meter

To assure the devices are experiencing identical conditions, the orifice meter, turbine meter, and SVP are placed in series, as shown in Figure 5, during the entire calibration. Before proving the "master meter," the flow and/or density conditions through the orifice meter, turbine meter, and SVP must approach steady state conditions.

In the "master meter" proving method, the turbine meter is calibrated on a mass basis under actual operating conditions by the SVP. Once the turbine meter has been proved, a comparison of the two meters' mass outputs forms the basis for determining the orifice meter factor as shown in Figure 6.

Proving Procedures

In order to assure accuracy, the operator must be certain the system approaches steady state conditions (flow rate, density, pressure, temperature), and that no leakage is occurring between the orifice, turbine, and the prover. If steady state conditions do not exist, accuracy and/or repeatability will be extremely difficult to obtain. The orifice meter's instrumentation is calibrated prior to proving.

Preparation.

1.1 Initiate flow through the SVP. Check for leakage on all critical valves, relief valves, and piston displacer seal.

1.2 Calibrate all instrumentation associated with the prover truck (pressures, temperatures, flow computers).

1.3 Check operation of assembled instrumentation by performing preliminary turbine meter proving runs.

1.4 If necessary, adjust hydraulic system to assure the prover does not act as a compressor or a brake. Adjust the hydraulic system if the piston displacer's differential pressure exceeds ± 6.89 kPa (1 psi).

1.5 Allow ethylene to circulate through the prover truck until the outlet temperature is stabilized.

Proving run.

2.1 Initiate the SVP proving sequence by commanding the prover computer/controller to obtain a VMFt. (Acceptable repeatability for the turbine meter is five consecutive runs in which the volumetric meter factor, VMFt, does not exceed a tolerance of 0.05%.) The computer will print a proving report when this criterion has been met.

2.2 Calculate the turbine meter's mass meter factor, MMFt.

2.3 Clear the turbine meter flow computer's internal mass prover counter.

2.4 Reset and program the master meter gate simulator to allow the master meter proving to occur for a minimum of one hour. Record the turbine meter's mass flow rate. (The simulator initiates a master meter proving after receiving five dry contacts from the orifice computer. The simulator stops the proving after approximately one hour. Example—for flow at 50,000 lb/hr, the gates would be set at 5 and 505.)

2.5 When the master meter run is complete, record the turbine meter's mass counts.

2.6 Calculate the orifice's mass meter factor (MMFo).

2.7 Repeat steps 2.1 thru 2.6 until two consecutive master meter runs satisfy repeatability criteria. (Acceptable repeatability for orifice meter is two consecutive runs in which the MMFo does not exceed a tolerance of 0.10%.)

2.8 Average the results of the master meter runs. Enter the average MMFo into the orifice flow computer for use in continuing operations.

Figure 7 is a typical orifice proving report.

Analysis of Data

Analysis of the data is contained in this section along with the calculation methods for reduction of the data.

Proving calculations. The prover base volume, PBV, is stated at 101.325 kPa (14.696 psia) and 15.6°C (60.0°F). Therefore, correction for the effect of operating pressure and temperature shall be applied to determine the prover's volume at operating conditions. These corrections are as follows:

Correction for Effect of Pressure on Steel Prover (Cpsp)

Cpsp = 1.0000 (since the prover is equipped with a double wall design)

Correction for Effect of Temperature on Steel Prover (Ctsp)

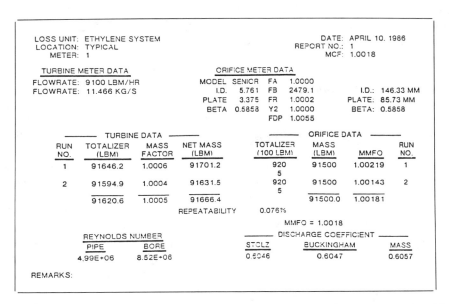

Figure 7. Shell Pipe Line Corporation orifice master meter proving report version 86.2A.

$$Ctsp = [1 + 0.000024(Tf - 15.6)] \quad \text{SI Units}$$
$$Ctsp = [1 + 0.0000134(Tf - 60)] \quad \text{IP Units}$$

The calibrated volume of the prover at operating conditions is:

$$PVtp = [PBV * Cpsp * Ctsp]$$

The mass of fluid in the prover's calibrated section at operating conditions is:

$$PMtp = [PVtp * DENp]$$

In the "master meter" proving method, the turbine meter is calibrated on a mass basis under actual operating conditions. The master meter's Mass Meter Factor, MMFt, can be calculated using the following equations:

$$TVtp = [(\text{Proving Run Pulses})/(\text{K Factor})]$$
$$TMtp = [TVtp * DENt]$$
$$VMFt = [PVtp/TVtp]$$
$$MMFt = [PMtp/TMtp]$$

The average mass meter factor, MMFt, for the turbine meter is calculated along with the repeatability for five consecutive runs. Acceptable repeatability is based on a maximum tolerance of 0.05% for the turbine's volumetric meter factor, VMFt. The master turbine meter is reproved at the start of each master meter proving run. This mass factor is applied forward.

In the "master meter" proving method, the orifice meter is proved on a mass basis under actual operating conditions. The average mass meter factor for the orifice meter, MMFo, is calculated along with the repeatability for two consecutive runs. Acceptable repeatability is based on a maximum deviation of 0.15%. An MMFo is calculated for each master meter proving run using the following equation:

$$MMFo = [(\text{Turbine Mass}) * (MMFt)]/[(\text{Orifice Mass})]$$

Discharge coefficients. When an orifice proving report was calculated, the empirically derived mass discharge coefficient was compared with two empirical predictions—the Stolz and Buckingham equations. The Mass Discharge Coefficient was calculated from the master meter proving results using the following formula:

$$\text{Mass } Cd = [(MMFo) * (\text{Buckingham's } Cd)]$$

Influence quantities. Orifice meter applications are based on steady flow of a homogeneous, single phase, Newtonian fluid with a velocity profile which follows the reference standard piping conditions. Departures from these reference piping conditions, the effect of the dynamic properties of the fluid, instrumentation effects, and operating conditions are termed influence quantities. Influence quantities are summarized in Figure 2.

Repeatability and reproducibility. Repeatability is the ability to obtain the same result when measuring the same quantity several times in succession. The turbine meter's VMFt has a required repeatability range of 0.05%. The orifice meter's MMFo has a required repeatability range of 0.10%. Repeatability has been expressed as a percentage on each orifice proving report.

Reproducibility is the ability to obtain the same result when measuring the same quantity under different conditions (i.e., time, observer, operating conditions, etc.). With respect to the turbine meter, reproducibility is not significant since the meter is proved immediately prior to proving an orifice meter. The orifice meter's MMFo reproducibility was demonstrated over a time period of several days at different flow rates and was generally within 0.10%.

Uncertainty of orifice calibrations. The generally accepted statistical method of the quadrature sum (square root of the sum of the squares) has been applied in deriving the overall measurement accuracy.

In calculating the uncertainties, the following assumptions were made:

- The universe is homogeneous (statistically controlled) with respect to MMFo
- The distribution is approximately Normal or Gaussian
- The sample is random
- A 95% confidence limit is required

The prover base volume, PBV, was established to within ±0.031% by water draw calibration using NBS certified standards.

When proved separately as a volume meter, the turbine meter's VMFt demonstrated a standard deviation of ±0.025% for five consecutive proving runs. An assigned uncertainty of ±0.05%, or two standard deviations, was used in calculating the overall uncertainty.

The flow computer has a stated uncertainty of 0.10%.

The density of ethylene was determined using API Chapter 11.3.2.1—Ethylene Density, using measured pressures and temperatures. The uncertainty of the equation of state is stated as ±0.20%.

Additional uncertainties when calculating density using an equation of state, no matter how accurate the equation of state might be, are errors resulting from the uncertainties of the measured temperature and pressure. Transmitter uncertainties of ±0.08°C (0.15°F) and ±6.89 kPa (1 psi) at metering conditions of 8 MPa (1,160 psia) and 20°C

(68°F) would result in a calculated density uncertainty of $\pm 0.15\%$.

Following is a summary of the master meter uncertainty equation:

	Tolerance (%)	Effect	Square of tol & eff
Prover calibration	0.03	1.0000	0.0009
Turbine repeatability	0.05	1.0000	0.0025
Turbine computer	0.10	1.0000	0.0100
Density computer	0.10	1.0000	0.0100
Equation of state	0.20	1.0000	0.0400
Density sensing	0.15	1.0000	0.0225
		Sum	0.0861

Square root of sum $\pm 0.2934\%$

When the orifice meters were proved on a mass basis, they demonstrated a standard deviation of $\pm 0.05\%$ for two consecutive proving runs. An assigned uncertainty of $\pm 0.10\%$, or two standard deviations, was used in calculating the overall uncertainty.

Following is a summary of the overall orifice meter uncertainty equation as a result of proving on a mass basis using the master meter method:

	Tolerance (%)	Effect	Square of tol & eff
Master turbine	0.2934	1.0000	0.0861
Master meter proving	0.10	1.0000	0.0100
		Sum	0.0961

Square root of sum $\pm 0.3100\%$

The uncertainties calculated are indicative at the time of proving and do not reflect additional uncertainties associated with measurements occurring beyond the proving period.

Many factors contribute an element of uncertainty to the flow rate indicated by the flow meter in relation to the actual or true flow. Orifice metering systems are made of an assembly of parts originating from different manufacturers. To determine the uncertainty with which the total assembly can measure, the tolerances on the different components and the way they affect the metering accuracy have to be considered. Since the orifice meter has been proved in situ, the practical working formula for the uncertainty can be expressed as follows:

$$\% = ([(\text{orifice calibration})^2] + 0.25[(\text{dp error})^2] + 0.25[(\text{density sensing error})^2])^{0.5}$$

Conclusions

The results of the orifice calibrations were rather interesting. Installations which were not in compliance with ANSI 2530 requirements were discovered, as well as human errors associated with flow computer programming.

The master meter method identified and measured errors associated with the following problems:

- Deformed orifice plates
- Eccentrically mounted orifice plates
- Orifice plates installed backwards
- Improperly programmed computers
- Effect of compressor pulsations

Additional information was no doubt masked by the influence quantities. From an application standpoint, identifying the source of error is not as important as measuring and correcting for the error.

Based upon the results, the majority of orifice meters have indicated a general understatement of measured quantities prior to proving.

In summary, our original goal was achieved—to develop an economically viable method for proving ethylene orifice meters under actual operating conditions. The master meter proving method has demonstrated to be a reliable measurement tool in identifying the quantifying measurement problems.

References

1. *Orifice Metering of Natural Gas and Other Related Hydrocarbon Fluids*, API MPMS Chapter 14.3, Second Edition, September 1985 (AGA Report No. 3), (GPA 8185-85), (ANSI/API 2530).
2. *Ethylene Density*, API MPMS Chapter 11.3.2.1, 1974.
3. *Shell Flow Meter Engineering Handbook*, Second Edition, McGraw-Hill Book Company (UK) Limited, 1985.
4. Miller, R. W., *Flow Measurement Engineering Handbook*, McGraw-Hill Book Company, 1983.
5. Stolz, J., *A Universal Equation for the Calculation of Discharge Coefficients of Orifice Plates*, Flow Measurements of Fluids, IMEKO Conference on Flow Measurement of Fluids, Groningen, The Netherlands, 1978.
6. Kolb, R. H., *Ethylene Mass Measurement System with Prover*, The Second Petroleum Measurement Seminar, 1980.
7. Pfrehm, R. H., "Improved Turbine Meter System Measures Ethylene Accurately," *Oil & Gas Journal*, April, 1981.

8. Bellinga, H., Kooi, A., Hoeks, C. P., van Laak, F. A. L., and Orbons, P. J., "Piston Prover Used to Calibrate Gas Meters," *Oil & Gas Journal*, August, 1985.

9. Boncher, F. S., Shipman, L. M., and Yen, L. C., *Hydrocarbon Processing*, 54 (2), 89 (1975).

10. *ASTM Manual on Presentation of Data and Control Chart Analysis*, ASTM Special Technical Publication 15D, 1976.

11. *Measurement Uncertainty for Fluid Flow in Closed Conduits*, ANSI/ASME MFC-2M-1983.

Notation

β	ratio of orifice plate bore to internal meter diameter
Cd	orifice discharge coefficient
Cpsp	correction for effect of pressure on prover
Ctsp	correction for effect of temperature on prover
d	orifice plate bore, in. (mm)
dp	differential pressure across orifice plate, in H_2O @ 60 (Pa)
D	meter internal diameter, in. (mm)
DENo	density of fluid at orifice, lbm/ft^3 (kg/m^3)
DENp	density of fluid at prover, lbm/ft^3 (kg/m^3)
DENt	density of fluid at turbine, lbm/ft^3 (kg/m^3)
Fa	orifice thermal expansion factor
Fb	basic orifice factor
Fdp	correction for static pressure on dp transmitter
Fr	Reynolds number factor
IP	Inch-Pound System of Units
MMFt	mass meter factor for turbine
MMFo	mass meter factor for orifice
PBV	prover base volume, ft^3 (m^3)
PMtp	prover mass, lbm (kg)
PVtp	prover volume, ft^3 (m^3)
qm	mass flow rate, lbm/s (kg/s)
Qm	mass flow rate, lbm/hr (kg/hr)
Rd	orifice bore Reynolds number
RD	pipe Reynolds number
SI	International System of Units
SVP	small volume prover
Tf	flowing temperature, °F (°C)
TMtp	turbine meter's mass, lbm (kg)
TVtp	turbine meter's volume, ft^3 (m^3)
VMFt	volumetric meter factor for turbine meter
Y2	expansion factor based on downstream static pressure

Source

Reprinted from *Oil & Gas Journal*, December 15, 1986 by permission and courtesy of PennWell Publishing Co. and James E. Gallagher, Shell Oil Company.

Gas or vapor flow measurement—orifice plate flange taps

HP 41CV Data Sheet

```
* * * * * * * * * *          ENTER DIFF RANGE INS H2O          * * * * * * * * * * *
                                  100.0000    ***
  ORIFICE    GAS            ENTER PIPE ID INS                              1.3160
  FLANGE    TAPS                  2.0670    ***
                             ENTER CP/CV
* * * * * * * * * *                1.3320    ***          ENTER REQD ORIFICE  INS
                             ENTER VISCOSITY cP                      1.3750  ***
 ENTER FLOW PPH                   0.0150    ***          CALC    BETA
     2,500.0000    ***                                   RATIO
 ENTER TEMP DEG F            CALC    BETA                         0.6652   ***
      366.0000    ***        RATIO                       CALC    FLOW
 ENTER PRSS PSIG                   0.6366   ***          PPH
      150.0000    ***        CALC ORIFICE                      2,731.2064    ***
 ENTER DENSITY PPCF           BORE    INS
        0.3640    ***              1.3160   ***          * * * * * * * * * * *
```

HP 41CV Register Loading

SIZE 080			
R00= 0.000000	R08= 0.000000	R17= 0.000000	
R01= 0.000000	R09= 0.000000	R18= 0.000000	
R02= 0.000000	R10= 0.000000	R19= 0.000000	
R03= 0.000000	R11= 0.000000	R20= 0.000000	
R04= 0.000000	R12= 0.000000	R21= 0.000000	
R05= 0.000000	R13= 0.000000	R22= 0.000000	
R06= 0.000000	R14= 0.000000	R23= 0.000000	
R07= 0.000000	R15= 0.000000	R24= 0.000000	
	R16= 0.000000	R25= "ORIFIC"	

R26= "E GAS"	R44= 0.000000	R62= "DIFF R"
R27= "FLANGE"	R45= 56.707100	R63= "ANGE I"
R28= " TAPS"	R46= 56.295000	R64= "NS H20"
R29= "FLOW "	R47= 0.000000	R65= "PIPE I"
R30= 0.000000	R48= 0.000000	R66= "D INS"
R31= 25.262728	R49= 0.000000	R67= "CP/CV"
R32= 57.295000	R50= "PPH "	R68= "VISCOS"
R33= 57.515200	R51= "TEMP D"	R69= "ITY cP"
R34= 57.535400	R52= "EG F "	R70= "BETA "
R35= 57.606100	R53= "PRSS P"	R71= "RATIO "
R36= 57.626364	R54= "SIG "	R72= " BORE "
R37= 57.656600	R55= "******"	R73= " INS "
R38= 57.670000	R56= "CALC "	R74= "REQD O"
R39= 57.686900	R57= "ENTER "	R75= "CALC O"
R40= 0.000000	R58= 0.000000	R76= "RIFICE"
R41= 56.707100	R59= 0.000000	R77= 0.000000
R42= 75.767273	R60= "DENSIT"	R78= 0.000000
R43= 57.747673	R61= "Y PPCF"	R79= 0.000000

HP 41CV Listing

SUBROUTINES

01♦LBL "FT GAS"
02 GTO A

PRINT LINE OF STARS

03♦LBL a
04 FC? 55
05 RTN
06 55.55
07 ADV
08 SF 12
09 XEQ b
10 ADV
11 RTN

PRINT LINES

12♦LBL d
13 SF 12
14 1
15 ST+ 58
16 RCL IND 58
17♦LBL b
18 STO 59
19 CLA
20♦LBL c
21 ARCL IND 59
22 RCL 59
23 FRC
24 100
25 *
26 STO 59
27 X≠0?
28 GTO c
29 FC? 55
30 GTO 01
31 PRA
32 CF 12
33 RTN
34♦LBL 01
35 CF 12
36 AVIEW
37 RTN

PRINT & STORE VARIABLES

38♦LBL e
39 1
40 ST+ 58
41 RCL IND 58
42 X=0?
43 GTO d
44 XEQ b
45 1
46 ST+ 01
47 RCL IND 01
48 TONE 9
49 FS? 00
50 GTO 01
51 STOP
52 STO IND 01
53 .7
54 STO 14
55 RCL IND 01
56♦LBL 01
57 XEQ J
58 GTO e

RE-ENTER VARIABLES

59♦LBL "REENTER"
60 ADV
61 1
62 +
63 STO 01
64 RCL IND 01
65 XEQ J
66 STOP
67 XEQ J
68 STO IND 01
69 ADV
70 SF 00
71 SF 02
72 STOP
73 GTO 02

BYPASS PRINTER

74♦LBL J
75 FC? 55
76 GTO 01
77 PRX
78 RTN
79♦LBL 01
80 VIEW X
81 RTN

ITERATE FOR β

82♦LBL B
83 .076

84 RCL 07
85 SQRT
86 /
87 .364
88 +
89 STO 13
90♦LBL F
91 .007
92 RCL 07
93 /
94 .5993
95 +
96 STO 12
97 .5
98 RCL 07
99 /
100 .07
101 +
102 RCL 14
103 PSE
104 -
105 0
106 X<>Y
107 X<=Y?
108 GTO 01
109 2.5
110 Y↑X
111 1.6
112 RCL 07
113 1/X
114 -
115 5
116 Y↑X
117 .4
118 *
119 *
120 ST+ 12
121♦LBL 01
122 .5
123 RCL 14
124 -
125 0
126 X<>Y
127 X<=Y?
128 GTO 01
129 1.5
130 Y↑X
131 .09034
132 RCL 07
133 /

```
134 +
135 *
136 ST- 12
137◆LBL 01
138 RCL 14
139 .7
140 -
141 0
142 X<>Y
143 X<=Y?
144 GTO 01
145 2.5
146 Y↑X
147 65
148 RCL 07
149 X↑2
150 /
151 3
152 +
153 *
154 ST+ 12
155◆LBL 01
156 RCL 12
157 2
158 *
159 RCL 14
160 *
161 RCL 14
162 5
163 Y↑X
164 6
165 *
166 RCL 13
167 *
168 +
169 STO 23
170 RCL 12
171 RCL 14
172 X↑2
173 *
174 RCL 13
175 RCL 14
176 6
177 Y↑X
178 *
179 +
180 830
181 RCL 14
182 5000
183 *
184 -
185 9000
186 RCL 14
187 X↑2
188 *
189 +
190 4200
191 RCL 14
192 X↑2
193 *
194 RCL 14
195 *
196 -
197 530
198 RCL 07
199 SQRT
200 /
201 +
202 STO 24
203 .000015
204 *
205 1
206 +
207 /
```

```
208 STO 11
209 FS? 01
210 GTO G
211 RCL 16
212 -
213 RCL 23
214 /
215 ST- 14
216 ABS
217 .000015
218 X<>Y
219 X>Y?
220 GTO F
221 RCL 14
222 RTN
```

CALC F_R

```
223◆LBL D
224 RCL 14
225 4
226 Y↑X
227 .35
228 *
229 .41
230 +
231 RCL 06
232 *
233 39.6
234 /
235 RCL 08
236 /
237 RCL 04
238 15
239 +
240 /
241 CHS
242 1
243 +
244 ST/ 16
245 RCL 14
246 RCL 07
247 *
248 X↑2
249 RCL 24
250 *
251 RCL 09
252 *
253 4.42
254 /
255 RCL 02
256 /
257 1
258 +
259 ST/ 16
260 RTN
```

MAIN PROGRAM ENTER DATA

```
261◆LBL A
262 1
263 STO 01
264 30
265 STO 58
266 XEQ a
267 XEQ d
268 XEQ a
269 XEQ e
270 CF 00
271◆LBL 02
272 41
273 STO 58
```

CALC $K_o\beta^2$

```
274 RCL 02
```

```
275 RCL 07
276 X↑2
277 /
278 358.5
279 RCL 03
280 .007
281 *
282 +
283 /
284 RCL 05
285 RCL 06
286 *
287 SQRT
288 /
289 STO 16
290 STO 10
```

CALC β

```
291 XEQ B
292 XEQ D
293 XEQ B
294 RCL 10
295 STO 16
296 XEQ D
297 XEQ B
298 XEQ J
```

CALC d

```
299 XEQ d
300 RCL 14
301 RCL 07
302 *
303 STO 19
304 GTO 01
```

ENTER REQD BORE

```
305◆LBL "EVEN SZ"
306 SF 01
307 14
308 STO 01
309 42
310 STO 58
311 XEQ e
```

CALC β

```
312 RCL 15
313 RCL 07
314 /
315 STO 14
316 XEQ J
```

CALC FLOW

```
317 GTO B
318◆LBL G
319 CF 01
320 XEQ d
321 RCL 11
322 RCL 16
323 /
324 RCL 02
325 *
326 STO 19
327◆LBL 01
328 XEQ J
329 FS? 00
330 STOP
331 XEQ a
332 CF 02
333 RCL 19
334 VIEW X
335 BEEP
336 ADV
337 STOP
338 .END.
```

Sequence of Calculations

1. Calculate $K_o\beta^2$ value.

$$K_o\beta^2 = \frac{W_m}{D^2(358.5 + .007T_f)\sqrt{\gamma_f h_m}}$$

2. Iterate to find uncorrected beta ratio.

$$K_o\beta^2 = \beta^2 \left[\left(\left(0.5993 + \frac{0.007}{D}\right) + \left(0.364 + \frac{0.076}{\sqrt{D}}\right)\beta^4\right.\right.$$

$$+ 0.4\left(1.6 - \frac{1}{D}\right)^5$$

$$\left(0.07 + \frac{0.5}{D} - \beta\right)^{5/2}\left(0.009 + \frac{0.034}{D}\right)$$

$$(0.5 - \beta)^{3/2} + \left(\frac{65}{D^2} + 3\right)$$

$$\left.(\beta - 0.7)^{5/2}\right] \div \left[1 + 0.000015\,(830\right.$$

$$\left.\left. - 5{,}000\beta + 9{,}000\beta^2 - 4{,}200\beta^3 + \frac{530}{\sqrt{D}}\right)\right]$$

Any negative factor, such as $(\beta - 0.7)^{5/2}$ when β is less than 0.7, shall be taken as equal to zero, and thus the whole term $(65/D^2 + 3)(\beta - 0.7)^{5/2}$ will drop out of the equation.

Let fx equal the equation in step 2—$K_o\beta^2$, and let:

$$f^1x = 2\left(\left(.5993 + \frac{.007}{D}\right) + 4\left(1.6 - \frac{1}{D}\right)^5\right.$$

$$\left(.07 + \frac{5}{D} - \beta\right)^{5/2} - \left(.009 + \frac{.034}{D}\right)$$

$$(0.5 - \beta)^{3/2} + \left(\frac{65}{D^2} + 3\right)(\beta - .7)^{5/2}\beta$$

$$\left. + 6\left(.364 + \frac{.076}{\sqrt{D}}\right)\beta^6\right.$$

Start with $\beta = 0.7$ and calculate fx/f^1x. If greater than 0.000015, subtract fx/f^1x from current value of β and repeat until the error is less than 0.000015.

3. Calculate the expansion factor.

$$Y = 1 - \frac{(.41 + .35\beta^4)h_m}{39.6\,kP_f}$$

4. Divide $K_o\beta^2$ by Y.
5. Calculate the Reynolds number correction factor.

$$F_r = 1 + \left(830 - 5{,}000\beta + 9{,}000\beta^3 - 4{,}200\beta^3 + \frac{530}{\sqrt{D}}\right)$$

$$\times \frac{(D\beta)^2\mu_{cp}}{4.42\,W_m}$$

6. Divide $K_o\beta^2$ by F_r.
7. Repeat step 2 to find corrected beta ratio.
8. Repeat steps 3, 4, 5, and 6 using corrected beta ratio.
9. Repeat step 2 for final beta ratio.
10. Calculate orifice bore.

   $d = BD$ in.

11. Calculate flow rate for an even-sized orifice bore.

$$\beta = \frac{d}{D}$$

Calculate $K_o\beta^2$ using the equation in step 2.

$$\text{new flow} = \frac{\text{new } K_o\beta^2}{\text{original } K_o\beta^2} \times \text{original flow, lb/hr}$$

Notation

D	inside diameter of pipe, in.
d	bore diameter of primary device, in.
F_r	Reynolds number correction factor for steam, vapor, or gas, dimensionless
h_m	maximum differential or range of differential gauge, in. of water, dry calibrated
K_o	coefficient of discharge, including velocity of approach at a hypothetical Reynolds number of infinity, dimensionless
k	ratio of specific heats C_p/C_r, dimensionless
P_f	flowing pressure, psia
T_f	flowing temperature, °F
W_m	maximum rate of flow, lb/hr
Y	expansion factor, dimensionless
β	diameter ratio d/D, dimensionless
γ_f	specific weight of flowing conditions, lb/ft³
μ_{cp}	absolute viscosity, centipoises

Source

Instrument Engineering Programs, Gulf Publishing Company, Houston, Texas.

Reference

1. Spink, L. K., *The Principles and Practice of Flowmeter Engineering*, The Foxboro Company.

Properties of gas and vapors

This program, designed for the HP-41CV calculator, will provide:

1. Compressibility factor Z, dimensionless.
2. Specific weight (or density), lb/ft³.
3. Mixture molecular weight M, dimensionless.
4. Mixture critical temperature T_c, °R.
5. Mixture critical pressure P_c, psia.
6. Mixture ratio of specific heats C_p/C_v, dimensionless.
7. Mixture normal boiling point latent heat of vaporization HV, Btu/lb.

Caution must be exercised when using the information in number 7. For mixtures having components with widely different values, the answer may be of limited value.

The program defaults to a compressibility factor and specific weight calculation if the first component volume % entry is 100. Any number of components can be handled by the program. It loops until 100% value is reached.

User instructions. Two programs are available. Program A calculates items 1–5 and program B calculates items 6–7.

Initiate desired program by pressing **A** or **B**.

Enter:

Flowing pressure, psig	Press **R/S**
Flowing temperature, °F	Press **R/S**
Component-number volume % (Note: When the volume % is 100, the program goes immediately to the calculation. When the volume % is not 100, it continues.)	Press **R/S**

Critical pressure, psia	Press **R/S**
Critical temperature, °R	Press **R/S**
Molecular weight, dimensionless	Press **R/S**

And, if program B is selected:

Ratio of specific heats C_p/C_v, dimensionless	Press **R/S**
Normal boiling point latent heat of vaporization, Btu/lb	Press **R/S**

Program calculates remaining volume %. If less than 100, it loops back to 3. If equal to 100, it continues and calculates.

- Mixture critical pressure, psia
- Mixture critical temperature, °R
- Mixture molecular weight, dimensionless

And, if program B was selected, it calculates:

- Mixture ratio of specific heats C_p/C_v, dimensionless
- Mixture normal boiling point latent heat of vaporization, Btu/lb

Both programs calculate:

- Compressibility factor Z, dimensionless
- Specific weight (or density), lb/ft³

See Table 1 for table of physical properties.

Table 1
Physical Properties

No.	Compound	Formula	Critical Constants		Molecular Weight (M)	Ratio of Specific Heats C_p/C_v at 14.7 PSIA and 60°F (k)	Heat of Vaporization, 14.696 PSIA at Boiling Point, Btu/lb (NBP ΔHV)
			Pressure, PSIA (P_c)	Temperature, °R (T_c)			
1	Methane	CH_4	667.8	343.4	16.043	1.311	219.22
2	Ethane	C_2H_6	707.8	550.0	30.070	1.193	210.41
3	Propane	C_3H_8	616.3	666.0	44.097	1.136	183.05
4	n-Butane	C_4H_{10}	550.7	765.0	58.124	1.094	165.65
5	Isobutane	C_4H_{10}	529.1	735.0	58.124	1.097	157.53
6	n-Pentane	C_5H_{12}	488.6	845.7	72.151	1.074	153.59
7	Isopentane	C_5H_{12}	490.4	829.1	72.151	1.075	147.13
8	Neopentane	C_5H_{12}	464.0	781.0	72.151	1.074	135.58
9	n-Hexane	C_6H_{14}	436.9	913.7	86.178	1.062	143.95
10	2-Methylpentane	C_6H_{14}	436.6	895.8	86.178	1.063	138.67
11	3-Methylpentane	C_6H_{14}	453.1	908.3	86.178	1.064	140.09
12	Neohexane	C_6H_{14}	446.8	860.1	86.178	1.064	131.24
13	2,3-Dimethylbutane	C_6H_{14}	453.5	900.3	86.178	1.065	136.08

Table 1
Physical Properties (Continued)

No.	Compound	Formula	Critical Constants Pressure, PSIA (P_c)	Critical Constants Temperature, °R (T_c)	Molecular Weight (M)	Ratio of Specific Heats C_p/C_v at 14.7 PSIA and 60°F (k)	Heat of Vaporization, 14.696 PSIA at Boiling Point, Btu/lb ($NBP\ \Delta HV$)
14	n-Heptane	C_7H_{16}	396.8	972.8	100.205	1.054	136.01
15	2-Methylhexane	C_7H_{16}	396.5	955.0	100.205	1.054	131.59
16	3-Methylhexane	C_7H_{16}	408.1	963.8	100.205	1.054	132.11
17	3-Ethylpentane	C_7H_{16}	419.3	973.5	100.205	1.054	132.83
18	2,2-Dimethylpentane	C_7H_{16}	402.2	937.2	100.205	1.053	125.13
19	2,4-Dimethylpentane	C_7H_{16}	396.9	936.0	100.205	1.053	126.58
20	3,3-Dimethylpentane	C_7H_{16}	427.2	965.9	100.205	1.053	127.21
21	Triptane	C_7H_{16}	428.4	956.4	100.205	1.055	124.21
22	n-Octane	C_8H_{18}	360.6	1024.2	114.232	1.046	129.53
23	Diisobutyl	C_8H_{18}	360.6	990.4	114.232	1.049	122.80
24	Isooctane	C_8H_{18}	372.4	979.5	114.232	1.049	116.71
25	n-Nonane	C_9H_{20}	332.0	1070.7	128.259	1.042	123.76
26	n-Decane	$C_{10}H_{22}$	304.0	1112.1	142.286	1.038	118.68
27	Cyclopentane	C_5H_{10}	653.8	921.5	70.135	1.117	167.34
28	Methylcyclopentane	C_6H_{12}	548.9	959.4	84.162	1.085	147.83
29	Cyclohexane	C_6H_{12}	591.0	996.7	84.162	1.089	153.00
30	Methylcyclohexane	C_7H_{14}	503.5	1030.3	98.189	1.068	136.30
31	Ethylene	C_2H_4	729.8	506.6	28.054	1.243	207.57
32	Propene	C_3H_6	669.0	656.9	42.081	1.154	188.18
33	1-Butene	C_4H_8	583.0	755.6	56.108	1.111	167.94
34	cis-2-Butene	C_4H_8	610.0	784.4	56.108	1.121	178.91
35	trans-2-Butene	C_4H_8	595.0	771.9	56.108	1.107	174.39
36	Isobutene	C_4H_8	580.0	752.6	56.108	1.106	169.48
37	1-Pentene	C_5H_{10}	590.0	836.9	70.135	1.084	154.46
38	1,2-Butadiene	C_4H_6	653.0	799.0	54.092	1.118	181.00
39	1,3-Butadiene	C_4H_6	628.0	766.0	54.092	1.120	174.00
40	Isoprene	C_5H_8	558.4	872.0	68.119	1.089	153.00
41	Acetylene	C_2H_2	890.4	555.3	26.038	1.238	—
42	Benzene	C_6H_6	710.4	1012.2	78.114	1.117	169.31
43	Toluene	C_7H_8	595.9	1065.6	92.141	1.091	154.84
44	Ethylbenzene	C_8H_{10}	523.5	1111.2	106.168	1.072	144.00
45	o-Xylene	C_8H_{10}	541.4	1135.0	106.168	1.069	149.10
46	m-Xylene	C_8H_{10}	513.6	1111.0	106.168	1.072	147.20
47	p-Xylene	C_8H_{10}	509.2	1109.6	106.168	1.072	144.52
48	Styrene	C_8H_8	580.0	1166.0	104.152	1.076	151.00
49	Isopropylbenzene	C_9H_{12}	465.4	1136.4	120.195	1.060	134.30
50	Methyl Alcohol	CH_4O	1174.2	923.0	32.042	1.237	473.0
51	Ethyl Alcohol	C_2H_6O	925.3	929.6	46.069	1.149	367.0
52	Carbon Monoxide	CO	507.0	240.0	28.010	1.403	92.7
53	Carbon Dioxide	CO_2	1071.0	547.9	44.010	1.293	238.2
54	Hydrogen Sulfide	H_2S	1306.0	672.7	34.076	1.323	235.6
55	Sulfur Dioxide	SO_2	1145.1	775.5	64.059	1.272	166.7
56	Ammonia	NH_3	1636.1	730.3	17.031	1.304	587.2
57	Air	N_2O_2	547.0	238.7	28.964	1.395	92.0
58	Hydrogen	H_2	188.1	60.2	2.016	1.408	193.9
59	Oxygen	O_2	736.9	278.9	31.999	1.401	91.6
60	Nitrogen	N_2	493.0	227.4	28.013	1.402	87.8
61	Chloride	Cl_2	1118.4	751.0	70.906	1.330	123.8
62	Water	H_2O	3208.0	1165.6	18.015	1.335	970.3
63	Helium	He	—	—	4.003	1.660	—
64	Hydrogen Chloride	HCl	1198.0	584.5	36.461	1.410	185.5

Note: C_p/C_v calculated by the formula $k = C_p/(C_p - 1.1987)/M$.

HP 41CV Data Sheet

```
*************          *************           ENTER CRIT  TEMP DEG R
                                                     500.0000000    ***
    *GAS*                *GAS*                 ENTER MW
 *PROPERTIES*         *PROPERTIES*                   20.00000000    ***
                                               ENTER CP/CV
*************          *************                  1.400000000    ***
                                               ENTER NBP dHV BTU/P
                      ENTER PRSS PSIG                 200.0000000    ***
                            100.0000000   ***  CALC MIXTURE
ENTER PRSS PSIG       ENTER TEMP DEG F         CRIT   PRSS
      100.0000000  ***       60.00000000   ***       200.0000000    ***
ENTER TEMP DEG F      ENTER % COMPNT NO 1.     CALC MIXTURE
      60.00000000 ***        25.00000000   ***  CRIT   TEMP
ENTER % COMPNT NO 1.  ENTER CRIT  PRSS PSIA           500.0000000    ***
      100.0000000  ***       200.0000000   ***  CALC MIXTURE
ENTER CRIT  PRSS PSIA ENTER CRIT  TEMP DEG R   MW
      200.0000000  ***       500.0000000   ***       20.00000000    ***
ENTER CRIT  TEMP DEG R ENTER MW                CALC MIXTURE
      500.0000000  ***       20.00000000   ***  CP/CV
ENTER MW              ENTER CP/CV                     1.400000000    ***
      20.00000000 ***         1.400000000  ***  CALC MIXTURE
CALC    Z            ENTER NBP dHV BTU/P        NBP dHV
      0.801732805  ***        200.0000000   ***       200.0000000    ***
CALC DENSITY          CALC % TO GO             CALC    Z
 PPCF                        75.00000000   ***        0.801732805    ***
      0.512814969  ***                         CALC DENSITY
                      ENTER % COMPNT NO 2.       PPCF
                             75.00000000   ***        0.512814969    ***
                      ENTER CRIT  PRSS PSIA
*************                200.0000000   ***  *************
```

HP 41CV Register Loading

SIZE 080			
	R26= "*PROPE"	R54= "% COMP"
R00= 0.000000	R27= "RTIES*"	R55= "******"
R01= 0.000000	R28= "CALC M"	R56= "CALC "
R02= 0.000000	R29= 24.252627	R57= "ENTER "
R03= 0.000000	R30= 57.505100	R58= 49.000000
R04= 0.000000	R31= 57.525300	R59= 0.000000
R05= 0.000000	R32= 0.000000	R60= 1.000000
R06= 0.000000	R33= 57.546260	R61= "V "
R07= 0.000000	R34= 0.000000	R62= "NT NO "
R08= 0.000000	R35= 57.635064	R63= "CRIT "
R09= 0.000000	R36= 57.635265	R64= "SIA "
R10= 0.000000	R37= 57.660000	R65= "EG R "
R11= 0.000000	R38= 0.000000	R66= "MW "
R12= 0.000000	R39= 57.670000	R67= "CP/CV "
R13= 0.000000	R40= 57.686970	R68= "NBP dH"
R14= 0.000000	R41= 0.000000	R69= "V BTU/"
R15= 0.000000	R42= 79.710000	R70= "P "
R16= 0.000000	R43= 28.726373	R71= " TO GO"
R17= 0.000000	R44= 28.726374	R72= "IXTURE"
R18= 0.000000	R45= 28.726600	R73= "PRSS "
R19= 0.000000	R46= 28.726700	R74= "TEMP "
R20= 0.000000	R47= 28.726861	R75= " Z "
R21= 0.000000	R48= 56.750000	R76= "CALC D"
R22= 0.000000	R49= 76.777800	R77= "ENSITY"
R23= 0.000000	R50= "PRSS P"	R78= " PPCF "
R24= " *GA"	R51= "SIG "	R79= "CALC %"
R25= "S* "	R52= "TEMP D"	
	R53= "EG F "	

HP 41CV Listing

SUBROUTINES

```
01◆LBL "GAS PRP"
02 GTO A
```

PRINT LINE OF STARS

```
03◆LBL a
04 FC? 55
05 RTN
06 55.55
07 ADV
08 SF 12
09 XEQ b
10 ADV
11 RTN
```

PRINT LINES

```
12◆LBL d
13 SF 12
14 1
15 ST+ 58
16 RCL IND 58
17◆LBL b
18 STO 59
19 CLA
20◆LBL c
21 FIX 0
22 ARCL IND 59
23 FIX 9
24 RCL 59
25 FRC
26 100
27 *
28 STO 59
29 X≠0?
30 GTO c
31 FC? 55
32 GTO 01
33 PRA
34 CF 12
35 RTN
36◆LBL 01
37 CF 12
38 AVIEW
39 RTN
```

PRINT & STORE VARIABLES

```
40◆LBL e
41 1
42 ST+ 58
43 RCL IND 58
44 X=0?
45 RTN
46 XEQ b
47 1
48 ST+ 01
49 RCL IND 01
50 TONE 9
51 FS? 00
52 GTO 01
53 STOP
54 FC? 01
55 GTO 02
56 XEQ J
57 RCL 04
58 *
59 100
60 /
61 ST+ IND 01
```

```
62 GTO e
63◆LBL 02
64 STO IND 01
65◆LBL 01
66 XEQ J
67 GTO e
```

RE-ENTER VARIABLES

```
68◆LBL "REENTER"
69 ADV
70 1
71 +
72 STO 01
73 RCL IND 01
74 XEQ J
75 STOP
76 XEQ J
77 STO IND 01
78 ADV
79 SF 00
80 SF 02
81 100
82 STO 04
83 STOP
84 GTO C
```

BYPASS PRINTER

```
85◆LBL J
86 FC? 55
87 GTO 01
88 PRX
89 RTN
90◆LBL 01
91 VIEW X
92 RTN
```

MAIN PROGRAM INCLUDES CP/CV & dHV

```
93◆LBL "CP/CV"
94 SF 03
```

ENTER DATA

```
95◆LBL A
96 1
97 STO 01
98 STO 21
99 STO 60
100 28
101 STO 58
102 FS? 00
103 GTO 01
104 0
105 STO 05
106 STO 06
107 STO 07
108 STO 08
109 STO 09
110 100
111 STO 20
112 STO 04
113◆LBL 01
114 XEQ a
115 XEQ d
116 XEQ a
117 XEQ e
118 CF 02
119◆LBL B
120 3
```

```
121 STO 01
122 32
123 STO 58
124 FS? 00
125 GTO G
126 XEQ e
127 RCL 04
128 ST- 20
129 100
130 -
131 X=0?
132 GTO G
133 SF 01
134 XEQ e
135 FS? 03
136 XEQ e
137 RCL 20
138 0
139 X≠Y?
140 GTO 01
141 CF 00
142 CF 01
```

PRINT MIXTURE VALUES

```
143 42
144 STO 58
145 XEQ d
146 RCL 05
147 XEQ J
148 XEQ d
149 RCL 06
150 XEQ J
151 XEQ d
152 RCL 07
153 XEQ J
154 FC? 03
155 GTO C
156 XEQ d
157 RCL 08
158 XEQ J
159 XEQ d
160 RCL 09
161 XEQ J
162 GTO C
163◆LBL 01
164 RCL 20
165 STO 04
166 41
167 STO 58
168 XEQ d
169 CF 01
170 RCL 20
171 XEQ J
172 1
173 ST+ 60
174 ADV
175 GTO B
176◆LBL G
177 34
178 STO 58
179 4
180 STO 01
181 XEQ e
```

CALC BP_f

```
182◆LBL C
183 47
184 STO 58
185 XEQ d
```

```
186 .0866403          217 *                  252 GTO F
187 RCL 06            218 -                  253 RCL 21
188 *                 219 RCL 12             254 XEQ J
189 RCL 05            220 RCL 11
190 /                 221 -
191 RCL 02            222 1                  CALC DENSITY
192 14.7              223 -
193 +                 224 RCL 11             255 XEQ d
194 *                 225 *                  256 RCL 02
195 RCL 03            226 STO 22             257 14.7
196 460               227 +                  258 +
197 +                 228 STO 23             259 RCL 03
198 /                 229 RCL 21             260 460
199 STO 11            230 3                  261 +
                      231 Y↑X                262 /
                      232 RCL 21             263 RCL 07
   CALC A²/B          233 X↑2                264 *
                      234 -                  265 RCL 21
200 RCL 06            235 RCL 22             266 /
201 RCL 03            236 RCL 21             267 10.73
202 460               237 *                  268 /
203 +                 238 +                  269 STO 13
204 /                 239 RCL 12             270 XEQ J
205 1.5               240 RCL 11             271 FS? 02
206 Y↑X               241 X↑2                272 STOP
207 4.93396           242 *                  273 XEQ a
208 *                 243 -                  274 CF 00
209 STO 12            244 RCL 23             275 CF 03
                      245 /                  276 CF 04
  ITERATE FOR Z       246 ST- 01             277 RCL 21
                      247 ABS                278 RCL 13
210♦LBL F             248 PSE                279 VIEW X
211 RCL 21            249 .00001             280 BEEP
212 X↑2               250 X<>Y               281 ADV
213 3                 251 X>Y?               282 STOP
214 *                                        283 END
215 RCL 21
216 2
```

Sequence of Calculations

1. Mixture values are calculated by the mixture law, which assumes that the property of the mixture is the sum of the partial fractions.

2. The compressibility factor is iterated from the Redlich-Kwong equation:

$$fx = Z^3 - Z^2 + BP_f \left[\left(\left(\frac{A^2}{B} \right) - BP_f - 1 \right) Z - \left(\frac{A^2}{B} \right) BP_f \right]^2$$

$$f^1x = 3Z^2 - 2Z + BP_f \left(\left(\frac{A^2}{B} \right) - BP_f - 1 \right)$$

where $A = \dfrac{.42748 T_c^{2.5}}{P_c T_f^{2.5}}$

$BP_f = .0866403 \dfrac{T_c P_f}{P_c T_f}$

$\dfrac{A^2}{B} = 4.93396 \left(\dfrac{T_c}{T_f} \right)^{1.5}$

Start with $Z = 1$ and calculate fx/f^1x. If greater than 0.00001, subtract fx/f^1x from the current value of Z and repeat until the error is less than 0.00001.

3. Calculate the specific weight:

$$\gamma_f = \frac{P_f}{T_f} \frac{M}{1.314Z} \text{ lb/ft}^3$$

Notation

A, B	partial calculations for Z
M	molecular weight, dimensionless
P_f	flowing pressure, atmospheres
P_c	critical pressure, atmospheres
T_f	flowing temperature, °K
T_c	critical temperature, °K
Z	compressibility, dimensionless
γ_f	specific weight, lb/ft³

Note: These units do not correspond with the calculator input units. The program converts as necessary.

Source

Instrument Engineering Programs, Gulf Publishing Company, Houston, Texas.

Determine required orifice diameter for any required differential when the present orifice and differential are known in gas measurement

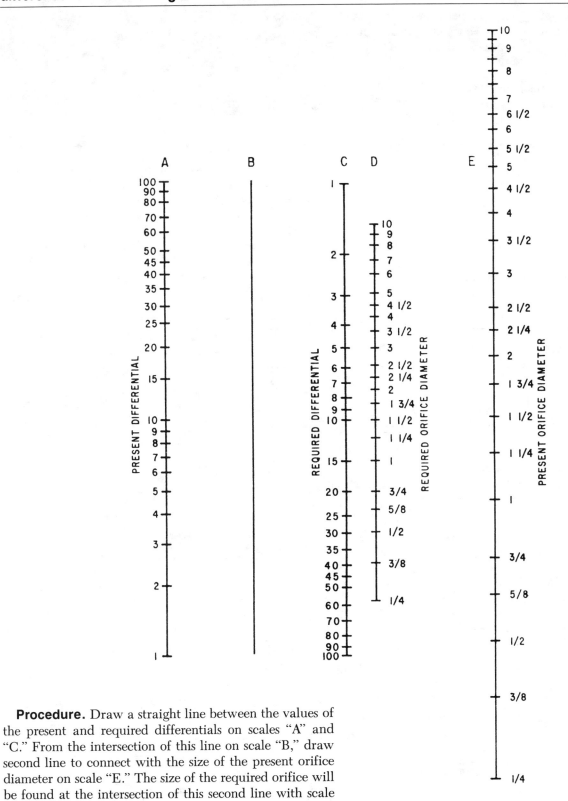

Procedure. Draw a straight line between the values of the present and required differentials on scales "A" and "C." From the intersection of this line on scale "B," draw second line to connect with the size of the present orifice diameter on scale "E." The size of the required orifice will be found at the intersection of this second line with scale "D."

Estimate the temperature drop across a regulator

The temperature drop that will be experienced when gas is passed through a regulator may be estimated as 1° drop for each atmosphere (15 psi) of pressure drop.

Example. Determine the temperature drop that will be experienced when reducing a gas from 750 psig to 600 psig.

Solution.

750 − 600 psig = 150 psig

$$\frac{150}{15} = 10 \text{ atmospheres}$$

$10 \times 1°/\text{atm} = 10°$ of drop

Estimate natural gas flow rates

This chart provides a good estimate of natural gas flow rates through 4-in. meter runs equipped with flange taps. No calculations are required to obtain a flow rate. Orifice plate size, and differential and static pressures from the orifice meter chart are used to make the estimate.

To use the Gas Flow Rate Estimator, enter the chart at the proper differential value at the indicated lower horizontal scale. Project this value vertically upward to the appropriate static pressure line, and then horizontally to the proper plate size line. Drop vertically from the plate size line to the lowermost horizontal rate scale at the base of the chart to obtain the estimated flow rate.

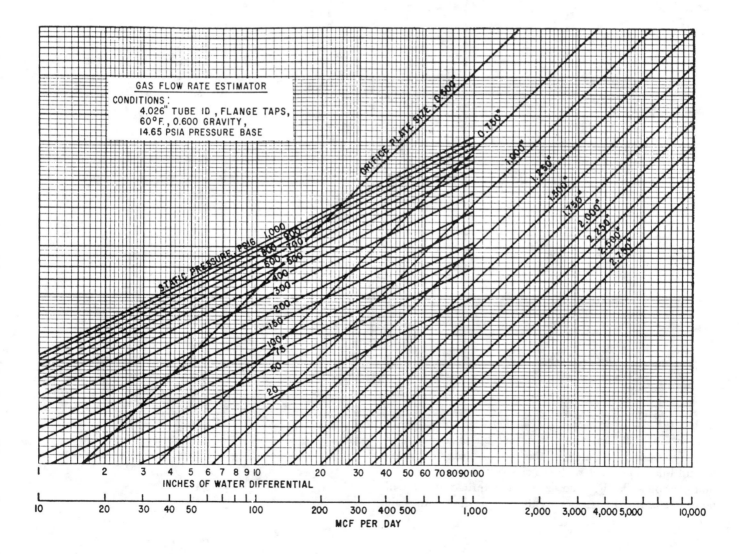

How to estimate the average pressure differential on the remaining meter runs of a parallel system when one or more runs are shut off

On gas systems where inferential meters are installed in parallel, it may at times be necessary to shut down one or more of the runs temporarily for repair or maintenance work. When this occurs, it is important to know beforehand what the pressure differential on the remaining runs will be so that the meters will not be over-ranged. The equation is:

Average differential for runs remaining in service (A) =

$$\left[\frac{(\text{No. of runs in service})\sqrt{(\text{Average differential})}}{(\text{No. of runs remaining in service})} \right]^2$$

Example. A measurement engineer wishes to shut off one meter run on a four-meter parallel system to inspect the orifice plate. The average differential on the four runs is 49 in. and his meters have a 100-in. range. He wishes to determine whether shutting down one meter will over-range the three remaining meters.

$$(\text{A}) = \left(\frac{4 \times \sqrt{49}}{3} \right)^2 = (9.3)^2 = 86.5 \text{ in.}$$

Since the average differential for the remaining runs is 86.5 in., the operator may safely shut the meter down as planned to inspect its orifice plate.

Sizing a gas metering run

Multiply the absolute pressure by the desired differential pressure (the differential in in. of water should not be greater than the absolute pressure in psia). Take the square root of the product. Divide this figure into the maximum flow rate in cubic ft/hr to obtain the approximate coefficient. Then divide the maximum coefficient by 250 and take the square root to determine the approximate inside diameter of the orifice in in. The metering run diameter should not be less than one-third larger than the orifice diameter.

Example. Find the orifice diameter and meter run for a line designed to carry a maximum flow of 2 MMcf/hr at a pressure of 600 psia. Use a differential pressure of 50 in. of water.

$$\sqrt{600 \times 50} = \sqrt{30,000} = 173.2$$

$$\frac{2,000,000}{173.2} = 11,540$$

$$\frac{11,540}{250} = 46.2$$

$\sqrt{46.2} = 6.8$ in., approximate orifice diameter

Meter run should not be less than 10 in. in diameter.

List of typical specifications for domestic and commercial natural gas

1. **Heating value:** Not less than 950 Btu's per std cubic ft.
2. **Purity:** The gas to be delivered shall be free from dust, gums, crude oils, hydrocarbons liquefiable at temperatures in excess of 15°F at 800 psig.
3. **Sulfur content:** The gas shall contain not more than one grain of hydrogen sulfide per 100 cubic ft of gas, and not more than 20 grains of total sulfur per 100 cubic ft of gas.
4. **CO$_2$ content:** The gas shall contain not more than 2% carbon dioxide.
5. **Moisture content:** The gas shall contain not more than 4 lb of water vapor per MMcf of gas measured at a pressure of 14.4 psi and a temperature of 60°F.
6. **Temperature:** Maximum temperature at point of delivery shall be 120°F.

Determine the number of purges for sample cylinders

To determine the proper number of times a sample cylinder should be pressured to ensure removal of the original gas, multiply the reciprocal of each filling pressure (in atmospheres) by each succeeding reciprocal. The final filling will determine what fraction of the original gas is still in the cylinder.

Example. The problem is to fill a gas sample for analysis to obtain an accuracy of 0.05%. Source pressure is 450 psia.

Solution. First filling to line pressure and emptying:

$$\frac{1}{450/15} = \frac{1}{30} \times 100 = 3.33\%$$

Second filling to line pressure and emptying:

$$\frac{1}{30} \times \frac{1}{30} \times 100 = 0.11\%$$

Third filling to line pressure:

$$\frac{1}{30} \times \frac{1}{30} \times \frac{1}{30} \times 100 = 0.0037\%$$

Find the British thermal units (Btu) when the specific gravity of a pipe line gas is known

Use the graph below prepared from the Headlee formula:

Btu @ 30 in. Hg and 60° saturated = 1,545 (specific gravity) + 140 − 16.33 (% N_2)

Example. Assume a gas having a gravity of 0.670 which contains 1.0% nitrogen. From the chart, read up from 0.670 gravity on the base line to the intercept with the curve marked 1% nitrogen. Reading horizontally across, the total Btu of the gas is 1,160 Btu at 30 in., 60° saturated.

Estimate for variations in measurement factors

Orifice measurement.
1. Each 10° change in flowing temperature will cause a 1% change in flow calculated.

2. Each 0.01-change in specific gravity of the gas measured will result in a 1% change in passing flow calculated.

3. Each 10-psi error in static pressure will make the following error in flow calculated:

at 100 psia—5%
at 500 psia—1%
at 1,000 psia—0.5%

4. Each 0.1 in. of differential pressure error will make the following error in flow calculated.

at 5 in. differential—1%
at 25 in. differential—0.2%
at 50 in. differential—0.1%
at 100 in. differential—0.05%

5. Supercompressibility affects volume measured approximately 0.5% per 100 psi pressure in flowing temperature range of 60–90°F.

Rules of measurement of gas by orifice meter

Where the static pressure remains constant, to reduce the flow by one-half, lower the differential reading to one-fourth. For example, if the differential is reading 64 in., to reduce the flow by one-half, reduce the differential to 16 in.

For better measurement, keep the beta ratio between 0.2 and 0.6.

Headers on meter stations should be twice the internal diameter of the meter runs they manifold together.

Measurement stations should be installed on the suction side of a compressor station rather than the discharge side where possible, to reduce effect of pulsation.

Use the same size orifice plate in meter runs that are manifolded together where possible; recordings can then be used to check against the other since their differentials should be approximately the same.

Make quick calculation of the gas flow through orifice meters

On meter runs up to and including 10-in., this can be accomplished by multiplying the coefficient for a 1-in. orifice times the actual orifice installed squared times the square root of the extension (the product of the differential in in. of water times the static pressure in lb/in.² absolute). AGA-3 coefficients for a 1-in. orifice in 2, 3, 4, 6, 8, and 10 in. standard orifice meter runs using flange taps should be committed to memory. Base conditions would be for a 0.60 specific gravity, 60°F and 14.73 psia pressure base. The answer would be in standard cubic ft/hr or day depending upon the coefficient used.

AGA Committee Report Number 3 coefficients at the conditions shown are:

2 in. meter run—1 in. orifice—271.4 cfh, 0.6 G, 60°F and 14.73 psia
3 in. meter run—1 in. orifice—262.8
4 in. meter run—1 in. orifice—260.8
6 in. meter run—1 in. orifice—259.3
8 in. meter run—1 in. orifice—258.7
10 in. meter run—1 in. orifice—258.5 (heavy)

Example. Find the hourly gas flow through an 8-in. meter run having a 2.00-in. orifice when the differential pen is recording 16 in. of water and the static pressure is 900 lb/in.² absolute.

Solution. From memory, the coefficient for an 8-in. meter run, 1-in. orifice is 258.7. Multiplying 258.7 times 4 (or 2 squared) times 120 (the square root of the product of 900 times 16) equals 124,176 cfh, the approximate flow in cubic ft of gas per hour at 0.6 gravity, 60°F temperature, and 14.73 psia base pressure.

How to measure high pressure gas

Accurate gas measurement is obtained only through a good maintenance and operating schedule

T. J. Lewis, Measurement Engineer, Columbia Gulf Transmission Co., Houston

The rapidly increasing value of natural gas, along with rising operating costs, has made accurate gas measurement vital for the natural gas industry.

Accurate gas measurement is obtained only through an efficient maintenance and operating schedule, adhered to rigidly, after a properly designed measuring station has been installed. Any error in the measurement of gas at high pressure and large volumes magnifies itself, so that there can be either a considerable gain or loss in revenue.

Field measurement at high pressure is defined as the measurement of gas at a specific point in a gas producing area at a pressure higher than that of contract base pressure.

Contracts

The gas purchase contract is of prime importance in determining the requirements for gas measurement at a given location.

Information such as gas analysis, gas/oil ratios, specific gravity determinations, flowing temperatures, pressures, etc., should be gathered before a gas purchase contract is written. This collection of data should be used to eliminate any assumptions regarding the requirements for performing accurate measurement. It also provides information for the engineering and accounting functions, so that they are made aware of their responsibilities and limitations in helping provide good measurement.

The contract should also state as clearly as possible all parameters of the gas measurement operation to prevent possible disputes by having a clear understanding between the buyer and the seller.

Design and planning

In the planning and designing of a natural gas measuring station, many items of equal importance must be considered if accurate measurement is to be obtained.

The design and fabrication of the orifice meter runs must meet the specifications of the AGA Gas Measurement Committee Report No. 3, latest revision. In addition to these requirements, consideration must be given to safety.

volume and quality of gas, term of contract, flexibility of station design, and physical location of the station. The most important point in this whole procedure is remembering that the measuring station is like a large cash register with money flowing both ways for the company.

A meter tube should be designed to allow easy internal inspection. Shutoff valves should be placed on each end of the meter, so that it may be depressured independently of any other piping. The instrumentation and/or control piping should be so located and supported to furnish maximum protection against accidental breakage.

Meter runs, piping, etc., should be designed as a Class III area facility. If possible, a properly vented walk-in type meter building should be used, with the total installation laid out with the safety of the operating personnel firmly in mind (Figure 1). As periodic orifice plate inspections are necessary, provisions should be made so that these inspections can be made without interruption of the gas flow. This can be done by installing a bypass line or by using a senior type orifice fitting.

The flow and temperature recording instruments should be located in a walk-in type meter building. This arrangement provides a year-round working area for chart changing, meter tests, maintenance work, etc.

Installation

The measuring or metering components in a measuring station are divided into two elements: primary and auxiliary.

The *primary element* of an orifice-type measuring station consists of the meter tube, the orifice plate fitting, the orifice plate, pressure taps and straightening vanes. While the primary element is being fabricated, its components should be checked against the specifications listed in AGA

Figure 1. Meter runs, piping, etc., should be designed as a Class III area facility. If possible, a properly vented walk-in type meter building should be used, with the total installation laid out with the safety of the operating personnel firmly in mind.

Gas Measurement Committee Report No. 3. Any discrepancies found then can be corrected before the meter tube is finally assembled and shipped to the field.

When the meter tube is being installed in the field, the orifice plate is checked to make sure it is centered in the meter tube, and that it meets AGA Report No. 3 specifications. Meter tubes are normally installed in a horizontal or vertical plane. Pressure taps in a vertical plane are preferred, since this allows the gage lines to be installed from the top of the orifice fitting. The pressure taps located on the bottom of the orifice fittings then can be used for installing drips or drains.

The *auxiliary element* consists of the meters (recording devices) used for measuring the static pressure, differential pressure, and temperature of the gas in the meter tube. Meters are placed as close as possible to the orifice or flange fitting. Gage lines are connected from the meter to the orifice fitting in a way that will best provide a drain to the meter tube. Liquids trapped in a gage line can cause significant errors in the recorded pressures.

There are two principal types of orifice meter gages: the mercury Manometer-type and the bellows-type meters. Each of these meters is suitable for field measurement, with the bellows-type meter becoming the favorite for wet gas measurement.

Installation of these elements in a measuring station should be under the supervision of experienced measurement personnel.

Operation and maintenance

After a measuring station has been properly installed, accurate measurement is obtainable only by conscientious care and maintenance by the field measurement technicians.

Any discrepancy noticed on the chart should be explained on the chart and a correction should be made immediately. Since immediate correction is not always possible, a report should then accompany the chart describing the condition of the meter and when the correction was completed. These reports, along with test and inspection reports, are necessary for the calculation of accurate volumes.

The accuracy of recording instruments depends on the men who test and operate the measuring station equipment. Good test equipment, properly maintained, is absolutely necessary in checking the calibration of temperature and pressure recording instruments and the quality of the gas being measured. The frequency of meter tests, gas quality tests, and meter tube inspections depends on the individual company and its contractual agreements with other companies.

Monthly calibration tests and inspections on the flow and temperature recorders are normally adequate for accu-

Figure 2. Relative effect of various factors on calculated flow volume (also Figure 3).

Figure 3. The above (also Figure 2) shows what happens to volume calculations when there are small errors in static pressure, differential pressure, temperature and specific gravity measurements.

rate measurement. In special cases, however, a weekly inspection may be required. Monthly calibration tests and inspections are vital to accurate measurement practices. Figures 2 and 3 show what happens to volume calculations when there are small errors in static pressure, differential pressure, temperature and specific gravity measurements.

Figure 4. Charts are shown being changed, and clocks wound and adjusted. Each measuring station has its own flow pattern, which is recognizable by experienced chart changers.

Since external tests and inspections do not reveal the internal condition of a meter tube, the meter tube and its components must be inspected internally on an annual basis for possible buildups and internal corrosion. This inspection ensures continuing accuracy.

Chart changing

Charts must be changed on time and clocks must be kept wound and adjusted (Figure 4).

Each measuring station has its own flow pattern, which is recognizable by experienced chart changers. Any change in this flow pattern indicates a possible error, which should be corrected if possible. If the correction cannot be made by the chart changer, he should notify the field measurement personnel who are qualified to perform required adjustment and/or repairs.

Each chart must be properly dated and identified along with field information necessary for calculating the required volumes. Automatic chart changers have become increasingly popular, especially in remote areas such as offshore Louisiana.

Periodic pulsation test should be made at each measuring station, particularly where a measuring station is located on the suction or discharge side of a compressor station. In this instance, pulsation dampeners must be installed between the compressor and the measuring station. Errors caused by pulsation can be very large, but they will not necessarily show in calibration tests.

A comprehensive maintenance program, faithfully followed, can detect and prevent costly errors before they occur. Leaking valves, gage lines, etc., will cause errors in an increase or decrease in company revenue. Such leaks should be corrected as soon as possible.

Wet gas measurement

The problems encountered in measuring wet gas are considerably different from dry gas measurement.

Wet gas contains entrained liquids consisting of water and hydrocarbon condensates which can cause hydrate formations in the orifice plate fitting, flow recorders and instrument lines. These hydrates are formed with the aid of free water under certain pressure and temperature conditions. Formation of hydrates can be reduced or eliminated in several different ways.

One successful solution to prevent hydrate formation is to use drip or condensate bottles in instrument and gage lines. Infrared gas heaters can be used in meter buildings for heating gage lines and flow recorders. This type of heater has become very popular lately, since it does not produce an open flame.

Errors caused by hydrate formations in the meter tube and around the orifice plate are common in wet gas measurement. The formation of these hydrates can be prevented by continuously injecting methanol into the gas stream ahead of the meter run. Experience and experimentation will determine the amount of methanol to be injected.

The use of methanol-containing inhibitors is not recommended as the inhibitor often solidifies in the meter tubes.

Four ways to calculate orifice flow in field

Rogers G. Thompson, Measurement Technician, Texas Eastern Transmission Corp., Shreveport, La.

Several methods are available for orifice flow field calculations.

The particular method used is determined by such factors as company policy, advance warning, time available, individual preference, accuracy required and the information and equipment available at the time of need. A good rule to follow is to calculate as precise a volume as possible at all times.

Any volume calculated manually will be considered a field calculation.

Most companies now use some type of computer system and rarely will a manually calculated volume match a computer volume. Field calculations are usually made under conditions which eliminate the use of adding machines and calculators. Accuracies can vary from an extremely accurate spot check to a very basic calculation. External factors usually determine what procedure to use.

Uses for the calculations determine the degree of accuracy required for each particular instance. Since time is usually at a premium it would be absurd to calculate to the

Table 1
Combined specific gravity and temperature correction
$$\text{factors—orifice meters combined factor} = \left(\sqrt{\frac{1}{G}} \sqrt{\frac{520}{460° \text{ F}}} \right)$$

Flow Temp. °F	Specific Gravity								
	.590	.591	.592	.593	.594	.595	.596	.597	.598
35	1.3343	1.3332	1.3321	1.3309	1.3298	1.3287	1.3276	1.3264	1.3254
36	1.3330	1.3319	1.3208	1.3296	1.3285	1.3274	1.3263	1.3251	1.3241
37	1.3317	1.3306	1.3295	1.3283	1.3272	1.3261	1.3250	1.3238	1.3228
38	1.3303	1.3292	1.3280	1.3269	1.3258	1.3247	1.3235	1.3224	1.3214
39	1.3290	1.3279	1.3267	1.3256	1.3245	1.3234	1.3222	1.3211	1.3201
40	1.3277	1.3266	1.3254	1.3243	1.3232	1.3221	1.3209	1.3198	1.3188
41	1.3264	1.3253	1.3241	1.3230	1.3219	1.3208	1.3197	1.3185	1.3175
42	1.3251	1.3240	1.3228	1.3217	1.3206	1.3195	1.3184	1.3172	1.3162
43	1.3238	1.3227	1.3215	1.3204	1.3193	1.3182	1.3171	1.3159	1.3149
44	1.3223	1.3212	1.3201	1.3190	1.3179	1.3168	1.3156	1.3145	1.3135
45	1.3210	1.3199	1.3188	1.3177	1.3166	1.3155	1.3143	1.3132	1.3122
46	1.3197	1.3186	1.3175	1.3164	1.3153	1.3142	1.3130	1.3119	1.3109
47	1.3184	1.3173	1.3162	1.3151	1.3140	1.3129	1.3118	1.3106	1.3096
48	1.3171	1.3160	1.3149	1.3138	1.3127	1.3116	1.3105	1.3093	1.3083
49	1.3158	1.3147	1.3136	1.3125	1.3114	1.3103	1.3092	1.3080	1.3070
50	1.3147	1.3135	1.3124	1.3113	1.3102	1.3091	1.3080	1.3069	1.3059
51	1.3134	1.3122	1.3111	1.3100	1.3089	1.3078	1.3067	1.3056	1.3046
52	1.3121	1.3109	1.3098	1.3087	1.3076	1.3065	1.3054	1.3043	1.3033
53	1.3108	1.3096	1.3085	1.3074	1.3063	1.3052	1.3041	1.3030	1.3020
54	1.3095	1.3083	1.3072	1.3061	1.3050	1.3039	1.3028	1.3017	1.3007
55	1.3081	1.3070	1.3059	1.3048	1.3037	1.3026	1.3015	1.3004	1.2994
56	1.3070	1.3059	1.3048	1.3037	1.3026	1.3015	1.3004	1.2992	1.2982
57	1.3057	1.3046	1.3035	1.3024	1.3013	1.3002	1.2991	1.2980	1.2970
58	1.3044	1.3033	1.3022	1.3011	1.3000	1.2989	1.2978	1.2967	1.2957
59	1.3032	1.3021	1.3010	1.2999	1.2988	1.2977	1.2966	1.2955	1.2945
60	1.3019	1.3008	1.2997	1.2986	1.2975	1.2964	1.2953	1.2942	1.2932
61	1.3006	1.2995	1.2984	1.2973	1.2962	1.2951	1.2940	1.2929	1.2919
62	1.2994	1.2983	1.2972	1.2961	1.2950	1.2939	1.2928	1.2917	1.2907
63	1.2982	1.2970	1.2959	1.2948	1.2937	1.2926	1.2915	1.2904	1.2894
64	1.2970	1.2959	1.2948	1.2937	1.2926	1.2915	1.2904	1.2893	1.2883
65	1.2957	1.2946	1.2935	1.2924	1.2913	1.2902	1.2891	1.2880	1.2870
66	1.2945	1.2934	1.2923	1.2912	1.2901	1.2890	1.2879	1.2868	1.2858
67	1.2932	1.2921	1.2910	1.2899	1.2888	1.2877	1.2866	1.2855	1.2845
68	1.2920	1.2909	1.2898	1.2887	1.2876	1.2865	1.2855	1.2844	1.2834
69	1.2908	1.2897	1.2887	1.2876	1.2865	1.2854	1.2843	1.2832	1.2822
70	1.2895	1.2884	1.2874	1.2863	1.2852	1.2841	1.2830	1.2819	1.2809
71	1.2884	1.2873	1.2862	1.2851	1.2840	1.2829	1.2818	1.2807	1.2798
72	1.2872	1.2861	1.2850	1.2839	1.2828	1.2818	1.2807	1.2796	1.2786
73	1.2859	1.2848	1.2837	1.2826	1.2815	1.2805	1.2794	1.2783	1.2773
74	1.2847	1.2836	1.2825	1.2815	1.2804	1.2793	1.2782	1.2771	1.2761
75	1.2835	1.2825	1.2814	1.2803	1.2792	1.2781	1.2770	1.2760	1.2750
76	1.2824	1.2813	1.2802	1.2791	1.2780	1.2770	1.2759	1.2748	1.2738
77	1.2811	1.2800	1.2789	1.2778	1.2767	1.2757	1.2746	1.2735	1.2725
78	1.2799	1.2788	1.2777	1.2767	1.2756	1.2745	1.2734	1.2723	1.2713
79	1.2787	1.2776	1.2766	1.2755	1.2744	1.2733	1.2722	1.2712	1.2702

cubic ft when calculating to the Mcf would suffice. A good example is automatic control of a compressor station where rates are given in terms of MMcf.

Field calculations are used for such things as determining purge gas volumes, relief valve losses, set point determination and sizing of plates. In general, volume, rate, differential, and plate size are the form variables that are determined by field calculations.

This discussion will be based on the AGA Committee Report No. 3 as it is the most commonly used. It is also the most complicated due to the large number of corrections applied.

Four categories have been devised to aid in discussing field calculations: the AGA formula, graphs, computers, and the quick method.

AGA Formula

In making field calculations a good rule to follow is—the more simple the better. Detailed calculations are difficult to perform under usual field conditions.

Another good rule to follow is cool and check. Always allow some time to elapse between calculating and checking. This can be from five minutes to five hours. Even better is to have someone else check your work. Following these rules will eliminate most errors.

The gas flow equation as stated in the AGA Committee Report No. 3 is:

$$Q_h = F_b\ F_r\ Y\ F_m\ F_{pb}\ F_{tb}\ F_m\ F_a\ F_l\ F_{tf}\ F_g\ F_{pv}\ \sqrt{h_w pf}$$

Obviously the most accurate method of volume calculation would use all of the above factors. Most field calculations will use some modified form of the AGA equation.

In order to present a logical discussion these correction factors are broken down into three groups—major, minor, and intermediate. This grouping is based upon the effect each factor has on the volume.

The major factors are the basic orifice factor F_b and the specific gravity factor F_g. These two factors have to be included in all computations.

The intermediate factors are the pressure base correction F_{pb}, flowing temperature correction F_{tf}, and supercompressibility correction F_{pv}. The value of these factors is variable. Usually they are determined by operating conditions at each specific location or are set by contract as in the case of pressure base.

Supercompressibility can become a large correction with transmission companies and producers where gas is measured at elevated pressures. For field calculation purposes, intermediate factors will be considered as either major or minor.

The F_r, Y, F_m, F_{tb}, F_a, and F_l corrections are minor factors for field calculation because of their minimal effect upon the final volume. Most of these factors are usually ignored.

Condensed Formula

As previously stated, the most accurate method of field calculation is the AGA formula using all corrections. However, this is usually not possible because of insufficient time, advance warning, or operating information.

If a little work is done prior to the time of use, the formula can be condensed to three or four calculations. However, some operating information must be available. The static pressure, differential, run and plate size, specific gravity, and gas composition must either be known or assumed.

Assumptions should be based on knowledge of normal operating procedures and characteristics of similar M&R stations. As an example, the inert composition of gas is assumed to be 0% nitrogen and 0% carbon dioxide until an analysis is performed. This procedure permits volume calculations until more definite information is available.

By using known and assumed data, an hourly coefficient composed of F_b, F_r, Y, F_m, F_{tb}, F_{pb}, F_a and F_l can be computed. The only corrections left to be made are F_g, F_{tf} and F_{pv}. By giving some thought, time and prior work the formula can be reduced to

$$Q_h = C\ F_g\ F_{tf}\ F_g\ \sqrt{h_w pf}$$

Tables are available which further reduce the amount of calculation that has to be done. Combined specific gravity and temperature tables are available from different sources. Also available are extension values for $\sqrt{h_w pf}$ or the individual values of $\sqrt{h_w}$ and $\sqrt{pf}$. These tables eliminate what is probably the hardest part of a field calculation.

When following these guidelines, calculations become fairly simple. One note of caution: always make computations in a legible and logical order. This makes everything fairly simple to check. As many errors are caused by illogical layout of calculations as any other factor.

All methods from this point on are an adaptation of the AGA Report No. 3 equation.

Flow Graphs

Usually the next most accurate method is the company developed flow graphs.

Although the graphs necessarily have to be based on average data, they are tailored to each particular company's

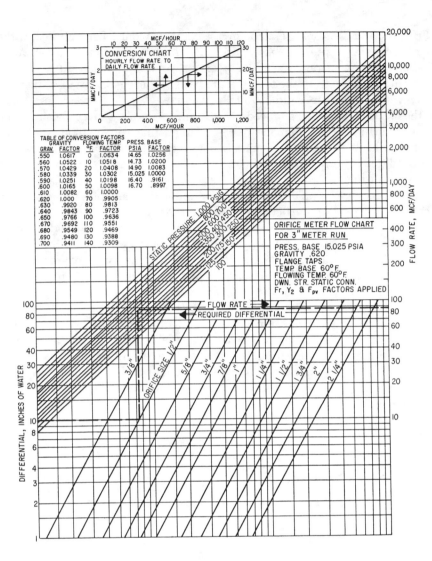

Figure 1. Example of orifice meter flow graph used by Texas Eastern.

operating conditions. Thus, you get more accurate results for your particular requirements. Corrections such as gravity can be made to the final figure if desired, but usually are not.

The graph has the advantage of giving considerable information quickly. All plate size and differential combinations are shown at a glance. Daily and hourly flow rates are also given. This is the fastest and easiest of all methods to use and accuracy is usually good enough to use under most conditions. No calculations are necessary under most conditions.

Most paper graphs are bulky, requiring a straight edge and space to use. Humidity will destroy them. However, replacements usually are available upon request. Oddly enough this is probably the least used of all methods.

Most companies use the AGA equation to develop their graphs. The newer factors Fa and Fl probably are not included in current data. Most factors are based on a particular company's average data.

Some factors are given a value of 1.000 and ignored. Supercompressibility corrections are usually by average specific gravity and inert composition.

Computers

Flow computers and slide rules are probably the most widely used method of field calculation. This is due to their compactness and simplicity of operation. Accuracy is generally in the acceptable range for field calculations.

Most manufacturers of measurement equipment furnish some type of flow computer either free or at a nominal cost. Construction usually is of paper and plastic.

All of these flow computers are based on the AGA Report No. 3 flow equation. Most of the minor corrections are not used, with the omissions listed in the instructions. Since computers are designed for use throughout the industry, they must be kept standard. Consequently they will not fit a particular set of conditions without making additional corrections.

Most flow computers incorporate Fb, Fg, Ftf, Ftb, and sometimes a generalized form of Fpv. The minor factors are usually ignored. The pressure base is usually set at 14.7 or 14.73, but some computers do incorporate a pressure base correction feature.

The biggest single drawback to computers is the fact that they must be general in scope and not slanted to one particular method of calculating volumes. Corrections are kept too general to fully correct for a specific set of conditions. There must be enough use of a computer for a person to maintain speed and accuracy of operation. If not, the instruction booklet will have to be studied each time the computer is used.

Flow computers are compact, easy to carry and store, and are very quick to use. As long as their limits are recognized they can do an excellent job.

Quick Method

Primarily, the quick method uses the major factors Fb and Fg only.

Inclusion of the intermediate factors is desirable if sufficient information is available. These intermediate corrections are made in a general way only. For example, if the pressure base is 15.025, subtract 2% of the volume. Algebraically, add 1% of the volume for each 10° variation from the temperature base. For each 100 psi of static pressure, add 8% of the volume.

This procedure will suffice for a lot of field calculation requirements. Keep in mind that this method is a rough calculation and is not intended to be used as a precise quantity.

Some mention must be made of the extension which probably causes more trouble than any other part of a volume calculation. In determining average differential and pressure, spot readings are the easiest to get and the most commonly used method. Sometimes predetermined values are used. Quite often differential and pressure are averaged over several hours. When averaging, care should be exercised for differentials under 25 in. and especially for differentials under 10 in. Due to the square root effect, errors are magnified in these ranges.

Extension books are available either in the form $\sqrt{h_w pf}$ or individual $\sqrt{h_w}$ and $\sqrt{pf}$ tables. Although the combined tables are bulky, the individual tables can be limited to a few small sheets. Use of these tables will improve accuracy and speed.

Practical maintenance tips for positive displacement meters

Here are the most common malfunction causes and points to watch on the widely used meters

J. A. Wilson, Meter Shop Manager, Northern District, Skelly Oil Co., Pampa, Texas

Thorough understanding of a meter's function is necessary to properly repair any mechanism within it.

There is a definite reason for each meter part, and each part has a definite relation to the other. The alignment of any one part with another has been precisely established for each type and size of meter.

The functions of a positive displacement meter are similar to a four-cylinder engine. For one revolution of the tangent, each chamber must fill and exhaust once.

One revolution of the tangent is 360°. Therefore, to make a smooth running meter with uniform displacement, one of the chambers must start to exhaust every 90°. Since the valves control the time at which the chamber starts to fill and exhaust, they must be coordinated, one with the other, so that this 90° relationship is maintained.

Similarly, the diaphragms must also be connected to assure this relationship, at least in their extreme positions. Both valves and diaphragms must be connected to coordinate both 90° relationships.

Following are practical tips for PD meter repair and testing.

PDM Repairs

The repairman cannot always carry parts for every type meter. Also, testers who only do an occasional repair job are not as efficient as men performing repairs daily. Therefore, most major repairs and all major overhauls are more efficiently handled in the meter shop.

Operating conditions in the field are various, and not always the best. Some of the operating problems in the field are gas saturated with water and hydrocarbons, iron oxide, sour gas and lubrication.

Shop repair. Meter shops, no matter how well equipped, cannot perform satisfactory meter repair without qualified and dedicated personnel.

When an experienced man disassembles a meter for repair, he inspects each part as he tears it down, considering each as a possible cause of the malfunction. When the trouble is found it is corrected before proceeding, if possible.

Some of the most common causes of malfunction and points to watch on the more widely used meters are:

Figure 1. Low pressure flow provers for PD meters can be used in the field or shop. The principle is essentially that of the orifice meter, except the orifice discs are calibrated for different flow rates instead of using various coefficients. Shown above is nomenclature for the Rockwell prover.

- Exhaust port gaskets should be inspected to make sure the seal is good. Some meters have metal spacers under the seal and must be in place when reassembled
- The clearance of the flag rods at the top of the meter should be checked. After a meter has been time checked and balanced, it will sometimes run rough. Look for a bright spot right over the top of the flag rods in the top cover. This is not a common problem, but is difficult to locate, and a few strokes with a file will take care of it
- The reassembly of the tangent gear between the valve gears of meters, as well as other gear-driven valves, should be centered carefully. There is usually a small amount of side play as well as end play on the tangent shaft bracket. A little extra care here will contribute to a quiet operating meter
- The body seal O-rings on the later models should be placed carefully in the slot when the top cover is put into place

Flag arms. Because new materials are being used in meters and the new aluminum-case meters are designed differently, there have been changes in some repair procedures. Some meters use an aluminum, long flag arm with a brass bushing cast in the center, so the flag arm can be soldered to the flag rod.

Most repairmen use a wet sponge to rapidly cool a solder joint on flag rods when flag arms are attached, but because of the different expansion coefficients of brass and alumi-num, the repairman must use extra precaution in cooling these solder connections.

Tests have proven that cooling the aluminum part of the flag arms too rapidly will cause fractures in the aluminum surrounding the brass bushing. Most of these fractured flag arms will go undetected and later break down in operation. However, if the repairman cools the solder joints by easing a wet sponge slowly up the flag rod, the joint is cooled from the inside out and no damage is done.

Valves and seats. Careful grinding of the valves and valve seats is very important. Number 80 grit may be used to clean up bad valves and seats, finishing with number 120 grit. Most repairmen use the figure eight motion in grinding, letting the weight of the glass block control the pressure on the valve seat. Use a very light touch on the valves. Where it is practical, both valve seats should be ground together. Grinding paper that is self adhesive is certainly recommended over grinding paper that has to be cemented to a glass block.

There are thousands of ironcase meters that have been in service for a great number of years. The valve seats on these meters should be carefully checked to make sure they are still bonded to the meter body. Remove and reseal the seats if there is any doubt about the strength of the bond. Use an old grinding block to true up the bottom of the valve seats and buff the meter case clean to ensure a good bond.

Clean all surplus cement from the intersurfaces of the valve seat opening. Be careful not to use screws that are

Crank — Adjustable Tangent

Partition, Table, Case and Channels

Grommet-Type Seals

Valve

Flag-Rods   Duramic Diaphragms

Figure 2. Thousands of ironcase PD meters, such as the American 25B and 35B shown above, have been in service for many years. Valve seats should be carefully checked to make sure they still are bonded to the meter body.

longer than standard in valve guides that screw into the valve seat. An overlength screw can jack the valve seat off the body and break the seal.

Watch the clearance on the new plastic parts when they are installed in old style meters. The raised reinforcing rib on some of these parts may pose clearance problems in many meters. Soldering acid will dissolve some plastic so if you must use acid core solder, be careful to keep it off the plastic.

Balance and time check. Take plenty of time to balance and time the valves and diaphragms to ensure a smooth operating meter. Use care in setting the tangent in proper relation to the valves. Time spent here will be regained manyfold when the meter is calibrated.

Most shops have a utility-type vacuum sweeper. This is an excellent means of rotating a meter before the top cover is installed. This gives the repairman a good chance to watch the valve action and make a final check on timing and balance, gage and depth of the undergear driving arm in relation to the tangent. Make sure the tangent wrist is engaged, but do not touch the short flag arm.

Installing a long pointer on the index driver dog will help proof test the meter since the circumference of the circle will be much longer than the test-hand on the index. It is a good practice to rotate the meter at least two rotations of the test-hand after the index is installed. This gives a final check to make certain the meter is operating smoothly. A vacuum cleaner can also be used to good advantage here.

Pan-type diaphragms can be rebuilt with a minimum amount of investment in jigs and other equipment. Many parts, such as crank brackets and flag arms which have pressed-in bushings, are easily repaired. Rebuilding parts can save up to 75% of the cost of a new item.

When it comes to repairing meters, "about right" isn't good enough. The meter repairman who wants to make his company money will strive to make all checks and adjustments as accurately as possible.

PDM Testing

Positive displacement meters may be tested by any of the following methods: (1) bell type meter prover, (2) low pressure flow prover, (3) critical flow prover, (4) by test meter. All of these methods when correctly used under proper conditions will give similar results.

Bell prover. Meter shop personnel are more familiar with the bell prover than with the other three types.

When a bell of sufficient size is available, large-capacity meters can be calibrated very rapidly in the shop. The use of a slope gage will certainly reveal if the repairman has assembled a well balanced, timed meter with a minimum of internal friction.

Bell provers should be in a temperature controlled area, and all meters should be at room temperature before testing. Normally both a 20% rate and capacity rate are run when proof testing flow rate on the meter. Two in. of water pressure on the bell is used instead of the 1½ in. used to test small domestic meters.

Low pressure flow prover. The low pressure flow prover is used both in the field and in the shop. It is very compact and portable, consisting of two pieces of aluminum pipe with straightening vanes in the upstream sections.

The principle is essentially that of the orifice meter, except the orifice discs are calibrated for different flow rates instead of using various coefficients. Many shops use the low pressure flow prover to test large capacity meters when a bell prover of sufficient size is not available. One or two blowers may be used to supply the necessary volume of air, or low pressure gas may be used where safe.

Critical flow prover. Critical flow is a condition occurring in the gas flow when the downstream pressure is approximately one-half the upstream pressure.

The critical flow prover consists of a piece of 2-in., extra heavy pipe approximately 18 in. long. One end is threaded to screw into a test connection placed as close to the meter outlet as possible. The other end has a special orifice plate holder. The critical flow prover can be used to test positive displacement meters operating on pressure in excess of 15 psi by passing gas or air through the meter and the prover, and discharging it into the air.

The orifice plates of a critical flow prover are stamped on the downstream side with the time (in seconds) required for one cubic ft of air to pass through the orifice at critical flow conditions. Plates used in critical flow provers have rounded edges on the upstream side, whereas orifice plates in a low pressure flow prover are sharp on the upstream edge.

It is vitally important that both the critical flow and low pressure orifice plate be handled with care to prevent damage, which would change the time stamped on the plate.

The stopwatch, which is important in timing prover tests, should be serviced and checked at regular intervals. Like many pieces of mechanical equipment, the stopwatch will operate for a long time before it completely gives out, and there is a very good possibility that during this time it will be operating inaccurately. Remember that tests taken with a defective timer are inaccurate.

Test meter method. Test meters have become popular in recent years with larger companies and contract testers. Continued development of rotary displacement has made the test meter a practical method for field use.

Although the shop tester normally runs only two rates—check rate and capacity rate—on a bell prover, most testers run three rates on the critical flow, low pressure and test meter methods. These rates are usually about 20, 50, and 100%.

Testing frequency is a much discussed subject. Sales through a large capacity meter represent a great deal of money; thus, it is important to regularly check the meter's accuracy. Some of the larger meters should be tested at 90-day intervals. Others of intermediate size could perhaps operate for two, five or seven years without being tested and at the end of the scheduled time, simply be changed out.

Many factors affect a testing and change-out schedule. The meter's mileage, its type of service, the type and amount of gas the meter must measure, and the length of time the meter is in service are important factors in setting up a testing and change-out schedule. Only after all of these factors have been considered, can an adequate testing and change-out program be established.

Sizing headers for meter stations

Calculate the velocity of gas in meter headers for proper sizing to achieve good measurement

Glenn E. Peck, Senior Engineer, Pipe Line Division, Bechtel Corp., San Francisco

Meter and regulator station design for natural gas transmission systems continues to grow more complex and demand more of the design engineer. Basic elements of today's station remain about the same but higher pressures and ever increasing capacities have replaced the positive displacement meter with multi-tube orifice meters and automatic chart changers. Buildings are often pre-fabricated structures designed specifically to house telemetering transmitters, supervisory and communications equipment in addition to basic recording instruments. The multi-run setting may have a computer giving the dispatcher instantaneous rate and totalized flow for the station. Electrical equipment may cost several times that of the entire station of a few years ago.

There is a sparsity of published criteria for sizing of orifice meter headers in the multi-run station. A discussion of factors that affect the sizing of the header and arrangement of component parts for satisfactory operating conditions should be in order.

We shall assume that the orifice tube and its component parts are designed in accordance with AGA Gas Measurement Report No. 3, that piping meets all the requirements of USASI B31.8 for pressure, materials, and class of construction, that all valves and fittings meet appropriate codes or standards, and that the differential pressure across the orifice plate will not exceed 100 in. of water.

The design goals may be listed as follows:

1. Commercially available equipment will be used that is specified to the closest reproducible economic tolerances.
2. The arrangement of the equipment shall conform to accepted practice for good measurement and allow fullest use of the available meter tubes.
3. Fittings and connections to the header shall minimize the flow turbulence in so far as is practical.

Meeting the above goals allows considerable latitude in the choice of components according to the judgment and experience of the designer without contradicting code, contract, or economic requirements.

In sizing the header a number of factors must be considered, namely:

- The number of orifice tubes which will eventually be connected to the header, or the ultimate flow capacity
- The setting layout, or arrangement, which governs the resistance flow paths the gas must take from entering the inlet header and dividing through the meter tubes before recombining in the outlet header and exiting
- The type of connections which the tubes make with the header

Station Layout

There are as many opinions of what constitutes good meter station layout as there are engineers working with gas measurement. Four of the more common arrangements are illustrated in Figures 1–4 and will be discussed relative to the flow-resistance paths offered.

Header Connections

Not many years ago, the header was a common piece of pipe with "orange-peel" ends and notched-miter nipple connections. The next step was a shaped nipple reinforced by a welding saddle or pad. Higher pressures brought forged steel tees into use. On larger sized tees the spacing between runs or tubes depended on the measurement for the tee and required a number of girth welds. With the introduction of the extruded nozzle header, the limitation of tees were eliminated for a generally more economical header.

The radius of curvature or flaring of the nozzle as extruded has superior flow characteristics over the reduced branch of a forged tee and both are vast improvements over the square cut opening for a shaped nipple. Some designers have gone a step further and reduced the 90° header nozzle by the use of a 45° or 60° lateral nipple to the axis of the header. This reduces the turbulence entering or leaving the header and could reduce the diameter of the header. This method can be much more costly for fabrication, is suspect for structural integrity at higher pressures and extreme ambient conditions, and goes beyond the configurations used in establishing the proportions, beta ratios, and coefficients for the orifice meter.

Runs Per Header

The number of runs, or meter tubes, for a given header is limited by the velocity criterion selected, the operating pressure, the size of fittings obtainable, and the relationship of the peak load to the normal load or capacity for which the meter setting is designed. The operating pressure has an important bearing on the number of orifice tubes used with a given header because the capacity of the tube is

Figure 1. The flow-resistant paths are identical for this layout and each of the two runs will furnish its full capacity with near identical differential readings on the meters. However, when additional capacity is required, an additional run will have a higher or longer resistance-flow path and will furnish a lower differential reading and capacity as compared with the first two runs.

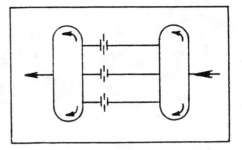

Figure 2. In this arrangement the middle tube has the lower resistance-flow path compared with the outer tubes and will affect the meter reading accordingly. The difference between meter readings will be more noticeable if the headers are undersized.

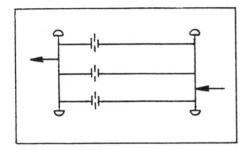

Figure 3. The "donut" header has been used frequently when large headers could be utilized. At higher pressures and capacities this layout places a high economic burden on the station and requires more space which can be a problem when a limited site is available.

Figure 4. This layout has been a favorite for many since the advent of the extruded or drawn nozzle header and because the resistance-flow paths are as nearly identical as is possible using commercially available equipment. It is limited to three or four meter tubes before the header begins to be "out-sized."

proportional to the square root of the pressure while the capacity (velocity) of the header is proportional to the pressure in atmospheres. At lower pressures the number of runs might be limited to three while at higher pressures the limit would be four runs. Size of fittings is generally less limiting than other factors unless the meter tube is greater than 12 in. in diameter. For the very large tube, the extruded nozzle header is required since reducing branch tees are not generally available. The relationship of the peakload to the sizing of the header will be discussed under the velocity sizing criterion.

Velocity Sizing

The following velocity criterion is suggested for sizing of headers for the sales orifice meter setting (expressed at ft per minute):

Target	Normal Limit	Maximum Limit
1,500	2,000	2,500

The relationship of the peak hour load to the average hour for the peak day designed for the station determines the range of velocity used in sizing the header. For example, the peak hour which is 10% of the peak day and one that is only 1% higher than the average hour for the peak day obviously demand that different limits be used for economic justification. The 10% peak hour suggests that for higher fluctuation in demand, a higher velocity be allowed in the header for the short time involved.

Keeping the velocity as low as possible within this criterion will provide the basis for good measurement and minimize the difference between adjacent meter readings. The way that gas enters and leaves the header connections to the meter tube is important in considering the velocity criterion which is chosen.

Sizing procedure. The normal procedure for sizing the station follows:

1. Take the peak day forecast or contract load and calculate the appropriate peak hour figure—provided the peak hour figure has not been fixed as a part of the forecast or contract. (Suggested period: 10 years, minimum.)

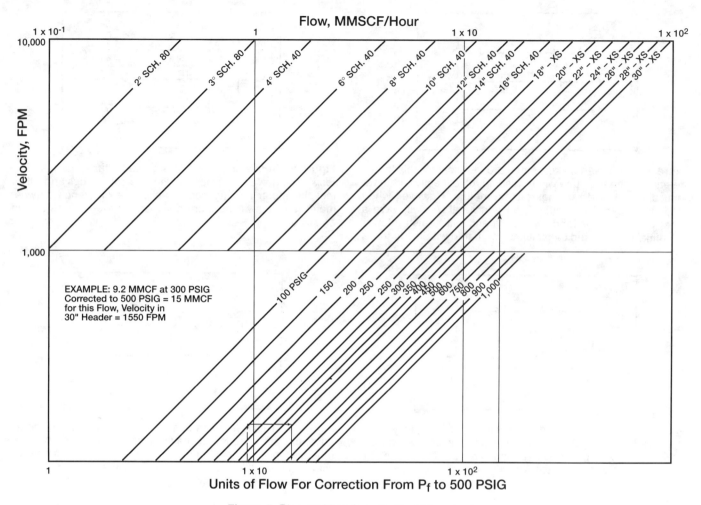

Figure 5. Pipe and header velocity sizing chart.

2. Divide the peak hour by the number of parallel tubes desired.

3. Use a flow rule or chart to size the orifice tube. Check the initial minimum hour against the minimum beta ratio and differential desirable since this may affect the size and number of tubes.

4. Use a chart, such as Figure 5, to calculate the velocity in the header for the proper criterion. Figure 5 has been found useful since the capacity can be traced up to the desirable velocity and the nearest appropriate size of header chosen.

Figure 5 was prepared for 500 psig or 514.65 psia at a temperature of 60°F in order that sizing of pipe or headers be made against velocity and capacity in Mscf/h on the upper part of the chart. The lower part of the chart is used to correct the flow at other pressures to the proportionate flow at 500 psig. Since the sizing procedure is an approximate one, it is felt that corrections for temperatures other than the standard condition and for compressibility are not necessary. Use of Figure 5 or a chart prepared to your own physical conditions will allow sizing of headers for the orifice meter setting quickly and accurately within the present state of the art.

16: Instrumentation

Types of control systems 454
Developments in Pipe Line Instrumentation.............456
Flow measurements 457
Proving devices... 459
Valves 460
Acoustic line break detectors.. 461
"Smart" pressure sensors 462
Densitometers 463
Pipe line samplers 464
Pipe line monitoring systems.. 465
Computer systems.. 466
SCADA systems 468
Cathodic protection.. 468
System design guidelines... 468
Future trends 469
**Choosing the Right Technology for Integrated
    SCADA Communications..470**

Types of control systems

Early pipe line systems depended upon manual control and the use of tanks to maintain a pipe line system in balance. On long pipe lines that required multiple pump stations, a surge tank was used to catch excess flow or make up the deficiency. Engine speed was varied to maintain the level in the surge tank at an acceptable level. This type of operation was fine as long as one grade of product was being pumped, but was not satisfactory for batching different products unless there was a tank available for each grade of product handled.

It was obvious that to build a tight line system, one that would not require surge tankage, some kind of automatic control system was needed to keep the booster stations in balance. Before the widespread use of centrifugal pumps, engine speed (for reciprocating pumps) was controlled to maintain the suction and discharge pressure within limits by sensing discharge and suction pressure and directing a control signal to either the engine governor or a control valve. Where control valves are used with reciprocating pumps, they are connected to recycle fluid from the discharge of the pump to the pump section.

When centrifugal pumps came into use, the same type of control system was used except that the final control element was a control valve installed in the pump discharge line as shown in Figure 1. The control system at origin pump stations normally controls discharge pressure or system flow rate. Intermediate booster stations require a control system that will control suction or discharge pressure

and in some cases, motor load. An examination of the system line friction and pump head curve will show that simultaneous control of suction and discharge pressure and high flow rate is not required since the pipe line system characteristics will not permit the coexistence of these events except in case of extreme system upset. As will be seen later, the control system is configured so that only one controller will be in control at a given time.

There are several modes of control available to the pipe line engineer. Mode of control is the manner in which control systems make corrections relative to the deviation of the controlled variable. The modes of control commonly used by pipe liners include:

- Proportional position
- Proportional plus reset
- Proportional-speed floating

Other modes such as proportional plus reset plus derivative are more commonly used in the process industries. This mode of control may, however, be used in pipe line measurement stations where it is necessary to control the temperature of a heater used to heat the product being pumped or measured.

In the proportional-position mode, there is a constant linear relation between the controlled value and the position of the final control element. With this type of control

Figure 1. Typical control scheme for a pipe line booster station.

system, the amount of correction will be the same regardless of the amount of deviation of the controlled variable. A proportional-position control system will not necessarily return the controlled variable to the exact value of the set point when an upset occurs. This may be satisfactory for some types of control systems, but not for pipe line pump station control.

The offset deviation noted in the proportional-position system can be overcome by adding reset to the controller. The reset action will continue to correct valve position until no deviation occurs between the controlled variable and controller setpoint.

Proportional-speed floating causes the final control element to change at a rate proportional to the amount of deviation between the setpoint and the measured variable. Correction to the final control element will continue until there is no deviation between setpoint and the measured variable. The amount of the correction signal is proportional to the amount of deviation between setpoint and the measured variable. This mode of control will produce an exact correction for any system load change.

Proportional-position and proportional-speed floating are the control modes most commonly used for pipe line control.

Figure 1 shows a typical control scheme for a single unit booster station. A minimum control system will include a suction pressure and discharge pressure controller. Other controllers such as motor current controllers and a ramp generator may be connected in the control loop.

The suction pressure controller is set to maintain the suction pressure above the low suction pressure shut down setting while the discharge pressure controller limits the station discharge pressure to a value which will not exceed the maximum allowable operating pressure. The output from each of the controllers is fed to a signal selector which will select either the high or low signal, depending upon the control action selected, and direct this signal to the control valve. Some control systems are designed so that a selector module may not be required, but they will include a means for selecting the proper signal. In a properly designed system, the transfer of control from one controller to another will be smooth and bumpless.

A ramp generator is used for ramping the control valve to a new position. The control valve may be closed when the station is at rest, and upon station startup the ramp generator will cause the valve to open slowly until one of the controllers is in control of the system or until the pump is "floating"—that is, operating at a condition where neither suction nor discharge pressure control is required. It may also be used to ramp the valve partially closed prior to unit shut down to minimize system upset. This feature is desirable in a multi-unit booster station since when one unit is shut down, the remaining operating units will be op-

erating momentarily at a no-flow condition. This condition will continue to exist until the system pressure decays to a value equal to the discharge pressure of the remaining units. Ramp speed should be adjustable so that it can be fine-tuned to the system requirements.

The control valve action, i.e., valve opens on increasing signal or valve closes on decreasing signal, will determine how the controller action is set. The suction pressure controller will be set opposite of the discharge controller.

Although now shown, the control system is usually equipped with a manual/automatic switch so that the control valve may be controlled in case one of the controllers fails or during system maintenance.

The controllers may also be equipped with remote setpoint control so that controller setpoint may be varied from a remote location to accommodate a changed condition in the system.

Figure 2 shows a control system for a multi-unit booster station. The control systems for single- and multi-unit stations are identical except that in a multi-unit station the station pressure transmitters are located so that one control system will suffice regardless of the number of units installed. It should also be noted that one control valve serves both units.

The control valve may be operated electrically, pneumatically, or hydraulically. Hydraulic systems offer the advantage of being able to operate the valve at a high rate of speed at virtually any system load. Most hydraulic operators are equipped with dual cylinders, and may be equipped with quad cylinders where the system loads are high. The selection of control valves is covered in Section 7.

Variable speed drives or controls are becoming an economically attractive choice for the operation of pumps. The use of these devices will virtually eliminate throttling excess head and the attendant energy costs. Few systems are operated at a constant rate and therefore some throttling is inevitable. Also, systems transporting multiple grades will require throttling to keep the system in balance as the batches move through the system.

Most control systems used for pipe line control are made up of standard control system elements. Some applications may, however, have unique characteristics that will require a system that is just a little bit different from a standard control system. Most of these types of systems can be put together using standard modules, or a programmable logic controller may be used.

In Section 13—Liquid Hydraulics, the discussion on hydraulic gradients illustrated a case where a hill required maintaining back pressure at the end of the line. It was shown that the amount of back pressure required varied with the system flow rate. A conventional back pressure controller and control valve installed at the end of the line and set for the worst case (maximum back pressure) would

SDP = Sta. Disch. Pr.
SFR = Sta. Flow Rate
SSP = Sta. Suc. Pr.
RTU = Remote Terminal Unit

Figure 2. Multi-unit booster station.

work; however, it would not be economical since more pressure than required would be maintained at the higher flow rates. The pressure at the hill could be telemetered to the end of the line where it could be used as the measured variable for the controller. The setpoint would be set to the amount of pressure required at the hill which in this case is 200 psig. This type of control system would not penalize system economics since it would always maintain the required 200 psig at the hill. However, it would require the installation and maintenance of a telemetry link.

Another option is to use a relationship controller which would not require telemetry of the hill pressure. The relationship controller would use the system flow rate to drive the setpoint of a pressure controller. The controller would need to be equipped with limit stops so that the setpoint could only be changed between the upper and lower limits. High system flow rates would drive the controller setpoint

to the minimum allowable value of back pressure and low system flow rates would drive the controller to the allowable maximum limit. Intermediate system flow rates would drive the setpoint to an intermediate pressure setting.

With today's technology and hardware, the control engineer is limited only by his imagination in solving simple as well as complex control problems. Programmable logic controllers (PLC) and self-tuning controllers are being used more and more in the control of pipe lines and pipe line stations. Control systems are being incorporated as a part of the PLC. With the addition of a modem and a suitable communication protocol, the PLC may also be used as the remote terminal unit (RTU) for the supervisory control and data acquisition system (SCADA).

The following article, "Developments in Pipe Line Instrumentation," by Dr. R. A. Furness, is an excellent discussion on the state-of-the-art in pipe line instrumentation.

DEVELOPMENTS IN PIPE LINE INSTRUMENTATION

R. A. Furness, Center for Flow Measurement, Department of Fluid Engineering and Instrumentation, Cranfield Institute of Technology

Abstract

Pipe lines are an integral part of the world's economy and literally billions of lb worth of fluids are moved each year in pipe lines of varying lengths and diameters. As the cost of some of these fluids and the price of moving them

has increased, so the need to measure the flows more accurately and control and operate the line more effectively has arisen. Instrumentation and control equipment have developed steadily in the past decade but not as fast as the computers and microprocessors that are now a part of most large scale pipe line systems. It is the interfacing of the new generation of digital and sometimes "intelligent" instrumentation with smaller and more powerful computers that has led to a quiet but rapid revolution in pipe line monitoring and control. This paper looks at the more significant developments from the many that have appeared in the past few years and attempts to project future trends in the industry for the next decade.

Introduction

Our need to transport fluids from the point of production to the area of end use has led in the past twenty-five years to a large increase in the number of pipe lines being designed and constructed. Many of these carry hazardous or toxic products, often through areas of high environmental sensitivity or close to centers of population. The cost of oil production rose sharply to over $35 per barrel before the recent sudden fall, and the cost of manufactured chemicals and industrial gases has also risen in the same period. With the need to safeguard the cost of the transported fluid and also the population and the environment close to the pipe line, on-line monitoring and in some cases 24-hour surveillance is now becoming routine in many systems throughout the world.

When one sees maps of the pipe line network in some parts of the world, the number and length of pipe lines is staggering. At the recent Aberdeen conference, one speaker gave just two of the many statistics that can be produced to show how vast the network is. One line runs from central Canada eastwards to the Toronto/Montreal area. This line alone has 54 million square meters of pipe surface to survey. Also in 1984 some 6 billion barrels of oil and 4.5 billion barrels of manufactured liquids were transported in pipe lines in the USA. The actual world volume of pipe line fluids is many times this amount and the value immense.

Pipe lines are not new, indeed almost 50% of the world's total network is 30 or more years old. These have been subject to corrosion and wear and some have carried fluids for which they were not designed and others have been operated outside design limits. Monitoring and control systems therefore are not just required for measurement purposes but also for integrity surveillance and on-line fault diagnostics. Some of the many developments that can be applied in modern day line operation are briefly reviewed and some design pointers for their integration into complete systems are given.

Flow measurements

Flow measurement is the most important process variable in the operation and control of pipe lines. Simple line flow balances are used to check for discrepancies on hourly, daily, or weekly bases. This may identify that a problem exists but flowmeters at the end of a line will not identify the location of the problem. For this, pressure transducers along the pipe line will be required together with the flowmeters to establish where the pressure gradient has deviated from that predicted from the basic equations of fluid flow. These equations use pressure/flow values at inlet and outlet to confirm the line is in balance. Such a simple system is described in more detail later in this paper.

Most flowmeters sold are of the differential pressure type and of these the predominant meter is the sharp-edged orifice plate. Many hundreds if not thousands of these are installed on pipe lines around the world to measure gas, liquid, and mixed fluid streams. With the arrival of the modern flow computer these types of flowmeter have almost received a new lease on life. One fact not appreciated though is the total measurement uncertainty of such devices. Frequently we read performance claims of $\pm 0.5\%$ and better for these pipe line metering systems. This is very difficult to believe when it must be remembered there is an uncertainty of $\pm 0.6\%$ on the data used to derive the flow coefficient in the basic calculation. When all the other variables that can affect the uncertainty of measurement are included, such as fluid composition changes, ambient and process temperature and pressure variations, errors in temperature and pressure measurement, A/D conversion and computational errors, to name but a few, then it is unreasonable to expect uncertainties much better than $\pm 1\%$ FSD. If wide flow rates are also experienced then the uncertainty of the low flow determination could be $\pm 5\%$ FSD or worse. This makes their use as accurate pipe line flow devices somewhat questionable, although they are very common in the metering of natural gas in this country.

Turbine flowmeters also find wide-spread use in pipe line applications, particularly hydrocarbons. They all suffer from bearing limitations, however, and this means they require regular calibration to ensure good performance. Recent developments have included twin rotor designs, where measurements of the speed of both rotors simultaneously has led to the meter becoming self diagnostic. These meters can now detect their own bearing wear and also deviations from initial calibration due to blade damage. The same on-board microprocessors can also check the integrity of the data generated by the turbine meter as well as providing alarm output for loss of signal and other problems.

Two different designs have been applied in pipe line systems with good success. The first design, from Quantum Dynamics and shown in Figure 1, has two rotors with two independent sets of bearings. The shaft and slave rotor ride on one set of bearings with the measurement rotor upstream of the slave motor riding on a second set of bearings on the rotating shaft.

Relative motion between the indicating turbine and the shaft therefore remains close to zero. This plus a drag-free RF pick-off results in an almost frictionless flowmeter.

Figure 1. Twin rotor turbine meter (courtesy Quantum Dynamics Inc).

Should the shaft or measuring motor bearings begin to wear, this will result in a change in the ratio of the rotor speeds for a fixed flow rate. Thus bearing wear can be detected on-line and action taken. This design of meter is also different from all other turbines because of the slender center body and axially short blades. Because friction is very low the fluid does not have to be accelerated past the rotor to obtain sufficient torque to overcome retarding forces. Thus pressure drop is also low and pressure and temperature measurements can be carried out at the point of volume flow rate determination. They have been shown to be outstanding meters for leak detection purposes on gas lines.

The Auto Adjust turbo-meter from Rockwell International is also self-checking but is in addition self-adjusting. The design shown in Figure 2 is quite different from the Quantum meter. The rotors are next to each other with the measuring rotor downstream of the main rotor. Constant performance is obtained because the sensor rotor detects deviations in blade angle of fluid leaving the main rotor.

Figure 2. Auto adjust gas pipeline meter (courtesy Rockwell International, Inc.).

With a very small angle, the sense motor runs at a much lower speed than the upstream rotor. Performance is then given by the difference in the speed of the two rotors rather than from a single rotor. When bearings begin to wear, the main rotor will spin slightly slower and so the exit angle will also change. This is sensed by the second rotor, which adjusts speed by an equal amount. Bearing wear is therefore automatically adjusted out. These meters have been installed in many gas transmission lines in North America with generally very good performance. The Quantum meter, on the other hand, is a newcomer to pipe line applications, but personal first hand experiences in a gas pipe line were favorable.

Pipe lines are generally operated on material balances (mass) but most flowmeters are volumetric devices. This means that for pure fluids measurements of temperature and pressure plus an appropriate equation of state are required to enable density to be determined. For mixtures of gas or liquids or indeed as an alternative to the above method on-line densitometers are used. The product of volume flow and density gives mass flow. Until recently pipe line mass flowmeters were not available but recently a number of Coriolis type direct mass meters have appeared. These have found slow but steady acceptance in industrial applications and are beginning to find applications in pipe line monitoring. One such design is shown in Figure 3 and recently four other designs were launched at the INTERKAMA trade show in West Germany. This type of meter is applicable to both liquids and gases and some have been used on liquid/liquid and gas/liquid mixtures but with mixed success. They are currently undergoing rigorous evaluation and analysis in many chemical, process, and small pipe line applications and indications are that they are standing up well in some applications but not others.

Figure 3. Micro motion direct mass flowmeter (courtesy Brooks Instruments).

Accuracy is around $\pm 0.5\%$ of reading or better and they do not need additional measurements to determine flowing density. Indeed by monitoring the output signal they can be used to indicate density as well as mass flow. The output is in digital and/or analog form making them compatible with many types of flow computer.

Direct mass measurement has one main advantage over volume measurement, since an integration of the signal over a period of time will indicate the pipe line fluid inventory directly. The disadvantage for pipe line operators is the current size range that is available, the largest being 150 mm bore. As most major pipe lines are in the range of 500–2,000 mm bore, these Coriolis meters must be installed in parallel to meet the larger applications. Mass metering is relatively new and the next decade should see a steady increase in sales and acceptance of these meters.

The ideal flowmeter is one which is installed on the outside of the line but gives the performance of the best flowmeters installed inside the line. Ultrasonic time-of-flight meters are slowly developing toward these apparently conflicting but demanding criteria. Multiple beam systems have been installed on many pipe lines, the most notable of which is the Alaskan oil pipe line. Some 23 meters each 4 ft in diameter look for gross imbalances along sections of this important installation. They permit the early detection of product spillage and also indicate the section of line giving the problem. This system has been operating since the late 1970s with millions of barrels of fluid being shipped through the line in that time. Experience of these meters has been very favorable in this application.

Ultrasonic meters, like most of the modern types of meter, were initially oversold but the main problem has been to convince users that the two types available, the Doppler and transit time meters, are completely different animals, that both use ultrasonics to infer flow. When ultrasonic meters are mentioned, people tend to think of Doppler types. Most of these are relatively crude flow indicating devices. The modern clamp-on transit time meter can indicate flow to around $\pm 2\%$ of reading or better depending on design, compared to the average Doppler meter which is a $\pm 5\%$ FSD device. The advantage of ultrasonic methods is the negligible headloss and the ability to install other portable or dedicated systems without line shutdown. It is surprising that more have not been installed in pipe line systems, but as discussed later in this paper, this is an example of the conservatism often shown in the application of new technology.

Flow measurement has really been transformed by the application of digital techniques to on-line signal analysis. Much more data than just flow rate is now available and with the latest systems, operators are taking advantage of the new generation of micro and flow computers to control and run their pipe lines. It is unusual these days not to install a local flow computer with each flowmeter, with temperature, pressure, and differential pressure signals being fed into the machine for local "number crunching."

Proving devices

The problem with all flowmeters is that they are affected to some degree by the fluid they are metering. For the least uncertainty of measurement, the flowmeter, whatever type is chosen, should be calibrated on-line under actual operating conditions. In the past this was performed with large and expensive pieces of hardware called ball provers. These had to be large to provide adequate volume within the calibrated section of pipe to ensure sufficient resolution and accuracy for the flowmeter they were intended to calibrate.

Figure 4. Modern compact meter prover (courtesy Brooks Instruments).

The latest generation of proving devices, called compact provers, are essentially time/volume calibrators that eliminate the ± 1 count error present on their "big brothers." It is physically much smaller but with the same or better performance. There are currently 10 different versions on the market, of which five are available in the UK. The prover, one design of which is shown in Figure 4, uses pneumatics to launch a piston, which is then driven by the fluid through a calibrated chamber, displaying a known volume. The key to their excellent performance is the data acquisition system illustrated in the lower part of Figure 4. As the piston passes through the calibrated section, pulses from a high frequency internal clock are counted to determine the time, A, to displace the calibrated volume D. Once the counting has started a second internal timer is gated to respond to the first input pulse from the flowmeter. This timer then records the time interval B to collect C whole flowmeter pulses.

During the proving run there were then C × A/B meter pulses and the meter factor is found by dividing this count by the volume displaced:

$$K = \frac{A}{B} \times \frac{C}{D}$$

Some designs are uni-directional and some bi-directional. In the uni-directional types pneumatics or hydraulics launch and retrieve the piston; in the bi-directional designs the process fluid is used to drive the piston. Other versions incorporate double-barrelled liners to remove the need for pressure compensation and also reduce the effect of temperature variations.

In one proving run there may be as few as 100 pulses collected, so timing error must be minimized. If the pulse output is frequency modulated, from a positive displacement bulk meter for example, then these few pulses may not represent a true statistical sample of the pulse train and multiple passes may be required. Alternatively, pulse interpolation techniques may be required, several of which are applicable to give the accuracy required. These are currently the subject of intense activity in standards committees who are producing a new general code of practice on meter provers.

The compact prover is better suited to offshore platform applications and to field use on operating pipe lines. Being portable it can be easily coupled in series with the meter to be proved. A suitable choice of materials allows proving on a vast range of fluids to be carried out. Operational experience has generally been favorable and several direct comparison tests with ball provers have shown the new generation of compact provers to be superior in most respects. Their acceptance though in custody transfer applications has been slower than had been expected, even though several countries have certified some designs for Weights and Measures approval. Some international calibration centers are now using this technology for routine traceable calibrations and confidence is growing as operational experience is accumulated.

By far the most economic method of measuring flows in pipe lines is to install pairs of flowmeters in custom-designed metering stations. By monitoring the ratio of the outputs, a measure of wear or drift is obtained. Provided the outputs of the meters remain within preset tight limits, uncertainty of flow rate measurement is low. The disadvantage of this on pipe line operation is that there are two meters to be maintained and checked and should faults develop the pipe line may have to be shut down more frequently, unless by-pass loops are provided when the metering station is designed. Turbine meter and PD meters with pulse outputs give the best performance, particularly if they are regularly proved with either a dedicated or portable prover of the type described.

For large bore lines, particularly gas lines, the proving of flow meters is very expensive if it can be performed at all. The majority of installations rely on orifice plates built to the standards. Recently a new high-pressure gas proving station was completed by British Gas at Bishop Auckland. Here turbine and other meters can be calibrated in actual operating conditions on natural gas. Such an installation has previously only been available in Holland where gas calibration standards have been developed to a high level

of confidence over many years of patient data collection. With the new British Gas facility we should see the ability to cross check our own standards against international standards.

Valves

Valves are an important part of any pipe line system. They come in many shapes and sizes and perform numerous functions. Some act as primary flow regulators, some as check valves for flow direction control, and some as pressure reducers for safety purposes. Developments have centered in two main areas, materials and actuators (both simple and intelligent). One of the development sources is again the North Sea, where the demands for better reliability, better in-field performance, and reduced cost have forced many manufacturers to look critically at designs. As one of the most expensive components in the valve is the body, effort has been concentrated in this area, both in terms of new materials to meet more demanding specifications and of quality and price. These last two items are critical to the survival of many companies and in a market sector that is growing slowly or even declining, as suggested by one leading instrument journal, valve body cost savings however small may mean the difference between life and death.

Slimline butterfly valves have recently been introduced by a number of manufacturers to meet the demand for lower cost reliable devices. These can be installed between existing pipe line flanges. They can be manufactured in a wide range of materials with special seal materials chosen for the most demanding of duties. With suitable actuators even valves as large as 60-in. diameter do not take up much space.

Other designs of control valve provide straight through flow patterns to minimize flow turbulence, vibration, and noise. Should a pipe line be fitted with acoustic line break detectors (see section following) line noise will be important in setting the sensitivity of the detectors and hence the type of valve fitted will be important to system operation. Other operational problems such as "hunting" in globe valves have prompted the designers to work more closely with the operators to define full specifications. Pipe line operators often run their systems with rapid transients present. Demand changes of both a positive and negative value can subject the control valves to very severe step changes in flow and pressure and initially many conventional valves could not withstand such treatment.

This is where intelligent sensors on pressure relief valves have come into their own. They are able to give better speed of response and meet some stringent operating criteria. Pressure may also rise gradually due to a steady rise in line temperature as well as the rapid transient caused by block valve closure or pump failure/surge. These transients travel at sonic velocity and even tenth of seconds advantage in operating the valves can affect safety on the line. Microprocessor-based detectors fitted to the valves can be used to sense line conditions and provide suitable inputs to the valve. In some advanced designs manufactured in the USA it is possible to program intelligence into the unit to begin closure/opening based on measurements of an anticipated condition such as pump surge further down the line. If the signals from one part of the line can be transmitted to other sectors, corrective action can be taken before safety blowoff valves are activated, releasing product into the environment.

Figure 5. Intelligent pressure relief valve.

A pipe line pressure relief system based on this principle is shown in Figure 5. A strain gauge sensor fitted on the suction side of the compressor is monitored by a microprocessor which linearizes and conditions the signal. The microprocessor operates a pair of solenoid valves acting the pilot mode with the line pressure acting as the actuating medium. By switching solenoids the relief valve can be opened, closed, or programmed to act as a regulator at the pressure trapped in the valve body. Such a system can provide adequate response to step changes on the order of 70 bar in 3 milliseconds. In this system the status of the pipe line pumps is provided as an additional input to the microprocessor to trigger action before the pressure surge occurs. Tests on large lines have shown such intelligence to be vital in the safe operation of systems when 5,000 hp pumps were suddenly lost.

Line block valves can also be actuated by output signals from line break detectors. This is important in isolating problem sections of line carrying toxic products. The wave detector will sense problems in seconds and the line can be closed to minimize the amount of toxic product released. That section of the line may well depressure but the release effects will be reduced. Such systems are fitted as anti-terrorist devices on lines in problem areas of the world such as the Middle East and Central Africa where all too often pipe lines are political targets.

The choice of materials for valve bodies and the styles of valve available is now very wide. A recent survey in *Processing* (January 1986) listed 15 different types supplied by 110 companies. These range from the cryogenic applications up to line temperatures in excess of 700°C. Selecting the correct valve for the duty is now the biggest problem facing the user and where it is necessary to take outputs into computers for safety and control purposes the choice becomes more difficult. In such a competitive market, developments are bound to come thick and fast and it will be interesting to watch the development of the next generation of intelligent pipe line valves.

Acoustic line break detectors

Problems with the pipe line are frequently accompanied by the generation of noise, either from problem valves (see above) or line breaks. If the flow becomes too high from a surge or a leak occurs through corrosion or third party intervention, then there are instances where vibrations in excess of 20 kHz are produced. These frequencies are in the ultrasonic range and can be detected with suitable transducers. Such devices have been made portable so that pipe line crews can clamp them at any point along the line to check for noise. By noting the signal strength, the source of the problem can be pinpointed. British Gas has used this principle to develop one of two types of pipe line monitoring tool that is described later.

A similar technique, though based on a different principle, is the acoustic "wavealert" monitor, though more correctly called a negative pressure wave detector. The central element is a piezo-electric sensor that gives an output when stressed. When a line rupture or leak occurs, there is a sudden drop in line pressure at the source of the problem followed by line repressurization a few milliseconds later. The rarefaction wave generated by these events moves away from the leak in both directions at the speed of sound. The pipewalls act as wave guides enabling this wave to travel for great distances, attentuating in amplitude as it does so. Sensors placed at known points along the line can detect the passage of this pressure transient. From the triggering of internal timing circuits and a knowledge of the prevailing acoustic velocity for the line, the leak location can be calculated.

Such devices are particularly useful for detecting large breaks in lines very rapidly. The transient wave moves at 1 mile per second in liquids and 1 mile in 5 seconds in gases. If the detectors are positioned at intervals of 2 to 5 miles in

Figure 6. Negative pressure wave 'wavealert' monitor (courtesy Spectra-Tek).

critical areas, then the break is detected in a few seconds. If the output signal is fed into a computer, then this can be used to trigger block valves to isolate the problem section of the line very easily. Figure 6 shows one further development where the instrument can be made to cancel line noise and use the full measuring capacity of the sensor for signal detection. There is, however, the practical problem in setting the background threshold correctly. Pipe line noise is generated by bends, valves, pumps, regulators, etc., and it is crucial to be able to remove this background noise to avoid unnecessary line shutdowns when the line noise exceeds the detector setting. The Reynolds number of the flow has been shown to be the important parameter in installations both in the USA and the UK. Experience has also shown that the installation of these devices is critical to reliable performance. It is important to have the sensor as close as possible to the line to avoid attenuation effects and improve sensitivity. Both this and the background line noise will influence the number of "false alarms" from the instrument.

ICI has been experimenting with strain gauge type sensors on their cross country ethylene line and these are showing more promise. These experiments will be closely monitored in the next few years to see whether we can improve the reliability and sensitivity of line break detectors.

'Smart' pressure sensors

One area where the impact of microprocessors has been felt is in the development of smart sensors, particularly pressure. Many physical effects respond directly or indirectly to pressure and as a result the first generation of "smart" process transducers appeared on the market some four years ago. Surprisingly, however, market penetration has been slow, mainly, I suspect, due to the fact that hand-picked conventional pressure and differential pressure cells

perform almost as well over the limited range often required. There is virtually no limit to the amount of intelligence and diagnostics that can be incorporated into pressure transducers, but so far only those which are reasonably linear or which have well defined characteristics have been commercially exploited. The need for linearity is now no longer there because all signal processing the digital and algorithms can be written to cope with any signal/pressure curve, provided it is repeatable. Even hysteresis can be taken account of.

It is in the communications area that the full potential of intelligent transducers is likely to be felt. Hand held field communicators enable the user to rerange, configure, and display process output values without the need to be close to the transmitters. On-line reranging is particularly useful when the pressure transducer is located in a very inaccessible place. It is possible to use two-way communication links to examine the outputs from a variety of intelligent transducers located at remote outstations, but the cost of such a feature should be carefully considered as this is likely to be expensive. Intelligent transducers are more expensive than the commercial instrumentation, but in certain applications cost is not a consideration.

The general features of the new generation of smart sensors in addition to those mentioned above includes:

- On-line temperature compensation
- Built in diagnostics
- Claimed accuracies of $\pm 0.1\%$ of span

Leading companies in the field include Honeywell, Bailey, and SattControl, but other established companies are working on similar devices. Most of these sensors use the properties of silicon plus the addition of micro electronics to give the performance. Silicon is an elastic drift free material and pressure sensing elements are diffused onto it. A cavity is usually etched on the reverse side to form the diaphragm. Where the line fluid is very aggressive a second insulating diaphragm is required to protect the silicon from corrosion. Generally speaking the cost of these sensors is coming down and it appears we can look forward to many more instruments appearing in the more difficult applications. Provided we can overcome the general conservatism about applying new technology this is one section of the instrumentation industry where micro electronics will have a great future influence, particularly when used with the computer and real time models. This will give the pipe line operator more confidence and more information on both the condition of the instrument and the accuracy of the measurement. Interfacing problems should also reduce.

Another type of pressure transducer that looks promising for future applications in pipe lines is the vibrating wire sensor. One such instrument from Foxboro has shown considerable reduction in size and manufacturing costs from other sensors. A tungsten wire encapsulated inside a cell is vibrated at its natural frequency. As pressure is applied to the silicon barrier diaphragms, this changes the tension in the wire and so alters the resonant frequency. This is sensed and amplified and gives the transducer excellent resolution characteristics. Pressure and temperature compensation is accomplished within the cell, and although the instrument is not strictly a "smart" transducer, it does possess basic characteristics which should be attractive to pipe line applications.

Densitometers

Continuous high accuracy on-line density measurement has become increasingly important in pipe line operations. The composition of natural gas and hydrocarbon streams are different from each production well and assumptions about constant composition cannot be made. In fiscal measurement and where the energy content of the stream is required, a knowledge of density is essential, particularly for gases.

Recently great strides in the development of rugged on-line densitometers have been made with the introduction of the vibrating element transducer. This type of instrument is fast becoming the industry standard for pipe line applications. One such instrument is shown in Figure 7. The operation is very simple, being based on the spring mass principle.

Figure 7. Pipeline densitometer (courtesy Solartron Inst and Control and Instrumentation).

A flat plate, cylinder, or fork is maintained at resonance by an electronic circuit. A fluid flowing over or through the vibrating element adds to the vibrating mass and causes a decrease in natural frequency. The higher the fluid density the lower the frequency and vice versa. By calibration using fluids of precisely known density, the relationship between frequency and density for each element can be determined very accurately. The characteristic equation for most vibrating element densitometers is of the form:

$$\rho = At^2 + Bt + C$$

where A, B, and C are calibration constants, t is the period of oscillation, and ρ the fluid density. Once the constants have been determined by the laboratory tests, the characteristic equation can be programmed into the system computer for on-line calculation of density. In some cases it may be possible to check the performance *in situ* by passing a fluid of known density through the transducer. This will establish if there has been any drift from the original manufacturer's calibration and will help to minimize the uncertainty of the operating system. The latest generation of instruments can approach accuracies of $\pm 0.1\%$ of reading under ideal conditions, and are typically $\pm 0.2\%$ of reading under actual operating conditions.

When used in conjunction with orifice meters, for example, the flow through the densitometer is maintained by taking fluid from a tapping downstream of the orifice and feeding this back at the downstream tapping of the plate. This method, known as pressure recovery, is very popular in pipe line applications, and a typical installation is shown in Figure 8. Other methods include pumping a small percentage of the stream flow through a flowmeter and densitometer before feeding back into the main stream.

Figure 8. Standard installation for pipeline density measurement (courtesy IP manual section 6.2 Nov 1983).

There are other types of fluid densitometer such as nuclear or force balance but these do not find the same widespread use on pipe lines as the vibrating element devices. The choice of transducer will depend on the application since fluid properties, flow rate, sensor element design, and materials of construction all influence the type of densitometer used. For example, if the fluid contains solid particles or is of a "dirty" nature, then straight pipe vibrating cylinder elements are more likely to give better long term performance. The actual elements are manufactured in a wide range of materials. Ni-Span-C is used for high accuracy metering and this material also has a low temperature coefficient. Stainless steel, on the other hand, has better corrosion characteristics and is cheaper to use. Materials such as hastelloy C-276 offers excellent resistance to the most corrosive of fluids.

It is probable that the use of these instruments will increase as general experience on pipe lines has been favorable. They must, however, be operated in single phase fluids as the presence of either droplets in gas lines or vapor bubbles in liquid lines will produce erroneous outputs of unknown magnitudes. The straight tube cylinder-type designs could be used in low second phase concentration applications if the densitometer is installed vertically with upward flow. This will ensure closer approximation to homogeneous flow. Experience has shown, however, that two-phase flows are significantly more troublesome to handle and the output of the transducer should be used with some caution.

Pipe line samplers

Oil lines, particularly crude oil lines, often contain significant amounts of water. This has to be monitored and allowed for when the oil is being metered for custody transfer or taxation purposes. It is a growing and important area of pipe line technology to install on-line samplers for the measurement of water content, and work is also underway to use statistical methods to estimate how many samples are representative of the composition of the flowing stream. There are currently two different basic methods being used on pipe lines to monitor the water content. These are the capacitance method and grab samplers.

The first method is based on the fact that the difference in the relative permittivities of water and oil makes a capacitive transducer sensitive to the water content flowing between two electrodes. In the past three years, commercial instruments using this method have appeared and have attracted a great deal of interest. particularly in North Sea applications where water content could be a problem.

They are basically also non-intrusive sensors with the potential for in-line installation. There are two basic designs, by-pass and main line types. By placing electrodes in a by-pass line from the main flow, a measure of water content in the main stream can be made. The accuracy of these meters, however, is highly dependent on the extent to which the flow in the by-pass loop is representative of the main stream. With the main line type this source of error is reduced but the problem of flow regime effects still remains. Both types are also affected by the excitation frequency, the salinity of the water content, the water droplet size, the thickness of the isolation material between electrode and fluid, and the homogeneity of the electric field.

Field experience has shown that meaningful measurements can only be made if the flow regime is constant and the best results are obtained when the water is evenly dispersed and in droplet form in the liquid. Such criteria often do not exist, with slugs of water passing down the pipe line at irregular intervals. One such commercial instrument is shown in Figure 9. The important feature of the design is the short distance between the electrodes which results in high sensitivity. The electrodes are intrusive and therefore liable to either coating or blockage effects if the oil is particularly dirty.

Grab samplers are devices which take small volumes of fluid from the line at regular intervals to build up a representative sample of the mixture over a period of time. An international standard is currently being reassessed in the light of operating experiences by many oil producers. The current practice states that 10,000 "grabs" of around 1 cc each should be taken and there are guidelines laid down for the sampling frequency and other criteria to be considered. Work at Cranfield and elsewhere has been evaluating this standard and some of the limitations have been identified.

Figure 9. Capacitance type water content transducer (courtesy Endress and Hauser).

Figure 10. Fast loop pipeline sampling system.

As with the capacitance devices above, these may be installed either in the main stream or in a by-pass loop. Often signals from flowmeters in the line are used to set the sampling frequency and the use of jet mixers, bends, pumps, and other devices that produce a more homogeneous stream is desirable to improve performance and accuracy.

Another variation, the fast loop sampler, is used for continuous monitoring of water content and such a system is shown in Figure 10. The main advantage of the fast loop sampler is that flow is continuously extracted and the sampler itself seems to have a smaller perturbing effect on the stream characteristics. A great deal of work is currently underway to examine the effect that the presence of the sampling probe has on the fluid dynamics of the stream. Early work has indicated that the flow streamlines around the probe head do have an influence on the droplet distribution and hence the oil/water ratio. This is developing technology and we should see increased activity in this area as some of the older North Sea fields begin to produce greater concentrations of water.

Pipe line monitoring systems

On-line inspection techniques have recently been developed to a very high level of sophistication. British Gas has spent some 12 years and over £100 M producing intelligent monitoring tools and the data analysis software to inspect line sizes from 8-in. to 42-in. bore. There is a demand worldwide for their expertise and with a potential 1 million miles plus of line to survey, they should be busy for years to come! These inspection tools are commonly called "pigs," but should be distinguished from the pipe line pigs used for cleaning, dewaxing, batching, filling, and de-watering of pipe lines. I feel the term *monitoring tool* should be used because these devices are very advanced indeed.

Two types of monitoring tool have been developed, one based on magnetic sensors to detect metal loss and the second based on ultrasonics to sense defects and cracks in pipe walls. The two used together give a complete picture of the

mechanical state of the pipe line. They are not continuous monitors like pressure, temperature, and flow transducers but are useful in supplementing the on-line measurements. If the lines are inspected at regular intervals, then changes can be identified and possible trouble spots pinpointed. The data can provide a very useful historical record to be used when maintenance, cleaning or line repairs are needed.

The magnetic pipe line inspection vehicle operates by magnetizing a short section of line as it travels down the pipe, driven by the process fluid. If a section of pipe has suffered metal loss it will cause a local distortion of the magnetic field. The data are recorded on 14 channel tape systems and on-board software reduces the amount of data saved to only that where significant signals are noted. The information from the clean or undamaged pipe is not saved. Once the inspection run is complete, the data are analyzed to identify the problem areas. The first offshore run was completed in early 1986 when the Rough and Morecambe gas field lines were surveyed. This run took around 5 hours. The method can be used to survey sections of line around 300 km long although a more typical single pass would be around 80 km (50 miles) (*Monitor*, No. 2, 1986).

The ultrasonic system is virtually identical except for the transducers and detectors. The sensors are ultrasonic operating in the megahertz region. They are housed in wheels which press against the pipe walls. They are pulsed from an on-board electronic control system and the reflected waves are stored in the same way as in the magnetic system. Both inspection vehicles travel at around three meters per second. If a line problem is detected by an imbalance between flowmeters for example, the location could be confirmed by the acoustic detectors on board the inspection vehicle. This could be arranged to alarm when the detector output reaches a maximum and the precise location can be given by radio transmitters mounted on board.

Other problems that can be identified include corrosion, fatigue, ground movement (causing pipe buckling) and third party damage. The inspection center at Cramlington has already undertaken a number of overseas surveys and has experience in gas, crude oil, kerosene and gasoline lines. The impressive part of these pieces of equipment is the engineering. They have to negotiate bends, valves, crossovers, etc. and still protect the relatively fragile electronic recording equipment from damage. They have been used both in on-shore and off-shore lines with equal success. This is one area of pipe line instrumentation development that I will be watching over the next two years as new on-line vehicles are developed for pipe coat-less monitoring.

Computer systems

It is in computer systems that the greatest amount of data can be gathered, analyzed and acted upon in the shortest amount of time. Most operators now have some form of computer-based monitoring system employing either commercially available or custom developed software to run the system. Programs can calculate the inventory of the line at any point in time and compare this with line measurements to provide independent cross checking of pipe line measurement data. These data can be transmitted by telemetry, radio, or telephone links to a central computer which continuously monitors the "health" of the line. The central operations room is now more likely to contain VDUs rather than banks of process indicators and enables the engineers to have rapid and accurate updates of conditions along the entire line or network. By changing programs and subroutines, a large number of tasks and functions can be performed easily, simultaneously and cost effectively.

The many functions that can be performed by computer based systems include:

- Leak detection and location
- Batch tracking of fluids
- Pig tracking
- On-line flow compensation
- Real time pipe line modeling
- Instrument data and malfunction checking
- Inventory and material balance accounting, etc.

Such systems have very rapid response, but are dependent on the accuracy of the input data to be effective. Most computer based systems consist of two major elements, namely:

1. A supervisory computer plus associated software.

2. A number of independent pipe line monitoring stations plus data transmission equipment for communication with the computer.

The complexity of the software, the number of input variables and the pipe line instrumentation selected, all influence the size and the choice of computer. Some systems are capable of running on the compact powerful micros such as the IBM AT or XT, while the more complex pipe line networks will require a main frame machine, possibly communicating with satellite micros installed close to the measurement stations. The overall design time varies with each system but should not be underestimated, especially if SCADA systems are being developed in-house (see section following).

At the most basic level the computer system can take flowmeter inputs to perform gross balances along the pipe line. Such a method is very useful in identifying small but consistent loss of product from corrosion pits in the pipe wall. This will arise if the integrated flowmeter outputs diverge with time when the flow in the line is maintained constant. If the line flow rate varies with time then imbalances are more difficult to detect, since the meters at each end may have different characteristics or their outputs may vary nonlinearly with flow rate. A loss of product will be identified simply as the difference between the steady state inventory of the system and the instantaneous inlet and outlet flows. Mathematically, this is:

$$\Delta V = V_{in} - V_{out} - V_1$$

where ΔV is the leakage volume, V_{in} and V_{out} are the metered inlet and outlet flows, and V_1 is the inventory of the pipe line. This last term can be calculated as the average of the integrated flowmeter signals, but is dependent on variations in temperature, pressure, and many other variables. It is the use of computers that enables this last term to be calculated on-line and as a function of pipe line elevation, pipe material characteristics, process variations, pipe line fitting loss data amongst the many variables. Temperature and pressure variations change both the density of the fluid and the actual dimensions of the line. Both will give rise to sources of errors in pipe line inventory if compensation is

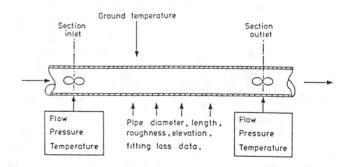

Figure 11. Input data for computer inventory calculation.

Crude/products system

Gas system

Locations of microcomputers*
Compressor station
Pump station
Line valve
Line sensors
Metering/measuring station
Direction of data flow

* A 16 or 32-bit microprocessor supporting a CRT (cathode-ray tube) screen and full-function keyboard.
† Data physically returned to station or communicated via modem.

Figure 12. Examples of modern computer-based-pipeline systems (courtesy Oil and Gas Journal, 1986).

not included in the calculations. Figure 11 shows typical input data requirements for simple on-line inventory calculations.

At the other extreme, computer-based systems can control entire networks with almost no need for the operators to intervene unless serious problems arise. Figure 12, taken from *Oil and Gas Journal*, March 3, 1986, shows two examples from the many hundreds that exist of computer based monitoring. The benefits from the addition of real time pipe line models cannot be overemphasized. Even when operating transients occur, some commercially available programs enable problems to be detected. An example is shown in Figure 13. Here a pipe line leak was induced following pump failure and restart, and although the line was operating under highly transient conditions, a material imbalance was detected. Real time modeling also allows density profiles to be calculated along the line. It is

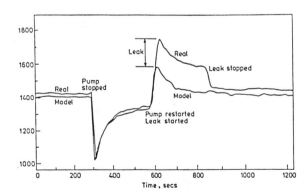

Figure 13. Example of transient real time modelling of pipelines (data from Ref 9).

fairly common practice to have amounts of different fluids in the same pipe line separated by a pig or sphere. From the pressure drop/flow characteristics, the real time can track the progress of these batches of fluids along the pipe.

SCADA systems

Real time computer modeling of pipe lines is a technique that uses the full data gathering capabilities of modern digital systems and the computational power of computers to give accurate "snapshots" of the line. These enable the process operator to act quickly when problems develop. The whole system is under the control of the SCADA (Supervisory Control And Data Acquisition) system. This suite of programs polls the data stations on the line in turn, processes the data, controls the running of any pipe line model, and activates alarm and other monitoring routines. Most modern SCADA systems now include checking routines to determine the validity of the data before any calculations are made. It has been found that the use of continuous quality and cross checking of data improves the sensitivity of the system. In addition to these basic functions, some of the more advanced SCADA systems might include predictive models to analyze "what if" operating scenarios, include automatic startup and shutdown procedures, or provide optimization routines for least-cost operating strategies.

There are a number of companies supplying standard SCADA packages, but the majority of systems use custom written master routines to enable the commercially available programs to be used. Each pipe line is different, in line characteristics, instrumentation installed, fluid properties, and need of the user to name but four requirements. Experience has shown that the adaptation of standard software modules to specific needs takes much longer than is first estimated and also gives rise to the greatest number of problems when pipe line monitoring and control systems are being commissioned.

It is very important for the software specialist to communicate directly with the instrument engineer who specified the hardware, to ensure proper integration of the various parts of the system. All too frequently these two key members of the design team do not meet until the project is well underway.

Cathodic protection

Protecting a pipe line from corrosion is an important and expensive part of building and running a pipe line system safely. Every year, operators spend millions of lb on pipe coatings, gas detectors, and sacrificial anodes to lengthen the effective life of pipe lines by reducing the effects of corrosion. Whenever a steel structure of any type comes into contact with the soil and/or water there is almost certainly going to be corrosion. The structure, in our case the pipe line, divides itself into anodes and cathodes with corrosion taking place at the anode. This chemical process is accompanied by the flow of electric current from one point on the line to another via the ground or water. There are two basic preventative methods, sacrificial anodes and impressed

currents (more commonly called cathodic protection). This second method is the more common and preferred method for long pipe lines to identify the problem spots.

If a voltage is applied to a section of line, a profile of the current flow can be drawn up by measuring the current flow at intermediate points. This gradient will show areas where current is escaping, indicting a pipe defect. Recently, techniques using AC current have been perfected where the magnitude of the current can be sensed inductively from above ground.

This method can be used by pipe line crews walking the line and compact lightweight handheld units store data on the mechanical condition of the line in lengths as short as five meters. On long cross-country lines the profile can be plotted at intervals of a few kilometers and from historical cross-comparisons, line changes can be identified.

The object of cathodic protection is to prevent current flow from the pipe. The operating principle is to apply a voltage in the opposite direction to the natural potential of the pipe. The pipe line voltage is made negative with respect to earth so that in theory the voltage and current flowing from the cathodic protection system balance the corrosion current that would otherwise flow. Recent developments include the use of solar panels and converters so that energy from the sun is used to run the system. One of the largest examples of this new technique is the Sarir to Tobruk in Libya where a 32-in. diameter, 319 mile pipe line is protected by such an installation.

Remote station monitoring is easily accomplished with the data being telemetered back to central control rooms many miles from the location of the actual installation. Continuous automatic control is often performed by means of thyristors. Such continuous monitoring is now routine on new lines that are being constructed and many older lines are being fitted with some form of cathodic protection system. In most cases it is probably more important for the older lines to be surveyed than the newer ones, especially where different fluids have been transported over the lifetime of the line.

System design guidelines

The aims and requirements of a project should be specified as fully as possible as early as possible when designing a pipe line monitoring system. Based on personal experience, the following points should be borne in mind. They are not in any special order but do form a useful checklist when starting system development.

1. Make the system as simple as is practical but do not underspecify the needs for the sake of simplicity.
2. Select instrumentation and pipe line equipment based on performance and not economic grounds. It

is better to install fewer high quality pieces of equipment than numerous poor ones.

3. Equipment compatibility is important. Use transducers, interface modules and other pieces or hardware that have standard communications protocol. Custom designed systems invariably put up development costs and do lead to endless sources of trouble and frustration.

4. When calibration and maintenance is due on the system do not put off. Wherever possible use computer software to cross check data to aid in the identification of problems. This will help to reduce operating and maintenance costs.

5. Where system availability is important install devices that are self checking, self diagnosing or dual systems where appropriate. The initial cost may be higher but the benefits may be more economic in the longer term.

6. Seek independent references, user experience or validation of the equipment chosen. Most pieces of hardware perform differently in real applications to the specification published under ideal conditions.

Future trends

Future trends are in one respect easy to assess but it is very difficult to estimate where we will be in ten years time. There is no doubt that the application of computer based technology is going to precede at an ever increasing speed to both the instrumentation and pipe line monitoring systems. There will be reliable and rugged intelligent flow, pressure, valve and other hardware components appearing and once the initial over the application has occurred, the second generation systems will give the desired performance, if previous history is anything to go on.

By far the biggest problem facing the whole spectrum of control and instrumentation is to develop standard protocols to facilitate communication between instruments. The situation at present means that most if not all monitoring computers have to have a multitude of interface boxes scattered through the system to enable some elements of the

system to "talk" to others. It is sincerely hoped that manufacturers will develop microcomputers that are able to communicate directly with each other, with full scale industrial process monitoring and control systems and with the individual process sensors and associated controllers. The more elements there are in the system, the greater the potential of line downtime. Some moves are being made in this area but it could be some time before we see commonality of process protocol.

The most rapid area of advancement has been in the microchip. Oil pipe line operators are among the many people who have been watching the development of 32-bit ma-

chines. The Intel 80386 chip for example, now contains over 275,000 transistors in one piece of silicon. This device has an operating speed of 4 million instructions per second and has the capability to address over 4 gigabytes of main memory. One other manufacturer is working on a computer which will perform 10 billion calculations per second, more than 10 times faster than any other computer on the current market. Such machines will permit easier real time modeling of pipe lines and will present the line operator with many extra pieces of data on the condition of his line. Of course complete new suites of software will be required for these fifth generation machines.

Conclusion

In such a short space, the paper cannot do justice or even mention the vast number of recent instrumentation developments that have had an impact on pipe line operation and control. The technology of measurement and control is proceeding at an ever increasing pace and it is now quite difficult to keep up with developments in one area. Pipe line instrumentation covers a very wide spectrum from pressure and surge control to intelligent instrumentation and advanced software. Certainly our "friend" the microchip has totally transformed our thinking in the past decade and the poor pipe line operator now needs specialized help in specifying a system to meet his needs. There are many options open, control using PLCs at sources, control using software in the process computer, and manual override if required. With the development of expert systems, systems can now be programmed to almost preempt problems before they arise and take appropriate corrective action. The flexibility in the industry is vast but one thing is sure, you cannot control unless you measure accurately, and all too frequently the primary elements of a system are under specified compared to the rest of the system. It is of very little use, for example, to pressure and temperature compensate transducer outputs when the basic measurement uncertainty is poor.

Source

Measurement + Control, Vol. 20, February 1987.

References

1. Adams, R. B., "Buckeye modernizes computer operations and SCADA systems," *Oil and Gas Journal*, 11, 55–59, August 1986.
2. Anderson, C. C., "The Ekofisk pipe line control system," *Pipes and Pipelines International*, 30, No. 3, 21–25, 1985.

3. Baker, R. C. and Hayes, E. R., "Multiphase measurement problems and techniques for crude oil production systems," *Petroleum Review*, 18–22, November 1985.

4. Chester, J. R. and Phillips, R. D., "The best pressure relief valves: simple or intelligent," *InTech*, 51–55, January 1985.

5. Covington, M. T., "Transient models permit quick leak identification," *Int Pipeline Industry*, August 1979.

6. Davenport, S. T., "Forecast of major international pipe line projects for the 1986–2000 period," *Pipes and Pipelines International*, Vol. 31, No. 1, 9–14, 1986.

7. Furness, R. A., "Modern pipe line monitoring techniques—Parts 1 and 2," *Pipes and Pipelines International*, Vol. 30, No. 3, 7–11, and Vol. 30, No. 5, 14–18, May and September 1985.

8. Kershaw, C., "Pig tracking and location—Parts 1 and 2," *Pipes and Pipelines International*, Vol. 31, No. 2, 9–15 and Vol. 31, No. 3, 13–16, 1986.

9. Stewart, T. L., "Operating experience using a computer model for pipe line leak detector," *Journal of Pipelines*, Vol. 3, 233–237, 1983.

10. Taffe, P., "Intelligent pigs crack pipe corrosion problems," *Processing*, 19–20, September 1986.

11. Tinham, B., "Micros and sensors make smart transmitters," *Control and Instrumentation*, 55–57, September 1986.

12. Williams, R. I., "Fundamentals of pipe line instrumentation, automation and supervisory control," *ISA Transactions*, Vol. 21, No. 1, 45–54, 1982.

CHOOSING THE RIGHT TECHNOLOGY FOR INTEGRATED SCADA COMMUNICATIONS
Service was consolidated for both current and projected computer environments

Vernon J. Sterba, Director, System and Telecommunications, Enron Gas Pipeline Group, Group Technical Services, Houston

Consolidation of overlapping telecommunication systems for Enron subsidiaries Florida Gas Transmission, Houston Pipeline, Northern Natural Gas and Transwestern Pipeline eliminated redundant facilities, and integrated information and communication required for upcoming open transportation services.

A wide-area communications study group (WAC) was formed to define and document operations/SCADA communications required for both current and projected computer environments. A functional and technical design would be developed and documented, as well as supporting costs, an implementation timetable, and equipment specifications.

The WAC project was to meet the following requirements:

- Pipe line monitoring and control (SCADA)
- Electronic flow measurement/custody transfer (EFM)
- Real-time pipe line modeling and optimization
- Compressor automation.

Since installing an advanced, integrated communication system required a considerable capital investment, the WAC team chose not to take a "SCADA-only" view, since this could miss opportunities to effectively develop the existing communications network.

Mainframe data and voice communications were considered for consolidation. This later became a key factor in technical decision making. The project emphasis was to identify where a cost-effective solution could simultaneously meet SCADA and other voice and data requirements.

The WAC team consisted of Enron members representing both organizational and technical experience. In addition, a telecommunications specialist served as project leader and expediter for formulating project methodology and imparting a "third party" perspective to team discussions.

WAC Methodology

The main sequence of tasks was as follows:

- Define existing communication systems
- Define communication requirements
- Analyze adequacy of existing systems
- Identify and evaluate technology alternatives
- Establish communication "direction"
- Develop functional and technical design
- Develop project implementation schedule.

This approach allowed the team to fulfill project objectives in a scheduled time frame. Next, an inventory of communication circuits and facilities was entered into a PC database. The database included the latitude and longitude of all facilities, type and name of facility, communication media, circuit number (where applicable), and monthly cost of leased communication.

The database was used to develop communication system maps by interfacing with a computer-automated design package. The maps identified redundant communication paths, duplication of facilities, high cost communication routes, and other details. Without the maps' visual assistance, it is unlikely the team could have quickly developed an understanding of the extensive and complex communication system.

Existing communication facilities included the following technology combinations:

- Multi-drop and point-to-point leased lines. These lines represented a major cost component; many redundant circuits were identified and some surprising cost anomalies were discovered. For example, the cost of a 40-mi hop from an RTU to the nearest microwave tower could cost more than a 300-mi hop from the RTU to Houston.
- Company-owned 2/6-GHz microwave system. The system provides voice and data communications in Texas inland and Gulf Coast areas, and incorporates voice circuits for toll free calling at 30 field offices along the microwave system.
- Radio telemetry. This system was used by Northern Natural Gas and Transwestern for gathering SCADA data at master control units (MTUs) or central data concentration points in major supply basins. The MTUs, in turn, relayed data to the SCADA master via leased phone lines.
- Meteor burst technology. The system "bounces" radio waves off the ionized trails of micro-meteorites entering the earth's atmosphere. Northern Natural Gas used this technology, as a prototype, for nine RTU sites.
- VSAT technology. The VSATs (very small aperture terminals) use satellites in orbit for relaying radio signals. Dish antennas of 1.8 to 2.4-meter diameter (6 to 8 ft) were required. In 1986, Transwestern Pipeline installed 11 satellite terminals to eliminate high construction costs involved with leased line service at remote RTU sites.

Analysis of Technology

Sending data to the SCADA master required both short and long-haul communication segments. The segments could be established by separate and independent technologies.

For example, to aggregate data from several RTUs at a central hub, use radio telemetry or local leased phone lines, the long-haul technology (central hub to SCADA master) could use either leased phone line, microwave, or VSAT. This requires a serial byte-oriented, SCADA protocol standard, which had been in place since 1987.

Table 1. Communication guidelines

Short-haul choices
1. Use radio, if available, in a radio hub arrangement.
2. Use leased lines if economically available, connected to a hub or the home office site.
3. Use VSAT only at remote sites when the above options are infeasible.

Long-haul choices
1. Use existing microwave, when available.
2. Use VSAT for sites with multiple applications.
3. Use single-application leased lines when the monthly cost is less than $450; otherwise, use VSAT.

To determine the best communication direction, various technology "mixes" were applied system-wide. This provided comparative technology costs for the entire communication system.

Amortized capital cost and O&M cost, including leased lines, were estimated for each design case. The cost analysis was conducted for both current and projected data density. The technology mixes considered were:

Case	Short-haul	Long-haul
1	radio	leased line
2	radio	microwave
3	radio	VSAT
4	leased line	leased line
5	leased line	microwave
6	leased line	VSAT
7	VSAT	VSAT

In both current and future data density environments, the combination of radio telemetry and long-haul leased lines (Case no. 1) yielded the minimum monthly cost. This was followed closely by radio telemetry and VSAT (Case no. 3). When VSAT was used also as a mainframe connection, Case no. 3 became the minimum cost option.

In all cases, radio telemetry was the most economical technology for the short-haul communication segment. Radio systems bypass local telephone-company leased lines to eliminate escalating costs and service problems found in remote areas. Radio is not always available, due to frequency availability.

For existing facilities, the microwave network was a clear choice for long-haul systems, with no capital investment required and a small incremental cost for adding circuits. However, when considering microwave expansion, the situation changed significantly. In the current data density environment, increased microwave coverage (Cases no. 2 and 5), did not justify capital costs involved, due to insufficient data traffic.

In the projected environment, higher data densities began to support the capital and O&M costs as microwave system capacity was fully utilized. This moved microwave technology from seventh to third most favorable system, compared to other technologies.

Generalizations shown in Table 1 apply to one system's economics and topography, and are not intended to suggest the same options should be chosen by other gas pipe line companies.

For example, the short-haul radio option was not viable in some cases. In one case studied, the cost of a high radio tower exceeded the cost of a VSAT system. Frequency availability was a problem as well.

In addition, the long-haul segment was affected by site-specific factors such as data density, leased-line installation and operating costs, and VSAT line-of-sight availability. VSATs require a clear view to the south.

The technical evaluation recommended combining short-haul radio telemetry with existing microwave or VSAT for the long-haul segment. Economics were especially favorable when the VSAT could provide mainframe connections, a situation already found at a majority of sites.

Another decision was whether to recommend C-band or Ku-band VSAT technology. The company was using C-band VSATs in the Transwestern Pipeline, but at the time had no experience with Ku-band VSATs.

C-band VSAT Advantages

- More cost effective for low data rates; C-band hardware is generally 25–50% less expensive.
- Capabilities meet existing SCADA transmission speed requirements.
- C-band SCADA communication services were available and had been used since 1986, a "known-quantity" advantage.

C-band VSAT Disadvantages

- Not amenable to moderate or high-speed data transmission rates, voice or video.
- Shares the radio wave spectrum with terrestrial microwave services, increasing the possibility of interference, a rare occurrence in rural areas, but a major concern in metropolitan areas.
- A single vendor was the main provider of VSAT transponder capacity.
- Existing C-band satellites have low utilization.
- C-band is older technology. The expected life span of a satellite, both C and Ku-band, is about seven years, usually set by fuel limitations for keeping the satellite on-station in geostationary orbit.
- Projected shortages of C-band satellite capacity were expected beyond 1991, as existing satellites reach life-span limits. However, the projected shortage did not develop.
- When conducting the WAC project, there were no launch plans for replacement C-band satellites.

Ku-band Advantages

- Growth trend in satellite communications is toward Ku-band, with future satellite launches planned.
- More suppliers are expected in the Ku-band marketplace.
- Ku-band VSATs can be used for high-speed applications such as voice, video, computer-to-computer data communications, and SCADA and mainframe workstations. Local networks can be extended into wide-area networks (WANs).
- Although the value of satellite voice communications is limited by the inherent 0.5-second delay, receive-only video and data communications will become more valuable in the future, if costs decrease.

- Television receive-only (TVRO) can be used to communicate with a dispersed workforce and provides cost-effective training when compared to increasing airfare costs.
- Computer-to-computer data communications will become necessary components as microcomputer installations increase at field locations.

Ku-band Disadvantages

- Cost usually exceeds C-band cost for one-application installations
- Can be affected by heavy rainfall, such as rainfall rates sufficient to restrict driving. This occurs because water droplets absorb the satellite frequency; however, this can be mitigated by proper antenna design and other equipment. Most users report actual outage rates less than statistically predicted.

VSAT Decisions

Because a distributed mainframe application was planned, the Ku-band scenario was cost-effective at a majority of sites. A secondary decision was made not to add more C-band VSAT sites. Also, because of the economics mentioned earlier, radio became the first choice for the short-haul option. The combined radio/VSAT or radio/microwave became the "radio hub."

Operating costs, for a Ku-band VSAT at each SCADA site, was $375 to $450 per month. The exact cost varied depending on vendor, data traffic costs, and capital amortization period. This estimated cost agreed with other published analyses.[1,2,3]

One disadvantage of VSAT technology was the delay introduced by equipment and related store-and-forward message handling. With one vendor, this delay ran as long as 12 seconds, reducing the effective circuit speed.

Do's and don'ts of VSAT technology installation:

- Do install bumper guards around VSATS in compressor station yards.
- Do install a security fence around VSATs in or near public areas and be prepared to install camouflage, such as shrubbery, to hide the VSAT from public view.
- Do allow extra time for installing VSATs mounted on building roofs, especially if the building is not company-owned.
- Do use dish heaters where ice storms are common
- Do install antennas high enough to avoid snow drifts in northern latitudes.
- Don't install dish heaters in snowy regions where ice storms are frequent.
- Don't install VSAT antennas just outside office buildings; this may obstruct view or reflect sunlight into windows.

Because of inherent VSAT delays, special features were mitigated into the SCADA system. A report-by-exception feature was implemented at data concentration (MTU) sites. Delay was reduced also by using single-hop rather than double-hop VSAT technology.

In a single-hop system (Fig. 1), the message path is:
• Outbound poll—SCADA master to vendor master earth station (MES) via leased lines
• Inbound response—VSAT at RTU to MES via space segment; MES to SCADA master via leased lines.

The double-hop system (Fig. 2), replaces the SCADA master to MES leased-line link with another pair of space-segment hops, increasing time delay, but still cost-effective for smaller systems.

A dial-up backup to key radio hubs and other critical VSAT sites was chosen. The SCADA system supports this failure mode by automatically dialing up failed VSAT sites.

Figure 1. Single-hop VSAT.

Figure 2. Double-hop VSAT.

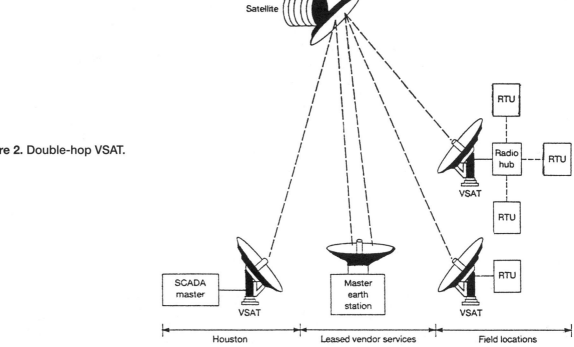

This is cost-effective, since the business line's monthly cost is low relative to other options. Dial-up online costs are high, but rarely needed.

Implementation

Because of very close bids and a WAC team recommendation to use multiple VSAT vendors, two vendors were awarded contracts to install Ku-band VSATs across the nation-wide 38,000-mi pipe line grid. Initial construction was completed in 1989.

Enron currently has 125 VSAT sites throughout the U.S. and has changed over to primarily Ku-band technology. Not all sites use the same VSAT options, as shown below:

Ku-band application area	Number of sites
SCADA	104
mainframe	60
TVRO	63
Other (ASCII/X.25)	14
Total number of sites	125

The television receive-only (TVRO) application currently is being installed. The X.25 application permits linkup between field and home office systems, including local area network, electronic mail, and remote access to centralized DEC VAX applications without any dial-up cost. Since many sites are in remote areas, the dial-up costs are often dwarfed by the costs and frustrations of poor quality telephone connections.

Acknowledgment

Based on a paper, "VSAT Communications Technology for SCADA System Communications," presented by the author at the AGA Distribution Conference, Nashville, Tenn., April 1991. The author received the Acker Medal Award 1991–92 for this technical paper, sponsored by Central Hudson Gas and Electric Corp., in recognition of excellent contribution to the gas industry.

Literature Cited

1. Sharma, R., "VSAT Network Economics: A Comparative Analysis," IEEE Communications, Feb. 1989.
2. Simo, E., "VSATs and the Pipeline Industry," Satellite Communications, Aug. 1988.
3. Coulter, J., "Design and Installation of a Gas Control SCADA Network using Satellite, Microwave and Radio Systems," Entelec News, Vol. 61, no. 2, Aug. 1988.

17: Leak Detection

Pipe Line Leak Detection Techniques 476
Causes and economic aspects of leaks................................ 476
Simple leak detection systems 477
Pig-based monitoring systems 478
Computer-based monitoring systems 478
Pipe line leak phenomena 479
Background philosophy of pipe line modeling.................... 479
Basic pipe line modeling equations 480
Impact of instrument accuracy.. 481
System design aspects and guidelines 482
Development of pipe line monitoring systems 483

Pipe line leak detection techniques

R. A. Furness, Cranfield Institute of Technology, Cranfield, Bedford, UK, and **J. D. van Reet,** Scientific Software Intercomp, Houston, Texas, USA

Summary

Pipe lines are now an integral part of the world's economic structure and literally billions of dollars worth of products are now moved annually in pipe lines. Both economic and environmental factors are influential in pipe line operation, and therefore integrity monitoring is vitally important in the control and operation of complex systems.

Leak detection systems range from simple, visual line walking and checking to complex arrangements of hardware and software. No one method is universally applicable and operating requirements dictate which method is the most cost effective. The aim of the paper is to review the basic techniques of leak detection that are currently in use. The advantages and disadvantages of each method are discussed and some indications of applicability are outlined.

Modern pipe line computer modeling and control is then reviewed in detail. These systems are the most flexible and versatile and are steadily being adopted. The influence of instrument accuracy on system design and performance is discussed and the basic modeling equations are reviewed.

Introduction

Our need to transport fluids from the point of production to the area of end use has led to a rapid increase in the number of pipe lines being designed and constructed. Many of these carry toxic or hazardous products, often close to centers of high population or through areas of high environmental sensitivity. With the need to safeguard these lines, on-line monitoring is becoming routine and in some cases 24 hour surveillance is mandatory. With the increase in world terrorism, the need for rapid and reliable pipe line measurement and control systems will increase.

The review begins with a summary of causes of leaks and the implications of the failure to detect them. Basic techniques are covered and the features of each are briefly discussed. The bulk of the paper deals with modern computer-based techniques. Basic flow equations are covered and the on-line dynamic calculations required are listed together with the input data required to enable the monitoring to be accomplished with the minimum of downtime. The latest systems are capable of resolving down to 1% of the maximum rated flow with a response time of a few minutes. Practical experience verifies this figure but the total costs of such a system could be high. The system therefore becomes a

compromise between response, performance, and alarm availability, and choice of instrumentation is crucial. The integration of good quality instruments with advanced real time models seems to be the current trend and the paper closes with some personal thoughts on future trends.

Causes and economic aspects of leaks

There are four main categories of pipe line failures. These are:

- Pipe line corrosion and wear
- Operation outside design limits
- Unintentional third party damage
- Intentional damage

Many pipe lines are operated for a number of years with no regard to any possible mechanical changes occurring in the line. Some of the products may be corrosive, the pipe line may be left partially full for periods of time, or atmospheric effects may cause external damage. These three reasons are responsible for pipe line corrosion and this may give rise to corrosion "pits" developing along the line. These are small in nature and could be responsible for material imbalances over a period of time. Very accurate flowmetering can be used to detect this as discussed in the next section. Abrasive fluids or dust-laden gas streams can give rise to pipe line wear. Again, this is a slow process, but should a weak spot develop (more often than not close to a change in direction or section) then a pipe break may occur very rapidly and totally unexpectedly.

Operation outside design guidelines is more common than is realized, as operators seek to use the line for as many fluids as possible. If the line is designed for a certain maximum temperature and pressure, then operation at higher pressure and/or higher temperature could lead to spontaneous failure. The problem could be compounded if the line has a large but unknown amount of corrosion.

Unintentional third party damage may occur if excavation or building occurs near buried lines. More often than not the right-of-ways are not clearly marked and lines are sometimes broken by bulldozers or similar plant machinery possibly with fatal results. An example occurred in West Virginia in 1984 when a 20-in. natural gas line was punctured by an excavator and product leaked slowly into a nearby supermarket during the night. The building was totally destroyed early next morning by a massive explosion caused by a staff member lighting a cigarette. Such an occurrence could have been avoided by some form of leak detection on the line.

Intentional damage unfortunately is on the increase and pipe lines carrying flammable or high value products make ideal targets. Alarm systems linked to block valves can help to minimize the amount of product released as a result of sabotage, so again certain lines are instrumented with the intention of reducing the effects of planned terrorism.

The costs of failure to detect leaks also fall into four main areas:

- Loss of life and property
- Direct cost of lost product and line downtime
- Environmental cleanup costs
- Possible fines and legal suits

These are all self explanatory, with the most costly of these being the last, although any of the four areas could be very expensive. The size of claims can run into many millions of dollars, so the cost of fitting and operating leak detection systems is often insignificant compared with the costs of failure of the line.

It is this background that is causing operators and designers to turn to on-line integrity monitoring systems and the next section of the paper looks at the more basic methods.

Simple leak detection systems

The most basic method of leak detection involves either walking, driving, or flying the pipe line right-of-way to look for evidence of discoloration of vegetation near the line or actually hear or see the leak. Often "unofficial" pipe line monitoring is performed by people living nearby who can inform the operator of a problem with the line.

The most cost effective way to detect leaks in non-flammable products is to simply add an odorant to the fluid. This requires some care in selection, as frequently the odorant has to be removed before the transported fluid can be used. Organic compounds make the most useful odorizers, especially when the fluid being carried has no natural smell of its own. A good example is carbon monoxide, a highly toxic but odorless gas which is often pumped in large quantities between chemical plants. Chemicals such as mercaptans (rotten egg smell) or Trimethylamine (rotten fish smell) can be added in small quantities to enable any leak to be located by smell. The disadvantage of such a method is that if the leak occurs in an area of no population the leak will go undetected unless the line is walked regularly by pipe line surveillance crews carrying suitable "sniffer" detectors. Thus to the apparent low costs of this method have to be added the costs of removing the odorant and maintaining staff to check the line

at frequent intervals. The location of a leak is also dependent on prevailing weather conditions. Strong winds may disperse the smell and atmospheric inversions may give an incorrect location of the leak and the uncertainty of relying on this method alone is high. Nevertheless, it is a useful method if used in conjunction with other techniques.

Simple line flow balances are frequently used to check for gross imbalances over hourly or daily bases. This method may identify that a leak is present but flowmeters at each end of the line will not identify the leak location. A line pressure measurement system will be required in conjunction with the flowmeters to establish that the pressure gradient has changed from the no-leak situation. The method is useful, however, in identifying the existence of corrosion pits as the outputs of the flowmeters at each end of the line will consistently diverge if flow in the line is maintained constant. If line flow rate varies with time, then imbalances are more difficult to detect, since the flowmeter outputs may vary nonlinearly with flow rate or may have different flow characteristics from each other.

A loss of product will be identified simply as the difference between the steady state inventory of the system and the instantaneous inlet and outlet flows. Mathematically this is:

$$\Delta V = V_{in} - V_{out} - V_1$$

where ΔV = leakage volume
   V_{in} = metered inlet flow
   V_{out} = metered outlet flow
   V_1 = pipe line fluid inventory

This last term can be calculated as the average of the integrated inlet and outlet flows in simple systems, but as will be seen later in this paper, the value of this term can be calculated more accurately and easily in real time as a function of several variables.

Another method is based on detecting the noise associated with or generated by a leak. There are many instances where fluid flow can generate vibrations at frequencies in excess of 20 kHz. These frequencies are in the ultrasonic range but can be detected with suitable transducers. The device can be made portable so that pipe line crews can clamp a transducer at any point along the line to check for noise. By noting the signal strength, the source of the leak can be pinpointed.

A similar technique, though based on a different principle, is the acoustic "wavealert" monitor, more correctly called a negative pressure wave detector. This is a piezoelectric sensor that gives an output when dynamically stressed. When a leak occurs there is a sudden drop in pressure at the leak followed by rapid line repressurization a few milliseconds later. The low pressure wave moves away from the leak in both directions at the speed of sound. The pipe walls act as a waveguide so that this rarefaction wave can travel for great distances, attenuating in amplitude as it does so. Sensors placed at distances along the line can be triggered as the wave passes and the location of the leak can be calculated from the line conditions and the internal timing devices in the instrument.

Such devices are particularly useful in identifying large breaks in lines very rapidly, since the transient wave typically moves 1 mile in 5 seconds in gases and almost 1 mile per second in liquids. Response is therefore on the order of a few seconds depending on the positioning of the transducers. Figure 1 shows one adaptation where the instrument can be made to cancel line noise and use the full measuring capability of the sensor for signal detection. There is the problem setting the background threshold correctly, as this may be affected by the location of the instrument in relation to bends, valves, pumps, regulators, etc. Experience has shown that the installation is also critical to reliable performance, and there is also a dependence on the Reynolds number of the flow. Both of these affect the number of "false alarms" from the instrument.

Figure 1. Negative pressure wave detector (acoustic monitor).

Pig based monitoring systems

Pipe line pigs are frequently used for pipe line commissioning, cleaning, filling, dewaxing, batching, and more recently pipe line monitoring. This last type of pig can be designed to carry a wide range of surveillance and monitoring equipment and can be used at regular intervals to check internal conditions rather than continuously monitoring the line. Data, however, can be built up over a period of time to provide a history of the line. This information can be used to predict or estimate when maintenance, line cleaning, or repairs are required. If a leak is detected, for example, by flow meter imbalance, the location can be

found by using a pig with acoustic equipment on board. This will alarm when the detection equipment output reaches a maximum and the precise location of the pig can be confirmed by radio transmitters also mounted on board.

Pigs require tracking because they may become stuck, at a point of debris build-up, for example. Pigging should be carried out at a steady speed, but occasionally the pig may stop and start, particularly in smaller lines. Information on when and where the pig stops is therefore important in interpreting the inspection records. Pig tracking is not new and many such proprietary systems exist. In the best systems, however, a picture of the line is often programmed in so that outputs from junctions, valves, cross-overs, and other geometries act as an aid to location. Pig tracking can make use of the acoustic methods discussed earlier. When the sealing cups at the front of the pig encounter a weld, vibrational or acoustic signals are generated. Each pipe line therefore has its characteristic sound pattern. When a crack occurs this pattern changes from the no-leak case and the location can be found from direct comparison. The technology has become so advanced that information on dents, buckles, ovality, weld penetration, expansion, and pipe line footage can be generated.

The equipment is often simple, consisting of sensor, conditioning, and amplifier circuits and suitable output and recording devices. Such a device developed by British Gas is shown in Figure 2. The range of detection is dependent on the pipe line diameter and the type of pig. Operational data have shown that light pigs in a 200 mm line can be detected at a range of 8 km, increasing to 80 km for a heavy pig in a 900 mm line. As the signals travel at acoustic velocity this means a signal from a pig at 80 km range will take 190 seconds to be picked up. Such technology is now becoming routine in both offshore gas and onshore liquid lines.

Figure 2. "Intelligent" pipe line monitoring pig. Courtesy British Gas Corp.

Computer-based monitoring systems

It is in computer-based systems that the greatest amount of data can be gathered, processed, analyzed, and acted upon in the shortest period of time. Programs can calculate the inventory of the line at any time and compare this with accurate measurements at any section in the system. The effects of pressure and temperature on line dimensions, for example, can be calculated to provide an accurate estimate of the mass of fluid in the line. Data from a wide range of instruments can be transmitted by telemetry, radio, or phone links to a central computer which monitors the "health" of the line continuously. By changing programs and subroutines, a vast amount of functions and tasks can be accomplished very easily and cost effectively.

The many functions that can be performed by computer-based systems include not only leak detection but also:

- Pig tracking
- Batch tracking of fluids
- Inventory accounting
- On-line flow compensation
- Instrument data and malfunction checking, etc.

Such systems have very rapid response and have the advantage of multiple inputs being required before leaks are declared. Thus some systems can run for short outage periods with no loss of integrity. They are often complex and costly to install but once the initial capital investment has been made running costs are low. The first section in the detailed review of such systems looks at the phenomena that need to be modeled when a leak occurs.

Pipe line leak phenomena

When a leak occurs in a pipe line the measured pressure downstream of the leak falls but the pressure at the same location is predicted to rise. The first is not difficult to understand as the line is depressurizing as mass leaves through the leak. The second effect can be explained as follows. The equations presented later in the paper predict pressure based on measured flow or flow based on measured pressure. As mass leaves the system through the leak hole, a reduced flow at the downstream end is compared to the inlet flow. This may not have changed and so to balance the system, the equations predict a downstream pressure rise. In physical terms the model thinks the line is "packing" and total system inventory is increasing. There is therefore a divergence between measured and modeled pressure.

The same is true of flow changes, but here the inlet flow could increase due to lower pipe flow resistance between meter and leak while the section outlet flow will fall as mass leaves through the leak instead of passing through the meter. Thus a real imbalance will result. The model however will show an inconsistency since the pressure comparison will indicate line packing and the flow comparison a line unpacking. If selected pressure and flow imbalance limits are exceeded then a leak is declared. The magnitude of the leak is predicted from the flow imbalance and the location from the pressure profile imbalance and the flow leak indicators. The impact of instrument accuracy on the predicted location is discussed later in the paper. It is vitally important to good leak sizing and location to have the best pipe line instrumentation possible to minimize uncertainty.

Background philosophy of pipe line modeling

Real time modeling is a technique that uses the full data gathering capabilities of modern digital systems and the computational power of small computers to give accurate "snapshots" of the pipe line. The whole system is under the control of a SCADA package of programs, which poll the data stations on the line, process the data, control the running of the transient pipe line model and activate the alarm and leak location routines. In addition to these basic software modules, more complex systems might include a predictive model to analyze "what if" operating scenarios, provide an optimization routine for least-cost operating strategies or include a separate man/machine interface for the model system.

The SCADA interface is responsible for acquiring the data from the SCADA system and relating them to the model representation of the line. As a point in the system where two large and independently developed systems join, this can be the source of many problems in the implementation of the real time modeling. Once the measurement data have been obtained, noise filtering and plausibility checking can be performed prior to running of the model. The model is the mathematical representation of the pipe line and will include such features as elevation data, diameters, valve and pump locations, changes of direction and the location or cross-overs and junctions. The model provides data on the flow conditions within the line at intervals between seconds and minutes, depending on operational needs.

With the data available from both the measurement system and the pipe line model, the real time applications modules are run. These are the leak detection and location routines in the context of integrity monitoring. The leak detection module functions by computing the difference between the modeled flows and pressures and the measured values at all points where measurements not already used as boundary conditions are located. Because the model accounts for normal transient operations, these differences will be small under normal conditions. When a leak is present, the differences become larger since the model system does not account for leakage. When these differences exceed preselected values, a leak alarm is declared. Sophisticated voting schemes which require multiple leak indicators to be in alarm for several consecutive time intervals are used to reduce false alarms while maintaining low thresholds. Often a simple pipe line balance of the type discussed earlier is used as a back-up to verify the transient model. Response characteristics are, however, much slower than the real time model.

Once the leak detection module declares a leak, the location routine is activated. The location is calculated from the magnitude and distribution of the leak indicators. As an example, in a straight pipe line with an upstream flow discrepancy and a downstream pressure discrepancy as leak indicators, it is an easy calculation to determine where the leak must be such that the leak flow when added to the modeled flow will produce the additional pressure drop observed at the downstream end. Solutions for pipe networks are more complicated and unique locations do not always exist. This might be the case with parallel looped lines, for example. In this case all the calculated locations should be checked.

The components of the real time modeling system work together to reduce the large volumes of raw data from the data acquisition system to a much smaller number of parameters and alarms that are more meaningful to pipe line operations. In the case of integrity monitoring, this means leak events that could not be detected by inspection of the measured data can be found and isolated quickly and reliably.

Basic pipe line modeling equations

The transient pipe line flow model is the heart of a pipe line modeling system. The model computes the state of the pipe line at each time interval for which data are available. The state of the pipe line is defined as a set of pressures, temperatures, flows, and densities that describe the fluids being transported at all points within the system. These quantities are found as the solution to a set of equations which describe the behavior of the pipe line system. These basic equations are the Continuity equation, the Momentum equation, the Energy equation, and an equation of state.

The continuity equation enforces the conservation of mass principle. Simply stated, it requires that the difference in mass flow into and out of a section of pipe line is equal to the rate of change of mass within the section. This can be expressed mathematically by the relation:

$$\frac{d\,(\rho A)}{dt} + \frac{d\,(\rho AV)}{dx} = 0$$

The momentum equation describes the force balance on the fluid within a section of pipe line. It requires that any unbalanced forces result in an acceleration of the fluid element. In mathematical form, this is:

$$\frac{dV}{dt} + V \times \frac{dV}{dx} + \frac{1}{\rho} \times \frac{dP}{dx} + g \times \frac{dH}{dx} + \frac{fV|V|}{2 \times D} = 0$$

The energy equation states that the difference in the energy flow into and out of a section equals the rate of change of energy within the section. The equation is:

$$\frac{dT}{dt} + V \times \frac{dT}{dx} + \frac{T}{\rho \times c} \times \frac{dP}{dT} \times \frac{dV}{dx} - \frac{f|V^3|}{2cD} + \frac{4U}{\rho cD}\,(T - T_g) = 0$$

These three are the basic one dimensional pipeflow equations and are present in one form or another in all transient pipe models. What is needed to solve them, however, is a relation between the pressure, density, and temperature for the fluid—an equation of state.

The state equation depends on the type of fluid being modeled, as no one equation fully describes the variety of products that are shipped in pipe lines. Some of the forms in use include a bulk modulus type of relation of the form:

$$\rho = \rho_o \left[1 + \frac{P - P_o}{B} + \alpha\,(T - T_o) \right]$$

This is normally used for liquids that can be regarded as incompressible. The bulk modulus B and the thermal expansion coefficient α can be constant or functions of temperature and/or pressure depending on the application. For light hydrocarbon gases a basic equation such as $P = \rho \times R \times Z \times T$ is appropriate, where Z (the compressibility) is a known function of temperature and pressure. For reasonable ranges of temperature and pressure a function of the form:

$$\frac{1}{Z} - 1 \cong \frac{P}{T^y}$$

may be adequate. For conditions where fluids are transported at or near the critical point, a more sophisticated correlation is required to obtain the required accuracy but there is still a large uncertainty in the true density under these operating conditions and they should be avoided wherever possible. Many real time systems have been installed on lines carrying ethylene, butane, propane, and LNG/LPG products and have used the Benedict-Webb-Rubin correlation as modified by Starling (commonly called the BWRS equation) with reasonable results. Alternatively, tables such as NX-19 or special correlations such as the NBS ethylene equation can be used, but this increases the complexity of the programming.

A further complication arises if the product is not uniform throughout the system. This can occur due to batching of fluids or from varying inlet conditions. The first is more common where different products are shipped in a common line. The properties are essentially discontinuous across the interface of the two fluids, but can be considered as uniform within batches. The basic problem here is to keep track of the location of the interface. Systems with continuous variations in inlet conditions occur in both liquid and gas systems. The variations can result from mixing of fluids of slightly different composition or from large variations in supply conditions.

The governing equations presented are non-linear partial differential equations which are not suitable for machine computation. They have to be solved by implicit or explicit finite difference techniques or the method or characteristics. Of these the implicit method seems the most appropriate, as the other two methods could give rise to mathematical instabilities if the wrong timestep or distance interval is used.

In order for the transient pipe line model to compute the state along the line at the end of each time interval, a set of initial conditions and a set of boundary conditions are required. The initial conditions specify the state at the beginning of the time interval and are normally the last set of data from the model. A steady state model must be used to generate an initial state when the model is started from rest, a so-called "cold start." In this case a period of time must elapse before the pipe line model truly represents the actual state of the line. This time period allows any transient conditions present and not represented by the steady state model to die out. Generally, less compressible systems will cold start faster than the more compressible ones, but the actual time for the transient model to be activated depends on the application. This may typically be on the order of 30 minutes for a gas pipe line.

The boundary conditions required by the model are taken from measured data along the line. For each point where fluids enter the system, its temperature, fluid type or composition, and either a flow or a pressure is required. For any equipment in the system that affects or controls the line, a suitable boundary condition must also be given. For a gas compressor for example, either its suction pressure, flow rate or discharge pressure must be specified. Additional measurements are used by the applications modules, generally by comparing their values to the corresponding model calculations.

Impact of instrument accuracy

The performance of real time pipe line monitoring systems is limited primarily by the accuracy of the instrumentation installed on the line. To estimate the performance of the leak detection and location routines, it is important to understand the effect of measurement uncertainty on the

model and the real time applications module. Measurement uncertainty is composed of bias and random components. The first is usually a fixed error between the indicated and true values but this could change with time as components wear. The second is temporal and possibly spatial fluctuation of the output about its mean value. Fortunately there are techniques that can be employed within the software to largely mitigate the effects of bias, but care should be exercised as this is not always the case.

Leak detection and location both use differences between the measured values and the modeled values to discern leak characteristics. The measured values could contain both bias and/or random errors as discussed. The model values, because they are driven by the measured values as boundary conditions, also include error terms which are less obvious. Because of these errors the differences in the model and measured values, or leak indicators, will not normally be zero but will fluctuate about some non-zero mean. This mean value is determined by observation during periods when no leak is present and is attributed to bias errors in the measuring system. By subtracting this from the leak indicators, the bias component can be eliminated. The leak detection then will be a function of the measurement precision errors.

Problems with this technique arise when dealing with pipe lines whose operations change substantially from time to time. An example would be a liquid line operating intermittently. By monitoring differences in this way, the instruments contributing to the error are not identified. Because fluid flow in a pipe line is governed by highly non-linear relationships, fixed errors in the boundary conditions can cause variable differences in the leak indicators when the pipe line operation changes. As an example, consider a steady state pipe line that is driven by pressure difference between upstream and downstream boundary points. The leak indicator is the difference in the measured and modeled flow. (Note that a transient model would have a flow difference at each end of the pipe line.) The true value of the data for this pipe line is an upstream pressure of 1,000 units with a pressure drop of 100 units for a flow of 100 units. The model of this system is then described by the equation:

$$P1 - P2 = 0.01 \ Q^2$$

If a 1% pressure error is introduced into the upstream pressure, the modeled flow becomes 105 and a difference of 5 units between measured and modeled flow would show up in the leak indicators. As long as the flow stays near this value, the error in the leak indicator will remain nearly constant. For instance, an actual flow of 80 units would result in a modeled flow of 86 units. Thus the effects of the bias error in the pressure measurement can be substantially mitigated by subtracting 5 units from the leak indicator. This is termed the leak indicator "offset."

Now assume the line is shut down. The 10 unit pressure measurement error causes the flow to compute a flow of 31 units. After applying the offset, the value of the leak indicator still remains at 25. In general, a line that undergoes large and rapid changes in operation will be affected by instrument bias errors. The more common case of lines that operate within relatively narrow bounds, or that change operations slowly so that the offset can be automatically adjusted, will only be affected by the precision error of the measurements.

Measurement errors impact leak detection by limiting the size of leak that can be detected by the monitoring system. The problem comes in finding a threshold value for each alarm in the system. For a simple system that operates within narrow bounds, this can be as simple as the offset previously discussed. The values of the leak indicators (now after the offset has been removed) can be observed during normal operation and the appropriate alarm values set. For more complicated systems, the thresholds can be set in a more rigorous way. The pipe line hydraulic equations can be used to determine the sensitivity of each leak indicator to the measurement at each boundary point. The error, whether bias or precision, of each boundary point can be estimated from knowledge of the transducers and the data gathering equipment installed on the pipe line. The error in the boundary instruments times the sensitivity of the leak indicator to the boundary point will give the threshold required to prevent normal noise in the measurement value from being interpreted as a leak. Using the root sum square as a result of combining the threshold for each boundary point with the error for the leak indicator's comparison measurement, a threshold for the leak indicator can be calculated on-line. This "auto tuning" of the leak indicator is found in the more advanced systems commercially available.

As an example, consider the steady state pipe line used in the earlier paragraph. The sensitivity of the modeled flow to the upstream and downstream pressures are:

$$\frac{dQ}{dP1} = -\frac{dQ}{dP2} = \frac{50}{Q}$$

Thus, at a flow of 100 units, the sensitivity of the flow to either pressure would be 0.5, with a decrease in the downstream pressure being equal to an increase in the upstream pressure. For a 1%, or 10 unit error in the pressure, a 5 unit error in the flow would be expected, which is consistent with the previous results. The threshold required for this system would then be:

$$TL = \left[\left(\frac{50}{Q} \times EP1 \right)^2 + \left(\frac{50}{Q} \times EP2 \right)^2 \right]^{1/2}$$

This shows that the required threshold for the system increases with decreasing flow, with leak detection being impossible at zero flow. The overly stringent requirement at zero flow is due to the simplified model used. Other than that, the results are representative of the manner in which the leak thresholds must be adjusted when large flow variations occur in a pipe line.

Measurement uncertainty affects leak location by increasing the uncertainty in the calculated leak position. Leaks are located in a line by discovering where a leak of a given size would need to be located to best match the observed discrepancies in the leak indicators. Consider our steady state model again. If there is a leak of 20 units halfway down the line so that the flow is 120 units before and 100 after the leak, then the pressure drop would be 122 units. This pressure drop would correspond to a modeled flow of 110. This would result in discrepancies in the leak indicator at both ends of the pipe at 10. The leak location for a pressure-pressure boundary condition is given by:

$$X = L \times \frac{Q2}{Q1 + Q2}$$

For the conditions given above, the correct location of half the pipe length is obtained. If, however, the flowmeters have a 2 unit error such that the upstream leak indicator is 12 and the downstream indicator is 8, the leak is located 40% down the pipe, and not halfway. In a 100 mile line this is an error of 10 miles which is very significant.

When evaluating the effect of measurement uncertainty on leak detection or location, it is useful to compare the uncertainty to the magnitude of the hydraulic events of interest. Thus a pressure transducer that is 1% accurate over a 1,500 psi span is only 30% accurate when the pressure drop between two closely spaced valve sites is 50 psi. This is because the flow is governed by pressure differences and not absolute pressures. Likewise a flow meter that is 2% accurate in comparison to its span is 70% accurate for sizing a 3% leak.

System design aspects and guidelines

The availability of leak alarm uptime depends heavily on the system design and the choice of hardware. Generally the more complex the system, the greater the risk of leak indicator loss, but the more accurate the location of the leaks. Design is therefore a compromise between cost, performance, and reliability. Consider a section of line shown in Figure 3. For the three simple alarms of flow imbalance, pressure imbalance, and acoustic alarms between stations A and B or B and C, the following components are needed:

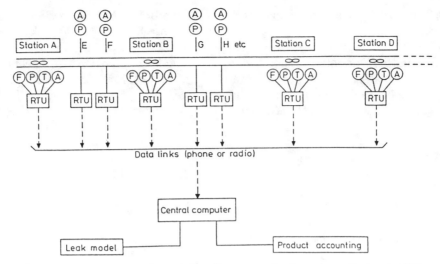

Figure 3. Schematic overview of pipe line computer-based monitoring system.

Mass flow: 2 flowmeters, 2 pressure sensors, 2 temperature sensors, 2 RTUs, 2 communications links,
   1 computer (11 elements)
Pressure: 4 pressure sensors, 4 RTUs, 4 communications
   links, and 1 computer (13 elements)
Acoustic: 2 acoustic monitors, 2 RTUs, 2 communications
   links, and 1 computer (7 elements)

If combined hybrid alarms of flow/acoustic, flow/pres-
sure, or pressure/acoustic are used then the number of components in the chain is increased. By summing the component availabilities for each element, an uptime for each
alarm can be estimated. For example, a flowmeter with a
failure of once in 3 years with a repair time of 4 hours has
an availability factor of 0.99985 (or 99.985%).

From such an analysis, the conclusion can be drawn that
the system should be made as simple as possible or instruments should be installed in duplicate to maximize alarm
uptime. Section 9 also showed that instruments should be
selected on performance and not on economic grounds. It is
better to install fewer high performance instruments than
numerous poor ones. Digital conversion also requires attention. Often 12 or 14 bit conversion is required to give the
necessary accuracy of data processing and usually double
precision computation is also required. The use of standard
outputs should be made wherever possible. Custom designed electronics invariably lead to problems. The best
guideline, however, is to seek users of pipe line monitoring
systems to ask their advice and experience with instruments
and system components. Independent validation of all information should be made wherever possible. Companies
that supply such complete systems usually have a client list

and it is worth spending time talking to these clients before
the final design specification is fixed.

With regard to instrumentation, flowmeters with the
highest accuracy are required for mass balance functions.
Suitable types include turbine and displacement meters
with pulse outputs. Orifice meters are not really suitable,
since the best accuracy that can be obtained from a well
maintained system is around ±1% of full scale. The correct choice of turbine by comparison is ±0.25% of reading
or better.

Development of pipe line monitoring systems

The speed of instrumentation development generally is
rather frightening. The impact of micro-electronics is still
being felt some ten years after they first appeared, and new
and improved transducers with on-board "intelligence" are
being sold in increasing numbers. At the same time the size
of computers is decreasing and the computating capability
is increasing. Software is also advancing rapidly and the
performance of modern flow monitoring systems is becoming dependent on the accuracy of the modeling equations.
Equations of state and the behavior of hydrocarbon mixtures are not particularly advanced or well understood,
and fundamental research is required before the next advance in this type of technology can proceed.

All of these points indicate that computer-based monitoring systems will become the standard technique of operating and controlling pipe lines in the future. Control algorithms can be integrated with the applications modules to
produce a semi-intelligent complete integrity monitoring
scheme. As experience in the design and operation of such

systems grows they will be applied with increasing confidence.

Future systems will use a combination of the new technology discussed. This could include internal monitoring pigs and advanced pipe line models, both run from a central control room. Thus the internal and external state of the line could be checked simultaneously. Such technology can enable safer and more economic operation of pipe lines to be carried out.

Conclusion

The paper cannot do justice in such a short space, to the complex and diverse subject of leak detection. Such systems have been in operation in many forms all over the world, but it is only recently that environmental as well as economic factors have influenced their development. Modern digital systems are transforming operation and design, with many parallel functions being possible with a single system. There is now a clear need to ensure complete integration of all components in the system to guarantee safer and more accurate pipe line management. Instrument selection is critical, as is the need to develop better thermodynamic models, for the next generation of systems to become more reliable and accurate.

References

1. Furness, R. A., "Modern Pipe Line Monitoring Techniques," *Pipes and Pipelines International*, May–June, 7–11 and September–October, 14–18, 1985.
2. Mailloux, R. L. and van Reet, J. D., "Real Time Transient Flow Modeling Applications," PSIG Annual Meeting, 1983.
3. Covington, M. T., "Pipe Line Rupture Detection and Control," ASME Paper 78-PET-54, 1978.
4. Bernard, G. G., "Energy Balance Derivation," CRC Bethany Internal Report, 1982.
5. Anon, "Pinpointing pigs in pipe lines," *R & D Digest*, 8, 14–15, British Gas Corporation, 1986.
6. Fujimori, N. and Sugaya, S., "A study of a leak detection based on in-out flow difference method," Proceedings of IMEKO Symposium on Flow Measurement and Control, pp. 205–209, Tokyo, Japan, 1979.
7. Stewart, T. L., "Operating experience using a computer model for pipe line leak detection," *Journal of Pipelines*, 3, 233–237, 1983.

Notation

A	cross-sectional area
B	fluid bulk modulus
c	specific heat at constant volume
D	pipe line diameter
EP1	uncertainty in upstream pressure measurement
EP2	uncertainty in downstream pressure measurement
EQ	uncertainty in flow measurement
g	gravitational acceleration
h	elevation
L	pipe length
P	pressure
P_o	reference or base pressure
P_1	section upstream pressure
P_2	section downstream pressure
Q	pipe line flow rate
Q_1	upstream measured vs. modeled flow discrepancy
Q_2	downstream measured vs. modeled flow discrepancy
R	gas constant
T	temperature
T_g	ground temperature
TL	leak detection threshold level
T_o	reference or base temperature
t	time
U	heat transfer coefficient
V	velocity
X	leak location
x	incremental distance along the pipe line
y	Z factor correlation coefficient
Z	gas compressibility factor
α	coefficient of thermal expansion
ρ	density
ρ_o	reference or base density

18: Tanks

Charts give vapor loss from internal floating-roof tanks 486
Estimating the contents of horizontal cylindrical tanks 488
How to gauge a horizontal cylindrical tank 489
Use nomograph to find tank capacity 489
Correct the volume of light fuels from actual temperature to a
  base of 60°F 491
Volume of liquid in vertical cylindrical tanks 491
Chart gives tank's vapor formation rate 491
Hand-held calculator program simplifies dike computations 492

485

Charts give vapor loss from internal floating-roof tanks

S. Sivaraman, Exxon Research & Engineering Co., Florham Park, N.J.

Nomographs, based on the guidelines presented in American Petroleum Institute (API) Publication No. 2519, have been constructed to estimate the average evaporation loss from internal floating-roof tanks.[1] Loss determined from the charts can be used to evaluate the economics of seal conversion and to reconcile refinery, petrochemical plant, and storage terminal losses.

The losses represent average standing losses only. They do not cover losses associated with the movement of product into or out of the tank.

The average standing evaporation loss from an internal floating-roof tank depends on:

- Vapor pressure of the product
- Type and condition of roof seal
- Tank diameter
- Type of fixed roof support

The nomographs (Figures 1–4) can estimate evaporation loss for product true vapor pressures (TVP) ranging from 1.5 to 14 psia, the most commonly used seals for average and tight fit conditions, tank diameters ranging from 50 to 250 ft, welded and bolted designs, and both self-supporting and column-supported fixed roof designs. The charts are purposely limited to tank diameters 250 ft and less because internal floating-roof tanks are generally below this diameter.

Typical values of the deck fitting loss factors presented as a function of tank diameters in the API Publication 2519, have been used in the preparation of these nomographs. In addition, for the calculations of the evaporation loss for the bolted deck design, a typical deck seam loss factor value of 0.2 has been assumed.

Table 1 gives the proper axis to use for various seal designs and fits.

Table 1
Selection of Seal Axis

Seal type	Seal axis Average fit	Tight fit
Vapor mounted primary seal only	H	G
Liquid mounted primary seal only	F	E
Vapor mounted primary seal plus secondary seal	D	C
Liquid mounted primary seal plus secondary seal	B	A

Use of these nomographs is illustrated by the following example.

Example. Determine the evaporation loss for an internal floating roof tank given the following:

- Tank diameter 200 ft

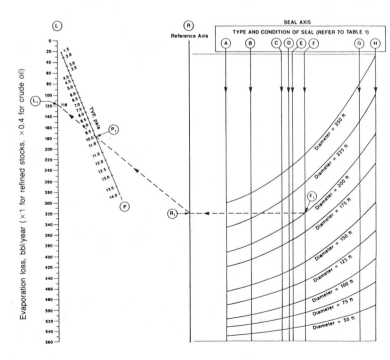

Figure 1. Loss from welded deck, self-supporting fixed roofs.

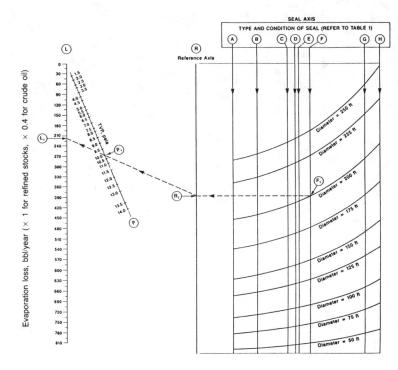

Figure 2. Welded deck, column-supported fixed roofs.

Figure 3. Bolted deck, self-supporting fixed roofs.

- Liquid mounted primary seal only and an average seal fit
- Product true vapor pressure of 10 psia
- Welded deck with self supporting fixed roof

Solution.

1. Use Figure 1 for the welded deck and self-supporting fixed roof.

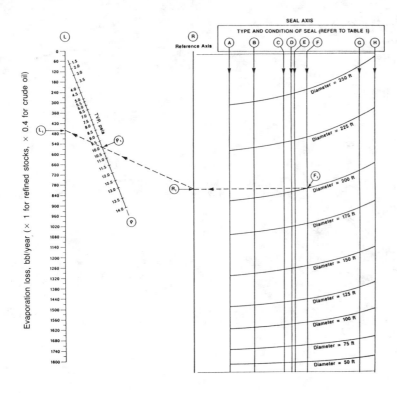

Figure 4. Bolted deck, column-supported fixed roofs.

2. From Table 1 select the seal axis. The seal axis for the example problem is F.
3. Locate the point of intersection F1 between the seal axis F and the tank diameter contour for the 200-ft diameter tank.
4. From the point F1 traverse horizontally to intersect the reference axis R at R1.
5. Locate the true vapor pressure point P1 corresponding to 10 psia on the pressure axis P.
6. Connect the point R1 on the reference axis R and the point P1 on the pressure axis P and extend it to intersect the evaporation loss axis L at L1.

Read the evaporation loss in bbl/year at L1. The average evaporation loss is 188 bbl/year for this example. The same example is shown in Figures 2, 3, and 4 for other deck designs and roof supports.

Source

Oil & Gas Journal, March 9, 1987.

Reference

1. "Evaporation Loss from Internal Floating-Roof Tanks," American Petroleum Institute Publication No. 2519.

Estimating the contents of horizontal cylindrical tanks

Horizontal cylindrical tanks are frequently used for water and fuel storage and in many cases it is important to be able to gage these vessels to determine the volume of liquid contained in them. However, it is normally much more difficult to establish a volume per inch scale for a horizontal tank than for one which is in a vertical position. The accompanying nomograph simplifies this problem.

To use the nomograph, it is necessary to gage the tank and determine the ratio of the depth of liquid in the tank to tank diameter. After this is found, draw a straight line from the point on the "ratio" scale through the known point on the "diameter of tank" scale and read the intercept on the "gallons per ft of length" scale. From this point, draw a

second line through the known point on the "length of tank" scale and read the intercept on the "gallons (or barrels) in total length" scale.

Example. Find the volume of liquid contained in a horizontal cylindrical tank 7 ft in diameter and 20 ft long when the depth of the liquid is 4 ft, 10.8 in.

The ratio of depth of liquid to tank diameter is:

58.8/84 = 0.70

Connect 0.70 on the ratio scale with 7 ft on the diameter scale and continue the straight line to obtain the intercept 215 on the gallons per ft of length scale. Draw a second line from the point 215 through the 20 on the length of tank scale and continue the line to obtain the intercept 4,300 on

the gallons in total length scale. Total liquid content of the tank is 4,300 gallons.

How to gauge a horizontal cylindrical tank

Express the depth in % of the diameter; then the result will be given in % of total capacity of the tank.

Rule 1. For depth up to 30; multiply the square root of the depth by the depth, and then by 0.155.

Example. Liquid depth is 16% of tank diameter

$16 \times 16 \times 0.155 = 4 \times 16 \times 0.155 = 9.9\%$

The correct answer is 10.3%; error is about .4%.

Rule 2. For depth between 30 and 50; subtract 10 from the depth, multiply by 1.25.

Example. Liquid depth is 44% of tank diameter

$(44 - 10) \times 1.25 = 34 \times 1.25 = 42.5\%$

The correct answer is 42.4%.

The maximum error, for depths less than 5% may be as great as 10%; gauging in this range is always difficult, and a very small slope can introduce a much larger error than this. When the depth is greater than 50%, apply the same rule to get the volume of the empty space above the fluid, and subtract.

Use nomograph to find tank capacity

This simple nomograph can be used to find the capacity of your vertical cylindrical tanks. Here's how it works:

Draw a straight line from the "height" scale through the "diameter" scale and to the first "capacity, barrels" scale.

Read directly the capacity of the tank in barrels. (Note: The "height" scale may be used to indicate the over-all height of the tank or the depth of liquid in the tank.)

Draw a second straight line connecting the two "capacity, barrels" scales at the same reading on each scale. Read the capacity of the tank in gallons and cubic ft on the proper scales.

The nomograph was constructed as follows:

1. The "height" scale is based on two log cycles per 10 in. with a range of 1 to 60 ft.
2. The "capacity, barrels" scale is based on four log cycles per 10 in. with a range of 20 to 150,000 barrels.
3. The "diameter" scale is based on three log cycles per 10 in. with a range of 4 to 150 ft.
4. The distance between the height and diameter scales is exactly two-thirds the distance between the height and "capacity, barrels" scale.
5. Determine points to locate the diameter scale from the following equation:

$$\text{Capacity, barrels} = 0.1399 \,(\text{diameter})^2 \,(\text{height}),$$
$$\text{units in ft}$$

6. The "capacity, gallons" scale is based on four log cycles per 10 in. The initial point on the scale is determined as follows:

   20 barrels × 42 gallons per barrel − 840 gallons

   The range of the scale is 900 to 6 million gal.

7. The "capacity, cubic ft" scale is based on four log cycles per 10 in. The initial point on the scale is determined as follows:

   20 barrels × 5.6146 cubic ft per barrel = 112.292 cubic ft

The range of the scale is 120 to 800,000 cubic ft.

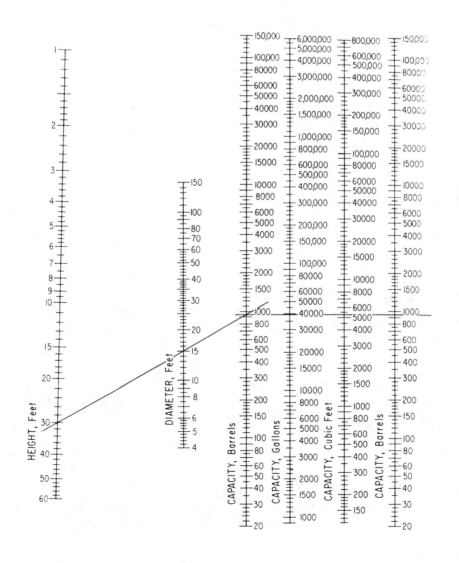

Correct the volume of light fuels from actual temperature to a base of 60°F.

To approximate quickly the volume of gasoline or other light liquid fuel at 60°F from a known volume at any temperature in the atmospheric range use the formula:

$$V_a - V_{60} = 0.0006 (T_a - 60) V_{60}$$

where V_a = volume at actual temperature
   V_{60} = volume corrected to 60°F
   T_a = actual temperature of fuel

Example. A tank contains 5,500 gallons of gasoline at 46°F. Correct the volume to a base of 60°F.

$(5,500 - V_{60}) = 0.0006 (46 - 60) V_{60}$
$(5,500 - V_{60}) = 0.0006 (- 14) V_{60}$
$5,500 = V_{60} - 0.0084 V_{60}$
$5,500 = 0.9916 V_{60}$
Volume at 60°F = 5,546.6 gallons

Shrinkage

To approximate the shrinkage or expansion obtain the difference between the actual volume measured and the corrected volume. In this case:

Shrinkage = 5,546.6 − 5,500 = 46.6 gallons

Volume of liquid in vertical cylindrical tanks

Measure the depth of the liquid and either the diameter or circumference of the tank; then the volume in:

Gallons = $0.0034 d^2h$ or $0.00034 c^2h$
Barrels = $0.000081 d^2h$ or $0.00082 c^2h$
Gallons = $5.88 D^2H$ or $0.595 C^2H$
Barrels = $0.140 D^2H$ or $0.0142 C^2H$

where   d = diameter, in.
   c = circumference, in.
   h = depth, in.
   D = diameter, ft
   C = circumference, ft
   H = depth, ft

If the circumference is measured on the outside, then three times the thickness of the tank wall should be sub-

tracted before using the formula. Naturally, these rules cannot supplant the results of accurate tank strapping, which take many other factors into account.

Example. How many gallons will a tank 12 ft in diameter and 16 ft high hold when full?

Gallons = $5.88 D^2H$
   = (5.88)(144)(16)
   = 13,548 gallons

Example. How many barrels will a tank 8 ft in diameter and 16 ft high hold when full?

Barrels = $0.140 D^2H$
   = (0.140)(64)(16)
   = 143 barrels

Chart gives tank's vapor formation rate

When sizing the vapor piping for a manifolded expansion-roof tank system, the rate of vapor formation must be known. While the rate of vapor formation can be computed by long-hand methods, the calculation is tedious and takes much valuable time.

Example. Determine the rate of formation of vapor in a 140,000 barrel capacity tank when it is filled at the rate of 8,000 barrels per hour.

Solution. Enter the chart on the left at a capacity of 140,000 barrels and draw a straight line through the filling rate of 8,000 barrels per hour on the right. At the intersection with the central scale read the vapor formation rate as 55,000 cubic ft per hour. The vapor piping for this tank would have to be designed for this formation rate if the maximum filling rate anticipated were 8,000 barrels per hour. But if a great filling rate was expected, the vapor formed would have to be computed for the higher rate.

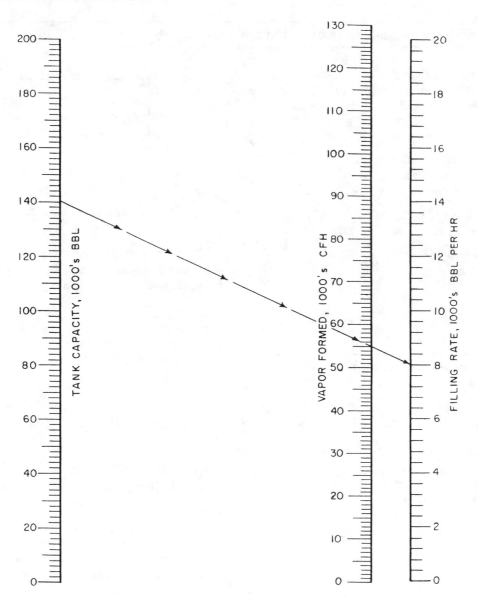

The chart could, of course, also be used for this computation.

This chart is based on the following equation:

$$V = \frac{\text{tank capacity, bbl}}{14.3} + \frac{\text{filling rate, bbl per hr}}{0.178}$$

where V = vapor formed, cubic ft per hr

Hand-held calculator program simplifies dike computations

Calculating height of earthen dikes around above-ground storage can be done easily with a program for a portable calculator

Frank E. Hangs, Sovereign Engineering Co., Houston

Earthen dikes are widely used all over the world to contain flammable volumes of above-ground storage. They perform two vital functions: to prevent loss of fluid into the environment and to reduce the likelihood of fire spreading from one tank to another.

Sizing dikes by conventional methods is a time-consuming, trial and error process. A complete assessment of the problem involves: applicable codes and regulations; land

Figure 1. Cross section of a typical dike.

area available; topography of the area; soil characteristics and the stipulated volume contained by dike and other dimensions of the dike section.

The following program for the HP-41CV hand-held calculator enables one to enter required data at a prompt and to calculate the height of the dike to retain the required volume of fluid, X-section of dike, width of the base and the cubic yards of earth required, quickly. When a printer is available, a record of the input and output (results) is made. Without a printer, the input and output items (all identified) can be displayed one at a time and advanced at will.

Many "what if" questions can be answered readily and different configurations compared as desirable. This is explained in detail in text and examples.

The Flammable and Combustible Liquids Code, as promulgated by the National Fire Protection Association, NFPA No. 30, is used as a basis for this program. Important stipulations are:

- Volume contained in dike area shall not be less than the full tank. (We have taken one tank per dike.)
- For crude petroleum with boilover characteristics, stored in fixed roof tanks, the contained volume above shall be calculated by deducting the volume of the tank below the height of the dike.
- Earthen dikes 3 ft or more in height shall have a flat section at the top not less than 2 ft wide.
- The slope of the earth wall shall be consistent with the natural angle of repose of the material of construction.
- The walls of the diked area shall be restricted to an average height of 6 ft above interior grade.

Dikes are constructed in circular, square or rectangular configurations. For the purposes of this program, the volumes contained in the dikes are calculated as invented frustums of a cone or pyramid. The dike volume (converted to barrels) is compared to the total volume (volume of tank or

```
EXAMPLE 1                    EXAMPLE 2

        XEQ "DIKE"                         SF 03
TV=?                          4.00 STO 07
       54,200.00    RUN                    XEQ A
TD=?                         DD=3TD?
         110.00    RUN              330.00    RUN
TH=?
          32.00    RUN       TV=54,200.00
RUN=1.5?                     TD=110.00
           1.50    RUN       TH=32.00
DWT=2??                      RUN=1.50
           2.00    RUN       DWT=2.00
DH INCR=?                    DH INCR=0.05
           0.05    RUN       DD=330.00
BOIL OVER?Y? N?              SQ=0.00
N                  RUN       LS=0.00
TRIAL DH=?                   SS=0.00
           3.00    RUN
A? B? C?                     DV=61,506.12
               XEQ P         DH=4.25
DD=3TD?                      X-SECT=35.59
          330.00    RUN      DWB=14.75
                             EARTH YD3=1,366.70
                             B O BBL=7,193.62
                             TOT BBL=61,393.62

TV=54,200.00                 EXAMPLE 3
TD=110.00
TH=32.00
RUN=1.50                                 CF 00
DWT=2.00                                 SF 01
DH INCR=0.05                             RCL 00
DD=330.00                                STO 01
SQ=0.00                         0.00 STO 24
LS=0.00                         4.50 STO 07
SS=0.00                                  XEQ A
                             DD=3TD?
                                    300.00    RUN
DV=54,521.90
DH=3.75
X-SECT=28.59
DWB=13.25                    TV=54,200.00
EARTH YD3=1,097.92           TD=110.00
B O BBL=0.00                 TH=32.00
TOT BBL=54,200.00            RUN=1.50
                             DWT=2.00
                             DH INCR=0.05
                             DD=300.00
                             SQ=0.00
                             LS=0.00
                             SS=0.00

                             DV=53,406.62
                             DH=4.50
                             X-SECT=39.38
                             DWB=15.50
                             EARTH YD3=1,374.45
                             B O BBL=0.00
                             TOT BBL=54,200.00
```

Figure 2. Examples of the dike computation program for the HP-41CV hand-held calculator.

DIKE PROGRAM

Legend and storage registers

		REG.
TV =	Tank vol. (bbl)	00
TOT BBL =	Total bbl TV + boilover (if needed)	01
TD =	Tank dia (ft)	02
TH =	Tank height (ft)	03
Run =	For angle of repose of dike earth expressed as bevel Rise/Run = 1/Run, 1/1.5 is widely used	04
DWT =	Dike width top—ft Min. 2 ft for dikes 3 ft and higher—NFPA No. 30.	05
DH INCR =	Dike height increment Suggest 0.10 ft for prelim. run; Use 0.05 ft (0.60-in.) to finalize	06
TRIAL DH =	Trial dike height (ft) Try 3 ft for tanks 10,000 bbl and larger	07
DD =	Dike dia (ft)	10
EARTH YD3 =	Dike vol. (cu yd)	20
SQ =	Square side (ft)	21
LS =	Long side (rectangle) ft	22
SS =	Short side (rectangle) ft	23
BO BBL =	Boilover barrels	24

Registers 8, 9, 13 and 17 not used.
Registers 11, 12, 14 and 15 scratch.

FLAGS

00—Boilover calculation
01—Single calculation
02—Circular dike
03—Square dike
04—Rectangular dike
21—Printer enables following: Allows data and results to be displayed one by one without printer. (Must be set each time calculator is turned on.)
Note: Flags 01 and 21 must be manually set. Flag 01—Clear manually—other flags set and cleared in program. See Example 3 for "short cut" exceptions.

PRINCIPAL LABELS

01—Calculates X-SECT and DWB; directs program to proper EARTH VOLUME routine, i.e., SQ ? etc.
03, 11 and 13 Bypass incrementation in a loop for a single calculation
06, 07 and 08 Calculates EARTH VOLUME for A, B, and C, respectively.
09 Summarizes data and results for display or printout.
A, B and C Subroutines for circular, square and rectangular dikes, respectively, calculating dike volumes for DH increments to converge dike volume and total volume.

a. Subroutine for calculating boilover volume.
b. Sets flag 00 for boilover calculations.

DV =	Dike vol. (bbl)	16
X-SECT =	Cross sect. dike (sq ft)	18
DWB =	Dike width base (ft)	19

FORMULA: For right truncated cone or pyramid:

$$V = \frac{H}{3}(A + \sqrt{AB} + B)$$

H = Height; A = Area of larger base,
B = Area of smaller base.

(See Fig. 1.)

Dia or side of top base = (DD, S, LS or SS) – DWT
Dia or side of lower base = (DD, S, LS or SS) – DWT – 2 (DH × Run.)

USER INSTRUCTIONS

1) Put dike program in calculator.
2) XEQ size 25 and set User Mode.
3) XEQ "Dike."
4) Key in data according to prompts and R/S. Notice Run = 1.5? and DWT = 2? If these values are acceptable, key in and R/S. DH INCR = ? Can be 0.10 ft for preliminary runs, otherwise, use 0.05 ft (0.60-in.).

Trial DH = ? Try 3 ft. If this is too much, machine will stop, STO smaller value in R07 and try again ("A," "B" or "C"). Likewise, if it is known that DH is much greater than 3—try larger value: This saves iterations!

A? Key in "A" for circular dikes (Be sure User Mode!)
B? Key in "B" for square dikes. (Be sure User Mode!)
C? Key in "C" for rectangular dikes. (Be sure User Mode!)
Boilover? Y? N?
This routine is available when needed. Usually answer is no. Press "N" and R/S.

Whichever key A, B, or C is pressed, a dia or side(s) will be called for; key in appropriate data and calculator will converge DV with total barrels for solution of DH (See Fig. 2).

Example 1: Illustrates a 54,200-bbl tank in a circular dike. Notice how data are entered according to prompts. A printer is a great convenience but is not indispensible. Without printer, SF 21 each time calculator is turned on, now data and results will be displayed one at time and advanced by pressing R/S key. Notice formating of data and results—zeros are shown for inactive dimensions or boilover volume.

Example 2: Demonstrates an approach, where most input data do not change: Calculate DH for above tank for boilover crude. (This avoids going through entering routine XEQ Dike.) SF 00 and a new trial DH try 4, STO 07. (Result of Example 1 is DH = 3.75 ft. It is evident dike will be higher to contain

larger volume.) Press "A," re-enter DD (330). Results: DH = 4.25 ft. larger X-SECT, MORE EARTH, a boilover volume is shown and included in total volume.

Example 3: Same tank, no boilover, what is DD for 4.5 ft. dike? CF 00 (Notice it appears in annunciator when set.) SF 01 (This is not a looping routine!) RCL 00 STO 01, 0.00 STD 24. STO 4.5 in 07. Press "A"; put some value for DD less than 330 (as above). Try 300 key in and R/S. One calculation will be made. Compare total volume and dike volume. Press "A" and key new DD. Repeat until satisfactory convergence is achieved.

The following table shows convergence for Example 3:

Total vol 54,200 bbl   DH = 4.5 ft.

Trial DD ft	DV bbl
300	53,406.62
305	55,255.73
302	54,142.48
302.5	54,727.24
302.3	54,253.30

It is evident that with a few iterations, one can home-in on a precise solution.

SUMMARY

For a new tank size, one should XEQ "Dike" and enter data according to prompts.

Subroutines "B" and "C" are similar to "A." One can build up a dike height for given center line distances (square or rectangle) with and without boilover.

Consider same tank as Example 1. (Let 3 = Trial DH STO 07.)

Square 300 ft (No B.O)
Square 300 ft B.O.

DH = 3.60
DV = 54,894.66
DH = 4.05
    DV = 61,473.78
    Total volume = 61,055.09

Rectangle LS 450: SS 200 (3.00 STO 07)
LS 450: SS 200 B.O.

H—3.60   DV = 54,655.16   DH = 4.10
    DV = 61,900.04
    Total volume = 61,139.72

One can "free-wheel" with "B" and "C" as for "A" for a fixed dike height. SF 01 and try different dike centerline distances. Compare DV and TOT. Val. and continue to desired convergence.

Warning: Be sure Flag 00 is clear when boilover routine is not required. When clearing Flag 00 STD 0.00 in 24 to prevent extraneous volume from being involved in program. BO VOL should be 0.00 when there is no boilover. Flat 01 must be clear for incrementing routines. Always STO new trial DH in 07 each run.

Note: XEQ 09 for printout or display of input and results in storage registers.

```
     PRP "DIKE"      51 SF 02        102 RCL 11       152 STO 22       204 /            255+LBL 06       306 "TD="
                     52 FS? 03        103 X↑2          153 STO 11       205 STO 16       256 RCL 18       307 ARCL 02
  01+LBL "DIKE"      53 CF 03         104 STO 14       154 "SS=?"       206 FS? 00       257 RCL 10       308 AVIEW
  02 "TV=?"          54 FS? 04        105 SF 03        155 PROMPT       207 XEQ a        258 *            309 "TH="
  03 PROMPT          55 CF 04         106 FS? 02       156 STO 23       208 RCL 01       259 PI           310 ARCL 03
  04 STO 01          56 FS? 01        107 CF 02        157 STO 12       209 X<=Y?        260 *            311 AVIEW
  05 STO 00          57 GTO 03        108 FS? 04       158 RCL 05       210 GTO 01       261 27           312 "RUN="
  06 "TD=?"                           109 CF 04        159 ST- 11       211 FS? 01       262 /            313 ARCL 04
  07 PROMPT          58+LBL 02        110 FS? 01       160 ST- 12       212 GTO 01       263 STO 20       314 AVIEW
  08 STO 02          59 RCL 06        111 GTO 11       161 RCL 11       213 GTO 12       264 CLX          315 "DWT="
  09 "TH=?"          60 ST+ 07                         162 RCL 12                        265 STO 21       316 ARCL 05
  10 PROMPT                           112+LBL 10       163 *            214+LBL a        266 STO 22       317 AVIEW
  11 STO 03          61+LBL 03        113 RCL 06       164 STO 14       215 RCL 02       267 STO 23       318 "DH INCR="
  12 "RUN=1.5?"      62 RCL 10        114 ST+ 07       165 SF 04        216 X↑2          268 GTO 09       319 ARCL 06
  13 PROMPT          63 STO 11                         166 FS? 02       217 7.1487                        320 AVIEW
  14 STO 04          64 RCL 05        115+LBL 11       167 CF 02        218 /            269+LBL 07       321 "DD="
  15 "DWT=2?"        65 ST- 11        116 RCL 21       168 FS? 03       219 RCL 07       270 RCL 21       322 ARCL 10
  16 PROMPT          66 RCL 07        117 STO 11       169 CF 03        220 *            271 4            323 AVIEW
  17 STO 05          67 RCL 04        118 RCL 05       170 FS? 01       221 STO 24       272 *            324 "SQ="
  18 "DH INCR=?"     68 *             119 ST- 11       171 GTO 13       222 RCL 00       273 RCL 18       325 ARCL 21
  19 PROMPT          69 2                             223 STO 01        274 *            326 AVIEW
  20 STO 06          70 *             120 RCL 07       172+LBL 12       224 RCL 24       275 27           327 "LS="
  21 "BOIL OVER?"    71 ST- 11        121 RCL 04       173 RCL 06       225 ST+ 01       276 /            328 ARCL 22
  22 "+Y? N?"        72 RCL 11        122 *            174 ST+ 07       226 RCL 16       277 STO 20       329 AVIEW
  23 AON             73 X↑2           123 2                             227 RTN          278 CLX          330 "SS="
  24 PROMPT          74 STO 15        124 *            175+LBL 13                        279 STO 10       331 ARCL 23
  25 ASTO Y          75 RCL 14        125 ST- 11       176 RCL 22       228+LBL b        280 STO 22       332 AVIEW
  26 AOFF            76 *             126 RCL 11       177 STO 11       229 SF 00         281 STO 23       333 ADV
  27 "Y"             77 SQRT          127 X↑2          178 RCL 23       230 GTO 14       282 GTO 09       334 ADV
  28 ASTO X          78 RCL 14        128 STO 15       179 STO 12                                         335 "DV="
  29 X=Y?            79 +             129 RCL 14       180 RCL 05       231+LBL 01        283+LBL 08       336 ARCL 16
  30 GTO b           80 RCL 15        130 *            181 ST- 11       232 RCL 07       284 RCL 22       337 AVIEW
  31 FS? 00          81 +             131 SQRT         182 ST- 12       233 X↑2          285 2            338 "DH="
  32 CF 00           82 RCL 07        132 RCL 14       183 RCL 07       234 RCL 04       286 *            339 ARCL 07
  33 CLX             83 *             133 +            184 RCL 04       235 *            287 RCL 23       340 AVIEW
  34 STO 24          84 21.4461       134 RCL 15       185 *            236 RCL 05       288 2            341 "X-SECT="
                     85 /             135 +            186 2            237 RCL 07       289 *            342 ARCL 18
  35+LBL 14          86 STO 16        136 RCL 07       187 *            238 *            290 +            343 AVIEW
  36 "TRIAL DH=?"    87 FS? 00        137 *            188 ST- 11       239 +            291 RCL 18       344 "DWB="
  37 PROMPT          88 XEQ a         138 16.8458      189 ST- 12       240 STO 18       292 *            345 ARCL 19
  38 STO 07          89 RCL 01        139 /            190 RCL 11       241 RCL 07       293 27           346 AVIEW
  39 "A? B? C?"      90 X<=Y?         140 STO 16       191 RCL 12       242 RCL 04       294 /            347 "EARTH YD3="
  40 PROMPT          91 GTO 01        141 FS? 00       192 *            243 *            295 STO 20       348 ARCL 20
                     92 FS? 01        142 XEQ a        193 STO 15       244 2            296 CLX          349 AVIEW
  41+LBL A           93 GTO 01        143 RCL 01       194 RCL 14       245 *            297 STO 10       350 "B O BBL="
  42 "DD=3TD?"       94 GTO 02        144 X<=Y?        195 *            246 RCL 05       298 STO 21       351 ARCL 24
  43 PROMPT                           145 GTO 01       196 SQRT         247 +            299 GTO 09       352 AVIEW
  44 STO 10          95+LBL B         146 FS? 01       197 RCL 14       248 STO 19                        353 "TOT BBL="
  45 STO 11          96 "SQ=?"        147 GTO 01       198 +            249 FS? 02       300+LBL 09       354 ARCL 01
  46 RCL 05          97 PROMPT        148 GTO 10       199 RCL 15       250 GTO 06       301 ADV          355 AVIEW
  47 ST- 11          98 STO 21                         200 +            251 FS? 03       302 ADV          356 .END.
  48 RCL 11          99 STO 11        149+LBL C        201 RCL 07       252 GTO 07       303 "TV="
  49 X↑2            100 RCL 05        150 "LS=?"       202 *            253 FS? 04       304 ARCL 00
  50 STO 14         101 ST- 11        151 PROMPT       203 16.8458      254 GTO 08       305 AVIEW
```

Example 3.

volume of tank plus boilover, if applicable). The calculations begin with given dike centerline, dike width at top, repose angle of soil and trial DH. As long as the dike volume is less than the total volume the program loops, incrementing DH for the next calculation. When the two volumes converge, calculations stop and input data and results are displayed or printed. The DH value, when the volumes converge, is the solution.

In some cases, it will be required to ascertain dike diameter or sides for a fixed dike height (DH). This is accomplished by storing DH Value in 07, setting Flag 01 (for single calculation). Press "A," "B" or "C" key in trial centerline distances. The results of any calculation give one an opportunity to compare total barrels with dike volume.

Then alter centerline distances to fit trend and continue. See Example 3.

This program is based on site being essentially level.

Source

Pipe Line Industry, August 1986.

19: Maintenance

How to Plan for Oil Pipe Line Spills 498
Regulatory requirements... 498
Contingency plan objectives.. 498
Related studies... 498
Planning concepts 499
Contingency response 500
Immediate response 501
Immediate response actions... 502
Flexible response actions... 502
Training 503

How to plan for oil pipe line spills

Here are ideas to consider in development of a comprehensive contingency plan

J. D. Sartor, vice president, and **R. W. Castle,** senior project engineer, Woodward-Clyde Consultants, Environmental Systems Division, San Francisco, California

Federal regulations and permit stipulations require the preparation of written plans and procedures for dealing with accidental spills of materials carried by liquids pipe lines. These requirements can be met through the preparation of comprehensive spill contingency plans.

Proper contingency planning involves analysis of the material(s) carried and the environments crossed by the pipe line. Based on company personnel and operating procedures, a response organization and necessary equipment are developed. The planning concept assumes that preplanned emergency actions are appropriate for protection of human life, property, and the environment.

In addition however, the planning must provide the flexibility to respond to unanticipated situations. The final planning element is the design of a comprehensive training program to ensure the safety and efficiency of the response teams.

Regulatory requirements

Historically, federal requirements pertaining to liquids pipe line construction and operation have contained no specific requirements or guidelines for the preparation of spill contingency plans.

In spite of this, such plans have been required in conjunction with various permits and right-of-way grants. Cases where contingency plans have been stipulated include the Trans Alaska, SOHIO West Coast to Midcontinent, Northern Tier, and the Northwest Alaska Gas pipe lines.

The National Environment Policy Act (NEPA) of January 1, 1970, and ensuing amendments have led to federal and state requirements for analysis of the potential for spills, their probable impact, and the degree to which spill impacts may be mitigated through the Environmental Impact Statement—Environmental Impact Report process. Spill contingency is often the major method of spill impact mitigation, and its association with the environmental review process is becoming increasingly common.

Effective July 15, 1980, the Department of Transportation (DOT) amended regulations for the transportation of liquids by pipe line. Title 49 of the Code of Federal Regulations, Part 195 (revised October 1, 1979) describes these regulations and amendments. Subpart—Operation and Maintenance—requires the preparation of written emergency plans and procedures for dealing with accidental release of commodity, operational failures, and natural disasters affecting facilities. The same regulations also require

assignment of personnel to emergency functions and the design and implementation of emergency procedures programs. These requirements summarize the essential elements of a spill contingency plan.

Contingency plan objectives

A rapid response capability is essential to adequately control and remove spilled material.

To provide this capability, adequate advance planning is mandatory. Such pre-spill planning should recognize the various types of spill situations which could occur and incorporate a response strategy for each.

These strategies should outline methods to be used, and describe the required types and amounts of equipment, materials, and manpower.

In addition to pre-spill planning for anticipated spill types and situations, a comprehensive plan must allow the flexibility to respond effectively to unanticipated spill situations.

Related studies

The development of a spill contingency plan is based on consideration and analysis of a wide variety of factors. Included are geographical elements (location of the spill, drainage characteristics, surface conditions, soil type, accessibility, trafficability), environmental elements (weather conditions, hydrology, etc.), and ecological elements (sensitive and vulnerable areas, rare and endangered species, etc.).

For special environments additional studies may be required. For example, a series of special studies on reclamation of spill affected areas was required for tundra during the implementation of the Alyeska pipe line. These studies included evaluation of plant types that could be used for rehabilitation, effects of spilled oil on native plants, effects of various types of cleanup measures, and residual effects of oil over different weathering periods.

Engineering elements necessary for contingency planning include pipe line pumping and drainage characteristics, valve placements, monitoring equipment, operating procedures, and communications and control systems. Typically much of this information is available through project-related design engineering and environmental studies at an early phase in project implementation.

In some cases, however, all of the required information may not be available and may require generation prior to the completion of the final spill plan. Early versions of spill contingency plans are considered "draft" plans and are submitted to satisfy the requirements of regulatory agencies. The "draft" plans are submitted with the stipulation that final spill plans are prepared and implemented prior to the pipe line becoming operational.

Planning concepts

Pipe lines characteristically extend considerable distances and encounter a variety of environmental conditions. The resulting spill planning must respond to these factors and is commonly either overly simplified or so complex that it is unusable in practice. To counter either tendency and yet provide sufficient information in a usable format, it is advantageous to design the contingency plan as a series of volumes, each specific to a particular need. Using this approach, the master volume or General Provi-

sions, contains organizational and operational information common to the entire pipe line system. It also forms a basis for personnel training.

The second volume or volumes delineate specific response actions and personnel for specific sections of the pipe line. Commonly the areas covered by these volumes are defined by major drainage patterns. If necessary as a result of drainage basin size or other unusual factors, each of these sections may be divided into further subdivisions until workable size field response plans are achieved.

Typical contents of successive volumes of a pipe line contingency plan are shown in Figure 1. Information pertinent to the entire system is contained only in the general provisions.

An effective contingency plan indicates the actions to be taken, their sequence, and timing in relation to other events. During or immediately following a spill, the following representative actions should be taken:

Figure 1. Contingency plan content.

a. Verify and locate the spill and notify the spill response team to initiate emergency procedures.
b. Control or limit the amount of material spilled. Exclude or prevent the spread of material to sensitive areas. Evaluate threats to the general public and act accordingly.
c. Determine the extent of the spill and predict its subsequent dispersion as a function of time.
d. Plan and direct overall operations; provide administrative support, liaison with government/local officials, and public information.
e. Remove and clean material from contaminated land areas, subsoils, shorelines, and water surfaces.
f. Determine procedures and site locations for the disposal of contaminated materials.
g. Document all phases of spill incident and subsequent cleanup activities.

Selection of the action to take at different times during a spill depends on: (a) the location and nature of the spill; (b) the quantity and type of material; (c) hydrology, topography, and soil types; (d) ice, weather, and currents. Supervision will start immediately and documentation must be conducted simultaneously with all other operations.

Spills require prompt action whether they are on land or on water. Detailed response plans must be prepared for immediate action to be taken wherever and whenever spills occur.

In general, the contingency plan describes the logical and sequential order of what-to-do actions. Some actions are taken in all situations. Other actions are contingent on and initiated only by special circumstances. The basic plan is supported by appendices which provide how-to-do-it instructions for basic actions. These appendices also provide information on regulations, sources of equipment, logistics, and general background.

Contingency Response

The nature of each individual response organization is dependent on the nature of the pipe line system and the area through which it crosses. In some remote areas such as Alaska, any effective response organization must rely heavily on internal resources. In more populated areas, a spill response organization can be designed more around the use of local resources.

Any response organization must, however, incorporate certain functions. Typically these functions may be grouped as management, advisory, support, and local field response. The management, advisory, and support roles generally perform staff functions and are usually staffed by company personnel. The local field response functions are normally staffed by local operating personnel stationed along the pipe line route. As necessary, these personnel may be supplemented by local contractors.

The assemblage of the above roles is known as the Spill Task Force (Figure 2).

Source

Pipe Line Industry, February 1982.

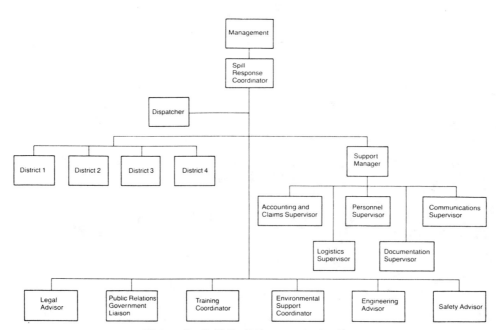

Figure 2. Spill Task Force organization.

How to plan for oil pipe line spills (Part 2)

What to consider for initial and followup actions for spills and task force training

J. D. Sartor, vice president and **R. W. Castle,** senior project engineer, Woodward-Clyde Consultants, Environmental Systems Division, San Francisco, California

Within the Spill Task Force, each district uses a two-level organization to provide fast and effective initial and followup response. A typical district organization is shown in Figure 1. The first level of response, the Immediate Response Team (IRT), reacts immediately to any spill within its area of concern. This team takes pre-planned response actions, limiting the spills' impacts until further actions can be activated.

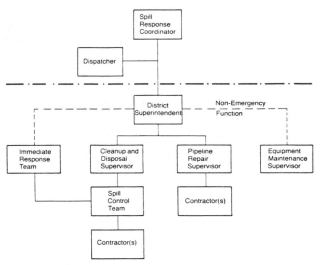

Figure 1. District Response Team organization.

The IRT is trained and equipped to handle minor spills without additional assistance. Should this team be unable to completely control a spill, it will take pre-planned control actions and notify the spill control coordinator. The spill coordinator will then activate company and contractor resources as necessary. The second level, the remainder of the District Response Team, is then activated as necessary to complete the control and recovery of the spill. For sufficiently large spills, the overall Spill Task Force may require activation.

Spills will normally be detected by or reported directly to the pipe line dispatcher, who will ensure proper communications within the Spill Task Force and who will shut down pumping operations if necessary. The dispatcher will ensure that the district superintendent responsible for the area in question is notified and that the appropriate Immediate Response Team has been activated. The dispatcher will then notify the spill coordinator who will activate additional resources as necessary after consulting with the local area superintendent. Figure 2 illustrates how the contingency response organization functions.

Immediate response

The composition of any individual Immediate Response Team will vary with the size and nature of the pipe line system. In many cases several company personnel will be available along a specific pipe line segment. With proper

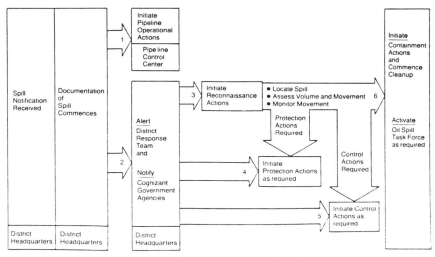

Figure 2. Immediate response action flow chart.

training and equipment however, even two or three man teams can be constructive in implementing immediate actions.

The duties of the Immediate Response Team may need to be taken simultaneously. These duties include:

- Verification and location of the leak
- Leak control
- Conducting prescribed initial actions to ensure the safety of the general public, contain the spill, and protect sensitive areas
- Assess the situation and provide initial inputs to the district supervisor and the spill coordinator
- Assume assigned positions in the spill task force when activated

Immediate Response Team positions can be manned 24 hours per day or on a 24-hour standby basis depending on the nature of the facility and available along-route personnel. Typically, the dispatcher position will be manned on a 24-hour basis. Individuals filling response positions are listed in the contingency plan by name and method of notification.

Immediate response actions

When a spill is detected by monitoring equipment or reported through some other mechanism, the dispatcher immediately notifies the IRT. A typical alert and notification procedure is shown in Figure 3. The primary functions of the IRT are to provide for public safety, contain the spill, and exclude the spill from sensitive or vulnerable areas. For small spills, the IRT may be capable of complete control and cleanup. For larger spills, they provide stop-gap measures until greater resources can be mobilized.

The type of response is dependent on the specific spill area and the nature of the material. The intention of all actions is to locate the leak and begin emergency actions immediately. To facilitate such actions, the contingency plan must provide predesignated control sites and actions, equipment recommendations, and sufficient information on the physical and ecologic characteristics of the threatened area to allow flexible and defendable response actions. Figure 4 depicts a presentation of pre-planned response actions and support information for a typical pipe line segment.

Directions of probable spill movement, access points, and potential control sites are indicated on the pipe line segment map. Supporting information is tabularized and includes predicted maximum spill sizes at selected points, location of equipment storage, recommendations for primary control actions under varying conditions (letter codes refer to a contingency plan appendix which details respective technique implementation), special features of concern during spill control, and general hydrologic characteristics. Ecologic characteristics of the area are provided in an additional supporting appendix.

Flexible response actions

While it is a basic premise in spill planning that pre-planned response actions are appropriate, it is also accepted that no two spills are ever identical.

As a result, spill contingency planning must provide the flexibility to respond to unpredicted situations. This can be accomplished through the design of decision guides which equate the types and ranges of environmental parameters to appropriate control and recovery techniques. A representative decision guide for the protection of inland waters is shown in Figure 5. A pre-planned decision making pro-

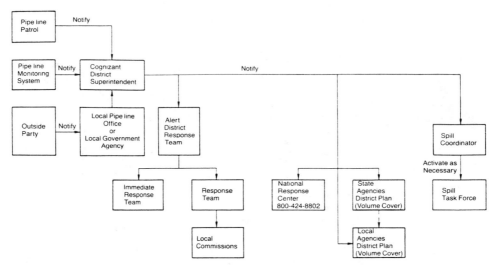

Figure 3. Alert and notification procedure.

PIPELINE MILEPOSTS   5.8 to 9.4   <u>COUNTY</u>   County C   <u>MAP REF.</u> 404-5

<u>MAXIMUM SPILL SIZE</u>   River 1: 1,682 bbl (16,148 bbl west of river)
Creek A:

<u>NEAREST EQUIPMENT SITE</u>   Site X:

SPILL CONTROL RECOMMENDATIONS

Location	Low Flow	High Flow	Terrestrial	Remarks
CS-1	F,G,H	G,H	-	Private property
CS-2	F,G,H	G,H	-	
CS-3	F,G,H	G,H	-	
CS-4	F,G	-	ABC	
CS-5	F,G,H	I,J,K,L	-	
CS-6	F,G,H	I,J,L	-	Limited access, if oil
CS-7	F,G,H	I,J,L	-	passes this point, contain in river at River 1
CS-8	F,G,H	G,H	-	
CS-9	F,G,H	G,H	-	Limited access, private property
CS-10	F,G,H	I,J,K	-	
CS-11	F,G,H	I,J,K	-	
CS-12	F,G,H	I,J	-	
CS-13	F,G,H	I,J	-	
CS-14	F,G,H	I,J	-	Private property

SPECIAL FEATURES/SPILL CONTROL PROCEDURES

1 Stream velocity during moderate to peak flow may preclude effective
booming. During lower flow periods bury boom skirt on banks and bars

- River 1 is a major salmon migration and spawning watercourse
- See Table C-29, Appendix C for an inventory of the Natural Resources
  of River 1.

WATER CROSSINGS/HYDROLOGIC CHARACTERISTICS

- Depth to water table generally exceeds 75 feet
- Irrigation ditch (mp 6.0)
- River 1 (mp 7.0)

Condition	Discharge	Velocity	Max. Width	Max. Depth	Occurrence
Approx. 100 yr. flow	8,800 cfs	8-9 fps	270-350 ft.	9-10 ft.	Dec. - Feb.
Low flow	70 cfs	3 fps	20 ft.±	2 ft.	Jun. - Aug.

- Unnamed Creek (mp 7.75):   perennial; water quality classification A
- Irrigation ditch (mp 7.9): underground
- Creek A (mp 8.6):   perennial; water quality classification A
- Unnamed Creek (mp 8.75):   perennial; water quality classification A

Figure 4. Typical pipe line segment information and response actions.

cess can be particularly useful in designing cleanup and disposal programs.

Training

Training of all spill response personnel, including simulated spill and equipment drills is critical. Accordingly, a comprehensive training program is an integral part of the contingency plan. The objectives of the training program are to:

- Maintain the plan as a fully operable and working document
- Inform Task Force members of their respective duties and standard communication procedures
- Allow Task Force members to become familiar with the use of all equipment and to update the plan to reflect the current state-of-the-art, including new equipment and procedures
- Modify the plan in the light of information gained from the field exercises
- Incorporate experience gained from response to spills into the plan

The training program is structured according to the level of responsibility of the participants and may include:

- Classroom instruction of field demonstrations
- Oil spill response drills to test notification, alert, and mobilization procedures
- Drills to practice specific containment measures and general containment and cleanup equipment and techniques

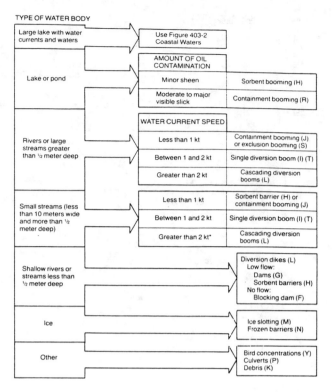

Figure 5. Decision guide for the protection of inland waters.

Conclusion

Accidental pipe line spills have and will continue to occur. Recognizing this fact, regulatory agencies have required emergency spill planning to minimize resulting impacts.

Comprehensive spill contingency planning satisfies these requirements and provides quick and effective response to environmental emergencies. Included in the preparation of such plans are the development of a response organization and response actions for each individual system. Also important in the development of a plan is the inclusion of a training and proficiency program.

(Based on a paper, Contingency Planning for Liquid Pipe Line Spills, presented by the authors as a contribution of the American Society of Mechanical Engineers at the Energy-Resources Conference and Exhibition, January 18–22, 1981, Houston, Texas.)

Source

Pipe Line Industry, March 1982.

20: Economics

Rule of thumb speeds payroll estimates 506
Rule of thumb estimates optimum time to keep construction
    equipment 507
How to estimate construction costs 509
Cost estimating strategies for pipe lines, stations, terminals.............. 512
Economics 519
Time value of money: concepts and formulas..................................... 524
Decision and evaluation criteria for investments and financial
    projects... 531
Accounting fundamentals 540
Estimate the cost of a pipe line in the U.S. (based on 1994 data) 545
How to compare the cost of operating an engine on diesel and
    natural gas... 545
How to estimate energy costs for different pipe line throughputs....... 545
Comparing fuel costs for diesel and electric prime movers 546
Nomograph for calculating scale-up of equipment or plant costs 546
Nomograph for calculating scale-up of tank costs 548
Nomograph for determining sum-of-years depreciation 549
Nomograph for estimating interest rate of return on investment
    ("profitability index").. 549
Nomograph for determining break-even point 551
Chart gives unit cost per brake horsepower of reciprocating
    compressors with various types of prime movers........................... 552
Chart shows influence on unit cost of numbers of reciprocating
    compressor units installed in one station 552
Chart gives unit cost per brake horsepower of centrifugal
    compressors with various types of prime movers........................... 553

Rule of thumb speeds payroll estimates

Hewlett Packard HP41-CV computation provides an estimate of the weekly expense of a "manpower spread" for pipe line construction in a given year

R. D. Green, Energy Center Fellow, University of Pennsylvania

Pipe line construction payroll expenses can be estimated in a rule-of-thumb computation that indexes 1982 survey data to any year of interest when an inflation figure is provided (software is provided in the appendix, p. 511).

An estimate of the weekly payroll expense of a "manpower spread" for pipe line construction in a given year is obtained as follows (Table 1):

$$P(y) = \left(\sum_{j=1}^{15} P_j E_j \right) e^{i(y-1982)}$$

where P_j = number of employees in category j
E_j = compensation expense per employee in category j
i = rate of inflation
y = year for the construction estimate

The present value (PV) of a stream of payroll expenses is found as follows:

TABLE 1—Compensation categories of manpower

Category* (j)	Position	Weekly compensation** (E_j)	Category* (j)	Position	Weekly compensation** (E_j)
I	Welding crew foreman	1,750	VII	Field office & overhead spreadman	1,100
II	Welding crew welders	1,670		Stringing foreman	
	Pipe crew foreman			Over-the-ditch, cleaning and priming crew foreman	
	Pipe crew stringer bead welder		VIII	Testing crew operator	1,050
	Pipe crew hot pass welders		IX	Field office & overhead asst. spreadman	1,000
	Tie-in welders				
III	Field office & overhead heavy-duty mechanics	1,250	X	R.O.W. clearing & fencing power saw operators	995
IV	Bending foreman	1,200		Pipe crew skid truck drivers	
	Bending engineer		XI	Field office & overhead lowboy truck driver	980
V	Field office & overhead fuel truck driver	1,175		Field office & overhead float truck driver	
	Field office & overhead grease truck driver			Pipe crew welder helpers	
VI	R.O.W. & fencing dozer operator	1,125		Tie-in welder helpers	
	R.O.W. grading foreman		XII	R.O.W. clearing & fencing laborers	960
	R.O.W. grading dozer operators			R.O.W. grading laborers	
	Ditching (earth) foreman			Ditching (earth) laborers	
	Ditching (earth) machine operators			Stringing truck drivers	
	Ditching (earth) backhoe operators			Stringing laborers	
	Road crossing road boring machine operator			Bending laborers	
	Road crossing backhoe operator			Pipe crew laborers	
	Stringing crane operator			Over-the-ditch, cleaning & priming crew laborers	
	Stringing sideboom operator			Tie-in laborers	
	Bending sideboom operator			Backfilling & clean-up laborers	
	Pipe crew side boom operator			Lower-in laborers	
	Pipe crew tack rig operators		XIII	R.O.W. clearing & fencing foreman	900
	Tie-in foreman		XIV	Field office & overhead office manager	800
	Tie-in sideboom operators			Field office & overhead payroll clerk	
	Lower in foreman		XV	Field office overhead night watchman	350
	Lower in sideboom operators				
	Testing crew foreman				
	Backfilling & clean-up foreman				
	Backfilling & clean-up dozer operators				

* Categories for hand-calculator program "CSTM".
** Average weekly compensation including overtime and bonuses, etc. (total compensation expense to employer). Figures are for professional pipeliners; compensation for local projects often handled by local or regional general contractors typically follow union published scales. This survey data is for 1982.

$$PV = \sum_{n=1}^{N} \frac{P(y)}{\left(1 + \dfrac{i}{52}\right)^n}$$

where N = number of weeks
      i = decimal "interest" rate

The above estimates are computed by the program CSTM (see appendix); the compensation data presented in Table 1 are incorporated into CSTM.

Estimate

If an estimate for P_j is needed, example manpower spreads for 8 and 30-in. cross-country pipe lines are provided in Table 2. In this case, an estimate of the number of weeks the spreads will be in operation is obtained as follows:

$$PD = \frac{L}{S \times d}$$

where PD = project duration in weeks
      L = length of the pipe line in miles
      S = average daily spread movement rate in miles per day
      d = number of days in the work week (e.g., d = 6 if the crews work 6 days per week)

TABLE 2—Typical manpower spreads

Category (j)	Numbers of employees in category j for 8 & 30-in. pipe line*	
	(P_j for 8-in. pipe line)	(P_j for 30-in. pipe line)
I	1	1
II	6	25
III	2	7
IV	1	3
V	1	5
VI	29	74
VII	3	4
VIII	1	3
IX	1	1
X	14	26
XI	11	36
XII	40	144
XIII	1	2
XIV	2	4
XV	1	1

* Cross-country pipe line construction.

For the manpower spreads of Table 2, S is probably in the range expressed by these inequalities:

1 mile/day $\leq$ S $\leq$ .35 miles/day for 8-in. size

Source

Pipe Line Industry, March 1985.

Rule of thumb estimates optimum time to keep construction equipment

Hewlett Packard HP41-CV computation estimates the optimum retention period for construction equipment

R. D. Green, Energy Center Fellow, University of Pennsylvania

Construction firm executives prefer cost-wise decisions concerning whether and when equipment will be replaced. The problem is extensive since, as depicted in Table 1, equipment spreads are both extensive and varied.

Equipment needed to construct a pipe line deteriorates with use, and maintenance costs rise over time. These maintenance costs might include, among other things, (1) funds for a preventive maintenance program; (2) funds to be set aside for costs incurred due to equipment failures; (3) replacement parts, and (4) labor expenses for repairs.

The problem of how to decide when to replace equipment is simplified with a "life cycle analysis" which compares (over an infinite time horizon) any number of alternatives of maintaining or replacing machines and then identifies the optimum choice. A "rule of thumb" approach to life cycle analysis for construction equipment can be performed by making some assumptions to simplify the usu-

ally complex calculations. Taking this approach, the assumptions and method are as follows.

Procedure

Let the original cost of the equipment item (e.g., a machine) be K. The problem is to find the number of years, N, that the machine should be kept such that the Net Present Worth (NPW) of operating costs is a minimum. This optimum number of years is called N* (read "N Star"). Suppose that the machine maintenance costs increase geometrically by a fraction, G, over the previous year and that the first year maintenance cost is C dollars. To simplify the equations, assume that all costs are paid as they are incurred; hence all costs are at present value. The relevant discount rate is R.

TABLE 1—Equipment spread

Typical spread—equipment type and number for 8-in. and 30-in. cross-country lines

	8"	30"		8"	30"
Field Office and Overhead			Sidebooms		3
Cars		2	Tow tractor		1
Pickups	2	7	Clean & prime machine		1
Welding rig		1	Coat & wrap machine		1
Fuel truck	1	1	Tar kettles		4
Grease truck	1	1	Winch truck		1
Utility truck	1	1	Flatbed truck		1
Tractor w/lowboy	1	1	Sets, cradles & strongbacks	2	2
Tractor w/float	1	1	Holiday detectors		2
Right of way, clear, burn, and fence			**Lower in**		
Pickups	3	2	Pickups		2
Dozers	2	2	Sidebooms		6
Winch truck		1	Clamshell		1
Flatbed truck	1	1	Dozer		1
Power saws	8	6	Trucks		2
Right of way, grade			Tar kettles		1
Pickup	2	1	Water pump		5
Dozers	4	3	**Tie in**		
Ditch (Earth)			Pickups	2	4
Pickups	2	7	Sidebooms	3	4
Ditching machines	1	2	Welders	1	4
Dozer w/ripper		1	Truck		1
Backhoes	3	4	Clamshell		1
Clamshell		1	Tar kettle		1
Ditch (rock)			Water pump		1
Sidebooms		2	**Testing**		
Wagon drills		2	Special truck	1	1
Compressors		2	Pumps, high pressure		2 sets
Trucks		2	**Utilities**		
Jackhammers		2	Pickup		1
Road crossings			Clamshell		1
Pickup	1	1	Clamps and cutters		1 set
Boring machine	1	1	Welders		2
Sideboom		1	**Backfill and clean up**		
Winch truck		1	Pickups	2	4
Tar kettles		1	Backfiller		1
Back hoe	1	1	Dozers	3	3
Stringing			High lift		1
Pickup	1	1	Farm tractor	1	1
Crane hoist	1	1	Dragline		1
Tractors w/pipe trailers	3	7	Winch truck		1
Sideboom	1	1	Flatbed truck		1
Tow tractors	1	3	**Other***		
Bending			Base station	1	3
Pickups	1	2	Two-way radios	11	33
Sideboom	1	1	Comfort station/toilet	3	9
Bending machine	1	1	Skids	5000	15000
Transit & level rod		1	Helicopters		
Pipe crew			Internal line-up clamp		
Pickup	1	1	External line-up clamp	5	15
Tractor welders	1	2	Tamper tractor		
Sidebooms		2	External plant coating machine		
Trucks		2	Vacuum-lift unit		
Set inside clamps		1	Laybarge		
Buffing rig	1	1	Motor patrol/grader		
Welding crew			Crane		
Pickup		1	Kitchen		
Welders	3	11	Trailer/sleeping		
Tow tractor		1	Medical station		
Crew bus		1	First aid kits	12	36
Coating crew					
Pickups	1	4			

* Survey data, 1982.

** Source: *A Manual of Pipeline Construction*, 3rd Ed., Dallas, Energy Communications, 1979, except for "other" category.

*** In "Other" category only: 30" data is estimated by multiplying 8" data by three.

Note: For more information on construction spreads, see *Cost Estimating Manual For Pipelines And Marine Structures*, John S. Page, Houston, Gulf Publishing Co., 1977.

The solution to this problem is to derive an equation that allows a comparison for machines with different lifetimes. This is done by finding the NPW of an infinite stream of constant scale replications. The analysis is performed by using the (algorithm) equations below to find an optimum replacement period for each alternative considered:

Find N^* such that:

$$f(N_i) - f(N_{i+1}) \leq 0 \text{ and } f(N_{i-1}) - f(N_i) > 0$$

where

$$f(N) = K + C \left(\frac{1 - \left(\frac{1+G}{1+R} \right)^N}{R - G} \times \frac{(1+R)^N}{(1+R)^N - 1} \right)$$

and

$$f(N+1) = K + C \left(\frac{1 - \left(\frac{1+G}{1+R} \right)^{N+1}}{R - G} \times \frac{(1+R)^{N+1}}{(1+R)^{N+1} - 1} \right)$$

and where
K = the initial cost of the machine
C = the first year maintenance cost
G = the fraction of increased maintenance costs (e.g., G = .20 when maintenance costs, C, are rising at 20% per year compounded annually)
R = the rate (e.g., R = .15 for a discount rate of 15%)

A stopping rule for operating the algorithm is to halt the computation when the value of N is no longer decreasing (e.g., the computation is over when, as N decreases, an N for which $f(N_i) - f(N_{i+1}) \leq 0$ is finally obtained). An example program and program run is given in the Appendix (written in the language of the Hewlett Packard HP41-CV). Figure 1 is a flowchart that is useful for interpreting the algorithm and translating the program into another computer language.

The procedures for choosing among the technical and managerial options (e.g., alternatives of repairing, replacing, rehabilitating, abandoning or decreasing the work load of machines) for managing the construction spread are: (1) Given i options, use the algorithm to identify the optimum replacement period N^* for each equipment replacement alternative (that is, given i alternatives, find all N^*_i); (2) For each N^*_i, there is an associated NPW_i. Compare these net present worths and identify the least expensive one (e.g., of several N^* values, each with an associated cost and equipment replacement scheme, the least expensive one is the smallest NPW_i); (3) Usually, the equipment replacement plan associated with this minimum NPW_i will be selected and then to implement the decision, that machine will be replaced every N^* years (where the N^*_i of interest is the one for which NPW_i is a minimum); (4) A simple sensitivity analysis may be performed by reworking

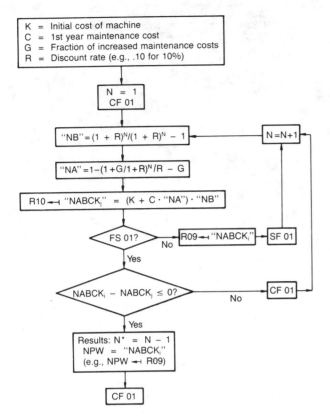

K = Initial cost of machine
C = 1st year maintenance cost
G = Fraction of increased maintenance costs
R = Discount rate (e.g., .10 for 10%)

N = 1
CF 01

"NB" = $(1 + R)^N/(1 + R)^N - 1$

"NA" = $1-(1+G/1+R)^N/R - G$

R10 ⊸ "NABCK$_i$" = $(K + C \cdot "NA") \cdot "NB"$

FS 01? → No → R09 ⊸ "NABCK$_i$" → SF 01

Yes

NABCK$_i$ − NABCK$_j$ ≤ 0? → No → CF 01

Yes

Results: N* = N − 1
NPW = "NABCK$_i$"
(e.g., NPW ⊸ R09)

CF 01

N = N+1

Figure 1. Flow chart for interpreting the algorithm and translating the program into another computer language.

items 1 through 3 for various values (e.g., for checking the sensitivity to interest rates, compute N* for various R values).

"Cutoff rates" may be used to decide when equipment failures are too high despite favorable cost curves for repairs since the pipe line construction firm must be reliable in serving customers, but these cutoff rates should be selected in recognition of the N* values for the various alternatives to avoid short-circuiting the whole decision process.

The result of the above procedure is a cost-wise strategy for the management of the equipment used in pipe line construction spreads.

If the assumptions of the life cycle analysis are modified to more accurately reflect the rising maintenance costs over time, then a new equation must be derived. If the analytical problem becomes unwieldy, an option is to solve the problem via a simulation; the saddle of the cost curve can then be found using a numerical searching method.

Another perspective is to view N* as the optimum retention period, e.g., the length of time that a machine should be kept if costs are to be minimized. Costs would be greater if the machine were kept shorter or longer than the ideal duration expressed by N*.

Source

Pipe Line Industry, August 1985.

Acknowledgments

Prof. Zandi, Civil and Urban Engineering Department, University of Pennsylvania; Dr. Pollack-Johnson, Department of Mathematics, Oberlin College, Ohio; and Mr. V. James Joyce, Joyce Western Corp., Andover, N.Y.

How to estimate construction costs

Two hand-held calculator programs divide pipe line costs into four component areas: right of way, material, labor, and miscellaneous

R. D. Green, Energy Center Fellow, University of Pennsylvania

Pipe line cost models are numerous and their uncertainty is associated with the degree of generality.

Costs vary with "distances from manufacturers, terrain, locality, labor costs, season of the year and other variables."[1] Hence, it is vital to the interests of construction firms to maintain historical records of costs and to update these records routinely. A well developed model may be used as a management tool during construction to identify *either* discrepancies between the model and "reality," or to encourage project superintendents to avoid cost over-runs and outperform previous time and motion constants.

A detailed account of cost estimating has been written by J. S. Page (1977);[2] the work is similar to the approach taken by contractors seeking to estimate costs of manpower, materials, etc., based on historical time frames and the like. Pipe line cost models that identify cost parameters and the inter-relationship of these parameters have been developed by I. Zandi (1982)[3] and K. Gimm and I. Zandi (1976).[4]

A rule of thumb calculation has also been computed (1964)[5] and it is updated here. Table 1, "Pipe line rule of thumb cost estimate," presents a means of estimating the cost based on costs of pipe lines in the U.S. The "1964" fig-

TABLE 1—Pipe line rule of thumb cost estimate
Cost = $D \times A \times$ Miles of pipe line

Year	Diameter of pipe, inches	Average cost US$ (inch diameter × mile)	Range US$ (in. dia. × mile) low	high	Average % inflation since 1964*
1964**	10	2,059	–	–	–
	16	2,849	–	–	–
	20	2,632	–	–	–
	24	3,325	–	–	–
	30	4,085	–	–	–
	36	4,260	–	–	–
1980***	8	16,920	6,953	24,125	–
	12	18,446	13,882	16,644	–
	16	10,823	9,200	12,368	8.3
	20	14,226	12,039	16,698	10.5
	24	14,738	8,184	28,263	9.3
	30	16,730	13,100	28,756	8.8
	36	25,468	19,087	33,929	11.2

* Compounded continuously; from $100\% \times \left(\ln \left(\frac{A\ 1980}{A\ 1964} \right) \right) \div 16.$

** Source is *Pipe Line Rules of Thumb Handbook*,[1] and they base these figures on 1964 data from the Federal Power Commission.

*** Based on averages of costs for onshore projects from FERC Construction Permit applications as reported in the *International Petroleum Encyclopaedia*.[5]

TABLE 2—Areal division of pipe line costs

Part I: By area, % of total pipe line costs over 10-year period:*

Size (inches)	ROW	Material	Labor	Misc.	Total
8	2.8	37.2	49.2	10.8	100.0
12	5.7	37.3	44.3	11.2	98.5
16	3.9	46.6	38.8	10.5	99.5
20	3.1	46.0	41.8	9.4	100.3
24	3.1	46.3	38.0	11.8	99.2
30	3.7	49.6	37.9	8.5	99.7
36	4.1	48.1	37.2	10.6	100.0
Average	3.8	44.4	41.0	10.4	99.6

Part II: Simplified version of Part I.

Size	ROW	Material	Labor	Misc.	Total
8	3.5	37.	49.	10.5	100.0
12	5.7	37.	44.3	10.5	97.5
16	3.5	47.	38.75	10.5	99.75
20	3.5	47.	38.75	10.5	99.75
24	3.5	47.	38.75	10.5	99.75
30	3.5	47.	38.75	10.5	99.75
36	3.5	47.	38.75	10.5	99.75
Average	3.8	44.1	41.0	10.5	99.46

Part III: Version used in hand-calculator program "CSTA."

(D inches)	ROW	Material	Labor	Misc.	Total
$D \leq 8$	3.5	37	49.	10.5	100.0
$8 < D \leq 12$	5.7	37	44.3	10.5	97.5
$12 < D \leq 36$	3.5	47	38.75	10.5	99.75
Average	–	–	–	–	–

* Tabulated from FERC based data presented in the *International Petroleum Encyclopaedia*.[5]

ures are listed for comparison only and are from the (1978) *Pipe Line Rules of Thumb Handbook*.[1]

Adjusting for inflation

The pipe line rule of thumb cost estimate is made more useful by including a means for adjusting the "1980" based formula of Table 1 to other years:

$$C(y) = DAme^{i}\ (y - 1980)$$

where $C(y)$ = cost of the pipe line in year y

D = diameter of the pipe line

A = average cost of pipe line construction per in.-diameter-mile

m = length of pipe line in miles

i = continuously compounded rate of inflation (e.g., % inflation from Table 1 divided by 100)

y = year for which the cost estimate is desired

Hand calculator program CST performs this cost calculation when the user inputs 1980 based data from Table 1; an example run is included in the appendix.

Estimate for cost areas

The cost of a pipe line may be divided into four component cost areas: Right of way (ROW), material, labor, miscellaneous.

An estimate of any one of these component costs (C_i) is obtained from the following:

$$C_i = a_i C(y)$$

where a_i = 10-year average proportion of total cost allocated for one of the four areas

$C(y)$ = total cost of the pipe line (or an estimate)

The a_i's are taken from Table 2, Part III, in the program CSTA (see appendix), according to the size (diameter) of the pipe line given the following divisions:

$D \geq 8$
$8 < D \leq 12$
$12 < D \leq 36$

Appendix—Software for Hewlett Packard HP41-CV

CSTA

Given both the cost and size of a pipe line, this program estimates the division of expenses between four categories: Right of way (ROW), material, labor and miscellaneous. Table 2 forms the basis of the program, and in Part III of the table it is apparent that the program user may elect to use pipe line sizes (diameters) for which no data is supplied. For example, a 10-inch line is viewed as a 12-inch line by the program.

```
PROGRAM "CSTA"
01 ▲ LBL "CSTA"
02 CF 01          Clear flags.
03 CF 02
04 "COST?"        User inputs pipe line construction cost
                  and diameter of line (D).
05 PROMPT
06 STO 03
07 "D?"
08 PROMPT
09 STO 04
10 8              Is D≤8?, 8<D≤12?, or 12<D≤36?
11 X( )Y
12 X( = Y?
13 SF 01          Set flags accordingly (1, 2, or neither).
14 FS? 01
15 GTO A
16 RCL 04
17 12
18 X( )Y?
19 X( = Y?
20 SF 02
21 FS? 02
22 GTO "AA"
23 ▲ LBL A        ROW cost calculation for D ≠ x
                  such that 8 < x ≤ 12.
24 RCL 03
25 .035
26 *
27 STO 08
28 GTO "AAA"
29 ▲ LBL "AA"     "ROW" cost calculation for D such
                  that 8<D≤12.
30 RCL 03
31 .057
32 *
33 STO 08
34 ▲ LBL "AAA"
35 "R.O.W. = "    output
36 PRA
37 RCL 08
38 PRX
39 FS? 01         "Material" calculation.
40 GTO B
41 FS? 02
42 GTO B
43 RCL 03
44 .47
45 *
```

CST

This program is to be used with the "1980" data of Table 1. The program estimates the cost of a pipe line from the equation,

$$\text{Cost} = D \times A \times \text{Miles} \times \exp\left(\frac{\% \text{ inflation}}{100} \times \# \text{ years}\right)$$

The exponential in this cost equation is merely a means for adjusting the "1980" figures to other years. User input is as follows: "Year" (e.g., 1983; the program also accepts years prior to the base year 1980), "INFLATION" (e.g., 8.3%; this will be compounded continuously), "D" (e.g., 16; diameter of pipe line from Table 1, "1980" category), "A" (e.g.,10,823; average cost from Table 1, "1980" category), "MILES" (e.g., 20.5; length of pipe line in miles).

Program output is as follows: "COST" estimate of pipe line in

U.S. dollars, based on U.S. data (e.g., 4,553,660); "IN YEAR", is the year for which the cost estimate has been computed (e.g., 1983).

```
PROGRAM "CST"
01 ▲ LBL "CST"
02    "YEAR?"       1980 is the base year; the program ac-
                    cepts any year prior to, after or includ-
                    ing 1980.
03    PROMPT
04    1980
05    -
06    STO 00
07    "INFLATION?"  Inflation rate (in percent) is any figure
                    selected by the user; it is continuously
                    compounded.
08    PROMPT
09    100
10    /
11    STO 01
12    "D?"          Diameter of pipe line.
13    PROMPT
14    "A?"          Average cost in units of US$ per (inch
                    diameter × mile).
15    PROMPT
16    *
17    "MILES?"      Length of pipe line in miles.
18    PROMPT
19    *
20    STO 02
21    RCL 01
22    RCL 00
23    *
24    E↑X           Compound factor is exp (rate × time).
25    RCL 02
26    *
27    STO 03
28    "COST = "     Cost estimate of pipe line in U.S. dol-
                    lars for a specified year. The cost esti-
                    mate is for construction and encom-
                    passes an average of material, labor,
                    right of way, and miscellaneous ex-
                    penses, as presented in Table 1.
29    PRA
30    RCL 03
31    PRX
32    "IN YEAR-"
33    PRA
34    RCL 00
35    1980
36    +
37    PRX
38    END
EXAMPLE RUN
```

```
46 STO 05
47 GTO C
48 ▲ LBL B
49 RCL 03
50 .37
51 *
52 STO 05
53 ▲LBL C
54 "MATERIAL = "   output
55 PRA
56 RCL 05
57 PRX
58 FS? 01          "Labor" calculation.
59 GTO D
60 FS? 02
61 GTO E
62 RCL 03
63 .3875
64 *
65 STO 06
66 GTO F
67 ▲ LBL D
68 RCL 03
69 .49
70 *
71 STO 06
72 GTO F
73 ▲ LBL E
74 RCL 03
75 .43
76 *
77 STO 06
78 ▲ LBL F
79 "LABOR = "      output
80 PRA
81 RCL 06
82 PRX
83 "MISC. = "      "Misc." calculation and output.
84 PRA
85 RCL 03
86 .105
87 *
88 STO 07
89 PRX
90 "TOTAL = "      "Total" = ROW + Material + Labor + Misc.
                   Due to assumptions of model (see Table 2), "Total"
                   might be different from input cost.
91 PRA
92 RCL 08
93 RCL 05
94 +
95 RCL 06
96 +
97 RCL 07
98 +
99 PRX             output
100 END
EXAMPLE RUN
```

	XEQ "CST"	
YEAR?		After program has been entered, user executes the program and responds to the prompts; after each data entry, the user presses the "RUN" button. User assignable keys may be defined to facilitate rapid program execution.
	1983	RUN
INFLATION?		
	8.3	RUN
D?		
	16	RUN
A?		
	10,823	RUN
MILES?		
	20.5	RUN
COST =		
	4,553,660.	···
IN YEAR =		
	1983	···

	XEQ "CSTA"
COST?	
5,605,238.00	RUN
D?	
8.0	RUN
R.O.W. =	
196,183.33	···
MATERIAL =	
2,073,938.06	···
LABOR =	
2,746,566.62	···
MISC. =	
588,549.99	···
TOTAL =	
5,605,238.00	···

The percentages listed in Table 2 are simple averages of pipe line construction cost data for a specific pipe line size:

$$a_i \times 100 = \sum_{y=1971}^{1980} \frac{A_y}{C_y}$$

where A_y = average pipe line cost for a given component cost area (e.g., ROW) in a given year

C_y = average total pipe line cost in a given year (e.g., sum of average costs for a given year in the four categories)

Source

Pipe Line Industry, September 1984.

References

1. *Pipe Line Rules of Thumb Handbook*, Gulf Publishing Company, 1978, p. 163.
2. *Cost Estimating Manual For Pipelines and Marine Structures*, Gulf Publishing Company, 1977.
3. "Coal Slurry Pipeline Cost Model," *Journal of Pipelines*, Vol. 2, No. 3, 1982, pp. 53–66.
4. "Freight Pipeline Cost Model," *Transportation of Solid Commodities Via Freight Pipeline*, Department of Transportation (DOT/RSPA/DPB-50/78/33-39), 1978.
5. "Pipelining Continues To Set New Records," *International Petroleum Encyclopaedia, 14th ed.*, Petroleum Publishing Co., 1981, pp. 368–373.

Cost estimating strategies for pipe lines, stations, terminals

Microcomputers and up-to-date estimating software are more efficient means for cost studies

R. D. Green, Energy Center Fellow, University of Pennsylvania, Folcroft, Pa.

Cost estimating strategies vary in the accuracy of representing cost factor data. Cost factors for pipe lines, stations and terminals are presented in Tables 1 and 2.

In the pipe line industry, the practice has been to assume that the most accurate cost estimates for constructing pipe lines, stations and terminals will come from a very detailed estimating procedure that sums the expected value or best guess for each cost factor involved. Such an assumption ignores more accurate cost estimating strategies that require many more computations and have therefore traditionally been avoided.

Historically, cost factor information has been organized in files, tables and ledgers. Estimating has been a lengthy procedure taking days, if not weeks, of work effort by a full staff of estimators. Now, with microcomputers and up-to-date estimating software, we have a more efficient means of storing, retrieving and handling data. It is time to reconsider the use of more accurate estimating strategies.

By accurate, I mean representing data in a form that identifies what is and is not known about the cost factor.

Current, future estimating practices

Currently, spreadsheet based cost models for pipe lines, stations and terminals sum the simply represented costs for the cost factors given in Tables 1 and 2. Future construction estimating practices will likely involve two changes from the current practices.

These changes will be a response to the two problems estimators face. One problem is how to speed the computational procedure that yields the final or grand total project cost estimate and the other is how best to represent cost factor data. This article considers these two problems.

Speeding computational procedures

Estimating computations are often lengthy and redundant. Estimating can be performed quickly either by simplifying the computations (often with an associated loss in accuracy) or by computerization in which case accuracy need not be sacrificed. Current and future construction estimating procedures involve the use of spreadsheet-based cost models. There are three basic types of spreadsheets.

- *Spreadsheets* are typically large ledger-like sheets of paper upon which are spread the cost factors and costs that go in to the cost estimating procedure in estimating the costs of a construction project.
- *Electronic spreadsheets* are a computerized version of the spreadsheet in which cost factors are listed and the estimator fills in the costs for the appropriate items and the computer automatically tallies the results.
- *Automated electronic spreadsheets* are an automated version of the electronic spreadsheet in the sense that much of the data, e.g., costs and cost factors, are provided by the computer rather than by having the estimator fill in the blanks with costs. The computer has a database to turn to for cost and material information, thus the estimator need not look up such information in compiling the estimate.

The time and effort required to produce routine estimates may be reduced by using an automated electronic

TABLE 1—Cost factors in pipe line estimate

Design engineering

Electrical & mechanical; engineering-supervisor; engineer, drafting; corrosion-engineer, drafting; planning engineering-supervisor; engineer; survey & drawing; aerial photography; stake out; as-built information & drawing; data processing-supervisor; programmer; travel; supplies.

Right of way, legal, environmental & safety

Acquisition labor; damages labor; permit/legal labor; environmental & safety labor; travel; field office rental; supplies; considerations-existing line; rights, new R/W rights; damages-damage payments, land; purchases, permit payments-permit fees; cost of deposits/bonds; first year rentals/leases.

Construction

Labor–electrical, mechanical; corrosion, inspection labor, expenses-regular, temporary; field office rental, equipment; rentals, supplies; inspection–corrosion-coating mill; pipe line, survey, mechanical-pipe mill; pipe line, X-ray, tuboscope, caliper pig; contract–pipe line installation–open area, moderate area, congested; install casing; supplementary work–install valves, purge line, caliper pig (contractor's labor), river crossing, hydrostatic test, gravitometer, station work, buildings, ground bed, rectifier, anodes, environmental requirements, performance bond; extra work–rock removal–blasting, air hammer; sand padding; contingencies.

Materials

Instrumentation–transmitters, gravitometers, miscellaneous; mainline pipe (size, grade, & wall), vent & marker, freight charges; casing pipe (size, grade & wall), freight charges; valves (size, series, type); fittings (description, size, series); valve operators; buildings, furnishings & fencing (size, type); coating (pipe size, type); cathodic protection–rectifiers; test stations; insulation joints; anodes (type); seals and insulators; data processing.

Overheads

Indirect labor (managers and above); travel; supplies; interest; fringe benefits.

TABLE 2—Cost factors in station and terminal estimate

Design engineering

Electrical & mechanical; engineering-supervisor; engineer, drafting; corrosion-engineer, drafting; planning engineering-supervisor; engineer; supplies; civil-engineer, drafting; surveying-property, topograph, aerial, soil borings; stake out/as-built-crew; draftsman; data processing-supervisor; programmer; travel; expenses; supplies.

Right of way, legal, environmental & safety

Acquisition/damages labor; permit/legal labor; environmental & safety; travel; field office rental; supplies; payments-considerations; damages; land purchase; title searches; permits & fees–first year rental/leases.

Construction

Labor-electrical, mechanical; corrosion, expenses–regular & temporary; office & equipment rental; supplies; inspection-mechanical; electrical; X-ray; soil testing; concrete testing; contracts; corrosion-anodes; test stations; deep well; groundbed; rectifier; set pole; insulating joint; painting (including sandblasting, primer, all coats); tanks; piping; pipe coating; excavation, machine, backfill; cable, substation-install; transformers; install switchgear; install fuse/circuit breakers; structure work; motors-set motor, install motor; control; starters; instrumentation-install transmitters, recorders, etc.; tank gauges; ticket printers; meter totalizers; electrical cabinets; valve operators; gravitometers; miscellaneous; conduit & wire; lighting-set poles, mount lights; install computer; other-concrete, excavation; modular building; mechanical-welds, valves, cuts & bevels; pipe, civil-excavation, backfill; concrete; stoning; fencing; gates; water well; septic system; building; oil/water separator; seeding; extra work–corrosion, electrical, mechanical, civil; electrical & mechanical-manpower; equipment.

Material

Motors; motor control center; starters; substation and transformers–transformer, substation structure, lightning arrestors, miscellaneous fittings; switchgear; instrumentation-general-gravitometers, transmitters, receivers, power supply, alarm annunciator, recorders, totalizers; ticket printers; switches-pressure, temperature, level; vibration monitor; cabinets; micro pumps; other & miscellaneous; instrumentation-tanks, tank gauges, transmitters, other; lighting-pole, lights; miscellaneous items; circuit breakers; wire & conduit; miscellaneous special equipment; mechanical-pipe (size, grade & wall), valves (size, series, type); fittings (description, size, series); valve operators; pumps & pump/motor combinations–spare rotating element; meters; prover loop; meter/prover skid; strainers; tanks-large storage tanks, other; communication-radios, other; fire protection-pipe, valves, fittings, fire extinguishers, foam proportioners, foam tank, foam; environmental control equipment-boat, spill boom, vapor recovery system; miscellaneous and special equipment; civil-buildings, building furnishings; oil water separator; structural steel supports; stone; special equipment; corrosion-coating, anodes, insulating joint; test stations; cable; rectifier; other; paint; data processing-computers, CRT displays, printers, disks, RTU; special equipment.

Overheads

Indirect labor-directors & managers of engineering, planning & analysis, construction, right of way, other; travel; supplies; interest; fringe benefits.

Administrative overheads

Administrative labor; travel expenses; supplies.

Indirect costs

Labor; materials, pipe, other.

---

spreadsheet. Given the tremendous data handling abilities of even low cost computers vis-a-vis manual computation, cost estimating may be performed using various strategies ranging from the traditional to the more complex.

A fully integrated cost estimating approach enables the estimator to rely on data that have as their basis historical, current and forecast aspects of the project. Since such an approach is conveniently packaged as a (high-speed) computerized estimating program, the estimator may base estimates on cost factor data in any of the strategies for representing data.

Three estimating strategies

The three basic strategies for obtaining an estimate differ in the way uncertainty is viewed and represented. Each strategy is characterized by the type of estimates that are utilized:

- A point estimate views factors in the estimate as quantifiable without regard to error, range and uncertainty (such estimates are rarely viewed as conclusive since managers often ask estimators, "How good are these numbers?").

- A range estimate is generally more sophisticated than the point estimate and may yield, for example, an "expected cost of pipe line construction, ranging as low as . . . and as high as. . . ."
- A probability estimate is the most sophisticated estimate since it does not ignore uncertainty and is based on an evaluation of probability distributions for (a) historical data, (b) current conditions, and (c) projections of future conditions.

Probability estimates may be depicted in simplified forms without regressing to a range or point estimating strategy. In particular, a probability distribution can be represented with some loss of accuracy as any of several simplified distributions (such as a histogram or a triangular distribution or some smooth distribution that is beta or normal).

One problem with point estimates is that these do not convey an awareness of any uncertainty behind the estimate. Point and range estimates can be misleading. As Hertz describes, the "estimated probability distribution paints the clearest picture of all possible outcomes,"[1] and he gives examples in the area of capital investment projects. Risk analysis is designed to provide more information than conventional techniques which fail, as Hespos and Strassman note, for not taking into account the "skewed factors," and for being "influenced by the subjective aspects of best guesses."[2]

To illustrate the subjective aspects, note that an estimator typically places the "best point estimate" in the low end if a high outcome is more serious than a low outcome. And the estimator's "best point estimate" is in the high end if a low outcome is more serious than a high outcome. The point estimate is a result of the estimator's judgments about uncertainties, outcomes, and risks or values of the consequences of outcomes. Probability estimates are more useful than point estimates since they more accurately represent what is and is not known about cost factors. There is less need or motivation to distort probability estimates (e.g., these are less vulnerable to subjective aspects).

Although the range estimate conveys the notion that a precise estimate is not available, it does not convey any probability of where the actual cost will fall. Range estimates are often misleading since the actual cost may fall not only within but outside of the range specified. Range estimates are usually defined by two point estimates (e.g., low and high).

The probability estimate conveys the most information and offers a means of expressing the uncertainty of the estimate.

Probability estimates of uncertain parameters are made by converting point forecasts to probabilistic forecasts. This is done generally according to either the historical records of the firm or the "guess-timates" provided by experienced personnel.

For example, where the future is statistically like the past (e.g., no identifiable characteristics such as cycles, trends, etc.), probability estimates for future cost factors can be based on historical relative frequency distributions that may have been "refined" by interpolation or smoothing. Concisely stated, the probability estimate conveys information concerning all that is and is not known about the estimated cost.

Therefore, it can visually represent the likeliness that the real value will be, for example, below a given cost level. This is especially important for construction contractors who engage in bidding to perform work on a project.

For the contract-engineering and construction firms, the variability in construction costs must be closely monitored in view of the competitive nature of the construction industry where profits are related to the accuracy of forecasting engineering and construction costs. As for pipe line operating firms that often have work performed by outside contractors, there is a need to pay attention to the variability in construction costs for both large and small projects.

The database ought to reflect the variability in cost factors so that uncertainties for large projects, which can impact the firm's capital position, are more clearly comprehended. Also, for an operating firm that has a large proportion of small construction projects relative to large projects, most of the data gathering opportunities will be missed if the management posture is one of ignoring variability for small projects.

Cost estimating strategies for pipe lines, stations, terminals (Part 2)

Developing a cost model for automated electronic spreadsheets may involve three estimating strategies

R. D. Green, Energy Center Fellow, University of Pennsylvania, Folcroft, Pa.

Developing a cost model (e.g., an automated electronic spreadsheet) for pipe lines, stations and terminals, which provides a probability estimate of cost, may involve the use of three estimating strategies. For example, the variability in cost for many cost factors may be so limited that it is feasible to use point estimates. This is often the case for material estimates.

Similarly, range estimates may be used where probability estimates will not add any significant information (e.g., where a probability distribution is approximately "mesa" shaped). But for the cost factors that exhibit great variability, it is essential to use probability estimates (estimated probability distributions).

Manpower (cost) estimates are often in this category. Whether or not historical data are available, probability estimates can be created. For example, based on the experience of an estimator, a three-point estimate can be obtained. Or if adequate data are available, histograms can be used to estimate the probability density functions (probability estimates). Some commonly used approaches for estimating the probability density functions are:

- The three-point estimate approach (which may involve any of several distributions, e.g., beta or the triangular distribution)
- The histogram approach (which may closely approximate the "smooth curve" of a probability distribution).

Three-point estimate approach

The three-point estimate, e.g., low (L), expected (E), and high (H), can be used to estimate the probability distribution for the cost factor. Given the simplifying assumption that the probability distribution for the cost factor is approximately beta, and that the standard deviation equals one-sixth of the range of reasonably possible costs for factors, then the estimated expected value of a given cost factor is:

$$C_e = (1/3) \times [2E + (1/2) \times (L + H)]$$

The variance is the square of the standard deviation:

$$s^2 = ((1/6) \times (H - L))^2$$

Given many cost factors that have been estimated in this way, the expected value of the final or grand total cost is:

$$G = \sum_{i=1}^{n} C_{e_i}$$

The variance for G is:

$$V_G = \sum_{i=1}^{n} s_i^2$$

The final or grand total cost may be viewed as a sum of many independent random variables. By invoking the general version of the central limit theorem, the probability distribution of such a sum is approximately normal. Given the expected value G and the variance V_G, it is then straightforward to find the probability that this normal random variable (grand total project cost, G) will be less than a given (e.g., "bid," B) price. This is done by finding the area under the standard normal curve from negative infinity to Z where:

$$Z = (B - G)/\sqrt{V_G}$$

That is, the estimated probability that the actual grand total cost is less than the given bid price is:

$$P = .5 + \int_{0}^{x=Z} 1/\sqrt{2\pi} - x^2/e^2 \, dx$$

This equation is solved by invoking Simpson's rule for several intervals and yields the following approximation (if more accuracy is needed, more intervals can be used):

$$P = .5 + \frac{\frac{Z}{6}}{3\sqrt{2\pi}} \times \left(1 + \left[\frac{4}{\frac{Z^2}{e^{72}}}\right] + \left[\frac{2}{\frac{Z^2}{e^{18}}}\right] + \left[\frac{4}{\frac{Z^2}{e^8}}\right] + \left[\frac{2}{\frac{2Z^2}{e^9}}\right] + \left[\frac{4}{\frac{25Z^2}{e^{72}}}\right] + \left[\frac{1}{\frac{Z^2}{e^2}}\right]\right)$$

Histogram approach

The histogram approach can be a more accurate means of representing the estimated probability distributions for cost factors (rather than using the three estimate approach) since bimodal and skewed characteristics, etc., can be portrayed. Depending on the mix of estimating strategies and the types of probability estimates involved, the final cost estimate will be obtained in part via a simulation by Monte Carlo sampling. Further details of this approach are beyond the scope of this article.

Observations and conclusion

The advantage of a well-developed electronic spreadsheet is that it can automatically maintain historically based data in the form of, for example, histograms that may be sampled later on to extract information. This supports future efforts to obtain probability estimates for cost

	A	B	C	D	E	F	G
1	CST						
2	DATA:	PIPE DIA.	AVG. COST	RANGE		AVG. % INFLATION	
3		D INCHES	A	LOW TO	HIGH	SINCE 1964	
4		8	$16920	$6953	$24125	8.3	
5		12	$18446	$13882	$16644	8.3	
6		16	$10823	$9200	$12368	8.3	
7		20	$14226	$12039	$16698	10.5	
8		24	$14738	$8184	$28263	9.3	
9		30	$16730	$13100	$28756	8.8	
10		36	$25468	$19087	$33929	11.2	
11	YEAR?	1983	expected	lo	hi		
12	D?	16.00					
13	MILES?	20.50					
14	A	$10823	GIVEN D=	16.00	24.00		
15	INFLATION	8.3					
16	_____	AVERAGE	LOW	HIGH			
17	COST=	$4,553,662.26	$3,870,802.25	$5,203,704.60			
18	_____	_____	_____	_____			
19	CSTA						
20	DATA:						
21		8	2.8	37.2	49.2	10.8	100.0
22		12	5.7	37.3	44.3	11.2	98.5
23		16	3.9	46.6	38.8	10.5	99.5
24		20	3.1	46.0	41.8	9.4	100.3
25		24	3.1	46.3	38.0	11.8	99.2
26		30	3.7	49.6	37.9	8.5	99.7
27		36	4.1	48.1	37.2	10.6	100.0
28	FOR COST=	$4,553,662.26	$3,870,802.25	$5,203,704.60			
29	FOR D =	16					
30	_____	POINT ESTIMATE	_____				
31	ROW	$178,485.25	GIVEN D=	16	24		
32	MATERIAL	$2,132,669.95					
33	LABOR	$1,775,699.44					
34	MISC.	$480,537.22					
35	TOTAL	$4,567,391.86					
36	_____	RANGE	ESTIMATE	_____			
37		LO	HI				
38	ROW	$151,719.89	$203,964.30				
39	MATERIAL	$1,812,858.13	$2,437,111.89				
40	LABOR	$1,509,418.36	$2,029,183.29				
41	MISC.	$408,476.62	$549,134.66				
42	TOTAL	$3,882,472.98	$5,219,394.13				
43	_____	PROBABILITY	ESTIMATE	_____			
44	B =	$6,000,000		Ce		s square	
45			ROW	$178,270.87		75,818,855	
46			MATERIAL	$2,130,108.30		10,824,798,700	
47			LABOR	$1,773,566.56		7,504,321,729	
48			MISC.	$479,960.03		549,574,535	
49			G =	$4,561,905.74	VG =	18,954,513,700	
50						Z =10.445543	
51							
52							0.2315
53						THE PROBABILITY	1.0000
54						OF THE ACTUAL	0.8789
55						GRAND TOTAL COST	0.0047
56						BEING LESS THAN	0.0000
57						A GIVEN COST "B"	0.0000
58						IS APPROXIMATELY	0.0000
59						EQUAL TO "P"	0.0000
60						WHERE	0.5000
61						P =	0.9361

Figure 1. Worksheet example of an electronic spreadsheet.

```
A1 CST
A2 DATA:
A11 YEAR?
A12 D?
A13 MILES?
A14 A
A15 INFLATION
A16 _____
A17 COST=
A18 _____
A19 CSTA
A20 DATA:
A21 8
A22 12
A23 16
A24 20
A25 24
A26 30
A27 36
A28 FOR COST=
A29 FOR D =
A30 _____
A31 ROW
A32 MATERIAL
A33 LABOR
A34 MISC.
A35 TOTAL
A36 _____
A38 ROW
A39 MATERIAL
A40 LABOR
A41 MISC.
A42 TOTAL
A43 _____PROBABILITY ESTIMATE_____
A44 B =
B2 PIPE DIA.
B3 D INCHES
B4 8
B5 12
B6 16
B7 20
B8 24
B9 30
B10 36
B11 1983
B12 16
B13 20.5
B14 @LOOKUP(D14,B4:F10,1)
B15 @LOOKUP(D14,B4:F10,4)
B16 AVERAGE
B17 B12*B13*B14*@EXP((B15/100)*(B11-1980))
B18 _____
B21 2.8
B22 5.7
B23 3.9
B24 3.1
B25 3.1
B26 3.7
B27 4.1
B28 B17
B29 B12
B30 POINT ESTIMATE
B31 @LOOKUP(D31,A21:A27,1)*B28/(@LOOKUP(D31,A21:A27,5))
B33 @LOOKUP(D31,A21:A27,3)*B28/(@LOOKUP(D31,A21:A27,5))
B34 @LOOKUP(D31,A21:A27,4)*B28/(@LOOKUP(D31,A21:A27,5))
B35 @SUM(B31,B34)
B36 _____ RANGE    ESTIMATE _____
B37 LO
B38 @LOOKUP(D31,A21:A27,1)*C17/(@LOOKUP(D31,A21:A27,5))
B39 @LOOKUP(D31,A21:A27,2)*C17/(@LOOKUP(D31,A21:A27,5))

B40 @LOOKUP(D31,A21:A27,3)*C17/(@LOOKUP(D31,A21:A27,5))
B41 @LOOKUP(D31,A21:A27,4)*C17/(@LOOKUP(D31,A21:A27,5))
B42 @SUM(B38,B41)
B43 _____PROBABILI
B44 1E+10
C2 AVG. COST
C3 A
C4 16920
C5 18446
C6 10823
C7 14226
C8 14738
C9 16730
C10 25468
C11 expected
C14 GIVEN D=
C16 LOW
C17 B12*B13*(@LOOKUP(D14,B4:B10,2))*@EXP((B15/100)*(B11-1980))
C18 _____
C21 37.2
C22 37.3
C23 46.6
C24 46
C25 46.3
C26 49.6
C27 48.1
C28 C17
C30 _____
C31 GIVEN D=
C36 ESTIMATE ____
C37 HI
C38 @LOOKUP(D31,A21:A27,1)*D17/(@LOOKUP(D31,A21:A27,5))
C39 @LOOKUP(D31,A21:A27,2)*D17/(@LOOKUP(D31,A21:A27,5))
C40 @LOOKUP(D31,A21:A27,3)*D17/(@LOOKUP(D31,A21:A27,5))
C41 @LOOKUP(D31,A21:A27,4)*D17/(@LOOKUP(D31,A21:A27,5))
C42 @SUM(C38,C41)
C43 TY ESTIMATE___
C45 ROW
C46 MATERIAL
C47 LABOR
C48 MISC.
C49 G =
D2        RANGE
D3    LOW TO
D4 6953
D5 13882
D6 9200
D7 12039
D8 8184
D9 13100
D10 19087
D11 lo
D14 @IF B12<=8 THEN 8 ELSE ( @IF B12<=12 THEN 12 ELSE
( @IF B12<=16 THEN 16 ELSE ( @IF B12<=20 THEN 20 ELSE E14)))
D16 HIGH
D17 B12*B13*(@LOOKUP(D14,B4:B10,3))*@EXP((B15/100)
*(B11-1980))
D18 _____
D21 49.2
D22 44.3
D24 41.8
D25 38
D26 37.9
D27 37.2
D28 D17
D29 @IF B29<=8 THEN 8 ELSE ( @IF B29<=12 THEN 12 ELSE
( @IF B29<=16 THEN 16 ELSE ( @IF B29<=20 THEN 20 ELSE E31)))
D36 _
D43 _____
D44 Ce

D45 (1/3)*(2*B31+(1/2)*(B38+C38))
D46 (1/3)*(2*B32+(1/2)*(B39+C39))
D47 (1/3)*(2*B33+(1/2)*(B40+C40))
D48 (1/3)*(2*B34+(1/2)*(B41+C41))
D49 @SUM(D45,D48)
E2 NGE
E3 HIGH
E4 24125
E5 16644
E6 12368
E7 16698
E8 20263
E9 28756
E10 33929
E11 hi
E14 @IF B12<=24 THEN 24 ELSE ( @IF B12<=30 THEN 30 ELSE 36)
E21 10.8
E22 11.2
E23 10.5
E24 9.4
E25 11.8
E26 8.5
E27 10.6
E31 @IF B29<=24 THEN 24 ELSE ( @IF B29<=30 THEN 30 ELSE 36)
E43 _____
E49 VG =
F2 AVG. % INFLATION
F3 SINCE 1964
F4 8.3
F5 8.3
F6 8.3
F7 10.5
F8 9.3
F9 8.8
F10 11.2
F21 100
F22 98.5
F23 99.5
F24 100.3
F25 99.2
F26 99.7
F27 100
F43 _____
F44 s square
F45 ((C38-B38)/6)*((C38-B38)/6)
F46 ((C39-B39)/6)*((C39-B39)/6)
F47 ((C40-B40)/6)*((C40-B40)/6)
F48 ((C41-B41)/6)*((C41-B41)/6)
F49 @SUM(F45,F48)
F50 Z =
F53 THE PROBABILITY
F54 OF THE ACTUAL
F55 GRAND TOTAL COST
F56 BEING LESS THAN
F57 A GIVEN COST "B"
F58 IS APPROXIMATELY
F60 WHERE
F61 P =
G43 _____
G50 (B44-D49)/@SQRT(F49)
G52 (G50/6)/(3*@SQRT(2*@PI))
G53 1/@EXP(0)
G54 4/@EXP(G50*G50/72)
G55 2/@EXP(G50*G50/18)
G56 4/@EXP(G50*G50/8)
G57 2/@EXP(2*G50*G50/9)
G58 4/@EXP(25*G50*G50/72)
G59 1/@EXP(G50*G50/2)
G60 0.5
G61 (G52*(@SUM(G53,G59)))+G60
```

Figure 2. Program listing of an electronic spreadsheet.

factors. The system may accommodate inflation rates that vary among the cost factors. These probability estimates enable the estimator to communicate information about the project risks and the uncertainty of the estimate so that the right risks can be taken by the executive or managerial component of pipe line construction and operating firms.

An implication for the future is related to the current "data policy" of the pipe line organization. In view of the estimating tools that are needed by the pipe line industry, data currently being gathered and preserved ought to reflect the variability of costs for the cost factors. Such data are important to the future of the pipe line organization.

Example spreadsheet

A simple automated electronic spreadsheet example is provided (with software). It illustrates the use of "lookup tables" as a means of accessing a "built-in" data base. In this example, a rule of thumb estimate[3] is used for estimating the probability distribution of pipe line construction costs in several cost areas. An overall distribution is obtained by using the procedure described above (three-point estimate approach). The estimated probability distribution for the grand total cost of the project is used in a step to assess the probability that the project cost is less than a given "bid" price. Given relatively few inputs (year of construction, pipe line diameter and length), and a limited database, this simple automated electronic spreadsheet finds point, range and probability estimates for several cost areas. The expected grand total cost G = \$4,561,905.74 is surrounded by considerable uncertainty. In a full-scale automated electronic spreadsheet, the variance surrounding the cost factors and expected grand total cost will be greatly reduced relative to this simple example.

A possible improvement upon this example is as follows. If more representative distributions for the cost factors are obtained and presented as histograms, then the estimated probability distribution for grand total cost will be obtained through a simulation that performs a summation involving random sampling of histograms (histogram approach).

Overall, the example provides an introductory automated electronic spreadsheet. Given some project-specific information, and the historical record held within the program, cost projections are effortlessly produced. A large-scale model would work in a similar way but with more detailed data to represent the many cost factors given in Figures 1 and 2.

Source

Pipe Line Industry, September 1986.

Acknowledgments

Mr. B. H. Basavaraj, Facilities Planning Supervisor, Buckeye Pipe Line Co., Emmaus, Pa.; Prof. I. Zandi, Civil and Urban Engineering Dept., University of Pennsylvania; Mr. Andy Anderson and Mr. K. N. Wheelwright, Weyher/Livsey Constructors, Inc., Salt Lake City, Utah.

References

1. Hertz, D. B., "Risk Analysis in Capital Investment," *Harvard Business Review*, Sept.–Oct. 1979, pp. 169–180.
2. Hespos, R. F. and Strassmann, P. A., "Stochastic Decision Trees for The Analysis of Investment Decisions," *Management Science*, August 1965, p. 11.
3. Green, R. D., "Rules of thumb, How to estimate construction costs," *Pipe Line Industry*, Vol. 61, No. 3, Sept. 1984, pp. 47–48.

APPENDIX

A worksheet and program listing for the example spreadsheet discussed above are provided in Figs. 1 and 2. The software and computer used is Syncalc on the Atari 800XL. (Note the similarities to VisiCalc and Lotus-123 electronic spreadsheets.) For industrial scale work, a more powerful system such as the Atari 520ST or IBM XT/AT computer offers an ideal system for developing a full-scale automated electronic spreadsheet for pipe line construction estimating. For more information on estimating, see *Cost Estimating Manual for Pipelines and Marine Structures* by John S. Page (Gulf Publishing Co., Houston, 1977).

Economics

The goal of all economic evaluations should be to eliminate all unprofitable projects. Some borderline, and perhaps a few good projects, will be killed along with the bad ones. However, it is much better to pass up several good projects than to build one loser.

Every company should develop a standard form for their economic calculations, with "a place for everything and everything in its place." As management becomes familiar with the form, it is easier to evaluate projects and choose among several based on their merits, rather than the presentation.

Components of an Economic Study

Most economic studies for processing plants will have revenue, expense, tax, and profit items to be developed. These are discussed in the following sections.

Revenues. The selling price for the major product and the quantity to be sold each year must be estimated. If the product is new, this can be a major study. In any case, input is needed from sales, transportation, research, and any other department or individual who can increase the accuracy of the revenue data. The operating costs may be off by 20% without fatal results, but a 20% error in the sales volume or price will have a much larger impact. Further, the operating costs are mostly within the control of the operator, while the sales price and volume can be impacted by circumstances totally beyond the control of the company.

Depending on the product and sales arrangement, the revenues calculation can take from a few man-hours to many man-months of effort. For instance, determining the price for a synthetic fuel from coal can be done in a very short time, based on the cost of service. However, if the price is tied to the price of natural gas or oil, the task becomes very difficult, if not impossible. On the other hand, determining the sales growth and selling price for a new product requires a great deal of analysis, speculation, market research, and luck, but projections can be made.

Most revenue projections should be on a plant net back basis with all transportation, sales expense, etc., deducted. This makes the later evaluation of the plant performance easier.

Operating Costs. The operating costs, as developed in the previous section, can be presented as a single line item in the economics form or broken down into several items.

Amortization. Items of expense incurred before start-up can be accumulated and amortized during the first five or ten years of plant operations. The tax laws and corporate policy on what profit to show can influence the number of years for amortization.

Such items as organization expense, unusual prestart-up expenses, and interest during construction, not capitalized, can be amortized. During plant operations this will be treated as an expense, but it will be a non-cash expense similar to depreciation, which is discussed later.

Land Lease Costs. If there are annual payments for land use, this item of expense is often kept separate from other operating costs, since it does not inflate at the same rate as other costs. If the land is purchased, it cannot be depreciated for tax purposes.

Interest. The interest for each year must be calculated for the type of loan expected. The interest rate can be obtained from the company's financial section, recent loan history, or the literature. The *Wall Street Journal* publishes the weekly interest rates for several borrowings. The company's interest rate will probably be the banks' prime rate, or slightly higher.

The calculation of interest is usually based on the outstanding loan amount. For example, semi-annual payments are expected, the amount of the first year's interest (I) would be:

$$I = \left(\frac{i}{2}\right)P + \left(\frac{i}{2}\right)(P - PP) \tag{1}$$

where   i = Annual interest rate
        P = Original principal amount
        PP = First principal payment

The equation for any annual payment would be:

$$I_y = \sum_{n=1}^{n=N} \left(\frac{i}{N}\right)(P_{y,n-1} - PP_{y,n-1}) \tag{2}$$

where I_y = Interest for any year
        $P_{y,n-1}$ = Principal or loan balance during the previous period
        n = Period for loan payment
        N = Total number of payments expected each year
        y = Year during the loan life
        $PP_{y,n-1}$ = Principal payment made at the end of the previous period

Debt Payments. The method of making principal and interest payments is determined by the terms of the loan agreement.

Loans can be scheduled to be repaid at a fixed amount each period, including principal and interest. This is sim-

ilar to most home mortgage payments. The size of level payments can be calculated using the formula:

$$I + PP = P \frac{\left(\frac{i}{n}\right)\left(1 + \frac{i}{n}\right)^{n \times y}}{\left[\left(1 + \frac{i}{n}\right)^{n \times y}\right] - 1} \tag{3}$$

In this type of loan, the principal payment is different each period, so the calculation of interest using Equation 2 must take into account the variable principal payment. Even though both interest and principal are paid together, they must be kept separate. Interest is an expense item for federal income taxes, while principal payments are not.

Alternatively, the debt repayment can be scheduled with uniform principal payments and varying total payments each period. The principal payment can be calculated by the formula:

$$PP_{y,n-1} = PP = \frac{P}{n \times y} \tag{4}$$

In some loans a grace period is allowed for the first six to twelve months of operation, where no principal payments are due. The economics must reflect this.

Depreciation. This is a noncash cost that accumulates money to rebuild the facilities after the project life is over. There are usually two depreciation rates. One is the corporation's book depreciation, which reflects the expected operating life of the project. The other is the tax depreciation, which is usually the maximum rate allowed by tax laws. To be correct, the tax depreciation should now be called "Accelerated Capital Cost Recovery System" (ACRS).

Economics can be shown using either depreciation rate. If book depreciation is used, a line for deferred income tax must be used. This is a cash available item that becomes negative when the tax depreciation becomes less than the book depreciation. At the end of the book life, the cumulative deferred income tax should be zero.

The 1984 allowable ACRS tax depreciation is shown in Table 1. Most chemical plants have a five-year tax life.

Federal Income Tax. The calculation of the federal income tax depends on the current tax laws. State income taxes are often included in this line item, since many are directly proportional to the federal taxes.

Table 1
ACRS Schedule in % Per Year of Depreciable Plant

Year	Years for Recovery			
	3	5	10	15
1	25	15	8	5
2	38	22	14	10
3	37	21	12	9
4		21	10	8
5		21	10	7
6			10	7
7			9	6
8			9	6
9			9	6
10			9	6
11–15				6

The federal tax laws are complicated and ever-changing. It is recommended that expert tax advice be obtained for most feasibility studies. If no current tax advice is available, many tax manuals are published each year that can be purchased or borrowed from a library.

Current tax laws allow deductions from revenues for sales expenses, operating costs, amortizations, lease costs, interest payments, depreciation, and depletion.

The net income before federal income tax (FIT) is calculated by subtracting all of the deductions from the gross revenues. If there is a net loss for any year, it can be carried forward and deducted the next year from the net income before FIT. If not all of the loss carried forward can be used in the following year, it can be carried forward for a total of seven years.

At times, tax laws allow a 10% investment tax credit for certain qualifying facilities. Thus, 10% of the qualifying investment can be subtracted from the first 85% of the FIT due. If the taxes paid in the first year are not enough to cover the investment tax credit, it can sometimes be carried forward for several years.

Exhaustible natural deposits and timber might qualify for a cost depletion allowance. Each year, depletion is equal to the property's basis, times the number of units sold that year, divided by the number of recoverable units in the deposit at the first of the year. The basis is the cost of the property, exclusive of the investment that can be depreciated, less depletion taken in previous years.

At times, mining and some oil and gas operations are entitled to an alternate depletion allowance calculated from a percentage of gross sales. A current tax manual should be consulted to get the percentage depletion allowance for the mineral being studied.

Percentage depletion for most minerals is limited to a percentage of the taxable income from the property receiving the depletion allowance. For certain oil and gas producers the percentage depletion can be limited to a percentage of the taxpayer's net income from all sources.

The procedure for a feasibility study is to calculate both the cost and percentage depletion allowances, then choose the larger. Remember, either type reduces the basis. However, percentage depletion can still be deducted after the basis reaches zero.

The cash taxes due are calculated by multiplying the book net income before tax, times the tax rate, then adjusting for deferred tax and tax credits, if any.

Economics Form

As stated at the start of this chapter, it is desirable to establish a form for reporting economics. Table 2 shows such a form. By following this form, the procedure can be seen for calculating the entire economics.

As can be seen in Table 2 all revenues less expenses associated with selling are summed in Row 17. All expenses including non-cash expenses such as depreciation, amortization, and depletion are summed in Row 30. The net profit before tax, Row 32, is obtained by subtracting Row 30 from Row 17 and making any inventory adjustment required. Row 34 is the cash taxes that are to be paid unless offset by investment or energy tax credits in Row 36. The deferred income tax is shown in Row 35. The deferred tax decreases the net profit after tax in the early years and increases the net profit after tax in later years. The impact on cash flow is just the other way around as discussed later. Row 37, profit after tax, is obtained as follows:

Row 32 (profit before tax)
−Row 34 (income tax)
−Row 35 (deferred tax)
+Row 36 (tax credits)
=Row 37 (profit after tax)

Row 33, loss carried forward, is subtracted from Row 32 before calculating the income tax.

The cash flow, Row 39, is calculated as follows:

Row 37 (profit after tax)
+Row 35 (deferred tax)
+Row 29 (depletion)
+Row 28 (amortization)
+Row 27 (depreciation)

−Row 38 (principal payments)
=Row 39 (cash flow)

If one of the rows is a negative number, such as deferred tax, the proper sign is used to obtain the correct answer. Any required inventory adjustment is also made.

Table 2
Economic calculation example

| Fiscal years | Periods/Year = 1 | |
	1992	1993
12 Sales revenue	239,093	248,657
13 Byproduct revenue	17,031	17,713
14 Advert'g & market'g	(1,638)	(1,703)
15 Shipping & freight	(1,310)	(1,363)
16 Other sales expense	(983)	(1,022)
17 Income from sales	252,194	262,282
18 Raw material cost	76,075	78,738
19 Operating labor	26,508	27,701
20 Maintenance labor	23,984	25,063
21 Maint'nce material	37,869	39,573
22 Utilities	25,246	26,382
23 Administr. & general	9,467	9,893
24 Unchanging expense	5,000	5,000
25 Temporary expense	0	0
26 Interest	7,614	7,188
27 Depreciation	5,500	5,500
28 Amortization	3,639	2,139
29 Depletion	12,621	19,541
30 Total Cost	233,524	246,719
31 Inventory adjust't	0	0
32 Profit before tax	18,671	15,563
33 Loss carried forw'd	12,621	4,031
34 Income tax	0	7,835
35 Deferred tax	2,783	(2,530)
36 Tax credits	0	7,835
37 Profit after tax	15,888	18,093
38 Principal payments	4,105	4,531
39 Cash flow	36,325	38,212

Economic Analysis

Once the cash flow has been calculated, various yardsticks can be applied to determine the desirability of the project. The years for payout (payback) is often used. This is obtained by adding the cash flow each year, from the first year of construction, until the sum of the negative and positive cash flows reaches zero. The payout is usually reported in years to the nearest tenth, i.e., 5.6 years.

Each company has its own cutoff period for payout, depending on the economy, the company position, alternate projects, etc. One company used five years as the maximum payout acceptable for several years. As conditions changed, this was reduced to three years, then one year. It was hard to find projects with one year payout after taxes.

Some companies feel they should earn interest on their equity as well as get a payout. To calculate this, the cash flows are discounted back to year 0 using some arbitrary interest (discount) rate, then the positive discounted cash flows are summed each year of operation until they equal the sum of the discounted negative cash flows. This yardstick also works but consistency must be maintained from one project to another. The discount (interest) rate and payout both become variables that must be set and perhaps changed from time to time depending on economic conditions inside and outside the company.

A simple return on equity can be used as a yardstick. Utilities are required by regulators to set their rates to obtain a given rate of return. This return is usually constant over the years and from project to project, so management must choose among projects on some basis other than profit.

In free enterprise projects, the return on equity will probably vary from year to year so an average would have to be used. However, use of an average would consider return in later years to be as valuable as income in the first year of operation, which is not true.

The best way to evaluate free enterprise projects is to use discounted cash flow (DCF) rate of return, sometimes called internal rate of return.

This calculation can be expressed as follows:

$$SDCF = \sum_{y=1}^{y=N} \frac{CF_y}{(1+i)^y} = 0 \tag{5}$$

where SDCF = Sum of the equity discounted cash flows
CF_y = Cash flow, both negative and positive for each year y
i = Discount rate in %/100 that results in the sum being zero
N = Total number of years

Since this calculation uses trial and error to find a discount rate at which all discounted negative and positive cash flows are equal, it can be tedious without a computer. Table 3 demonstrates this calculation for a simple set of cash flows. The DCF rate of return results in a single number that reflects the time value of money. It is certainly one of the yardsticks that should be used to measure projects.

Sensitivity Analysis

Once the economic analysis has been completed, the project should be analyzed for unexpected as well as expected impacts on the economics. This is usually done through a set of "what if" calculations that test the project's sensitivity to missed estimates and changing economic environment. As a minimum, the DCF rate of return should be calculated for ±10% variations in capital, operating expenses, and sales volume and price.

Table 3
Discounted cash flow calculations

Discount Rate (%)	Discounted Cash Flows			
Year	0	20	30	25.9
1	−20	−17	−15	−16
2	−20	−14	−12	−12
3	−10	−6	−5	−5
4	10	5	3	4
5	15	6	4	4
6	20	7	4	5
7	25	7	4	5
8	30	7	4	5
9	40	8	4	5
10	50	8	4	5
Totals	+140	+11	−5	00

The impact of loss of tax credits, loss carry forward and other tax benefits should be studied. Such things as start-up problems that increase the first year's operating costs and reduce the production should also be studied. With a good economic computer program, such as described earlier, all of these sensitivity analyses can be made in an afternoon.

Additional reading on economics can be found in the References.

References

1. Perry, R. H. and Chilton, C. H., *Chemical Engineers Handbook,* Fifth Edition, McGraw-Hill, Inc., New York, section 25, p. 3.
2. Constantinescu, S., "Payback Formula Accounts for Inflation," *Chemical Engineering,* Dec. 13, 1982, p. 109.
3. Horwitz, B. A., "How Does Construction Time Affect Return?" *Chemical Engineering,* Sept. 21, 1981, p. 158.
4. Kasner, E., "Breakeven Analysis Evaluates Investment Alternatives," *Chemical Engineering,* Feb. 26, 1979, pp. 117–118.
5. Cason, R. L., "Go Broke While Showing A Profit," *Hydrocarbon Processing,* July, 1975, pp. 179–190.
6. Reul, R. I., "Which Investment Appraisal Technique Should You Use," *Chemical Engineering,* April 22, 1968, pp. 212–218.
7. Branan, C. and Mills, J., *The Process Engineer's Pocket Handbook, Vol. 3,* Gulf Publishing Co., 1984.

TIME VALUE OF MONEY: CONCEPTS AND FORMULAS

The value of money is not constant, but changes with time. A dollar received today is worth more than a dollar received a month from now, for two reasons. First, the dollar received today can be invested immediately and earn interest. Second, the purchasing power of each dollar will decrease during times of inflation, and consequently a dollar received today will likely purchase more than a dollar next month. Therefore, in the capital budgeting and decision-making process for any engineering or financial project, it is important to understand the concepts of *present value, future value,* and *interest.* Present value is the "current worth" of a dollar amount to be received or paid out in the future. Future value is the "future worth" of an amount invested today at some future time. Interest is the *cost* of borrowing money or the *return* from lending money. Interest rates vary with time and are dependent upon both risk and inflation. The rate of interest charged, and the rate of return expected, will be higher for any project with considerable risk than for a "safe" investment or project. Similarly, banks and investors will also demand a higher rate of return during periods of monetary inflation, and interest rates will be adjusted to reflect the effects of inflation.

Simple Interest vs. Compound Interest

Simple interest is interest that is only earned on the original principal borrowed or loaned. Simple interest is calculated as follows:

I = (P) (i) (n)

where: I = simple interest
P = principal (money) borrowed or loaned
i = interest rate per time period (usually years)
n = number of time periods (usually years)

Compound interest is interest that is earned on both the principal and interest. When interest is compounded, interest is earned each time period on the original principal *and* on interest accumulated from preceding time periods. Exhibits 1 and 2 illustrate the difference between simple and compound interest, and show what a dramatic effect interest compounding can have on an investment after a few years.

Exhibit 1
Comparison Between $5,000 Invested for 3 Years at 12% Simple and 12% Compound Interest

	Simple Interest		
End of Year	Interest Earned	Cumulative Interest Earned	Investment Balance
0	$0.00	$0.00	$5,000.00
1	$5,000.00 × .12 = $600.00	$600.00	$5,600.00
2	$5,000.00 × .12 = $600.00	$1,200.00	$6,200.00
3	$5,000.00 × .12 = $600.00	$1,800.00	$6,800.00

	Compound Interest (Annual Compounding)		
End of Year	Interest Earned	Cumulative Interest Earned	Investment Balance
0	$0.00	$0.00	$5,000.00
1	$5,000.00 × .12 = $600.00	$600.00	$5,600.00
2	$5,600.00 × .12 = $672.00	$1,272.00	$6,272.00
3	$6,272.00 × .12 = $752.74	$2,024.64	$7,024.64

Exhibit 2
Comparison of Simple vs. Compound Interest

Nominal Interest Rate vs. Effective Annual Interest Rate

Almost all interest rates in the financial world now involve compound rates of interest, rather than simple interest. Compound interest rates can be quoted either as a *nominal rate* of interest, or as an *effective annual rate* of interest. The nominal rate is the stated annual interest rate compounded periodically (at the stated compounding time interval). The effective annual rate is the rate that produces the same final value as the nominal rate, when compounded only once per year. When the compounding period for a stated nominal rate is one year, then the nominal rate *is the same* as the effective annual rate of interest. The following formula can be used to convert nominal interest rates to effective annual rates:

$$i_{eff} = \left(1 - \frac{i_{nom}}{c}\right)^c - 1$$

where: i_{eff} = effective annual rate
i_{nom} = nominal rate
c = number of compounding periods per year

Note: interest rates are expressed as fractions (.12) rather than percentage (12%).

For *continuous* compounding, natural logarithms may be used to convert to an effective annual rate:

$$i_{eff} = e^{i_{cont}} - 1$$

where: i_{cont} = continuous compounding rate

Note: interest rates are expressed as fractions (.12) rather than percentage (12%).

Present Value of a Single Cash Flow To Be Received in the Future

How much is an amount of money to be received in the future worth today? Stated differently, how much money would you invest today to receive this cash flow in the future? To answer these questions, you need to know the *present value* of the amount to be received in the future. The present value (PV) can easily be calculated by *discounting* the future amount by an appropriate compound interest rate. The interest rate used to determine PV is often referred to as the *discount rate, hurdle rate,* or *opportunity cost of capital.* It represents the opportunity cost (rate of return) foregone by making this particular investment rather than other alternatives of comparable risk.

The formula for calculating present value is:

$$PV = \frac{FV}{(1+r)^n}$$

where: PV = present value
FV = future value

r = discount rate (per compounding period)
n = number of compounding periods

Example. What is the present value of an investment that guarantees to pay you $100,000 three years from now? The first important step is to understand what risks are involved in this investment, and then choose a discount rate equal to the rate of return on investments of comparable risk. Let's say you decide that 9% effective annual rate of interest (9% compounded annually) is an appropriate discount rate.

$$PV = \frac{\$100,000}{(1+.09)^3} = \$77,218$$

Thus, you would be willing to invest $77,218 today in this investment to receive $100,000 in three years.

Future Value of a Single Investment

Rearranging the relationship between PV and FV, we have a simple formula to allow us to calculate the future value of an amount of money invested today:

$$FV = PV (1 + r)^n$$

where: FV = future value
  PV = present value
  r = rate of return (per compounding period)
  n = number of compounding periods

Example. What is the future value of a single $10,000 lump sum invested at a rate of return of 10% compounded monthly (10.47 effective annual rate) for 5 years? Note that monthly compounding is used in this example.

$$FV = \$10,000 (1 + .10/12)^{60} = \$16,453$$

The $10,000 investment grows to $16,453 in five years.

The Importance of Cash Flow Diagrams

Many financial problems are not nearly as simple as the two previous examples, which involved only one *cash outflow* and one *cash inflow*. To help analyze financial problems, it is extremely useful to diagram cash flows, since most problems are considerably more complex than these previous examples, and may have multiple, repetitive, or irregular cash flows. The *cash flow diagram* is an important tool which will help you understand an investment project and the timing of its cash flows, and help with calculating its PV or FV. The cash flow diagram shown represents the earlier example (in which PV was calculated for an investment that grew to $100,000 after annual compounding at 9% for 3 years).

The cash flow diagram is divided into equal time periods, which correspond to the compounding (or payment) periods of the cash flows. Note that this diagram is also presented from the investor's (lender's) point of view, since the initial investment is a cash outflow (negative) and the final payoff is a cash inflow (positive) to the investor. The dia-

gram could also have been drawn from the borrower's point of view, in which case the PV would be a cash inflow and therefore positive. Before drawing a cash flow diagram, you must first choose which viewpoint (lender or borrower) the diagram will represent. It is then extremely important to give each cash flow the correct sign (positive or negative), based upon whichever point of view is chosen. Other examples of cash flow diagrams will be presented in additional sample problems in this chapter.

Analyzing and Valuing Investments/Projects with Multiple or Irregular Cash Flows

Many investments and financial projects have multiple, periodic, and irregular cash flows. A cash flow diagram will help in understanding the cash flows and with their analysis. The value of the investment or project (in current dollars) may then be found by discounting the cash flows to

determine their present value. Each cash flow must be discounted individually back to time zero, and then summed to obtain the overall PV of the investment or project. This *discounted cash flow formula* may be written as:

$$PV = \sum_{i=1}^{n} \frac{C_i}{(1+r)^i}$$

where: PV = present value
  C_i = cash flow at time i
  n = number of periods
  r = discount rate of return (opportunity cost of capital)

Example. How much would you be willing to invest today in a project that will have generated positive annual cash flows of $1,000, $1,500, $2,400, and $1,600, respectively, at the end of each of the first 4 years? (Assume a 10% annual rate of return can be earned on other projects of similar risk.)

Step 1: Cash flow diagram:

Step 2: Discount cash flows into current dollars (present value):

$$PV = \frac{1,000}{(1+.1)^1} + \frac{1,500}{(1+.1)^2} + \frac{2,400}{(1+.1)^3} + \frac{1,600}{(1+.1)^4}$$
$$= \$5,044.74$$

Thus, you should be willing to invest a maximum of $5,044.74 in this project.

Perpetuities

A *perpetuity* is an asset that provides a periodic fixed amount of cash flow in perpetuity (forever). Endowments and trust funds are often set up as perpetuities, so that they pay out a fixed monetary amount each year to the beneficiary, without depleting the original principal. For valuation purposes, some physical assets such as an oil well, natural gas field, or other long-lived asset can also be treated as a perpetuity, if the asset can provide a steady cash flow stream for many years to come. The cash flow diagram for a perpetuity is represented here:

The value of a perpetuity can be found by this simple formula:

$$PV_{perpetuity} = \frac{C}{r}$$

where: C = cash flow per time period
  r = discount rate of return (per time period)

A *growing perpetuity* is a cash flow stream that grows at a constant rate. The formula to value a growing perpetuity expands the basic perpetuity formula to include a growth rate for the cash flow stream:

$$PV = \frac{C_1}{r-g}$$

where: C_1 = initial cash flow
  r = discount rate of return (per time period)
  g = growth rate of cash flows (per time period)

Example. An existing oil well has been producing an average of 10,000 barrels per year for several years. Recent estimates by geologists show that this well's oil field is not large enough to justify additional drilling, but can maintain the existing well's production rate for hundreds of years to come. Assume that the recent price of oil has averaged $20 per barrel, oil prices are historically estimated to grow 3% a year forever, and your discount rate is 10%. Also assume that maintenance costs are insignificant, and the land has no value other than for the oil. What is the current monetary value of this oil field and well, if the owner were to sell it?

Solution:

$$PV = \frac{(10,000 \times 20)}{.10 - .03} = \$2,857,143$$

The value of the oil field and well should be $2,857,143 based upon these assumptions.

Future Value of a Periodic Series of Investments

Both individual investors and companies often set aside a fixed quantity of money every month, quarter, or year, for some specific purpose in the future. This quantity of money then grows with compound interest until it is withdrawn and used at some future date. A cash flow diagram for a periodic investment is shown.

C (fixed investment each time period)

It can easily be calculated how much these periodic, fixed investments each time period will accumulate with compounding interest to be in the future. In other words, we want to determine the future value (FV), which can be calculated using the following formula:

$$FV = \frac{C}{r}[(1+r)^t - 1]$$

where: C = cash flow invested each time period
r = fractional compound interest rate (each time period)
t = number of time periods (length of investment)

Example. A self-employed contract engineer has no company-paid retirement plan, so he decides to invest $2,000 per year in his own individual retirement account (IRA), which will allow the money to accumulate tax-free until it is withdrawn in the future. If he invests this amount every year for 25 years, and the investment earns 14% a year, how much will the investment be worth in 25 years?

Solution:

$$FV = \frac{\$2,000}{.14}[(1+.14)^{25} - 1] = \$363,742$$

The periodic investment of $2,000 each year over 25 years will grow to $363,742 if a 14% rate of return is achieved. This example illustrates the "power of compounding." A total of only $50,000 was invested over the 25-year period, yet the future value grew to over 7 times this amount.

Annuities, Loans, and Leases

An *annuity* is an asset that pays a constant periodic sum of money for a fixed length of time. Bank loans, mortgages, and many leasing contracts are similar to an annuity in that they usually require the borrower to pay back the lender a constant, periodic sum of money (principal and interest) for a fixed length of time. Loans, mortgages, and leases can therefore be considered "reverse" annuities, as shown by the cash flow diagrams.

Cash Flow Diagrams: Annuity

Cash Flow Diagrams: Reverse Annuity

The relationship between the value (PV) of the annuity/loan/lease, its periodic cash flows, and the interest or discount rate is:

$$PV = \frac{C}{r}\left[1 - \frac{1}{(1+r)^t}\right]$$

where: C = cash flow received/paid each time period
  r = interest or discount rate (per time period)
  t = number of time periods (time length of the annuity)

This formula may be used to determine the PV of the annuity, or conversely, to determine the periodic payouts or payments when the PV is known. It should be noted that the future value (FV) at time "t" will be zero. In other words, all the principal of the annuity will have been expended, or all the principal of the loan or lease will have been paid off, at the final time period. Financial calculators are also pre-programmed with this formula, and can simplify and expedite annuity, loan, and other cash flow problem calculations.

Example. Oil pipeline equipment is to be purchased and installed on a working pipeline in a remote location. This new equipment will have an operational life of 5 years and will require routine maintenance once per month at a fixed cost of $100. How much would the oil company have to deposit now with a bank at the remote location to set up an annuity to pay the $100 a month maintenance expense for 5 years to a local contractor? Assume annual interest rates are 9%.

Solution:

$$PV = \frac{100}{(.09/12)}\left[1 - \frac{1}{(1+.09/12)^{60}}\right] = \$4,817.34$$

Example. Apex Corp. borrows $200,000 to finance the purchase of tools for a new line of gears it's producing. If the terms of the loan are annual payments over 4 years at 14% annual interest, *how much are the payments?*

$$\$200,000 = \frac{C}{.14}\left[1 - \frac{1}{(1+.14)^4}\right]$$

Solution:

C = $68,640.96 per year

Gradients (Payouts/Payments with Constant Growth Rates)

A *gradient* is similar to an annuity, but its periodic payouts increase at a constant rate of growth, as illustrated by its cash flow diagram. A gradient is also similar to a growing perpetuity, except that it has a fixed time length. The cash flow diagram for a gradient loan or lease is simply the inverse of this cash flow diagram, since the loan/lease payments will be negative from the borrower's point of view. The PV of a gradient may be found by:

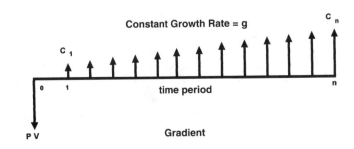

$$PV = \frac{C_1(1-x^n)}{(1+r)(1-x)} \quad \text{when } g \neq r$$

or

$$PV = \frac{C_1 n}{(1+r)} \quad \text{when } g = r$$

where: C_1 = initial cash flow at the end of first time period
r = rate of return (discount rate)
n = number of time periods
$x = (1+g)/(1+r)$
g = fractional growth rate of cash flow per period

Analyzing Complex Investments and Cash Flow Problems

No matter how complex a financial project or investment is, the cash flows can always be diagrammed and split apart into separate cash flows if necessary. Using the discounted cash flow (DCF) formula, each cash flow can be discounted individually back to time zero to obtain its present value (PV), and then summed to determine the overall valuation of the project or investment. It may also be possible to simplify things if you can identify a series of cash flows in the complex investment. Cash flow series can be discounted using either annuity or perpetuity formulas, and then added together with the PV of any singular cash flow.

When analyzing any complex cash flow problem, it is important to remember the basic time value of money (TVM) concept (i.e., *the value of money changes with time*). Therefore, if you wish to add or compare any two cash flows, they both must be from the same time period. Cash flows in the future must be discounted to their present value (PV) before they can be added or compared with current cash flows. Alternatively, current cash flows can be compounded into the future (FV) and then added or compared with future cash flows.

Example. An automotive components company has been asked by a large automobile manufacturing company to increase its production rate of components for one particular car model. Sales of this car model have been much higher than expected. In order to increase its production rate, the components company will have to purchase additional specialized machinery, but this machinery will easily fit into its existing facility. The automobile manufacturer has offered to sign a contract to purchase all the additional output that can be produced for 4 years, which will increase annual net revenues by $300,000 without any additional administrative overhead or personnel expense. The added machinery will cost $450,000 and will require maintenance every 6 months at an estimated cost of $20,000.

The machinery will be sold at the end of the 4-year period, with an estimated salvage value of $80,000. *If the company requires a minimum rate of return of 20% on all investments, should it sign the contract, purchase the machinery, and increase its production rate?* (Depreciation and tax effects are not considered in this example.)

Step 1: From the point of view of the automotive components company, the cash flow diagram will be:

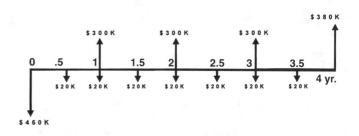

Step 2: Separate the overall cash flow diagram into individual components and series, and then determine the PV of each future cash flow or series of flows by discounting at the required annual rate of return of 20%:

$$PV = \frac{\$20,000}{.20/2}\left[1 - \frac{1}{(1+.20/2)^7}\right] = \$97,368 \text{ (negative)}$$

ANNUITY

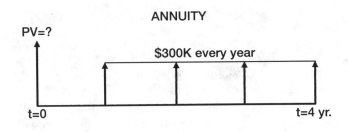

$$PV = \frac{\$300,000}{.2}\left[1 - \frac{1}{(1+.2)^4}\right] = \$776,620 \quad \text{(positive)}$$

SINGLE CASH FLOW FROM SALVAGE VALUE

$$PV = \frac{\$80,000}{(1+.2)^4} = \$38,580 \text{ (positive)}$$

Step 3: Sum the PV of all cash flows to obtain the project's overall PV from all *future* cash flows:

$$PV = \$776,620 + \$38,580 - \$97,368$$
$$= \$717,832 \text{ (positive)}$$

Step 4: Compare the PV of the *future* cash flows with the project's *initial* cash flow (initial investment):

The $717,832 PV of the project's future cash inflows (positive) is clearly greater than the $450,000 (negative) initial investment. Thus, this project has a positive *net present value* (NPV) because the PV of the future cash flows are greater than the initial investment. A positive NPV means that the rate of return on the investment is greater than the rate (20%) which was used to discount the cash flows.

Conclusion: The company should buy the additional machinery and increase its production rate.

DECISION AND EVALUATION CRITERIA FOR INVESTMENTS AND FINANCIAL PROJECTS

How should you decide whether or not to make an investment or go forward with a financial project? If you also have a number of alternative investment options, how do you decide which is the best alternative? How do you evaluate an investment's performance some time period after the investment has been made and the project is up-and-running? There are a number of analysis techniques that can help with these questions. Four separate methods are presented here, although each method varies in its effectiveness and complexity. There are also additional methods and techniques that can be found in other engineering economics, finance, and accounting textbooks. However, most modern books and courses in these fields of study recommend the *net present value* (NPV) method as the most effective and accurate technique of evaluating potential investments and financial projects. Accounting measures such as *rate of return* (ROR) on an investment or *return on equity* (ROE) are useful to evaluate a project's financial results over a specific time period, or to compare results against competitors.

Payback Method

The payback method is a simple technique that determines the number of years before the cash flow from an investment or project "pays back" the initial investment. Obviously, the shorter the length of time it takes a project to pay off the initial investment, the more lucrative the project.

Example. A company wants to add a new product line and is evaluating two types of manufacturing equipment to produce this new line. The first equipment type costs $700,000 and can produce enough product to result in an estimated $225,000 in additional after-tax cash flow per year.

The second type of equipment is more expensive at $900,000, but has a higher output that can bring an estimated $420,000 per year in after tax cash flows to the company. Which type of equipment should the company buy?

Solution:

Equipment Type 1
Payback Period = $700,000/$225,000 per year
= 3.11 years

Equipment Type 2
Payback Period = $900,000/$420,000 per year
= 2.14 years

Based upon this payback analysis, equipment type 2 should be purchased.

The advantage of the payback method is its simplicity, as shown in the example. However, there are several disadvantages that need to be mentioned. First, the payback method does not evaluate any cash flows that occur after the payback date, and therefore ignores long-term results and salvage values. Secondly, the payback method does not consider the time value of money (TVM), but assumes that each dollar of cash flow received in the second year and beyond is equal to that received the first year. Thirdly, the payback method has no way to take into account and evaluate any risk factors involved with the project.

Accounting Rate of Return (ROR) Method

The *accounting rate of return* for an existing company or project can be easily calculated directly from the company's or project's quarterly or annual financial accounting statements. If a potential (rather than existing) project is to be evaluated, financial results will have to be projected. Since ROR is an accounting measure, it is calculated for the same time period as the company or project's income statement. The rate of return on an investment is simply the net income generated during this time period, divided by the book value of the investment at the start of the time period.

$$\text{R.O.R.} = \frac{\text{Net Income}}{\text{Net Investment (book value)}}$$

Net income is an accounting term which is defined as "revenues minus expenses plus gains minus losses" (i.e., the "bottom line" profit or loss on the income statement). Net investment is the "book value" of the investment, which is the original investment amount (or purchase price) less any depreciation in value of the investment.

Example. Company XYZ invested $500,000 in plant and equipment four years ago. At the end of year 3, the company's financial statements show that the plant and equipment's value had depreciated $100,000, and had a listed book value of $400,000. If company XYZ has a net income of $60,000 during its 4th year of operation, what is its ROR for this time period?

Solution:

$$\text{R.O.R.} = \frac{60,000}{400,000} = 0.15 \quad (15\%)$$

Like the payback method, the advantage of the accounting ROR is that it is a simple calculation. It is especially simple for an existing company or project when the accounting statements are already completed and don't have to be projected. One disadvantage of ROR is that it doesn't consider TVM when it is used for more than one time period. It also uses accounting incomes rather than cash flows, which are not the same (see Accounting Fundamentals in this chapter). Another disadvantage is that ROR is a very subjective method, because the ROR is likely to vary considerably each time period due to revenue and expense fluctuations, accumulating depreciation, and declining book values. (Some textbooks recommend averaging ROR over a number of time periods to compensate for fluctuations and declining book values, but an "averaged" ROR still does not consider TVM.)

In general, accounting measurements such as ROR are best used for evaluating the performance of an *existing* company over a specific annual or quarterly time period, or for comparing one company's performance against another. Net present value (NPV) and internal rate of return (IRR) are superior methods for evaluating *potential* investments and financial projects, and in making the final decision whether to go ahead with them.

Internal Rate of Return (IRR) Method

Internal rate of return (IRR) is an investment profitability measure which is closely related to net present value (NPV). The IRR of an investment is that rate of return which, when used to discount an investment's future cash flows, makes the NPV of an investment equal zero. In other words, when the future cash flows of an investment or project are discounted using the IRR, their PV will exactly equal the initial investment amount. Therefore the IRR is an extremely useful quantity to know when you are evaluating a potential investment project. The IRR tells you the exact rate of return that will be earned on the original investment (or overall project with any additional investments) if the projected future cash flows occur. Similarly, when analyzing past investments, the IRR tells you the exact rate of return that was earned on the overall investment. The IRR decision rule for whether or not to go ahead with any potential investment or project being considered is simple: *If the IRR exceeds your opportunity cost of capital (rate of return that can be earned elsewhere), you should accept the project.*

The following example will make it clear how the IRR of an investment is calculated. It is highly recommended that a cash flow diagram be drawn first before trying to calculate the IRR. Particular attention should be paid to make sure cash inflows are drawn as positive (from the investor's point of view) and cash outflows are drawn as negative. It is also recommended that a financial calculator (or computer) be used to save time, otherwise an iteration procedure will have to be used to solve for the IRR.

Example. Acme Tool and Die Company is considering a project that will require the purchase of special machinery to produce precision molds for a major customer, Gigantic Motors. The project will last only four years, after which Gigantic Motors plans to manufacture its own precision molds. The initial investment for the special machinery is $450,000, and the machinery will also require a partial overhaul in three years at an estimated cost of $70,000. Projected after-tax cash flows resulting from the project at the end of each of the four years are $140,000, $165,000, $185,000, and $150,000, respectively. These projections are based upon a purchase agreement that Gigantic is prepared to sign, which outlines the quantities of molds it will buy. Expected salvage value from the machinery at the end of the four-year period is $80,000.

Should Acme proceed with this project, if its opportunity cost of capital is 23%?

Step 1: Cash flow diagram:

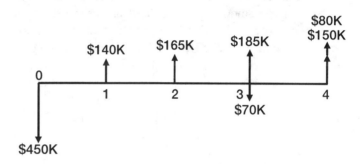

Step 2: Write an equation which sets the original investment equal to the PV of future cash flows, using the discounted cash flow formula. Solve for r, which is the IRR. (This is a simple process with a financial calculator or computer, but otherwise requires iteration with any other calculator).

$$450,000 = \frac{140,000}{(1+r)} + \frac{165,000}{(1+r)^2} + \frac{(185,000 - 70,000)}{(1+r)^3}$$
$$+ \frac{(150,000 + 80,000)}{(1+r)^4}$$

Solving for r (IRR):

IRR = 0.1537 (15.37%)

Step 3: Apply decision rule to results: Since the IRR for this project is 15.37%, and Acme's required rate of return (opportunity cost of capital) is 23%, Acme should not proceed with this project.

IRR has a number of distinct advantages (as does NPV) over both the payback and ROR methods as a decision-making tool for evaluating potential investments. The first advantage is that the IRR calculation takes into account TVM. Secondly, IRR considers all cash flows throughout the life of the project (including salvage values), rather than looking at only one time period's results like accounting ROR does, or the years until the initial investment is paid off like the payback method. Therefore, IRR is an objective criterion, rather than a subjective criterion, for making decisions about investment projects. Thirdly, IRR uses actual cash flows rather than accounting incomes like the ROR method. Finally, since IRR is calculated for the entire life of the project, it provides a basis for evaluating whether the overall risk of the project is justified by the IRR, since the IRR may be compared with the IRR of projects of similar risk and the opportunity cost of capital.

The IRR method has several disadvantages compared to the NPV method, though only one disadvantage is mentioned here for purposes of brevity. Further information about potential problems with the IRR method (compared to NPV) may be obtained from most finance textbooks. One major problem with IRR is the possibility of obtaining multiple rates of return (multiple "roots") when solving for the IRR of an investment. This can occur in the unusual case where cash flows change erratically from positive to negative (in large quantities or for sustained time periods) *more than once* during the life of the investment. The graph in the following example illustrates how multiple IRR roots can occur for investments with these types of cash flows. The second graph is for comparison purposes, and typifies the IRR calculation and graph for most investments. (Normally, NPV declines with increasing discount rate, thus giving only one IRR "root".) When multiple "roots" are obtained using the IRR method, it is best to switch to the NPV method and calculate NPV by discounting the cash flows with the opportunity cost of capital.

Example. A project under consideration requires an initial investment/cash flow of $3,000 (negative) and then has cash flows of $26,000 (positive) and $30,000 (negative) at the end of the following two years. A graph of NPV versus discount rate shows that two different IRR "roots" exist. (IRR is the point on the curve that intersects the hor-

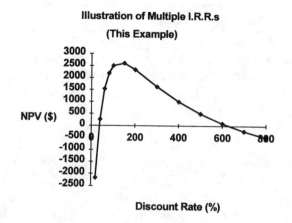

Illustration of Multiple I.R.R.s
(This Example)

Illustration of Single I.R.R. Root
(Previous Example - Acme Tool & Die)

izontal axis at NPV = 0.) The second graph is from the previous example, showing that most "normal" investments have only one IRR root.

The IRR for this example is both 37.1% and 629.6%, as NPV equals zero at these two discount rates.

Net Present Value (NPV) Method

Net present value (NPV), as its name suggests, calculates the net amount that the discounted cash flows of an investment exceed the initial investment. Using the discounted cash flow (DCF) formula, the future cash flows are discounted by the rate of return offered by comparable investment alternatives (i.e., the opportunity cost of capital), and then summed and added to the initial investment amount (which is usually negative).

$$NPV = C_0 + \sum_{i=1}^{n} \frac{C_i}{(1+r)^i}$$

where: C_o = initial cash flow (negative for a cash "outflow")
C_i = cash flow in time period i
n = number of time periods
r = opportunity cost of capital/discount rate

Though similar to the IRR method, NPV does not calculate an investment's exact rate of return, but instead calculates the exact dollar amount that an investment exceeds, or fails to meet, the *expected* rate of return. In other words, if an investment provides a rate of return exactly equal to the opportunity cost of capital, then the NPV of this investment will be zero because the discounted future cash flows equal the initial investment. Thus, NPV provides an excellent decision criterion for investments. An investment with a positive NPV should be accepted, since it provides a rate of return above the opportunity cost of capital. By the same reasoning, an investment with a negative NPV should be rejected. NPV also does *not* suffer from any of the drawbacks of the payback, accounting ROR, or IRR methods. Because of this, NPV is the method most recommended by financial experts for making investment decisions. IRR is still useful to determine the exact rate of return for an investment, but NPV has none of the problems that IRR may have with unusual investments (such as the "multiple root" problem illustrated previously).

The following example shows how NPV is used to decide between investment alternatives.

Example. A company must choose between two alternative manufacturing projects, which require different amounts of capital investment and have different cash flow patterns. The company's opportunity cost of capital for projects of similar risk is 15%. Which project should it choose?

Project	C_0	C_1	C_2	C_3	C_4
A	−$125,000	$31,000	$43,000	$48,000	$51,000
B	−$210,000	$71,000	$74,000	$76,000	$79,000

Solving with the NPV formula:

$$NPV(A) = -\$4,809$$

$$NPV(B) = \$2,833$$

Conclusion: The company should choose project B. Despite a higher initial cost than A, Project B earns $2,833 more than its required rate of return of 15%. Incidentally, IRR also could have been used, since future cash flows are all positive, and therefore no multiple roots exist. Project A's IRR is 13.23%; Project B's IRR is 15.65%.

SENSITIVITY ANALYSIS

The previous sections of this chapter show how to analyze investments and financial projects. These analysis methods can help quantify whether or not an investment is worthwhile, and also help choose between alternative investments. However, no matter what method is used to analyze a financial project, that method, and the decision about the project, will ultimately rely upon the *projections* of the project's future cash flows. It is, therefore, extremely important that the future cash flows of any financial project or investment under consideration be forecast as accurately and realistically as possible. Unfortunately, no matter how precise and realistic you try to be with your projections of cash flows, the future is never certain, and actual results may vary from what is projected. Thus, it is often useful to perform a *sensitivity analysis* on the investment or financial project.

Sensitivity analysis is a procedure used to describe analytically the effects of uncertainty on one or more of the parameters involved in the analysis of a financial project. The objective of a sensitivity analysis is to provide the decision maker with quantitative information about what financial effects will be caused by variations from what was projected. A sensitivity analysis, once performed, may influence the decision about the financial project, or at least show the decision maker which parameter is the most critical to the financial success of the project. The NPV method lends itself nicely to sensitivity analysis, since the discount rate is already fixed at the opportunity cost of capital. One parameter at a time can be varied in steps (while the other parameters are held constant) to see how much effect this variance has on NPV. A financial calculator or computer spreadsheet program will greatly expedite the multiple, repetitive calculations involved in this analysis. Plotting the results graphically will help show the sensitivity of NPV to changes in each variable, as illustrated in the following example.

Example. A-1 Engineering Co. is contemplating a new project to manufacture sheet metal parts for the aircraft industry. Analysis of this project shows that a $100,000 investment will be required up front to purchase additional machinery. The project is expected to run for 4 years, resulting in estimated after-tax cash flows of $30,000 per year. Salvage value of the machinery at the end of the 4 years is estimated at $32,000. A-1's opportunity cost of capital is 15%. Perform a sensitivity analysis to determine how variations in these estimates will affect the NPV of the project.

Step 1: Cash flow diagram:

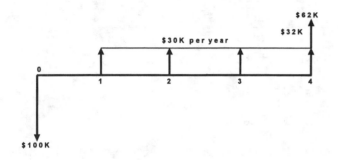

Step 2: Calculate baseline NPV:

$$NPV = -100,000 + \frac{30,000}{(1+.15)^1} + \frac{30,000}{(1+.15)^2} + \frac{30,000}{(1+.15)^3}$$
$$+ \frac{62,000}{(1+.15)^4} = \$3,945.45$$

Step 3: Change salvage value, annual cash flow revenues, and initial investment amount in steps. Calculate NPV for each step change.

Step 4: Plot values to illustrate the NPV sensitivity to changes in each variable.

Change in NPV ($) Caused by % Change in Parameter Value ($)

Conclusion: The graph shows that this project's NPV is equally sensitive to changes in the required initial investment amount (inversely related), and the annual revenues (after-tax cash flow) from the project. This is easily seen from the slope of the lines in the graph. The graph also shows that NPV is not affected nearly as much by changes in salvage value, primarily because the revenues from salvage value occur at the end of the fourth year, and are substantially discounted at the 15% cost of capital. In conclusion, this sensitivity analysis shows A-1 Machinery that it cannot accept anything more than a 5% increase in the initial cost of the machinery, or a 5% decrease in the annual cash flows. If these parameters change more than this amount, NPV will be less than zero and the project will not be profitable. Salvage value, however, is not nearly as critical, as it will take more than a 20% decrease from its estimated value before the project is unprofitable.

DECISION TREE ANALYSIS OF INVESTMENTS AND FINANCIAL PROJECTS

Financial managers often use *decision trees* to help with the analysis of large projects involving sequential decisions and variable outcomes over time. Decision trees are useful to managers because they graphically portray a large, complicated problem in terms of a series of smaller problems and decision branches. Decision trees reduce abstract thinking about a project by producing a logical diagram which shows the decision options available and the corresponding results of choosing each option. Once all the possible outcomes of a project are diagrammed in a decision tree, objective analysis and decisions about the project can be made.

Decision trees also allow *probability theory* to be used, in conjunction with NPV, to analyze financial projects and

investments. It is often possible to estimate the probability of success or failure for a new venture or project, based upon either historical data or business experience. If the new venture is a success, then the projected cash flows will be considerably higher than what they will be if the project is not successful. The projected cash flows for both of these possible outcomes (success or failure) can then be multiplied by their probability estimates and added to determine an *expected* outcome for the project. This expected outcome is the *average* payoff that can be expected (for multiple projects of the same type) based upon these probability estimates. Of course, the *actual* payoff for a single financial project will either be one or the other (success or failure), and not this average payoff. However, this average expected payoff is a useful number for decision making, since it incorporates the probability of both outcomes.

If there is a substantial time lapse between the time the decision is made and the time of the payoff, then the time value of money must also be considered. The NPV method can be used to discount the expected payoff to its present value and to see if the payoff exceeds the initial investment amount. The standard NPV decision rule applies: *Accept only projects that have positive NPVs.*

The following example illustrates the use of probability estimates (and NPV) in a simple decision tree:

Example. A ticket scalper is considering an investment of $10,000 to purchase tickets to the finals of a major outdoor tennis tournament. He must order the tickets one year in advance to get choice seats at list price. He plans to resell the tickets at the gate on the day of the finals, which is on a Sunday. Past experience tells him that he can sell the tickets and double his money, but only if the weather on the day of the event is good. If the weather is bad, he knows from past experience that he will only be able to sell the tickets at an average of 70% of his purchase price, even if the event is canceled and rescheduled for the following day (Monday). Historical data from the *Farmer's Almanac* shows that there is a 20% probability of rain at this time of year. His opportunity cost of capital is 15%. Should the ticket scalper accept this financial project and purchase the tickets?

Step 1: Construct the decision tree (very simple in this example).

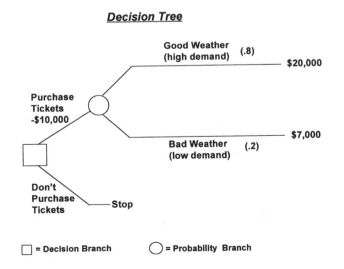

Decision Tree

□ = Decision Branch   ◯ = Probability Branch

Step 2: Calculate the expected payoff from purchasing the tickets:

Expected payoff = (probability of high demand) × (payoff with high demand) + (probability of low demand) × (payoff with low demand)

= (.8 × 20,000) + (.2 × 7,000) = $17,400

Step 3: Calculate NPV of expected payoff:

$$NPV = -10,000 + \frac{\$17,400}{(1+.15)^1} = \$5,130$$

Remember that the cash flow from selling the tickets occurs one year after the purchase. NPV is used to discount the cash flow to its present value and see if it exceeds the initial investment.

Conclusion: The ticket scalper should buy the tickets, because this project has a positive NPV.

———

The preceding example was relatively simple. Most financial projects in engineering and manufacturing applications are considerably more complex than this example, therefore, it is useful to have a procedure for setting up a decision tree and analyzing a complex project.

Here is a procedure that can be used:

1. Identify the decisions to be made and alternatives available throughout the expected life of the project.
2. Draw the decision tree, with branches drawn first for each alternative at every decision point in the project. Secondly, draw "probability branches" for each possible outcome from each decision alternative.
3. Estimate the probability of each possible outcome. (Probabilities at each "probability branch" should add up to 100%.)
4. Estimate the cash flow (payoff) for each possible outcome.
5. Analyze the alternatives, starting with the most distant decision point and working backwards. Remember the time value of money and, if needed, use the NPV method to discount the expected cash flows to their present values. Determine if the PV of each alternative's expected payoff exceeds the initial investment (i.e., positive NPV). Choose the alternative with the highest NPV.

The following example illustrates the use of this procedure:

Example. Space-Age Products believes there is considerable demand in the marketplace for an electric version of its barbecue grill, which is the company's main product. Currently, all major brands of grills produced by Space-Age and its competitors are either propane gas grills or charcoal grills. Recent innovations by Space-Age have allowed the development of an electric grill that can cook and sear meat as well as any gas or charcoal grill, but without the problems of an open flame or the hassle of purchasing propane gas bottles or bags of charcoal.

Based upon test marketing, Space-Age believes the probability of the demand for its new grill being high in its first year of production will be 60%, with a 40% probability of low demand. If demand is high the first year, Space-Age estimates the probability of high demand in subsequent years to be 80%. If the demand the first year is low, it estimates the probability of high demand in subsequent years at only 30%.

Space-Age cannot decide whether to set up a large production line to produce 50,000 grills a month, or to take a more conservative approach with a small production line that can produce 25,000 grills a month. If demand is high, Space-Age estimates it should be able to sell all the grills the large production line can make. However, the large production line puts considerably more investment capital at risk, since it costs $350,000 versus only $200,000 for the small line. The smaller production line can be expanded with a second production line after the first year (if demand is high) to double the production rate, at a cost of an additional $200,000.

Should Space-Age initially invest in a large or small production line, or none at all? The opportunity cost of capital for Space-Age is 15%, and cash flows it projects for each outcome are shown in the decision tree. Cash flows shown at the end of year 2 are the PV of the cash flows of that and all subsequent years.

Steps 1–4 of procedure: (Construct decision tree and label with probabilities and cash flow projections).

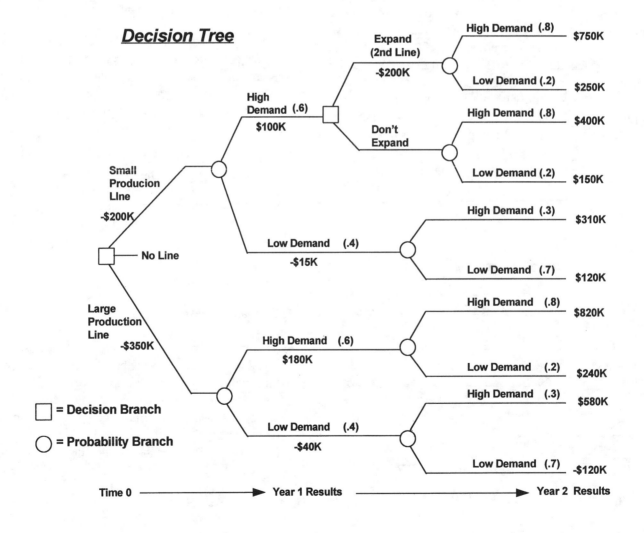

Decision Tree

= Decision Branch	
= Probability Branch	

Time 0 → Year 1 Results → Year 2 Results

Step 5: Analysis of alternatives:

(a) Start the analysis by working backwards from the R/H side of the decision tree. The only decision that Space-Age needs to make at the start of year 2 is deciding whether or not to expand and install a second production line if demand is high the first year (if the initial decision was for a small production line).

Expected payoff of expanding with a second small production line (at the end of year 2):

$$= (.8 \times \$750 \text{ K}) + (.2 \times \$250 \text{ K}) = \$650,000$$

Expected payoff without the expansion (at the end of year 2):

$$= (.8 \times \$400 \text{ K}) + (.2 \times \$150 \text{ K}) = \$350,000$$

To decide if expansion is worthwhile, the NPVs of each alternative (at end of year 1) must be compared:

$$NPV \text{ (with 2nd line)} = -\$200 \text{ K} + \frac{\$650 \text{ K}}{(1+.15)^1}$$
$$= \$365,217$$

$$NPV \text{ (without 2nd line)} = \$0 + \frac{\$350 \text{ K}}{(1+.15)^1}$$
$$= \$304,348$$

These NPV amounts show that Space-Age *should clearly expand with a second small production line* (if its initial decision at time 0 was to open a small production line and the demand was high the first year). Now that this decision is made, the $365,217 NPV of this expansion decision will be used to work backwards in time and calculate the overall NPV of choosing the small production line.

(b) The expected payoff (at the end of year 2) if the demand is low the first year with the small production line:

$$= (.3 \times \$310 \text{ K}) + (.7 \times \$120 \text{ K}) = \$177,000$$

This expected payoff must be discounted so that it can be added to the projected cash flow of year 1:

$$PV = \frac{\$177,000}{(1+.15)} = \$153,913$$

(c) The expected payoff (at the end of year 1) for the small production line can now be determined:

$$= [.6 \times (\$100,000 + \$365,217)] + [.4 \times (-\$15,000 + \$153,913)] = \$334,695$$

Notice that the cash flow for year 1 is the sum of the first year's projected cash flow plus the discounted expected payoff from year 2.

(d) Calculate the NPV (at time 0) of the small production line option:

$$NPV = \$200,000 + \frac{\$334,695}{(1+.15)} = \$91,039$$

(e) Calculate the NPV (at time 0) of the large production line option. (Since there are no decisions at the end of year 1, there is no need to determine intermediate NPVs, and the probabilities from each branch of each year can be multiplied through in the NPV calculation as follows:

$$NPV = -\$350 \text{ K} + \frac{(.6)(\$180 \text{ K}) + (.4)(-\$40 \text{ K})}{(1+.15)^1}$$
$$+ \frac{(.6)[(.8)(\$820 \text{ K}) + (.2)(\$240\text{K})] + (.4)[(.3)(\$580 \text{ K}) + (.7)(-\$120 \text{ K})]}{(1+.15)^2}$$

Solving: NPV = $76,616

(f) *Conclusion:* Both the NPVs of the large and small production lines are positive, and thus both production lines would be profitable based upon the expected payoffs. However, the NPV of the small production line is higher than the NPV of the large production line ($91,039 vs. $76,616). Therefore, Space-Age's decision should be to set up a small production line, and then expand with a second line if the demand for electric grills is high in the first year of production.

ACCOUNTING FUNDAMENTALS

Regardless of whether a person is managing the finances of a small engineering project or running a large corporation, it is important to have a basic understanding of accounting. Likewise, it is also impossible to study engineering economics without overlapping into the field of accounting. This section, therefore, provides a brief overview of accounting fundamentals and terminology.

Accounting is the process of measuring, recording, and communicating information about the economic activities of accounting entities (financial projects, partnerships, companies, corporations, etc.). Public corporations, and most other types of companies, issue financial statements from their accounting data on a periodic basis (usually quarterly and yearly). These financial statements are necessary because they provide information to outside investors and creditors, and assist a company's own management with decision making. Four primary types of financial statements are normally issued:

- Balance sheet
- Income statement
- Statement of changes in retained earnings
- Statement of cash flows

The *balance sheet* provides information (at one particular instant in time) about financial resources of a company, and claims against those resources by creditors and owners. The balance sheet's format reflects the *fundamental accounting equation:*

Assets = Liabilities + Owner's Equity

Assets are items of monetary value that a company possesses, such as plant, land, equipment, cash, accounts receivable, and inventory. *Liabilities* are what the company owes its creditors, or conversely, the claims the creditors have against the company's assets if the company were to go bankrupt. Typical liabilities include outstanding loan balances, accounts payable, and income tax payable. *Owner's equity* is the monetary amount of the owner's claim against the company's assets. (Owner's equity is also referred to as stockholder's equity for corporations that issue and sell stock to raise capital.) Owner's equity comes from two sources: (1) the original cash investment the owner used to start the business, and (2) the amount of profits made by the company that are not distributed (via *dividends)* back to the owners. These profits left in the company are called *retained earnings.*

There are numerous asset, liability, and owner's equity *accounts* that are used by accountants to keep track of a company's economic resources. These accounts are shown in the following "balance sheet" illustration, which is a typical balance sheet for a small manufacturing corporation.

Notice in this illustration that the balance sheet's total assets "balance" with the sum of liabilities and owner's equity, as required by the fundamental accounting equation. There are also a number of additional *revenue* and *expense* accounts that are maintained, but not shown in the balance sheet. The balances of these revenue and expense accounts are closed out (zeroed) when the financial statements are prepared, and their amounts are carried over into the income statement. The *income statement* is the financial statement that reports the profit (or loss) of a company over a specific period of time (quarterly or annually). The income statement reflects another basic and fundamental accounting relationship:

Revenues – Expenses = Profit (or Loss)

A sample "income statement" for our typical small corporation is illustrated.

The income statement simply subtracts the company's expenses from its revenues to arrive at its *net income* for the period. Large companies and corporations usually separate their income statements into several sections, such as income from operations, income or loss from financial activities and discontinuing operations, and income or loss from extraordinary items and changes in accounting principles.

Companies frequently show an expense for depreciation on their income statements. *Depreciation* is an accounting method used to allocate the cost of asset "consumption" over the life of the asset. The value of all assets (building, equipment, patents, etc.) depreciate with time and use, and depreciation provides a method to recover investment capi-

XYZ Company
Balance Sheet
As of March 31, 1994

__ASSETS__

Current Assets:

Cash	$25,000.00	
Accounts Receivable	$268,000.00	
Inventory	$163,000.00	
Total Current Assets		$456,000.00

Property, Plant, & Equipment:

Building	$600,000.00		
less Accumulated Depreciation	($250,000.00)	$350,000.00	
Equipment	$300,000.00		
less Accumulated Depreciation	($125,000.00)	$175,000.00	
Land		$100,000.00	
Total Property, Plant, & Equipment			$625,000.00
Total Assets			**$1,081,000.00**

__Liabilities__

Current Liabilties:

Accounts Payable	$275,000.00	
Salaries Payable	$195,000.00	
Interest Payable	$13,000.00	
Income Tax Payable	$36,000.00	
Total Current Liabilities		$519,000.00

Long-Term Liabilities:

Notes Payable		$260,000.00
Total Liabilities		$779,000.00

__Owner's (Stockholder's) Equity__

Capital Stock	$200,000.00	
Retained Earnings	$102,000.00	
Total Owner's Equity		$302,000.00
Total Liabilities & Owner's Equity		**$1,081,000.00**

$$\text{Depreciation Expense} = \frac{\text{Cost} - \text{Salvage Value}}{\text{Expected Life of Asset}}$$

tal by expensing the portion of each asset "consumed" during each income statement's time period. Depreciation is a *noncash expense,* since no cash flow actually occurs to pay for this expense. The *book value* of an asset is shown in the balance sheet, and is simply the purchase price of the asset minus its accumulated depreciation. There are several methods used to calculate depreciation, but the easiest and most frequently used method is *straight-line depreciation,* which is calculated with the following formula:

Straight-line depreciation allocates the same amount of depreciation expense each year throughout the life of the asset. *Salvage value* (residual value) is the amount that a company can sell (or trade in) an asset for at the end of its useful life. "Accelerated" depreciation methods (such as sum-of-the-years'-digits and declining balance methods) are available and may sometimes be used to expense the cost of an asset faster than straight-line depreciation. Howev-

```
┌──────────────────────────────────────────────────────────────────┐
│                        XYZ Company                                 │
│                      Income Statement                              │
│                For the Quarter Ended March 31, 1994                │
│                                                                    │
│  Sales Revenues                                      $2,450,000.00 │
│  Less: Cost of Goods sold                           ($1,375,000.00)│
│  Gross Margin                                        $1,075,000.00 │
│  Less: Operating Expenses                                          │
│     Selling expenses              $295,000.00                      │
│     Salaries expense              $526,000.00                      │
│     Insurance expense              $32,000.00                      │
│     Property Taxes                 $27,000.00                      │
│     Depreciation, Equipment        $35,000.00                      │
│     Depreciation, Building         $70,000.00         ($985,000.00)│
│  Income from Operations                                 $90,000.00 │
│  Other Income                                                      │
│     Sale of Assets                 $16,000.00                      │
│     Interest Revenue                $2,000.00           $18,000.00  │
│                                                        $108,000.00 │
│  Less: Interest Expense                                ($22,000.00)│
│  Income before Taxes                                    $86,000.00 │
│  Less: Income Tax Expense                              ($34,400.00)│
│  Net Income                                             $51,600.00 │
└──────────────────────────────────────────────────────────────────┘
```

```
┌──────────────────────────────────────────────────────────┐
│                      XYZ Company                           │
│         Statement of Changes in Retained Earnings          │
│            For the Quarter Ended March 31, 1994            │
│                                                            │
│   Retained Earnings, Dec. 31, 1993          $50,400.00     │
│   Net Income, Quarter 1, 1994               $51,600.00     │
│   Retained Earnings, March 31, 1994        $102,000.00     │
└──────────────────────────────────────────────────────────┘
```

er, the use of accelerated depreciation methods is governed by the tax laws applicable to various types of assets. A recent accounting textbook and tax laws can be consulted for further information about the use of accelerated depreciation methods.

The profit (or loss) for the time period of the income statement is carried over as retained earnings and added to (or subtracted from) owner's equity, in the "statement of changes in retained earnings" illustration.

If the company paid out any dividends to its stockholders during this time period, that amount would be shown as a deduction to retained earnings in this financial statement. This new retained earnings amount is also reflected as the retained earnings account balance in the current balance sheet.

The last financial statement is the *statement of cash flows*, which summarizes the cash inflows and outflows of the company (or other accounting entity) for the same time period as the income statement. This statement's purpose is to show where the company has acquired cash (inflows) and to what activities its cash has been utilized (outflows) during this time period. The statement divides and classifies cash flows into three categories: (1) cash provided by or used by operating activities; (2) cash provided by or used by investing activities; and (3) cash provided by or used by financing activities. Cash flow from operations is the cash generated (or lost) from the normal day-to-day operations of a company, such as producing and selling goods or services. Cash flow from investing activities includes selling or purchasing long-term assets, and sales or purchases of investments. Cash flow from financing activities covers the issuance of long-term debt or stock to acquire capital, payment of dividends to stockholders, and repayment of principal on long-term debt.

The statement of cash flows complements the income statement, because although income and cash flow are related, their amounts are seldom equal. Cash flow is also a better measure of a company's performance than net income, because net income relies upon arbitrary expense allocations (such as depreciation expense). Net income can also be manipulated by a company (for example, by reducing R&D spending, reducing inventory levels, or changing accounting methods to make its "bottom line" look better in the short-term. Therefore, analysis of a company's cash flows and cash

flow trends is frequently the best method to evaluate a company's performance. Additionally, cash flow rather than net income should be used to determine the rate of return earned on the initial investment, or to determine the value of a company or financial project. (Value is determined by discounting cash flows at the required rate of return.)

The statement of cash flows can be prepared directly by reporting a company's gross cash inflows and outflows over a time period, or indirectly by adjusting a company's net income to net cash flow for the same time period. Adjustments to net income are required for (1) changes in assets and liability account balances, and (2) the effects of non-cash revenues and expenses in the income statement.

For example, in comparing XYZ Company's current balance sheet with the previous quarter, the following information is obtained about current assets and liabilities:

- Cash increased by $10,000
- Accounts receivable decreased by $5,000
- Inventory increased by $15,000
- Accounts payable increased by $25,000
- Salaries payable increased by $5,000

Additionally, the balance sheets show that the company has paid off $166,600 of long-term debt this quarter. This in-

---

XYZ Company
Statement of Cash Flows
For the Quarter Ended March 31, 1994

Net Income		$51,600.00	
Adjustments to reconcile net income to cash flow from op. activities:			
Decrease in accounts receivable	$5,000.00		
Increase in inventory	($15,000.00)		
Increase in accounts payable	$25,000.00		
Decrease in salaries payable	$5,000.00		
Depreciation expense, equipment	$35,000.00		
Depreciation expense, building	$70,000.00		
Total Adjustments:		$125,000.00	
Net cash flow from operating activities:			$176,600.00
Cash flows from financing activities:			
Retirement of long-term debt (notes payable)	($166,600.00)		
Net cash flow from financing activities:			($166,600.00)
Net Increase in cash:			$10,000.00
Cash account balance, Dec. 31, 1993			$15,000.00
Cash account balance, March 31, 1994			$25,000.00

formation from the balance sheets, along with the noncash expenses (depreciation) from the current income statement, are used to prepare the "statement of cash flows." The example demonstrates the value of the statement of cash flows because it shows precisely the sources and uses of XYZ Company's cash. Additionally, it reconciles the cash account balance from the previous quarter to the amount on the current quarter's balance sheet.

REFERENCES AND RECOMMENDED READING

Engineering Economics

1. Newman, D. G., *Engineering Economic Analysis.* San Jose, CA: Engineering Press, 1976.
2. Canada, J. R. and White, J. A., *Capital Investment Decision Analysis for Management and Engineering.* Englewood Cliffs, NJ: Prentice-Hall, 1980.
3. Park, W. R., *Cost Engineering Analysis: A Guide to Economic Evaluation of Engineering Projects,* 2nd ed. New York: Wiley, 1984.
4. Taylor, G. A., *Managerial and Engineering Economy: Economic Decision-Making,* 3rd ed. New York: Van Nostrand, 1980.
5. Riggs, J. L., *Engineering Economics.* New York: McGraw-Hill, 1977.
6. Barish, N., *Economic Analysis for Engineering and Managerial Decision Making.* New York: McGraw-Hill, 1962.

Finance

1. Brealey, R. A. and Myers, S. C., *Principles of Corporate Finance,* 3rd ed. New York: McGraw-Hill, 1988.
2. Kroeger, H. E., *Using Discounted Cash Flow Effectively.* Homewood, IL: Dow Jones-Irwin, 1984.

Accounting

1. Chasteen, L.G., Flaherty, R.E., and O'Conner, M.C. *Intermediate Accounting,* 3rd ed. New York: McGraw-Hill, 1989.
2. Eskew, R.K. and Jensen, D.L., *Financial Accounting,* 3rd ed. New York: Random House, 1989.

Additional References

Owner's Manual from any Hewlett-Packard, Texas Instruments, or other make business/financial calculator.

Note: A business or financial calculator is an *absolute must* for those serious about analyzing investments and financial projects on anything more than an occasional basis. In addition to performing simple PV, FV, and mortgage or loan payment calculations, modern financial calculators can instantly calculate NPVs and IRRs (including multiple roots) for the most complex cash flow problems. Many calculators can also tabulate loan amortization schedules and depreciation schedules, and perform statistical analysis on data. The owner's manuals from these calculators are excellent resources and give numerous examples of how to solve and analyze various investment problems.

Estimate the cost of a pipe line in the U.S. (based on 1994 data)

The following cost data is based on average costs for pipelines constructed on and offshore in the U.S. Cost data is based on FERC permit filings. These costs are average and consideration will need to be given to the location, terrain, season of the year, length of line, environmental issues and other variables before using these data.

Line size inches	Onshore ($/dia.in./mi.)	Offshore ($/dia.in./mi.)
8	25,321	
10	81,515	
12	21,160	

Line size inches	Onshore ($/dia.in./mi.)	Offshore ($/dia.in./mi.)
16	22,382	
20	54,688	
24	29,243	
26	29,923	
30	40,022	49,664
36	27,277	47,037
42	42,193	
48	26,305	

Example. To find the cost of 100 miles of 36-in. diameter pipeline (onshore construction), 27,277x100x36 = $98,197,200.

How to compare the cost of operating an engine on diesel and natural gas

Rule. For a rough estimate of the cost of diesel fuel per hour of operation, multiply: (.05) (engine horsepower) (cost per gallon of diesel fuel). To find the cost of natural gas per hour's operation, multiply: (.008) (engine horsepower) (cost of natural gas per Mcf).

Example. What would it cost per hour to operate a 500 horsepower engine on diesel fuel? On natural gas? Diesel fuel costs 10 cents per gallon and natural gas costs 20 cents per Mcf.

Solution.

Cost/hour for diesel fuel operation = (.05)(500)(.10)
= $2.50

(A more accurate answer can be obtained by substituting .0533 for .05. This would make the cost of diesel fuel operation $2.67 per hour.)

Cost/hour for natural gas = (.008)(500)(.20) = $0.80

Example. Compare the cost of operating the engine on straight diesel fuel with dual fuel consisting of 90% natural gas and 10% diesel pilot fuel, if the conditions are the same as above.

Solution.

Cost per hour = (.10)(2.50) + (.90)(.80) = $0.97

Generally speaking, 1,000 cubic ft of natural gas has a Btu content of 1,000,000 and a gallon of diesel fuel has a Btu content of 150,000. Thus 1 Mcf of natural gas is equivalent to $6\frac{2}{3}$ gallons of diesel fuel. To arrive at the figures above, the engine was assumed to require an average of 8,000 Btu per brake horsepower hour. This is an arbitrary figure, which is about the average for engines in good mechanical condition operating under high percentage of load. Some pipe line companies are equipped with engines that consume about 7,000 Btu/bhp; others run as high as 11,000 Btu/bhp. Condition of the engine and the loading account for most of the difference.

If an engine used an average of 7,000 Btu/bhp in developing its full horsepower of say, 1,000, naturally it would use 7,000,000 Btu per hour. This would require 7 Mcf of natural gas or about 46.7 gallons of diesel fuel.

How to estimate energy costs for different pipe line throughputs

The rule of thumb given below will apply to liquid pipe lines where the flow is in the turbulent region only. It will serve to estimate energy requirements for pumping through pipe lines between increments of +15% and decrements of −15%.

Rule. Multiply the normal throughput in barrels times the cost of energy per barrel. Then multiply the increment or decrement (provided it is within a range of +15% or −15%) times the cost of energy per barrel at normal throughput times 3. Then add or subtract the latter amount from the cost of normal throughput.

Or, restated in simpler form:

Cost of energy per barrel at normal throughput = Y

Cost of energy per barrel of increment or decrement within a range of $\pm 15\%$ = 3Y

Example. Assume energy requirements for pumping 2,000 barrels per hour through a 12-in. pipe line are $.01 per barrel. What are energy requirements for pumping 2,200 barrels per hour?

Energy cost for 2,000 b/h:

$2,000 \times .01 = \$20$ per hour

Energy cost for additional 200 b/h:

$200 \times .01 \ (3) = \$6$ per hour

Energy cost for 2,200 b/h = $26 per hour

Comparing fuel costs for diesel and electric prime movers

Here's a formula which will give you a quick comparison of fuel costs of a pump station—for diesel engine or electric motor.

Just find out which of the expressions below, C_e or C_d is the smaller:

$$C_e = \frac{0.7457}{n} \times E$$

$$C_d = 0.065 \ D + 0.04$$

where C_e = power consumption cost of electric motors, cents/bhp-hr

C_d = fuel and lube oil cost of diesel engines, cents/bhp-hr

E = electric power rate, cents/kW-hr

D = diesel fuel cost, cents/gal

n = efficiency of electric motor

Example. Let's see which is more economical: running a 1,000 bhp unit on diesel or electric driver.

The data are:

$n = 95\%$

$E = 1¢/kW\text{-}hr$

$D = 10¢/gal$

So, we have:

$$C_e = \frac{0.7457}{0.95} \times 1 = 0.785$$

$$C_d = 0.065 \times 10 + 0.04 = 0.690$$

Then, for a 1,000 bhp unit the estimated operating cost per hour would be:

motor: $7.85/hr

engine: $6.90/hr

Keep in mind that this rule deals only with costs of power and fuel consumption. It does not take into consideration such expenses as operating personnel, repairs and renewals, depreciation, overhead, and so on.

Nomograph for calculating scale-up of equipment or plant costs

In correlating the cost of equipment or complete plants as a function of size, it has been shown that very often the cost varies as a fractional power of the ratio of capacities (or equipment sizes), as follows:

$$\left|\frac{C_2}{C_1}\right|^n = \frac{I_2}{I_1}$$

where C_1 and C_2 = the capacities (or sizes) of the equipment or plants in question

I_1 and I_2 = the cost or fixed investment of the known unit and the new unit, respectively

n = the exponent that applies to the particular case at hand

The nomograph enables one to calculate the capacity ratio raised to the fractional power n and the equipment or plant cost if the size for one capacity level is known.

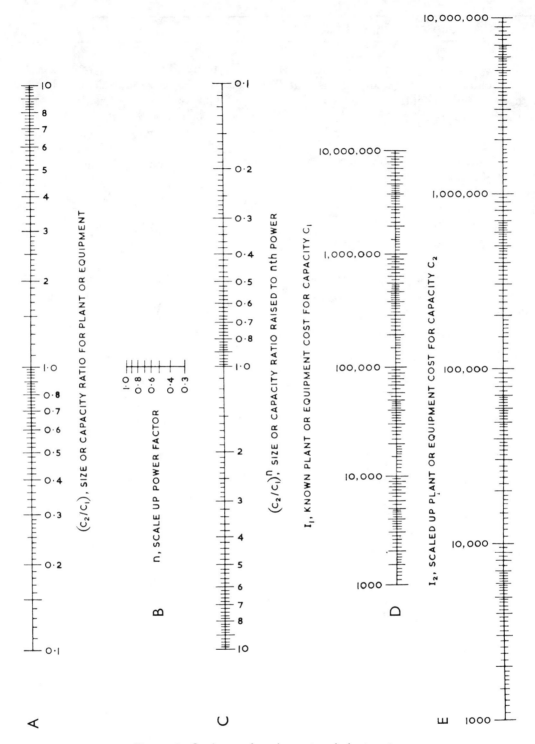

Figure 1. Scale-up of equipment and plant costs.

Example. A rotary drum filter costs $20,000 when its filtering area is 50 ft². Assuming the cost varies as the 0.6 power of the capacity, how much will a 100 ft² filter cost?

The capacity ratio is

$$C_2/C_1 = 100/50 = 2$$

Enter 2 on the capacity ratio scale (A) and align with 0.6 to the intersection point on the middle scale with $20,000 on the (D) scale and read I_2 = $30,000 on the (E) scale.

Source

Kuong, J. F., *Brit. Chem. Eng.*, 7, No. 7, 546 (1962).

Nomograph for calculating scale-up of tank costs

The nomograph enables one to estimate the cost of a steel container of given volume when one knows the cost and volume of another or standard container.

The chart is based on the equation:[1]

$$C = cV_s^{0.3}V^{0.7}$$

where   C = cost of container in question, dollars
      c = cost of standard container, dollars per unit volume
     V_s = volume of standard container
     V = volume of container in question

V_s and V are expressed in the same units: gallons cubic ft, barrels.

Based on Dickson's suggestions,[1] the chart can be used for drums of the same diameter and internal pressure, and for spheres of the same internal pressure.

Example. If a 30,000-barrel tank costs $36,000, what is the cost of a 15,000-barrel tank? For the standard tank, c = 36,000/30,000 or $1.2 per barrel. Follow the key; connect 30,000 on the V_s scale with 15,000 on the V-scale and note the intersection with the α-axis. Connect this point with 1.2 on the c-scale and read the desired cost of approximately $22,000 on the C-scale.

Source

Davis, D. S., *Petrol Refiner*, 37, No. 4200 (1958).

Reference

1. Dickson, R. A., *Chem. Eng.*, 54, (8) 122 (1947).

Figure 1. Scale-up of tank costs.

Nomograph for determining sum-of-years depreciation

In estimating annual depreciation of fixed capital assets, the sum-of-the-years-digits depreciation method is commonly used. This method allows for the effect of accelerated depreciation in the earlier years of a project. It is based on the number of years of service remaining each year and on the sum of the arithmetic series of the years from 1 to N. The depreciation factor for each year is calculated by dividing the number of remaining service years by the sum of the digits corresponding to the total service years. The depreciation charge can then be obtained by multiplying the depreciation factor each year by the total fixed asset cost.

The nomograph was constructed from the following equation:

$$D = \frac{N_D}{\sum_1^N N_i} \times V = \frac{N_D}{1 + 2 + 3 + \ldots N} \times V$$

where   D = annual depreciation, dollars
N = number of years of service of the depreciable item
N_D = number of years of service remaining each year (N − 0, N − 1, … N − (N − 1)) for the first, second and Nth year of depreciation)
N_i = digit corresponding to the ith year

Example. The value of a piece of equipment is $5,200. Its salvage value at the end of 10 years is $200. What is the annual depreciation for the first, second and third years?

Solution. The depreciable amount is 5,200 − 200 = $5,000. The first, second and third years correspond to 10, 9, and 8 years of remaining service life, respectively.

Connect	With	Mark or Read
55 for 10 yrs	N_D = 10 (1st yr)	R
R	V = 5,000	D = $910

Repeat the procedure for N_D = 9 and 8 and obtain D = 820 and 720 for the second and third years of depreciation, respectively.

Figure 1. Sum-of-years depreciation.

If the total depreciable value V is larger than the range of the scale, divide by a factor that brings it within the range of the scale, then read the D value, but remembering that the result must be multiplied by the same factor.

Source

Kuong, J. F., *Chem. Eng.*, 67, No. 23, 237 (1960).

Nomograph for estimating interest rate of return on investment ("profitability index")

Several methods are used for evaluating the profitability of a given project. A method for estimating profitability which has received considerable emphasis in the past few years is the "interest rate of return," (Figure 1), also known as "profitability index" or "discount cash flow" method. This profitability method takes into account the time value

Figure 1. Interest rate of return.

of money. Mathematically, interest rate of return may be defined as that rate at which the company's outstanding investment is repaid by proceeds from the project.[1]

The nomograph was developed for the case in which the yearly profits (P) (or the difference between income and costs) are equal through the life of a project. The nomograph is based on the following assumptions:

1. Depreciation method is sum of the years-digits.
2. Project economic life is equal to depreciable life (T).
3. Construction time is zero, i.e., total investment occurs instantaneously at time zero.
4. There is no salvage value at the end of the project life.
5. There is no start-up expense, (S = 0).

Where these conditions apply, the nomograph gives the interest rate of return R, as a function of P and T, fixed investment (F), and working capital (W).

Example 1. *Basic Profitability.* For the preliminary evaluation of a project the following estimates are available:

Plant capacity and sales	10,000,000 lb/yr
Average price	$0.20/lb
Unit cost (at capacity)	$0.10/lb
Overhead 15% of price	$0.03/lb
Project life 10 years (T)	
Depreciation is sum of the years digits	
Plant investment (F)	$1,000,000
Working capital (W)	$200,000

It may be assumed that no start-up expense is incurred. Estimate the interest rate of return if the project is undertaken. The reference or base case is not to invest.

First compute P, the yearly before-tax and depreciation profit.

Gross income: 10,000,000 × 0.20 = $2,000,000
Less overhead @ 15%: 10,000,000 × 0.03 = $300,000
Less operating cost: 10,000,000 × 0.10 = $1,000,000
P, profit before tax and depreciation = $700,000

Compute:

$$\frac{P \times T}{F} = \frac{700,000 \times 10}{1,000,000} = 7$$

Compute:

$$\frac{W}{F} = \frac{200,000}{1,000,000} = 0.2$$

Connect	With	Read
$\dfrac{P \times T}{F} = 7$	$\dfrac{W}{F} = 0.2$	$R \times T = 330$

Since T = 10, the interest rate of return for the project is 33%/yr.

This is equivalent to saying that the outlay of capital of $1,000,000 in plant facilities plus 200,000 in working capital when invested in the project would generate enough cash to pay back the total investment plus interest at the rate of 33%/yr.

Example 2. *Comparison of alternatives.* Consider the following alternatives for replacing a pump:

- *Alternative A.* Replace existing pump with one of the same type at a cost of $3,000, working capital of $50 and maintenance expense of $300/yr.

- *Alternative B.* Install a new-type pump costing $3,500 with working capital of $100 and maintenance costs of $100/yr. Economic and depreciable life for both pumps is 10 years.

The saving per year for using pump B vs. pump A is

$300 − 100 = $200/yr

to justify an increment of investment of $500 ($3,500 − 3,000). What is the profitability index or interest rate of return on the differential investments?

Calculate:

$$\frac{P \times T}{F} = \frac{200 \times 10}{500} = 4$$

$$\frac{100(W)}{F} = \frac{50}{500} = 0.10$$

Connect	With	Read
$\dfrac{P \times T}{F} = 4$	0.1 on $\dfrac{W}{F} = 0.1$ scale	$R \times T = 210$

Since T = 10, R = 21%/yr = profitability index for project.

If 21% is considered an adequate rate of return, after consideration of risk, then pump B would be the preferred one.

Source

Greist, W. H., Bates, A. G., and Weaver, J. B., *Ind Eng. Chem.*, 53, 52A (1961).

References

1. Dean, J., *Harvard Bus. Rev.*, 32, 120–130 Jan., Feb. (1954).
2. Martin, J. C., in *Chemical Process Economics in Practice*, Hur, J. J., Ed., Reinhold Publishing Corp., New York, 1956.
3. Weaver, J. B., and Reilly, R. J., *Chem. Eng. Progr.*, 52, 405 (1956).

Nomograph for determining break-even point

The nomograph gives the break-even point for a plant as percent of design capacity when the unit fixed cost, Cf, the sales price, S, and the unit variable cost, Cv are known. It is based on the following equation:

$$BE = 100 \cdot Cf/(S - Cv)$$

The nomograph can also be used to determine the sensitivity of the break-even point to any combination of fixed variable costs or sales price or to determine the maximum fixed cost that can be absorbed for a given break-even point.

Example. Determine the break-even point when:

Design output	10,000 lb/yr
Unit fixed cost	$0.045/lb
Unit variable cost	$0.075/lb
Sales price	$0.200/lb

Calculate Δ = 0.200 − 0.075 = 0.125 or 12.5¢/lb
Align Cf = 4.5¢ with Δ = 12.5 and obtain BE = 36% of plant capacity.

Source

Kuong, J. F., *Petro/Chem. Engineer*, 35, No. 1, 44 (1963).

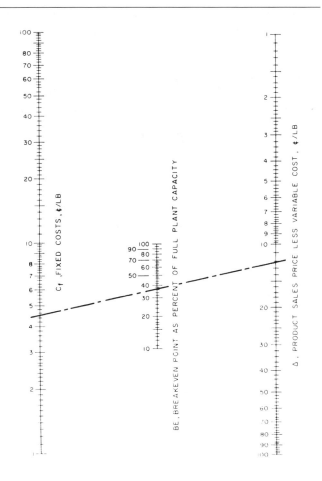

Figure 1. Break-even point.

Chart gives unit cost per brake horsepower of reciprocating compressors with various types of prime movers

Chart shows influence on unit cost of numbers of reciprocating compressor units installed in one station

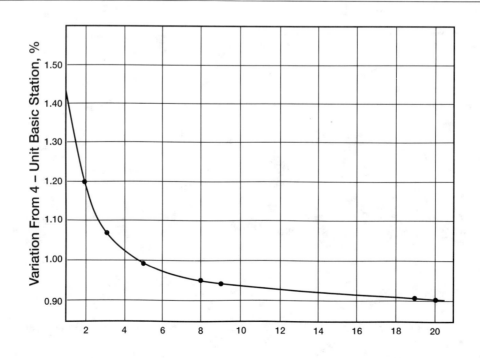

Chart gives unit cost per brake horsepower of centrifugal compressors with various types of prime movers

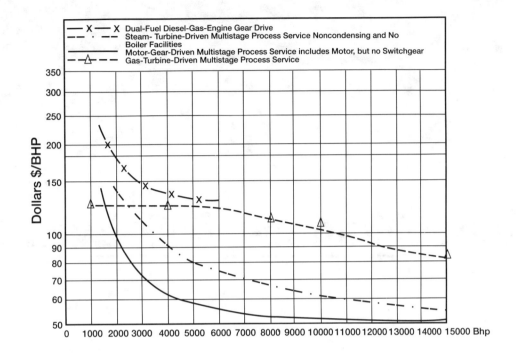

21: Rehabilitation—Risk Evaluation

When does a pipe line need revalidation? The influence of defect growth rates and inspection criteria on an operator's maintenance program.. 555
Risk evaluation 564

When does a pipe line need revalidation? The influence of defect growth rates and inspection criteria on an operator's maintenance program

by John F. Kiefner and Patrick H. Vieth, Kiefner and Associates, Inc. Worthington, Ohio

Introduction

Buried, high-pressure pipelines are subjected to loads and environmental effects which may cause them to become degraded with the passage of time. All pipeline operators are well aware of this, and the prudent operators have active programs to mitigate deterioration and to remediate defective pipe. The efficiency of an operator's revalidation program can be optimized if the necessary responses can be made when needed but not until then. The authors have assisted several pipeline operators in establishing timely intervention programs to assure continuing pipeline serviceability. A variety of techniques are used depending on the nature of the pipeline and the perceived problems. Some of the basic techniques are described in this document.

Background

Pipelines may suffer degradation from a variety of causes, including corrosion, mechanical damage, fatigue, and stress-corrosion cracking. The appropriate remedies for these problems are well known but other than the routine patrolling of the rights-of-way and monitoring of cathodic-protection potentials and rectifier currents, such remedies are too expensive to be applied on a regular or routine basis. By that we mean that most operators cannot afford to routinely and periodically utilize in-line inspection and/or hydrostatic testing to revalidate their pipelines. Usually, these techniques are invoked when some special circumstances exist. The special circumstances may be the existence of excessive amounts of low pipe-to-soil potential readings, the occurrence of leaks or ruptures, or just an intuitive feeling that it is time to check the condition of a pipeline. Alternatively, as we are finding, more and more operators are coming to depend on more sophisticated models to determine when intervention is needed. The types of models we have used are described.

General Concepts

Failure Pressure Versus Defect Size

Pipeline integrity is usually defined in terms of pressure-carrying capacity. For a given diameter, wall thickness, and grade of material, one can expect that a sound piece of line pipe will be able to sustain an internal pressure level of at least 100 percent of its specified minimum yield strength (SMYS). And, that pressure level can readily be calculated by means of the Barlow formula. For example, if one has a 16-inch O.D. by 0.250-inch wall thickness X52 pipeline, the pressure level corresponding to 100 percent of SMYS is 1,625 psig. For comparison, the 72 percent of SMYS maximum operating pressure level is 1,170 psig. Thus, one has the expectation that a new piece of this particular pipe has a safety margin against failure of 1,625/1,170 or 1.39. In fact, if it is free of defects, it will have a failure pressure of at least 1,938 psig. The latter is based upon research conducted by the pipeline research Committee of the American Gas Association[1].

Realistically, not all pipe is defect free, otherwise there would never be pre-service hydrostatic test breaks. But, after a preservice test to a pressure level of 90 to 100 percent of SMYS, one can expect a pipeline to perform satisfactorily at an operating stress level up to 72 percent of SMYS, that is, until or unless it becomes degraded in service by corrosion, mechanical damage, fatigue, or stress-corrosion cracking.

Thanks to extensive research by the previously mentioned Pipeline Research Committee of A.G.A.[1], an experimentally validated model exists for calculating the effects of a longitudinally oriented part-through-the-wall defect on the pressure-carrying capacity of the pipe. For any given piece of pipe, the model can be used to generate the relationships between failure pressure and flaw size. For our example, the 16-inch O.D. by 0.250-inch wall thickness X52 material, the relationships for selected depths of flaws are shown in Figure 1. Each of the nine parallel curves represent the failure-pressure-versus-flaw-length relationship for a particular depth (d) to wall thickness (t) ratio. The curve which cuts across the others is the dividing line between leaks and ruptures. Defects with failure at pressures

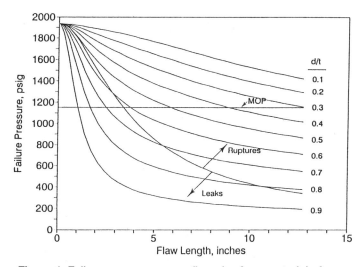

Figure 1. Failure pressure versus flaw size for a material of normal toughness (16 inch by 0.250 inch, x52, normal toughness).

above and to the right of this curve will fail as ruptures whereas those which have failure pressures below and to the left of this curve will fail as leaks.

A horizontal line is drawn on Figure 1 to represent the maximum operating pressure (MOP) of this pipe material, 1,170 psig, corresponding to a hoop stress level of 72 percent of SMYS. All flaws with length and d/t combinations which lie below this line will fail at pressure levels below the MOP. All flaws which lie above this line could exist in the pipeline in service. For example, the curve representing a d/t of 0.5 crosses the line at a flaw length of 5.8 inches. Flaws of that length which are more than half way through the wall will fail at the MOP; flaws of that length which are less than half way through will not.

The family of curves shown in Figure 1 is quite useful in the kinds of models we use to determine the time-dependent degradation of a pipeline. Two features of the curves should be kept in mind as the discussion proceeds. First, these curves represent failure pressures that are achievable with steadily increasing pressure over a relatively short period of time (minutes). Because of the phenomenon of time-dependent growth, it is possible to observe failures at pressure levels of 5 to 10 percent lower than these curves predict if pressure levels 5 to 10 percent below the predicted levels are held long enough.

Secondly, these curves are affected by the ductile-fracture toughness of the material. The curves shown in Figure 1 are characteristic of a garden-variety X52 line-pipe material having a Charpy V-notch upper-shelf energy (full-size specimens) of 25 ft-lb. The set of curves shown in Figure 2 was generated for a hypothetical material of the same pipe geometry and grade but with a Charpy energy of 250 ft-lb (a material of optimum toughness). Note that for a given defect size the material of optimum toughness has a higher failure pressure. To illustrate, it is recalled from Figure 1 that a 5.8-inch-long flaw half way through the wall fails at 1,170 psig. In contrast, the same 5.8-

inch-long flaw with a d/t of 0.5 in the very tough material has a failure pressure of 1,210 psig. It is noted that blunt flaws such as corrosion-caused metal loss tend to behave as illustrated in Figure 2 rather than Figure 1 regardless of the toughness of the material. This is because the material in a corrosion pit is strained over a large "gage length" unlike the situation of a crack-like flaw. For the latter, toughness becomes a very important parameter.

Representing the other extreme, Figure 3 presents the failure-pressure-versus-flaw-size relationship for a very low toughness material such as the ERW bond line in an older, low-frequency welded material. Even though the diameter, thickness, and grade are the same as that represented in Figures 1 and 2, the critical flaw sizes are much smaller.

Benefits of Hydrostatic Testing and In-Line Inspection

Another useful concept in evaluating pipelines for revalidation arises from the unique failure-pressure-versus-flaw-size relationships of the type described in Figures 1, 2, and 3. That is the manner in which those relationships illustrate the benefits of hydrostatic testing and in-line inspection. For illustrative purposes, let us consider the 25 ft-lb material for which Figure 1 gives the failure-pressure-versus-flaw-size relationship. This relationship is replotted in Figure 4 but Figure 4 contains additional information. It contains horizontal lines at 1,170 psig (maximum operating pressure), at 1,463 psig (90 percent of SMYS), and at 1,625 psig (100 percent of SMYS). As illustrated previously, no flaws having failure pressures below 1,170 psig can exist if the pipeline is operating at 72 percent of SMYS otherwise they would fail (i.e., their particular lengths and d/t values would lie below the horizontal line at 1,170 psig). Similarly, if the pressure in the pipeline is raised to 1,463 psig

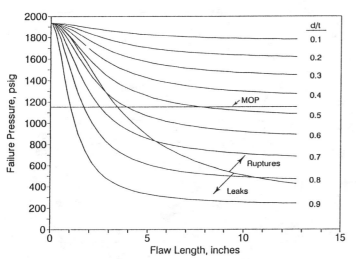

Figure 2. Failure pressure versus flaw size for a material of optimum toughness or for blunt defects (16 inch by 0.250 inch, x52, optimum toughness or blunt defects).

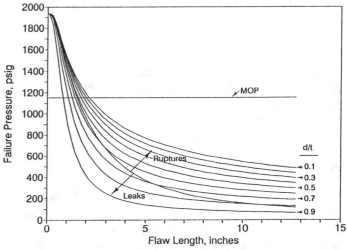

Figure 3. Failure pressure versus flaw size for a material of low toughness (16 inch by 0.250 inch, x52, low toughness).

Figure 4. Sizes of flaws located by hydrostatic testing (all types except gouge and dent combinations) (16 inch by 0.250 inch, x52, normal toughness).

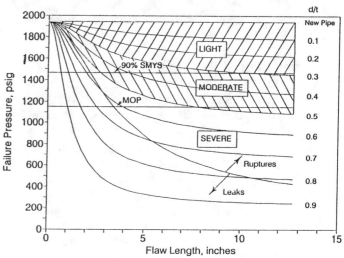

Figure 5. Sizes of flaws located by in-line inspection (corrosion) (16 inch by 0.250 inch, x52, blunt defects).

(90 percent of SMYS), any flaws with length and d/t combinations lying above the line at 1,170 psig but below that at 1,463 psig will fail and be eliminated. If no failures occur, the absence of flaws in this size range is proven. And, if the pressure in the pipeline is raised to 1,625 psig (100 percent of SMYS), the absence of additional still smaller flaws will be proven. Hence, the benefit of a hydrostatic test is expressible as the ratio of test pressure to operating pressure. The higher the ratio, the more beneficial the test, because the flaws that can remain will be smaller.

With respect to in-line inspection, most pipeliners are aware that the conventional magnetic-flux leakage tools are reasonably capable of sorting metal-loss anomalies into categories by depth of light, moderate, and severe. Lights generally are those with depths of less than 30 percent of the wall thickness (d/t < 0.3). Moderate are those with depths of 30 to 50 percent through the wall thickness (d/t = 0.3 to d/t = 0.5). Severes have depths greater than 50 percent of the wall thickness (d/t > 0.5). Recalling that the behavior of blunt metal-loss flaws is best represented by the optimum toughness relationships of Figure 2, one can see in Figure 5 (based on Figure 2) how in-line inspection data can be interpreted in terms of the effects on pressure-carrying capacity. The top cross-hatched area corresponds to light; the lower cross-hatched area corresponds to moderate, and the unshaded area corresponds to severe. Note that for long flaws, the failure pressures corresponding to moderates become low enough to intersect the MOP level. This illustrates the need to excavate and examine moderates in a program of conventional in-line inspection. With the use of the more expensive, high-resolution in-line inspection tools, it is possible to define flaw length and, hence, to avoid many excavations.

Figure 5 provides an important conceptual comparison between in-line inspection and hydrostatic testing. With the aid of this figure, one can see where each has its advantages and disadvantages. The hydrostatic test represented by the 90-percent SMYS line provides a demonstration of the immediate pressure-carrying capacity up to a pressure level of 1,463 psig. The use of conventional magnetic-flux tools cannot provide this kind of assurance unless all severe and moderate anomalies are excavated, leaving only the lights. However, the hydrostatic test does not locate any of the anomalies that do not fail. Hence, very deep pits can survive the test and develop leaks shortly thereafter (as illustrated by the amount of unshaded area—severe anomalies—lying above the horizontal line at 1,463 psig). At least with in-line inspection, one can locate and remove or repair the anomalies (especially the deep ones). After having removed or repaired all moderates and severes, one can have a high degree of confidence that the pipeline will not fail or leak for a long time as the result of corrosion-caused metal loss. The same cannot be said after a hydrostatic test. Very short but deep pits could have survived the test and may become leaks (not ruptures) within a short time after the test.

On the other hand, because none of the existing in-line inspection tools is capable of finding crack-like flaws, hydrostatic testing is the only means to assure the integrity of a pipeline that may have sustained cracking such as fatigue-crack growth or stress-corrosion cracking.

Determining When to Revalidate

Having established the concepts illustrated previously, we can now proceed to the determination of when to revalidate a pipeline. This is done by different means depending on the type of anomaly of concern. First, let us consider corrosion-caused metal loss.

Corrosion-Caused Metal Loss

A key parameter for scheduling the revalidation of an externally or internally corroded pipeline is the worst-case corrosion rate. A pipeline operator's corrosion control personnel may be able to estimate such a rate. Alternatively, one can obtain a reasonable estimate if prior corrosion leaks have occurred. If leaks have occurred, the decision to revalidate may be obvious without the use of a model. Nevertheless, having some idea of the corrosion rate is useful for scheduling the "next" revalidation. The rate can be estimated by dividing the nominal wall thickness of the pipeline by the number of years between the time of the original installation and the time of the first leak. This involves assuming a constant rate of corrosion over the life of the pipeline. For any pit that is not a leak, the corrosion rate will be less; it will be proportional to the d/t ratio of the pit. For our example pipeline (0.250-inch wall thickness), if we assume that it first developed a leak after 25 years of service, its worst-case corrosion rate is 0.25/25 or 0.010 inch per year (10 mils per year). If we postulate that there is a pit on this pipeline that is 80 percent through the wall, the corrosion rate for that pit is 0.8 x 0.01 = 0.008 inch per year (8 mils per year). With this kind of a corrosion-rate rationale, we can utilize a figure like Figure 2 to plan the revalidation of a corroded pipeline.

The family of curves represented in Figure 2 is reproduced in Figure 6 for the purpose of examining the effects of time-dependent metal loss. In Figure 6, the pipeline when it was installed is represented by the horizontal line at 1,938 psig (d/t = 0). If this pipeline is losing metal to corrosion, the effect of the metal loss is to lower the failure pressure*. In terms of Figure 6, one can see that as metal loss proceeds, the horizontal line originally at 1,938 psig would move downward as indicated by the arrows until it would eventually reach the 1,170 psig operating level, and a failure would occur. It is important to note that this occurs at a d/t of 0.75 for a 2-inch-long flaw (Arrow 1) and at a d/t of 0.47 for a 10-inch-long flaw (Arrow 2). If the company's corrosion specialist has determined that the worst-case corrosion rate is 8 mils per year, then 75 percent of the wall thickness (250 mils) will be penetrated in less than 23 years, and 47 percent of the wall thickness will be penetrated in less than 14 years. If the type of metal loss that is occurring is long (10 inches or more along the axis of the pipe), this pipe line will require revalidation in less than 14 years. On the other hand, if experience shows that the corrosion is occurring as isolated pits

---

* The ASME B31G method for evaluating corroded pipe is based on the same empirical equation which is used to derive the curves of Figure 6. The only difference is that the flaws represented in Figure 6 have been idealized as being rectangular in profile.

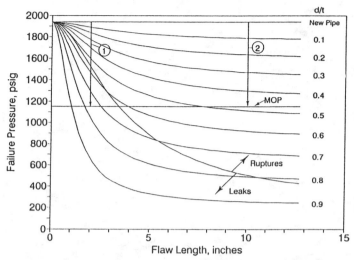

Figure 6. Determination of need for revalidation of a pipeline affected by corrosion (16 inch by 0.250 inch, x52, blunt flaws).

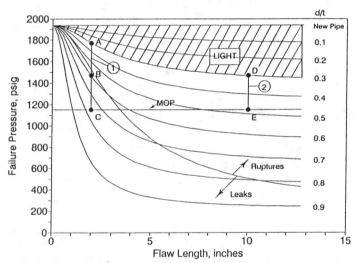

Figure 7. Determination of time to next revalidation after in-line inspection or hydrostatic test (corrosion) (16 inch by 0.250 inch, x52, blunt flaws).

(less than 2 inches in axial extent), the time for revalidation would be less than 23 years.

Perhaps a more practical case is that corresponding to a pipeline which has either been hydrostatically tested or has been repaired following in-line inspection. Now the question becomes: when will revalidation be needed again? The question can be answered in terms of Figure 7. Let us consider two alternatives simultaneously. In one case, let us assume that our 16-inch O.D. by 0.250-inch wall thickness pipeline has been inspected by a conventional magnetic-flux leakage tool and that all anomalies except the lights have been repaired (see the cross-hatched area in Figure 7). In the other case, let us assume that the pipeline has been hydrostatically tested to 1,463 psig (90 percent of SMYS).

With respect to a 2-inch-long isolated pit (Arrow 1), the deepest pit remaining after the in-line inspection is represented by Point A (d/t = 0.3, pit depth = 75 mils). From the standpoint of the hydrostatic test, the deepest pit of that can exist with a 2-inch length is represented by Point B (d/t = 0.6, pit depth = 150 mils). To cause a leak in service (and it will be a leak rather than a rupture if the length remains less than about 3 inches), the pit must grow to Point C (d/t = 0.76, pit depth = 190 mils). If the corrosion rate remains at 8 mils per year, the time to grow from A to C is (190 - 75)/8 = 14.4 years, and the time to grow from B to C is (190 - 150)/8 = 5 years. This shows again the advantage of in-line inspection over hydrostatic testing with respect to dealing with pitting corrosion.

With respect to a long corrosion anomaly (Arrow 2), the deepest possible remaining anomaly is the same whether the present revalidation was by in-line inspection or by hydrostatic test. The anomaly must grow from Point D to Point E to fail in service (and it will fail as a rupture rather than a leak because it is well above the leak/rupture dividing line). To do so, d/t must change from 0.3 to 0.47. This is a change in wall thickness of 42.5 mils and will take place in 5.3 years at a corrosion rate of 8 mils per year. From this example it appears that if the corrosion is occurring over long regions as opposed to isolated pits, the advantage of in-line inspection over hydrostatic testing is eroded. This is not quite correct for two reasons. First, the high-resolution tools give reasonably accurate information on the length of the anomaly as well as the depth. Hence, one can focus the repairs on the basis of remaining pressure-carrying capacity. Second, even with the use of conventional tools, the advantage of locating the corrosion and having the option to examine and repair it provides benefits that do not accrue with hydrostatic testing.

Returning to our analysis, we see that for our example, a revalidation interval of less than 5 years is suggested. An appropriate factor of safety should be applied to this interval, so if this were a real case, the pipeline operator might arbitrarily decide to take action after 2 or 3 years. This probably is an excessively conservative example. Hopefully, with adequate corrosion control, a lower corrosion rate than 8 mils per year can be assured. If so, the revalidation interval would be longer than shown in this example. However, the principles are the same regardless of the corrosion rate.

Fatigue Crack Growth

Model for Crack Growth

As we have seen in terms of Figure 4, if a pipeline is hydrostatically tested to a given pressure, defects of a certain size which are present at the time of the test will fail. Smaller defects would be expected to survive the test but would not be expected to grow further at a static pressure level much below the test pressure level. However, defects of size a_T limited by pressure level PT in Figure 8 could grow during cyclic-pressure service

a. Failure pressure as a function of crack size.

b. Crack growth with time.

Figure 8. Effect of crack growth on pipe line integrity.

until they reach the size as. At that point, failure will occur at the maximum operating pressure (MOP). If a defect as large as at remains after a hydrostatic test and if the pipeline is subjected to sufficient pressure fluctuations, the defect will fail after time t_1.

It should be obvious from Figure 8 that flaw enlargement by fatigue-crack growth will erode the effective safety margin provided by a hydrostatic test. This situation may be addressed by analyzing the effects of the service-pressure cycles in terms of the size of the margin between test pressure and operating pressure and assessing when it may be necessary to conduct a subsequent hydrostatic test in the event that flaws which survive the first test become enlarged.

For this purpose, one may use a crack-growth model developed explicitly for pipelines[2]. The model is based upon linear-elastic fracture mechanics. It assumes that crack growth takes place under fluctuating stress levels according to the relationship

$$da/dN = C\Delta K^n \qquad (1)$$

where

da/dN   is the increment of crack advance per cycle of loading is the range of stress intensity factor, a function of the change in stress and the initial crack size.

ΔK   is the range of stress intensity factor, a function of the change in stress and the initial crack size.

C and n are material properties determined through crack-growth-rate tests. In the use of the model, the initial crack size is fixed by the hydrostatic test. A defect which just barely survives the test is assumed to be the initial crack. The stress range is determined by the maximum and minimum pressure levels. The stress-intensity factor which provides the cracking-driving force is calculated via a Newman-Raju algorithm[3]. The model is capable of handling irregular-size pressure variations. For example, we have analyzed pressure spectra as complex as that shown in Figure 9. Equation (1) is numerically summed over a number of cycles until the crack becomes large enough to fail in service. The failure criterion for flaw size at failure is that used to generate Figures 1, 2, and 3.

To demonstrate the importance of a high test-pressure-to-operating-pressure ratio in a situation where service pressure cycles can cause flaw growth, one can consider an example based on an actual cyclic pressure test. The results of a full-scale pipe experiment involving crack growth are described in Reference 4 (Figure 47 of Reference 4) and are shown in Figure 10. In this experiment, a 2-inch-long part-through (0.095-inch-deep) flaw was placed in the fusion line of the ERW seam of a 12/-3/4-inch O.D. by 0.250-inch wall thickness X52 material. It was then subjected to constant pressure cycles from 0 to 1,020 psig until it grew through the wall and failed (after about 8,000 cycles). The actual failure pressure levels are not important, so for purposes of illustration, we can consider the implications of Figure 10 for a material having an ERW seam toughness equivalent to 2 ft-lb of Charpy V-notch impact energy. If this were a 2-ft-lb Charpy-energy material, the flaw would have had a failure pressure level of 1,300 psig after growing to a depth of 0.1 inch as indicated by the 1,300 psi "test-pressure" line. Similarly, it would have had failure pressure levels of 1,250 psig and 1,200 psig after growing to depths of 0.12 inch and 0.13 inch, respectively. These values were calculated on the basis of the failure-pressure flaw-size relationship of the types shown in Figures 1-3 but for this particular pipe geometry.

Consider the following scenarios based on Figure 10. First, as service cycles are applied and the flaw has grown from 0.095

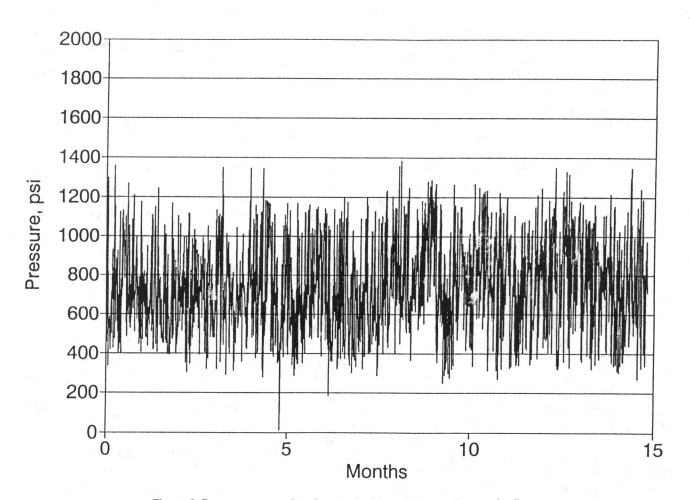

Figure 9. Pressure versus time for a typical petroleum products pipeline.

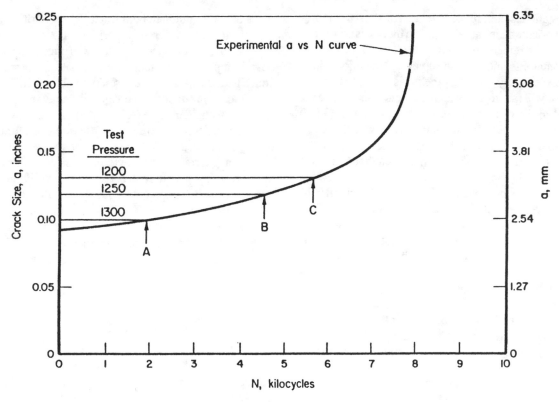

Figure 10. Example of an actual a versus N relationship used to compare effects of test pressure.

inch to 0.1 inch (N = 2,300 cycles corresponding to Point A), the remaining flaw is exactly the size that would just survive a test to 1,300 psig. Since the number of cycles to failure is 8,000, the flaw has a remaining life of 5,700 cycles. Second, after the flaw has grown to 0.12 inch (N = 4,800 cycles corresponding to Point B), the remaining flaw is exactly the size that would just survive a test to 1,250 psig. It has a remaining life of 3,200 cycles. Third, after the flaw has grown to 0.13 inch (N = 6,000 cycles corresponding to Point C), the remaining flaw is exactly the size that would just survive a test to 1,200 psig. It has a remaining life of 2,000 cycles. What this shows is the effectiveness of maximizing the test-pressure-to-operating pressure ratio. In terms of test pressure versus cycles to failure, the results would be as follows:

Test Pressure psig	Cycles to Failure After the Test	Relative Time to Failure
1,200	2,000	1.0
1,250	3,200	1.6
1,300	5,700	2.9

This exercise illustrates the valuable gain that can be made by small increases in test pressure. For a 4-percent increase over the base case (1,200 psig), the gain in time to failure is 60 percent. For an 8 percent increase, the gain is nearly 300 percent. The above-demonstrated effect of test-pressure-to-

operating-pressure ratio on service life provides one of the strongest incentives for high-pressure testing.

Calculating a Retest Interval

In the example shown above, let us assume that the pipeline sees 200 cycles of pressure from 0 to 1,020 psig per year. Let us also assume that it was tested to 1,200 psig. Such a test probably does not give an adequate margin of safety and, in fact, it does not meet the D.O.T.'s minimum requirement of 1.25 x MOP for a liquid pipeline (Part 195). But, it serves as a useful example. As the example shows, the worst-case flaw that might survive the test could be expected to fail in 2,000 cycles or 10 years at 200 cycles per year. Prudence dictates that one use a factor of safety as was suggested in our previous discussion on corrosion. The recommended factor of safety is 2, and it is based upon an examination of the "a" versus "N" relationship of Figure 10.

Shown in Figure 11 is a replot of the "a" versus "N" relationship of Figure 10. Early in the life of a crack, the growth is slow. It accelerates, however, as the crack grows deeper because the stress intensity factor driving the crack is a function of both applied stress and crack size. The implications of this behavior are illustrated as follows. Consider the crack represented by the curve in Figure 11. This crack is initially 0.095 inches in depth and will grow to failure in 8,000 cycles of the service pressure.

The hydrostatic test and retest level of 1,200 psig is represented by the horizontal line at "a" = 0.13. This line is arbitrarily divided into four increments represented by Points 1, 2, 3, 4, and 5. The horizontal line intersects the a versus N relationship at about 5,500 cycles. Any time the pipeline is tested after 5,500 cycles, the flaw will fail in the test. Suppose one chooses a retest interval equal to the distance from Point 1 to Point 3 (5,000 cycles). If the pipeline is tested at Point 3, the flaw will not fail. But, it will also not survive until the next test at Point 5 (10,000 cycles). Such a retest interval is too long even though no test failure occurred in the first test.

Now consider what happens if the test interval is cut in half. The retests are then conducted at Points 2, 3, and 4. the flaw fails during the test at Point 4. Obviously, this is a somewhat contrived example. However, the point is that if the model is reasonably accurate, the times to failure it predicts should be adjusted by a substantial factor of safety to establish an appropriate retest interval. Because of this, we normally recommend a factor of safety of 2. In our example, this would correspond to a retest interval of only 5 years. For a higher test pressure, say 1,300 psig, the interval could be 4,700/(2 x 200) = 14 years where the 5,700 cycles comes from our previous analysis based on Figure 10.

Mechanical Damage

Although mechanical damage is frequently inflicted on pipelines and can cause severe degradation of their pressure-carrying capacities, we have not applied any revalidation model to this type of defect for three reasons. First, the state of analysis of mechanical-damage flaws is not as advanced as that for other types of flaws although both Maxey[5] and British Gas[6] have made attempts to derive algorithms for dents and gouges. Second, the time-dependent nature of mechanical damage is nearly impossible to model. Since its occurrence is often concealed, the "zero" time is seldom known. The rate at which it degrades may depend on creep-like behavior and/or environmental effects. Third, we feel that the best action with respect to mechanical damage is preventative action, patrolling the rights-of-way, participating in one-call systems, conducting public education programs, and providing on-site locating of pipelines at excavation sites. In fact, our primary activity in this respect has been the development of ranking algorithms to calculate relative risk numbers for damage-prevention activities.

It should be mentioned that both in-line inspection and hydrostatic testing are viable remedies for mechanical damage. The problem is that since the existence of concealed damage is difficult to detect short of doing either of these, it is difficult to know when or even if such remedies are needed.

Stress-Corrosion Cracking

Perhaps the hardest type of defect to deal with is stress-corrosion cracking. With this type of cracking, the main problem is that no reliable model exists to define the crack-growth rate. Furthermore, it is suspected that the rate is highly variable with changes in environmental conditions. However, unlike the case for mechanical damage, it is possible to establish a "zero" time based on the date of installation. And, if the cracks exist, it is almost certain that they will grow.

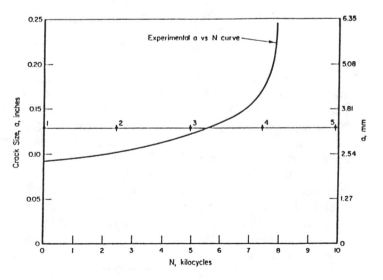

Figure 11. Example of a versus N relationship used to compare effects of retest interval.

Figure 12. Concept of retesting when defect growth rate is unknown.

A scheme used by several pipeline operators to control stress-corrosion cracking is periodic hydrostatic retesting. In the absence of a crack-growth-rate model, the retest interval has to be set arbitrarily at first and corrected as required by the results of the test. A generalized scheme for setting retest intervals to deal with defect growth at unknown rates is illustrated in Figure 12. First, an initial test is made in which defects may or may not be discovered. But, in either case, the maximum size flaw remaining is fixed by the test pressure, P_f in Figure 12, and the margin of effectiveness is proportional to P_f/P_o where P_o is the MOP. It is assumed that the defects will grow with the passage of time. Retest 1 is made after some arbitrary period of time short enough to assure that none of the remaining flaws can grow to a size that will fail in service. If this is done correctly, the growing flaws, 1, 2, and 3 in Figure 12, will fail in the test instead of in service at the times suggested by the dashed curves. If no service failures have occurred before Retest 1, then the first interval was adequate (based on the assumption that the crack-growth rate is fairly constant on the average). Depending on the outcome of Retest 1, the pipeline operator may decide to increase the interval between retests. If no test failures occur or if they occur only at levels at or near the test pressure, then it makes sense to consider a longer interval. If Retest 2 is then conducted before any flaws grow to failure in service, the new interval is apparently valid and can be lengthened depending again on the failure pressure levels of flaws in Retest 2. In this manner, the pipeline operator is able to control the phenomenon by preventing service failures, and in effect, determine the average flaw-growth rates.

It should be emphasized that the scheme portrayed in Figure 12 is valid only if the crack-growth rate is reasonably constant. For example, such a scheme would not be appropriate for fatigue cracks because of the shape of the "a" versus "N" curve as shown in Figures 10 and 11. Where the growth of the flaw accelerates rapidly, it is necessary to determine the actual growth rate. The setting of retests based upon arbitrarily assumed intervals could easily be defeated by the crack being in a rapid stage of growth as was illustrated in the discussion of Figure 11.

Summary

We have presented concepts which are frequently used by us and others to attempt to determine when it is time to be concerned about the integrity of a pipeline. The models we use rely on well-established principles and require the use of data which is usually readily available. The models we have discussed can be used to assess revalidation needs for pipelines affected by corrosion, fatigue-crack growth, and stress-corrosion cracking. By means we alluded to but did not discuss, the risks associated with mechanical-damage defects can also be evaluated. While there is never an absolute certainty about the correctness of these types of models, we recommend them to our clients to provide guidance for establishing revalidation policies. We also recommend that they be used in conjunction with sound engineering judgement and generous factors of safety.

REFERENCES

(1) Maxey, W. A., Kiefner, J. F., Eiber, R. J., and Duffy, A. R., "Ductile Fracture Initiation, Propagation, and Arrest in Cylindrical Vessels", Fracture Toughness Proceedings of 1971 National Symposium on Fracture Mechanics, Part II, ASTM STP 514, American Society for Testing and Materials, pp 70-81 (1972).

(2) Kiefner, J. F., and Forte, T. P., "Model Predicts Hydrostatic Retest Intervals", Oil and Gas Journal (January 7 and January 14, 1985).

(3) Newman, J. C., and Raju, I. S., "An Empirical Stress-Intensity Factor Equation for Surface Crack", Engineering Fracture Mechanics, Vol. 15, No. 1-2 (1981).

(4) Mayfield, M. E., and Maxey, W. A., ERW Weld Zone Characteristics, NG-18 Report No. 130 to American gas Association, Catalog No. L51427 (June 18, 1982).

(5) Maxey, W. A., "Outside Force Defect Behavior", NG-18 Report No. 162, Line Pipe Research Supervisory Committee, Pipeline Research Committee, American Gas Association (August 15, 1986).

(6) Hopkins, P., Jones, D. G., and Clyne, A. J., "The Significance of Dents and Defects in Transmission Pipelines", Conference on Pipework Engineering and Operations, I Mech E., London, Paper C376/049 (February 1989).

Reprinted with Permission of Systems Integrity and Specialty Pipeline Inspection & Engineering, Inc., organizers of the International Pipeline Rehabilitation Seminar. Paper presented at the 1992 International Pipeline Rehabilitation Seminar, Houston, Texas January 28-31, 1992

Risk Evaluation

by W. Kent Muhlbauer, Dow Pipeline Company, Houston, Texas

In order to make rational decisions in operating and maintaining pipelines, we must understand the pipeline risk factor. This includes the nature of the hazard, the probability of an accident and the potential consequences. Risk assessment is not necessarily a complex process, but is a process that may be set up and functioning in a relatively short time by just one evaluator. A suggested format and methods for assigning points to various categories is discussed in detail in *Pipeline Risk Management Manual,* Gulf Publishing Company.

The point schedules shown in the manual may be used or they may be customized based on the pipeline operator's experience. More important items should have a greater numerical impact.

Beginning Risk Management

Building a risk management tool takes four steps:

1. Sectioning the pipeline. Breaking the pipeline system(s) into sections. The size of the sections depends upon how often conditions change and upon the cost of data gathering/ maintenance vs the benefit of increased accuracy.
2. Customizing. Deciding on a list of risk contributors and risk reducers and the relative importance of each item on the list.
3. Data gathering. Building the data base by completing an evaluation for each pipeline section.
4. Maintenance. Identifying when

and how changes in risk items occur; updating the data base to reflect the changes.

Steps 2 and 3 will most likely be the most costly part of the process. These steps can be time-consuming not only in the hands-on aspects, but also in obtaining the necessary consensus from all key players. The initial consensus often makes the difference between a widely accepted and a partially resisted system. Time and resources spent in Steps 2 and 3 should be viewed as initial investments in a successful risk management tool. Steps 1 and 4 are necessary costs in keeping the tool useful, but should not be very costly to perform. The risk assessment technique described in Pipeline Risk Management Manual should become a useful tool for pipeline operators and managers. By providing timely answers to sometimes complicated questions, the tool should become a constant reference point for decision making.

The main thrust of this risk assessment is the risk exposure to the public and how that risk can be effectively managed.

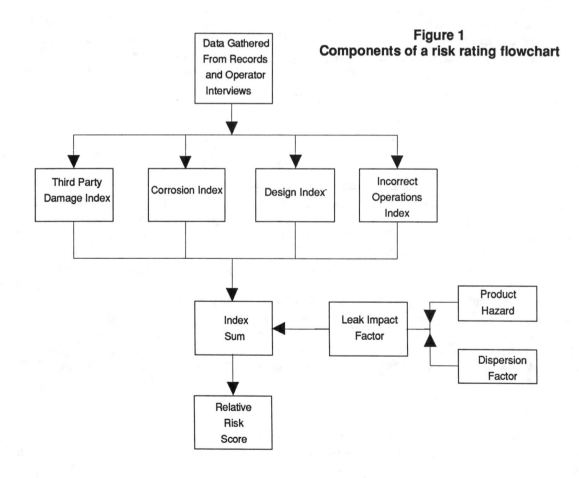

**Figure 1
Components of a risk rating flowchart**

Risk evaluation at a glance

A flow chart is shown in Figure 1 to illustrate the application of risk management.

Relative Risk Rating = (Index Sum)÷Leak Impact Factor)
= [(Third Party Index)
+ (Corrosion Index)
+ (Design Index)
+ (Incorrect Operations Index)]
÷ Leak Impact Factor

Each index roughly corresponds to a category that has historically been a cause of pipeline failures. Specific items are identified in each index as items which may increase or decrease the risk.

Third Party Index

A. Minimum Depth of Cover	0 - 20 pts	20%
B. Activity Level	0 - 20 pts	20%
C. Aboveground Facilities	0 - 10 pts	10%
D. One-Call System	0 - 15 pts	15%
E. Public Education	0 - 15 pts	15%
F. Right-of-Way Condition	0 - 5 pts	5%
G. Patrol Frequency	0 - 15 pts	15%
	0 - 100 pts	100%

Corrosion Index

Corrosion Index = (Atmospheric Corrosion)		20%
+(Internal Corrosion)		20%
+ (Buried Metal Corrosion)		60%
		100%

A. Atmospheric Corrosion	
1. Facilitiess	0 - 5 pts
2. Atmosphere	0 - 10 pts
3. Coating/Inspection	0 - 5 pts
	0 - 20 pts 20%

B. Internal Corrosion	
1. Product Corrosivity	0 - 10 pts
2. Internal Protection	0 - 10 pts
	0 - 20 pts 20%

C. Buried Metal Corrosion	
1. Cathodic Protection	0 - 8 pts
2. Coating Condition	0 - 10 pts
3. Soil Corrosivity	0 - 4 pts
4. Age of System	0 - 3 pts
5. Other Metals	0 - 4 pts
6. AC Induced Current	0 - 4 pts

7. Mechanical Corrosion	0 - 5 pts	
8. Test Leads	0 - 6 pts	
9. Close Interval Survey	0 - 8 pts	
10. Internal Inspection Tool	0 - 8 pts	
	0 - 60 pts	60%

Design Index

A. Pipe Safety Factor	0 - 20 pts	20%
B. System Safety Factor	0 - 20 pts	20%
C. Fatigue	0 - 15 pts	15%
D. Surge Potential	0 - 10 pts	10%
E. System Hydrostatic Test	0 - 25 pts	25%
F. Soil Movements	0 - 10 pts	10%
	0 - 100 pts	100%

Incorrect Operations Index (Human Error Potential)

A. Design		
1. Hazard Identification	0 - 4 pts	
2. MAOP Potential	0 - 12 pts	
3. Safety Systems	0 - 10 pts	
4. Material Selection	0 - 2 pts	
5. Checks	0 - 2 pts	
	0 - 30 pts	30%

B. Construction		
1. Inspection	0 - 10 pts	
2. Materials	0 - 2 pts	
3. Joining	0 - 2 pts	
4. Backfill	0 - 2 pts	
5. Handling	0 - 2 pts	
6. Coating	0 - 2 pts	
	0 - 20 pts	20%

C. Operation		
1. Procedures	0 - 7 pts	
2. SCADA/Communications	0 - 5 pts	
3. Drug-testing	0 - 2 pts	
4. Safety Programs	0 - 2 pts	
5. Surveys	0 - 2 pts	
6. Training	0 - 10 pts	
7. Mechanical Error Preventors	0 - 7 pts	
	0 - 35 pts	35%

D. Maintenance		
1. Documentation	0 - 2 pts	
2. Schedule	0 - 3 pts	
3. Procedures	0 - 10 pts	
	0 - 15 pts	15%
Incorrect Operations Index	0 - 100 pts	100%

Total Index Sum	**0 - 400 pts**

Leak Impact Factor

A. Product Hazard (Acute + Chronic Hazards) 0 - 22 pts

    1. Acute Hazards
        a. Flammability   0 - 4
        b. Reactivity   0 - 4
        c. Toxicity   0 - 4
        Total Acute   0 - 12
    2. Chronic Hazards   0 - 10

B. Dispersion Factor (Spill Score) ÷ (Population Score) 0 - 6

    1. Liquid Spill or Vapor Spill 0 - 6

    2. Population Density   0 - 4

Leak Impact Factor = (Product Hazard) ÷ (Dispersion Factor)

Relative Risk Score = (Index sum) ÷ (Leak Impact Factor)

        = 0 - 400 points

A complete discussion of Risk Assessment and Management can be found in *Pipeline Risk Management Manual*, published by Gulf Publishing Company.

Source: *Pipeline Risk Management Manual*, Gulf Publishing Company.

22: Conversion Factors

Units of measurement from one system to another 568
Viscosity—equivalents of absolute viscosity 577
General liquid density nomograph.. 577
Chart gives specific gravity/temperature relationship for
    petroleum oils 579
Weight density and specific gravity of various liquids 579
Metric conversions—metric to English, English to metric 580
Temperature conversion—centigrade to Fahrenheit, Fahrenheit to
    centigrade.. 581
Viscosity—equivalents of kinematic viscosity.................................. 582
Viscosity—equivalents of kinematic and Saybolt Universal
    Viscosity 582
Viscosity—equivalents of kinematic and Saybolt Furol Viscosity
    at 122°F... 583
Viscosity—general conversions 584
Pressure conversion chart 585
A simple method to determine square root.. 585
SI data 586
Energy conversion chart 587
Flow conversion chart.. 587
Conversions involving different types of fuel 588
Weight of water per cubic ft at various temperatures........................ 593
Engineering constants.. 593
Mensuration units 594
Minutes to decimal hours conversion table 594
How to compare costs of gas and alternate fuels............................... 595
Typical characteristics of fuel oils... 595

Units of measurement from one system to another

TO CONVERT	INTO	MULTIPLY BY
A		
Abcoulomb	Statcoulombs	2.998 E+10
Acre	Sq. chain (Gunters)	10
Acre	Rods	160
Acre	Square links (Gunters)	1 E+05
Acre	Hectare or sq. hectometer	.4047
acre	section	640
acres	sq feet	43,560.0
acres	sq meters	4,047.
acres	sq miles	1.562 E-03
acres	sq yards	4,840.
acre-feet	cu feet	43,560.0
acre-feet	gallons	3.259 E+05
acre-feet	cu. meters	1.233489 E+03
amperes/sq cm	amps/sq in.	6.452
amperes/sq cm	amps/sq meter	1 E-04
amperes/sq in.	amps/sq cm	0.1550
amperes/sq in.	amps/sq meter	1,550.0
amperes/sq meter	amps/sq cm	E-04
amperes/sq meter	amps/sq in.	6.452 E-04
ampere-hours	coulombs	3,600.0
ampere-hours	faradays	0.03731
ampere-turns	gilberts	1.257
ampere-turns/cm	amp-turns/in.	2.540
ampere-turns/cm	amp-turns/meter	100.0
ampere-turns/cm	gilberts/cms	1.257
ampere-turns/in.	amp-turns/cm	0.3937
ampere-turns/in.	amp-turtis/meter	39.37
ampere-turns/in.	gilberts/cms	0.4950
ampere-turns/meter	amp/turns/cm	0.01
ampere-turns/meter	amp-turns/in.	0.0254
ampere-turns/meter	gilberts/cms	0.01257
Angstrom unit	Inch	3937 E-09
Angstrom unit	Meter	1 E-10
Angstrom unit	Micron or (Mu)	1 E-04
Are	Acre (US)	0.02471
Ares	sq. yards	119.60
ares	acres	0.02471
ares	sq meters	100.0
Astronomical Unit	Kilometers	1.495 E+08
Atmospheres	Ton/sq.inch	.007348
atmospheres	cms of mercury	76.0
atmospheres	ft of water (at 4°C)	33.90
atmospheres	in. of mercury (at O°C)	29.92
atmospheres	kgs/sq cm	1.0333
atmospheres	kgs/sq meter	10,332.
atmospheres	pounds/sq in.	14.70
atmospheres	tons/sq ft	1.058
B		
Barrels (U.S., dry)	cu.inches	7056.
Barrels (U.S., drY)	quarts (dry)	105.0
Barrels (U.S., liquid)	gallons	31.5
barrels (oil)	gallons (oil)	42.0
barrels/hr	cu. meters/hr	1.589873 E-01
barrels/day	cu. meters/day	1.589873 E-01

TO CONVERT	INTO	MULTIPLY BY
barrels/day	cu. meters/hr	6.624471 E-03
barrels/day	cu. meters/year	5.803036 E+01
barrels/ft	cu. meters/meter	5.216119 E-01
barrels/mi	cu. meters/km	9.879013 E-02
barrels	cu. inches	9702
barrels	cu. feet	5.6150
barrels	cu. meters	0.15899
barrels/day	ton/annum	58.0304
barrels/ton	cu. meters/ton	0.17525
bars	atmospheres	0.9869
bars	dynes/sq cm	1 E+06
bars	kgs/sq meter	1.020 E+04
bars	pounds/sq ft	2,089.
bars	pounds/sq in.	14.50
Baryl	Dyne/sq. cm	1.000
board feet	cu. inches	144
Bolt (US Cloth)	Meters	36.576
Btu	Liter-Atmosphere	10.409
Btu	ergs	1.0550 E+10
Btu	foot-lbs	778.3
Btu	gram-calories	252.0
Btu	horsepower-hrs	3.931 E-04
Btu	joules	1,054.8
Btu	kilograin-calories	0.2520
Btu	kilograin-meters	107.5
Btu	kilowatt-hrs	2.928 E-04
Btu/bhp-hour	watt/kilowatt	0.39301
Btu/cu. ft.	megajoules/cu. meter	3.725895 E-02
Btu/hr	foot-pounds/sec	0.2162
Btu / hr	gram-cal/sec	0.0700
Btu/hr	horsepower-hrs	3.929 E-04
Btu/hr	watts	0.2931
Btu/lb	megajoules/kg	2.326000 E-03
Btu/min	foot-lbs/sec	12.96
Btu/min	horsepower	0.02356
Btu/min	kilowatts	0.01757
Btu/min	watts	17.57
Btu/sq ft/min	watts/sq in.	0.1221
Btu/UK gal	megajoules/cu. meter	2.320800 E-01
Btu/US gal	mega joules/cu. meter	2.787163 E+04
Bucket (Br. dry)	Cubic Cm.	1.818 E+04
bushels	cu ft	1.2445
bushels	cu in.	2,150.4
bushels	cu meters	0.03524
bushels	laters	35.24
bushels	pecks	4.0
bushels	pints (dry)	64.0
bushels	quarts (dry)	32.0
C		
Calories, gram (mean)	B.T.U. (mean)	3.9685 E-03
Candle/sq- cm	Lamberts	3,142
Candle/sq. inch	Lamberts	.4870
centares (centiares)	sq meters	1.0
Centigrade	Fahrenheit	(°Cx9/5)+32
centigrams	grams	0.01
Centiliter	Ounce fluid (US)	.3382

TO CONVERT	INTO	MULTIPLY BY	TO CONVERT	INTO	MULTIPLY BY
Centiliter	Cubic inch	.6103	cubic feet	cu meters	0.02832
Centiliter	drams	2,705	cubic feet	cu yards	0.03704
centiliters	liters	0.01	cubic feet	gallons (U.S. liq.)	7.48052
centimeters	feet	3.281 E-02	cubic feet	liters	28.32
centimeters	inches	0.3937	cubic feet	pints (U.S. liq.)	59.84
centimeters	kilometers	1 E-05	cubic feet	quarts (U.S. liq.)	29.92
centimeters	meters	0.01	cubic feet/min	cu cms/sec	472.0
centimeters	miles	6.214 E-06	cubic feet/min	gallons/sec	0.1247
centimeters	millimeters	10.0	cubic feet/min	liters/sec	0.4720
centimeters	mils	393.7	cubic feet/min	pounds of water/min	62.43
centimeters	yards	1.094 E-02	cubic feet/sec	million gals/day	0.646317
centimeter-dynes	cm-grams	1.020 E-03	cubic feet/sec	gallons/min	448.831
centimeter-dynes	meter-kgs	1.020 E-08	cubic inches	cucms	16.39
centimeter-dynes	pound-feet	7.376 E-08	cubic inches	cu feet	5.787 E-04
centimeter-grams	cm-dynes	980.7	cubic inches	cu meters	1.639 E-05
centimeter-grams	meter-kgs	1 E-05	cubic inches	cu yards	2.143 E-05
centimeter-grams	pound-feet	7.233 E-05	cubic inches	gallons	4.329 E-03
centimeters of mercury	atmospheres	0.01316	cubic inches	liters	0.01639
centimeters of mercury	feet of water	0.4461	cubic inches	mil-feet	1.061 E+05
centimeters of mercury	kgs/sq meter	136.0	cubic inches	pints (U.S. liq.)	0.03463
centimeters of mercury	pounds/sq ft	27.85	cubic inches	quarts (U.S. liq.)	0.01732
centimeters of mercury	pounds/sqin.	0.1934	cubic meters	bushels (dry)	28.38
centimeters/sec	feet/min	1.969	cubic meters	cucms	1 E+06
centimeters/sec	feet/sec	0.03281	cubic meters	cu feet	35.31
centimeters/sec	kilometers/hr	0.036	cubic meters	cu inches	61,023.0
centimeters/sec	knots	0.1943	cubic meters	cu yards	1.308
centimeters/sec	meters/min	0.6	cubic meters	gallons (U.S. liq.)	264.2
centimeters/sec	miles/hr	0.02237	cubic meters	liters	1,000.0
centimeters/sec	miles/min	3.728 E-04	cubic meters	pints (U.S. liq.)	2,113.0
centimeters/sec/sec	feet/sec/sec	0.03281	cubic meters	quarts (U.S. liq.)	1,057
centimeters/sec/sec	kms/hr/sec	0.036	cubic meters/ton	bbls/US ton	5.70602
centimeters/sec/sec	meters/sec/sec	0.01	cubic yards	cucms	7.646 E+05
centimeters/sec/sec	miles/hr/sec	0.02237	cubic yards	cu feet	27.0
centipoise	pascal-second	0.001	cubic yards	cuinches	46,656.0
Chain	Inches	792.00	cubic yards	cu meters	0.7646
Chain	meters	20.12	cubic yards	gallons (U.S. liq.)	202.0
Chains (surveyors'			cubic yards	liters	764.6
or Gunter's)	yards	22.00	cubic yards	pints (U.S. liq.)	1,615.9
circular mils	sq cms	5.067 E-06	cubic yards	quarts (U.S. liq.)	807.9
circular mils	sq mils	0.7854	cubic yards/min	cubic ft/sec	0.45
Circumference	Radians	6.283	cubic yards/min	gallons/sec	3.367
circular mils	sqinches	7.854 E-07	cubic yards/min	liters/sec	12.74
Cords	cord feet	8			
Cord feet	cu. feet	16		**D**	
Coulomb	Statcoulombs	2.998 E+09			
coulombs	faradays	1.036 E-05	Dalton	Gram	1.650 E-24
coulombs/sq cm	coulombs/sq in.	64.52	days	seconds	86,400.0
coulombs/sq cm	coulombs/sq meter	1 E+04	decigrams	grams	0.1
coulombs/sq in.	coulombs/sq cm	0.1550	deciliters	liters	0.1
coulombs/sq in.	coulombs/sq meter	1,550.	decimeters	meters	0.1
coulombs/sq meter	coulombs/sq cm	1 E-04	degrees (angle)	quadrants	0.01111
coulombs/sq meter	coulombs/sq in.	6.452 E-04	degrees (angle)	radians	0.01745
cubic centimeters	cu feet	3.531 E-05	degrees (angle)	seconds	3,600.0
cubic centimeters	cu inches	0.06102	degrees/sec	radians/sec	0.01745
cubic centimeters	cu meters	1 E-06	degrees/sec	revolutions/min	0.1667
cubic centimeters	cu yards	1.308 E-06	degrees/sec	revolutions/sec	2.778 E-3
cubic centimeters	gallons (U. S. liq.)	2.642 E-04	dekagrams	grams	10.0
cubic centimeters	liters	0.001	dekaliters	liters	10.0
cubic centimeters	pints (U.S. liq.)	2.113 E-03	dekameters	meters	10.0
cubic centimeters	quarts (U.S. liq.)	1.057 E-03	Drams (apothecaries'		
cubic feet	bushels (dry)	0.8036	or troy)	ounces (avoirdupois)	0.1371429
cubic feet	cu cms	28,320.0	Drams (apothecaries'		
cubic feet	cu inches	1,728.0	or troy)	ounces (troy)	0.125

TO CONVERT	INTO	MULTIPLY BY
fluid or apoth.)	cubic cm.	3.6967
drams	grams	1.7718
drams	grains	27.3437
drams	ounces	0.0625
Dyne/cm	Erg/sq. millimeter	.01
Dyne/sq. cm.	Atmospheres	9.869 E-07
Dyne/sq. cm.	Inch of Mercury at 0°C	2.953 E-05
Dyne/sq. cm.	Inch of Water at 4°C	4.015 E-04
dynes	grams	1.020 E-03
dynes	joules/cm	1 E-07
dynes	joules/meter (newtons)	1 E-05
dynes	kilograms	1.020 E-06
dynes	poundals	7.233 E-05
dynes	pounds	2.248 E-06
dynes/sq cm	bars	1 E-06

E

TO CONVERT	INTO	MULTIPLY BY
Ell	Cm.	114.30
Ell	Inches	45
Em, Pica	Inch	.167
Em, Pica	Cm.	.4233
Erg/sec	Dyne - cm/sec	1.000
ergs	Btu	9.480 E-11
ergs	dyne-centimeters	1.0
ergs	foot-pounds	7.367 E-08
ergs	gram-calories	0.2389 E-07
ergs	gram-cms	1.020 E-03
ergs	horsepower-hrs	3.7250 E-14
ergs	joules	1 E-07
ergs	kg-calories	2.389 E-11
ergs	kg-meters	1.020 E-08
ergs	ki lowatt-hrs	0.2778 E-13
ergs	watt-hours	0.2778 E-10
ergs/sec	Btu/min	5,688 E-09
ergs/sec	ft-lbs/min	4.427 E-06
ergs/sec	ft-lbs/sec	7.3756 E-08
ergs/sec	horsepower	1.341 E-10
ergs/sec	kg-calories/min	1.433 E-09
ergs/sec	kilowatts	1 E-10

F

TO CONVERT	INTO	MULTIPLY BY
farads	microfarads	1 E+06
Faraday/sec	Ampere (absolute)	9.6500 E+04
faradays	ampere-hours	26.80
faradays	coulombs	9.649 E+04
Fathom	Meter	1.828804
fathoms	feet	6.0
Feet	centimeters	30.48
feet	kilometers	3.048 E-04
feet	meters	0.3048
feet	miles (naut.)	1.645 E-04
feet	miles (stat.)	1.894 E-04
feet	millimeters	304.8
feet	mils	1.2 E+04
feet of water	atmospheres	0.02950
feet of water	in. of mercury	0.8826
feet of water	kgs/sq cm	0.03048
feet of water	kgs/sq meter	304.8
feet of water	pounds/sq ft	62.43

TO CONVERT	INTO	MULTIPLY BY
feet of water	pounds/sqin.	0.4335
feet/min	cms/sec	0.5080
feet/min	feet/sec	0.01667
feet/min	kms/hr	0.01829
feet/min	meters/min	0.3048
feet/min	miles/hr	0.01136
feet/sec	cms/sec	30.48
feet/sec	kms/hr	1.097
feet/sec	knots	0.5921
feet/sec	meters/min	18.29
feet/sec	miles/hr	0.6818
feet/sec	miles/min	0.01136
feet/sec/sec	cms/sec/sec	30.48
feet/sec/sec	kms/hr/sec	1.097
feet/sec/sec	meters/sec/sec	0.3048
feet/sec/sec	miles/hr/sec	0.6818
feet/100 feet	per cent grade	1.0
Foot-candle	Lumen/sq. meter	10.764
foot-pounds	Btu	1.286 E-03
foot-pounds	ergs	1.356 E+07
foot-pounds	grain-calories	0.3238
foot-pounds	hp-hrs	5.050 E-07
foot-pounds	joules	1.356
foot-pounds	kg-calories	3.24 E-04
foot-pounds	kg-meters	0.1383
foot-pounds	ki lowatt-hrs	3.766 E-07
foot-pounds/min	Btu/min	1.286 E-03
foot-pounds/min	foot-pounds/sec	0.01667
foot-pounds/min	horsepower	3.030 E-05
foot-pounds/min	kg-calories/min	3.24 E-04
foot-pounds/min	kilowatts	2.260 E-05
foot-pounds/sec	Btu/hr	4.6263
foot-pounds/sec	Btu/min	0.07717
foot-pounds/sec	horsepower	1.818 E-03
foot-pounds/sec	kg-calories/min	0.01945
foot-pounds/sec	kilowatts	1.356 E-03
Furlongs	miles (U.S.)	0.125
furlongs	rods	40.0
furlongs	f eet	660.0

G

TO CONVERT	INTO	MULTIPLY BY
Gallons	cu cms	3,785.0
gallons	cu feet	0.1337
gallons	cu inches	231.0
gallons (US)	cu meters	3.785412 E-03
gallons (UK)	cu. meters	4.546092 E-03
gallons	cu yards	4.951 E-03
gallons	liters	3.785
gallons (liq. Br. Imp.)	gallons (U.S.liq.)	1.20095
gallons (U.S.)	gallons (imp.)	0.83267
gallons of water	pounds of water	8.3453
gallons/min	cu ft/sec	2.228 E-03
gallons/min	liters/sec	0.06308
gallons/min	cu ft/hr	8.0208
gausses	lines/sq in.	6.452
gausses	webers/sq cm	1 E-08
gausses	webers/sq in.	6.452 E-08
gausses	webers/sqmeter	1 E-04
gilberts	ampere-turns	0.7958
gilberts/cms	amp-turns/cm	0.7958
gilberts/cms	amp-turns/in	2.021
gilberts/cms	amp-turns/meter	79.58

TO CONVERT	INTO	MULTIPLY BY
Gills (British)	cubic cm.	142.07
gills	ltters	0.1183
gills	pints (liq.)	0.25
Grade	Radian	.01571
Grains	drams (avoirdupois)	0.03657143
grains (troy)	grains (avdp)	1.0
grains (troy)	grams	0.06480
grains (troy)	ounces (avdp)	2.0833 E-03
grains (troy)	pennyweight (troy)	0.04167
grains/U.S. gal	parts/million	17.118
grains/U.S. gal	pounds/million gal	142.86
grains/Imp. gal	parts/million	14.286
grams	dynes	980.7
grams	grains	15.43
grams	joules/cm	9.807 E-05
grams	joules/meter (newtons)	9.807 E-03
grams	kilograms	0.001
grams	milligrams	1,000.
grams	ounces (avdp)	0.03527
grams	ounces (troy)	0.03215
grams	poundals	0.07093
grams	pounds	2.205 E-03
grams/cm	pounds/inch	5.600 E-03
grams/cu cm	pounds/cu ft	62.43
grams/cu cm	pounds/cuin	0.03613
grams/cu cm	pounds/mil-foot	3.405 E-07
grams/liter	grains/gal	58.417
grams/liter	pounds/ 1,000 gal	8.345
grams/ liter	pounds/cu ft	0.062427
grams/liter	parts/million	1,000.0
grams/sq cm	pounds/sq ft	2.0481
gram-calories	Btu	3.9683 E-03
gram-calories	ergs	4.1868 E+07
gram-calories	foot-pounds	3.0880
gram-calories	horsepower-hrs	1.5596 E-06
gram-calories	kilowatt-hrs	1.1630 E-06
gram-calories	watt-hrs	1.1630 E-06
gram-calories/sec	Btu/hr	14.286
gram-centimeters	Btu	9.297 E-08
gram-centimeters	ergs	980.7
gram-centimeters	joules	9.807 E-05
gram-centimeters	kg-cal	2.343 E-08
gram-centimeters	kg-meters	1 E-05

H

TO CONVERT	INTO	MULTIPLY BY
Hand	Cm.	10.16
hectares	acres	2.471
hectares	sq feet	1.076 E+05
hectograms	granis	100.0
hectoliters	liters	100.0
hectometers	meters	100.0
hectowatts	watts	100.0
henries	millihenries	1,000.0
Hogsheads (British)	cubic ft.	10.114
Hogsheads (U.S.)	cubic ft.	8.42184
Hogsheads (U.S.)	gallons (U.S.)	63
horsepower	Btu/min	42.44
horsepower	foot-lbs/min	33,000.
horsepower	foot-lbs/sec	550.0
horsepower (metric) (542.5 ft lb/sec)	horsepower (550 ft lb/sec)	0.9863

TO CONVERT	INTO	MULTIPLY BY
horsepower (550 ft lb/sec)	tiorsepower (metric) (542.5 ft lb/sec)	1.014
horsepower	kg-calories/min	10.68
horsepower	kilowatts	0.7457
horsepower	watts	745.7
horsepower (boiler)	Btu/hr	33.479
horsepower (boiler)	kilowatts	9.803
horsepower-hrs	Btu	2,547.
horsepower-hrs	Pegs	2.6845 E+13
horsepower-hrs	foot-lbs	1.98 E+06
horsepower-hrs	gram-calories	641,190.
horsepower-hrs	joules	2.684 E+06
horsepower-hrs	kg-calories	641.1
horsepower-hrs	kg-meters	2.737 E+05
horsepower-hrs	kilowatt-hrs	0.7457
hours	days	4.167 E-02
hours	weeks	5.952 E-03
Hundredweights (long)	pounds	112
Hundredweights (long)	tons (long)	0.05
Hundredweights (short)	ounces (avoirdupois)	1600
Hundredweights (short)	pounds	100
Hundredweights (short)	tons (metric)	0.0453592
Hundredweights (short)	tons (long)	0.0446429
hydraulic horsepower	kilowatts	7.460430 E-01

I

TO CONVERT	INTO	MULTIPLY BY
inches	centimeters	2.540
inches	meters	2.540 E-02
inches	miles	1.578 E-05
inches	millimeters	25.40
inches	mils	1,000.0
inches	yards	2.778 E-02
inches of mercury	atmospheres	0.03342
inches of mercury	feet of water	1.133
inches of mercury	kgs/sq cm	0.03453
inches of mercury	kgs/sq meter	345.3
inches of mercury	pounds/sq ft	70.73
inches of mercury	pounds/sqin.	0.4912
inches of mercury @ 60°F	kiloPascal	3.376850 E+00
inches of mercury @ 32°F	kiloPascal	3.386380 E+00
inches of water @ 4°C	atmospheres	2.458 E-03
inches of water @ 4°C	inches of mercury	0.07355
inches of water @ 4°C	kgs/sq cm	2.540 E-03
inches of water @ 4°C	ounces/sq in.	0.5781
inches of water @ 4°C	pounds/sq ft	5,204
inches of water @ 4°C	pounds/sq in.	0.03613
inches of water @ 39.2°F	kiloPascal	2.490820 E-01
inches of water @ 60°F	kiloPascal	2.488400 E-01
international Ampere	Ampere (absolute)	.9998
international Volt	Volts (absolute)	1.0003
International volt	Joules (absolute)	1.593 E-19
International volt	Joules	9.654 E+04

J

TO CONVERT	INTO	MULTIPLY BY
joules	Btu	9.480 E-04
joules	ergs	1 E+07
joules	foot-pounds	0.7376
joules	kg-calories	2.389 E-04
joules	kg-meters	0.1020
joules	watt-hrs	2.778 E-04
joules/cm	grams	1.020 E+04

TO CONVERT	INTO	MULTIPLY BY	TO CONVERT	INTO	MULTIPLY BY
joules/cm	dynes	1 E+07	kilometers/hr/sec	meters/sec/sec	0.2778
joules/cm	joules/meter (newtons)	100.0	kilometers/hr/sec	miles/hr/sec	0.6214
joules/cm	poundals	723.3	kilopascal	pound force/inch²	0.14504
,joules/cm	pounds	22.48	kilowatts	Btu/min	56.92
			kilowatts	foot-lbs/min	4.426 E+04
	K		kilowatts	foot-lbs/sec	737.6
			kilowatts	horsepower	1.341
kilograms	dynes	980,665.	kilowatts	kg-calories/min	14.34
kilograms	grams	1,000.0	kilowatts	watts	1,000.0
kilograms	joules/cm	0.09807	kilowatt-hrs	Btu	3,413.
kilograms	joules/meter (newtons)	9.807	kilowatt-hrs	ergs	3.600 E+13
kilograms	poundals	70.93	kilowatt-hrs	foot-lbs	2.655 E+06
kilograms	pounds	2,205	kilowatt-hrs	gram-calories	859,850.
kilograms	tons (long)	9.842 E-04	kilowatt-hrs	horsepower-hrs	1.341
kilograms	tons (short)	1.102 E-03	kilowatt-hrs	joules	3.6 E+06
kilograms/cu meter	grams/cu cm	0.001	kilowatt-hrs	kg-calories	860.5
kilograms/cu meter	pounds/cu ft	0.06243	kilowatt-hrs	kg-meters	3.671 E+05
kilograms/cu meter	pounds/cuin.	3.613 E-05	kilowatt-hrs	pounds of water evaporated from and at 212° F.	3.53
kilograms/cu meter	pounds/mil-foot	3.405 E-10			
kilograms/meter	pounds/ft	0.6720			
Kilogram/sq. cm.	Dynes	980,665	kilowatt-hrs	pounds of water raised from 62° to 212° F.	22.75
kilograms/sq. cm	atmospheres	0.9678			
kilograms/sq. cm	feet of water	32.81	knots	feet/hr	6,080.
kilograms/sq cm	inches of mercury	28.96	knots	kilometers/hr	1.8532
kilograms/sq cm	pounds/sq ft	2,048.	knots	nautical miles/hr	1.0
kilograms /sq cm	pounds/sq in.	14.22	knots	statute miles/hr	1.151
kilograms/sq meter	atmospheres	9.678 E-05	knots	yards/hr	2,027.
kilograms/sq meter	bars	98.0 E-06	knots	feet/sec	1.689
kilograms/sq meter	feet of water	3.281 E-03			
kilograms/sq meter	inches of mercury	2.896 E-03		**L**	
kilograms/sq meter	pounds/sq ft	0.2048			
kilograms/sq meter	pounds/sqin.	1.422 E-03	league	miles (approx.)	3.0
kilograms/sq mm	kgs/sq meter	1 E+06	Light year	Miles	5.9 E+12
kilograin-calories	Btu	3.968	Light year	Kilometers	9.46091E+12
kilograin-calories	foot-pounds	3,088.	lines/sq cm	gausses	1.0
kilograin-calories	hp-hrs	1.560 E-03	lines/sq in.	gausses	0.1550
kilograin-calories	joules	4.186.	lines/sq in.	webers/sq cm	1.550 E-09
kilograin-calories	kg-meters	426.9	lines/sq in.	webers/sq in.	1 E-08
kilograin-calories	kilojoules	4,186	lines/sq in.	webers/sq meter	1.550 E-05
kilogram-calories	kilowatt-hrs	1.163 E-03	links (engineer's)	inches	12.0
kilogram meters	Btu	9.294 E-03	links (surveyor's)	inches	7.92
kilogram meters	ergs	9.804 E+07	liters	bushels (U.S. dry)	0.02838
kilogramm meters	foot-pounds	7.233	liters	cu cm	1,000.0
kilogram meters	joules	9.804	liters	cu feet	0.03531
kilogram meters	kg-calories	2.342 E-03	liters	cu inches	61.02
kilogram meters	kilowatt-hrs	2.723 E-06	liters	cu meters	0.001
kilolines	maxwells	1,000.0	liters	cu yards	1.308 E-03
kiloliters	liters	1,000.0	liters	gallons (U.S. liq.)	0.2642
kilometers	centimeters	1 E+05	liters	pints (U.S. liq.)	2.113
kilometers	feet	3,281.	liters	quarts (U.S. liq.)	1.057
kilometers	inches	3.937 E+04	liters/min	cu ft/sec	5.886 E-04
kilometers	meters	1,000.0	liters/min	gals/sec	4.403 E-03
kilometers	miles	0.6214	lumens/sq ft	foot-candles	1.0
kilometers	millimeters	1 E+06	Lumen	Spherical candle power	.07958
kilometers	yards	1,094.	Lumen	Watt	.001496
kilometers/hr	cms/sec	27.78	Lumen/sq. ft.	Lumen/sq. meter	10.76
kilometers/hr	feet/min	54.68	lux	foot-candies	0.0929
kilometers/hr	feet/sec	0.9113			
kilometers/hr	knots	0.5396		**M**	
kilometers/hr	meters/min	16.67			
kilometers/hr	miles/hr	0.6214	maxwells	kilolines	0.001
kilometers/hr/sec	cms/sec/sec	27.78	maxwells	webers	1 E-08
kilometers/hr/sec	ft/sec/sec	0.9113			

TO CONVERT	INTO	MULTIPLY BY	TO CONVERT BY	INTO	MULTIPLY
megalines	maxwells	1 E+06	miles/min	kms/min	1.609
megohms	microhms	1 E+12	miles/min	knots/min	0.8684
megohms	ohms	1 E+06	miles/min	miles/hr	60.0
meters	centimeters	100.0	mil-feet	cu inches	9.425 E-06
meters	feet	3.281	milliers	kilograms	1,000.
meters	inches	39.37	Millimicrons	meters	1 E-09
meters	kilometers	0.001	Milligrams	grains	0.01543236
meters	miles (naut.)	5.396 E-04	milligrams	grams	0.001
meters	miles (stat.)	6.214 E-04	milligrams/liter	parts/million	1.0
meters	millimeters	1,000.0	millihenries	henries	0.001
meters	yards	1,094	milliliters	liters	0.001
meters	varas	1,179	millimeters	centimeters	0.1
meters/min	cms/sec	1,667	millimeters	feet	3.281 E-03
meters/min	feet/min	3.281	millimeters	inches	0.03937
meters/min	feet/sec	0.05468	millimeters	kilometers	1 E-06
meters/min	kms/hr	0.06	millimeters	meters	0.001
meters/min	knots	0.03238	millimeters	miles	6.214 E-07
meters/min	miles/hr	0.03728	millimeters	mils	39.37
meters/sec	feet/min	196.8	millimeters	yards	1.094 E-03
meters/sec	feet/sec	3,281	million gals/day	cu ft/sec	1.54723
meters/sec	kilometers/hr	3.6	mils	centimeters	2.540 E-03
meters/sec	kilometers/min	0.06	mils	feet	8.333 E-05
meters/sec	miles/hr	2.237	mils	inches	0.001
meters/sec	miles/min	0.03728	mils	kilometers	2.540 E-08
meters/sec/sec	cms/sec/sec	100.0	mils	yards	2.778 E-05
meters/sec/sec	ft/sec/sec	3.281	miner's inches	cu ft/min	1.5
meters/sec/sec	kms/hr/sec	3.6	Minims (British)	cubic cm.	0.059192
meters/sec/sec	miles/hr/sec	2.237	Minims (U.S., fluid)	cubic cm.	0.061612
meter-kilograms	cm-dynes	9.807 E+07	minutes (angles)	degrees	0.01667
meter-kilograms	cm-grams	1 E+05	minutes (angles)	quadrants	1.852 E-04
meter-kilograms	pound-feet	7.233	minutes (angles)	radians	2.909 E-04
microfarad	farads	1 E-06	minutes (angles)	seconds	60.0
micrograms	grams	1 E-06	myriagrams	kilograms	10.0
microhms	megohms	1 E-12	myriameters	kilometers	10.0
microhms	ohms	1 E-06	myriawatts	kilowatts	10.0
microliters	liters	1 E-06			
Microns	meters	1 E-06			
miles (naut.)	feet	6,080.27			
miles (naut.)	kilometers	1.853		**N**	
miles (naut.)	meters	1,853.			
miles (naut.)	miles (statute)	1.1516	nepers	decibels	8.686
miles (naut.)	yards	2,027.	Newton	Dynes	1 E+05
miles (statute)	centimeters	1.609 E+05			
miles (statute)	feet	5,280.			
miles (statute)	inches	6.336 E+04		**O**	
miles (statute)	kilometers	1.609			
miles (statute)	meters	1,609.	OHM (International)	OHM (absolute)	1.0005
miles (statute)	miles (naut.)	0.8684	ohms	megohms	1 E-06
miles (statute)	yards	1,760.	ohms	microhms	1 E+06
miles/hr	cms/sec	44.70	ounces	drams	16.0
miles/hr	feet/min	88.	ounces	grains	437.5
miles/hr	feet/sec	1.467	ounces	grams	28.349527
miles/hr	kms/hr	1.609	ounces	pounds	0.0625
miles/hr	kms/min	0.02682	ounces	ounces (troy)	0.9115
miles/hr	knots	0.8684	ounces	tons (long)	2.790 E-05
miles/hr	meters/min	26.82	ounces	tons (metric)	2.835 E-05
miles/hr	miles/min	0.1667	ounces (fluid)	cu inches	1.805
miles/hr/sec	cms/sec/sec	44.70	ounces (fluid)	cu. cm	29.5735
miles/hr/sec	feet/sec/sec	1.467	ounces (fluid)	liters	0.02957
miles/hr/sec	kms/hr/sec	1.609	ounces (troy)	grains	480.0
miles/hr/sec	meters/sec/sec	0.4470	ounces (troy)	grams	31.103481
miles/min	cms/sec	2,682.	ounces (troy)	ounces (avdp.)	1.09714
miles/min	feet/sec	88.	ounces (troy)	pennyweights (troy)	20.0

TO CONVERT	INTO	MULTIPLY BY	TO CONVERT	INTO	MULTIPLY BY
ounces (troy)	pounds (troy)	0.08333	pounds (troy)	pounds (avdp.)	0.822857
ounce/sq. inch	Dynes/sq. cm.	4309	pounds (troy)	tons (long)	3.6735 E-04
ounces/sq. in.	pounds/sqin.	0.0625	pounds (troy)	tons (metric)	3.7324 E-04
			pounds (troy)	tons (short)	4.1143 E-04
			pounds/US gal (gas)	kilograms/cu. meter	1.601846 E+01
	P		pounds/US gal (liquid)	kilograms/cu. meter	1.198264 E+02
			pounds/UK gal (liquid)	kilograms/cu. meter	9.977633 E+01
Parsec	Miles	19 E+12	pounds/cu. foot	kilograms/cu. meter	1.601846 E+01
Parsec	Kilometers	3.084 E+13	pounds/barrel	kilograms/cu. meter	2.853010 E+00
Pascal-second	centipoise	1,000.0	pounds of water	cu. feet	0.01602
parts/million	grains/U.S. gal	0.0584	pounds of water	cu. inches	27.68
parts/million	grains/Imp. gal	0.07016	pounds of water	gallons	0.1198
parts/million	pounds/million gal	8.345	pounds of water/min	cu. ft/sec	2.670 E-04
Pecks (British)	cubic inches	554.6	pound-feet	cm-dynes	1.356 E+07
Pecks (British)	liters	9.091901	pound-feet	cm-grams	13,825.
Pecks (U.S.)	bushels	0.25	pound-feet	meter-kgs	0.1383
Pecks (U.S.)	cubic inches	537.605	pounds/cu ft	grams/cu. cm	0.01602
Pecks (U.S.)	liters	8.809582	pounds/cu ft	kgs/cu. meter	16.02
Pecks (U.S.)	quarts (dry)	8	pounds/cu ft	pounds/cu. in.	5.787 E-04
pennyweights (troy)	grains	24.0	pounds/cu ft	pounds/mil-foot	5.456 E-09
pennyweights (troy)	ounces (troy)	0.05	pounds/cu in.	gms/cu. cm	27.68
pennyweights (troy)	grams	1.55517	pounds/cu. in.	kgs/cu. meter	2.768 E+04
pennyweights (troy)	pounds (troy)	4.1667 E-03	pounds/cu. in.	pounds/cu. ft	1,728.
pints (dry)	cu inches	33.60	pounds/cu. in.	pounds/mil-foot	9.425 E-06
pints (dry)	cu. cms.	550.611	pounds/ft	kgs/meter	1.488
pints (ljq.)	cu cms.	473.2	pounds/in.	gms/cm	178.6
pints (liq.)	cu feet	0.01671	pounds/mil-foot	gms/cu cm	2.306 E+06
pints (liq.)	cu inches	28.87	pounds/sq ft	atmospheres	4.725 E-04
pints (liq.)	cu meters	4.732 E-04	pounds/sq ft	feet of water	0.01602
pints (liq.)	cu. yards	6.189 E-04	pounds/sq ft	inches of mercury	0.01414
pints (liq.)	gallons	0.125	pounds/sq ft	kgs/sq meter	4.882
pints (liq.)	liters	0.4732	pounds/sq ft	pounds/sqin.	6.944 E-03
pints (liq.)	quarts (liq.)	0.5	pounds/sq in.	atmospheres	0.06804
Planck's quantum	Erg -second	6.624 E-27	pounds/sq in.	feet of water	2.307
Poise	Gram/cm. sec.	1.00	pounds/sq in.	inches of mercury	2.036
Pounds (avoirdupois)	ounces (troy)	14.5833	pounds/sq i n.	kgs/sq meter	703.1
poundals	dynes	13,826.	pounds/sq in.	pounds/sq ft	144.0
poundals	grams	14.10			
poundals	joules/cm	1.383 E-03			
poundals	joules/meter (newtons)	0.1383		**Q**	
poundals	kilograms	0.01410			
poundals	pounds	0.03108	quadrants (angle)	degrees	90.0
pounds	drams	256.	quadrants (angle)	minutes	5,400.0
pounds	dynes	44.4823 E+04	quadrants (angle)	radians	1.571
pounds	grains	7,000.	quadrants (angle)	seconds	3.24 E+05
pounds	grams	453.5924	quarts (dry)	cu inches	67.20
pounds	joules/cm	0.04448	quarts (liq.)	cu cms	946.4
pounds	joules/meter (newtons)	4.448	quarts (liq.)	cu feet	0.03342
pounds	kilograms	0.4536	quarts (liq.)	cu inches	57.75
pounds	ounces	16.0	quarts (liq.)	cu meters	9.464 E-04
pounds	ounces (troy)	14.5833	quarts (liq.)	cu yards	1.238 E-03
Pounds	poundals	32.17	quarts (liq.)	gallons	0.25
pounds	pounds (troy)	1.21528	quarts (liq.)	liters	0.9463
pounds	tons (short)	0.0005			
pounds/sq. in./ft	kiloPascal/meter	2.262059 E+01		**R**	
pounds/sq. in/mile	kiloPascal/kilometer	4.284203 E+00			
pounds (troy)	grains	5,760.	radians	degrees	57.30
pounds (troy)	grams	373.24177	radians	minutes	3,438.
pounds (troy)	ounces (avdp.)	13.1657	radians	quadrants	0.6366
pounds (troy)	ounces (troy)	12.0	radians	seconds	2.063 E+05
pounds (troy)	pennyweights (troy)	240.0	radians/sec	degrees/sec	57.30
			radians/sec	revolutions/min	9.549
			radians/sec	revolutions/sec	0.1592

TO CONVERT	INTO	MULTIPLY BY
radians/sec/sec	revs/min/min	573.0
radians/sec/sec	revs/min/sec	9.549
radians/sec/sec	revs/sec/sec	0.1592
revolutions	degrees	360.0
revolutions	quadrants	4.0
revolutions	radians	6.283
revolutions/min	degrees/sec	6.0
revolutions/min	radians/sec	0.1047
revolutions/min	revs/sec	0.01667
revolutions/min/min	radians/sec/sec	1.745 E-03
revolutions/min/min	revs/min/sec	0.01667
revolutions/min/min	revs/sec/sec	2.778 E-04
revolutions/sec	degrees/sec	360.0
revolutions/sec	radians/sec	6.283
revolutions/sec	revs/min	60.0
revolutions/sec/sec	radians/sec/sec	6.283
revolutions/sec/sec	revs/min/min	3,600.0
revolutions/sec/sec	revs/min/sec	60.0
Rod	Chain (Gunters)	0.25
Rod	Meters	5.029
Rods (Surveyors' meas.)	yards	5.5
rods	feet	16.5

S

TO CONVERT	INTO	MULTIPLY BY
Scruples	grains	20
seconds (angle)	degrees	2.778 E-04
seconds (angle)	minutes	0.01667
seconds (angle)	quadrants	3.087 E-06
seconds (angle)	radians	4.848 E-06
Slug	Kilogram	14.59
Slug	Pounds	32.17
Sphere	Steradians	12.57
square centimeters	circular mils	1.973 E+05
square centimeters	sq feet	1.076 E-03
square centimeters	sq inches	0.1550
square centimeters	sq meters	0.0001
square centimeters	sq miles	3.861 E-11
square centimeters	sq millimeters	100.0
square centimeters	sq yards	1.196 E-04
square feet	acres	2.296 E-05
square feet	circular mils	1.833 E+08
square feet	sq cms	929.0
square feet	sq inches	144.0
square feet	sq meters	0.09290
square feet	sq miles	3.587 E-08
square feet	sq millimeters	9.290 E+04
square feet	sq yards	0.1111
square inches	circular mils	1.273 E+06
square inches	sq cms	6.452
square inches	sq feet	6.944 E-03
square inches	sq millimeters	645.2
square inches	sq mils	1 E+06
square inches	sq yards	7.716 E-04
square kilometers	acres	247.1
square kilometers	sq cms	1 E+10
square kilometers	sq ft	10.76 E+06
square kilometers	sq inches	1.550 E+09
square kilometers	sq meters	1 E+06
square kilometers	sq miles	0.3861
square kilometers	sq yards	1.196 E+06
square meters	acres	2.471 E-04
square meters	sq cms	1 E+04

TO CONVERT	INTO	MULTIPLY BY
square meters	sq feet	10.76
square meters	sq inches	1,550.
square meters	sq miles	3.861 E-07
square meters	sq millimeters	1 E+06
square meters	sq yards	1.196
square miles	acres	640.0
square miles	sq feet	27.88 E+06
square miles	sq kms	2.590
square miles	sq meters	2.590 E+06
square miles	sq yards	3.098 E+06
square millimeters	circular mils	1,973.
square millimeters	sq cms	0.01
square millimeters	sq feet	1.076 E-05
square millimeters	sq inches	1.550 E-03
square mils	circular mils	1.273
square mils	sq cms	6.452 E-06
square mils	sq inches	1 E-06
square yards	acres	2.066 E-04
square yards	sq cms	8,361.
square yards	sq feet	9.0
square yards	sq inches	1,296.
square yards	sq meters	0.8361
square yards	sq miles	3.228 E-07
square yards	sq millimeters	8.361 E+05

T

TO CONVERT	INTO	MULTIPLY BY
temperature (°C) + 273	absolute temperature (°C)	1.0
temperature (°C) + 17.78	temperature (°F)	1.8
temperature (°F) + 460	absolute temperature(°F)	1.0
temperature (°F) -32	temperature (°C)	5/9
temperature (°R)	temperature (°K)	5/9
therm/US gal.	megajoules/cu. meter	2.787163 E+04
tons (long)	kilograms	1,016.
tons (long)	pounds	2,240.
tons (long)	tons (short)	1.120
tons (metric)	kilograms	1,000.
tons (metric)	pounds	2,205.
tons (short)	kilograms	907.1848
tons (short)	ounces	32,000.
tons (short)	ounces (troy)	29,166.66
tons (short)	pounds	2,000.
tons (short)	pounds (troy)	2,430.56
tons (short)	tons (long)	0.89287
tons (short)	tons (metric)	0.9078
tons (UK)	kilograms/sec	2.822353 E-01
tons (US)	kilograms/sec	2.519958 E-01
tons (short)/sq ft	kgs/sq meter	9,765.
tons (short)/sq ft	pounds/sqin.	2,000.
tons of water/24 hrs	pounds of water/hr	83.333
tons of water/24 hrs	gallons/-min	0.16643
tons of water/24 hrs	cu ft/hr	1.3349
tons of refrigeration	kilowatts	3.516853 E+00

V

TO CONVERT	INTO	MULTIPLY BY
Volt/inch	Volt/cm.	0 .39370
Volt (absolute)	Statvolts	0.003336

W

TO CONVERT	INTO	MULTIPLY BY
watts	Btu/hr	3.4129
watts	Btu/min	0.05688
watts	ergs/sec	107.
watts	foot-lbs/min	44.27
watts	foot-lbs/sec	0.7373
watts	horsepower	1.341 E-03
watts	horsepower (metric)	1.360 E-03
watts	kg-calories/min	0.01433
watts	kilowatts	0.001
watts (Abs.)	B.T.U. (mean)/min.	0.056884
watts (Abs.)	joules/sec.	1
watt-hours	Btu	3.413
watt-hours	ergs	3.60 E+10
watt-hours	foot-pounds	2,656.
watt-hours	gram-calories	859.85
watt-hours	horsepower-hrs	1.341 E-03
watt-hours	kilogram-calories	0.8605
watt-hours	kilogram-meters	367.2
watt-hours	kilowatt-hrs	0,001
Watt (International)	Watt (absolute)	1.0002
webers	maxwells	1 E+08
webers	kilolines	1 E+05
webers/sq in.	gausses	1.550 E+07
webers/sq in.	lines/sq in.	1 E+08
webers/sq in.	webers/sq cm	0.1550
webers/sq in.	webers/sq meter	1,550.
webers/sq meter	gausses	1 E+04
webers/sq meter	lines/sq in.	6.452 E+04
webers/sq meter	webers/sq cm	1 E-04
webers/sq meter	webers/sq in.	6.452 E-04

Y

TO CONVERT	INTO	MULTIPLY BY
yards	centimeters	91.44
yards	kilometers	9.144 E-04
yards	meters	0.9144
yards	miles (naut.)	4.934 E-04
yards	miles (stat.)	5.682 E-04
yards	millimeters	914.4
year (common)	days	365
year (common)	hours	8760
year (leap)	days	366
year (leap)	hours	8784

Viscosity—equivalents of absolute viscosity

Absolute or Dynamic Viscosity		Centipoise	Poise Gram/Cm Sec Dyne Sec/Cm²	Slugs/Ft Sec *Pound$_f$ Sec/Ft²	†Pound$_m$/Ft Sec Poundal Sec/Ft²
		(μ)	$(100\ \mu)$	(μ'_e)	(μ_e)
Centipoise	(μ)	1	0.01	2.09 (10^{-5})	6.72 (10^{-4})
Poise Gram/Cm Sec Dyne Sec/Cm²	$(100\ \mu)$	100	1	2.09 (10^{-3})	0.0672
Slugs/Ft Sec *Pound$_f$ Sec/Ft²	(μ'_e)	47 900	479	1	g or 32.2
†Pound$_m$/Ft Sec Poundal Sec/Ft²	(μ_e)	1487	14.87	$\frac{1}{g}$ or .0311	1

*Pound$_f$ = Pound of Force   †Pound$_m$ = Pound of Mass

To convert absolute or dynamic viscosity from one set of units to another, locate the given set of units in the left hand column and multiply the numerical value by the factor shown horizontally to the right under the set of units desired.

As an example, suppose a given absolute viscosity of 2 poise is to be converted to slugs/foot second. By referring to the table, we find the conversion factor to be 2.09 (10^{-3}). Then, 2 (poise) times 2.09 (10^{-3}) = 4.18 (10^{-3}) = 0.00418 slugs/foot second.

Courtesy Crane Co. (Technical Paper #410).

General liquid density nomograph

X AND Y VALUES FOR DENSITY NOMOGRAPH

Compound	X	Y	Compound	X	Y	Compound	X	Y
Acetic Acid	40.6	93.5	Ethyl chloride	42.7	62.4	Methyl sulfide	31.9	57.4
Acetone	26.1	47.8	Ethylene	17.0	3.5	n-Nonane	16.2	36.5
Acetonitrile	21.8	44.9	Ethyl ether	22.6	35.8	n-Octadecane	16.2	46.5
Acetylene	20.8	10.1	Ethyl formate	37.6	68.4	n-Octane	12.7	32.5
Ammonia	22.4	24.6	Ethyl propionate	32.1	63.9	n-Pentadecane	15.8	44.2
Isoamyl alcohol	20.5	52.0	Ethyl propyl ether	20.0	37.0	n-Pentane	12.6	22.6
Ammine	33.5	92.5	Ethyl auindo	25.7	55.3	n-Nonadecane	14.9	47.0
Benzene	32.7	63.0	Fluorobenzene	41.9	87.6	Isopentane	13.5	22.5
n-Butyric acid	31.3	78.7	n-Heptadecane	15.6	45.7	Phenol	36.7	103.8
Isobutane	13.7	16.5	n-Heptane	12.6	29.8	Phosphine	28.0	22.1
Isobutyric acid	31.5	75.9	n-Hexadecane	15.8	45.0	Propane	14.2	12.2
Carbon dioxide	78.6	45.4	n-Hexane	13.5	27.0	Propionic acid	35.0	83.5
Chlorobenzene	41.7	105.0	Methanethiol	37.3	59.5	Piperidene	27.5	60.0
Cyclohexane	19.6	44.0	Methyl acetate	40.1	70.3	Propionitrile	20.1	44.6
n-Decane	16.0	38.2	Methyl alcohol	25.8	49.1	Propyl acetate	33.0	65.5
n-Dedecane	14.3	41.4	Methyl n-butyrate	31.5	65.5	Propyl alcohol	23.8	50.8
Diethylamine	17.8	33.5	Methyl isobutyrate	33.0	64.1	Propyl formate	33.8	66.7
n-Elconane	14.8	47.5	Methyl chloride	52.3	62.9	n-Tetradecane	15.8	43.3
Ethane	10.8	4.4	Methyl ether	27.2	30.1	n-Tridecane	15.3	42.4
Ethanethiol	32.0	55.5	Methyl ethyl ether	25.0	34.4	Triethylamine	17.9	37.0
Ethyl acetate	35.0	95.0	Methyl formate	46.4	74.6	n-Undecane	14.4	39.2
Ethyl alcohol	24.2	48.6	Methyl propionate	36.5	68.3			

Reprinted by permission of Tubular Exchanger Manufacturers Association (TEMA).

GENERAL DENSITY NOMOGRAPH

Density Conversions

To convert the numerical value of a property expressed in one of the units in the left-hand column of the table to the numerical value expressed in one of the units in the top row of the table, multiply the former value by the factor in the block common to both units.

Units→ ↓	$\frac{g}{cu\ cm}$	$\frac{g}{ml}$	$\frac{lb}{cu\ in.}$	$\frac{lb}{cu\ ft}$	$\frac{lb}{gal\ (U.S.)}$
$\frac{g}{cu\ cm}$	1	1.000028	0.036128	62.428	8.3455
$\frac{g}{ml}$	0.99997	1	0.036126	62.427	8.3452
$\frac{lb}{cu\ in.}$	27.680	27.681	1	1,728	231
$\frac{lb}{cu\ ft}$	0.016018	0.016019	5.7870×10^{-4}	1	0.13368
$\frac{lb}{gal\ (U.S.)}$	0.11983	0.11983	4.3290×10^{-3}	7.4805	1

Chart gives specific gravity/temperature relationship for petroleum oils

C₂H₆ = Ethane
C₃H₈ = Propane   iC₄H₁₀ = Isobutane
C₄H₁₀ = Butane   iC₅H₁₂ = Isopentane

Example: The specific gravity of an oil at 60 F is 0.85. The specific gravity at 100 F = 0.83.

Courtesy Crane Co. (Technical Paper #410).

To find the weight density of a petroleum oil at its flowing temperature when the specific gravity at 60 F/60 F is known, multiply the specific gravity of the oil at flowing temperature (see chart above) by 62.4, the density of water at 60 F.

Weight Density and Specific Gravity of Various Liquids

Liquid	Temp. *t* Deg. Fahr.	Weight Density *ρ* Lbs. per Cu. Ft.	Specific Gravity *S*	Liquid	Temp. *t* Deg. Fahr.	Weight Density *ρ* Lbs. per Cu. Ft.	Specific Gravity *S*	
Acetone	60	49.4	0.792	Mercury	20	849.74	...	*Liquid at 60 F referred to water at 60 F.
Ammonia, Saturated	10	40.9	...	Mercury	40	848.03	...	
Benzene	32	56.1	...	Mercury	60	846.32	13.570	†Milk has a weight density of 64.2 to 64.6.
Brine, 10% Ca Cl	32	68.05	...	Mercury	80	844.62	...	
Brine, 10% Na Cl	32	67.24	...	Mercury	100	842.93	...	‡100 Viscosity Index.
Bunkers C Fuel Max.	60	63.25	1.014	Milk	...	†	...	
Carbon Disulphide	32	80.6	...	Olive Oil	59	57.3	0.919	
Distillate	60	52.99	0.850	Pentane	59	38.9	0.624	
Fuel 3 Max.	60	56.02	0.898	SAE 10 Lube‡	60	54.64	0.876	
Fuel 5 Min.	60	60.23	0.966	SAE 30 Lube‡	60	56.02	0.898	
Fuel 5 Max.	60	61.92	0.993	SAE 70 Lube‡	60	57.12	0.916	
Fuel 6 Min.	60	61.92	0.993	Salt Creek Crude	60	52.56	0.843	
Gasoline	60	46.81	0.751	32.6° API Crude	60	53.77	0.862	
Gasoline, Natural	60	42.42	0.680	35.6° API Crude	60	52.81	0.847	
Kerosene	60	50.85	0.815	40° API Crude	60	51.45	0.825	
M. C. Residuum	60	58.32	0.935	48° API Crude	60	49.16	0.788	

Courtesy Crane Co. (Technical Paper #410).

Metric conversions—metric to English, English to metric

Length

To convert from	To	Multiply by	To convert from	To	Multiply by
mm	inches	.03937	inches	mm	25.40
cm	inches	.3937	inches	cm	2.540
meters	inches	39.37	inches	meters	.0254
meters	feet	3.281	feet	meters	.3048
meters	yards	1.0936	feet	km	.0003048
km	feet	3280.8	yards	meters	.9144
km	yards	1093.6	yards	km	.0009144
km	miles	.6214	miles	km	1.609

Area

To convert from	To	Multiply by	To convert from	To	Multiply by
sq mm	sq inches	.00155	sq inches	sq mm	645.2
sq cm	sq inches	.155	sq inches	sq cm	6.452
sq meters	sq feet	10.764	sq feet	sq meters	.09290
sq meters	sq yards	1.196	sq yards	sq meters	.8361
sq km	sq miles	.3861	sq miles	sq km	2.590
hectares	acres	2.471	acres	hectares	.4047

Volume

To convert from	To	Multiply by	To convert from	To	Multiply by
cu cm	cu inches	.06102	cu inches	cu cm	16.387
cu cm	fl ounces	.03381	cu inches	liters	.01639
cu meters	cu feet	35.314	cu feet	cu meters	.02832
cu meters	cu yards	1.308	cu feet	liters	28.317
cu meters	US gal	264.2	cu yards	cu meters	.7646
liters	cu inches	61.023	fl ounces	cu cm	29.57
liters	cu feet	.03531	US gal	cu meters	.003785
liters	US gal	.2642	US gal	liters	3.785

Weight

To convert from	To	Multiply by	To convert from	To	Multiply by
grams	grains	15.432	grains	grams	.0648
grams	ounces†	.0353	ounces†	grams	28.350
kg	ounces†	35.27	ounces†	kg	.02835
kg	pounds†	2.2046	pounds†	kg	.4536
kg	US tons	.001102	pounds†	tonnes	.000454
kg	long tons	.000984	US tons	kg	907.2
tonnes	pounds†	2204.6	US tons	tonnes	.9072
tonnes	US tons	1.1023	long tons	kg	1016
tonnes	long tons	.9842	long tons	tonnes	1.0160

Unit weight

To convert from	To	Multiply by	To convert from	To	Multiply by
gr/sq cm	lb/sq in	.01422	lb/ft	kg/m	1.4881
gr/cu cm	lb/cu in	.0361	lb/sq in	gr/sq cm	70.31
kg/sq cm	lb/sq in	14.22	lb/sq in	kg/sq cm	.07031
kg/cu m	lb/cu ft	.0624	lb/cu in	gr/cu cm	27.68
kg/m	lb/ft	.6720	lb/cu ft	kg/cu m	16.018

Unit volume

To convert from	To	Multiply by	To convert from	To	Multiply by
liters/min	US gpm	.2642	US gpm	liters/min	3.785
liters/min	cfm	.03531	US gpm	liters/hr	227.1
liters/hr	US gpm	.0044	US gpm	cu m/hr	.2271
cu m/min	cfm	35.314	cfm	liters/min	28.317
cu m/hr	cfm	.5886	cfm	cu m/min	.02832
cu m/hr	US gpm	4.4028	cfm	cu m/hr	1.6992

Power

To convert from	To	Multiply by	To convert from	To	Multiply by
watts	ft-lb/sec	.7376	ft-lb/sec	watts	1.356
watts	hp	.00134	hp	watts	745.7
kw	hp	1.3410	hp	kw	.7457
cheval-vap	hp	.9863	hp	cheval-vap	1.0139

Heat

To convert from	To	Multiply by	To convert from	To	Multiply by
gr-cal	Btu	.003969	Btu	gr-cal	252
kg-cal	Btu	3.9693	Btu	kg-cal	.252
kg-cal/kg	Btu/lb	1.800	Btu/lb	kg-cal/kg	.5556
gr-cal/sq cm	Btu/sq ft	3.687	Btu/sq ft	gr-cal/sq cm	.2713
kg-cal/cu m	Btu/cu ft	.1124	Btu/cu ft	kg-cal/cu m	8.899

Work or energy

To convert from	To	Multiply by	To convert from	To	Multiply by
joule	ft-lb	.7376	ft-lb	joule	1.356
meter-kg	ft-lb	7.2330	ft-lb	meter-kg	.1383
gr-cal	ft-lb	3.087	ft-lb	gr-cal	.3239
kg-cal	ft-lb	3087	ft-lb	kg-cal	.000324
hp-hr	ft-lb	1,980,000	ft-lb	hp-hr	5.05×10^{-7}
kwhr	ft-lb	2,655,000	ft-lb	kwhr	3.766×10^{-7}
Btu	ft-lb	778.0	ft-lb	Btu	.001285

†avoirdupois pounds and ounces.

Courtesy Cameron Hydraulic Data Book, *Ingersoll-Rand Company.*

Temperature conversion—centigrade to Fahrenheit or Fahrenheit to centigrade

NOTE: The numbers in boldface refer to the temperature in degrees, either Centigrade or Fahrenheit, which it is desired to convert into the other scale. If converting from Fahrenheit to Centigrade degrees, the equivalent temperature will be found in the left column; while if converting from degrees Centigrade to degrees Fahrenheit, the answer will be found in the column on the right.

Centigrade		Fahrenheit	Centigrade		Fahrenheit	Centigrade		Fahrenheit	Centigrade		Fahrenheit
−73.3	−100	−148.0	2.8	37	98.6	33.3	92	197.6	293	560	1040
−67.8	−90	−130.0	3.3	38	100.4	33.9	93	199.4	299	570	1058
−62.2	−80	−112.0	3.9	39	102.2	34.4	94	201.2	304	580	1076
−59.4	−75	−103.0	4.4	40	104.0	35.0	95	203.0	310	590	1094
−56.7	−70	−94.0	5.0	41	105.8	35.6	96	204.8	316	600	1112
−53.9	−65	−85.0	5.6	42	107.6	36.1	97	206.6	321	610	1130
−51.1	−60	−76.0	6.1	43	109.4	36.7	98	208.4	327	620	1148
−48.3	−55	−67.0	6.7	44	111.2	37.2	99	210.2	332	630	1166
−45.6	−50	−58.0	7.2	45	113.0	37.8	100	212.0	338	640	1184
−42.8	−45	−49.0	7.8	46	114.8	43	110	230	343	650	1202
−40.0	−40	−40.0	8.3	47	116.6	49	120	248	349	660	1220
−37.2	−35	−31.0	8.9	48	118.4	54	130	266	354	670	1238
−34.4	−30	−22.0	9.4	49	120.2	60	140	284	360	680	1256
−31.7	−25	−13.0	10.0	50	122.0	66	150	302	366	690	1274
−28.9	−20	−4.0	10.6	51	123.8	71	160	320	371	700	1292
−26.1	−15	5.0	11.1	52	125.6	77	170	338	377	710	1310
−23.3	−10	14.0	11.7	53	127.4	82	180	356	382	720	1328
−20.6	−5	23.0	12.2	54	129.2	88	190	374	388	730	1346
−17.8	0	32.0	12.8	55	131.0	93	200	392	393	740	1364
−17.2	1	33.8	13.3	56	132.8	99	210	410	399	750	1382
−16.7	2	35.6	13.9	57	134.6	100	212	414	404	760	1400
−16.1	3	37.4	14.4	58	136.4	104	220	428	410	770	1418
−15.6	4	39.2	15.0	59	138.2	110	230	446	416	780	1436
−15.0	5	41.0	15.6	60	140.0	116	240	464	421	790	1454
−14.4	6	42.8	16.1	61	141.8	121	250	482	427	800	1472
−13.9	7	44.6	16.7	62	143.6	127	260	500	432	810	1490
−13.3	8	46.4	17.2	63	145.4	132	270	518	438	820	1508
−12.8	9	48.2	17.8	64	147.2	138	280	536	443	830	1526
−12.2	10	50.0	18.3	65	149.0	143	290	554	449	840	1544
−11.7	11	51.8	18.9	66	150.8	149	300	572	454	850	1562
−11.1	12	53.6	19.4	67	152.6	154	310	590	460	860	1580
−10.6	13	55.4	20.0	68	154.4	160	320	608	466	870	1598
−10.0	14	57.2	20.6	69	156.2	166	330	626	471	880	1616
−9.4	15	59.0	21.1	70	158.0	171	340	644	477	890	1634
−8.9	16	60.8	21.7	71	159.8	177	350	662	482	900	1652
−8.3	17	62.6	22.2	72	161.6	182	360	680	488	910	1670
−7.8	18	64.4	22.8	73	163.4	188	370	698	493	920	1688
−7.2	19	66.2	23.3	74	165.2	193	380	716	499	930	1706
−6.7	20	68.0	23.9	75	167.0	199	390	734	504	940	1724
−6.1	21	69.8	24.4	76	168.8	204	400	752	510	950	1742
−5.6	22	71.6	25.0	77	170.6	210	410	770	516	960	1760
−5.0	23	73.4	25.6	78	172.4	216	420	788	521	970	1778
−4.4	24	75.2	26.1	79	174.2	221	430	806	527	980	1796
−3.9	25	77.0	26.7	80	176.0	227	440	824	532	990	1814
−3.3	26	78.8	27.2	81	177.8	232	450	842	538	1000	1832
−2.8	27	80.6	27.8	82	179.6	238	460	860	566	1050	1922
−2.2	28	82.4	28.3	83	181.4	243	470	878	593	1100	2012
−1.7	29	84.2	28.9	84	183.2	249	480	896	621	1150	2102
−1.1	30	86.0	29.4	85	185.0	254	490	914	649	1200	2192
−0.6	31	87.8	30.0	86	186.8	260	500	932	677	1250	2282
0.0	32	89.6	30.6	87	188.6	266	510	950	704	1300	2372
0.6	33	91.4	31.1	88	190.4	271	520	968	732	1350	2462
1.1	34	93.2	31.7	89	192.2	277	530	986	760	1400	2552
1.7	35	95.0	32.2	90	194.0	282	540	1004	788	1450	2642
2.2	36	96.8	32.8	91	195.8	288	550	1022	816	1500	2732

The formulas at the right may also be used for converting Centigrade or Fahrenheit degrees into the other scale.

$$\text{Degrees Cent., } C° = \frac{5}{9}(F° + 40) - 40 \qquad \text{Degrees Fahr., } F° = \frac{9}{5}(C° + 40) - 40$$

Courtesy Natural Gas Processors Suppliers Association.

Viscosity—equivalents of kinematic viscosity

Kinematic Viscosity		Centistokes	Stokes $\dfrac{Cm^2}{Sec}$	$\dfrac{Ft^2}{Sec}$
		(ν)	$(100\,\nu)$	(ν')
Centistokes	(ν)	1	0.01	1.076 (10^{-5})
Stokes $\dfrac{Cm^2}{Sec}$	$(100\,\nu)$	100	1	1.076 (10^{-3})
$\dfrac{Ft^2}{Sec}$	(ν')	92 900	929	1

To convert kinematic viscosity from one set of units to another, locate the given set of units in the left hand column and multiply the numerical value by the factor shown horizontally to the right, under the set of units desired.

As an example, suppose a given kinematic viscosity of 0.5 square foot/second is to be converted to centistokes. By referring to the table, we find the conversion factor to be 92,900. Then, 0.5 (sq ft/sec) times 92,900 = 46,450 centistokes.

Courtesy Crane Co. and American Society for Testing and Materials.

Viscosity—equivalents of kinematic and Saybolt Universal Viscosity

Kinematic Viscosity, Centistokes ν	Equivalent Saybolt Universal Viscosity, Sec		Kinematic Viscosity, Centistokes ν	Equivalent Saybolt Universal Viscosity, Sec	
	At 100°F Basic Values	At 210°F		At 100°F Basic Values	At 210°F
1.83	32.01	32.23	180.0	834.2	
2.0	32.62	32.85	200.0	926.9	
4.0	39.14	39.41	220.0	1019.6	
6.0	45.56	45.88	240.0	1112.3	
8.0	52.09	52.45	260.0	1205.0	
10.0	58.91	59.32	280.0	1297.7	
15.0	77.39	77.93	300.0	1390.4	
20.0	97.77	98.45	320.0	1483.1	
25.0	119.3	120.1	340.0	1575.8	
30.0	141.3	142.3	360.0	1668.5	Saybolt Seconds
35.0	163.7	164.9	380.0	1761.2	equal
40.0	186.3	187.6			Centistokes
45.0	209.1	210.5	400.0	1853.9	times 4.6673
50.0	232.1	233.8	420.0	1946.6	
55.0	255.2	257.0	440.0	2039.3	
60.0	278.3	280.2	460.0	2132.0	
65.0	301.4	303.5	480.0	2224.7	
70.0	324.4	326.7	500.0	2317.4	
75.0	347.6	350.0		Saybolt Seconds	
80.0	370.8	373.4	Over 500	equal	
85.0	393.9	396.7		Centistokes	
90.0	417.1	420.0		times 4.6347	
95.0	440.3	443.4			
100.0	463.5	466.7			
120.0	556.2	560.1			
140.0	648.9	653.4			
160.0	741.6				

Note: To obtain the Saybolt Universal viscosity equivalent to a kinematic viscosity determined at t, multiply the equivalent Saybolt Universal viscosity at 100°F by $1 + (t - 100)\,0.000\,064$. For example, 10 ν at 210°F are equivalent to 58.91 multiplied by 1.0070 or 59.32 see Saybolt Universal at 210°F.

Courtesy Crane Co. (Technical Paper #410).

Viscosity—equivalents of kinematic and Saybolt Furol Viscosity at 122°F

Kinematic Viscosity, Centistokes v	Equivalent Saybolt Furol Viscosity, Sec	
	At 122°F Basic Values	At 210°F
48	25.3	
50	26.1	25.2
60	30.6	29.8
70	35.1	34.4
80	39.6	39.0
90	44.1	43.7
100	48.6	48.3
125	60.1	60.1
150	71.7	71.8
175	83.8	83.7
200	95.0	95.6
225	106.7	107.5
250	118.4	119.4
275	130.1	131.4
300	141.8	143.5
325	153.6	155.5
350	165.3	167.6
375	177.0	179.7
400	188.8	191.8
425	200.6	204.0
450	212.4	216.1
475	224.1	228.3
500	235.9	240.5
525	247.7	252.8
550	259.5	265.0
575	271.3	277.2
600	283.1	289.5
625	294.9	301.8
650	306.7	314.1
675	318.4	326.4
700	330.2	338.7
725	342.0	351.0
750	353.8	363.4
775	365.5	375.7
800	377.4	388.1
825	389.2	400.5
850	400.9	412.9
875	412.7	425.3
900	424.5	437.7
925	436.3	450.1
950	448.1	462.5
975	459.9	474.9
1000	471.7	487.4
1025	483.5	499.8
1050	495.2	512.3
1075	507.0	524.8
1100	518.8	537.2
1125	530.6	549.7
1150	542.4	562.2
1175	554.2	574.7
1200	566.0	587.2
1225	577.8	599.7
1250	589.5	612.2
1275	601.3	624.8
1300	613.1	637.3
Over 1300	*	†

*Over 1300 Centistokes at 122°F:
Saybolt Fluid Sec = Centistokes × 0.4717

†Over 1300 Centistokes at 210°F
Log (Saybolt Furol Sec − 2.87) = 1.0276 [Log (Centistokes)] − 0.3975
Courtesy Crane Co. and American Society for Testing and Materials.

Viscosity—general conversions

APPROXIMATE CONVERSION TABLE

SECONDS SAYBOLT UNIVERSAL	SECONDS SAYBOLT FUROL	ENGLER DEGREES	ENGLER TIME	BARBEY	REDWOOD STANDARD	REDWOOD ADMIRALTY	KINEMATIC VISCOSITY STOKES
35		1.18	60	2,800	32		0.026
50		1.60	82	880	44		0.074
75		2.30	102	460	65		0.141
100	15	3.00	153	320	88		0.202
150	19	4.40	230	205	128		0.318
200	23	5.90	305	148	170	18	0.431
250	28	7.60	375	118	212·	23	0.543
300	33	8.90	450	98	254	27	0.651
400	42	11.80	550	72	338	36	0.876
500	52	14.50	750	59	423	45	1.10
600	61	17.50	900	48	518	53	1.32
700	71	20.6	1050	41	592	62	1.54
800	81	23	1,200	36.5	677	71	1.76
900	91	27	1,300	32.0	762	78	1.98
1,000	100	29	1,500	29.5	846	87	2.20
1,500	150	42	2,300	19.5	1,270	135	3.30
2,000	200	59	3,000	14.5	1,695	175	4.40
2,500	250	73	3,750	11.5	2,120	230	5.50
3,000	300	87	4,500	9.6	2,540	260	6.60
4,000	400	117	6,000	7.4	3,380	350	8.80
5,000	500	145	7,500	6.0	4,230	435	11.00
6,000	600	175	9,000	5.2	5,080	530	13.20
7,000	700	205	10,500	4.1	5,925	610	15.40
8,000	800	230	12,000	3.7	6,770	700	17.60
9,000	900	260	13,500	3.2	7,620	780	19.8
10,000	1,000	290	15,000	2.9	8,460	870	22
20,000	2,000	590	30,000	1.4	16,920	1,760	44
40,000	4,000	1,170	60,000		33,850	3,600	88
60,000	6,000	1,750	90,000		50,800	5,300	132
80,000	8,000	2,300	120,000		67,700	7,000	176
100,000	10,000	2,900	150,000		84,600	8,700	220
200,000	20,000	5,900	300,000		169,200	17,600	440
400,000	40,000	11,700	600,000		338,500	36,000	880
600,000	60,000	17,500	900,000		508,000	53,000	1,320
800,000	80,000	23,000	1,200,000		677,000	70,000	1,760
1,000,000	100,000	29,000	1,500,000		846,000	87,000	2,200

ABSOLUTE VISCOSITY is the force required to move a unit plane surface over another plane surface at a unit velocity when surfaces are separated by a layer of fluid of unit thickness

KINEMATIC VISCOSITY is the absolute viscosity divided by the density of the fluid.

SAE VISCOSITY NUMBERS are arbitrary numbers which have been assigned to classify lubricating oils and do not have any functional relationship with other systems.

VISCOSITY INDEX is an arbitrary system relating the effect of temperature change to viscosity. It is the slope of a plot of viscosity versus temperature for a given fluid.

1 Stoke	= 1 Poise/Density in Grams	1 Centistoke	= 1/100 Stoke	1 Centipoise	= 0.0000209 Slugs/(Ft)(Sec)
1 Centistoke	= 1.076 x 10⁻⁵ Ft²/Sec	1 Centipoise	= 0.000672 Lbs/(Ft)(Sec)	1 Centipoise	= 2.42 Lbs/(Ft)(Hr)
1 Poise	= 1 Gram/(Cm)(Sec)	1 Centipoise	= 0.000672 (Poundals)(Sec)/Ft²	1 Slug/(Ft)(Sec)	= 1/0.002089 Poise
1 Centipoise	= 1/100 Poise	1 Centipoise	= 0.0000209 (Lbs)(Sec)/Ft²	1 Ft²/Sec	= 92903 Centistokes

Pressure conversion chart

This line chart provides an easy method for converting units of pressure. Draw a line perpendicular to the scale lines through a known value of flow and read the equivalent value on any of the other scales. For values off the scale, known values can be read in multiples of 10 or 100.

Example. Known value—120 psi is equivalent to:

278 ft of water
1,920 ounces per in.^{2}
17,400 lb/ft^2
3,300 in. of water
242 in. of mercury
6,180 mm of mercury

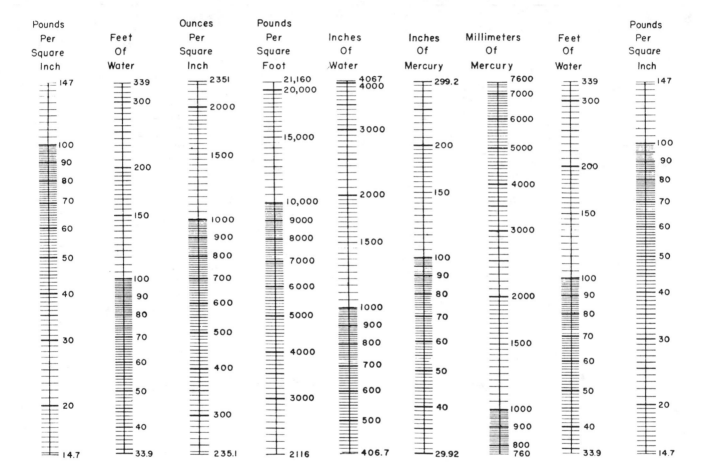

A simple method to determine square root

Problem. Determine the square root of the number 6,280.

Solution.

1. *Guess the square root to an even number.* For the example, use 80 since (80)^2 equals 6,400.

2. *Divide the number used into the number of which the square is desired* and carry the quotient to only two figures. Thus

$$6,280 \div 80 = 78$$

3. *Average this quotient with the divisor.*

80 + 78 = 158, and 158 ÷ 2 = 79

4. *Divide this figure into the number* and carry quotient to four or five figures, if desired.

6,280 ÷ 79 = 79.5

5. *Average this quotient with #3* for the square root.

79 + 79.5 = 158.5, and 158.5 ÷ 2 = 79.25

6. *To check answer*, multiply answer by itself.

79.25 × 79.25 = 6,280.5

SI data*

Examples of SI Derived Units

	SI Unit	
Quantity	Name	Symbol
Area	square metre	m^2
Volume	cubic metre	m^3
Speed, velocity	metre per second	m/s
Acceleration	metre per second squared	m/s^2
Density	kilogram per cubic metre	kg/m^3
Concentration (of amount of substance)	mole per cubic metre	mol/m^3
Specific volume	cubic metre per kilogram	m^3/kg
Luminance	candela per square metre	cd/m^2

SI Prefixes

Factor	Prefix	Symbol
10^{18}	exa	E
10^{15}	peta	P
10^{12}	tera	T
10^{9}	giga	G
10^{6}	mega	M
10^{3}	kilo	k
10^{2}	hecto	h
10^{1}	deka	da
10^{-1}	deci	d
10^{-2}	centi	c
10^{-3}	milli	m
10^{-6}	micro	μ
10^{-9}	nano	n
10^{-12}	pico	p
10^{-15}	femto	f
10^{-18}	atto	a

Examples of SI Derived Units With Special Names

	SI Unit			
Quantity	Name	Symbol	Expression in Terms of Other Units	Expression in Terms of SI Base Units
Frequency	hertz	Hz		s^{-1}
Force	newton	N		$m \cdot kg \cdot s^{-2}$
Pressure	pascal	Pa	N/m^2	$m^{-1} \cdot kg \cdot s^{-2}$
Energy, work, quantity of heat	joule	J	$N \cdot m$	$m^2 \cdot kg \cdot s^{-2}$
Power, radiant flux	watt.	W	J/s	$m^2 \cdot kg \cdot s^{-3}$
Electric potential, potential differences, electromotive force	volt	V	W/A	$m^2 \cdot kg \cdot s^{-3} \cdot A^{-1}$
Electric resistance	ohm	Ω	V/A	$m^2 \cdot kg \cdot s^{-3} \cdot A^{-2}$
Conductance	siemens	S	A/V	$m^{-2} \cdot kg^{-1} \cdot s^3 \cdot A^2$

* Reprinted from API *Manual of Petroleum Measurement Standards*, Chapter 15, ''Guidelines for the Use of the International System of Units (SI) in the Petroleum and Allied Industries,'' with permission of American Petroleum Institute.

Energy conversion chart

This nomograph provides an easy method for determining the fuel requirements for various types of fuel. Since the Btu content per unit volume of natural gas and crude oil vary over a limited but significant range of values, nomographs for the energy value of several gases and oils are included rather than using an average value.

Example. 30 cubic ft of a gas with 1,100 Btu per cubic ft yields 33,000 Btu. Conversion of 33,000 Btu to the other units of energy is obtained by drawing a straight line connecting the values 33,000 on two end Btu scales. Using the right-hand nomograph, an oil with 11,000 Btu per gallon (about 4.6 million Btu per barrel) will produce 33,000 Btu from 0.3 gallon (about 0.0072 barrel).

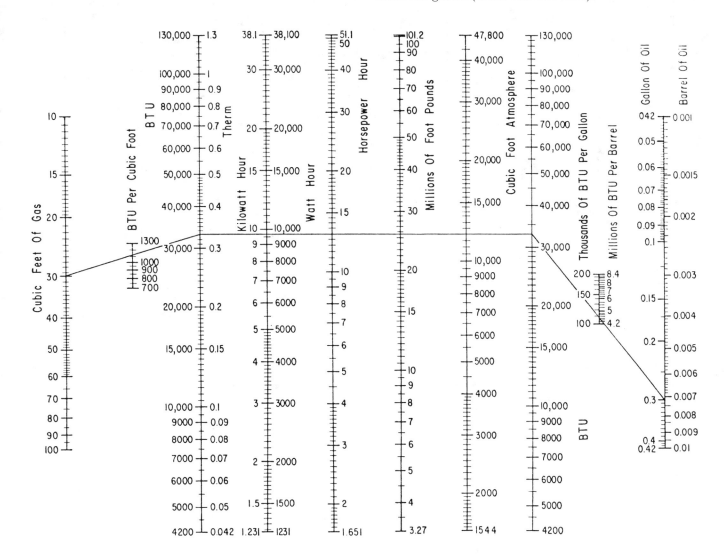

Flow conversion chart

This line chart provides an easy method for converting units of volume flow. Simply draw a line perpendicular to the scale lines through a known value of flow and read the equivalent value on any of the other scales.

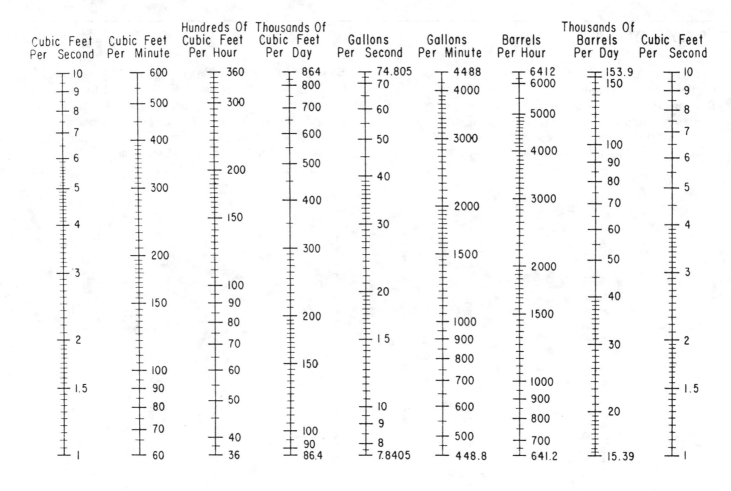

Cubic Feet Per Second	Cubic Feet Per Minute	Hundreds Of Cubic Feet Per Hour	Thousands Of Cubic Feet Per Day	Gallons Per Second	Gallons Per Minute	Barrels Per Hour	Thousands Of Barrels Per Day	Cubic Feet Per Second
10	600	360	864	74.805	4488	6412	153.9	10
9			800	70		6000	150	9
8	500	300	700	60	4000			8
7	400		600	50		5000		7
6		200	500		3000	4000	100	6
5	300			40			90	5
		150	400		2000	3000	80	
4				30			70	4
	200		300		1500		60	
3		100		20		2000	50	3
	150	90	200		1000	1500	40	
2		80 70		15	900 800		30	2
	100	60	150		700	1000		
1.5	90 80	50	100	10 9	600	900 800	20	1.5
	70	40	90	8	500	700		
1	60	36	86.4	7.8405	448.8	641.2	15.39	1

Conversions involving different types of fuel

Natural Gas

The calorific value of natural gas varies according to its composition, and virtually no two fields produce gas of identical quality; the composition of gas from any one field can vary over its producing life. Thus there is no such thing as a "standard" natural gas, nor a corresponding "standard" calorific value. However, when referring to natural gas in hypothetical cases it is quite common to indicate an "average" gross calorific value of 1,000 Btu/ft³ (the equivalent of which is taken, for purposes here, to be 9,400 kcal/Nm³ gross; see note on Table 3). Although in the U.S. most natural gas transmitted is at or slightly above this heat value, this does not generally apply in other countries. As an important example, Groningen (Netherlands) gas, because of a high nitrogen content, has a gross calorific value no greater than 898 Btu/ft³ measured at 60°F, 30 in Hg, dry (7,980 kcal/m³ₛ or 8,420 kcal/Nm³).

When precise values are required, reference should be made to the proprietors of the gas field concerned or to published data.

Manufactured Gases

These are derived from primary energy sources by processes involving chemical reaction, e.g., gases produced from coal and naphtha, also propane, butane, etc. Gas as piped to consumers can comprise both manufactured gases and natural gas used for enrichment. Calorific values of the component gases can thus vary widely depending on origin, but the resulting blend of what is known as *town gas* has a fairly constant heating value which, depending on locality, is usually about half that of natural gas. Exact figures can only come from the distribution company concerned.

Solid Fuels

The calorific value of solid fuels varies widely between countries, grades, qualities, and even from seam to seam. It is particularly important, therefore, to use local data wherever possible. Measurements in Btu/lb or kcal/kg will normally include moisture and ash content as received.

N.B. Btu/lb × 0.5556 = kcal/kg

kcal/kg × 1.8 = Btu/lb

Where actual values are not known, an average gross calorific value of 12,400 Btu/lb (6,890 kcal/kg) is considered reasonable for hard coal grades of good quality. This also serves as the basis of a common unit, "coal equiva-

Table 1
Petroleum fuels volume/weight/calorific value factors

Hydrocarbons	Specific Gravity	Weight/Volume bbl/ton	Gross Calorific Value Btu/Imp. gal	Btu/lb	kcal/kg	10^6 Btu/bbl	10^6 kcal/bbl	b.o.e./bbl
Propane	0.501	12.790	. . . . . .	21,500	11,943	3.795	0.956	0.654
Butane	0.579	11.060	. . . . . .	21,140	11,742	4.315	1.088	0.744
Motor Gasoline	0.733	8.740	148,000	20,160	11,200	5.176	1.304	0.892
Kerosene	0.780	8.213	156,000	20,000	11,110	5.456	1.377	0.941
Vaporizing Oil	0.823	7.783	163,000	19,800	11,000	5.701	1.442	0.983
Gas Oil	0.840	7.625	164,000	19,500	10,835	5.736	1.445	0.988
Fuel Oils	0.850	7.536	165,660	19,490	10,830	5.794	1.460	0.999
	0.860	7.448	166,840	19,400	10,780	5.835	1.470	1.006
	0.870	7.362	168,080	19,320	10,730	5.878	1.481	1.013
	0.880	7.278	169,310	19,240	10,690	5.921	1.492	1.021
	0.890	7.196	170,440	19,150	10,640	5.961	1.502	1.028
	0.900	7.116	171,640	19,070	10,590	6.002	1.513	1.035
	0.910	7.038	172,730	18,980	10,540	6.041	1.522	1.042
	0.920	6.961	173,800	18,890	10,490	6.078	1.532	1.048
	0.930	6.887	174,860	18,800	10,440	6.115	1.541	1.054
	0.940	6.813	175,890	18,710	10,390	6.151	1.550	1.061
	0.950	6.741	176,910	18,620	10,340	6.187	1.559	1.067
	0.960	6.671	177,910	18,530	10,290	6.222	1.568	1.073
	0.970	6.602	178,890	18,440	10,240	6.256	1.577	1.079
	0.980	6.535	179,860	18,350	10,190	6.290	1.585	1.084
	0.990	6.469	180,800	18,260	10,140	6.323	1.593	1.090
	1.000	6.404	181,730	18,170	10,090	6.356	1.602	1.096

Sources:
W.P.C. Publication "Technical Data on Fuel" (Editor H. M. Spiers) 6th Edition 1961; all products except fuel oil. Latter from H. Hyams, "Petroleum Cargoes, Measurement and Sampling, 1966."

Notes:
1. Btu/gal × 34.9726 = Btu/bbl
2. Btu/bbl × 0.252 = kcal/bbl
3. b.o.e./bbl = $\dfrac{10^6 \text{ Btu/bbl}}{5.8}$
4. kcal/kg × 1.8 = Btu/lb
5. To convert 10^6 Btu/bbl to ft³ natural gas (1000 Btu), multiply by 10^3
6. To convert 10^6 kcal/bbl to m³ natural gas (9400 kcal), multiply by 106.4

Table 2
Conversion factors for calorific values of gases under different conditions of measurement

A \ B	Btu/ft³ @ 60°F 30 inHg dry	Btu/ft³ @ 60°F 30 inHg saturated	Btu/ft³ @ 60°F 760 mmHg (60 °F) dry	kcal/m³ @ 0°C 760 mmHg dry	kcal/m³ @ 0°C 760 mmHg saturated	kcal/m³ @ 15°C 760 mmHg dry	kcal/m³ @ 15°C 760 mmHg saturated
Btu/ft³ @ 60°F 30 inHg dry	1	0.9826	0.9946	9.381	9.325	8.893	8.743
Btu/ft³ @ 60°F 30 inHg saturated	1.0177	1	1.0122	9.547	9.490	9.050	8.898
Btu/ft³ @ 60°F 760 mmHg (60 °F) dry	1.0054	0.9879	1	9.431	9.375	8.940	8.790
kcal/m³ @ 0°C 760 mmHg dry	0.10659	0.10474	0.10602	1	0.9940	0.9480	0.9320
kcal/m³ @ 0°C 760 mmHg saturated	0.10724	0.10537	0.10666	1.0060	1	0.9537	0.9376
kcal/m³ @ 15°C 760 mmHg dry	0.11245	0.11049	0.11184	1.0548	1.0486	1	0.9832
kcal/m³ @ 15°C 760 mmHg saturated	0.11437	0.11238	0.11375	1.0729	1.0665	1.0171	1

Notes

1. For equal conditions of temperature, pressure and water vapor content, the following factors apply:

$$1 \, Btu/ft^3 = 8.8992 \ kcal/m^3$$
$$1 \, kcal/m^3 = 0.11237 \ Btu/ft^3$$

2. To convert one unit of A to its equivalent heat value in terms of alternative units/conditions B, multiply by the appropriate factor.

lent," to which the many grades of solid fuel and other forms of energy may be reduced for the purposes of aggregation. Standard fuel for the purposes of European Economic Community statistics is considered to have a heat value of 7,000 kcal/kg.

Electricity

The basic unit of electric power is the *watt* (W), with the *kilowatt-hour* (kWh) as the derivative of most importance.

1 kilowatt-hour (kWh) = 3,412.14 Btu
                     = 859.845 kcal
                     = 3,600 kJ

See note to Table 4.

Liquid Fuels

In the case of liquid fuels, the actual calorific value of the particular fuel under consideration should be used where possible for conversions to other units, for example when calculating the natural gas equivalent of some specific application. For more general conversions, refer to Table 1.

Standard Reference Fuels

To convert different types and grades of fuel to a common basis for the purposes of presentation or comparison, it is the practice to express them in terms of some standard unit of a defined calorific value, for example:

Barrel of Oil Equivalent (b.o.e.), with a heat content of 5.8×10^6 Btu gross.

Tons of Oil Equivalent (t.o.e.), simply a hypothetical "ton of oil" with an average heating value of 43×10^6 Btu gross.

Oil equivalents may, if required, also be expressed in terms of standard fuel oil equivalent with a calorific value of 41.4×10^6 Btu per ton gross, i.e., 18,500 Btu/lb $\times$ 2,240.

Table 3
Heat value conversions and natural gas equivalents of various fuel units

	10^6 Btu gross	10^6 kcal gross	bbl oil equiv.	tons oil equiv.	metric t.c.e.	ft³ natural gas	m³ natural gas	10^6 kWh
1 barrel oil equivalent	5.8	1.462	1	0.1349	0.21217	5,800	155.50	0.0017
1 metric tonne coal equivalent	27.337	6.888	4.713	0.6357	1	27,337	732.90	0.0080
1 ton oil equivalent	43	10.836	7.414	1	1.5730	43,000	1152.82	0.0126
1 ton fuel oil equivalent	41.4	10.433	7.138	0.9628	1.5144	41,400	1109.92	0.0121

Notes
The natural gas equivalents above and in Tables 1 (notes) and 4 are based on gas with a gross heat content of 1,000 Btu/ft³ measured dry at 60° F and 30 inHg, which within 0.2% is equal to 9,400 kcal/m³ gross measured dry at O°C and 760 mmHg. On this basis, 37.3 cubic feet = 1 cubic meter.

The abbreviation m.t.c.e. is often used for metric tonnes of coal equivalent or million tons coal equivalent. Neither practice is sound as small m is the symbol for meter and not for metric or million. Instead, the ton unit used should be stated in full and the quantity expressed by the use of exponents; for example, 36×10^6 metric t.c.e.

Conversions above and in Table 4 are based on gross calorific values in line with normal statistical practice both inside and outside the gas industry. Net calorific values may however be more appropriate in situations where it is desired to compare or convert fuels in terms of useful heat content, due allowance having been made for changes in end-use efficiency.

Metric Tonne Coal Equivalent (t.c.e.), with a heat content of 27.337×10^6 Btu gross, i.e., 12,400 Btu/lb × 2,204.62.

The heat values of these different units and conversions between them are given in Table 3, together with natural gas equivalents.

Table 4

Conversion for daily/annual rates of energy consumption (gross heat basis)

	1×10^6 metric t.c.e. per year	1×10^6 ft^3 nat. gas per day	1×10^9 m^3 nat. gas per year	bbl daily oil equiv.	1×10^{12} (US trillion) ft^3 nat. gas per year	1×10^6 kWh per year	1×10^6 thermies per year
1×10^6 metric t.c.e. per year =	1	74.896	0.7329	12,913	0.027337	8,012	6,891
100×10^6 ft^3 nat. gas per day =	1.3352	100	0.9785	17,241	0.0365	10,697	9,201
1×10^9 m^3 nat. gas per year =	1.3645	102.2	1	17,620	0.0373	10,932	9,403
1×10^4 bbl daily oil equiv. =	0.7744	58	0.5675	1×10^4	0.0212	6,204	5,336
1×10^{12} (US trillion) ft^3 nat. gas per year =	36,580	2740	26,808	472,367	1	293,071	252,074
1×10^6 kWh per year =	0.1248×10^{-3}	0.9348×10^{-2}	0.9147×10^{-4}	1.6118	0.341×10^{-5}	1	0.8601
1×10^6 thermies per year =	0.1451×10^{-3}	0.1087×10^{-1}	0.1064×10^{-3}	1.8739	0.397×10^{-5}	1.1626	1

Notes

The table is good for natural gas conversions on the basis of a gas of 1,000 Btu/ft^3 gross (9,400 kcal/Nm^3), see note to Table 3 for conditions of measurement. Corrections may be applied as follows:

Gases with other calorific values: apply a correction based on the ratio of the calorific values; e.g. to use for Groningen gas, multiply the horizontal figures shown against 1×10^9 m^3 per year by 8420 ÷ 9400 i.e. 0.896 and the corresponding column by 1.116

Gases measured at other conditions of temperature and pressure: apply a correction based on the appropriate factor shown in Table 2; e.g. for gas measured saturated at 60°F and 30₆inHg, multiply the horizontal figures shown against 100×10^6 ft^3 per day by 0.9826 and the corresponding column by 1.0177

Calorific value and conditions of measurement different: apply a combined correction; e.g. to use for Gronigen gas measured dry at 15°C and 760 mmHg (m³$_s$), multiply the horizontal figures shown against 1×10^9 m³ by 0.948 × 8420 ÷ 9400 i.e. 0.849 and the corresponding column by the reciprocal of this number, i.e. 1.178

To use the table for *1×10^6 therms per year*, multiply U.S. trillion row by 10^{-4} and the corresponding column by 10^4

To use the table for *1×10^6 tons oil equivalent*, multiply the horizontal figures shown against 1×10^6 metric t.c.e. per year by 1.573 and the corresponding column by 0.6357

The factors relating to kilowatt-hours are based on the heat content of the electricity when used and not to the corresponding power station intakes of coal, oil or natural gas, which would need to reflect station generating efficiencies and load factor.

Weight of water per cubic ft at various temperatures

Temperature °F.	Weight per Cubic Foot, Lbs.	Temperature °F.	Weight per Cubic Foot, Lbs.	Temperature °F.	Weight per Cubic Foot, Lbs.	Temperature °F.	Weight per Cubic Foot, Lbs.	Temperature °F.	Weight per Cubic Foot, Lbs.	Temperature °F.	Weight per Cubic Foot, Lbs.
32	62.42	150	61.18	260	58.55	380	54.36	500	48.7		
40	62.42	160	60.98	270	58.26	390	53.94	510	48.1		
50	62.41	170	60.77	280	57.96	400	53.5	520	47.6		
60	62.37	180	60.55	290	57.65	410	53.0	530	47.0		
70	62.31	190	60.32	300	57.33	420	52.6	540	46.3		
80	62.23	200	60.12	310	57.00	430	52.2	550	45.6		
90	62.13	210	59.88	320	56.66	440	51.7	560	44.9		
100	62.02	212	59.83	330	56.30	450	51.2	570	44.1		
110	61.89	220	59.63	340	55.94	460	50.7	580	43.3		
120	61.74	230	59.37	350	55.57	470	50.2	590	42.6		
130	61.56	240	59.11	360	55.18	480	49.7	600	41.8		
140	61.37	250	58.83	370	54.78	490	49.2	. . .	. . .		

From "National Pipe Standards" The National Tube Co.

Engineering constants

Atmosphere:
Theoret. air to burn 1 lb. carbon = 11.52 lbs.
Vol. of 1 lb air at 32 deg. F and 14.7 lbs. per sq. in. = 12.387 cu. ft.
Vol. of 1 lb. air at 62 deg. F and 14.7 lb. per sq. in. = 13.144 cu. ft.
14.223 lb. per sq. in. = 1 Kg. per sq. cm. = 1 metric atm.
28.8 = equivalent mol. wgt. of air
30 in. Hg. at 62 deg. = atm. press
0.07608 lb. = wgt. 1 cu. ft. air at 62 deg. F and 14.7 lb. per sq. in.
0.08073 lb. = wgt. 1 cu. ft. air at 32 deg. F and 14.7 lb. per sq. in.

Heat:
8 lb. oxygen required to burn 1 lb. hydrogen (H)
777.5 ft. lb. = 1 B. T. U.
62,000 B. T. U. = cal. val. hydrogen (H)
14,600 B. T. U. per lb. cal. val. of carbon (C)
1.8 F deg. = 1 C deg.
2547 B. T. U. per hr. = 1 h.p.
−270 deg. C = absolute zero
3,415 B. T. U. = 1 kw. hr.
34.56 lb. = wt. air to burn 1 lb. hydrogen (H)
3.9683 B. T. U. = 1 Kg. calorie
4,050 B. T. U. = cal. val. of sulphur (S)
4.32 lb. = wgt. air to burn 1 lb. sulphur (S)

Water:
Freezing point = 0 deg. C, 32 deg. F
Boiling point, atm. press = 100 deg. C, 212 deg. F
Water power available from 1 cu. ft.-sec. falling 1 ft. = 0.1134 h.p.
1.134 ft. at 62 deg. F = 1 in. Hg. at 62 deg F
2.309 ft. at 62 deg. F = 1 lb. per sq. in.
27.71 in. at 62 deg. F = 1 lb. per sq. in.
33.947 ft. at 62 deg. F = atm. press.
5.196 lb. per sq. ft. = 1 in. at 62 deg. F
59.76 lb. = wgt. 1 cu. ft. at 212 deg. F
62.355 lb. = wgt. 1 cu. ft. at 62 deg. F

1 in. at 62 deg. F = 0.0735 in. Hg. at 62 deg. F
8.3356 lb. = wgt. 1 gal. at 62 deg. F
970.4 B.T.U. = latent heat of evap. at 212 deg. F

Steam:
Total heat sat. at atm. press = 1150.4 B.T.U.
0.47 B.T.U. per lb. deg. F = approx. specific heat of superheated steam at atm. press.

Surfaces:
1,273,239 circular mils = 1 sq. in.
1.4142 = square root of 2
144 square ins. = 1 sq. ft.
1728 cu. in. = 1 cu. ft.
1.7321 = square root of 3
2.54 cm = 1 in.
57.296 deg. = 1 radian (angle)
0.062428 lb. per cu. ft. = 1 Kg. per cu. meter
area of circle = 0.7854 (= 3.1416 ÷ 4) × diameter squared

Miscellaneous:
2,116.3 lb. per sq. ft. = atm. press.
2.3026 x log 10a log ea
2.7183 = c = base hyperbolic logs
288,000 B.T.U. per 24 hrs. = 1 ton of refrigeration
3.1416 = π = ratio of circumference of circle to diameter = ratio area of circle to square of radius
550 ft.-lb. per sec. = 1 h.p.
58.349 grains per gal. = 1 gram per liter
61.023 cu. in. = 1 liter
745.7 watts = 1 h.p.
8,760 hr. = 1 year of 365 days
88 ft. per sec. (min.) = 1 mile per min. (hr.)
2.666 lb. = wgt. oxygen required to burn 1 lb. carbon

Mensuration units

Diameter of a circle × 3.1416 = Circumference
Radius of a circle × 6.283185 = Circumference

Square of the radius of a circle × 3.1416 = Area
Square of the diameter of a circle × 0.7854 = Area
Square of the circumference of a circle × 0.07985 = Area
Half the circumference of a circle × half its diameter = Area

Circumference of a circle × 0.159155 = Radius
Square root of the area of a circle × 0.56419 = Radius

Circumference of a circle × 0.31831 = Diameter
Square root of the area of a circle × 1.12838 = Diameter

Diameter of a circle × 0.86 = Side of an inscribed equilateral triangle
Diameter of a circle × 0.7071 = Side of an inscribed square
Circumference of a circle × 0.225 = Side of an inscribed square
Circumference of a circle × 0.282 = Side of an equal square
Diameter of a circle × 0.8862 = Side of an equal square

Base of a triangle × ½ the altitude = Area
Multiplying both diameters and .7854 together = Area of an ellipse

Surface of sphere × ⅙ of its diameter = Solidity

Circumference of a sphere × its diameter = Surface
Square of the diameter of a sphere × 3.1416 = Surface
Square of the circumference of a sphere × 0.3183 = Surface

Cube of the diameter of a sphere × 0.5236 = Solidity
Cube of the radius of a sphere × 4.1888 = Solidity
Cube of the circumference of a sphere × 0.016887 = Solidity

Square root of the surface of a sphere × 0.56419 = Diameter
Square root of the surface of a sphere × 1.772454 = Circumference

Cube root of the solidity of a sphere × 1.2407 = Diameter
Cube root of the solidity of a sphere × 3.8978 = Circumference

Radius of a sphere × 1.1547 = Side of inscribed cube
Square root of (⅓ of the square of) the diameter of a sphere = Side of inscribed cube

Area of its base × ⅓ of its altitude = Solidity of a cone or pyramid, whether round, square or triangular

Area of one of its side × 6 = Surface of cube
Altitude of trapezoid × ½ the sum of its parallel sides = Area

Minutes to decimal hours conversion table

Minutes	Hours	Minutes	Hours
1	.017	31	.517
2	.034	32	.534
3	.050	33	.550
4	.067	34	.567
5	.084	35	.584
6	.100	36	.600
7	.117	37	.617
8	.135	38	.634
9	.150	39	.650
10	.167	40	.667
11	.184	41	.684
12	.200	42	.700
13	.217	43	.717
14	.232	44	.734
15	.250	45	.750
16	.267	46	.767
17	.284	47	.784
18	.300	48	.800
19	.317	49	.817
20	.334	50	.834
21	.350	51	.850
22	.368	52	.867
23	.384	53	.884
24	.400	54	.900
25	.417	55	.917
26	.434	56	.934
27	.450	57	.950
28	.467	58	.967
29	.484	59	.984
30	.500	60	1.000

How to compare costs of gas and alternate fuels

Growing shortage of natural gas prompts consideration of practical replacements, including coal, heavy fuel oil, middle distillates and LPG

Conversion from natural gas to other fuels has progressed from possibility to reality for many companies and will become necessary for many others in months and years ahead.

Fuels considered practical replacements for gas include coal, heavy fuel oils, middle distillates (such as kerosine-type turbo fuel and burner fuel oils) and liquefied petroleum gas (LPG).

To compare relative costs of fuels, the quoted prices normally are converted to the price per million British thermal units (Btu). With HHV being the higher (gross) heating value of the fuel, the following equations may be used.

Coal:

$$\text{Cents/million Btu} = \frac{\text{dollars/ton} \times 100 \times 10^6}{\text{lb/ton} \times \text{HHV}}$$

Heavy fuel oils:

$$\text{Cents/million Btu} = \frac{\text{dollars/barrel} \times 100 \times 10^6}{\text{gallons/barrel} \times \text{HHV}}$$

Middle distillates and LPG:

$$\text{Cents/million Btu} = \frac{\text{cents/gallon} \times 10^6}{\text{HHV}}$$

Natural gas:

$$\text{Cents/million Btu} = \frac{\text{cents/Mcf} \times 10^6}{\text{HHV} \times 1,000}$$

The approximate heat contents (HHV) of the fuels are as follows:

Coal:

Anthracite	13,900 Btu/lb
Bituminous	14,000 Btu/lb
Sub-bituminous	12,600 Btu/lb
Lignite	11,000 Btu/lb

Heavy fuel oils and middle distillates:

Kerosine	134,000 Btu/gal
No. 2 burner fuel oil	140,000 Btu/gal
No. 4 heavy fuel oil	144,000 Btu/gal
No. 5 heavy fuel oil	150,000 Btu/gal
No. 6 heavy fuel oil, 2.7% sulfur	152,000 Btu/gal
No. 6 heavy fuel oil, 0.3% sulfur	143,800 Btu/gal

Gas:

Natural	1,000 Btu/cubic ft
Liquefied butane	103,300 Btu/gal
Liquefied propane	91,600 Btu/gal

Typical characteristics of fuel oils

	Kerosine	No. 2 burner fuel
Gravity, °API	41.4	36.3
Pounds per gallon	6.814	7.022
Flash point, °F	133	165
Btu/gal.	134,000	140,000
Viscosity, SSU at 100°F	31.5	35.0
Pour point, °F	−30	0
Sulfur, wt. %	0.04	0.18
Carbon residue, wt. %	0.01	0.05
BS&W, %	—	1.30[2]

[1]Saybolt Furol Viscosity at 122°F
[2]mg/100 ml

LPG

Vapor pressure, psig at 100°F	138
Distillation, 95% point, °F	−40
Residue, vol. %	0.05
Oil-ring test	none

Index

Above/below ground transition, stress in, 94–97
Accounting fundamentals, 540–544
Air, moisture content of, 138–139
Air receiver, size of, 28
Annuities, 528–529
Annular sections, pressure loss in, 323
Anodes
  graphite, 194
  magnesium, 191, 193, 195
    life of, 196
  sacrificial, group installation of, 195
ANSI forged steel flanges, 20
Anti-freeze chart, 13
Arc length, 24
Area, calculating, 3–4
Basic formulas, 2–4
Batch injection rate, determining, 326
Belts, 11–12
Block and tackle, advantages of, 6
Break-even point, 551
Brush pigs
  run with gas, 135
  run with liquid, 135
Bulk modulus, 297, 299
Cables, 5–11
  estimating strength of, 5
  ground bed, calculating voltage drop in, 192
  rectifier, most economical size, 192
Capacity increases, 360
Carbon steel, color and temperature, 17
Cash flow, 526–527, 530–531
Cathodic protection
  for pipe lines, 179–187
  of heat exchanger shells, 194
  troubleshooting systems, 178–179, 201
Cathodic protection rectifiers
  efficiency of, 191
  monthly power bill for, 194
Cement, sacks of, needed to fill form, 35

Centrifugal compressors. *See* Compressors, centrifugal.
Centrifugal pumps. *See* Pumps, centrifugal.
Coated lines, current requirements for, 196
Coating defects, locating, 205–208
Coatings, 168–210
  enamel, NAPCA specifications for, 171–174
  for concrete, 169–171
  for pipe, 209
    coal-tar enamel per mile, 175
    coefficient of friction, 176–178
    primer per mile, 174
    tape for, 209–210
    wrapping per mile, 175
  standard applied weights for NAPCA
    specifications, 172–173
Compressibility factor
  generalized, 250
  for water, 132
Compressible fluids, velocity of in pipe, 301
Compressor cylinders, diagnosing problems with, 251
Compressor horsepower, 261–262
Compressor installation, fuel requirements for, 264
Compressor station, determining size of fuel line, 262–263
Compressors, 242–264
Compressors, centrifugal
  costs of, 553
  nomograph for estimating performance, 256–260
  performance calculations, 252–255
Compressors, gas, 243
Compressors, reciprocating, 244–247
  costs of, 552
Concrete
  coating thickness, 169–171
  mixing and finishing, 35
Concrete work, 35
Conductors, electric power, 113
Constants, engineering, 593
Construction, 29–69
Construction costs, estimating. *See* Costs, construction.

Construction equipment, optimum time to keep, 507–509
Control systems, 454–456
Control valves. *See* Valves, control.
Conversion factors, 567–595
Conversions, 568–576
  energy, 587
  flow, 587–588
  fuel, 588–590
  metric, 580
  minutes to decimal hours, 594
  pressures, 585
  temperature, 581
  viscosity, 577
Corrosion, 168–210
  controlling at compressor stations, 202–203
  metal loss in pipe lines, 558–559
Corrosion index, 565
Cost comparison, for engine operation, 546
Cost estimating
  construction costs, 509–512
  for pipe lines, 545
  for pipe lines, stations, terminals, 512–518
Costs
  break-even point, 551
  comparison between gas and alternate fuels, 595
  energy, for pipe line throughputs, 545
  for centrifugal compressors, 553
  for reciprocating compressors, 552
  of fuel for prime movers, 546
  profitability index, 549–551
  scale-up
    of equipment, 546–547
    of plants, 546–547
    of tanks, 548
  sum-of-years depreciation, 549
Cylinders, number of purges for, 438
Decision tree analysis, 536–539
Design, polypipe, 98–100
Dewatering, 134
Dike computations, with calculator, 492–496
Ditch
  cubic yards of excavation in mile of, 34
  shrinking and expansion of excavated soil from, 34
Ditching, 34
  labor/equipment considerations, 34
Drag-reducing agents, 339–356
  effects of active, passive drag on, 350–356
  outweighing looping or boosting, 344–350
Drag reduction, basics of, 339–344
Drums, 5, 11
  capacity of, 11
Economics, 519–544

Elbows, odd-angle, 24
Electric generators. *See* Generators, electric.
Electric motors. *See* Motors, electric.
Electric power conductors. *See* Conductors,
  electric power.
Electrical design, 101–119
Electrical formulas, 111
  what to remember, 114
Electricity, conversions for, 590–591
Electrodes, reference, 189
Ells, three-diameter, dimension of, 97
Enamel. *See* Coatings, for pipe.
Energy conversion chart, 587
Energy costs. *See* Costs, energy.
Energy measurement equipment, selecting, 227–232
Engine, internal combustion, fuel requirements
  for, 263–264
Engine cooling, water requirements for, 263
Engineering constants, 593
Equipment costs, 546–547
Excavations, sloping sides of, 21
Fatigue crack growth, 559–561
Flanged fittings, identifying series number of, 97
Flanges
  ANSI forged steel, 20
  raised face weld neck, bolting dimensions for, 18
Floodlighting, 117–119
Flow conversion chart, 587–588
Friction, pull out, 6
Fuel oils, typical characteristics of, 595
Fuels
  liquid, conversions for, 591
  solid, conversions for, 589
  standard reference, conversions for, 591–592
Full-load currents
  single-phase transformers, 111
  three-phase transformers, 112
Future value, of an investment, 526, 528
Gas, 211–292
  BTUs when specific gravity is known, 438
  compression of, 242–264
    determining horsepower needed, 248–250
  density and specific volume of, 226
  energy conversion data for estimating, 272
  glossary of common industry terms, 239–241
  high pressure, how to measure, 439–443
  hydraulics, 265–291
    pipe line calculations, 266–267
  in pipe lines
    in a given section, 270
    lost through puncture of line, 278
    pressure of, 279

velocity of, 277
weight of, 279
manufactured, conversions for, 588–589
measurement of, by orifice meter, 439
natural
and reserves terminology, 238
conversions for, 588
flow-rates, 436
required to compress, 261
specifications for, 437
viscosity of, 280
properties of, 214–225, 430–434
Gas blow-down system, throat pressure, 277–278
Gas compressors. *See* Compressors, gas.
Gas flow, 278, 280–291
calculating for pipe lines, 278
multi-phase, 284–288
Gas laws, 212–214
Gas lines
comparative capacities of, 271
equivalent length factors, 270–271
leakage from, 271
size of, 272
time for shut-in test, 273
Gas metering run, sizing, 437
Gas properties, 214–225, 430–434
Generators, electric, sizing of, 105
Geometry, 3–4
Gradients
financial, 529–530
hydraulic, 317–320
Ground bed resistance, 190–191
Head loss due to friction, 358
Hoses, marine, 310–314
Hydraulic gradients, 317–320
Hydrocarbons, physical properties of, 218–225
Hydrostatic testing, 120–132
for pipe lines, 121–127
using charts to estimate pressure change, 129–132
volume required at test pressure, 128–129
volume required to fill test section, 127
Incorrect operations index, 565
Induction motors. *See* Motors, induction.
Injection rate, batch, 326
Instrument accuracy, impact of, 481–482
Instrumentation, 453–474
pipe line, developments in, 456–470
Interest, calculating, 524–526
Investments, evaluation criteria, 531–535
Knockout dimensions, 106
Land descriptions, 25–27
Leak detection, 475–484

Leak detection systems
computer-based, 478–479
design aspects and guidelines, 482–483
development of, 483–484
pig-based, 478
simple, 477–478
Leak impact factors, 566
Leaks
causes and economic aspects of, 476–477
estimating, 307
in pipe line systems, 203–204
Lighting, flood, 117–119
Line loss, 275
Liquid density nomograph, 577–578
Liquid discharge, rate of, from pipe, 356
Liquid flow, estimating, 358
Liquid fuels, volume of, 491
Liquid gravity and density conversion chart, 295
Liquid hydrocarbons, 295–296
Liquids, 292–372
hydraulic, 309–372
lines, pressure drop in, 335–338
velocity of, in pipe, 301–303
volume of, in tanks, 491
Magnesium rods, 194
Maintenance, 497–504
Marine hose data, 310–314
Material in a roll, estimating length of, 13
Mathematical calculations
area, 3
surface area, 4
volume, 4
Measurement, 404–452
by orifice meter, 419–426
of gas, 439
of carbon dioxide, 413–419
of gas, 419–452
of light hydrocarbons, 412
of liquid, 408–411
of vapor flow, 426–429
units of, from one system to another, 568–576
Measurement factors, variations in, 438–439
Mensuration units, 594
Mercury, reuniting separated, 23
Meters, positive displacement, maintenance tips for, 446–449
Meter stations, sizing headers for, 450–452
Metes and bounds descriptions, 26
Methane, 233–237
pure, conversion table for, 237
Metric conversions, 580
Microwave hops, on ground level, 114–115

Monitoring systems. *See* Leak detection systems.
Motor controller sizes, 112
Motors
    electric, selection of, 102–103
    induction, quick determination of hp/ampere, 115
Multi-phase flow, 284–288
Nails, common, 17
NAPCA specifications for enamel coatings, 171–172
National Electric Code tables, 107–110
NEMA enclosure types, 104–105
Net present value, of an investment, 534–535
NFPA hazardous locations, 103–104
Oil, blended
    API gravity of, 294
    gravity of, 297
    viscosity of, 297
Oil, crude
    estimating lines, pressure drop, and spacing, 359
    pressure and temperature loss, 323–332
    pressure in transmission lines, 359
    sizing lines for capacity increases, 360
    viscosity of, 293
Oil spills, planning for, 498–504
Oils, petroleum, specific gravity/temperature
        relationships, 579
Orifice diameters, 435
Orifice flows, how to calculate in field, 443–446
Orifice meters, measurement of gas by, 439
Orifice plate flange taps, 408, 426–429
Panhandle A, 267–268
Payroll estimates, 506–507
Perpetuities, 527–528
Petrochemical pipe lines, commissioning, 140–142
Pipe bends
    calculation of, 79–80
    calculating maximum on cold pipe, 41–42
    computations for, 36–41
    determining degrees of, 35
    determining length of, 42, 78
    to fit ditches, 36
Pipe contents, calculating with computer, 85–88
Pipe design, 71–97
Pipe-equivalent factors, 268–270
Pipe laying, 35–47
Pipe line drying, 133–145
Pipe line flow measurement, 405–408
Pipe line instrumentation. *See* Instrumentation, pipe line.
Pipe line leak detection, 476–484
Pipe line modeling
    background philosophy of, 479–480
    basic equations, 480–481
Pipe line oil spills, planning for, 498–504

Pipe line revalidation, 555–563
    determining when to proceed, 557–558
Pipe line systems, leaks in, 203–204
Pipe line throughputs, 358, 545–546
Pipe lines
    above/below ground transition, stress in, 93–97
    advances in protection of, 204–209
    bare and concrete coated steel, buoyance of, 91
    carbon steel, allowable working pressure of, 92–93
    cleaning of
        with brush pig, 135
        with chemicals, 136
        with sand blasting, 135–136
    commissioning petrochemical, 140–142
    contraction or expansion of, 89
    corrosion-caused metal loss in, 558–559
    cost estimating for, 512–518
    cross-country, 69
    determining length of when subtended by
            angle, 42–43
    dewatering of, 134
    drying of, 136–137
    equivalent length of, 320
    estimating weight in metric tons/km, 89–90
    fatigue crack growth in, 559–561
    fill of, 306
    fitting in a ditch, 45–47
    flow of fuel gases in, 276
    gas in, 279
    HVDC effects on, 197–201
    hydrostatic testing of, 556–557
    in-line inspection of, 556–557
    looped systems, 320, 322
    lowering one still in service, 48–51
    measuring currents on, 194
    mechanical damage to, 562
    properties of, 76–77
    puncture of, 321
    rehabilitation of, 554–566
    schedule number of, 90
    series systems, 321
    sizing for water flow, 357
    span between supports, 44–45, 90
    stress-corrosion cracking of, 562–563
    vacuum drying of, 143–145
    velocity of gas in, 277–278
    volume of, 306
    welding of, 53–69
Pipe lowering, 48–51
Pipe size
    capacity increase vs. investment increase, 273
    increasing throughput volumes of natural gas, 274

Pipe supports, spacing of, 81–82
Pipe taps, drill sizes for, 17
Pipe threads
  American standard taper, 83
  British standard taper, 84
  normal engagement to make tight joints, 85
Pipe velocity, calculating with computer, 85–88
Pipe wall, stress in due to internal pressure, 93–94
Pipe weight
  calculating from outside diameter and wall thickness, 90
  calculating with computer, 85–88
Pipelay table
  on supports, 44
  underground, 43
Piping materials, weight of, 92
Plant costs, 546–547
Polypipe design data, 98–100
Positive displacement meters. *See* Meters, positive displacement.
Power conductors, electric, most economical size for, 113
Pressure, differential, average, 437
Pressure conversion chart, 585
Pressure drop method, 271
Primer. *See* Coatings, for pipe.
Profitability index, 549–551
Project scoping data, 30–31
  worksheet for major facilities, 30–31
Pumping stations, 116
Pumping units, electric motor horsepower chart, 115
Pumps, 373–403
  centrifugal, 350–387
    conversion factors and formulas, 353
    discharge at various speeds, 387
    estimating head for, 387
  discharge pressure, 387
  horsepower requirement of, 390
  motors for, 390
  reciprocating, 387, 388–389, 391–398
  rotary, 399–403
  specific speed of, 389–390
Purges, number of, for sample cylinders, 438
Rates of return, economic, 531–534
Reciprocating compressors. *See* Compressors, reciprocating.
Rectifier, voltage rating of, 196
Rectifier power, 190–191
Rehabilitation of pipe lines. *See* Pipe lines, rehabilitation of.
Relief valves. *See* Valves, relief.
Reynolds number, 289–291, 333–335
Right-of-way, 31–33
  clearing and grading of, 31–32
  crop acreage included in, 31

estimating man-hours
  for clearing trees and brush, 33
  for removing trees in, 32
  for removing tree stumps in, 32
Risk evaluation, 564–566
Rope, 5–11
  manila, working strength of, 5
  wire, 5
    causes of field troubles of, 9–11
    measuring diameter of, 9
    safe loads for, 6
    strength and weight, 9
Rubber hose, pressure drop calculations for, 315–317
Sand blasting, internal 135–136
SCADA communications, 470–474
Screws, machine, tap drills and clearance drills for, 16
Sensitivity analysis, 535–536
Shaft, 11–13
  estimating horsepower transmitted by, 13
Shaft key, calculating stress in, 12
Shafting
  properties of, 15–16
  round steel, weight in pounds, 14
Sheaves, 5
SI data, 586
Slurries, viscosity of, 299–300
Specific gravity, 297
Specific gravity/temperature relationships for petroleum oils, 579
Square roots, determining, 585–586
Steel fitting dimensions, 19
Steel flanges, 20
Stress in guy wires, 7–8
Sum-of-years depreciation, 549
Surface area, calculating, 4
Tank costs, 548
Tank farm line sizing, 370–372
Tanks, 485–496
  floating-roof, vapor loss from, 486–488
  horizontal cylindrical
    estimating contents of, 488–489
    gauging, 489
    volume of liquid in, 491
  nomograph to find capacity, 489–490
  vapor formation rate for, 491–492
Tape. *See* Coatings, for pipe.
Temperature
  drop across a regulator, 436
  predicting subsurface ranges, 356–357
  protection, glycol requirements for, 14
Temperature conversion, 581
Throat pressure, in gas blow-down system, 277–278

Throughputs. *See* Pipe line throughputs.
Time conversions, 594
Tools and appliances, typical wattages for, 106
Transient pressure analysis, 361–369
Transmission lines, 359
Trench shoring, minimum requirements, 22
Tubular goods, formulas and constants for problem
    solving, 88–89
Ultrasonic flow equations, 304–305
Units of measurement, 568–576
Utilities, underground, color codes for locating, 21
Vacuum drying. *See* Pipe lines, vacuum drying of.
V-belt, length of, 11–12
Valves
    control, 146–167
    selection, 156–157
    sizing for gas and vapor, 163–167
    sizing for throughput, 151–155
    sizing formula for, 147–150
    variable orifice rotary, 162–163
Valves, relief, 158–161
Vapors, properties of, 430–434
Velocity, calculating, 277
Velocity heads, 308
Viscosity
    equivalents of absolute, 577

equivalents of kinematic, 582
equivalents of kinematic and Saybolt Furol, 583
equivalents of kinematic and Saybolt Universal, 582
general conversions, 584
Viscosity units, 297
Voltage drop
    circuits using 600 V. copper conductors in steel
        conduit, 113
    in ground bed cable, 192
Volume, calculating, 4
Water
    resistivity of fresh, 193
    weight of, 593
Weight, calculating, 279
Weld, lb of electrodes required per, 62
Welding, 51–53. *See also* Pipe lines, welding of.
Welding criteria
    for safe and effective repair, 63–69
    table for selecting an appropriate repair method, 65–66
Welding rod, needed for mile of schedule 40 pipe line, 62
Welding terminology, 68
Welds, average per hour per welder, 62
Wire, resistance, 23
Wires, copper, resistance and weight of, 112
Wires, guy, stress in, 7–8
Wrapping, per mile of pipe, 175